WEATHER AND
CLIMATE MODIFICATION

"Little Bear to plane. Little Bear to plane. Start seeding."

Drawing by Dana Fradon; © 1953
The New Yorker Magazine, Inc.

WEATHER AND CLIMATE MODIFICATION

Edited by
W. N. Hess

National Oceanic and Atmospheric Administration,
Environmental Research Laboratories

A WILEY-INTERSCIENCE PUBLICATION

JOHN WILEY & SONS, New York • London • Sydney • Toronto

Copyright © 1974, by John Wiley & Sons, Inc.

Library of Congress Cataloging in Publication Data:

Hess, Wilmot N
  Weather and climate modification.

  "A Wiley-Interscience publication."
  Includes bibliographies.
  1. Weather control. I. Title.

QC928.H47   551.6′8   73-21829
ISBN 0-471-37453-9

Printed in the United States of America

10 9 8 7 6 5 4 3 2 1

Preface

Weather Modification is a rapidly changing subject. According to the Annual Report of the Interdepartmental Committee for Atmospheric Sciences for fiscal year 1973, $20 million is spent annually on weather modification in the United States for research and development. In addition, about $2 million is spent annually on operations. There are several hundred degreed scientists and engineers in the United States whose major occupation is weather modification. Very clearly major portions of this book could not been written five years ago or would have been written very differently. Similar changes will certainly take place in the next five years. Why then should one bother to write a book on weather modification when parts of it may soon be out of date? I think there are at least two answers:

1. The idea for this book started when I, a nuclear physicist, became involved in weather modification activities—trying to understand the field and having to manage some of the work. In starting to read the literature on the subject, I rapidly became overwhelmed by an unfamilar language, a great amount of detail, and obvious disagreement on major problems by senior researchers in the field. What to do? Write a book! Try to treat each major subject in the field by a chapter written by an outstanding student of that subject, and have it written in a language that I, a nonmeteorologist, could understand. In doing this, hopefully the book would be useful to many other nonspecialists who are also interested in weather modification. So this book is specifically written for the nonmeteorologist, but hopefully it might also find some use in meteorological areas.

2. There is a 28-year history of weather modification now since the revolutionary dry ice seeding experiments were conducted by Vincent Schaefer and Irving Langmuir at General Electric. The subject of weather modification has gone through some wild oscillations since that time. There was a major upswing in the late 1940s and early 1950s after Schaefer's discovery that dry ice would produce snow in a box filled with supercooled water drops. This led to considerable enthusiasm for seeding anything in sight and hopefully modifying all kinds of weather systems. The initial enthusiasm about large controllable effects led to a group of practitioners of the art, not all of whom were well based technically, and maybe a few of whom bordered on charlatanism. This period passed as the people who paid the bills found that the extravagant claims could not be backed up very often.

After some time, when it became obvious that the job wasn't as simple as that, a reaction set in and the subject was rather in disrepute among many professional meteorologists and others. But for the last 10 years or more, a strong research activity has developed in universities and in government agencies that is starting to answer the basic questions. I do mean *starting*. Serious long-range research to prove causality and to demonstrate the magnitude of the effects by statistically designed experiments using crossover randomization, etc., is beginning. Commercial operators are now, in the main, meteorologically trained, technically sophisticated groups, and are not given to extravagant claims. However, the subject is not yet in good shape. In very few locations and meteorological settings have significant controllable effects of weather modification been *proved* to the satisfaction of scientists who think seriously about the subject.

Much more work remains to be done to answer the important questions of causality and magnitude of effects. But now most of the "initial transient" has damped out and the subject is on a pretty steady growth curve of knowledge and respectability. Because of this present reasonable state and also because of the lengthy and rather controversial past, now is a good time to write a summary of the subject. However, it is clear that there is a long road ahead before most professional meteorologists would agree on the answers to questions like the following:

1. Is there a significant downwind effect from cloud seeding?
2. Can we effectively clear warm fogs?
3. Can cloud seeding produce severe weather, like tornadoes?
4. Can we increase precipitation from winter storms in the Great Plains?
5. Can we decrease hail and lightning in thunderstorms?
6. Can we steer hurricanes by seeding them?
7. Can we decrease rainfall under some seeding conditions?
8. Is it possible to understand long-term climatic changes now?

It seems now that we are able to answer some useful questions. For example, I think it is safe to say now that we (Dr. Joanne Simpson, that is) can modify cumulus clouds in Florida on nondistrubed days by heavy seeding with silver iodide. I think it is also safe to say that we know how to clear cold fog from airports and do it operationally quite well now.

This book presents technical discussions of the major fields of weather modification. The reader will find the subject is not monolithic, and different opinions of what is important and what is proved true are held by different authors. This is what the real world is like and we have not tried to edit it out.

In putting together a book like this, one thing is certain—you can never satisfy all the readers. We have designed this book to try to appeal to one class of reader, although a rather large class. We aimed the book at intelligent laymen, such as senior-level college students in the sciences or engineering, or working scientists or engineers who don't know much meteorology. Instructions to the authors said they should write a little above the technical level of *Scientific American* (but not too much above) and that they could use some mathematics, but that the physical meaning of the subject should come through even if one were to skip the math. A problem common to all books by a group of authors is that the different chapters are not written at the same level, or with the same mathematical detail, attention to referencing, or crispness and clarity. We were very fortunate to have an excellent editorial review committee chaired by Professor George Benton and including Mr. Eugene Bollay and Professor Patrick Squires, who read all the chapters and suggested revisions to almost all authors in order to bring the chapters more nearly in line. This effort produced major improvements, but of course the chapters are still not of the same level and the personalities of the different authors show through, as they should. Readers must understand that they will need to shift gears sometimes in going from one chapter to another.

Very many people help to put together a book like this. First, I would like to thank all the authors for getting in their contributions only slightly late and with only modest prodding. Second, my very special thanks to all the people who reviewed chapters of the book: Dr. William Aron, Professor Louis J. Battan, Mr. Eugene Bollay, Mr. Glenn Brier, Professor Marx Brook, Dr. Robert Bushnell, Professor Horace Byers, Dr. William R. Cotton, Dr. Robert M. Cunningham, Dr. Arnett S. Dennis, Mr. Hugh Dolan, Dr. Earl G. Droessler, Dr. Robert Elliott, Dr. Neal Frank, Dr. Tetsuya Fujita, Dr. R. E. Hallgren, Dr. Peter Hobbs, Professor Charles Hosler, Professor Henry G. Houghton, Mr. Raud E. Johnson, Dr. Jim Jiusto, Professor Robert Kates, Dr. Ed Kessler, Professor Wayne Larson, Dr. Noel E. LaSeur, Professor R. Lavoie, Dr. Chuck Leith, Dr. Roland List, Professor Julius London, Dr. Robert Long, Dr. Howard Mason, Mr. Don Moore, Dr. Bruce Morton, Dr. Harry Orville, Mr. Donald H. Pack, Mr. Allen Pearson, Mr. Bryon Phillips, Dr. Richard S. Schleusner, Dr. Joanne Simpson, Dr. Robert H. Simpson, Dr. Joseph Smagorinsky, Dr. Theodore Smith, Dr. Jerome Spar, Dr. Patrick Squires, Professor Gene Summers, Dr. Donald L. Veal, Dr. Bernard Vonnegut, Dr. James R. Wail, and Dr. Helmut Weickmann. Mr. Charles Chappell helped substantially in reviewing the book and also in preparing the subject index.

Finally, my very special thanks to all the people who helped put the book together. My secretary, Ms. Fridel Settle, did a great job in assembling the material and keeping after the authors. Ms. Robin Smith was always willing to do the necessary Xeroxing and collating. Mr. Don Valentine did his usual excellent job on all the illustrations. Ms. Janice Cavaliere edited the entire manuscript with great skill.

W. N. Hess

Boulder, Colorado
September 1973

Contents

A. HISTORY   **1**

   1. History of Weather Modification—Horace R. Byers   **3**

The Period Prior to 1946, 4   Langmuir, Schaefer, and Vonnegut, 9   Project
Cirrus, 14   The Weather Bureau Cloud Physics Project, 16   Hawaii and
Honduras, 17   More on Project Cirrus, 18   Commercial Seeding, 21   The
Australian Experiments, 23   The Painful Search for the "Truth", 24   The
Scientists Look at Their Field, 33   Recent and Ongoing Projects, 37
References, 39

   2. Experience of the Private Sector—Robert D. Elliott   **45**

Present-Day Cloud Seeding in the Private Sector, 45   Evolution of a
Rationale, 51   Interaction with the Federal Government, 70   Interaction
with the Public, 80   References, 87

B. BACKGROUND   **91**

   3. The Meteorological Background for Weather Modification—Morris   **93**
      Neiburger and Helmut K. Weickmann

Scales of Weather Phenomena, 93   The Formation of Clouds, 99   The
Formation of Precipitation, 103   Size Distribution of Raindrops, 127
Energetics of the Atmosphere, 130   References, 133

   4. Weather Modification Instruments and Their Use—R. E. Ruskin and **136**
      W. D. Scott

Perspective, 136   Priorities in Instrumentation, 138   Instrumentation
Platforms, 142   Standard Measurements, 144   Sampling in Clouds, 162
Measurements of Cloud Particles, 165   The Characteristics of Aerosols, 181
Remote Measurements, 189   The Measurement of Electrical Param-
eters, 192   Aircraft Systems for Delivery of Seeding Agents, 193
Systems Design, 195   Ground-Based Instrumentation, 199   Future of
Instrumentation, 199   Acknowledgments, 201   References, 201

5. Design and Evaluation of Weather Modification Experiments—   **206**
Glenn W. Brier

Design, 207   Analysis, 210   Some Problem Areas, 217   Proposed Direction
of Research, 220   Summary and Appraisal, 223   References, 223

C. PRECIPITATION MANAGEMENT   **227**

6. Cumulus Clouds and Their Modification—Joanne Simpson and   **229**
Arnett S. Dennis

The Dynamics and Physics of Individual Cumuli, 233   Cumulus Models, 237
Cumulus Groups and Interactions, 240   Design and Evaluation of Cumulus
Modification Experiments, 241   Cumulus Modification to Enhance Coales-
cence, 242   Modification of Supercooled Cumuli by Artificial Glaciation—
Methods and Basic Principles, 245   Seeding Supercooled Cumuli—Static
Approach, 246   Dynamic Seeding of Isolated Supercooled Cumuli, 247
Cumulonimbus Mergers, Dynamic Seeding of Many Cumuli Over a Target Area,
and Applications, 262   Are There Other Ways Than Seeding to Modify
Cumuli? 271   Extended Area and Persistent Effects. Potential Impacts of
Cumulus Modification, 274   Outlook, 278   Acknowledgments, 279
References, 279

7. Weather Modification for Augmenting Orographic Precipitation—Lewis   **282**
O. Grant and Archie M. Kahan

The Requirement and Potential, 282   The Nature of Orographic Clouds, 283
The Physical Basis for Seeding Orographic Clouds, 288   Results of Field
Experiments, 294   The Technology for Seeding "Cold" Orographic Clouds,
307   References, 315

8. The Mitigation of Great Lakes Storms—Helmut K. Weickmann   **318**

Structure and Properties of Great Lakes Cloud Systems, 318   Summary of
Seeding Experiments, 344   Conclusion, 351   References, 353

9. Fog—Bernard A. Silverman and Alan I. Weinstein   **355**

Morphology of Fog, 357   Fog Modification Concepts, 362   General Requirements of Fog Dispersal, 364   Supercooled Fog Dispersal, 365   Warm Fog
Dissipation, 369   Ice Fog Elimination, 381   Outlook, 381   References, 382

D. PROGRAMS IN OTHER COUNTRIES   **385**

10. Modification of Meteorological Processes—Y. K. Fedorov   **387**

Early Soviet Weather Modification Efforts, 389   Recent Soviet Work, 391
Changing the Climate, 397   Conclusion, 400   References, 401   Appendix:
Annotated Soviet Bibliography, 402

11. **Progress of Hail Suppression Work in the USSR** —G. K. Sulakvelidze,   **410**
B. I. Kiziriya, and V. V. Tsykunov

Concept of the Mechanism of Hail Formation, 411   Modes of Intervention in
Hail Processes and Some Supplements to Work on Modeling a Hail Cloud,
415   Basic Technological Means for Modification of Hail Processes, 418
Organization and Development of Hail Suppression Operations, 421   Some
Special Aspects of Hail Suppression Operations, 423   Assessing the Results
of Modification, 424   Conclusion, 428   References, 429

12. **Cloud Seeding in Australia** —E. J. Smith   **432**

Experiments with Dry Ice, 432   The Use of Silver Iodide, 434   Silver Iodide
Experiments on Single Clouds, 437   Area Experiments, 438   Early Experiments, 438   The Experiment in Tasmania, 442   Operational Cloud
Seeding, 443   Assessment of Operations, 444   Cloud Physics Research,
445   Future Experiments, 451   References, 451

13. **Rain Stimulation and Cloud Physics in Israel** —A. Gagin and J. Neumann   **454**

Rain Stimulation Experiments in Israel and Israel's Water Potential, 454
Statistical Aspects of the Experiment: Design, Results, and Problems, 458
Cloud Physics Aspects of the Israeli Cloud Seeding Experiment, 468
Acknowledgments, 492   References, 492

E. **SEVERE STORMS**   **495**

14. **Hurricane Modification** —R. Cecil Gentry   **497**

Structure and Energetics of Hurricanes Related to Modification Experi-
ments, 500   History of Hurricane Modification Experiments, 505   The
Stormfury Experiment, 508   Future Prospects, Special Problem Areas, and
Plans, 516   Acknowledgments, 519   References, 520

15. **Computer Simulation of Hurricane Development and Structure** —   **522**
Stanley L. Rosenthal

The Theoretical Problem and a Strategy for Its Solution, 523   A Theory for
Hurricane Development, 527   Quantitative Aspects of CISK: The Problem of
Cumulus Parameterization, 531   A Survey of Results from Circularly Symmetric Hurricane Models, 532   A Three-Dimensional Model of the Hurricane, 545   The Future, 549   Acknowledgments, 550   References, 550

16. **Tornadoes** —Robert Davies-Jones and Edwin Kessler   **552**

Tornado Characteristics, 552   Tornado Climatology, 554   Tornado Prediction, 555   Tornado Warning, 558   Relevant Aspects of Thunderstorm
Circulation, 558   Observations, 562   Experimental Models, 577   Mathematical Modeling, 581   Outlook for Modification, 590   References, 591

17.   Lightning Modification—George Dawson (Section I), Donald M. Fuquay   596
      (Sections II and III), and Heinz W. Kasemir (Sections IV and V)

An Introduction to Atmospheric Electricity, 596   Lightning Damage and
Lightning Modification Caused by Cloud Seeding, 604   Lightning Suppression by Chaff Seeding and Triggered Lightning, 612   References, 628

F.   CLIMATIC CHANGE   631

18.   Global Atmospheric Modeling and the Numerical Simulation of   632
      Climate—Joseph Smagorinsky

Basic Parameters of a Model, 636   The Governing Laws, 638   The Nature of
the Energy Source, 644   The Scope of the Atmospheric Response, 648
Conducting Numerical Experiments, 652   Experiments in the Simulation
of Climate, 656   Models for Long-Term Prediction, 673   Acknowledgments, 685   References, 685

19.   Inadvertent Large-Scale Weather Modification—Lester Machta and   687
      Kosta Telegadas

Carbon Dioxide, 688   Particles, 700   Cloudiness, 710   Land Usage and
Heat, 715   Climatic Prediction, 718   Summary and Conclusions, 722
Acknowledgment, 723   References, 723

20.   Inadvertent Atmospheric Modification Through Urbanization—Helmut   726
      Landsberg

Urban Effects on Radiation and Energy Balance, 727   The Urban Temperature Field, 731   Some Secondary Effects of the Urban Heat Island,
735,   The Urban Wind Field, 738   Urban Pollution, 744   Some Meteorological Effects of Urban Pollution, 748   Urban Effects on Precipitation, 752
Conclusions, 759   Acknowledgment, 760   References, 760

G.   OTHER PROBLEMS   765

21.   Weather Modification Litigation and Statutes—Ray J. Davis   767

The Lawsuits, 767   The State Statutes, 773   The Federal Bills, 779 The
Unanswered Questions, 782   References, 786

22.   Sociological Aspects of Weather Modification—J. Eugene Haas   787

The Sociological Significance of Weather Modification, 787   Various Perspectives on Weather Modification, 793   Weather Modification and Cultural Lag, 800   Political and Economic Implications, 801   How Shall the
Interests of the Unorganized be Protected? 804   How Shall the Interests

of the Minority in a Weather Modification Area be Protected? 806   What Factors are Associated with Conflict over Planned Weather Modification Efforts? 808   Acknowledgments, 810   References, 810

INDEX   **813**

WEATHER AND
CLIMATE MODIFICATION

A

HISTORY

A weather modification student will find that not all people agree in detail on weather modification history.

The two chapters on history in this book—Chapter 1 by Horace Byers and Chapter 2 by Robert Elliott—are interesting to read side by side because of their dissimilarity. We wondered if we should try to bring them more into agreement to change them and decided not to do so. Part of history is in the eye of the beholder and these two different versions are both valid and represent two different group's opinions. These two views largely reflect the strong differences in opinion that existed in about 1950 between the practitioners and most of the meteorological profession.

1 | History of Weather Modification

HORACE R. BYERS

Weather modification, like weather forecasting, is based on sound physical principles that cannot be applied precisely in the open atmosphere because several processes are interacting together in a manner difficult to predict. Weather modification has the further difficulty in that its manipulations are superimposed upon natural processes acting, perhaps indistinguishably, to the same or opposite effect. Added to this is the serious difficulty in learning the details of how the natural systems work, including an assessment of the extent and frequency of deficiencies or excesses within cloud systems that could be altered by treatment. Therefore it should not be surprising that the history of weather modification is one of painfully slow progress. Yet it has been an exciting endeavor, supported by important discoveries and widening knowledge in its background field of cloud physics.

In writing a history in a limited number of words or pages, the historian must carefully select, on as objective a basis as possible, those developments that appear to constitute the main thread of

Horace R. Byers is Distinguished Professor of Meteorology, Texas A & M University, College Station, Texas.

the narrative, leaving out the majority of happenings which, although perhaps significant, have contributed in a minor way to the general march of events. This is particularly risky in writing about science, since a contribution that may today seem obscure might emerge later as a development of great importance in a heretofore unrecognized aspect of the science. It is also awkward to sit in judgment over colleagues who are still active in the field. Modern weather modification had its origin in the United States, and more effort has been devoted to the subject in this country than elsewhere. The author apologizes to his non-American colleagues if the emphasis on United States contributions is deemed to be too strong. The editor has wisely included in this volume some chapters by other nationals, which should serve to provide international balance in the total treatment.

In this historical survey the 20-year period from roughly 1947 to 1967 will be emphasized.[84] Before 1946–1947 the physical principles applied today were scarcely known. The period since 1968 is represented by discussions of ongoing programs in other chapters. The years selected for emphasis here marked the development of the theory and applica-

3

tions of cloud physics from a primitive state to a point where regular weather enhancement programs by government agencies were started in several countries of the world. It was a period characterized by overoptimism on the one hand and conservatism on the other, with claims and counterclaims sometimes exceeding the bounds of objective science.

The discussion in this chapter is mainly on the history of *cloud* modification. Such topics as climate modification, inadvertent weather modification, and various agricultural practices such as orchard heating and fanning or use of shelter belts for mitigating weather severities are not touched upon.

A history of cloud physics and the other scientific and technical fields underlying the development of weather modification cannot be included without distracting the reader from the main purpose of this treatise and destroying the unity of the subject matter. Only those aspects of basic science that bear directly on the development of weather modification are treated.[110] (See Chapter 3.)

THE PERIOD PRIOR TO 1946

To bring on needed rain, primitive man used incantations, dances, various rituals, and prayers to one or more deities. Praying for rain is still practiced in civilized as well as primitive societies. Superstitions, some of which were believed to have a scientific basis, have developed from time to time. For example, in the first century A.D., Plutarch declared, "It is a matter of current observation that extraordinary rains pretty generally fall after great battles."[73] In an attempt to rationalize cause and effect, some people in those times suggested that the noise of battle touched off the rains. After the introduction of gunpowder this theme was reemphasized, and through the literature even to the present day, one can find records of rain ascribed to battles and explosives. W. J. Humphreys [82, 83] tells of the popularity of this idea over the centuries. An explanation for this belief lies in the fact that preparations for a battle usually are made in periods of good weather, and by the time the battle is over, rain is due to occur in accordance with the natural periodicity of extratropical disturbances.[158] In many parts of Europe and North America there is a high probability that rain will occur every 3 to 5 days. Pseudoscientific folklore even in modern times goes beyond these theories. During World War II many intelligent people were heard to expound the belief that observed unusual weather conditions (the weather always seems to be "unusual") were caused by the gases emitted from munitions factories. After the explosion of the first nuclear weapons and following the various tests in the Pacific and elsewhere, many citizens were certain that local and global weather was being adversely affected. These notions became so widespread that the U.S. Weather Bureau repeatedly was called upon to comment. The beliefs persisted despite official statements pointing out fallacies in the reasoning.

Early Attempts at Weather Modification

Probably the first rain-inducement proposal having a reasonable scientific basis was that made by James P. Espy,[56] author of the well-known meteorological classic *Philosophy of Storms*. In the April 5, 1839 issue of the *National Gazette and Literary Register* of Philadelphia, Espy proposed that large fires be built to generate updrafts that, in a humid atmosphere, would support cumulus clouds leading to rain. There is no record of field trails of this scheme until about 115 years later when experiments were started in

France and Africa with some success, as is discussed in a later section.

In the United States the next important excursion into the technology of rain-making came in the 1890's. It was a return to the idea that "it always rains after a battle." Edward Powers, in a book *War and Weather*, first published in 1871,[118] cited Civil War experience in which action after action was followed by rain. His theory won advocates in Congress and among some engineers and scientists, until finally in 1890–1891 Congress appropriated $10,000, of which $9,000 was to be spent in field experiments. After some tests of balloons and rockets for carrying explosives were conducted near Utica, New York, and Washington, D.C., full-scale experiments were carried out at Midland, Texas. Robert Dyrenforth, a former Commissioner of Patents and an enthusiast for the theory, who put forth the hypothesis that the explosions caused "airquakes" that started the rain,[73] supervised the tests under the aegis of the Forestry Division of the Agriculture Department.[105] He gave optimistic and promising reports of results, but criticisms by meteorologists and other objective scientists prevailed, and in 1893 the Secretary of Agriculture asked for no further funds to support the activity. However, the publicity and public discussion produced a rash of rain-makers who, armed with crude pyrotechnics and a convincing sales pitch, became a conspicuous part of the American rural scene, plying their questionable trade for a decade or more.

While American impetuosity and interest in "scientific progress" led to pseudoscientific gullibility, similar beliefs prevailed in rural areas of Europe. There, many of the farmers were less interested in possible scientific explanations than in outright mysticism. Sorcery, superstitions, and good-luck tokens, many of them related to practices of an entirely local character, were reported. In writing about hail-suppression endeavors in Italy, Gori et al.[65] state: "Exorcism, ritual fires, antihail brooms, 'aerostats' hoisted to 'discharge the cloud's electricity,' church-bell ringing and the use of cannons (some of them 1.4 meters in diameter) are known to have been widely spread for the purpose of hail suppression since remote times until quite recently."

In 1921 Dr. E. Leon Chaffee, professor of physics at Harvard University and a pioneer in electronics and vacuum-tube physics, developed the concept, suggested to him rather casually by an acquaintance, Mr. L. F. Warren, that cloud particles could be made to coalesce by sprinkling charged particles such as sand over the clouds.[109] Dr. Chaffee became enthusiastic about the idea and developed in his laboratory a nozzle for charging sand and dispersing it from an airplane. The nozzle could deliver sand grains having surface gradients of the order of 1000 V/cm. Flight experiments were carried out in August and September of 1924 at Aberdeen, Maryland with an airplane scattering the sand particles in the clear air above clouds having tops at 5000 to 10,000 ft. Dr. Chaffee reported "success" in the reverse sense, in that several clouds were observed to *dissipate* after treatment. The tests were well publicized in newspapers and scientific news journals, and this author, then a freshman at the University of California, recalls that his physics professors were enthusiastic about the idea. Chaffee's results probably would not endure the type of statistical scrutiny to which experiments of this kind are subject today.

Experiments by the Dutchman August W. Veraart in 1930[151] are often referred to because he used solid carbon dioxide as a seeding material just as in the more modern science of weather modification. However, it is probably only a coincidence that he used CO_2. He had no clear notion of the spontaneous development of ice crystals through cooling of super-cooled clouds by solid CO_2. Apparently he did not always work with supercooled clouds; he had some vague ideas about changing the thermal structure of cloud air, modifying temperature inversions, and creating electrical effects. His claims of success were not accepted by the scientific community.[23]

The first partially successful dissipation of fogs at temperatures above freezing was demonstrated in the 1930s in the fog studies of the Massachusetts Institute of Technology under the direction of Henry G. Houghton.[78] Sprays of water-absorbing solutions, especially calcium chloride, were used to clear the fog from a small area at the M.I.T. field station at Round Hill (South Dartmouth), Massachusetts. Houghton and his colleague, W. H. Radford, also designed an apparatus for dispensing fine particles of initially dry hygroscopic material. Their work predated by 30 years some of the methods used today.

Epsy's 1839 suggestion that convective rains be induced by burning large fires was examined again in 1938 by Görög and Rovo′ of Hungary.[66] From their experience with fires on lakes of oil, they made computations of the engineering, physical, and meteorological factors involved, and concluded that the method was practical for stimulating rain under certain conditions.

Although we are upsetting the chronology of this historical narrative by doing so, we can point out here that the fire method was revived in an ingenious series of experiments by the Frenchman Henri Dessens[49, 50] of the Observatoire de Puy de Dôme and the University of Toulouse in the 1950s and 1960s. In 1955 Dessens was commissioned by a large Belgian plantation organization in what was then the Congo to study possibilities of artificial stimulation of rain. While

there he noted that brush fires in the savanna portions of Africa frequently produced convective clouds which developed showers. In the following dry season, an associate of Dessens, Dr. Guy Soulage,[140] conducted a series of observations of cumulus formations over controlled brush fires, with mostly negative results as far as precipitation was concerned. Henri Dessens and his son Jean then conceived the idea of a controlled heat generator burning up to 100 metric tons of fuel in 30 min.[50] A cluster of oil burners consuming 1000 liters (approximately 1 ton) of fuel per minute was set up at Lannemezan, in southwestern France near the Pyrenees. The first trials were carried out in 1961. In one case the convection generated was strong enough to produce a small tornado.[53]

Early Discoveries on Nucleation

Lord Kelvin's landmark paper[149] showing the elevation of the vapor tension when a liquid surface is deformed into a convex curvative was published in the Proceedings of the Royal Society of Edinburgh in 1870. Further analysis indicated that free droplets of pure water could be in equilibrium with their vapor only at very high supersaturations, corresponding to relative humidities of a few hundred percent in the case of the smallest embryo droplets. In 1875 the Frenchman Coulier[46] published results of experiments demonstrating that particles floating in the air served as nuclei around which condensation could occur with little or no supersaturation. This work was followed by the remarkable research by the Scottish physicist John Aitken through nearly 35 years of painstaking effort.[3] With his portable expansion chamber, still known as the Aitken nuclei counter, he surveyed the particle content of the lower atmosphere. His equipment produced supersaturations at which the most numerous particles, those between 0.01 and 0.1 μm radius, are active. Today these are generally spoken of as Aitken particles. They are capable of serving as condensation nuclei for clouds and fogs, but it is now apparent that nuclei larger than predominantly measured in the Aitken counter are sufficiently numerous in the lower half of the troposphere to lay first claim to the available water. Concurrently with Aitken's work, C. T. R. Wilson[166] was carrying out his classical cloud-chamber experiments, in which he showed, among other things, that the ever-present small ions can serve as nuclei at quite high supersaturations and that in the absence of ions or any other particles, the supersaturations necessary for condensation to begin were of the order of 500 percent.

Aitken was the first to imply that there were two types of nuclei, those with an affinity for water vapor on which condensation begins before water saturation occurs, called hygroscopic particles, and nonhygroscopic nuclei that require some degree of supersaturation in order to serve as centers of condensation. Some years later, in 1921 and 1926, the Swedish chemist-meteorologist Hilding Köhler[89] demonstrated the importance of sea-salt nuclei and developed the theory of condensation on hygroscopic nuclei.

Following this pioneering work there was carried out in Germany, at the Taunus Observatory near Frankfurt, the series of measurements of numbers and sizes of nuclei, first under Helmut Landsberg, published in 1934,[94, 95] and subsequently by Christian Junge.[85, 86] The latter, like John Aitken before him, has become the world authority on this subject.[87] H. L. Wright[167] carried out a similar series of experiments at the Kew Observatory in London. The curves of number concentrations as a function of size obtained from the measurements

around Frankfurt augmented by data from other parts of the world have become the standard of reference for condensation nuclei everywhere. The first measurements on record of the vertical distribution of nuclei in the free atmosphere were those taken by A. Wigand[165] in 1919 in manned balloon flights in which Aitken nuclei were counted at heights up to about 6 km.

With the new knowledge of condensation on hygroscopic nuclei, A. Schmauss and Wigand in 1929 published their book *Die Atmosphäre als Kolloid,*[130] in which they expressed ideas about the growth of drops, referring to "colloidal stability" and "colloidal instability" in clouds.[152]

Supercooling and the Bergeron Hypothesis

The importance of the ice phase in precipitation was, at the beginning of the 1930s, yet to be clarified. Since the turn of the century meteorologists were aware of the abundance of droplets in clouds at temperatures well below the nominal freezing point. For example, A. Berson[25] before 1900 made scientific balloon flights in which he observed supercooled water, and in 1908 A. Wagner[157] reported liquid droplets observed at the Sonnblick (3106 m) at subfreezing temperatures, and A. Wegener in his 1911 book *Thermodynamik der Atmosphäre*[159] discussed liquid droplets observed to temperatures of -20 to $-30°C$. In the years before 1940 very little was known about the microphysics of nucleation of ice crystals in clouds, but it was obvious that something (lack of suitable nuclei?) was holding back the process. A few careful observers of clouds, notably C. Dauzère[48] and M. Guilbert[68] in France and Daniel Hull[81] in the United States, had noted that appreciable rains fell only after the upper parts of clouds were visibly glaciated. In Germany, G. Stüve,[143] in 1931 discussed evidence of the first occurrence of ice crystals commonly at temperatures of −10 to −20°C.

The physical process of the growth of ice crystals in the presence of water droplets was clearly set forth in Wegener's 1911 book where it is stated, in Bergeron's translation,[24] "The vapor tension will adjust itself to a value in between the saturation values over ice and over water. The effect of this must be that condensation will take place continually on the ice, whereas at the same time liquid water evaporates, and this process must go on until the liquid phase is entirely consumed." It was 22 years later that Tor Bergeron[24] put the observed facts together and promulgated his famous theory of the initiation of precipitation in mixed liquid and ice clouds.

Bergeron presented his theory in 1933 at the Fifth General Assembly of the International Union of Geodesy and Geophysics held at Lisbon. It was not until 1935 that the proceedings containing Bergeron's paper were published. Acknowledging the physical process described by A. Wegener in 1911, Bergeron discussed the existence of mixed clouds in the atmosphere in his own special English as follows:

Berson[25] Wegener[159] Douglas[54] and others have observed fogs consisting of droplets down to temperatures of −20° or even −30°C. Theoretically, i.e., under undisturbed conditions, *crystallization* will demand more special properties of the nuclei than condensation (solid particles showing the angles of the hexagonal crystal system), i.e., will not take place in lack of such nuclei. In the air there will, however, probably be a small amount of such particles, which can gradually get into action as sublimation nuclei, as the temperature falls. Moreover, disturbed droplets—and according to Kohler[89] those larger

than a certain size ("Gefriergrösse")—will crystallize easier. With falling temperature the smallest size or disturbance necessary for crystallization of already formed drops will diminish, because the crystallization forces grow. Thus to every temperature < 0°C will correspond a certain probability of crystallization resp. a certain frequency of crystals within a supercooled water cloud (though above −10°C this frequency will be very small). Thus, within an ascending supercooled water cloud by and by a few scattered elements may spontaneously solidify without any relative velocities of cloud elements at all.

Then, invoking Wegener's diffusion-growth theory, Bergeron tried a rough calculation as to how long it would take for the liquid to be consumed by the ice in the case of 4.2×10^{-6}g of liquid water per cm^3 dispersed in 1000 droplets of diameter 20 μm in the presence of one ice crystal. He calculated that the complete diffusion to the crystal would be achieved in about 10 to 20 min.

In reading Bergeron's paper one has the impression that he did not at the time realize fully the great importance of his precipitation theory, for the latter half of his presentation degenerated into an advocacy of a change in the international cloud classification. Others were aware of the significance of the concept, and among these was W. Findeisen.[57] He expanded on Bergeron's ideas and in 1938 published a clearer statement of the theory with a number of new interpretations. As a result, the Bergeron theory came to be known in many quarters as the Bergeron-Findeisen theory. (B. J. Mason in his book *The Physics of Clouds*[106] refers to the Wegener-Bergeron mechanism but speaks of the Bergeron hypothesis.) Findeisen considered at some length the manner in which ice crystals could form, favoring direct crystallization from the vapor (sublimation) in preference to freezing of droplets. He also speculated as to what might be the nuclei for this process, mentioning quartz crystals as possible nuclei. Findeisen foresaw the possibility of initiating the mechanism by introducing suitable nuclei.

In a 1939 paper[58] Findeisen categorically stated that all rain of consequence originates as snow or hail, although Bergeron had previously conceded that in the tropics "warm rains" were common. Meteorologists were skeptical of the absolute requirement of the ice-crystal process to produce rain but could not present documentation from airplane flights. By this time World War II had restricted international communication among meteorologists, but in Germany Peppler[117] presented aerological evidence. These results were published in papers appearing in 1940. Growth by collison-coalescence in "warm" clouds, especially in low latitudes or in summer, was found to occur, but the Bergeron process remained as the mainstay of precipitation mechanisms.

Collision-Coalescense of Drops

Henry G. Houghton,[77] in a paper published in 1938, summarized the problem of going from an observed distribution of sizes of condensation nuclei to a spectrum of droplet sizes in a cloud. He pointed out that the initial growth of droplets is of such a nature that a narrow spectrum of sizes is to be expected. He then showed how precipitation could occur with, as well as without, the occurence of the ice phase. Growth to raindrop size was considered possible by collision of a few drops large enough to have an appreciable fall velocity with respect to the numerous small cloud droplets. He noted that these larger drops could be formed by condensation on "giant" hygroscopic nuclei which are present in the atmosphere in relatively small but nevertheless significant numbers. He emphasized the relative ease with which precipitation could be started by the Bergeron process but gave credence to the development of rain by the collison-coalescence process. It is interesting to note that he did not speculate about the effects of the breaking of large drops suspended in an updraft to create the "chain reaction" that later became the basis for one of Langmuir's hypotheses.[98] Nor did he attempt a theoretical examination of the collection efficiency (ratio of liquid water collected to the total in the volume swept out by the falling sphere).

A lucid assessment of condensation and precipitation processes was presented by G. C. Simpson[133] in 1942. He also saw no grounds for Findeisen's rejection of the possibility of formation of "warm" rains by the collision-coalescence process. He again pointed out that the only requirements were (a) that there exist, initially, droplets in a wide range of sizes so that the coagulation process could start, and (b) that the cloud attain sufficient depth to provide adequate liquid water in the total path of the collector drops. He spoke of drops reaching breakup size only in thunderstorms.

LANGMUIR, SCHAEFER, AND VONNEGUT

Scientific weather modification had its beginnings in 1946. To most meteorologists the discoveries came out of a clear sky because they were made by persons new to the field of meteorology, although Dr. Irving Langmuir, the leader of the new developments, was known as one of America's foremost Nobel laureates. Actually he and his laboratory assistant, Vincent J. Schaefer, had been working for three years or more on problems related to cloud physics, first in the development of smoke screens, where fundamental studies of particle sizes and light scattering were made, then on precipitation static, followed by icing research.[97] In the last two research efforts they had started a series of observations in the clouds that sweep the summit of Mount Washington in New Hampshire. In speaking of this early work with Schaefer, Langmuir stated in a speech before the National Academy of Sciences in 1947[96]:

In all this work, on the problems I tell you about tonight, Mr. Schaefer and I have been associated. Vincent Schaefer is a good deal younger than I am. He came to the laboratory, working in the machine shop, and was a very skillful mechanic. He made a lot of fine apparatus for us. Later, he wanted to do research work. He worked on surface chemistry problems with us, did excellently at it, has published work on it, and has been working with me on these war jobs as they came along. He likes mountain climbing and I like mountain climbing; we like outdoor life in general.

"You will see that such things are very important, and that they are closely related to our subsequent interest in weather control. Many things we are now doing have originated from that common background. Nobody could have planned that for us. If we had not been naturally inclined that way, no one would have prevailed on us to go up on Mount Washington in the wintertime to carry on the research that occurred as a natural development.

That Langmuir was not familiar with existing literature and had not talked to knowledgeable meteorologists (he rediscovered several meteorological facts during these years) was evidenced by observations related in the same address. "We found, much to our surprise," he said, "that anything exposed on the summit of Mount Washington during the winter when clouds were there, as they are most of the time, immediately becomes covered with ice. This was due to the presence of supercooled water droplets. We had heard about rime, we had seen it, but we did not know that it occur-

red all the time at a place like Mount Washington. There are very few winter days without it. This puzzled us a lot."

At Mount Washington Langmuir and Schaefer found work in progress on icing research and the measurement of the liquid-water content of the clouds. David L. Arenberg,[8] while serving in graduate research assistant positions at Harvard and at the Massachusetts Institute of Technology, had developed a method of collecting drops of an assortment of sizes based on collection efficiency as a function of size. Following a suggestion by Albrecht,[4] Arenberg formed a set of seven cylindrical collectors of various diameters rotating on a single wooden spindle. In effect, the drops sorted themselves upon the collectors according to their kinetic energy. During the winter of 1943–1944 Langmuir and Schaefer made studies of the rate of growth of rime on the rotating cylinders. In 1945, under an Army contract, they expanded on the Mount Washington multicylinder data to investigate the trajectories of spherical particles carried by air flowing around cylinders, spheres, and ribbons. The results, representing the joint efforts of Langmuir and Dr. Katherine B. Blodgett, and published in a 1945 report to the Army[101] marked an important step forward in providing the beginning of a series of studies of collection efficiencies of raindrops by a number of authors since that time.

Reminiscing in 1947 (in the NAS talk) about the Mount Washington experience, Langmuir said:

The thing that struck us most is that, if there are any snow crystals in the cloud, they will be growing and falling. In the winter-time, if you see any stratus clouds from which no snow is falling, even though the temperatures in the clouds are below freezing, there just simply are no crystals there in any reasonable number. Such clouds consist of water droplets. They can be supercooled to very low temperatures.

It is interesting to note that up to this point Langmuir neither appeared to be aware of what had been known in meteorology for more than 25 years nor did he seem to know about Bergeron's 1933 paper. But, having observed this phenomeon, Langmuir and Schaefer *did something* about it. That is what set them apart from the rest of us and brought them fame.

Schaefer's Great Discovery

It was early summer in 1946; Langmuir reported[96]:

"Schaefer and I decided we must make some careful experiments in the laboratory to try to duplicate these conditions. I then went to California for three or four months. When I came back, Schaefer, I found, had made some beautiful experiments. He had taken a home freezing unit,* which is used for food storage, about four cubic feet capacity and lined it with black velvet, directed a beam of light into it, in order to see what happened in the chamber. He then breathed into it and the moisture condensed and formed fog particles which were just like ordinary cloud particles, although the temperature was about $-23°C$. No ice crystals formed. He tried many different substances dusted into the box to get ice crystals to form, and practically never got any. He got just enough to convince him that, if he did get them, he could see them. However, the number was totally insignificant.

Finally, one day† the temperature of the chamber was not low enough. He wanted it somewhat lower so he took a big piece of dry ice and put it in the chamber to lower the temperature. In an instant, the air was just full of ice crystals, millions of them.

He then took the dry ice out, and the ice crystals persisted for a while. Then he found that even the tiniest piece of dry ice would fill the cloud with crystals. Then he took a needle which he dipped into liquid air and he let that pass once across the box. The result again was the production of hundreds of millions of ice crystals by just one brief contact with the needle. The effect rapidly spread through the entire box. It is a wonderful experiment . . . . The effects to be seen are wonderful to look at and it is a simple matter to duplicate all the natural conditions of an actual cloud in the sky.

Well, that discovery changed the whole situation. What was discovered first was that the dry ice had no direct effect on the super-cooled cloud but rather its temperature was the important thing. The needle, instead of being cooled in liquid air, could be cooled with dry ice. Instead of cooling it with dry ice, anything can be used having a temperature less than $-40°C$.

Finally further study showed that there is a quite critical temperature of about $-39°C$, the temperature of freezing of mercury, where a spontaneous reaction occurs to produce natural nuclei. This can be proved quite conclusively by covering the box with a sheet of glass and continuing to supply moisture into a cold chamber having a temperature of $-40°C$. The ice crystals continue to separate out, fall on the bottom of the chamber, and the new crystals form from moisture produced by evaporation of the ice.

If, on the other hand, you take the box and cool it to $-20°C$ and let air in from the laboratory, sometimes you find a few crystals, but if a cover is applied, they grow and snow out and there will be no more, unless you stir in some new air.

Schaefer made replicas of the snow crystals in Formvar softened in ethelene dichloride, a technique that he had developed five years earlier and had applied to his hobby-interest in studying the beauties and remarkable symmetries of snowflakes. The crystals that formed in his freezing box were like those of the natural atmosphere.

---

* Units of this kind were just appearing on the market. In the postwar swords-to-plowshares economy they were in great demand but in extremely short supply. Newspapers contained allegations that "influence peddlers" in Washington were making gifts of them (and mink coats) to officials in the high circles of government.

† An unusually hot day, July 12, 1946, according to Schaefer in "The early history of weather modification," an address before the American Association for the Advancement of Science, in New York, December 30, 1967.

Schaefer submitted a scientific article on his experiments to *Science,* where it was published in the November 15, 1946 issue.[125] At the conclusion of the article he wrote:

It is planned to attempt in the near future a large-scale conversion of supercooled clouds in the atmosphere to ice crystal clouds, by scattering small fragments of dry ice into the cloud from a plane. It is believed that such an operation is practical and economically feasible and that extensive cloud systems can be modified in this way.

That last statement was more than timely, for on November 13, 1946, two days before the appearance of that issue of *Science*, Schaefer made his historic flight, representing man's first scientific seeding of a supercooled cloud. From the cockpit of a small airplane piloted by Curtis Talbot, he scattered 3 lb of dry ice along a line about 3 mi long over a cloud above Mt. Greylock east of Schenectady, New York. The cloud was of a stratified character at 14,000 ft where the temperature was $-20°C$. Within about 5 min the whole cloud appeared to have been turned into snow, which fell about 2000 ft below the cloud before evaporating.[127]

The *General Electric Review* for November, 1952[6, 75] quotes the following from Schaefer's laboratory notebook: "While still in the cloud as we saw the glinting crystals all over, I turned to Curt and we shook hands and I said, 'We did it.' "

Vonnegut's Silver Iodide Experiments

During these fall months of 1946 Dr. Bernard Vonnegut, who had worked on nucleation problems in connection with aircraft icing research at the Massachusetts Institute of Technology, was busy in the General Electric laboratories in Schenectady working on a variety of nucleation processes, but after Schaefer's laboratory experiment, turned his attention again to ice nucleation. Aware of the possibility of nucleation of crystal growth on substances of matching crystalline structure, he looked in the handbook of X-ray crystallography for substances that were close to the ice crystal in lattice structure and that were insoluble in water. Silver iodide and lead iodide seemed to fulfill the requirements best. He scattered some silver iodide powder into the supercooled fog of Schaefer's freezer box and was disappointed in finding no results. Powdered lead iodide produced a very slight effect in starting the formation of ice crystals. Schaefer was away on a trip, and when he came back he was greatly interested in what Vonnegut was doing. Schaefer tried iodoform, which produced the desired effect in a rather unsatisfactory way, and also found that iodine itself gave a slight tendency toward icing.

A few weeks later Vonnegut began trying the introduction of metallic aerosol particles formed by creating sparks between various electrodes. Then he took a silver coin from his pocket and sent sparks to it. The results were spectacular. The supercooled cloud in the chamber was transformed to ice crystals as readily as it had been with dry ice. The effect did not last long, and in the ensuing days Vonnegut was puzzled to find that the results could not be repeated. It was finally realized that there had been some traces of iodine in the box, and the sparking to the silver had created pure silver iodide. He then found that the silver iodide he had taken from a reagent bottle on the laboratory shelf for the first experiment was of very low purity, consisting of about 50 percent sodium nitrate. With a more nearly pure reagent of silver iodide, he found that the powder did the trick.

In a private communication to this author, Dr. Vonnegut writes:

. . . now that I look at my notebook . . . I find that my finding that a smoke of silver iodide is indeed a good nucleus effective at temperatures only a few degrees below freezing was made on November 14th, the day after Vince's historic flight.

In a recent conversation with this writer, Vonnegut stated that at the time he had no thought of its being used on any scale outside the laboratory. He said he was not looking for another way to seed clouds. But the application of silver iodide to cloud modification was soon on its way,[153, 154] and Vonnegut played a leading role in the development of means for generating silver iodide smokes.[155, 156] Thus was developed man's first successful attempt at artificial nucleation of supercooled clouds. Dry ice could make crystals of ice form by lowering the temperature to that necessary for nucleation on whatever might be available as nuclei, or even in the absence of any nuclei, while silver iodide provided a nucleus that was much more efficient than those occurring naturally.[99] It was subsequently found that in the presence of silver iodide particles, ice was nucleated at a threshold temperature as high as $-4°C$—depending on the degree of supersaturation—as compared with observed temperature thresholds for natural freezing or crystallization nuclei of -15 to $-20°C$. (Fourteen years later Motoi Kumai[91, 92] determined that the predominant natural nuclei for snow crystals were clay-mineral particles, mostly of the kaolin group, active at about $-15°C$ or lower.)

Spreading the Gospel

Immediately following Schaefer's November 13, 1946 flight, newspapers reported the results, and during the ensuing months and on into following years glowing reports in the popular press quoted Langmuir in making extremely optimistic predictions of the benefits to be derived from "weather control." At the annual meeting of the American Meteorological Society in December, 1946 in Boston, Langmuir reported on cloud seeding activities of November and December.[128] Although the talk was not published, he was reported to have spoken in general terms about changing the arid Southwest into farmland, asserted that the December 20–21 snowstorm in New York State and northern New England had been largely triggered by seeding which was performed on December 20, and implied that although the U.S. Weather Bureau had predicted the snow, its characteristics were derived from the effects of seeding. He touched on a theme that he repeatedly used, even in scientifically published reports, stating that forecasting can never be made with great accuracy because of "triggering" reactions that can be produced artificially. In his well-known 1948 paper in the *Journal of Meteorology*[98] he stated, " . . . it becomes apparent that important changes in the whole weather map can be brought about by events which are not at present being considered by meteorologists." Published statements of this type, and even more challenging informal comments, started years of heated arguments with members of the meteorological profession. The Weather Bureau, which even then was in the midst of a difficult task of assuring the public that atomic explosions were not changing and could not change large-scale weather, had the responsibility of guiding public policy in such matters, and was forced into the unpleasant position of trying to restrain Langmuir who, because of his high standing in the scientific community, had strong support from scientists and the general public alike. Until his death in 1957 Langmuir was not restrained in his enthusiasm for tremendous effects of weather control.

PROJECT CIRRUS

The General Electric Company supported their researchers in the studies in 1946, even though its business was not in the field of meteorology. The obvious importance of the discoveries was sufficient justification for fostering their development. However, it was realized that since the atmosphere, with its largess and its vagaries, was the concern of every citizen of the world, modification of its processes was more properly a field for government research. Accordingly, a show of interest on the part of the U.S. Army Signal Corps was welcomed, and a contract for support from that agency was negotiated in February of 1947. Later the Office of Naval Research participated in the support and the U.S. Air Force furnished flight support. Finally the U.S. Weather Bureau participated in a consultative capacity. The whole activity, with field experiments being conducted by the government agencies and with laboratory work and general advice by General Electric, was designated as *Project Cirrus*.

The most pronounced effect produced by Project Cirrus and subsequently substantiated by a number of tests by others, was the clearing of paths through supercooled stratus cloud layers by means of seeding from an airplane with dry ice or with silver iodide. When such clouds were not too thick, the snow that was artificially nucleated swept all the visible particles out of the cloud. It was shown that eddy diffusion spread the effect from the narrow wake of a single airplane to a path more than 1 mi wide. In one of the first flights of Project Cirrus, on April 7, 1947, the supercooled particles in stratus clouds were removed using only 12 lb of

dry ice distributed along a 14-mi line.[97] In later flights, even more spectacular results were achieved, documented by good photography.

Although Project Cirrus made some studies of cumulus clouds in the Schenectady area during the summer of 1947, it was not until the following year in New Mexico that important cumulus-cloud seeding experiments were conducted. The most significant event of 1947 was an attempt to modify a hurricane by seeding. On October 13 a flight was made into the vicinity of a hurricane located about 350 mi east of Jacksonville, Florida. One of the spiral rain bands, called a "squall line" in the report, was seeded with 80 lb of dry ice dispensed along a 100-mi track. Project Cirrus flight personnel reported from visual observations that there was a pronounced modification of the cloud deck after seeding. Shortly after the seeding, the hurricane, which had been traveling northeastward, changed its course and headed almost straight westward to strike the coast of South Carolina and Georgia. There was much speculation concerning the possibility that the seeding was responsible for changing the path of the hurricane; however, a somewhat similarly erratic path was noted in the records of a hurricane 41 years earlier. This attempt was the first to modify a hurricane. (See Chapter 14.) Because of the frightening possibilities of legal responsibility for damages caused over the land, the erratic behavior of this hurricane taught the lesson that hurricanes having the possibility of reaching land should not be the subjects of experimentation. Langmuir discussed this hurricane seeding in his address 1 month later before the National Academy of Sciences.[96] Meteorologists were nettled by his statements concerning the uniqueness of the hurricane rain band. "The squall line is not a formal feature of a hurricane," he said. "When it started, when it ended, what happened to it, nobody knows. The Weather Bureau has no record whatever." Apparently Langmuir was unaware that spiral rain bands or squall lines, as he called them, were known as identifying features of hurricanes. In a paper presented on December 3, 1946, and published in abstract form, with photograph, Harry Wexler[164] had shown the radar portrayal of such bands in a hurricane. A more complete discussion was contained in his article published by the New York Academy of Sciences in September 1947, but presented before that Society several months earlier.[163] Langmuir had rediscovered the hurricane rain band which radar meteorologists had seen during the war and talked about publicly as soon as security was lifted. The first published radar photographs showing the characteristic bands were in an article by R. H. Maynard in the December, 1945 *Journal of Meteorology*.[108]

Langmuir the Indomitable

Although General Electric Company's legal counsel strongly admonished Langmuir not to relate the erratic course of the hurricane to the seeding,[119] he nevertheless mentioned it in informal discussions. His talk ended with the statement: "The stakes are large and, with increased knowledge, I think we should be able to abolish the evil effects of these hurricanes." Statements that were more positive in their optimism and giving promise of great benefit to mankind were made by Langmuir throughout the remainder of his career, appearing mainly in newspaper and magazine interviews. In Chapter 3 of the Project Cirrus Final Report,[97] Langmuir wrote: "The chance that a given hurricane between latitudes 20 and 40°N will change its direction in any given six-hour period is

only about one in 110. Therefore, the fact that the 1947 hurricane *did* change its direction within six hours after seeding has a significance factor of the order of 100, so that there is considerable evidence that seeding hurricanes does tend to change their behavior." A page or two later, he referred to criticisms by Dr. F. W. Reichelderfer, Chief of the Weather Bureau, and wrote: "I pointed out to him, however, that the larger the storm, and the more energy that is stored in it, the easier it should be at the proper stage in its development to get widespread effects. To assume that a hurricane could not be successfully modified by even a single pellet of dry ice is like assuming that a very large forest could not be set on fire by such a small thing as a single match."

The U.S. Weather Bureau was beset by interests eagerly expecting the Government to push ahead with massive weather control efforts. The Bureau and many meteorologists thus were placed in an uncomfortable defensive position, while many other scientists, impressed by Langmuir's prestige and convincing arguments, were ranged against them.

During that summer of 1947 I had two long visits with Langmuir. First Langmuir came to visit the Thunderstorm Project being conducted in Wilmington, Ohio. He was with Dr. Schaefer, and the two were on their way home after visiting forest areas in Montana at the request of the U.S. Forest Service. Dr. Langmuir described some careful visual observations he had made from the ground that convinced him that glaciation (ice formation) in cumulus clouds occurred only after they had reached a height where the temperature was around −40°C. It was difficult to argue without a reference library at hand or without the completed analyses of the Thunderstorm Project flights. It is interesting to note that aircraft measurements documenting the temperature range of supercooling in various convective clouds in various regions of the United States were not available until 10 or 20 years later, notably in the University of Chicago flights.[30, 31]

The second meeting that summer was in Dr. Langmuir's study at his home in Schenectady. We were warmly received and listened to his accounts of the effectiveness of cloud seeding. The most vivid impression I came away with was of the stack of copies of the newspaper *The Arizona Republic,* and some other newspapers, piled on the floor and on an empty chair, which carried headline front-page accounts of his recent visit to the Southwest. These articles quoted Dr. Langmuir in extremely glowing statements about rain enhancement by seeding. Langmuir was proud of this coverage of his Southwest visit and pointed to these newspapers during the conversation.

On another occasion in New York 2 or 3 years later a lunch at the old Astor Hotel was arranged with Langmuir and Carl-Gustaf Rossby, one of the greatest and most articulate meteorologists of our time. Rossby gave a lucid discussion of how the atmosphere works, emphasizing the scales of atmospheric phenomena. Langmuir expressed interest and an understanding of some of the problems of fluid dynamics, but when it came to discussing the atmosphere generally, I had the impression that Langmuir was hardly listening. The discussion did not change his views one iota, as later comments showed.

These personal reminiscences are brought up at this point in order to make understandable the great controversies that raged between Langmuir and the Weather Bureau, and to some extent the meteorological profession as a whole. Langmuir made some great contributions to weather modification, and perhaps if he had comprehended fully the magnitude

of atmospheric phenomena he might have been discouraged from the start.

THE WEATHER BUREAU CLOUD PHYSICS PROJECT

With the tempo of publicity mounting and claims of spectacular results of cloud seeding emanating from Project Cirrus, the U.S. Weather Bureau in August of 1947 made the decision to carry out a project of its own to test the efficacy of seeding. It was felt that if the results were really spectacular, no elaborate design of the tests would be necessary to prove statistical significance. One criterion decided upon was that if at approximately the same time precipitation was occurring naturally from nearby unseeded clouds, those seeded might be considered to have been able to generate precipitation if left to themselves. In other words, the results would be considered as positive only if seeded clouds were unique in producing precipitation in the area. The project was in operation for nearly 2 years, and included flight operations in Ohio, California, and the Gulf States.[41-45]

Substantiating the Project Cirrus findings in one aspect, namely, the striking visual changes occurring in the seeded portions of clouds, the Weather Bureau Cloud Physics Project nevertheless found very little to justify optimism about spectacular precipitation effects. They found that supercooled stratus cloud layers could be cleared along wide swaths by dry ice seeding but noted that the effect was often occurring at the same time that a general dissipation of the cloud layer was beginning. With regard to cumulus clouds in Ohio, the report concluded: " . . . the artificial modification of cumuliform clouds is of doubtful economic importance for the production of rain. Dissipation rather than new development was the general rule. There is no indication that seeding will initiate self-propagating storms, and therefore the only precipitation that can be extracted from a cloud is that contained within the cloud itself. The methods are certainly not promising for the relief of drought."

In clouds over the Gulf States it was noted that ice crystals usually formed naturally just a few thousand feet above the freezing level, often when the temperature at the tops of the clouds was not below $-6°C$.

These results had virtually no effect in moderating the enthusiasm for rain-making by cloud seeding, except in the hard core of the meteorological profession.* Commercial rainmakers were now in full swing in many parts of the United States and abroad. As the use of silver iodide from ground generators obviated the necessity for airplane delivery of the seeding agents and, of course, at much lower cost, these endeavors were becoming quite profitable. Many of the commercial operations were poor imitations of Langmuir's and Schaefer's experiments. Some of the more reliable commercial operators continue in business to this day, and have done much to push forward the frontiers of knowledge of weather modification (see Chapter 2).

HAWAII AND HONDURAS

In October 1947 Dr. Luna B. Leopold and Maurice H. Halstead[103] of the Pineapple Research Institute in Hawaii reported to Dr. Langmuir a seeding flight in which they dropped dry ice in a cumulus cloud that was everywhere warmer than freezing. Subsequently, rain

* Langmuir criticized the Weather Bureau Project on several grounds in "Widespread control of weather by silver iodide seeding," *GE-RL-1263*, July 1955; and also in *Collected Works*, Vol. 11.

fell from it. Langmuir surmised that the cloud actually was seeded by water which coated the dry-ice pellets, and he then developed his chain-reaction theory of the development of precipitation by collision-coalescence of drops as had been postulated by others, for example, H. G. Houghton[77] and G. C. Simpson[133] several years earlier. Langmuir, however, introduced results on collection efficiencies as they had been calculated by him and Dr. Katharine Blodgett in connection with the Mount Washington studies.[101] He also considered the consequences of having the drops grow to such a size that they would break up, the smaller drops thus formed being carried upward again to repeat the cycle. This he called the chain reaction, analyzed in his well-known 1948 paper in the *Journal of Meteorology*.[98] He now suggested that warm clouds could be seeded with water drops to produce rain. Thus was started the possibility of enhancing the production of rain in warm clouds by introducing water drops near their tops.[28] The use of giant hygroscopic nuclei for the same purpose was suggested later, to start the process in a manner originally deemed possible in natural clouds in Houghton's 1938 treatment.[77]

In 1948 and 1949 Langmuir went to Honduras to study tropical clouds and familiarize himself with work being done by Joe Silverthorne, a commercial cloud seeder, working for the United Fruit Company. Silverthorne was testing the possibility of controlling rainfall, and particularly he was trying to stop the blow-downs that result from winds associated with thunderstorms, destroying on occasion large stands of fruit trees. Langmuir reported in detail on two clouds seeded on a day he flew with Silverthorne.[97] He claimed credit for causing these clouds to dissipate by dropping, in one case, a single pellet of dry ice and, in the other, two pellets.

In a 1950 paper in *Science*[102] Langmuir suggested two different procedures for treating cumulus clouds, one to cause heavy rain and self-propagating storms, and another to dissipate the cloud. For rain enhancement he advocated using a single pellet of dry ice, postulating that although the falling pellet would produce a concentration of nucleating ice along its path, the cloud as a whole would have parts that were underseeded. "Under these conditions," Langmuir stated, "large snowflakes will form at all altitudes above the freezing level, and this is more apt to produce self-propagating storms and may lead to heavy rain." To dissipate a cloud, he advocated dropping relatively large amounts of dry ice into the cloud to cause overseeding. In this case so many nuclei are produced that the snowflakes that form may even outnumber the original droplets, and they tend not to fall out of the cloud. The heat of fusion generated by the conversion to ice crystals makes the top float off, according to Langmuir, so that it separates from the lower part and the cloud dissipates. Thus the idea of overseeding was another new concept introduced by Langmuir which even today puzzles those engaged in cloud seeding because it is difficult to assess quantitatively.* He maintained that one or two pellets of dry ice introduced just above the freezing level are enough "to transform a cloud into an efficient rain producer."

In this *Science* article, based on a paper presented before the American Meteorological Society on January 25, 1950 in New York, Langmuir figuratively threw the gauntlet before the Weather Bureau when he said: "The control of a

* In a 1955 paper Roscoe R. Braham and John R. Sievers[33] presented data on aircraft penetrations of cumulus clouds after seeding with dry ice at a rate of 9 to 26 lb/mi showing that supercooled droplets were present, thus with no evidence of overseeding in 21 of 24 dry ice seeded clouds.

system of cumulus clouds requires knowlecge, skill, and experience. Failure to consider the importance of the type of seeding, the place, and the time, and also the failure to select the best available clouds, explain why the Cloud Physics Project of the U.S. Weather Bureau was not able to obtain 'rainfall of economic importance'. "

MORE ON PROJECT CIRRUS

The next area of operations for Dr. Langmuir and Project Cirrus was in New Mexico. Langmuir wanted to work in an area where storms "originate" rather than in a region such as upstate New York, which is "traversed" by storms. Also, his colleague Schaefer knew of the activities of Dr. E. J. Workman of the New Mexico Institute of Mining and Technology in studying thunderstorms, observing them by radar, and performing experiments in the laboratory on the effects of ice in thunderstorm electrification.

New Mexico Cumulus

In October of 1948, members of the project spent 3 days in Albuquerque, quickly made arrangements with Dr. Workman and his group for radar tracking and photography, and made two cumulus-cloud seeding flights. The second one, on October 14, produced one of Langmuir's classical examples of phenomenal results and the starting of self-propagating storms by dry ice seeding of cumulus. He reported that rainfall was produced over an area of more than 40,000 mi^2 as a result of the seeding—about a quarter of the area of the state of New Mexico. He declared: "The odds in favor of this conclusion as compared to the assumption that the rain was due to natural causes are many millions to one."[97]

Impressed by the results of this dry ice seeding, the Project Cirrus group returned to New Mexico the following summer, conducting 10 flights between July 13 and July 22, 1949. This time, according to the *History of Project Cirrus*, compiled by Barrington S. Havens,[75] Langmuir became aware of the fact that Dr. Vonnegut was dispensing silver iodide smoke from a ground generator in the area. Langmuir thought that this contribution to the general seeding might explain some of the discrepancies he had observed. Furthermore, he reconsidered the seeding activities in New Mexico the preceding year when Vonnegut was also operating his generator. He felt that it was reasonable to conclude that the similar widespread effects produced in October, 1948 were the result of the silver iodide seeding which was done at that time, rather than of the dry ice seeding, which had been the previous interpretation.

Again Langmuir reported spectacular results[97] and, goaded by previous challenges to his 1948 interpretations, developed his "new probability theory" to distinguish the natural from the artificially induced storms. His statistical methods and conclusions were challenged, especially by the U.S. Weather Bureau. For example, after a paper on the subject in *Science*,[102] F. Hall et al.,[71] in a communication to the editor, wrote that Langmuir mistakenly assumed that the rainfall values for adjacent subareas could be treated as independent random numbers. The distribution of rainfall is never purely random; adjacent areas are not independent of one another. Hall also presented data tending to nullify the assumed significance of the similarity of rainfall patterns on two seeded days. Langmuir was undaunted and proceeded to develop his probability

theory further, arriving at apparent improvements. But he was to come out with even more astonishing correlations between rainfall and seeding with silver iodide.

In analyzing the rainfall associated with the 1949 seedings, Langmuir noted that Dr. Vonnegut only operated the ground silver iodide generator on certain days. Langmuir detected a response in the form of rainfall occurrences related to the times the generator was in operation. This led him to his controversial experiment in periodic seeding, to which he devoted the rest of his career.

Periodic Seeding and Periodic Weather

In the periodic experiment, the silver iodide generator was operated by Project Cirrus in New Mexico so as to introduce a 7-day periodicity in the behavior of the various weather elements. The periodic seedings were begun in December, 1949, and already in the early part of 1950 unusual weekly periodicities in the weather were observed over a large part of the country. Langmuir was led to conclude that the effects of artificial nucleation were more widespread than just local cloud modification and extended to other elements such as temperature and pressure for thousands of miles away.

Langmuir presented his first results on periodic seeding and weather at the October 12, 1950 meeting of the National Academy of Sciences, published a few months later in the *Bulletin of the American Meteorological Society*.[100] It was apparent from the data that the 7 day periodicities in April, 1950 were unusually striking, not only in rainfall but also in such rainfall-related elements as 850-mb temperatures at Chicago, Omaha, Columbia, Missouri, Oklahoma City, Buffalo, St. Cloud, Minnesota,

Nashville, Washington, D.C., and Charleston, South Carolina.

Meteorologists pointed out, however, that such periodicities are not uncommon, although these could, indeed, be the most striking they had seen. But it was argued that a thorough examination of past records ought to be made. Langmuir's statistical methods also were criticized.[70]

The periodic seeding was discontinued in July, 1951 during great floods in Kansas and adjacent states. These floods, characterized by excessive rains in both June and July, were declared by Verne Alexander, North Central Area Hydrologic Engineer of the U.S. Weather Bureau, as the most devastating in the nation's history.[5, 7] They occurred in northern and eastern Kansas, somewhat astride the Kansas River which joins the Missouri at Kansas City. The devastation to the industrial area in Kansas City on both the Kansas and Missouri sides was estimated at a billion dollars in physical damage and an equal loss caused by interruption of business and production. This was a good time to stop the periodic seeding. Another reason for discontinuance was the background of silver iodide increasingly being generated by commercial operators, particularly in California.

The Weather Bureau started immediately to see if similar periodicities in the weather had occurred naturally in the past. A study was completed and a report written in 1952, but it did not appear in published form until 1955.[35] Glenn W. Brier, author of the article, stated in a footnote simply, " . . . the original plans for its publication did not materialize." Using records from 1899 to 1951 Brier showed that a 7-day component in the harmonic analysis of the data appeared with some frequency, although seldom as marked as in Langmuir's periodic seeding months. Details of Brier's analysis cannot be given here, but the evidence appears just as good for the occurrence of a natural periodicity as for an artificially controlled one.

In two earlier papers, Hawkins[76] and Brier[36] discussed a remarkable 7-day periodicity in May, 1952, several months after the periodic seeding stopped. Brier's harmonic analysis showed on the 700-mb chart for the Northern Hemisphere, high correlations in Asia and Iceland during a 28-day period. Similar hemispheric patterns were shown in the 1955 paper.

End of Project Cirrus and Langmuir's Final Words

Part II of the Final Report of Project Cirrus appeared in May, 1953. Entitled "Analysis of the Effects of Periodic Seeding of the Atmosphere with Silver Iodide," it covered 340 pages in multicopy format. In regular printed form in Volume 11 of Langmuir's *Collected Works*, it was 276 pages long. The first three chapters briefly describe the discoveries and activities through 1949, and the remaining nine chapters are devoted to the periodic seeding in New Mexico and its effects. Although this report would appear to have been his great work on the subject, Langmuir did not stop there, but continued to work on analytical aspects of the problem after the close of Project Cirrus. His final paper on the subject was a GE Research Laboratory Report (No. 55-RL-1236) entitled "Widespread Control of Weather by Silver Iodide Seeding," issued in July, 1955. In an appendix he wrote a critique on some ongoing experiments on seeding and statistical methods for evaluation.

Langmuir wrote one more paper before his death in 1957 at the age of 76. The paper was called "Freedom—The Opportunity to Profit from the Unexpected." This was a fitting philosophical close to the career of the man who, while in his sixties, climbed Mount Washington with Vincent Schaefer in the middle of winter and from that experience developed with his colleagues a new vista of the mysterious atmosphere.

Before closing this account of Project Cirrus, mention should be made of a somewhat peripheral activity concerning the possibility of suppressing lightning. From 1947 to the close of Project Cirrus, Vincent Schaefer made regular visits to U.S. Forest Service installations in the northern Rockies in response to speculations that lightning could be suppressed by cloud seeding.[126] Project Skyfire, aimed at lightning suppression, was started. In the formative period some of Vonnegut's early developments of silver iodide smoke generators were tried.[154-156] The project was carried on through the 1960s with some reports of success, although a number of uncertainties remained.[11, 12, 62]

Project Cirrus in Retrospect

The two most important discoveries for the development of cloud-physics and weather modification were made in the General Electric Research Laboratory before Project Cirrus was organized. Schaefer had shown that a tiny pellet of dry ice or a similar very cold object could transform a large volume of supercooled cloud into ice crystals. Vonnegut found an artificial ice nucleus which had a much higher threshold temperature for nucleation than any naturally occurring substances. In either dry ice or silver iodide treatment it was found that once the initial ice crystals are formed, glaciation spreads through large cloud volumes. The effect of dry ice on natural clouds had already been tested out by the end of 1946. The marked effect in clearing stratus decks was demonstrated shortly after Project Cirrus got under way. But beyond that the effects of seeding became more a program of advocacy than of objective proof. Although some spectacular visible

developments were seen after seeding. Project Cirrus failed to demonstrate that seeding of cumulus clouds increases rainfall, that seeding initiates self-propagating storms, that the atmosphere responds periodically to periodic seeding, or that a hurricane could be deflected in its path by seeding. This is not to say that meteorologists and others do not believe any of these things are possible; reasonable proofs of *some* of the effects appear now to have been achieved. The solid proofs of effects were not obtained by the project.

Much of the difficulty came from Langmuir's failure to use accepted statistical methods of verification.[104] He developed his own statistics in which experienced statisticians found weaknesses. It was also noted that in other aspects he worked out his own analytical tools. For example, in a rather complicated way, he arrived at an expression for the condensation adiabatic rate in ascent of moist air, apparently without knowledge of the development of this theory beginning in the last century.

Langmuir developed a reasonable theory of growth of raindrops by coalescence in warm clouds, using data of collision efficiency that he and Dr. Katharine B. Blodgett[101] had analyzed theoretically some few years earlier.

This writer has heard colleagues say that the overly enthusiastic advocacy impeded the orderly progress of the science, but it probably can also be said that without this pushing, governments would not have put their resources behind cloud-physics research the way they did.

COMMERCIAL SEEDING

News media were advised immediately of Schaefer's first cloud seeding success in 1946. Public interest in the following weeks and months rivaled that engendered by the announcement of the discovery of nuclear fission 15 months earlier. Langmuir and Schaefer were in great demand to speak before and consult with groups of water users, farmers and ranchers, city officials, Federal Government program directors, and scientific societies. Nearly all these contacts, carrying with them the hope of spectacular achievement in weather control, resulted in some new cloud seeding effort, by either commercial operators, industrial organizations, water districts, or groups of farmers.[37] As might be expected, there were many unscrupulous or incompetent "rainmakers" whose activities flourished for a short time.[69] Some of the commercial operators are still in business today, apparently having proved their competence through the years.*

The development of generators of silver iodide smoke which could be set up on the ground without the great cost of aircraft flights permitted the operation of rain-enhancement programs with great ease. It was not until several years later that investigations were made to find out if the air currents actually carried smokes from ground generators into the subfreezing portions of clouds.[34, 47] As might be expected, it was found that in many cases, especially in mountain regions, the ground-generated materials were carried high enough; in other cases very little vertical transport occurred.

The development of commercial weather-modification endeavors is covered in Chapter 2. Some of these activities were noteworthy from a historical point of view.

In 1949–1950 the City of New York found its water supply becoming alarmingly low. Ten years earlier the city had realized that the water system had to be expanded to take care of unusual

* For various assessments on the state of the art through the years since 1946 the reader is referred to References 9, 59, 61, and 107.

drought situations, but with World War II limitations on big projects, it was decided to take the risk and let the new development come at a more propitious time from the viewpoint of construction of public works. It was a bad risk, for the long-continued drought came sooner than might be expected. Hearing of Langmuir's work, the city decided to try augmenting the water supply by cloud seeding. On Langmuir's recommendation, they engaged Dr. Wallace E. Howell, who had worked with Langmuir at the Mount Washington Observatory on icing research.[79] The historical interest in Howell's activities on the New York watershed stems mainly from the fact that the entire citizenry of the city, the world capital of news media, became involved in discussion of them. The project was also of interest because it was the first case in which legal action was taken against cloud seeding by persons whose businesses (resorts, etc.) slumped during rains. Although rains came, and the reservoirs eventually were adequately filled, Wallace could not claim that his activities were responsible for the end of the drought.[114]

Another commercial activity that drew a great deal of attention was the Santa Barbara project in California. It was likewise aimed at increasing the water supply which, in this case, was derived from rains and snows in the mountains north and northeast of the city. The work is discussed in detail in Chapter 2 by Robert G. Elliott, who was in charge of the undertaking. The unique feature of it was that it was one of the principal cloud seeding programs evaluated by the California State Water Resources Board, which organization assigned the statistical evaluation to the Statistical Laboratory at the University of California in Berkeley.[40] Under the noted statistician Dr. Jerzey Neyman[116] a number of statistical tests were made. The project was also unique in that it is probably the only cloud seeding contract in which the clients permitted randomization, that is, the seeding of only some of the storms, randomly selected.

In order to obtain a complete perspective of commercial activities, it is necessary for the reader to refer to Chapter 2. Comparing the views expressed there with those of other authors in this volume, one must have the impression that the academic and commercial workers in this area have different perceptions of the development and present status of weather modification. It is particularly noteworthy that those in the commercial sector place considerable emphasis on the hearings before the U.S. Congress on the subject.

THE AUSTRALIAN EXPERIMENTS

Australia has extensive deserts, and even in the more favored places where crops are grown the drought risks are high. Furthermore, her mountains are not high enough or extensive enough to produce high rates of discharge in the rivers that come down to the plains or to store water in a mountain snowpack. The most useful mountain range, the Snowy Mountains, slopes abruptly to the moist southeast coast, while a fertile plain extends widely on the drier west side. The Murray River and its tributaries help to water the plain, but it is neither a Nile nor a San Joaquin. Diversion of water to this river system and at the same time the generation of electrical energy was the task of the Snowy Mountains Hydro-Electric Authority (SMHEA), but more water was needed to ensure a continuously optimum yield of irrigation water and hydroelectric energy.

The Commonwealth Scientific and Industrial Research Organization (CSIRO) had the resources and manpower to which the SMHEA could turn for help. In fact, Australian scientists were already deeply interested in the possibilities of rain enhancement by seeding. Less than three months after Schaefer's first dry ice seeding of a natural cloud, E. B. Kraus and P. Squires[90] of the Australian Weather Bureau seeded a stratocumulus cloud, on February 5, 1947, near Sydney while the effect was being observed by radar.

Under the direction of the noted radio astronomer, Dr. E. G. Bowen,[26, 27] the CSIRO organized what could probably be called the world's most outstanding scientific group dealing with cloud physics and weather modification. It is probably correct to say that no other organization contributed more to practical cloud physics during the period approximately from 1950 to 1965.

In 1949 two papers by E. J. Smith[134] and by P. Squires and Smith[142] described showers produced in Australia in 15 of 20 cumulus clouds treated with dry ice. The precipitation was monitored by radar. Other clouds within 25 miles of the seeded ones remained unaffected. These and other experiments were reviewed by Bowen in a 1952 paper.[26] Although seeding operations have been carried out in various places in Australia, the Snowy Mountains project has attracted the most attention.[2] The objective was to produce a significant increase in precipitation over the mountains by silver iodide seeding. The experiment was conducted for 5 years, from 1955 to 1959 inclusive, concentrating on the colder part of the Southern Hemisphere year, from March 15 to December 1. The silver iodide was dispensed from an airplane flying along a line upwind from the target area. The target area of 35 mi² was in the center of the estimated seeded zone totaling 1100 mi.² A control area of 750 mi² was separated from the seeded area by a neutral zone approximately 10 mi wide. The cloud systems were associated with moving cyclones interspersed with ridges of high pressure so that there was usually a lapse of a few days between each seedable system. Operations were divided into seeded and unseeded periods each of not less than 8 days' duration, with the division being made on the basis of a set of random numbers. The controller who decided on the beginning and ending of each activity related to a cloud system was unaware of whether he was dealing with a seeded situation or not.

First reports of the experiment indicated that it was a success in the sense that the precipitation over the target area appeared to have been increased by the seeding, but the final report, published in 1963, was less positive. The authors, E. J. Smith, E. E., Adderley, and D. T. Walsh[136] concluded: "The marginal significance of the difference between seeded and unseeded period precipitations, together with the subsidiary analyses, lead to an overall result of the experiment which, while encouraging, is inconclusive."

At the same time, Smith, Adderley, and F. D. Bethwaite[137] reported on a cloud seeding experiment in South Australia, near Adelaide, conducted during the winter months of 1957, 1958, and 1959. It was found that much of the rain was caused by the coalescence process and therefore the silver iodide technique used could not have been helpful. Two separate areas were treated during alternate periods, for a total of 40 seeded periods in each. The conclusion was that there was no evidence that cloud seeding influenced the mean precipitation.

This same type of target-control crossover design was used in two other areas, New England and the Warragamba Catchment, again with results of doubtful significance.[135] Undaunted, the Australians have continued up to the present time with cloud seeding experiments, apparently with improved designs and better knowledge of the natural processes involved in each area.[139]

An interesting side issue arising from the Australian research which set the world community of atmospheric and planetary scientists astir was Bowen's hypothesis that meteoritic dust served to nucleate clouds and stimulate rain at periods related to meteor showers.[29] In examining many years of rainfall records in various parts of the world he noted a tendency for episodes of heavy rain to occur on certain calendar dates, that is, in certain parts of the earth's orbit. He deduced that the rain maxima followed by about 30 days the earth's passage into meteoritic clouds or meteor showers. It was reasoned that 30 days was about the time required for the smaller particles to settle through the upper atmosphere and downward into the tops of tropospheric clouds where the nucleating action would occur. It was concluded that meteoritic dust particles are active in the atmosphere as ice nuclei.

Against the onslaughts of meteorologists, planetary physicists, and statisticians, the Bowen hypothesis did not stand up well. Little is heard of the idea today, although it was a startling theme when it was introduced, and it occupied much time and energy in efforts to substantiate or disprove it.

THE PAINFUL SEARCH FOR THE "TRUTH"

The 10-year period from about 1953 to 1963 was one of more elaborate testing of seeding effects than had previously been tried. Several prominent statisticians had begun to show an interest in the subject in 1951–1952, and for the first time some attention was given to a rational experimental design.[39] In the United States the Weather Bureau, the Army, the Navy, and the Air Force started various phases of the work. Together, these agencies in 1951 appointed an Advisory Group under the chairmanship of Dr. Sverre Petterssen, Professor of Meteorology at the University of Chicago.

The United States
1953–1954 Experiments

Under the advice and guidance of this committee, six projects were started as follows: (1) seeding of extratropical cyclones, carried out by New York University under the sponsorship of the Office of Naval Research; (2) seeding of migratory cloud systems associated with fronts and cyclones, conducted by the U.S. Weather Bureau; (3) treatment of convective clouds, a project of the University of Chicago supported by the Air Force; (4) dissipation of cold stratus and fog, performed by the Army Signal Corps Engineering Laboratories; (5) studies of the physics of ice fogs by Stanford Research Institute under Air Force auspices; (6) a special warm stratus and fog treatment system by A. D. Little, Inc., under contract from the U.S. Army. Only the first four were directly pertinent to the testing and evaluation of seeding effects of the kind advocated by Langmuir and others.

The field experiments were conducted in 1953 and 1954, and the reports were published in a Meteorological Monograph (Vol. 2, No. 11) of the American Meteorological Society in 1957.

In the introduction to the Monograph containing the reports, Chairman Petterssen explains the extratropical cyclone seeding in the following words: "The purpose of this project was to ascertain whether or not it would be possible to modify the development and behavior of extratropical cyclones by artificial nucleation, with the further aim of eluci-

dating Langmuir's contention that the larger-scale motion systems can be modified by seeding." Dr. Jerome Spar[141] professor of meteorology at New York University was in charge of this activity, which was designated *Project Scud*. The intensifying East Coast cyclones, usually involving the typical Cape Hatteras secondary, were the targets. Navy airplanes dispensing dry ice were flown in a specified pattern in the storm area while silver iodide smoke generators located on the ground were operated at 17 points from southern Florida to Long Island, New York. In order to provide a useful statistical test, randomization was introduced by tossing coins to determine if a predicted suitable situation should be seeded or not.

With the data obtained, statistical analysis was applied to the precipitation and the pressure changes in the seeded and the control (nonseeded) cases. The conclusion was that " ... the seeding in this experiment failed to produce any effects which were large enough to be detected against the background of natural meteorological variance."

The Weather Bureau project in Western Washington was operated on the background of cloud systems that sweep into the area from the Pacific with great frequency during the winter months in this rainy climatic area. Unlike the systems treated in Project Scud, the storminess in the area is most often associated with degenerating cyclonic systems. But also, in contrast to the East Coast experiment, the project in Washington dealt with cloud systems over a limited area—about 25,000 mi². Dry ice seeding from aircraft was used exclusively during the tests. There were approximately 100 stations in the area provided with recording rain gauges, and cloud and precipitation evaluations were also made with 0.85-cm and 5-cm radars. From the aircraft the usual visual changes in clouds as a result of the seeding were observed, at least in the more stratified systems.

Randomization was accomplished by dividing the storm systems into two groups, one seeded and the other not, in accordance with instructions contained in sealed envelopes arranged serially according to random numbers. Although physical processes in the clouds were examined in such a way as to add much to our understanding of how the precipitation process operates in winter cloud systems, the statistical results were inconclusive. Six different evaluation systems were tried on the data, none of which produced significant results, either positive or negative, although one of the methods was suggestive in the sense of increased precipitation. In his final report, Ferguson Hall,[72] who directed the project, wrote: "In general, the variation of the results with different methods of analysis, as well as the rather wide confidence limits imposed by the natural variability of rainfall, both point to the fact that field tests of limited duration will not define the effects of seeding within the narrow limits that are undoubtedly desired, and that tests carried forward over a considerable period of time will be needed to establish cloud seeding effects to an acceptable degree of precision."

This project came in for quite a lot of criticism from commercial seeders and some others. It was conducted in the West, which some commercial operators considered as "our" territory. Furthermore, it was carried out by the Weather Bureau, whose organization was already accused, before the project started, of seeking a negative result and, to paraphrase—oh well, it was just a backward, negative, bureaucratic organization, anyway. In writing the report, Hall went farther than the results would justify in trying to avoid giving a negative impression. "The evaluations do not necessarily furnish information on what the effects might have been with more or less intense seeding activity, rate of release of dry ice, etc," he wrote. "Also it might be speculated that the seeding increased the rainfall on some occasions and decreased it on others."

The project on cumulus clouds was assigned to a well-developed, essentially permanent cloud physics laboratory at the University of Chicago.[31] At that time under the leadership of this writer, with such well-known cloud physicists as Roscoe R. Braham, Jr., Louis J. Battan, James P. Lodge, Guy Goyer, and the late James E. McDonald as associates, the group had been formed as part of a more general large-scale scientific organization known as the Chicago Midway Laboratory, supported by the Department of Defense. As a result, it was one of the better equipped and more completely staffed cloud physics laboratories, surpassed at that time perhaps only by the Australian group of the Commonwealth Scientific and Industrial Research Organization (CSIRO) and the group at the Imperial College of Science and Technology in London, As a result of this special competence, numerous investigations relating to the physics of clouds and cloud processes were incorporated into the project in addition to the seeding tests.

In order to obtain as many cumulus clouds as possible for testing, it was decided to work in the Caribbean area during the winter months and in the Middle West in summer. It was realized that the clouds of the tropical seas initiated rain by the warm-cloud process, and that the well-known trade-wind cumulus might be amenable to water-spray treatment, as advocated by Langmuir from the Hawaiian experience, or by seeding with large hygroscopic nuclei. It was thought that the cumulus clouds of the Middle West, on the other hand, would operate through the ice-crystal process in initiating precipitation.

In the Caribbean tests, water was allowed to flow out from a large opening in a 450-gal tank carried in the bomb bay of an Air Force B-17. Since it was determined that half of the trade-wind cumulus initiated rain when their tops reached 9000 ft, clouds were selected mostly in the range of 6000 to 9000 ft, summit altitude, provided they were not already producing a detectable radar echo. The airborne nose radar, scanning laterally in a vertical plane, was used to determine the presence of precipitation particles as the airplane circled the cloud. Two clouds, one treated and the other not, were examined on each flight. From random numbers, sealed instructions were used by the controller of the seeding to determine whether the first cloud selected or the second one would be seeded. The scientist-controller of the flight sat in an isolated bubble on top of the airplane and had no way of knowing whether or not the first of his selected clouds was seeded. In this way the possibility of bias in choosing the second cloud of the pair was avoided. The water was dumped during the penetration of the cloud at a height about 1000 ft below the top. The statistical analysis indicated that the treatment was successful in that many more treated clouds had radar echoes than the untreated.[32] The statistical details are not discussed here, but the simplest positive statement coming from the analysis was to the effect that the probability of occurrence of an echo in a cloud was doubled (from 0.23 to 0.48) by treatment. The amount of water falling out of the clouds was not measured, but it could be estimated that the amount of water producing the radar echo was many times the amount dumped in. (The water falling from the airplane did not in itself produce an echo).

During the summers of 1953 and 1954

dry ice seeding of cumulus clouds in an area of southern Illinois and Missouri was carried out by the Chicago group. Again the clouds were treated in pairs on a randomized seed no-seed basis. Drought prevailed in the area during the two summers, so the number of cloud pairs suitable for study was not large. The dry ice was dispensed at temperatures in the clouds ranging from -1 to $-7°C$. Only clouds showing no echo initially were chosen for study. There were 27 pairs of clouds treated with dry ice, in 14 of which no echo was found in either the treated or untreated cloud. In seven pairs the treated cloud showed an echo while the untreated did not, in one pair both clouds gave a radar return, and in five pairs the untreated cloud showed an echo without one appearing in the seeded cloud. Leaving out the cases when the two clouds behaved the same, one finds a score of 7 to 5 in favor of the seeded cloud, as far as production of an echo was concerned. However, the sample was deemed much too small to be significant.

Seventeen pairs of cumulus in the Central United States were treated with waterspray. In 11 of these, neither cloud produced an echo. In terms of one cloud producing an echo while the other did not, the score was 3 to 2 in favor of the seeded cloud of the pair. There was one case in which both clouds produced an echo. These results, of course, were not significant.

As might be expected, the seeding experiments with subcooled stratus clouds conducted by the Army Signal Corps Engineering Laboratories,[160] substantiated the results of Project Cirrus for this type of cloud, with some minor exceptions. A number of new relationships with regard to seeding rates, spread of glaciating effect, relative results from thick versus thin stratus, formation of new clouds after seeding, and the question of overseeding were determined. In this as well as in the cumulus project, it became apparent that earlier concerns about over-seeding had been exaggerated.

In retrospect it seems surprising that the results of this series of statistically controlled experiments did not receive more attention than was the case. For one thing, it was published in the American Meteorological Society Monographs wich, unfortunately, have a somewhat limited circulation. Furthermore, publication delays caused much of the material to be out of date when it appeared. The Petterssen committee held all manuscripts to be submitted at the same time, which meant a 1-year delay for some of the material, and on top of that one of the contributions was returned for revision, adding another 7 months to the publication lag. It is also interesting to note that at the time of its appearance, in the late summer of 1957, all U.S. Government supported cloud-physics groups, along with their colleagues in other branches of science, were facing the most serious reduction in Department of Defense and other support in the history of the science. It took about 3 years before signs of recovery, especially under the support of the National Science Foundation, began to take effect. The 1960s proved to be the great period of development in the United States. But that is getting ahead of the story.

Arizona Mountain Cumulus

After completion of the 1953–1954 tests, the University of Chicago group co-operated with the newly formed Institute of Atmospheric Physics at the University of Arizona in seeding tests on summer cumulus and, to some extent, winter storms in Southern Arizona. Early airborne results were again not extensive enough to be conclusive, but enough progress was made to warrant the

inauguration of a series of silver iodide seeding tests by the University of Arizona under support of the National Science Foundation. Program I, started in the summer of 1957 and continuing through the summer of 1960, involved randomization on pairs of suitable days.[13 – 15] One suitable day was allowed between members of a pair. If more than one day interrupted the pair by reason of being unsuitable, it was discarded and a new envelope of the randomized series of instructions was opened. The target was the Santa Catalina Mountains, covering an area of about 15 by 20 mi and with elevations ranging up to 9200 ft above sea level. Seeding was from an airplane flying 2 to 4 hr along a path 15 to 20 mi long perpendicular to the airflow about 30 mi upwind from the foothills of the mountains. The silver iodide was dispensed while flying at the altitude of the $-6°C$ temperature.

Analysis of the precipitation data for the first two years showed greater rainfall on the seeded than on the nonseeded days, but in the next two years, the nonseeded days came out with considerably more rainfall, so that for the four summers, 1957 to 1960 combined, the result showed a ratio of 0.70 for the ratio of precipitation on seeded days to that of nonseeded days.[16, 17] It so happened that on August 17, 1959, a day when there was no seeding, a torrential rain occurred. If the pair August 17 and 18 was left out of the statistics, the ratio of seeded to nonseeded precipitation still came out less than one (0.93).

Because of questions raised, a new effort, Program II, was operated in 1961, 1962, and 1964.[22] The moisture requirement for the prediction of a suitable day was increased in order to eliminate some no-rain days such as had occurred in Program I. Another new feature was the decision to dispense the silver iodide from an airplane at an altitude about

1000 to 2000 ft below the cloud-base altitude. It was also decided to have the airplane fly the same pattern on nonseeded days as on seeded days. This peculiar requirement was for the purpose of ensuring that no day, seeded or unseeded, was eliminated because of the inability of the airplane to fly because of mechanical or other difficulties. In order to obtain greater representativeness of the data, the number of rain gauges was increased for this second program.

Only one of the three years, 1964, showed more rain on seeded than on nonseeded days. For the three years in Program II the ratio of precipitation on seeded days to that on nonseeded days was 0.70. A number of analyses of the data were tried, but none of them supported the hypothesis that airborne silver iodide seeding of these clouds increased rainfall or influenced its areal extent over the mountain range where the convective clouds formed.[18]

Perhaps the explanation of the failure to increase precipitation lay in the finding, from radar studies, that the initiation of rain did not require the ice-crystal process. Radar echoes characteristically appeared at temperatures indicating that precipitation had been formed by the coalescence mechanism.[19]

Today one must be precise in using the words "Arizona project" because there has been another project going on, partly simultaneously, in the mountains of Northern Arizona around Flagstaff.[161, 162] Here the plain is at an elevation of 5000 to 6000 ft as contrasted with the much lower elevation in southern Arizona and the summer convective-orographic clouds appeared to be different from those around Tucson. Meteorologists of the private, commercial sector were mainly the ones involved in the Flagstaff activities, although occasional Air Force support and participation by academic meteorologists was arranged. In the early phases the emphasis was on the gathering of cloud-physics information, but in the late 1960s a cumulus-cloud seeding project was started by Meteorology Research, Inc., under sponsorship of the U.S. Bureau of Reclamation (see Chapter 6).

Project Whitetop and Other Tests

The most extensive and, in some respects, the most sophisticated experiment in weather modification was Project Whitetop, a 5-year program of seeding of summer convective clouds in the Middle West, specifically in south-central Missouri. The field program and the analysis was under the direction of Dr. Roscoe R. Braham, Jr. of the University of Chicago.[30] In addition to tests of seeding effects, it included a well-organized study of the physical processes in the clouds, using an airplane instrumented and equipped for cloud-physics measurements, radar coverage, stereophotography, ground measurements of ice nuclei, and studies of various characteristics of collected ice particles.

The project was operated in the five summers, 1960 to 1964, inclusive. The purpose was to settle the question, once and for all, whether or not summer convective clouds, which produce most of the significant rain in the nation's Middle Western breadbasket, could be treated with silver iodide to initiate or to enhance precipitation. The research area was within a circle of 60-mi radius centered at the radar site. Although the area is spoken of as part of the Ozarks, it consists of a rolling plain with relief of more than 400 ft above or below the average plain at only a few widely scattered places. Thus, unlike the Arizona site, it was not a region giving rise to mountain cumulus.

Each operating day a seeding line was

chosen, roughly perpendicular to the wind and extending for 30 mi along the upwind side of the research area. Three seeding airplanes flew individual 10-mi seeding segments along this track, back and forth, for 6 hr. The flights were at the altitude of the average cloud bases, usually about 4000 ft above terrain, with continuous dispensing of silver iodide smoke from generators that had been tested in a wind tunnel to produce 10^{14} to 10^{16} active ice nuclei per gram of silver iodide at $-17°C$. The total output in the seeding operation was 2700 g/hr.

The selection of operational days, suitable for clouds, was based on criteria of total moisture in nearby radiosonde observations in the early morning and winds at 4000 ft. A unique feature of the project was determining as precisely as possible the spread of the silver iodide smoke plume from the seeding line, accomplished by taking pilot-balloon observations of the winds every two hours from 10 a.m. until midnight.

Randomization was provided by opening a sealed envelope, after the day was declared operational, containing "seed" or "no-seed" instructions. The envelopes were in a dated series based on random numbers. The envelopes for nonoperational dates were not used. The design of the experiment thus was such that it gave comparative data in-plume, outside-plume, seeded day, and non-seeded day.

The effects on precipitation were assessed by the range-height indicating radar and by an augmented rain gauge network. The radar could pinpoint the rain, even if falling between rain gauges, but it could not give values of the amount of rain. It was most valuable in supplying information about the height at which the rain formed, thus indicating the probability of coalescence rather than ice initiation in certain cases. The rain gauge data provided the most useful information for statistical analysis.

The final statistical analysis showed that in the presence of silver iodide nuclei, the rainfall was *less* than in the unseeded areas or situations. Never before had weather-modification data been subject to more different kinds of statistical analysis with such great interest on the part of statisticians, interspersed with occasional misinterpretations.

The best answer to the negative results appears in the physical data from the cloud-physics airplane flights. Most of the Missouri clouds produced raindrops by the coalescence process below the freezing line, and these drops were carried in the updrafts and frozen as ice pellets at surprisingly high subfreezing temperatures (-5 to $-10°C$). Furthermore, in-flight collections at these relatively warm subfreezing levels repeatedly revealed 10^4 to 10^5 ice particles/m^3 in size ranges of 10 to 300 μm under natural (unseeded) conditions. These are already the concentrations one attempts to achieve in seeding operations. Thus the addition of silver iodide could have no effect but to overseed. The ever-present ice pellets presumably operate in some manner to promote the formation of great numbers of smaller ice particles (see Braham[32] and Koenig,[88] where ice splinters from the pellets are suggested, although now much doubted).

Other important United States experiments, discussed in other chapters of this volume, deserve notice in compiling a history. A randomized seeding experiment in the high Rocky Mountains of Colorado aimed at increasing the snowfall from winter orographic clouds was begun in 1960 under the direction of Lewis O. Grant, professor of atmospheric science at the Colorado State University. This experiment looms large in the history of weather modification not only because it marked the beginning of intensive operational projects but also because positive effects were reported. All this work is described in detail by Professor Grant in Chapter 7.

Another series of experiments that has attracted widespread attention is the Florida cumulus seeding under the direction of Dr. Joanne Simpson of the U.S. National Oceanic and Atmospheric Administration (NOAA). These activities are described by her in Chapter 7. Finally, attention is called to the hurricane seeding program (Project Stormfury) of NOAA and the U.S. Navy, described by the director, Mr. R. Cecil Gentry, in Chapter 14. These undertakings take an important place in the history of weather modification and must be noted here with emphasis.

Experiments in Other Countries

In many countries of the world weather modification projects have been carried out, in some cases by government agencies and in others by contract with commercial operators.[10, 120] Some of them, because of their thoroughness or uniqueness have attracted a great deal of attention internationally. The Australian experiments have already been discussed. Some others are worthy of mention.

During the 1960s some interesting tests were carried out in Israel.[63, 64] A randomized crossover (alternating target and control area) experiment was conducted, involving airborne silver iodide seeding of clouds, mostly cumulus. The two selected areas were separated by a buffer zone about 20 mi wide. The seeding airplane flew at an altitude just below the cloud base at a distance upwind corresponding to a 1/2-hr fetch. The area designated for seeding on each day with suitable clouds was determined on the basis of a ran-

domized instruction prepared in advance. Over the first 5 1/2 yrs the statistical analysis indicated a rainfall increase of 18 percent.

In Switzerland, a project on the southern slopes of the Alps called Grossversuch III was conducted.[148] It was intended as a randomized experiment in hail suppression, using ground-based silver iodide generators. The statistical analysis revealed that the frequency of hail was greater on seeded than on unseeded days, but the average rainfall on seeded days was 21 percent higher than on unseeded days.[131]

Scientists of the Soviet Union have been active in the field of cloud-physics and cloud dynamics, following closely the developments in the West and contributing significantly. Battan[20, 21] has published summaries and bibliographies of their work. In the area of weather modification Soviet activities were little known until 1965 when Sulakvelidze et al.[144 - 146] of the High Mountain Geophysical Institute of the Hydrometeorological Service published results of the 1964–1965 Caucasus Anti-Hail Expedition. This project and its results are discussed in the next section.

Hail Suppression

Although the United States and Australia appear to have taken the lead in weather modification work aimed at precipitation enhancement, the suppression of hail was tried mainly elsewhere. Switzerland, France, Italy, the Soviet Union, Argentina, Bulgaria, Yugoslavia, Kenya, and Canada were early starters in this activity. The principal scientific basis for seeding to suppress hail is the hypothesis that by furnishing to the cloud a superabundance of ice-crystal nuclei, the bulk of the available water will be used up in the formation of a tremendous number of snow crystals so that the amount of supercooled water, which causes growth of hailstones by riming, can be reduced greatly.

In the vineyard and orchard regions of France, Italy, and, to some extent, Switzerland and Germany, the growers did not await the proof of hypotheses or the development of the science of cloud-physics but instead were swept away by the promise of breaking up hailstorms by the sound effects of exploding rockets. During the early 1950s a Frenchman, General F. L. Ruby[121, 122] invented inexpensive rockets for the use of growers in combatting hail. A phenomenally successful business was developed; it seemed that every prosperous grower in France and Italy had rockets to shoot at menacing looking clouds, The psychological effect of thus being able to shoot at the enemy must have been very satisfying.

Early in the 1950s an organization was formed in the southwest of France under the scientific guidance of H. Dessens of the Observatory of Puy du Dôme, University of Clermont-Ferrand. An area of 7,000,000 ha ultimately was treated with 240 ground-based silver iodide generators. In 1968 Dessens gave a report on the results, showing no convincing evidence of an effect.[52] Similar disappointing results were reported for the Argentina experiment in the grape-growing region of Mendoza.[67] The Swiss results of a negative nature have already been mentioned.[123, 131, 148]

An interesting experiment with explosives was conducted in Kenya, where 10,-000 rockets, each containing 800 g of TNT, were exploded to treat about 150 hailstorms in a small area of 1500 ha.[124] It was speculated that the acoustic shock waves would shatter the hailstones or produce large numbers of ice crystals by the temperature decrease in a sudden adiabatic expansion of the air. The results were disappointing. Subsequently, a commercial operator was employed for airborne silver iodide seeding.

The Caucasus hail suppression experiments of the Russians, mentioned previously, attracted much attention in the world cloud-physics community. In research preliminary to the 1964–1965 expedition, Sulakvelidze and colleagues[144 – 146] determined that from the radar echo intensities the "concentration zone" of the greatest amount of water in the convective clouds could be determined. The experiment was designed around a method of delivering silver iodide directly in that zone. The method consisted of shooting antiaircraft shells and rockets carrying an explosive and lead iodide warhead into this zone of each cloud. (Lead iodide was used instead of silver iodide because of its lower cost and its having a nucleating threshold temperature very close to that of silver iodide.) Three zones were chosen for protection, each surrounded by a control area. There were about 300,000 ha under crop in the protected areas. Comparisons were made with crop damage in the protected area and the control sector as well as comparisons during 4 yr before start of the project with results during the 1964 and 1965 protection experiments. In the protected cultivated area the number of hectares of crops damaged was 20/year before the protection and only 15.8 and 12.8 ha, respectively, in 1964 and 1965. In the control sectors, the number of hectares damaged was greater in 1964 and 1965 than in the 1959–1963 average year. In other words, while the area of damaged crops was decreased in the protected areas, there was an increase in area damage in the control sectors. There has been some skepticism about the results, but certainly the report of

Sulakvelidze et al. stands out as one of the important events in the history of weather modification.

As a result of this work, hail suppression activity is now carried out on a large scale in the Soviet Union. The operations are discussed by Sulakvelidze himself in Chapter 11.

Schleusener[129] in 1968 presented a summary of hail suppresion undertakings in the United States. Most of them were commercial operations that did not produce data of significant value. A few research-type (randomized) projects had only recently started. The most complete, well-designed project, conducted near Rapid City, South Dakota, 1966–1967, showed no statistically significant differences in hailfalls between target and control areas.

Fog Dissipation

Beginning in 1962–1963 experiments have been conducted at Orly Airport in Paris to clear supercooled fog from the runways by means of sprays of liquid propane. Although the propane is not as cold as dry ice, it is cold enough to cause spontaneous nucleation of supercooled fog. It boils at $-44.5°C$ at atmospheric pressure. The Orly experiments, carried out under the guidance of R. Serpolay of the Clermont-Ferrand University group, soon became operational. The method has generally proved successful there and in various U.S. Air Force installations. In fact, the dissipation of cold (supercooled) fog is now operational at many locations around the world, including some in the Soviet Union and in North America.

Warm fogs, that is, those that are not supercooled, are more common than supercooled ones over most of the inhabited globe, and effort has been directed at dissipating them. By 1970 there had not been much improvement over the methods introduced by Houghton and Radford in the 1930s[78] but various heating techniques, helicopter downwash, and some proprietary chemicals have been tried. At Orly Airport heat and turbulence in the wake of stationary jet engines has replaced propane methods.

THE SCIENTISTS LOOK AT THEIR FIELD

As the early wave of overoptimism formed, the American Meteorological Society took note of events by forming a Committee on Cloud Physics and Weather Modification. As the years have gone on this committee has continued, with a rotating membership. It has issued repeatedly statements of "cautious optimism" in the early period decrying the claims of spectacular results. Later, as the science settled down to more modest enthusiasm, the statements have appeared less embittered than formerly, but the general tone has been about the same: a plea for more research, better knowledge of physical processes, and more careful evaluations. In recent years more attention to legal and socioeconomic problems has been urged.

The "Confidential" Report

Early in 1950, when the controversy over some of the results reported by Project Cirrus was at its height, it was agreed by all concerned that an independent scientific review of the work and the resulting claims of spectacular results should be undertaken. A review committee was appointed. Since Cirrus was a project of the Department of Defense, the committee was organized under the jurisdiction of Dr. H. E. Landsberg, then chairman of the Panel on Meteorology of the Research and Development Board of the Department of Defense. Dr. Bernard Haurwitz, then Chairman of the Department of Meteorology at New York University, was appointed chairman of the committee which was to investigate rainmaking results and report to the Department of Defense for appropriate action. The other committee members were Gardner Emmons, of the New York University Department of Meteorology, Dr. Hurd C. Willett, Professor of Meteorology, and Dr. George P. Wadsworth, Professor of Mathematics at Massachusetts Institute of Technology.

When the committee was appointed, it was thought that its report would be distributed or published widely in order to explain to the public what the real prospects of weather control were, but when the report was received in the late spring of 1950 it was classified as "confidential." Dr. F. W. Reichelderfer, Chief of the Weather Bureau, felt more than frustrated, because he had intended to use the report to inform the public of the extravagance of some of Langmuir's claims. The committee members themselves were annoyed at the security classification. Finally the Department of Defense agreed to let the report be published by the American Meteorological Society without any reference to the Department of Defense. Thus the report appeared in the *AMS Bulletin*[74] in the guise of a report requested by the president of the Society, so it had hardly any more significance than the regular reports of the Society's Committee on Cloud Physics and Weather Modification. After pointing out the need for more physical knowledge of cloud behavior and better statistical evaluation, the report concluded with the following paragraph:

It is the considered opinion of this committee that the possibility of artificially producing

any useful amounts of rain has not been demonstrated so far, if the available evidence is interpreted by any acceptable scientific standards.

If Reichelderfer was incensed by the Department of Defense act of classifying the report for "security reasons," Langmuir's infuriation by the contents of the report matched these feelings. From that time on he seemed to stand alone with his statistics.

The Orville Committee

A committee that had great impact, especially in the area of commercial activities, was the U.S. Advisory Committee on Weather Control. It was created by an act of Congress, signed into law by President Eisenhower on August 13, 1953. Captain Howard T. Orville (U.S.N., Ret.) who formerly had served as head of the Navy weather service, president of the American Meterological Society, and a member of the steering committee of Project Cirrus, was appointed chairman of the committee. The law provided for representation on the committee from the departments of Agriculture; Commerce; Defense; Interior; Health, Education, and Welfare; and from the National Science Foundation. Members from the private sector were Lewis W. Douglas, former ambassador to Great Britain and Arizona rancher and banker; Alfred M. Eberle, dean of agriculture at South Dakota A & M College; Joseph J. George, chief meteorologist of Eastern Air Lines; and Kenneth C. Spengler, executive secretary of the American Meteorological Society.

The committee, which came to be known as the Orville Committee, was charged with assaying the state of the science and recommending public policy. At its first meeting the committee amplified this charge by stating that it had been "given the job of finding out whether the U.S. Government should 'experiment with, engage in, or regulate activities designed to control weather conditions'." The committee carried on its work from early 1954 until its report was completed in December of 1956.[150] In his early public pronouncements and interviews, Chairman Orville struck a note of advocacy rather than objectivity* but the final report was temperate. The committee visited all the major weather modification experiments and operations in the country, holding hearings, conferences, and reading reports.

The temperate nature of the final report could partially be ascribed to the reception some of the preliminary statements received at a conference on "The Scientific Basis of Weather Modification Studies" in April, 1956 at the University of Arizona. The conference afforded an opportunity to stage a preview of tentative conclusions of the committee in the presence of outstanding scientists in the fields of cloud physics and statistics. A warm debate developed between Dr. K. A. Brownlee, a University of Chicago statistician, and committee members, and especially Mr. H. C. S. Thom, statistical consultant to the committee. The controversy continued for weeks after the meeting, apparently with some errors of analysis or interpretation on both sides. It seemed that this was the first serious confrontation of the committee with scientists demanding high standards of proof, and some of the comments in the proceedings of the meeting[111] make interesting reading.

The final report was in two volumes. The first contained a slightly optimistic evaluation of the field in general, and the

second volume described various seeding experiments and operations. It was the first compendium of commercial operations. (By printing error, one detailed report slipped into the first volume.)

From wintertime seeding with silver iodide, the committee found that in mountainous areas in the western United States precipitation increases were possible, but that the same statistical tests applied to nonmountainous areas failed to reveal an effect, that no negative effects were found, and that not enough data were available to evaluate hail suppression projects. The report emphasized the meagerness of our knowledge and advocated greater effort in research. The second volume of the report is a good reference source of commercial operations going on at that time, although the committee did not pass judgment on any of them.

In accordance with the recommendations of the Orville Committee and other interested groups, the U.S. Congress passed and the President signed Public Law 85–510 (1958) which charged the National Science Foundation with responsibility to "initiate and support a program of study, research and evaluation in the field of weather modification" and "to report annually to the Congress and the President thereon." This act not only placed emphasis on research into the scientific problems associated with weather modification, but it also assigned to the NSF the leading responsibility within the Federal Government. In the succeeding ten years other agencies developed research and operational interests in weather modification, notably the Bureau of Reclamation in the Department of the Interior, the National Weather Service, and the Department of Defense. Recognizing these broader objectives of federal agencies, in 1968 the Congress passed and the President signed Public Law 90–407 which encouraged ef-

* Orville had the great respect and backing of those engaged in weather modification. The Proceedings of the First National Conference on Weather Modification, 1968, were dedicated to him, with a full-page picture of him as the frontispiece.

forts directed toward a variety of agency missions in addition to scientific research and which left to the Executive Branch the responsibility of developing a coordinated program. In effect this means that any agency may submit plans for a weather-modification activity which will be discussed in an interagency conference and put through the usual budgetary proposal process.

The National Academcy Speaks

Neither the report of the Orville Committee nor the results of the series of government-sponsored tests reported through the "Petterssen Committee" produced an authoritative statement as to how an experiment to test the efficacy of cloud seeding should be run.[160] The National Science Foundation, given by the 1958 law the responsibility of carrying out the weather modification investigations, needed guidelines, especially from the point of view of experimental design and statistical evaluation. Dr. Earl G. Droessler, then program director for Atmospheric Sciences in the Foundation, discussed with Dr. Paul E. Klopsteg, at that time Chairman of the Committee on Atmospheric Sciences of the National Academy of Sciences, the possibility of having a conference of meteorologists, mathematicians, and statisticians to examine the progress made and the needs to be met in weather-modification experiments. The result was a conference on May 1–3, 1959, organized by the Division of Mathematics of the National Academy of Sciences and sponsored by the National Science Foundation. Held at a lodge on the Skyline Drive in Shenandoah National Park, Virginia, it was called "The Skyline Conference on the Design and Conduct of Experiments in Weather Modification." There were 30 participants, of whom 17 were mete-

orologists and 13 were statisticians although two or three were qualified in both areas.

After remarking on the nonscientific nature of many commercial weather modification operations and dismissing as "unsubstantiated" Langmuir's claim of pronounced correspondence between the changes in the seeding schedule and the rainfall patterns in the Project Cirrus periodic seeding, the report of the conference[112] made a strong plea for careful statistical design of weather modification experiments. It pointed out the need for long-term experimental programs, the standardization of design for purposes of replication, the need for more basic research in cloud physics, and the requirement for cooperation between meteorologists and statisticians. Field surveys prior to formal testing were advocated in order to determine the seedability and natural habit of various types of clouds in various regions. Although phrased in general terms, the report contained a good list of "commandments" for would-be weather modifiers.

Two reports were published by the National Academy of Sciences Panel on Weather and Climate Modification of the Committee on Atmospheric Sciences. Dr. Gordon J. F. MacDonald, noted geophysicist, was chairman of the panel. The first of the two reports entitled "Scientific Problems of Weather Modification," 1964,[113] was based on four 2-day meetings of meteorologists and aeronomers talking about what might be done in weather modification, including for the first time in published form a lengthy discussion of the general circulation of the atmosphere, together with some speculations about modifying the circulation and the climate by some "triggering" mechanism. There is also a section on contamination or modification of the upper atmosphere. Otherwise the report was little more than a sketchy review of its subject.

The second report of the panel entitled "Weather and Climate Modification Problems and Prospects," published in 1966[114] in two volumes, was a lengthy assessment of the state of the art. Again it was based on a series of conferences of the 14 members of the panel, adding up to a total of 34 days of meetings and discussions with more than 50 invited individuals. The first volume contains general statements and recommendations and Volume II contains nearly 200 pages of discussion of experiments or unrandomized commercial operations on which the conclusions are based.

Very guarded phrases are used in the conclusions, such as, "There is increasing but somewhat ambiguous statistical evidence that precipitation from some types of cloud and storm systems can be modestly increased or redistributed by seeding techniques," and "The available evidence, though not conclusive, indicates that artificial nucleation techniques, under certain meteorological conditions, may be used to modify the space or time distribution of precipitation."

One gets the impression that the members of the panel wanted to express optimism concerning the future of weather modification, but had to hedge their statements in order to preserve their self respect. This is a very charitable judgment compared with the scathing review of the report by the University of Chicago statistician K. A. Brownlee.[38]

The second volume of the report contains an extensive study of commercial operations. Four long-period snowpack enhancement projects in the mountains of the Far West were examined in detail and the conclusion was reached that stream flow increases of 7 to 18 percent occurred in seeded years as compared with historical stream flow records. The evaluation also suggested

that a dozen or so short-term seeding projects in the eastern United States had increased the precipitation by 10 to 20 percent.

In 1971 the NAS-NRC Committee on Atmospheric Sciences published a general survey of their field under the title, "The Atmospheric Sciences and Man's Needs—Priorities for the Future."[115] Chapter 4 of the report deals with weather and climate modification. The general statements support the 1966 Committee findings, but several new issues are brought out that relate to public policy. Economic, legal, environmental, and decision-making problems are discussed, as are also questions of the roles of Government, industry, and the universities; the international ramifications; and the design criteria for operational programs. The committee published in 1973 a revised edition of the 1966 report, containing a broader coverage of the problem but reaching conclusions similar to those of 1966.

RECENT AND ONGOING PROJECTS

To round out a history of weather modification, several recent and ongoing projects must be included. Those receiving the greatest attention, namely, the Florida cumulus, the winter orographic precipitation enhancement, the Great Lakes snow, fog modification, hurricane modification, and lightning-suppression projects are treated in detail in Chapters 6, 7, 8, 9, 13, 14, and 17, respectively. Another recent development is the recognition of various possibilities of inadvertent weather modification, discussed in Dr. Machta's Chapter 19 and Dr. Lansberg's Chapter 20.

One of the developments beginning in the middle 1960s has been the use of numerical models of cumulus dynamics to predict the expected behavior as a basis for evaluating the observed effects after seeding. As of this writing numerical modeling, as indicated in various chapters of this book, has become the third support in the triangle of testing physical hypotheses—prediction, experiment (treatment), and observation.

Another development goes back to Langmuir's Honduras experiments, where in one case the bouyancy is increased by high efficient conversion from supercooled droplets to ice (overseeding?) and where in another case the seeding is directed at enhancing the growth of existing cloud particles without creating new ones to compete for the water. In the new jargon of weather modification, the former is called "dynamic seeding" and the latter "microphysical" seeding. A differentiation is also made between the use of hygroscopic materials[60] in warm cumulus to produce the early drop growth that will lead to a Langmuir "chain reaction," as contrasted with seeding with silver iodide, to start the Bergeron-Wegener process among supercooled droplets. A variety of considerations of this type are brought up by some of the writers in this book.

Finally a new development of the 1960s was the consideration of social, legal, and other human aspects of weather modification. Books have been written on these subjects, notably "Human Dimensions of Weather Modification," edited by W. R. Derrick Sewell[132] and "Controlling the Weather—A Study of Law and Regulatory Procedures" edited by Howard J. Taubenfeld.[147] The report prepared by the Legislative Reference Service of the Library of Congress by Hartman[73] for the committee on commerce of the United States Senate in 1966 is rich in material on the subject.

Recent studies also have drawn attention to apparent effects outside target areas, especially for considerable distances downwind. In the Santa Barbara Project, Elliott and Brown[55] reported striking downwind effects at distances as great as 100 to 130 mi. Adderley[1] has presented data of downwind effects up to 200 mi in Australian experiments. There was also evidence of downwind effects on Project Whitetop, as reported by Braham,[30] but more recently with the use of a denser network of precipitation stations, Huff[80] finds that the evidence for the downwind effects is too weak to draw definite conclusions.

As a means of determining the influence natural variability can have on the assessment of seeding effects, desk-top weather-modification "games" have been carried out using observed data from dense precipitation networks with make-believe randomized seeding programs. From the unusually dense networks of the Illinois State Water Survey, Huff[80] has shown that conclusions about seeding effects can be very misleading. Taking a hypothetical 5-year seeding experiment for 1960–1964 (the same period as Project Whitetop) he finds that a rainfall increase in "seeded" summer convective storms of 50 percent due entirely to natural rainfall variability can be expected in approximately 14 percent of the experiments in the Little Egypt (Southern Illinois) network. A 20 percent increase resulting strictly from natural variability will occur in 30 percent of the 5-year experiments in this area. From Illinois data he showed that if a 10 percent increase due to seeding is to be detected with a 95 percent statistical confidence limit, the experiment would have to run for a period of 11 to 28 years, assuming the most efficient experimental design. The length of period would depend on the type of rainfall investigated, the longest period being for summer cold fronts and airmass showers and the shortest for summer thunderstorms. If a 20 percent increase had been induced, the period of experiment required to substantiate it against the background of natural variability would range from about 3 to 7 years in the most efficient design.

Huff's data would indicate that in the lack of experimental design or in the inefficient designs of commercial operations, up to 50 years might be required to show the 10 to 20 percent increases claimed, if they actually did occur.

Studies of this kind illustrate why progress in weather modification has been slow and controversial. The elusive nature of the natural phenomena has compounded the problem so much that cloud seeding experiments often become exercises in frustration. Until we can predict what the atmosphere, or a storm, or a cloud, intends to do, we must rely on an astute use of statistics, often involving many years of experimentation.

REFERENCES

1. Adderley, E. E., Rainfall increases downwind from cloud seeding projects in Australia, *Proceedings of the 1st National Conference on Weather Modification*, American Meteorological Society, Boston, 1968, pp. 42–46.

2. Adderley, E. E. and S. Twomey, An experiment on artificial stimulation of precipitation in the Snowy Mountains regions of Australia, *Tellus* 10, 275–280, 1958.

3. Aitken, J., *Collected Scientific Papers of J. Aitken*, C. G. Knott, Ed., Cambridge University Press, Cambridge, 1923.

4. Albrecht, F., Theoretische Untersuchungen über die Ablagerung von Staub aus strömender Luft und ihre Anwendung auf die Theorie der Staubfilter, *Phys. Z.* 32, 48–56, 1931.

5. Alexander, V., The greatest flood of history, *Weatherwise* 4, 110–111, 1951.

6. Anonymous, "Project Cirrus"—The story of cloud seeding, *Gen. Elec. Rev.* 55, 8–26, Nov. 1952.

7. Anonymous, Flood disaster in Kansas, *Weatherwise* 4, 79, 1951.

8. Arenberg, D. L. and P. J. Harney, The Mt. Washington icing research program, *Bull. Am. Meteorol. Soc,* 22, 61–63, 1941.

9. Bannon, J. K., Artificial stimulation of rain formation, *Meteorol. Mag.* 76, 169–174, 1947.

10. Bannon, J. K., South African rain-making experiments, *Weather* 4, 155, 1949.

11. Barrows, J. S., Weather modification and the prevention of lightning caused forest fires, *Human Dimensions of Weather Modification*, W. R. D. Sewell, Ed., University of Chicago Dept. of Geography Research Paper No. 105, pp. 169–182, 1966.

12. Barrows, J. S., V. J. Schaefer, and P. B. McCready, Project Skyfire: A progress report on lightning fire and atmospheric research, *U.S. Forest Serv. Forest and Range Expt. Sta., Ogden, Utah, Res. Paper No. 35*, 1954.

13. Battan, L. J., Relationship between cloud base and initial radar echo, *J. Appl. Meteorol.* 2, 333–336, 1963.

14. Battan, L. J. and A. R. Kassander, Jr., Design of a program of randomized seeding of orographic cumuli, *J. Meteorol.* 17, 583–590, 1960.

15. Battan, L. J. and A. R. Kassander, Jr., Artificial nucleation of orographic cumulus clouds, *Physics of Precipitation, Geophys. Monogr. (Am. Geophys. Union)* 5, 409–411, 1960.

16. Battan, L. J. and A. R. Kassander, Jr., Evaluation of effects of airborne silver iodide seeding on convective clouds, *Univ. Arizona Inst. Atmos. Phys. Sci. Rept.* 18, 1962.

17. Battan, L. J., A. R. Kassander, Jr., and L. S. Sims, Randomized seeding of orographic cumuli, 1957, Pt. II, *Univ. Arizona. Inst. Atmos. Phys. Sci. Rept.* 9, 1958; 10, 1959.

18. Battan, L. J. and A. R. Kassander, Jr., Summary of results of a randomized cloud seeding project in Arizona, *Proceedings of the 5th Berkeley Symposium on Mathematical Statistics and Probability*, Vol. 5, University of California Press, Berkeley, 1967, p. 29.

19. Battan, L. J., Relationship between cloud base and initial radar echo, *J. Appl. Meteorol.* 2, 333–336, 1963.

20. Battan, L. J., Soviet research in radar meteorology and cloud modification, *Bull. Am. Meteorol. Soc.* 42, 755–764 1961.

21. Battan, L. J., A survey of recent cloud physics research in the Soviet Union, *Bull. Am. Meteorol. Soc.* 44, 755–771 1963.

22. Battan, L. J., Silver-iodide seeding and rainfall from convective clouds, *J. Appl. Meteorol.* 5, 669–683, 1966.

23. Berg, A., Bemerkungen über das Verfahren des Herrn A. Veraart (Rijswijk) zur Erzeugung künstlicher Regen, *Z. Angew. Meteorol.* **46**, 161–171, 1930.

24. Bergeron, T., On the physics of cloud and precipitation, *Proc.—Verbaux Assoc. Meteorol. Intl. Union Geodesy Geophys., 5th General Assembly, Lisbon, 1933*, Paris, 1935, pp. 156–178.

25. Berson, A., *Wissenschaftliche Luftfahrten* II, Braunschweig, 1900, p. 192.

26. Bowen, E. G., Australian experiments on artificial stimulation of rainfall, *Weather* **7**, 204, 1952.

27. Bowen, E. G., Australian experiments in artificial rain-making, *Bull. Am. Meteorol. Soc.* **33**, 144–246, 1952.

28. Bowen, E. G., A new method of stimulating clouds to produce rain and hail, *Quart. J. Royal Meteorol. Soc.* **78**, 37–45 1952.

29. Bowen, E. G., The relation between rainfall and meteor showers, *J. Meteorol.* **13**, 142–151, 1956.

30. Braham, R. R., Jr., Project Whitetop, Reports, University of Chicago Dept. of Geophysical Sciences, Cloud Physics Laboratory (with various other authors), 1965 to 1971.

31. Braham, R. R., Jr., L. J. Battan, and H. R. Byers, Artificial nucleation of cumulus clouds, *Cloud and Weather Modification Meteorol. Monogr.* **2**(11), 47–85, 1957.

32. Braham, R. R., Jr., What is the role of ice in summer rain showers?, *J. Atmos. Sci.* **21**, 640–645, 1964.

33. Braham, R. R., Jr., and J. R. Sievers, Overseeding of cumulus clouds, in *Artificial Stimulation of Rain*, Pergamon, New York, 1957, pp. 250–266.

34. Braham, R. R., Jr., B. K. Seely, and W. D. Crozier, A technique for tagging and tracing air parcels, *Trans. Am. Geophys. Union* **33**, 825–833 1952.

35. Brier, G. W., Seven-day periodicities in certain meteorological parameters during the period 1899–1951, *Bull. Am. Meteorol. Soc.* **36**, 265–277, 1955.

36. Brier, G. W., Seven-day periodicities in May 1952, *Bull. Am. Meteorol. Soc.* **35**, 118–121, 1954.

37. Brier, G. W. and I. Egner, An analysis of the results of the 1951 cloud seeding operations in Arizona, *Bull. Am. Meteorol. Soc.* **33**, 208–210, 1952.

38. Brownlee, K. A., Review of "Weather and climate modification problems and prospects," *J. Am. Statist. Assoc.* **62**, 690–694, 1967.

39. Byers, H. R., A rational approach to rain-making, *Proc. Indian Acad. Sci., Sect. A* **37**, 237–247, 1953 (Ramanathan 60th Birthday Issue).

40. California Department of Water Resources, Santa Barbara weather modification project, Interim Report to Board of Directors, 1960.

41. Coons, R. D., and R. Gunn, Relation of artificial cloud modification to the production of precipitation, *Compendium of Meteorology,* American Meteorological Society, Boston, 1951, pp. 235–241.

42. Coons, R. D., R. C. Gentry, and R. Gunn, First partial report on the artificial production of precipitation—stratiform clouds—Ohio 1948, *Bull. Am. Meteorol. Soc.* **29**, 266–269, 1948. Also in *U.S. Weather Bur. Res. Paper* **30**, 1948.

43. Coons, R. D., E. L. Jones, and R. Gunn, (1949), Second partial report on the artificial production of precipitation—cumulus clouds—Ohio 1948, *Bull. Am. Meteorol. Soc.* **29**, 544–546, 1948. Also in *U.S. Weather Bur. Res. Paper* **31**, 1949.

44. Coons, R. D., E. L. Jones, and R. Gunn, Third partial report on the artificial production of precipitation—orographic clouds—California 1949, *Bull. Am. Meteorol. Soc.* **30**, 255–256, 1949. Also in *U.S. Weather Bur. Res. Paper* **32**, 1949.

45. Coons, R. D., E. L. Jones, and R. Gunn, Fourth partial report on the artificial production of precipitation—cumulus clouds—Gulf States 1949, *Bull. Am. Meteorol. Soc.* **30**, 289–292, 1949. Also in *U.S. Weather Bur. Res. Paper* **33**, 1949.

46. Coulier, P. J., Note sur une nouvelle propriété de l'air, *J. Pharm. Chim.* **22**, 165, 1875.

47. Crozier, W. D., and B. K. Seely, Concentration distribution in aerosol plumes 3 to 22 miles from a point source, *Trans. Am. Geophys. Union* **36**, 46–52, 1954.

48. Dauzère, C., La foudre, la grêle et la pluie, *Rev. Sci.* **68**, 711–717, 1930.

49. Dessens, H., Essai de formation artificielle de cumulus par utilization exclusive de l'énergie solaire, *Bull. Obs. Puy de Dome* (4), 114–125, 1956.

50. Dessens, H. and J. Dessens, La formation ar-

tificielle de grands cumulus producteurs de pluie, *Compt. Rend.* **243**(11), 814–817, 1956.

51. Dessens, H., Expérience de pluie provoquée par ensemencement in-interrompu a partir du sol. *Proceedings of the International Conference on Cloud Physics*, Toronto, 719–723, 1968.

52. Dessens, J., Expérience de suppression de la grêle dans le sudouest de la France, *Proceedings of the International Conference on Cloud Physics*, Toronto, 773–777, 1968.

53. Dessens, J., Man-made tornadoes, *Nature*, **193**(4810); 13–14, 1962.

54. Douglas, C. K. M., The appearance of the sun and the moon through a cloud, *Meteorol. Mag.*, Jan. 1929.

55. Elliott, R. D., and K. J. Brown, The Santa Barbara II project—downwind effects, *International Conference on Weather Modification, Canberra, Australia, American Meteorological Society, Boston, 1971, pp. 179–184.*

56. *Espy, J. P., Artificial rains, National Gazette and Literary Rev.* (Philadelphia) **19**, 5798, 1839, Reprinted in J. P. Espy, *Philosophy of Storms*, Boston, 1841, pp 492–500.

57. Findeisen, W., Die kolloidmeteorologischen Vorgänge bei der Niederschlagsbildung, *Meteorol. Z.* **55**, 121–133, 1938.

58. Findeisen, W., Zur Frage der Regentropfbildung in reinen Wasserwolken, *Meteorol. Z.*, **56**, 365, 1939.

59. Fletcher, J. E., Weather modification, *Science* **158**, 276–277, 1967.

60. Fournier d'Albe, E. M., Artificial cloud nucleation with sodium chloride, *in* Atmospheric chemistry of chlorine and sulfur compounds, *Geophys. Monogr. (Am. Geophys. Union)* **3**, J. P. Lodge, Jr., Ed., 104–105, 1959.

61. Fraser, D., Production of ice-crystal clouds by seeding, *Nature*, **164**, 179–180, 1949.

62. Fuquay, D. M., Generator technology for cloud seeding, *Am. Soc. Civil Engrs. J. IR*, **86**, 79–91, 1960.

63. Gabriel, K. R., The Israeli artificial rain stimulation experiment. Statistical evaluation, *Proceedings of the Berkeley Symposium on Mathematical Statistics and Probability*, Vol. 5, 1967, p. 91.

64. Gabriel, K. R., Y. Avichal, and R. Steinberg, A statistical investigation of persistence in the Israeli artificial rainfall stimulation ex-

periment, *J. Appl. Meteorol.* **6**, 323–325, 1967.

65. Gori, E. G., G. Musso, and H. M. Papee, On the hailstorms of Asti Province, Part II—Past hail endeavours, *Geofis. Meteorol. Boll. Soc. Ital. Geofis. Meteorol.* **20**, 24–27, 1971.

66. Görög, and A. Rovó, Attempt at artificial production of rain [translated title], *Idöjaras* (Budapest) **14**, 2–9, 1938.

67. Grandoso, H. N. and J. V. Iribarne, Evaluation of the first three years in a hail prevention experiment in Mendoza (Argentina), *Z. Angew. Math. Phys.* **14**, 549–553, 1963.

68. Guilbert, G., On the formation of rain and the origin of cirrus, *Compt. Rend.* **173**, 999 (1921).

69. Halacy, D. S., Jr., *The Weather Changers*, Harper and Row, New York, 1968, p. 246.

70. Hall, F., An evaluation of the technique of cloud seeding to date, *Trans. N.Y. Acad. Sci. Ser. II* **14**(1), 45–50, Nov. 1951.

71. Hall, F., G. Emmons, and B. Haurwitz, et al., Dr. Langmuir's article on precipitation control, *Science* **113**, 189–192, 1951.

72. Hall, F., The Weather Bureau ACN project, *Cloud and Weather Modification, Am. Meteorol. Soc. Monogr.* **2**(11), 24–46, 1957.

73. Hartman, L. M., Report on weather modification and control to Committee on Commerce, United States Senate, 89th Congress, *Senate Report No. 1139*, U.S. Gov't. Printing Office, Washington, D.C., 1966.

74. Haurwitz, B., G. Emmons, G. Wadsworth, and H. C. Willett, Correspondence: On the results of recent experiments in the artificial production of precipitation, *Bull. Am. Meteorol. Soc.* **31**, 346–347, 1950.

75. Havens, B. S., History of Project Cirrus, *Gen. Elec. Res. Lab. Rep. RL-756*, 1952.

76. Hawkins, H. F., Jr., The weather and circulation of May 1952, *Monthly Weather Rev.* **80**, 82–87, 1952.

77. Houghton, H. G., Problems connected with the condensation and precipitation processes in the atmosphere, *Bull. Am. Meteorol. Soc.* **19**, 152–159, 1938.

78. Houghton, H. G. and W. H. Radford, On the local dissipation of natural fog, *Mass. Inst. Tech. Woods Hole Oceanogr. Inst. Papers Phys. Oceanogr. Meteorol.*, **6**(3), 1938.

79. Howell, W. E. More rain for New Yorkers?, *Weatherwise* **3**, 27–29, 1950.

80. Huff, F. A., Evaluation of precipitation

records in weather modification experiments, *Advances in Geophysics*, H. Landsberg and J. Van Mieghem, Eds., Vol. 15, 1971, pp. 60–135.

81. Hull, D., Paper presented before the Indiana Academy of Science Dec. 2, 1927, printed by Notre Dame University South Bend, Ind. Discussed in H. R. Byers, Correspondence, *Bull. Am. Meteorol. Soc.* **28,** 158, 1947.

82. Humphreys, W. J., *Rainmaking and Other Weather Vagaries*, Williams and Wilkins, Baltimore, 1926.

83. Humphreys, W. J., *Ways of the Weather*, Jaques Cattell Press, Lancaster, Pa., 1942, pp. 323–335.

84. Huschke, R. E., A brief history of weather modification since 1946, *Bull. Am. Meteorol. Soc.* **44,** 425–429, 1963.

85. Junge, C. E., Übersättigungsmessungen an atmosphärische Kondensationskernen, *Gerlands Beitr. Geophys.* **46,** 108–129, 1935.

86. Junge, C. E., Zur Frage der Kernwirksamkeit des Staubes, *Meteorol. Z.* **53,** 186–188, 1936.

87. Junge, C. E., *Air Chemistry and Radioactivity*, Academic, New York, 1963, Press 382 p.

88. Koenig, L. R., The glaciating behavior of small cumulonimbus clouds, *J. Atmos. Sci.* **20,** 29–47, 1963.

89. Köhler, H., Zur Thermodynamik der Kondensation an hygroskopischen Kernen und Bemerkungen über das Zusammenfliessen der Tropfen, *Medd. Statens Meteorol.-Hydrograf. Anstalt* **2**(5), 1925.

90. Kraus, E. B., and P. Squires, Experiments on the stimulation of clouds to produce rain, *Nature* **159,** 489, 1947.

91. Kumai, M., Electron-microscope study of snow crystals, *J. Meteorol.* **8,** 151–160, 1951.

92. Kumai, M., Snow crystals and the identification of the nuclei in the northern United States of America, *J. Meteorol.* **18,** 139–150, 1961.

93. Kumai, M., Fog modification on the Greenland ice cap, *Proceedings of the National Conference on Weather Modification,* American Meteorological Society, Boston, 1968, pp. 414–422.

94. Landsberg, H., Atmospheric condensation nuclei, *Ergeb. Kosm. Phys.* **3,** 155–212, 1938.

95. Landsberg, H., Zahlen der Kondensationskernen auf den Taunus Observatorium, *Bioklimat. Beibl.* **1,** 125, 1934.

96. Langmuir, I., The growth of particles in smokes and clouds and the production of snow from super-cooled clouds, *Proc. Am. Philos. Soc.* **92,** 167–198, 1948.

97. Langmuir, I., *Collected works of Langmuir*, Vols. 10 and 11, G. Suits and H. E. Way, Eds., Pergamon Press, New York, 1961.

98. Langmuir, I., The production of rain by a chain reaction in cumulus clouds at temperatures above freezing, *J. Meteorol.* **5,** 175–192, 1948.

99. Langmuir, I., Cloud seeding by means of dry ice, silver iodide, and sodium chloride, *Trans. N.Y. Acad. Sci.* **14**(1), 40–47, 1951.

100. Langmuir, I., A seven-day periodicity in weather in United States during April 1950, *Bull. Am. Meteorol. Soc.* **31,** 386–387, 1950.

101. Langmuir, I. and K. B. Blodgett, Mathematical investigation of water droplet trajectories, Report to Army, Dec. 1944 and July 1945. Reissued in 1949 as *Gen. Elec. Res. Lab. Rept. RL-225,* 1949.

102. Langmuir, I., Control of precipitation from cumulus clouds by various seeding techniques, *Science* **112,** 35–41, 1950.

103. Leopold, L. B., and M. H. Halstead, First trials of the Schaefer-Langmuir cloud-seeding technique in Hawaii, *Bull. Am. Meteorol. Soc.,* **29,** 525–534, 1948.

104. Lewis, W., Correspondence: on a seven-day periodicity, *Bull. Am. Meteorol. Soc.* **32,** 192, 1951.

105. Mallen, J., The artificial production of rain, *Sci. Am. Ser. 2* **65,** 144–145, 1891.

106. Mason, B. J., *The Physics of Clouds*, 2nd ed., Clarendon Press, Oxford, 1971, p. 284.

107. Mason, B. J., Recent developments in the physics of rain and rain-making, *Weather* **14,** 81–97, 1959.

108. Maynard, R. H., Radar and weather, *J. Meteorol.* **2,** 214–226, 1945.

109. McDonald, J. E., An historical note on an early cloud-modification experiment, *Bull. Am. Meteorol. Soc.* **42,** 195, 1961.

110. McDonald, J. E., The physics of cloud modification, *Advances in Geophysics*, H. E. Landsberg and J. Van Mieghem, Eds., Academic Press, New York, Vol. 5, 1958, pp. 223–301.

111. McDonalds, J. E., Ed., Proceedings of the conference on the scientific basis of weather modification studies, Tucson, Institute of Atmospheric Physics, University of Arizona, 1956 (mimeographed).

112. National Academy of Sciences, Report of the Skyline conference on the design and conduct of experiments in weather modification, *NAS-NRC Publ.* **742**, 1959.

113. National Academy of Sciences, Scientific problems of weather modification—a report of the panel on weather and climate modification, Committee on Atmospheric Sciences, *NAS-NRC Publ.* **1236**, 1964.

114. National Academy of Sciences, Weather and climate modification problems and prospects. Vol. I: summary and recommendations; Vol. II: research and development, final report of the panel on weather and climate modification, Committee on Atmospheric Sciences, *NAS-NRC Publ.* **1350**, 1966.

115. National Academy of Sciences, The atmospheric sciences and man's needs: Priorities for the future—a report of the Committee on Atmospheric Sciences, *NAS-NRC Publ.*, unnumbered, 1971.

116. Neyman, J., E. L. Scott, and M. Vasilevskis, Statistical evaluation of the Santa Barbara randomized cloud seeding experiment, *Bull. Am. Meteorol. Soc.* **41**, 531–547, 1960.

117. Peppler, W., Unterkühlte Wasserwolken and Eiswolken, *Forsch. Erfah. Ber., Reichsamt Wetterdienst* **1**, 289, 1940.

118. Powers, E., *War and the Weather, or Artificial Production of Rain*, S. C. Griggs Co., Chicago, 1871, 171 pp.

119. Rosenfeld, A., The quintescence of Irving Langmuir: a biography, in *Collected Works*, Vol. 12, G. Suits and H. E. Way, Eds., Pergamon Press, New York, 1961, p. 203.

120. Roy, A. K., Rain-making and its possibilities in India, *Indian J. Meteorol. Geophys.* **2**, 241–249, 1951.

121. Ruby, F. L., Étude sur l'action des fusées paragrêles explosives et à iodure d'argent, *Agr. Prat.* **115**, 419–422, 1951.

122. Ruby, F. L., Rapport général sur la défense contre la grêle en France, 1947–48, *Centre Natl. Défense Contre la Grêle, Rapp.* Lyoñ, 1949.

123. Sänger, R., The Swiss randomized hail suppression experiment, *in* "Physics of Precipitation," H. Weickmann, Ed., *Geophys. Monogr. Am. Geophys. Union* **5**, 388–395, 1960.

124. Sansom, H. W., A four-year hail suppression experiment using explosive rockets, *Proceedings of the International Conference on Cloud Physics*, Toronto, 1968, pp. 768–772.

125. Schaefer, V. J., The production of ice crystals in a cloud of supercooled water droplets, *Science* **104**, 457–459, 1946.

126. Schaefer, V. J., The possibilities of modifying lightning storms in the northern Rockies, *Station Paper No. 19, Northern Rocky Mountain Forest and Range Experiment Station*, Missoula, Mont, Jan. 1949, 18 pp.

127. Schaefer, V. J., The producton of clouds containing supercooled water droplets or ice crystals under laboratory conditions, *Bull. Am. Meteorol. Soc.* **29**, 175–182, 1948.

128. Schaefer, V. J. and I. Langmuir, The modification produced in clouds of supercooled water droplets by the introduction of ice nuclei, paper presented before American Meteorological Society annual meeting, Dec. 28, 1946.

129. Schleussener, R. A., Hailfall damage suppression by cloud seeding—a review of recent experience, *Proceedings of the 1st National Conference on Weather Modification*, American Meteorological Society, Boston, 1968, pp. 484–493.

130. Schmauss, A., and A. Wigand, *Die Atmosphäre als Kolloid*, Braunschweig, 1929.

131. Schmid, P., On "Grossversuch III," a randomized hail experiment in Switzerland, *Proceedings of the 5th Berkeley Symposium on Mathematical Statistics and Probability*, Vol. 5, University of California Press, Berkeley, 1967, p. 141.

132. Sewell, W. R. D., Ed., Human dimensions of weather modification, *Univ. Chicago Dept. Geogr. Res. Paper* **105**, 1966.

133. Simpson, G. C. On the formation of cloud and rain, *Quart. J. Royal Meteorol. Soc.* **67**, 99–133, 1941.

134. Smith, E. J., Experiments in seeding cumuliform cloud layers with dry ice, *Australian J. Sci. Res., A* **2**, 78–91, 1949.

135. Smith, E. J., Cloud seeding experiments in Australia, *Proceedings of the Berkeley Symposium Mathematical Statistics and Probability*, Vol. 5, University of California Press, Berkeley, 1967, p. 161.

136. Smith, E. J., E. E. Adderley, and D. T. Walsh, A cloud-seeding experiment in the Snowy Mountains, Australia, *J. Appl. Meteorol.* **2**, 324–332, 1963.

137. Smith, E. J., E. E. Adderley, and F. D. Bethwaite, A cloud-seeding experiment in South Australia, *J. Appl. Meteorol.* **2**, 565–568, 1963.

138. Smith, E. J. and K. J. Heffernan, Airborne

measurements of the concentration of natural and artificial freezing nuclei, *Quart. J. Royal Meteorol. Soc.* **80,** 182–197, 1954.

139. Smith, E. J., E. E. Adderley, L. Veitch, and E. Turton, A cloud-seeding experiment in Tasmania, *Proceedings of the International Conference on Weather Modification, Canberra, Australia,* American Meteorological Society, Boston, 1971, pp. 90–96.

140. Soulage, G., Nuages équatoriaux naturels et essais de formation artificielle de cumulus précipitants, *Bull. Obs. Puy de Dôme* (1):9–24, 1957.

141. Spar, J., Project Scud, *Cloud and Weather Modification, Am. Meteorol. Soc. Monogr.* **2**(11), 5–23, 1957.

142. Squires, P., and E. J. Smith, The artificial stimulation of precipitation by means of dry ice, *Australian J. Sci. Res., A* **2,** 232–245, 1949.

143. Stüve, G., Zur Kenntnis der Kristallisation des Wasserdampfes aus der Luft, *Gerlands Beitr. Geophys.* Köppen-Band I) **32,** 326–335 (1931).

144. Sulakvelidze, B. K., Findings of the Caucasus anti-hail expedition of 1965, Israel Program for Scientific Translations, Jerusalem, 1967, published by National Science Foundation, procurable through U.S. Dept. Commerce Clearinghouse for Federal Scientific and Technical Information, Springfield, Va.

145. Sulakvelidze, B. K., N. Sh. Bibilashvi, and V. E. Lepcheva, Formation of precipitation and modification of hail process, *ibid.,* 1967.

146. Sulakvelidze, B. K., On the principles of hail control method applying in the USSR, *Proceedings of the International Conference on Cloud Physics,* Toronto, 1968, pp. 796–803.

147. Taubenfeld, H. J., Ed., *Controlling the Weather,* Dunellen, New York, 1970, 275 pp.

148. Thams, J. C., et al., Die Ergebnisse des Grossversuches III zur Bekampfung des Hagels in Tessin in den Jahren 1957–1963, *Meteorol. Zentralanst. Veroeff.* (Zurich) **3,** 1966.

149. Thomson, W. (Lord Kelvin) On the equilibrium of vapour at a curved surface, *Proc. Roy. Soc. Edinburgh* **7,** 63, 1870.

150. U.S. Advisory Committee on Weather Control, *Final Report,* Vols. 1 and 2, 1957.

151. Veraart, W., *Meer Zonnenschijn in het nevelig Noorden: meer Regen in de Tropen,* Seyf-fardt's Boek en Musiekhandel, Amsterdam, 1931, 32 pp.

152. Volmer, M., and A. Weber, Keimbildung in übersättigten Gebilden, *Z. Phys. Chem. A* **119,** 277, 1926.

153. Vonnegut, B., Nucleation of supercooled water clouds by silver iodide smokes, *Chem Revs.* **44,** 177–289, 1949.

154. Vonnegut, B., Techniques for generating silver iodide smokes, *J. Colloid Sci.* **5,** 37–48, 1950.

155. Vonnegut, B., and K. Maynard, Spray-nozzle type silver-iodide smoke generator for airplane use, *Bull. Am. Meteorol. Soc.* **33,** 420–428, 1952.

156. Vonnegut, B., and C. B. Moore, Preliminary attempts to influence convective electrification in cumulus clouds by introduction of space charge into the lower atmosphere, A.D. Little, Inc., 1958.

157. Wagner, A., Untersuchungen der Wolkenelemente auf dem Hohen Sonnblick (3106 m.), *Sitzber. Math. Naturw. Klasse Akad. Wiss. Vienna* **117,** 1908.

158. Ward, R. de C., Artificial rain: a review of the subject to the close of 1889, *Am. Meteorol. J.* **8,** 484–493, 1892.

159. Wegener, A., *Thermodynamik der Atmosphäre,* Barth, Leipzig, 1911.

160. Weickmann, H. K., A realistic appraisal of weather control, *Z. Angew. Math. Phys.* **14,** 528–543, 1963.

161. Weinstein, A. I., and P. B. McCready, Jr., An isolated cumulus cloud modification project, *J. Appl. Meteorol.* **8,** 936–947, 1969.

162. Weinstein, A. L., Case study of a seeded storm system, *J. Rech. Atmos.* **4,** 161–171, 1970.

163. Wexler, H., Structure of hurricanes as determined by radar, *Ann. N.Y. Acad. Sci.* **48**(8), 821–844, 1947.

164. Wexler, H., Radar storm detection, *Trans. Am. Geophys. Union,* **28,** 70, 1947.

165. Wigand, A., Die vertikale Verteilung der Kondensationskerne in der freien Atmosphäre, *Ann. Physik* **59,** 689–742, 1919.

166. Wilson, C. T. R., Condensation of water vapour in the presence of dust-free air and other gases, *Phil. Trans. Roy. Soc. London Ser. A* 265, 1897.

167. Wright, H. L., The size of atmospheric nuclei, *Proc. Phys. Soc., London* **48,** 675–689, 1936.

2 | Experience of the Private Sector

ROBERT D. ELLIOTT

PRESENT-DAY CLOUD SEEDING IN THE PRIVATE SECTOR

Scientific weather modification dates from the time of the discoveries of the famous Nobel laureate, Dr. Irving Langmuir, Dr. Vincent Schaefer, and their co-workers at the General Electric Company Research Laboratories, Schenectady, New York. Most of this work was conducted under the title "Project Cirrus," which was supported during the late 1940s and early 1950s by the Office of Naval Research and the Signal Corps. (This era and the antecedent scientific developments are treated in detail in Chapter 1.)

Their discoveries followed earlier work carried out on Mount Washington in New Hampshire, which had to do with icing tests. The General Electric scientists were intrigued by the large amount of supercooled cloud droplets that blew upslope to the mountain crest and were converted to rime ice upon striking objects. Going back to the earlier studies of T. Bergeron and W. Findeisen,[4, 25] they became aware of the natural deficit of ice-forming nuclei needed to convert su-

percooled cloud water droplets to precipitation-size ice particles. They began speculating about how this natural deficit could be remedied by artificial means. They had been working on devices for generating smoke particles and knew how to produce large numbers in the submicroscopic range. It would be necessary to produce large numbers of artificial nuclei in order to have an appreciable effect on a cloud volume. But before they found smokes which were effective as ice nuclei, Vincent Schaefer discovered that he could transform a cloud of supercooled droplets, produced by breathing into a freezer, into minute ice crystals, which grew rapidly to sufficient size to fall to the bottom of the box. This he did by dropping dry ice pellets into the box. The crystals were easily visible as they twinkled in the light of a flashlight.

The crystals formed because the air near the falling dry ice pellets was chilled to below $-40°C$. Schaefer established that at $-40°C$, supercooled water droplets froze spontaneously, without any need for nuclei. Once frozen, their vapor pressure was well below that of the ambient air, which in this case was just saturated with respect to water, but not with respect to ice. Hence water vapor deposited (sublimated) upon the ice crystals

Robert D. Elliott is President of North American Weather Consultants, Santa Barbara Municipal Airport, Goleta, California.

45

caused them to grow. The removal of water vapor from the chamber in turn lowered the vapor pressure of the ambient air so that it was lower than that over the supercooled droplets, and they evaporated, supplying more vapor for deposition on the ice crystals.

Schaefer next flew over stratocumulus clouds which were not precipitating and were not at a temperature near $-20°C$. He seeded them with pea-sized dry ice pellets. Very quickly there was a transformation of the cloud to snow, which fell from the base, leaving a hole along the line of the seeding. This occurred in November, 1946 at a location in western Massachusetts.

Within a year Dr. Bernard Vonnegut, of the laboratory staff, discovered that silver iodide smoke particles served as an excellent ice-forming nuclei when introduced into a laboratory cloud at $-5°C$ or lower. Subsequently, other nucleating agents were found by Vonnegut and others, but none had as warm a threshold as did AgI. It has been found that under some circumstances, AgI is effective at -3 to $-4°C$.

Some of the very first cloud seeding performed after the GE discoveries was carried out by crop dusters who experimented in cloud seeding on their own or in service to a farm group. Within several years a more systematic type of cloud seeding came into being as individuals and small professional firms possessing scientific know-how in the field of meteorology and cloud physics started seeding, still somewhat on an experimental basis, for a variety of clients. A few of the early projects are still in existence today.

Besides providing extra water where needed and at the same time accumulating data about seeding effects, these private projects have served as a testing ground for various trial modes of seeding and for different approaches to operational routines. For example, improvements in the reliability and versatility of the silver iodide smoke generator has been a contribution of the private sector. The experience of the private sector has provided a history of practical solutions, in the arena of the free market and within the framework of a pluralistic society, to the problems arising from the demand for expanding water resources for industry and agriculture.

Private cloud seeding activity in the United States has been carried on at a moderately uniform level since the early 1950s. There are about 30 projects annually (excluding fog clearing), most of which are in the rain or snowpack enhancing categories. In a typical year they are distributed as follows: a dozen in the west coast states, half a dozen in the Rocky Mountain-Great Basin area, half a dozen in the Great Plains, and half a dozen in the rest of the United States. Hail suppression projects have been fewer than five per year through the 1960s but are increasing in number.

It may come as a surprise to some that the total number of seeding projects in the country has remained relatively constant through many years. There is a widespread impression of a wildly fluctuating market for the cloud seeder but this impression is the result of large fluctuations in publicity rather than in seeding. Some projects draw publicity like a lightning rod does lightning. Others are by nature "quiet." There are periods when the "noisy" projects predominate, which are followed by periods of "quiet" projects. At the start of extensive seeding in the early 1950s there was, of course, a sharp increase and it was accompanied by great publicity as the drought had started in the Great Plains. Dr. Irving P. Krick and his Water Resources Development Corporation were deeply involved in cloud seeding in this region. It was the country's first drought alleviation attempt. At about the same time Dr. Wallace Howell was working for the City of New York to increase precipitation and runoff into the city's dangerously depleted reservoirs. He drew national attention, especially when the city was sued by resort owners who felt any additional rain drove their business away.

Subsequent to this, during the middle and latter 1950s, seeding diminished, as did the drought, in the dry farm and ranching areas, but increased in the form of watershed seeding for utilities and for irrigation districts. Decisions to start seeding projects were made internally, not at public meetings, by people who eschewed publicity. Therefore, private cloud seeding appeared to the news readers to diminish in extent, although it had not.

The 30 or so annual seeding projects in the United States during the 1950s and 1960s (excluding fog clearing) were carried out, for the most part, by five firms. In addition to these, there was an equal number of individuals or groups that started on a project or two, which usually lasted only a few years. The five firms had on their staffs people skilled in meteorology, cloud physics, and enough engineering to properly install and maintain the necessary ground or air systems.

In the East, Dr. Wallace Howell, a product of Harvard and MIT, who had conducted the original project for the City of New York, formed Howell Associates Inc., of Lexington, Massachusetts. This firm has conducted many seeding projects in the eastern United States and Canada, in the Carribbean area, and in South America.

In the West initial activity developed under Dr. Irving Krick. While Krick was head of the Department of Meteorology at the California Institute of Technology, the department was asked to monitor some aerial dry ice seeding being conducted over Mt. San Jacinto, in 1947.

The results looked interesting. After Krick left the department and formed his own company, some seeding projects were carried out for ranchers in San Diego County, California, in Mexico, and for the Salt River Valley water users in Arizona in 1948 and 1949. In 1950, he moved to Denver and there formed the Water Resources Development Corporation which started the seeding activity over the Great Plains and elsewhere in the West. Expansion of the work to other countries followed.

In 1950, Eugene Bollay and I, both former students of Krick, and I having worked on early cloud seeding experiments under Krick, formed North American Weather Consultants. This company expanded its activities into the western United States and Canada, and became involved primarily in the seeding of mountain watersheds.

Weather Modification Company, a California company, was formed about the same time by Bert O'Hanlon and John Battle, the latter also being a former student of Krick's at Cal Tech. This company conducted projects primarily in the western United States.

Atmospherics Inc., another California company, was formed by Thomas Henderson in 1960. Henderson had been involved in cloud seeding operations while serving as a hydrographer and snow surveyor with the California Electric Power Company in the late 1940s and continued work thereafter on the Kings River Conservancy district project. Atmospherics Inc. has been involved in both winter orographic seeding projects and summer seeding in the Great Plains, especially in connection with hail suppression. In Africa, the company has been conducting hail suppression work for tea growers.

Mention should be made of still another company founded in the early days by Charles Barnes of Phoenix, Arizona. It was active in the late 1940s and early 1950s. It was later taken over by Richard Merrill of Taft, California. A company called Weather Engineers was founded by Boyd Quate in the 1950s and was active in the early days, conducting seeding operations for farm groups.

During the 1960s and continuing into the 1970s, several aviation oriented operators with limited theoretical but considerable practical experience have gone into summer cumulus cloud seeding in the northern Great Plains. They contract with farm groups to increase rainfall for crops and range land, or to suppress hail.

For much of the 1950s and 1960s, there occurred annually six to nine utility type watershed seeding projects, mostly in the West. Many of these projects were long enduring. For example, the Southern California Edison project in the upper San Joaquin River basin in the Sierra Nevada range has been operated continuously every winter season since the 1950–1951 season. The farm project in the Big Bend area, State of Washington, has had an equally long and continuous duration. As a rule, utility company projects tend to run several years in a row when demand exceeds power resources. When new generating units are brought into full operation, with full reservoirs, the seeding may be cancelled for a period of years until demand once again catches up with, then exceeds, supply. Droughts always stimulate activity.

Hail suppression is undertaken for farmers in the Great Plains where it is very destructive of grain crops. It is also being undertaken in fruit growing areas where even small hail marks fruit and reduces its market value. In Africa hail suppression is practiced in the high-elevation tea growing regions where hail is a serious problem.[30, 31] It is also used in tobacco growing regions. Hail suppression requires the use of an abundance of nuclei at the right time and place to prevent large natural growth on a few embryos. The use of aircraft backed by radar surveillance, and close meteorological scrutiny, is a key feature of most such projects.

An interesting early commercial hail suppression project was started in 1958 in northeastern Colorado by Weather Engineers Inc. This project was operated by the Weather Modification Company in 1959, at which time it involved five seeding aircraft and roughly 125 ground-based generators which made it the largest cloud seeding project up to that time. Professor Richard Schleusner (now head of the Institute of Atmospheric Sciences at South Dakota School of Mines, then at Colorado State University) examined the results and made a presentation at the International Hail Conference held in Verona, Italy, in 1960. This project stimulated the interest of scientists and provided the historical roots leading to the establishment of the present large National Hail Research project in northeastern Colorado.

Other more experimental applications of seeding lying within the realm of the experience of the private sector include the amelioration of conditions favoring the production of strong gusty winds in thunderstorms. These blow down banana trees in tropical areas. Overseeding plays a role in the technique for accomplishing this.

The lumber industry has made some use of cloud seeding in the west to enhance summer precipitation and thus to maintain a higher fuel moisture index to avoid early shutdown of logging operations. The seeding of cumulus clouds over forest fire areas to limit or put out the fires is an occasional use of seeding. There have been instances of success in this endeavor, and it is beginning to be employed routinely in Alaska.

During the 1960s the use of cloud

seeding to clear airport fog had been brought into being as an operational procedure under the guidance of the airlines, with Boynton Beckwith, a United Airlines meteorologist, playing a leading role in getting the practice started.[3] This procedure is applied extensively only to cold fog (subfreezing) and hence is used only in the more northerly or higher-elevation airports of the United States. About 15 such projects are underway in the United States each winter, with the contracts being given by the airlines to local aircraft operators to carry out the necessary seeding flights. Usually, dry ice is dropped into the fog by a low-flying plane. This fog clearing program has been successful enough to make it well worth the airlines' investment in it.

The U.S. Air Force has conducted, partly under contract to the private sector, extensive engineering development studies of fog clearing techniques and now uses them operationally.[11] Experimentation is also being carried out in seeding cold fogs with propane, a method used on an operational basis at Orly, the Paris Airport. Research and development are underway at present in clearing warm fog (the other 90 percent of the fog problem) at airports by seeding it with various types of hygroscopic particles to induce coalescence of fog droplets into fewer large drops, which leads to an increase in visibility.

All the principal cloud seeding firms also contract for seeding in other parts of the world, especially in Central and South America, in Africa, the near East, and in Europe. Several of them employ their talents in types of work other than cloud seeding. For example, they engage in studies of air pollution meteorology for new industrial plants, and in specialized weather forecasting for industry. They have become involved in one way or another with government-sponsored research projects in weather modification.

In this involvement they have made use of their wealth of practical experience and observation to guide their research work.

Several private meteorologically oriented firms, not in the business of industrial cloud seeding, have become involved in weather modification research under government contract. In particular, Meteorology Research Inc., of Altadena, California, under the leadership of Drs. Paul MacCready, Jr., and Theodore Smith, has accomplished useful research into summer cumulus activity in the Flagstaff, Arizona, area and has developed and marketed a number of unique cloud physics instruments. E. Bollay Associates, later part of EG&G, conducted interesting experiments in orographic seeding in Colorado and also produced specialized cloud physics instruments. Weather Science Inc. is one of the new companies on the scene. Its president, Dr. Ray Booker, is a meteorologist, engineer, and pilot. He has brought his talents to bear on the various problems of collecting, storing, and reducing by machine methods atmospheric data collected by aircraft. Western Scientific Services Inc., a Colorado company headed by Gerald Price, has developed the engineering of data collection in remote and difficult areas of the mountainous west.

The annual expenditure on cloud seeding by private concerns or local government agencies under contract to private cloud seeding firms from the 1950s to date has run about $1 or 2 million a year. The annual expenditure by the Federal Government on weather modification was somewhat less than this during the 1950s; however, during the 1960s expenditure increased steadily, first as the National Science Foundation became active, then in the mid-1960s as the Bureau of Reclamation program got underway. The total Federal expenditure rose from about $3 million/year at the

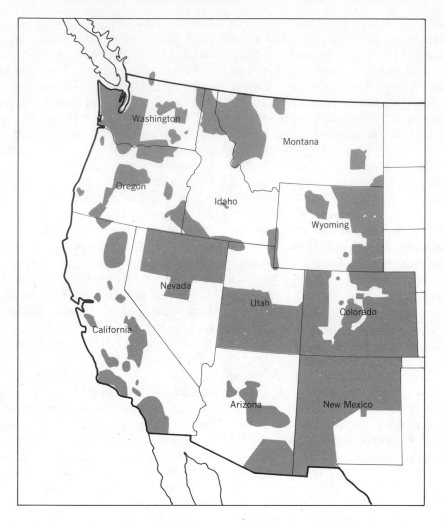

Figure 2.1 Extent of area in western United States in which cloud seeding was practiced at some time in the years 1948–1964.

start of the decade to about $10 million at the end and is currently running above this.

The extent of the area in the western United States which has, at one time or anther, been under contract for cloud seeding during the years 1948–1964 is shown in Figure 2.1. The extent is somewhat startling, but it must be remembered that only a fraction of this total area has been seeded in a given season. In the years since 1964 not many

new areas have been added. Also, in a number of cases, the seeding was concentrated on certain portions of the indicated project area, such as high mountain ranges.

Weather modification projects supported by private ultilities, by farmers, by irrigation districts, and by local governments did not favor the expenditure of funds on any but a few items that could be called "research." Their thinking was largely water oriented. A purely water oriented seeding project can be defined as one in which the full effort is focused upon the artificial nucleation of all clouds deemed suitable for cloud seeding. No effort is devoted to gathering evidence of the effectiveness of the seeding, the object being only to optimize the production of added precipitation on the basis of current seeding methodology as applied by a competent professional practitioner. The basis for deciding to use such an approach is largely economic, and runs along these lines: "Seeding promises to enhance precipitation by 10 percent. According to our calculations this is worth $500,000 in additional water over one winter season. The project cost is one-tenth of this, $50,000. In the face of the large benefit/cost ratio, we must accept the risk of failure of the seeding method. We shall leave the long-term reduction of this risk to organizations whose purpose it is to engage in research. Our purpose is to enhance our water supplies and thereby our own productivity."

A purely information oriented seeding project, on the other hand, can be defined as one in which the full effort is focused upon seeding in such a way as to optimize the production of scientific information, under the guidance of scientists of various disciplines. In view of the large number of variables involved in the atmosphere, it is customary to seed half the cases only (storms, or other seeding and observing intervals), reserving the non-seeded cases for comparison purposes. The seeded cases are selected by random process in order to avoid any subconscious selection bias; furthermore, considerable investment in a rain gauge network and in other types of observational instrumentation is the rule. The argument for this approach runs along these lines: "A project of this type will provide the basic data needed further to elucidate the mechanisms involved in natural and artificially stimulated precipitation processes. Our purpose is to add to the fund of scientific knowledge."

These two approaches, as pure strategies, could be considered mutually exclusive. In practice, a useful increase in precipitation occurs in some information oriented projects, and most water oriented seeding projects provide information having some research value. In information oriented projects, it quite often becomes necessary, because of the impact on weather, for accommodations to be made such as not seeding during harvest or at other agriculturally inconvenient times. On the other hand, farmers and others in the area affected usually would prefer that all cases be seeded. Sometimes a compromise is made and the ratio of seeded to non-seeded cases is set at 2:1 instead of 1:1.

There have been differences of opinion about the propriety of conducting water-oriented versus information oriented weather modification programs. In earlier years strong arguments were raised in the more traditional scientific circles against any water oriented projects, insisting that research should be completed before operations are commenced. However, even so eminent a scientist as Irving Langmuir thought some operational seeding was merited. He pointed out that a vast chemical industry existed years before the nature of the chemical bond was understood. Both types of projects have been carried out since the late 1940s, and coexistence continues today.

EVOLUTION OF A RATIONALE

Precipitation production is a complex process involving mechanisms ranging from the microphysical scale up to the cyclonic scale. Meteorologists began posing questions that were unanswerable:

what was the optimum concentration of artificial nucleant required, and was it possible to overseed a cloud and reduce precipitation? How could the desired concentration of nucleant be generated and dispersed? How did the release of latent heat alter the dynamics of the cloud? Were there any larger-scale effects extending well beyond the seeded cloud? Was there a compensating reduction in precipitation downwind of the seeded cloud? What were the relative roles of the Bergeron ice nucleation process and the cloud droplet coalescence process in the generation or precipitation? How can an evaluation of the effects of seeding be made against the highly variable background of natural precipitation?

The development of a cloud seeding rationale from Schaefer's first primitive effort in his 1946 field experiment to the present sophisticated ones is largely a history of how successively more refined answers to these questions have evolved.

The problem of determining the optimum concentration of nucleant was initially answered by the Project Cirrus group as being that required to convert the available supercooled cloud droplets to ice. In 1949, the Swedish atmospheric scientist, Tor Bergeron, the coauthor of the Bergeron-Fendeisen theory of rain production in the 1930s, provided a more detailed answer.[5] He made the assessment that a ratio of one ice crystal (which could be formed on an artificial nucleus) to 1000 supercooled water droplets was correct for enhancing precipitation. Since there are about 10^8 cloud droplets/m^3, 10^5 nuclei/m^3 would be required. This would ensure a thousandfold growth in mass, tenfold in radius. Thus a 10-μm cloud droplet could become a 100-μ ice sphere. The particle terminal velocities would then be sufficiently large to ensure fallout.

Another interesting observation was that one ice crystal for every supercooled cloud droplet would convert the cloud to small ice crystals, but this would not lead to precipitation. The result would be overseeding.

Later estimates of the required number tended to be considerably less. Nature, under precipitation conditions, may supply around 10^3 nuclei/m^3, so theoretically, any concentration higher than this might lead to precipitation enhancement. However, the concentration of effective natural nuclei varies sharply with temperature. Most often they are in a concentration of $10^3/m^3$ at $-21°C$, but they tend to increase by an order of magnitude with a 4 deg drop in temperature, and decrease by an order of magnitude with every 4 deg rise in temperature. The crucial temperature is that of the coldest part of the cloud, normally that at its top.

At Colorado State University's Climax field experiment, P. O. Mielke and L. W. Grant[44] found seeding from the mountain slope with a silver iodide-sodium iodide complex at the rate of 20 g/hr did not overseed until the cloud top was colder than $-23°C$. In Santa Barbara County field tests carried out by the author's company indicated similar results, although the clouds in Santa Barbara storms consist of cumuliform clouds embedded in stratiform decks in contrast to the stratiform cap clouds which predominate at Climax. The mode of seeding was different also, being accomplished by a ground-based pyrotechnic device.[16]

In Australia, on the other hand, cloud tops are warmer and so the overseeding event was not observed there. However, positive effects were found only in seeded clouds whose tops were colder than $-10°C$.[55] There is further evidence from other areas to substantiate the notion that in seeding with artificial ice-forming nuclei to increase precipitation, there is a favorable cloud top temperature "window" somewhere between about -10 and $-23°C$.

An unusual effect has been found to occur in clouds whose temperature is in the warmer range. Ice crystals are found to be more abundant by orders of magnitude than would be expected from the numbers of effective natural nuclei found by sampling the air in a cold box set at that temperature. Apparently at these warmer temperatures there is some natural mechanism for increasing the supply of ice crystals. Perhaps it involves shattering of ice crystals formed initially on natural nuclei. The effect is not present in clouds with cold tops (see Chapters 6 and 7 for more details).

The whole matter of optimum concentration is complicated also by the fact that a cloud is seldom static, even when it is rather stratiform in appearance. There are areas of upward motion, that is, updrafts, where cloud droplets are being formed by condensation of water vapor on condensation nuclei (not to be confused with freezing or sublimation nuclei. In some air masses there are 10 times as many cloud droplets formed as in others, because of more condensation nuclei. Also, there are downdrafts, sometimes on the outside of the cloud but also sometimes within, wherein total or partial evaporation of cloud droplets occurs. In such nonstatic conditions one must consider the rate of introduction of nuclei versus the rate of cloud droplet production. It should be sufficient to convert the cloud droplets generated in the updrafts to precipitation particles before they can get into a downdraft and evaporate. In small unseeded thunderstorms, fully 80 percent of the cloud droplets generated eventually evaporate. Its precipitation efficiency can be said to be 20 percent.[7] Giant thunderstorms in squall lines are generally from 50 to nearly 100 percent efficient.

In the case of orographic clouds over a mountain barrier, the air is lifted and condenses on the upwind slope. To the lee of the range are the downdrafts and evaporation. Over small mountain barriers,[20] the efficiency was found to be about 20 percent in a field investigation carried out by the author's firm under National Science Foundation auspices. Over large barriers and in strong storms, it may be 50 percent, although in major storms it may approach 100 percent. The natural precipitation rate is usually too slow to convert all the water condensed in updrafts into precipitation. Much is lost by evaporation. The question is, under particular circumstances, what rate of introduction of artificial nuclei into the ascending airstreams will do most to reduce the evaporative loss? Rather detailed numerical models of the whole process were constructed in order to arrive at satisfactory answers for varying conditions. It is not possible in this chapter to go into the details required to explain the models. Reference should be made to the chapters on orographic cloud seeding and on cumulus modification.

The matter of generation and dispersal brings into play the manner in which natural atmospheric mixing processes are relied upon to dilute the effluent from a typical generator down to an acceptable value. A typical continuous smoke generator such as that used on orographic seeding projects emits around 10^{16} smoke particles/of AgI burned. But only 1 in 10 are effective as ice-forming nuclei at $-20°C$ and one in 100,000 at $-10°C$. With a consumption rate of AgI of 10g/hr, the generator effluent at 10 m downwind of the generator, in a good wind, will have a concentration of nuclei effective at $-20°C$ of about 3×10^{11} crystals/m³. Of those effective at $-10°C$, the concentration would be about 3×10^7/m³. According to Bergeron's criterion, the effluent would overseed the cloud near the generator in either case.

But suppose the generator is located on a mountain slope, or in a low-flying airplane, at about 1 km or so below the $-5°C$ level, where it can be entrained into convective updrafts. Under usual conditions the smoke will be diluted by turbulence from that in the immediate vicinity of the generator by a factor of 1000 to 1,000,000 by the time it has risen to these colder temperatures. This brings the concentration down to a more acceptable value. In the absence of convection, but with the smoke blowing up a mountain slope, turbulent mixing still produces a dilution by about the correct factor.

This implies that the nucleant should not be injected into a cloud at a cold temperature. Injection of such a high concentration directly into cloud would overseed it, leading to the production of a swarm of small ice crystals, not snow. However, the trail of small ice crystals formed downwind of such a generator spreads as it mixes into surrounding supercooled cloud droplets. The concentration is thus reduced somewhere to the proper value for enhancing precipitation.

At this point, it can only be stated that a view of seeding based upon the notion that the nucleant should be injected at a given concentration into a cloud at a given temperature level and a given location, and that all the desired effects will occur right there, is a gross oversimplification. It is important to know about the transport and diffusion of nuclei and ice crystals formed by nucleation before optimum dosage rates and location of nuclei sources can be formulated.

Over complex mountainous terrain it has been found that nucleant smoke rises and spreads rapidly. That the plume top can rise to 1 1/2 km, 5 km downwind, has been demonstrated in the Climax area by aerial plume sampling and in wind tunnel modeling.[48]

Complex valley and ridge terrain patterns also influence the manner in which the air flows over the mountain. T. J. Henderson[28] conducted a research program in the Sierra Nevada where he was carrying out operational seeding at the time. He traced the AgI smoke plume by means of a portable cold box on the ground and in the air. He found a very intricate pattern of flow in the foothills. In general, there is a speed up of air as it flows over a mountain barrier, and this has an appreciable effect on transport rates. There are several numerical models in existence which predict the basic transmountain flow pattern. Required inputs to such models are air mass thermal structure and wind data obtained by balloon soundings taken upwind of the orographic barrier. This is one component of larger models of the entire orographic seeding pattern.

Various cloud seeders within the private sphere have developed what may be termed early conceptual models for organizing their procedures in the conduct of orographic cloud seeding. Components of such models would include the transport and dispersion of nucleant, the nucleation process at various temperature levels, and the growth and fallout of ice crystals. One such model which included these features in numerical form was developed for seeding in the San Gabriel watershed by the author's company for the Los Angeles County Flood Control District. Subsequently, the author has developed a more detailed orographic seeding model suitable for rapid computerized calculation.[15] Grant[26] developed an orographic seeding model which explains the results obtained in the Climax experiment. In connection with the Park Range project conducted for the Bureau of Reclamation, another computerized orographic model was developed.[50]

Generators located on the ground in mountain valleys may at times be under a local inversion which will effectively eliminate the rise of the nucleant up the slope. Generators are, wherever possible, placed well above any possible inversions. This practice has led through the years to locating generators at higher and more remote locations. In the early days of orographic seeding, it was the practice of the author's company to install couples, serving as generator operators, at various high-level sites where suitably located summer cabins could serve for winter occupancy. During the 6 or 8 months that they were snowed in, communication with them was carried on by radio. Surprisingly, it was not difficult to find couples willing to do this work since many people find this type of isolated life a challenge. In spite of their enthusiasm, various sorts of problems arose as a result of the deep snow accumulations as represented by Figure 2.2. One man skied out in midwinter, leaving his wife forever.

Fortunately, we were able to bring a replacement couple in, and her out, in one snowcat trip. A couple ran out of fuel for the radio motor generator—they had used too much to run their hi-fi. Another man put his extra time to good use. He wrote his doctoral dissertation on cotton culture.

In the interests of avoiding human problems, attention was focused upon developing remote radio-controlled generators and the first test installations were made in the late 1950s in the back country of the upper San Joaquin River, above Big Creek. The Southern California Edison Company microwave system was used to telemeter signals to and from the test generator. The generator was designed to open its gas and liquid valves in response to a command signal, ignite the mixture ejected

Figure 2.2   Seasonal snow accumulation: (*a*) fall; (*b*) early winter; (*c*) late winter; (*d*) spring, with only the radio antenna showing above the roof.

Figure 2.3 Radio-controlled silver iodide smoke generator. The generator is located on Swan Peak in Montana at 9500 ft altitude. The generator is mounted above the silver iodide acetone containing tank atop the left-hand tower. Its bonnet is in *up* position for maintenance. Electronic gear is stored in the round top hut, which is entered in winter down through the Santa Claus chimney. The tall object on the tower to the right contains the antenna. The object is a plastic shroud which minimizes ice loading. The basic energy comes from propane tanks, not shown. In addition, batteries are charged by means of solar panels mounted on the mast in back of the Santa Claus chimney.

from the burning head, and on being interrogated, return a signal indicating that it was operating properly. In the event of a blowout in the strong mountain winds, it would relight itself. After turnoff, it had to purge its own pipes of remaining fluid to prevent clogging. Tanks sufficient to hold a season's supply of propane and nucleating agent were essential. The unit was mounted on a tower in order to clear the deep snow.

The development was successful and such units are in use now in several projects in western United States. It adds much to the flexibility of ground seeding operations. Figure 2.3 shows one such installation in rugged terrain in Montana. Servicing of this particular installation can be performed only by helicopter.

In actual operations, there are a number of commonly practiced seeding modes. One mode is to seed continuously into the airstream from a ground nucleant generator at a fixed point. This is most commonly used in mountainous terrain. Orographic upslope flow, often combined with entrainment into convection, is the means for transport up to nucleation level, with the required dilution mixing occurring along the way. If the generator is located over flat country at temperatures well above freezing (as in summer), entrainment into convection is the only possible means for attaining the nucleation level, which may be 10,000 ft or more above.

Two types of nucleant have been used with ground-based generators. One is the AgI and NaI complex, the other AgI and NH_4I, which does not form a permanent complex. The two have quite different nucleating properties.[13] Generator outputs are generally 6 to 20 g/hr, but much higher output ground generators are occasionally employed.

A third type of nucleating device is the pyrotechnic flare. This has been used successfully in the air as well as on the ground. Its results can be grouped with those of the AgI-NH_4I type, although the output is much higher. Flares are ignited at successive time intervals and an output of several hundred grams per hour is not uncommon. Flares of various types were used experimentally in the private sector from the earliest times. However, their use was not placed upon a firm basis until the 1960s when the Earth and Planetary Sciences Division of the Naval Weapons Test Center developed pyrotechnic devices for all types of use, including hurricane seeding.[51, 52]

A second mode is to seed into single cumuli by flying into the updraft at cloud base, or circling beneath it. A third is to fly into and seed an ascending turret at the −5 or −10°C level. High-output pyrotechnic flares are commonly employed in these modes.

Still another mode is to seed continuously from an airplane flying in a line just upwind of the clouds to be seeded (patrol seeding). The flight line is usually between cloud base and the −5° level. Generator output is usually several hundred grams per hour (ram type generator). The reader should refer to the cumulus modification chapter for more details. Pyrotechnic flares may also be dropped down into promising cumulus towers.

It should be noted that cumulus clouds act as giant vacuum sweepers. They process a huge amount of air, much of which is gathered into the base from an extensive subcloud region. When silver iodide smoke, or other tracer material, is released in this lower region by aircraft or at ground level, the tracer is characteristically found concentrated inside the cumuli, with little or none between clouds. However, in seeding cumuli in this manner, if the distance to cloud base is great, the nucleant may be distributed into many clouds, with a large dilution.

Furthermore, with only partial cloud cover, photolytic decay of silver iodide occurs on exposure to sunlight and this can render the seeding ineffective.

An interesting way of ensuring the ascent of a nucleating agent from the ground level in rough terrain is by employing small sounding balloons to carry dry ice aloft. The dry ice, in pea-sized pellets, is suspended beneath the balloon in a hair net. Releases are made from certain ground positions every 15 min, the position depending upon the wind pattern. The author's company carried out such a project for the Pacific Gas and Electric Company over a small watershed in the Sierra Nevada.

The General Electric Laboratories suggested, or did preliminary testing on, a surprising number of generator types. It remained, to a considerable extent, to the private sector to develop their capability to function reliably in the real world of storms, rough terrain, and airplanes. Howell improved upon the string type generator and used it extensively. Krick developed and used an impregnated coke type generator. Later he made extensive use of an electric-arc type generator.

One way of determining where the silver iodide particles end up is to sample the precipitation in and around the target area. Naturally, the few grams of finely divided silver iodide, emitted from a generator over a period of minutes or an hour, are widely dispersed and concentrations found at ground level in the precipitation are very low. However, there are special techniques (spectro-photometric and neutron activation) by which amounts of silver as small as one part in 10^{12} can be detected. This is near the natural background count found in most areas. The method requires collection of samples of precipitation which must be frozen and shipped to a laboratory equipped to perform the necessary analysis.

In the experience of the author's company, concentrations of 10^{-9} to 10^{-10} of silver/g of water were common in the target areas of the Hungry Horse project in Montana, while upwind the values were near background (10^{-11} to 10^{-12}). The finding of silver in the target is proof that the targeting was good, but it is not evidence of a precipitation increase. In fact, with overseeding there could be a decrease in precipitation. The indirect dynamic consequences of cloud seeding, to be discussed below, can also produce precipitation increases where no silver is involved.

The question of toxicity of the effluent from a generator has been covered in detail by W. J. Douglas[14] of the Division of Atmospheric Water Resources Management of the U.S. Bureau of Reclamation. He found that the effluent produces no smoke concentrations which could be classed by public health service standards as toxic. The arthor's firm made computations of particle concentration in the plume over the Hungry Horse target. They were about $0.0001~\mu g/m^3$. The Environmental Protection Agency has set a primary ambient air standard for particulates at $75~\mu g/m^3$ on an annual basis. The values of silver found in the Hungry Horse snowpack are 1/50 to 1/500 of the limiting value set by the Public Health Service for silver in water. The secret of seeding is the production of a large effect from a small amount of material. Hence pollution is not a problem.

Dynamic effects of seeding refer to the possibility of producing a more profound effect than the release of droplets from a given cloud. A by-product of the transformation of cloud droplets to ice particles is the release of heat of sublimation. Bergeron[5] enunciated the principle that stimulation of an updraft would occur as a result of this heat release; the buoyancy of the updraft is increased, and the air accelerates upward. An increased updraft means a faster condensation of cloud water and this added condensation could in turn be subjected to seeding. He termed this "double release."

A stimulated updraft, and enhanced buoyancy, could lead to a rise in the cloud top. The cloud tops were observed to rise following seeding by a number of experimenters in the 1940s; the Australians observed this effect in 1947.[37] Some early cloud seeding reports by Krick's group on projects in southwestern United States and in northern Mexico in the late 1940s comment on dynamic rise in cloud tops, and expanding rain areas, following the seeding of cumuliform clouds with dry ice. Not only were updrafts beng increased, but the added buoyancy was carrying the updraft through inhibiting stable layers of inversions at the old cloud top level. The volume of condensed water in the cloud at any given time was thus enhanced.

Although the dynamic effect in the seeding of cumulus clouds was known for years, it is only more recently that it has been carefully studied and seeding procedures developed to exploit the effect. Numerical models now predict, from balloon sounding data, what increase in updraft and rise in height can be expected through seeding and give indications of the seeding-produced increase in precipitation. The models show that under typical conditions seeding may increase the temperature of the updraft by only a third of a degree and yet the updraft can be doubled and the top raised by a kilometer or more. The principal working models have been developed by Dr. Joanne Simpson of the National Oceanographic and Atmospheric Administration (NOAA) Laboratories and her staff[54] and Drs. A. J. Weinstein, P. B. MacCready, and L. G. Davis.[60-62] All three of the latter have been active in the private field although Davis and Weinstein did much of the model development as doctoral dissertations at Pennsylvania State University. (Reference should be made to the Chapter 6 for more technical details on the very interesting matter of cumulus cloud models.)

An interesting application of seeding involving the alteration of buoyancy has been its use to prevent, or reduce, the incidence of strong wind gusts generated by thunderstorms. They are the result of heavy rains chilling the lower layers of air through evaporation, thus producing a negative buoyancy with a resultant downdraft. These winds produce great damage in tropical banana-growing lands. Howell, in the late 1950s, conducted such a project near Santa Marta, Columbia, in which he seeded to get showers started earlier than the usual time of the day.[41] In this way he hoped to dissipate the total energy of the system over a longer period. There might be as much total wind movement but it would be less concentrated in time. Also, the early production of high and intermediate clouds as a by-product would reduce sunlight and surface heating. The seeding was intensive, with typically 250 generator-hr of operation in one day. The evaluation of results indicated a 20 percent reduction in damage. Another similar project, planned and evaluated by Loren Crow, a consulting meteorologist, was conducted in Panama by Boyd Quate, of Weather Engineering.

Possible larger-scale effects of seeding, that is, larger in scale than the immediate effects on the particular group of clouds seeded, are now considered. The popular question along these lines is: does seeding rob the area lying downwind from the target area of precipitation they might otherwise receive? Several articles published in law reviews in the late 1940s and early 1950s proposed, or rather assumed, that this was so, and then

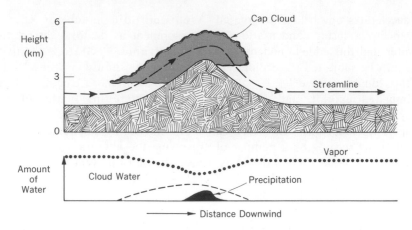

Figure 2.4 Schematic diagram of airflow, cloud, and precipitation over an orographic barrier for a stable cap cloud case.

proceeded to apply the laws of riparian rights in analogy to the removal of water from a stream. Another variant of the question is: does seeding in the target area produce heavier rain at great distances downwind, perhaps even a thousand miles away?

The matter of very large scale effects of seeding (as much as 2000 mi downwind) was advanced by Langmuir, who had seeded with a ground generator from Socorro, New Mexico every seventh day over a period of several years (1949–1952). He examined rainfall[40] and other weather statistics throughout central and eastern United States, and claimed to have discovered a response to this periodic seeding. His findings were disputed by other scientists. Before he completed his tests, other seeding in the United States became widespread enough so that he felt any effect he produced might be masked from then on (see Chapter 1). His contentions were listened to by Senators Anderson and Case, and were partly responsible for their introducing legislation in Congress in the early 1950s to establish some kind of Federal weather control. Extensive hearings were held in 1951, and are discussed in more detail in the following paragraphs.

To convey to the reader the current understanding about the scales of seeding effects, reference will be made to a series of sketches. Figure 2.4 depicts an orographic cap cloud, under stable air mass conditions. Air ascends on the upwind side (arrow) and condenses into a cap cloud. The cloud remains fixed as the air blows through it, although the individual droplets move with the airstream. In the descending motion on the lee side, evaporation occurs. Since some cloud water is converted to precipitation, which falls from the cloud onto the mountain, there is less water to evaporate on the lee side than condenses on the windward side, and the ceiling is higher there. In many cap clouds there is no precipitation whatsoever, and the ceiling is as high on the lee side as on the windward. The lower portion of Figure 2.4 depicts in idealized form the total amount of water contained in the air mass lying above a point at the surface (corresponding to the ground point in the upper figure) in each of the three forms: cloud droplets, precipitating particles, and water vapor. By far the largest amount is in vapor form. Water in the form of cloud droplets increases on the ascending portion of the barrier because of condensation, and decreases on the descending portion because of precipitation loss and evaporation. There is a net loss of water in vapor form. In the absence of precipitation water, there would be no net loss of vapor. It should be noted that the precipitation rate at ground level is not the same as the precipitation water. Precipitation water is what would be measured by suitable sensors flown at various levels through the clouds. Because of the drift of precipitation particles as they fall to ground level, precipitation rates are highest just downwind of the peak in the precipitation water curve.

In a typical precipitating cap cloud about 20 percent of the water vapor in the upwind air mass (which we shall assume is cloud free) condenses. Of this, about 20 percent falls out as precipitation. Therefore, $100(0.20 \times 0.20) = 4\%$ of the water vapor is removed. If cloud seeding increases the precipitation by 10 percent, then the water vapor is depleted an additional $100(0.10 \times 0.04) = 0.4\%$. a relatively small figure.

The argument is quite valid and has been used for many years to explain that a rather trivial reduction in total water would occur in the area downwind of a target area. Such a small deficit would soon be reduced to nought through mixing with the air mass to either side. However, when there is a dynamic effect, the picture is more complex.

Let us consider the orographic case in which the air rising up the slope in unstable and cumuliform clouds are present, buried within a stratiform layer. Figure 2.5 depicts this situation, using the same format as Figure 2.4. In this case the streamline of the broad airflow over the mountain has superimposed upon it little transient eddies (also depicted by arrows) which are the updrafts in convection cells. There is more condensation because of this added motion, and also more precipitation. However, a considerable amount of the convection- produced extra condensation evaporates between cumulus clouds, especially near their tops. Seeding has two effects; it hastens the conversion of cloud droplets to precipitation and it creates new buoyancy, new convective updrafts, and therefore new condensate. In a sense the effect is to heighten the mountain barrier by superimposing a "convective moun-

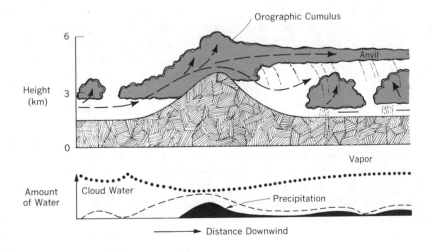

Figure 2.5 Schematic diagram of airflow, cloud, and precipitation over an orographic barrier for the case of convection embedded within stratiform cloud.

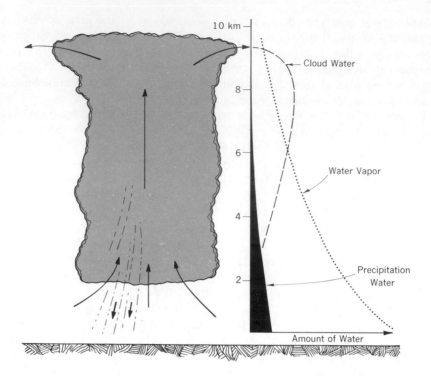

Figure 2.6   Schematic diagram of airflow, cloud, and precipitation in an isolated convection cell.

tain" on top of it. Therefore, it adds to the precipitation in two ways, and there is a little more net loss in water vapor in the downwind region. However, this ignores one other effect which is significant in this orographic situation, and also in isolated cumulus seeding over flat country.

When updrafts are enhanced there will be an increase in inflow near the base of convection and an increase in outflow near the top. This is required by mass continuity. From the energy point of view, the added buoyancy represents added potential energy, which is converted to kinetic energy in the form of horizontal inflow and outflow motion, as well as of updraft motion. Therefore, the dynamic effect of seeding includes this mechanism: it acts as a pump which gathers together the moisture-laden lower air mass, lifts it up, condenses it, and

ejects air somewhat reduced in total water content if precipitation has occurred. This can be understood better by reference to Figure 2.6, the simple convection cell uncomplicated by orographic lifting. The inset figure to the right depicts the usual distribution of total water in an air mass. About 90 percent of it is confined to the lower half of the mass of the atmosphere.

At the end of one pumping "cycle" the air ejected at the top is more moist than the upper environment. The air near the base of convection is unchanged in moisture content, having been replaced by air drawn in laterally. At the same time, precipitation has been enhanced. In the seeding of an unstable air mass, the correct analogy for the lawyers to use is a pump which draws water up from a limitless ocean of moisture, not a diversion of water from a stream.

There has been a burst of activity in recent years in the analysis of data in the area well downwind of seeding projects. A National Academy of Sciences report in 1966[56] made veiled reference to unexpected increases and to decreases in precipitation far downwind of Project Whitetop. Statistical analyses of precipitation in the area several hundred miles downwind of a group of long-term commercial seeding projects by my company[10] did indicate some positive effects up to 100 mi downwind and a similar analysis in Australia indicated positive effects even further downwind.[1] But the real test of such effects comes from looking at the areas downwind from long-term randomized seeding tests. Recent studies[17, 18] of the results of long-term randomized seeding projects have clarified the meaning of a number of heretofore partially understood or misunderstood data from the field. In particular, the analyses of data have been extended 150 to 200 mi downwind from two long-term randomized seeding projects, one being the Santa Barbara project (conducted by my firm) and the other the Climax, Colorado, project (conducted by Colorado State University under Professor L. O. Grant). In both cases, positive precipitation effects were found to extend somewhat downwind of the conventional and originally (in the case of Climax) evaluated target area. In the Santa Barbara case, the area of effect model was employed to compute, on an extended grid, where the seeding-enhanced fallout could reach ground level. It was found that the model did indeed predict the full downwind extent of the effect, although it failed with respect to certain features of the observed effect. In the Climax case, the available evidence suggests that an excess of ice crystals were ejected from the top of the cloud seeded over Climax and drifted to the east downwind as an artificial cirrus anvil. Some oversized ice crystals falling from this cirrus then seeded lower-level clouds. It is known that a few extra large crystals can fall great distances downward in sufficient concentration to seed lower-level cumulus clouds.[9]

The data indicate that in about 10 percent of the cases when seedable conditions occurred at Climax, there were relatively low clouds in an Arctic air mass over the Great Plains. With top temperatures near $-10°C$ (tops at 12,000 ft, lower than the 18,000 to 20,000-ft tops over the Rockies) they were excellent candidates for seeding and responded to the falling crystals. This process is illustrated in the right-hand part of Figure 2.5. This effect was detected mainly about 100 to 150 mi downwind of Climax (see Chapter 7).

In the Whitetop randomized seeding project, a University of Chicago group under Professor Roscoe Braham seeded cumuli over Missouri during four summer seasons. They employed the patrol method of aerial seeding in a line upwind of the target area, expecting the nucleant to drift downwind and become entrained in cumulus clouds there. The data analysis[8] appears to indicate that positive effects occurred only in the cases where radar tops were in the 20,000 to 40,000 ft range. This is about the -10 to $-45°C$ cloud top temperature range. In the case of tops higher than 40,000 ft, marked decreases occurred. In the lower than 20,000 ft category, there were decreases, but they were not statistically significant. The net result was decrease in precipitation (see Chapter 6). Although this approach to analysis is complicated by not including the effects of the seeding on the height of cloud tops, nevertheless it conforms in general to the responses found elsewhere, and now predicted by models, on the basis of cloud top temperature. In a separate study which extended up to 200 mi downwind, a net decrease was said to have occurred,[46] but the reality of this was contested.[53]

Finally, it is known that a large thunderstorm produces a slow high-level descending motion (subsidence) over an area extending some 60 mi from its center. This suppresses surrounding cloud development. Updraft enhancement in the seeding of convective systems which are naturally large can conceivably produce a suppressing effect in the outside area. The nature of this effect is being investigated, largely through the activities of the South Dakota School of Mines, Institute of Atmospheric Sciences.

With regard to the role of the coalescence process versus the Bergeron ice nucleation process, it can be stated that this is fairly well in hand at present. Originally, the focus of weather modification was on the ice process. As time went on the Commonwealth Scientific Industrial Research Organization in Australia and other scientific groups became interested in the manner in which some clouds produce rain even though their entire mass is at temperatures above freezing. Attention soon became focused on this problem, sometimes almost to the exclusion of the ice process. Models were developed to predict how fast the collision and coalescence of cloud droplets might lead to the formation of precipitation size particles. It became evident that this was highly dependent upon the nature of the natural condensation nuclei present.

There are many more condensation nuclei available over land than over seas. Hence there are more cloud droplets to be found in continental clouds, more than $500/cm^3$, as against 20 to $100/cm^3$ in maritime clouds. The water is shared over many cloud droplets in continental clouds and consequently the continental cloud droplets are smaller than maritime ones. The chances of growth by collision and coalescence are best when there are a few large droplets (50 to 150 μ) among the small ones, as is the case in maritime clouds.

The addition of any small condensation nuclei might reduce the chance of rain, particularly in the case of continental clouds. Some confusion arose because some cloud seeding generators add condensation nuclei (NaI particles), as well as ice-forming or sublimation nuclei (AgI particles) to the atmosphere. For example, *Science Newsletter* for May 4, 1957 contained an article reporting that Dr. Ross Gunn of the Weather Bureau had found that cleaner air had a better chance of rain from warm clouds. Contaminating particles could lead to a decrease in rainfall. The article then quoted Dr. Francis Reichelderfer, Chief of the Weather Bureau, as calling for discontinuance of "hit or miss cloud seeding." The article suggested that Gunn's results contradicted the idea of adding particles to increase the rainfall as cloud seeders had done. By failing to make the distinction between condensation and ice-forming nuclei, a badly garbled picture emerged as to what cloud seeding constituted. The output of ice-forming nuclei from the conventional generator is truly enormous, but the output of condensation nuclei is about the same as that from a home furnace.

It would appear that seeding with water drops could generate more precipitation because the drops in falling through a cloud would collide with and collect many cloud droplets. The University of Chicago found that this technique did indeed work, and Howell subsequently carried out a number of water seeding projects in tropical areas (see Chapter 6).

Salt particles (NaCl, or NH_4NO_3 and/or urea) of about 5 to 20 μ size, if injected into clouds, grow rapidly because of their hygroscopicity, growing to 40 to 100 μ very quickly. Subsequently, precipitation develops as the larger grown particles start colliding and coalescing with the smaller cloud droplets. This method of seeding has been practiced by Howell in the tropics as a supplement to AgI seeding.[36] Tests of the effects of seeding by brine spray from the ground under monsoon conditions were conducted at New Delhi using ground-based spray rigs.[6]

The problem in connection with water or salt seeding is that it lacks the multiplying effect of silver iodide seeding. The salt particles are over a million times larger in mass than AgI particles. They are in concentrations of 100 times less. The logistics of return of precipitation per ton of water or salt delivered is low. However, in 1948 Langmuir pointed out a process by which a chain reaction could multiply the effect enormously.[38] Raindrops are known to break up when they grow to a diameter of somewhere between 5 and 10 mm. They become hydrodynamically unstable and break into perhaps 10 droplets. Their terminal velocity at breakup point is about 8 m/sec. Such a drop may find itself suspended at a fixed level in an 8 m/sec updraft. This could happen if the peak of the updraft were over 8 m/sec and lay below the level of the drop. After breakup, the little drops move upward toward lower updraft velocities, growing by coalescence on the way up until they become large enough to fall back. They stop again at the 8 m/sec level, then break up again, etc. If salt seeding enhances the initial growth of a few large drops, then under these circumstances the overall effect can be greatly multiplied. The extra rain thus generated falls out after the updraft dies away, usually in about 20 min.

It turns out that summer cumuli produce precipitation by a mixture of the coalescence and ice processes. Hail embryos are apparently the oversized drops produced originally by the coa-lescence process and then frozen. There now exist computer models which predict these processes with some veracity and which can distinguish between the results when using various types of salts with spectra of different size.

In the seeding of summer cumuli, the stage of the development of the cloud is important. The heating of the ground by the sun in turn heats the lower layers of the air mass and creates instability, with the rising thermals carrying some air up above the condensation level and producing clouds. Depending upon the stability of the air at higher levels, the clouds may either top off at levels too low for precipitation to form naturally, or may go on developing to levels where full precipitating systems develop. If the first condition prevails, then in clouds whose tops are not going above the freezing level, ice nucleation seeding would make no difference. If the tops rose to above the $-5°C$ level, then seeding would start precipitation that would not have occurred naturally. Dynamic effects could lead to a multiplication of the precipitation production.

In the second case, seeding could lead to an enhancement of the natural rain production. However, premature seeding might start precipitation but reverse or restrict the natural growth of cloud later in the day. Thus it might advance the time or precipitation without necessarily increasing the total precipitation.

W. E. Howell has presented a detailed appraisal of the importance of stage of development in connection with the seeding of tropical cumuli.[32]

As to the evaluation of results of seeding against an enormous and highly varying background of natural precipitation, experience has shown that a fully information oriented field project cannot be expected to produce results of statistical significance in less than 3 years, and most require 5 years or more.

Most of the early experiments by government agencies, many of which were deemed initially to have shown the lack of value of seeding to increase precipitation, were run for too short a period to prove what it was contended they proved. What they did prove was that more experimentation was needed to discern the effects for which the tests were presumably designed.

In most cases it became clear that basing the analysis wholly on randomly selected seed/no seed precipitation in the target area led to a very insensitive test. The use of control areas helped, but even this could be improved by stratifying the data according to cloud top temperature and other observed physical variables. After sorting out the results in this way, it became clear that in the prior testing "lemons had been mixed with apples," confusing the results. Once a separation was made, rather clear-cut results appeared.

Another way to sensitize the testing procedure is to work with shorter time intervals (or smaller observational units). For example, many individual cumuli can be tested quickly, building up a large number of cases in the sample in short order. In Santa Barbara it was found possible to treat bands of convective activity passing through the test area as separate units, rather than whole storms. In 4 years more than 100 band units were tested, whereas in the same period there were many fewer seedable storms.

The private sector pioneered in the evaluation of project results where the non-seeded sample consisted of historical data. A statistical relationship, usually in the form of a regression equation, would be developed between precipitation in the target area and that in a control area, on the basis of historical precipitation records. This relationship could then be used during the seeded period to predict what precipitation should be expected in the target on the basis of the observed control area precipitation (see Chapter 5 for more details). A difference between observed and expected target area precipitation could then be attributed to the seeding, assuming the target was affected by the seeding, but not the control. The prediction of target area precipitation could then be attributed to the seeding, assuming the target was affected by the seeding, but not the control. The prediction of target area precipitation has an error, and statistical tests give the probability that the observed difference is simply caused by this error. There remain, of course, some doubts about the consistency of the historical records, and this is why randomization is practiced in purely information oriented projects.

This regression method has been applied to storm, daily, monthly, and seasonal precipitation. It has also been applied to April 1 snowpack water content as measured on the snow courses. Snow course measurements have been made at monthly intervals during the winter at hundreds of snow courses in the West for the last 20 to 40 years and these records provide a fund of useful information. The maximum snow accumulation usually occurs about April 1, and accurate forecasts of seasonal stream flow, so important to the farmer, are based upon these data. Seasonal stream flow itself has been used for evaluation by comparing that issuing from the target watershed to that from some nearby watershed.

One of the earliest evaluations of the results of seeding is the private sector was published in 1953.[27] The California Electric Power Company had conducted seeding on the eastern slopes of the Sierra Nevada for several years starting in 1948. W. F. Hall of the Weather Bureau joined with T. J. Henderson (of California Electric, at that time) to produce an evaluation indicating that runoff had indeed been increased.

In 1954, an evaluation of 3 years of orographic seeding on three projects was published by Elliott and Strickler[23] of North American Weather Consultants. About a 20 percent increase in April 1 snowpack was claimed. Later, in 1966, Elliott and Walser[24] published an evaluation for three long-term projects which indicated increases in runoff. Two of the projects covered were simply long-term continuations of those covered in the original article. An evaluation of 15 yr of seeding on a Sierra Nevada project was reported in 1967[22] and of ten years of seeding on still another Sierra project in 1966.[29] The evaluation of results of a long-term project in the California coastal mountains of Santa Clara County was published in 1966.[12]

The efforts at evaluating California cloud seeding should be mentioned. One of the earliest independent evaluations of weather modification was conducted by the California State Water Resources Board.[59] The California reports lists individual projects conducted from the 1947–1948 season to 1955. There were a number of small projects in the earlier years performed by agricultural air services for farmers, consisting mainly of aerial dry ice seeding. Gradually more systematic seeding was taken over by such weather modifiers as Water Resources Development Corporation (I. P. Krick), Precipitation Control Company of Phoenix (C. L. Barnes), North American Weather Consultants (Bollay and Elliott), and Weather Modification Company (O'Hanlon and Battle).

A statistical evaluation was performed on two projects, Santa Barbara County and Carrizo Plain. The former employed ground-based AgI generators, the latter aerial seeding with AgI generators. The evaluation was based upon comparison of seeded storm precipitation with nonseeded storm precipitation from the historical records. Storms were categorized into three types corresponding to the basic steering flow: meridional storms, intermediate storms, and zonal storms. The Santa Barbara data indicated significant increases in the intermediate storm cases, and Carrizo Plain significant negative effect in some types in certain years, and no effect in other cases.

The complexity of the precipitation mechanisms, natural or artificially modified, was recognized. Additional evaluations were made of a project in the southern Sierras. A surprisingly comprehensive analysis of the economics of a 10 percent precipitation increase was made for a typical California watershed and for a marginal farming zone in the northern Great Plains.

This report lay the groundwork for future cooperation between the state, the University of California, Santa Barbara County, and North American Weather Consultants in the SBA I project, 1957–1960.

Howell and his associate, Lopez, published a number of articles covering the results of cloud seeding in South America.[34, 36, 42] These articles contain very interesting accounts of the local peculiarities in cloud generating areas of the Andes and explain how special seeding problems were met. One of the projects[34] had been carried on for 12 years at the time of the article. Howell also published results for silver iodide seeding in a group of projects during the drought period in northeastern United States.[35]

Seeding of summer cumulus by the Mexican Power and Light Company over the Necaxa basin has been reported on by E. Perez-Siliceo.[49] This project had been carried on a daily randomization schedule for 12 years at the time of reporting.

There have, of course, been innumerable evaluations performed by the private sector which did not appear in the scientific literature. Most of these are recorded in evaluation and operations reports to the clients. These reports are written on a professional level but receive only limited distribution since they are the property of the client. However, it is interesting to note that permission has been granted through the years by the industries concerned to send copies to interested congressional committees and to government bodies investigating weather modification. Krick alone submitted copies of about 200 of his reports to Senate hearings on weather modification, Bill S.23, 2nd Session, 89th Congress.

The private sector has also been involved in the conduct of purely information oriented projects. Meteorology Research Inc. and its work on summer cumuli at Flagstaff have already been mentioned.[43] A series of experiments and studies have also been conducted in the Santa Barbara County area by my company. The first Santa Barbara randomized seeding project (1957–1960), besides my company, involved an effort by the State of California and the University of California's statistical laboratory. Support was rendered by the National Science Foundation, the State of California, the Counties of Santa Barbara and Ventura, the Weather Bureau, and the U.S. Forest Service. The results were provocative, there being overall considerably more precipitation in the seeded cases. However, the statistical significance of the results remained in doubt, partly as a result of the inclusion of Ventura County into the project after its commencement.[47] It was realized that physical measurements other than rainfall were required fully to evaluate a project. In the subsequent storm study, conducted by the author's company under a National Science Foundation contract, an extensive data collection (sequential rawinsondes at several locations, aerial cloud physics collection, and an extensive rain gauge network) provided a basis for examining in detail the structure of frontal systems responsible for winter precipitation production in the area.[21] It was found at this time that most of the precipitation and most of the updrafts and supercooled cloud forms most suitable for cloud seeding occurred within relatively narrow traveling bands of convection clouds buried within the general cloud mass and spaced out about 50 mi apart.[19]

In the second Santa Barbara randomized seeding project (1967 to the present) being conducted by my company under contract to the Earth and Planetary Sciences Division of the Naval Weapons Center, China Lake, California, it was possible to design the project in such a way as to use the convection bands themselves as units to be seeded, or not seeded, on a random schedule. Additional sounding and weather radar information were also obtained. With this finer resolution and more complete collection of physical data in addition to band rainfall amounts, it was possible to arrive at quite significant results.[16] A stratification of the data by cloud top temperature clearly sorted out the effects of overseeding at colder temperatures and explained some of the previously unexplained effects found in the first Santa Barbara project.

Another randomized test project conducted within the private sphere was that carried out by Pacific Gas and Electric Company in the Lake Almanor drainage basin of the Sierra Nevada range during the 1970s. They were able to sort out the types of storms within which seeding was most effective.[45]

The detailed problems of evaluation of results of seeding cannot be dwelt upon here. There have been a number of false conclusions drawn in the past about seeding effectiveness, or ineffectiveness. For the most part these erroneous conclusions have stemmed from the examination of the results of a project whose duration has been too short and whose data collections have been too limited to warrant the drawing of any firm conclusions. Initially there was a strong tendency to assume that results found in one area, using a certain mode of seeding, pertained to all cloud seeding conducted in any area and by any mode. In the course of time and shortcomings of this oversimplified view became apparent.

Personal convictions as to the effectiveness of cloud seeding are apt to come to the scientific experimenter in a different and quicker way. The beginnings of conviction most often occur as a result of direct visual observation of effects, an effect which is found to occur repeatedly. The seeding of clouds by aircraft can be most convincing, as Schaefer discovered. In the author's case conviction developed following the seeding and observation of two separate fronts which passed Pasadena in April, 1950. The generator was located on the roof of the office building. During the first frontal passage the wind blew northeastward up the front face of the east-west oriented San Gabriel range, with its crest line 5 to 7 mi away. Heavy clouds lay over the mountains, with lighter clouds over the valley. Light intermittent rain fell at the seeding site. The front was rather weak. Within 20 min after seeding commenced, the sky downwind grew darker, and it was quite evident that heavy rain was falling from cloud base. The entire mountain area was darkened. L. O. Grant, who later became a professor at Colorado State University and planned and carried out the Climax experiments, had driven some 10 to 15 mi down into San Gabriel valley to get a panoramic view of the phenomenon. He reported in by telephone that heavy dark clouds concentrated in the downwind area, and nowhere else. Later, an examination of recording rainfall records from the mountain stations showed much heavier amounts in the downwind area. The seeding of the second front was almost a repeat of the first. Twice in a row these extraordinary observations were made.

Previously, Grant had been greatly impressed by his observations made in aerial seeding of cumuliform cloud with dry ice over San Diego County, in Arizona, and in Chihuahua, Mexico, during 1948 and 1949. He reported that some clouds responded to the seeding not only with intensified rain, or new rain, but with a lateral spreading of the rainfall area and sometimes with a spectacular rise in cloud top. Subsequent analysis of rainfall observations seemed to verify the visual observations. Further ground seeding tests in the State of Washington and in New Mexico in 1950 showed similar effects with observed precipitation patterns difficult to duplicate in the records.

Analytical knowledge about the various microphysical and dynamic processes in clouds and their interactions has been synthesized into numerical models whose predictions bear a reasonable facsimile to observations. One might ask the question: why not use the empirical knowledge so laboriously built up as a sole basis for designing and conducting seeding projects of the future? Are not statistical-empirical models adequate to the task? The answer is that the numerical models offer several advantages which are absent in empirical models. First, they serve as a means for transferring knowledge about seeding effects from areas where we have empirical knowledge to an area with a different climate, different terrain, and different cloud forms, and for which we have no empirical knowledge about seeding effects but where we do possess knowledge about the weather parameters (air-mass temperature, stability, moisture, wind) that are inputs to the numerical models. The numerical models provide a means for synthesizing our general theoretical and empirical knowledge in such a way as to avoid repeating the whole elaborate procedure of experimental test seeding in each new region before starting operational seeding. Certainly the private sector has accepted the approach of applying theoretical concepts and models to many strange meteorological situations around the globe.

Secondly, computerized numerical models offer a means for testing the effects of varying modes of treatment, dosage rates, etc., without going into elaborate field tests. This sensitivity testing can also be applied to naturally varying meteorological inputs, so that one does not have to wait for a long period to make an observation of seeding effects under certain unusual weather patterns.

Some numerical models can also be used in a practical way as a basis for operational decision making. They are becoming an important tool in cloud seeding.

INTERACTION WITH THE FEDERAL GOVERNMENT

Weather modification in the private sector has from the beginning interacted with the activities of Federal government legislative bodies, agencies, commissions, and committees. In order to convey to the reader the true flavor of the private sector, some of this interaction is covered in this section.

The Weather Bureau commenced a series of field tests in the summer of 1947, which continued through 1948. It was called the cloud physics project. Aerial seeding of cumuli was carried out near Wilmington, Ohio, and in the Gulf states. Ground seeding was tried briefly in the Sierra Nevada mountains of California. Some response to seeding was observed but it was concluded that whenever rain

followed seeding, there was always rain without 30 mi anyway, so it could have been a natural development. It was also concluded, without any real analysis of the economics involved, that any rain which had been produced was of no economic significance. These conclusions were at variance with those of the cloud seeding enthusiasts.

At about this time Dr. Francis Reichelderfer, Chief of the Weather Bureau, became convinced that Langmuir's claims were exaggerated, and that unscrupulous commercial seeders were reaping fortunes, or were about to, by using his claims. In the interest of protecting the public, the Weather Bureau issued public statements designed to caution potential customers of cloud seeders. The statements pointed out that there is actually only a few hundredths to a tenth of an inch or so of water in any given cloud. This is the most that seeding might remove. The term "fraudulent rainmaking" appeared in official documents, although who the rainmakers were was not defined.

Langmuir's assessment of seeding effectiveness, on the other hand, increased as time went on. As a result of seeding in New Mexico in October, 1948,[39] he claimed that 0.37 in. was produced in one seeding over an area of 4000 mi^2. This was, according to Langmuir, 10^8 tons of water (later revised upward to 8×10^8 tons). Odds against this occurring by chance were a million to one, he claimed. The Weather Bureau challenged this, and there were debates at various National Academy of Science and American Meteorological Society meetings over this and other points during 1949 and 1950. When the Weather Bureau representative pointed out the numerous meteorological reasons why the New Mexico rain could have developed naturally, Langmuir countered by asking, if they knew this much about it, why had not the Bureau forecast the rain? Langmuir later asserted that large-scale weather control might eventually provide the solution to the problem of weather forecasting.

Claims made by Langmuir that disturbed the Weather Bureau and some other government meteorologists included these items: clearing winter fogs over airports, reducing damage from hailstorms, controlling forest fires, enhancing the water supply in arid areas, and forcing snowstorms to the lee of the Great Lakes to drop their moisture short of the cities. In retrospect, it is interesting to note that fog clearing is now a standard operation at airports in the colder part of the country; hailstorm abatement is practiced in several parts of the world, and apparently with some success; controlling forest fires is now being carried out operationally in Alaska. In forest fire control the object is to initiate rain at the right time and place, and not necessarily to increase rainfall. Watershed seeding is widely practiced, and lake storm snow redistribution is under extensive field investigation by National Oceanic and Atmospheric Administration (NOAA), the Federal agency within which the Weather Bureau (now the National Weather Service) is the largest component.

The point is that Langmuir had little or no conclusive evidence at the time that these things could be done. But the power of his scientific intuition has certainly stood the test of time.

Other government agencies such as U.S. Forest Service and U.S. Bureau of Reclamation were anxious to get started on seeding trials. They expressed resentment, as did Langmuir, at the Weather Bureau's seeming negative attitude, a stigma which after subsequent congressional hearings on weather modification, seemed to become permanently attached to it. Reichelderfer, in order to gain scientific clarification, asked the Academy

of Science to appoint a committee of investigation. The Haurwitz Committee was appointed and it reported, in 1950, that the production of rain in useful amounts had not been demonstrated.

But Langmuir had even bigger claims in store for the harried Weather Bureau. In December 1949, he started a periodic seeding trial wherein a single AgI generator near Socorro, New Mexico, burned a kg of AgI/during 1 day each week. He continued this through 1952. Already in October, 1950 he began making public statements to the effect that he was finding that precipitation patterns more than 2000 mi downwind were showing a 7-day periodicity. This led to a great deal of concern in many quarters, including the halls of Congress. By 1951 Langmuir was able to make a case for large-scale effects. It did not convince the Weather Bureau and other meteorologists who visited him and went over his data, although they agreed that there was indeed a 7-day periodicity in precipitation over a large segment of the country, and that it did change phase about the time Langmuir changed the day on which he seeded. However, it was argued that such transitory and changing periodicities had developed in the past, and evidence was produced that this was indeed the case.

The widespread publicity and sensationalism of reporting accompanying the activities of Langmuir and his Weather Bureau "opposition," Richelderfer on the one hand, and Krick and Howell on the other hand, drew the attention of Congress to the matter of weather modification, which seemed to them to be fraught with implications. In particular, Senators O'Mahoney, Anderson, and Case introduced bills early in 1951, which would recognize the national importance of weather modification and set up Federal control and a lead agency. Actually, there had been a previous bill, the Simpson bill introduced in 1948, which would have provided the Weather Bureau with considerable funding for studying weather modification. The Weather Bureau was cool toward it, and it did not pass. In addition, Anderson had introduced a bill in 1950 which did not pass.

S222 (Senator Anderson) proposed the formation of an independent weather control commission consisting of the Secretary of Commerce and four members appointed from private life. S798 (Senator Case) authorized the Secretary of Agriculture to conduct research and experiments in controlling or producing precipitation in moisture-deficient areas. S5 (O'Mahoney bill) provided for research into demonstration of economical production of useful water from saline water or from the atmosphere, under the leadership of the Department of Interior. It provided for allocation of $25 million for this purpose.

There was considerable testimony presented by the Weather Bureau and General Electric staff members. The private sector was also represented. Dr. Wallace Howell (fresh from his experiences in New York) expressed his opinion that any possible widespread effect has nothing to do with the seeding agent. He felt that "Federal legislation would be of great assistance to those rain-making groups that do try to use scientific methods. . . ." He "claimed" a 14 percent increase in his New York seeding. He went into some detail in explaining his evaluation methods, which are similar to ones employed even today, but which were advanced for that time.

Dr. Irving Krick, President of the Water Resources Development Corporation (recently moved from Pasadena to Denver) outlined his early work. He discerned no effects more than a few hundred miles away. He considered S222 premature. He pointed out how his organization suspended seeding whenever

rain would not benefit certain agriculturists, such as during fruit blooming or harvest periods. Senator Anderson brought up the matter of the conflict of interest between the wheat farmers of the Horse Heavens area (in Washington) who wanted rain, and nearby cherry growers who wanted none, and who had hired someone who would "reduce" rain. Krick said that this problem could easily be resolved by modifying seeding schedules.

The American Meteorological Society's statement on weather modification of May 3, 1951 was entered in testimony. This is the first AMS statement, and has subsequently been modified. It tended to emphasize the small scale and magnitude of seeding effects. It discounted Langmuir's large-scale effect. It emphasized the need for research in cloud physics.

The upshot of the hearings and other legislative activities of the various senators and their staffs was an airing of the whole matter of weather modification. However, no bills were made into law. The airing seemed to have the following results: it indicated that there was no evidence of a large-scale effect in the minds of most scientists; that it was premature to institute federal control; and that more research was needed.

In 1953 Senator Case introduced legislation (co-sponsored by Anderson and other senators) which was passed and became Public Law 256. It created the Advisory Committee on Weather Control (ACWC). Retired Navy Captain and former chief of Navy Aerology, Howard T. Orville, was appointed its chairman by President Eisenhower. The committee itself consisted of the Secretaries (or their designees) of the Departments of Agriculture, Commerce, Defense, Interior, and Health, Education and Welfare, the Director of the National Science Foundation (or his designee), and four other members drawn from private life. The committee was authorized to make "a complete study and evaluation of public and private experiments in Weather Control for the purpose of determining the extent to which the United States should experiment with, engage in, or regulate activities designed to control weather conditions."

The committee had a scientific staff headed by retired Admiral Frederic Berrey, another former Naval aerologist. It conducted a series of studies and evaluations, and held many meetings at the scenes of industrial cloud seeding activities, a practice not followed since by those writing government reports on weather modification.

Most of the scientific, information oriented tests of cloud seeding that had been conducted in the United States up to the time of the committee's activities had employed methods for seeding that were significantly different from those employed by the private sector. The private sector had turned largely to the use of silver iodide smoke generators, and as a rule they were ground based. The older cloud physics projects of the late 1940s and early 1950s had used aircraft and dry ice. The artificial cloud nucleation (ACN) program of the early 1950s, which was being wound up at the time the Committee came into being, had also employed aircraft and dry ice. It was a large-scale project in the southwestern portion of Washington State. It had originally been intended to test methods in use in the private sector, particularly as applied to winter orographic storms. Although theoretically a well-designed project having a good observational network including an extensive rain gauge network and weather radar, the percentage of the seedable precipitation regimes which were operated was extraordinarily low because of flying problems, and insufficient data were gathered upon which to draw valid conclusions. Thus the Weather Bureau

discovered what the private sector had already learned, namely, that in winter orographic storms the aviation of the day was not reliable.

The ACWC turned to the private sector to see whether it could base a scientific evaluation on the data from the numerous projects which had been conducted up to that time. After a diligent review of sources, the Committee based its evaluation of seeding primarily on data from twelve commercial cloud seeding projects. The projects were broken down into storm units, 427 in all. Measured rainfall (U.S. Weather Bureau) data from 5516 unseeded storms during the preseeding period were used as a basis for comparison. Eleven of the projects were in the United States, one in France. In one project, stream discharge (USGS figures) was used as a basis for measurement. With one exception, the projects all occurred in the fall, winter, and spring seasons. The projects were categorized three ways: orographic, semiorographic, and nonorographic. North American Weather Consultants contributed three snowpack projects: Big Creek in the southern Sierra Nevada, Coeur d'Alene in the Bitteroots of Idaho, and the Mokelume-Stanislaus in the Central Sierra. A fourth project was a rain-enhancing project in Santa Barbara County. W. E. Howell Associates conducted three of the projects, one in eastern Kentucky, one (Mohawk) in upstate New York, and one in South Carolina. Weather Modification Company conducted two of the projects. Three of the projects were conducted by Irving P. Krick Associates. They were the Tri-Counties project in the plains of central Oregon, the Dallas project over the watershed providing domestic water supply to the City of Dallas, Texas and the Tignes, France project in a mountainous watershed where hydroelectric generation was the user.

The statistical evaluation[2] indicated that the average increase in precipitation due to seeding was 10 to 15 percent in the winter-type storm in the mountains of western United States. These all fell into the orographic or semiorographic categories. For the nonorographic projects, their evaluation did not detect any increase or decrease. There was some question as to whether their method for testing was sufficiently sensitive in this latter instance. Most of the nonorographic projects involved the seeding of convective clouds where the background variability of precipitation is great.

The Committee also concluded that there was no evidence of a negative effect in any of these projects which all had as their goal increasing the precipitation; the data on hail prevention projects was too scanty to base any conclusion on it.

The Committee conducted under contract a variety of field experiments which came under the heading "physical evaluation program." These involved the tracking of silver iodide smoke plumes, the study of convective cloud motions, and nuclei measurement studies. On the basis of these and their evaluation studies, they concluded that the conventional ground-based AgI generator is a valid technique for seeding clouds. Adequate concentrations occur up to 30 mi from the source. Furthermore, they found no evidence of serious photolytic deactivation of the nuclei plume.

The Committee looked into the state laws then in existence. They felt the approaches followed by the various states were neither a scientific nor a legal solution to the uncertainties of weather modification, but that the concept of regulation was salutory.

Subsequently, the Committee was criticized by statisticians because the projects they evaluated were not randomized. It was claimed that they could have been biased in their selection of not-seeded storms from the historical sample.

The ACWC findings lent support to S86, 85th Congress, a bill entitled the Water, Cloud Modification Research Act of 1957. It designated the National Science Foundation (NSF) as the institute for supporting and coordinating research projects in weather modification. This bill, very important to weather modification, was passed and a new era of open-minded, federally supported investigation of weather modification at all levels ranging from theoretical through laboratory to elaborate field tests was commenced.

The bulk of the support went to universities; however, some members of the private sector were able to obtain research contracts. For example, the Santa Barbara randomized seeding project 1957–1960 was supported in part by NSF funding, and the subsequent Santa Barbara storm study (1960–1964) conducted by NAWC's research affiliate, Aerometric Research Inc., was fully funded by the NSF. The results of these two projects provided a sound basis for a later project in the Santa Barbara area, 1967–1971, where the testing was based upon the seeding of mesoscale systems. Initial financial support was also rendered to Meterology Research Inc. in their very interesting and productive work on the physical aspects of seeding summer cumuli over Flagstaff, Arizona.

Two other rather important field projects which were supported was a series of summer cumulus seeding in southeast Arizona conducted by Louis Battan of the University of Arizona and the Whitetop project conducted on summer cumuli in Missouri by Roscoe Braham of the University of Chicago. In addition, there were innumerable other projects, especially those involving laboratory studies or theoretical aspects. The NSF rapidly became the foremost federal funder of weather modification research studies. It was not until the late 1960s that it was overtaken by the Bureau of

Reclamation's Atmospheric Water Resources Research group. The NSF remains active in its support of weather modification research and is now focusing attention on the hail suppression area.

The Office of Atmospheric Water Resources Research of the Bureau of Reclamation (now the Division of Atmospheric Water Resources Management) had become a major government center for weather modification activity by the mid-1960s. Its head, Dr. Archie Kahan, is a well-known meteorologist who had a brief exposure to weather modification in the private sector earlier in his career. The Bureau has funded numerous research projects in weather modification. Most of these have been at western universities. The private sector has been involved in several. For example, Meteorology Research Inc.'s recent work at Flagstaff was largely supported by the Bureau.

The Weather Bureau is now called the National Weather Service. In the 1960s it became part of the Environmental Science Services Administration (ESSA), the function of which moved to the U.S. Commerce Department's new National Oceanic and Atmospheric Administration (NOAA) in 1970. Within NOAA the Environmental Research Laboratories are engaged in research in atmospheric and other geophysical processes. The field tests of seeding to ameliorate heavy snowstorms on the lee side of the Great Lakes, already mentioned, are being conducted by one of these, the Atmospheric Physics and Chemistry Laboratory, under the direction of Dr. Helmut Weickmann. The extensive development and testing of seeding techniques and methods in Florida under Dr. Joanne Simpson are also a NOAA operation, in the Experimental Meteorology Laboratory. NOAA is currently interested in the whole field of weather modification research and is planning to expand its field test activities, making full use of the capabilities within the private sector in carrying out the mission.

It should be mentioned that other national governments besides ours were interested in weather modification in the 1950s. In Australia, a continent often shy of water, the government's Commonwealth Scientific and Industrial Research Organization produced, under the direction of Dr. E. G. Bowen, a great deal of useful cloud physics work in the lab and in the field, and had started on some long-term field test programs. Other foreign countries gave less, but significant, attention to weather modification.

Two more efforts were made in the Senate to accelerate weather modification. Senator Engle introduced S943 on February 5, 1959, and S152 on January 14, 1963. Both were cosponsored by Senator Anderson. The first would set up a weather modification program under the Department of Interior and National Science Foundation to seed the Colorado River Basin. The second would set up a National Weather Council to accelerate weather research. Neither passed.

In addition to the research activity subsequent to the ACWC report[2] the private sector experienced continued demand for seeding, with a widening market overseas in Israel, North Africa, and Central and South America. There was a setback in seeding for utilities, brought on by the start of hearings on the Feather River suit (to be covered later), but otherwise things were going well. Especially, there was good liaison and exchange of information, in the eyes of the private field, between the scientists active in weather modification research and the private field.

Up to the early 1960s, weather modification experiments had been conducted by several government agencies other than the Weather Bureau. One field project of more than usual interest was the U.S. Forest Service's Project Skyfire, directed toward finding means for alleviating or eliminating the type of lightning responsible for producing forest fires. In 1962, the Bureau of Reclamation announced its entry into the field. Its interest was in conducting research in atmospheric water resources with an eye to ultimately increasing stream flow in the Colorado River Basin and other major basins. This created some concern among the other government agencies because of its potential scope.

In November, 1963, the National Academy of Sciences' Committee on Atmospheric Sciences appointed a panel on Weather and Climate Modification to undertake a deliberate review of the present status of activities in this field and its potential for the future. A report entitled "Scientific Problems of Weather Modification" was issued in 1964 and, although considered to be preliminary in nature, was distributed widely including release at a press conference. The report started with the assertion that the "commercial operators" who had exploited the findings of Langmuir and Schaefer for business interests had contributed but little to advancing the state of the art. To support this concention (contrary to the findings of the Advisory Committee on Weather Control), it was stated that "it has not been demonstrated that precipitation from winter orographic storms can be increased significantly by seeding . . . ." By way of destroying any possible remaining confidence, the argument that satisfied customers came back for more seeding was blasted by the assertion, "Many reputable firms and state executive bodies subscribe large sums of money to support these operations. There can be no answer to such an argument, based as it is on faith and hope of economic gain, except to point out that the theories and predictions of astrology could be substantiated in a like manner."

This report was received with shock and amazement by those scientists who had by now become part of the weather modification community, being involved in either research or operations, or in a combination of the two. The press took a different view. A headline in the Newark Sunday Star-Ledger (November 22, 1964) was, "So rainmakers are phony after all." Dr. Seitz, President of the Academy, ordered the panel to look deeper into the matter. The panel then obtained the services of Professor James E. McDonald of the Department of Atmospheric Sciences, University of Arizona. Professor McDonald now encountered the same sort of a problem as the ACWC had encountered a decade before. The various government and university conducted tests of cloud seeding had not employed the methods which were most commonly employed in the private sector, and could not therefore be used as a basis for assessing the efficiency of the bulk of the seeding which was being conducted in the United States. The two-large-scale randomized seeding tests had been those operated in southeast Arizona by L. J. Battan of the University of Arizona, and in Missouri by R. R. Braham of the University of Chicago. Both projects involved summer cumulus seeding, and both employed the aerial seeding mode (patrol) with high-output silver iodide-sodium iodide-acetone type generators.

Howell, Krick, and others during the same time had used ground-based generators in summer convection. There was indeed some aerial seeding in this country, and the Australians were using it consistently. With respect to winter orographic seeding, the exhaustive and illuminating tests carried out by L. O. Grant at Colorado State University had not been completely analyzed and reported upon as yet. Therefore the academy panel, as had the ACWC, turned to the private sector for information and data to serve as a basis for an evaluation of seeding effects.

McDonald reviewed a large amount of

project data that were made available by the private sector to the panel. He made a very thorough evaluation of a number of the projects, using as basic data government-published reports on precipitation and stream flow from target and control areas, and making statistical comparisons of seeded and historical non-seeded data by the regression methods.

He examined four long-period (8 to 14 years 41 project seasons) snowpack seeding operations in the west, three of which had been operated by North American Weather Consultants, and one by Atmospherics Inc. Instead of using storm precipitation data as the ACWC had, which had opened them up to statistical criticism of possible hidden selection bias, he used seeded-years seasonal stream flow and all historical stream flow records. He found stream flow increases of 7 to 18 percent, significant at levels ranging from .002 to .04. Two of the projects were also evaluated by the ACWC in 1957; they had added more years of seeding in the meantime. It is one of the ironies of this field that the early government test programs from which so little was learned had simply been run for too short a period. And yet the term "hit or miss commercial seeding" came into being in those days.

In addition, McDonald examined 14 short-term seeding projects largely from nonorographic areas in eastern United States. They ranged in length from 1 to 5 months, and were seeded by ground generators. They were projects operated by Dr. Wallace Howell's group. In spite of the short duration of these projects, two were significant at the .04 level and four at the .07 level or better. Careful studies of the way in which the seeding months were picked, for operational purposes, and how this might bias the evaluation one way or another, were made by independent agencies. No possible bias could be found.

These figures were published in the panel's final report.[56] The panel had evidently been surprised at the results, especially for the flatlands projects. The ACWC had previously found no effect in this latter type of seeding. The panel concluded that their analysis of the 41 project seasons of orographic seeding supported the conclusions of the ACWC. They also indicated that according to their analysis, the seeding over flatlands had increased precipitation about 10 to 20 percent in the target areas.

In spite of avoiding the errors in approach which had been attributed to the ACWC by some statisticians, the Academy report duly received criticism from the same people.

The panel stated categorically "that there is, at present, no known way deliberately to reduce predictable changes in the very large scale feature of climate or atmospheric general circulation." They did not specify what these terms meant, but presumably Langmuir's periodic effects would be large enough to qualify. However, the panel went on to point out the hazards inherent in attempting to manipulate weather on a large scale without first developing a comprehensive theory and numerical simulation models. They also made a cause for developing a better understanding of energy-exchange processes in the earth-air boundary layer. The question of inadvertent atmospheric modification through the burning of carboniferous fuels and other human activities was introduced. The need to study atmospheric water budgets on all scales was stressed. The panel emphasized that the tropics in general, and specifically hurricanes, need field investigations.

Resumes of federal spending on weather modification and meteorology in general were presented. The federal support of research in meteorology had been running around $100 million/year. That for weather modification rose from $2.75 million in 1963 to $4.97 million in 1965. The estimate for the USSR was $20 million/year. The panel stated that the United States' effort had been dissipated over many small projects of insubstantial size and that large field studies should be implemented, and that the level of financial support should be raised from $5 million in 1965 to $30 million in 1970 (it was about $10 million in 1970).

The report of the Academy has not been accorded the "implementation" given the ACWC report. The latter was translated almost immediately into congressional action when Public Law 85-50 put weather modification research into the hands of the National Science Foundation. The Academy report proposed that a lead agency be designated for a concerted and larger-scale attack, and for playing a crucial role in interagency coordination and in ultimate control of the private sector. This has not occurred, nor has the larger program recommended ($30 million by 1970) developed. There was indeed a considerable activity in the Congress, but most of it occurred prior to the issuance of the final NAS report.

During the period when the NAS/NRC panel on weather and climate control was active, other government agencies were busy studying weather modification.[58] The U.S. Weather Bureau issued a report in July, 1965 that covered the technical aspects of weather modification, the socioeconomic, legal, and legislative aspects, and the direction that weather modification research might take.

The National Science Foundation activated a special commission on weather modification in October 1963. This was chaired by Dr. A. R. Chamberlain, an engineer, and Vice President of Colorado State University. A report[57] was issued in December, 1965. In this report, little attention was devoted to evaluation, there being an acceptance of what other government reports had found or were finding.

A considerable amount of attention was focused on ecological, social, and legal implications. It was pointed out that much remains to be learned of how man responds to the normal variability in weather conditions, and of how he uses weather prediction in decision making. Legal problems such as responsibility for damages need study as does the whole area of international relationships. Inadvertent weather modification needs thorough exploration.

In 1964, Senator Anderson introduced S1020, directing the Secretary of Interior to conduct a seeding program in five areas of the United States to increase usable precipitation. It was not reported out of the Commerce Commitee. Later, in January, 1965, he introduced a bill, S23, then subsequently in February, 1966, S2875. The interesting thing about the latter bill was that it authorized $155 million for the next 3 years for a comprehensive program of scientific and engineering research, experiments, and tests and operations for increasing the yield of water from atmospheric sources. The Department of the Interior was to be responsible. S23 was similar, but authorized $20 million. Still another bill, S2916 introduced by Magnuson, would establish a national weather modification program under the Department of Commerce. Commerce would also regulate commercial seeding.

At various places and times several members of the private sector delivered testimony at hearings on S23 and S2875. Wallace Howell presented three points concerning a national weather modification program: (1) the importance of a pluralistic approach to a national program of weather modification; (2) the importance of private initiative in reducing weather modification to useful practice; and (3) the importance of separating the regulation of weather modification from its operation. Howell noted the following.

Unless the scientific objectives now in view fail of ultimate realization, we are already committed unavoidably to an undertaking of which the operational, economic, and social magnitude will tremendously overshadow the seminal role of pure science. Science alone is not a sufficient guide. If in testing the scientific foundations we neglect the other foundations of this undertaking, we risk our best chance to build a well-balanced and effective national program.

In the history of science, the pure and applied aspects have always cross-fertilized each other. In weather modification, experience with applications is an important source of ideas and intuitions for basic research. Both of these involve potential direct impact on the social and political communities to an unprecedented degree, for the weather is everybody's. The exploration of applications and impacts is an important as exploration of pure science itself.

The author, in his testimony, introduced into the records the statement of position of the Weather Control Research Association (see details on this organization in following paragraphs). This position on weather modification regulation had been hammered out at several meetings of this organization over the past year. It was their opinion that regulation and licensing of weather modification should be in the hands of an independent commission. They also were of the opinion that the federal policy should be to encourage private industries and local government to continue their weather modification work.

Dr. Booker of Weather Science Inc., Norman, Oklahoma, specifically recommended that responsibility for regulation of weather modification be separate from any agency which conducts the activity. He expressed the opinion that any national program should make use of existing capabilities in weather modification.

Drs. Vincent Schaefer and B. Vonnegut also made presentations, bringing the Senators up to date on seeding technology and larger implications.

The Weather Bureau (then in ESSA) stuck to the previously voiced argument that weather modification progress is directly linked to progress in forecasting, so it should lie within the ESSA regime. Later on Dr. Mordy of the Desert Research Institute, University of Nevada, said he was bothered by this dependence, and also the alleged dependence upon a global observing system (also emphasized by ESSA). "Cloud seeding has advanced from its beginning in 1946 to its present level of achievement without requiring similar progress in weather forecasting," he said. He agreed that general progress in meteorology would help weather modification, but that it was a distortion to say nothing can be done in weather modification until we can more accurately predict the weather.

Dr. Thomas Malone, Second Vice President and head of Research Department, The Travelers Insurance Company, suggested separation of powers in regulation.

Eugene Bollay, President, Bollay Associates, Boulder, Colorado, felt that the time had come for a coordinated national program. He felt the NSF should be assigned the responsibility for regulation. Other agencies should be assigned specific missions.

Two flying farmers, Bill Fisher and Wilbur Brewer, had been conducting cloud seeding operations during summers in Bowman and Slope Counties, North Dakota. They had obtained financial support from farmers in the area. They gave very down-to-earth testimony as to the value to the farmer of added moisture and the methods (adapted from the products of other's research and development) they employed in their aerial seeding. They told of the helpful advice received from Dr. Richard Schleusener and staff at the South Dakota School of Mines.

Marion Bruce, Chairman of South Dakota Weather Control Commission reviewed the activities of the Control Commission and of seeding in South Dakota. The interest in weather modification had been long enduring in the State of South Dakota and Marion Bruce was active from its beginnings.

None of the bills which were the subject of so many hearings in Washington, D.C., and in various cities and towns of the West, were made into law. In a sense, there was a shifting of the interest away from the thought of spending large sums on a national weather modification program toward the matter of federal regulation of weather modification.

In the private sector, further attention was given to this at meetings of the Weather Control Research Association, which had renamed itself the Weather Modification Association. As a result of action on the part of individuals within this organization, a bill (S1182) was introduced into the Senate in 1969 by Senator Young of North Dakota and a companion bill (HR9055) into the House by Congressman Kleppe. These would establish a national "Weather Modification Commission" composed of members drawn from the Federal Government, states, colleges and universities, and private industry. The commission would undertake a comprehensive investigation of weather modification activities and the type of regulation required and come up with a plan for a federally sponsored permanent commission within two years.

Senator Dominick once again, in 1969, introduced a bill (S2826) to develop a practical weather modification program for the upper Colorado River Basin. Over $6 million would be appropriated to cover expenses for the program through 1974.

INTERACTION WITH THE PUBLIC

A large part of the growth in private sector weather modification has occurred in response to farm area determination to exploit the new technology. This support continues today as strong as in the early days, and at present is most in evidence in the Great Plains region. However, there are those who for one reason or another express themselves, usually in the local newspaper, against the practice of weather modification, and who attribute unexplained and unpleasant weather happenings to it. One lady in Colorado was convinced that the silver iodide from cloud seeding caused the then-current pine tree blight in the Rockies. A man in coastal British Columbia complained that the above-normal precipitation they had been receiving in Vancouver was due to the seeding, even though the target area was a remote mountain region. He pointed out that people came to British Columbia to enjoy its climate and asked, "By what right do a handful of avaricious power moguls dare to meddle with and disrupt factors of nature that endow our habitations?"

But the expressions of concern and discontent at the possible short-term personal inconveniences which weather modification might impose have usually been offset by a concern for the community as a whole. A storekeeper might say, "Extra rain harms me but if it helps the farmers it helps the whole community." On the other hand, weather incidents harmful to the community sometimes occurred during seeding operations and, rightly or wrongly, they were associated with the seeding. Concerns based upon false premises could often be corrected through discussions by the weather modifier at public meetings, service club programs, and other such forums.

In some cases, problems arose because people demanded an operational cloud seeding project where it would interfere with an on-going research project. A case arose in the second Santa Barbara research project area where following a dry year water users approached North American Weather Consultants, the conductors of the research program, asking for a water oriented project. They were persuaded to forego such a project until completion of the research program, from which they were receiving full benefits of seeding on half the occasions anyway. In another case, in the early Santa Barbara test seeding project, a group of farmers in adjacent Ventura County wanted a seeding program of their own during the second year of the project. We tried to induce them to join in with the Santa Barbara randomized project. They insisted on a fully operational program of their own and indicated they would engage another cloud seeder if we were not interested. We finally convinced them that they should at least agree to establishing a buffer zone between the two counties, and seeding was carried out for them on a continuous basis for a year. By the next year we had induced them to join the Santa Barbara project on a randomized research project basis.

The practice of cloud seeding has encountered some interesting local problems abroad. Dr. Manuel Lopez tells of the situation he encountered in the late 1950s while working for Howell Associates in Cuba. Every time he serviced a nucleating generator located at a remote mountainous site he had to make very sure that he got official permission from both the Cuban government and the followers of Castro who controlled the area.

In Iran, the author's firm discovered that anything left in a remote area was taken to be a present given by Allah to any person who happened to wander by. It was necessary to place a stockade around all generators.

In California, the pioneer of watershed seeding, the California Electric Power Company found itself embarrassed to the point of dropping its seeding program. During a rate hearing it had publicized its seeding to show how progressive it was. The state utilities commission took it at its word and reduced the requested rate increase by an amount reflecting the total value of the water produced by seeding, minus the cost of the seeding.

In certain cases, complaints reached the point where injunctions and law suits were brought, as for example the injunction against the City of New York in 1950 already mentioned. The chapter on law should be referred to for details, but a few highlights of suits involving the private sector are given here.

One suit was brought against Krick in the Federal Court for the Western District of Oklahoma in 1956. The plaintiff complained that a heavy rain, during seeding operations, damaged his store. He lost, as the defendant proved the seeding could not have affected his area.

In Texas, in 1959, on the other hand, a group of ranchers obtained an injunction against hail suppression seeding being done for a group of cotton farmers. They merely had some old-timers relate in court how the clouds behaved differently now than they had ever done in the days before seeding. This convinced the judge that he should issue an injunction against the seeding.

Later, threats of injunctions elsewhere to prevent hail suppression activities were cooled at the thought of those for whom the injunction was imposed having to pay for any hail damages which might result from the lack of the suspended hail suppression program.

In 1955 in California, a terrible storm occurred from December 19 through December 24 which brought much rain, including rain on top of old snow, to the mountains and foothills of the Sierra Nevada range. A flood occurred which topped levees in Sutter County. It became known as the Yuba City flood. Suits were brought against the State of California for its alleged mismanagement of the levee system, and against the U.S. Engineers. Somewhat later, it was discovered that seeding had been conducted by the author's firm for the Pacific Gas and Electric Company in an adjoining watershed. Pacific Gas and Electric Company (PGE) and North American Weather Consultants were sued for $13 million, along with the other defendants, even though no seeding had been conducted during this particular storm because of its predicted flood potential. The pretrial period dragged on for 5 years and it was not settled in favor of PGE and NAWC until 1963. The judge ruled that the seeding effects had been confined safely above the dam in the seeded watershed, and that there had been no effect of seeding on the flooded watershed.

In the meantime, the insurance industry reacted sharply. Rates went up and utility companies backed out of watershed seeding. North American Weather Consultants' watershed cloud seeding business was abruptly cut to a third, and recovered only slowly thereafter, even though the case was settled for the defendants. The suit is reported to have cost PGE some $250,000 to defend itself.

The belief that the major technical projects of man are responsible for profound changes in weather and climate is deep-seated. The H bomb has been blamed for unusual weather on a worldwide basis. A recent example of such unfounded belief occurred in some counties of Pennsylvania, Maryland, and West Virginia in the 1960s where cloud seeding was blamed for a drought which was actually caused by a large-scale shift in storm tracks and covered all of the northeastern United States. In 1957, a Blue Ridge Weather Modification Association was formed, and contracted with the Weather Modification Company of San Jose to seed in an attempt to

reduce hail. Although some farmers in the area had feared that rainfall might be reduced, and the summer of 1957 was dry, the rainfall was ample during the next 4 years of the project and initial opposition died out. but in 1962, the northeastern drought commenced and intense opposition developed against the seeding company and the orchardists themselves, who received anonymous threats. One orchard was vandalized.

A bill to outlaw seeding was introduced in the Pennsylvania State Legislature in 1963, but died in committee. A bill, HR8708, was introduced into the U.S. Congress which would outlaw cloud seeding by air unless permission was obtained from all landowners in the target area. It was not passed.

In 1963 the contract let to Weather Modification Company in 1957[33] was transferred to W. E. Howell Associates, Inc. Meetings were held with the farmers to discuss the matter. The year 1964 was also dry. The drought extended over the entire northeastern United States. In 1965 legislation to prohibit cloud seeding was sponsored by the Natural Weather Association in the legislatures of West Virginia, Maryland, and Pennsylvania.

Hearings were held before the committee of Agriculture and Natural Resources of the State of Maryland's Senate in March, 1965, on a bill to stop all seeding for 2 years. A number of the legislators pointed out that a controversy was raging in the farm counties, the farmers believing that the drought was caused by the seeding and the orchardists believing that this was not so. Mr. Donald Spickles, President of the Natural Weather Association of Maryland, said that in view of doubts by competent scientists (the 1964 preliminary NAS report was just out), the emergency legislation calling for the 2-year moratorium on weather modification should be enacted. Many others testified in a similar vein.

On the other hand, John R. Martin, a farmer and orchardist member of the Blue Ridge Weather Modification Association (BRWMA) cited a Weather Bureau response to a letter of inquiry from Fred Glaize, President of the BRWMA, which stated that: "We know of no scientifically established proof either from our own cloud seeding experiments or from the published results of tests carried to completion by other government and university groups that provides an acceptable basis for suggesting that rainfall is prevented or diminished by the release of chemical reagents." Dr. Wallace Howell testified at length, presenting the scientific basis for hail suppression and exposing the fallacy of many of the arguments relating the drought to cloud seeding. The moratorium bill was passed and weather modification was made illegal in Maryland for 2 years.

Ordinances against cloud seeding were passed in several counties in south-central Pennsylvania. The Howell Associates initiated a legal test of one of these ordinances and of an injunction and a damage claim, on the basis of their being patently beyond the justifiable need of remedy and an invasion of individual liberty without sufficient basis. The judge before whom the test of the county ordinance was tried died before handing down a decision. Howell allowed it to pass to his successor, rather than re-trying it. The decision was to uphold the Ayr county ordinance against having seeding equipment in the county, but to deny the injunction and damage claim. The decision was based upon the judge's finding that AgI was a poisonous substance and that the town Board of Health was within its rights in proscribing it. However, Howell says that no evidence was presented in court as to its toxicity.

For years now, no seeding has been carried on in Pennsylvania. Yet the belief that it is continuing holds strong. Every time there is a dry spell, there is a spate of letters to the editors of farm area newspapers condemning the authorities for not stopping the seeders who are at it again. The belief has been expressed that road contractors hire the seeders secretly to reduce precipitation. The U.S. Army was accused of carrying out secret seeding tests in the area. It was stated that seeding planes have been "positively" identified. At a farmer's meeting in Pennsylvania in 1968, a state Senator proposed that the farmers gather evidence of illegal seeding activity in the area. Others at the meeting supported his suggestion. An attendee from a nearby college discussed the possibility that government operations had caused a drought in Cuba and had rained out the poor people's march on Washington. Some participants at the meeting went so far as to suggest shooting down the planes. The shooting suggestion was emphatically vetoed. In spite of the stand against shooting at planes, the State Department of Agriculture had to call off an aerial survey of oak wilt disease in the area, when farmers started shooting at the low-flying planes.

Subsequently, the drought was broken. There were years of wet weather which followed. This prompted the development of a new theory about rainmaking. According to Paul Hoke, head of the Tri-State Natural Weather Association, Inc., "There's no question that during a dry season, cloud seeding aggravates conditions to produce a drought, and during a wet cycle, it triggers even more rain and possibly floods. It's a disaster either way."

It is contended that cloud seeders pollute the air. One man claimed that he received serious iodine burns on his arms after brushing against trees in the area. One legislator claimed the seeding is having a toxic effect on flora and fauna and that the reproduction abilities of eagles have been weakened.

In Quebec in the mid-1960s, excessive rainfall produced floods, eroded soils, damaged crops, and reduced tourist trade. There happened to be a cloud seeding project over a target located within the province. More than 60,000 women signed a petition requesting government intervention to stop the seeding. The cloud seeders were threatened. A government report explained the excessive rainfall was due to natural causes. It was claimed clandestine experiments to ease New York's water shortage were being conducted 240 mi away on New York's border. Several doctors found the health level of children and old people in the area was declining due to lack of sunshine. Most of this was going on two years after the real rain-making experiments, conducted under Canadian Government auspices, were stopped.

An interesting cloud seeding project was carried out by my firm for the Bonneville Power Authority but under contract to the Bureau of Reclamation, for the five winter seasons from 1967–1968 through 1970–1971. It required seeding to increase snowpack over the watershed above Hungry Horse dam. This would increase the water available for hydroelectric generation. More such water meant less interruption of industrial power which in turn meant less chance of job losses in Montana, Idaho, Washington, and Oregon. The benefit cost ratio would be very large.

At first, local opposition to the program was sharp. It was argued that the elk population in the nearby Bob Marshall Wilderness area would be reduced. An extra 10 percent in snowpack could destroy the browse needed for wintering over. It seems that an important income in the area derived from hunters who came to shoot elk in the fall. Seventy five to 100 elk are known to be

killed in the wilderness each fall and hunters spend about $100/day in the area while trying to get their elk. Seeding was regarded as a possible threat to the hunting industry. However, a program of explaining and teaching about cloud seeding quieted fears and 5 years went by with seeding, and with better elk herds than ever.

Now we are in a new era. Attention has been refocused upon ecology, and for the first time the Montana Water Board, the licensing and permit agency for weather modification in Montana, saw fit to hold a public meeting prior to issuing a permit for the coming season. The controversy was rekindled. It was now claimed that it is against the Wilderness Act to seed to affect clouds over a wilderness area. Seeding is an unwanted sign of the hand of man. The beneficiaries of the extra water produced by seeding are mainly industries and jobholders in Oregon, Washington, and Idaho. A local newspaper came out against this "economic exploitation" of the wilderness area. Hearings were held. A permit was conditionally issued, then withdrawn.

Several professional associations have played important roles in weather modification and each to a certain extent has served as a home for the weather modification community. The American Meteorological Society (AMS) was, from the first, the logical home. It has been inevitably dominated by the character of the membership. For most of them, professional meteorology was forecasting, and usually forecasting as employees of the Weather Bureau or of the armed services. There was an elite of university professors who played a role exceeding their numbers in importance. They lent the society more of a scientific than a professional character. The AMS was urged by Reichelderfer to look into Langmuir's claims. In 1951, the first AMS statement of weather modification was issued. The AMS had developed a policy of issuing statements for transmission to and use by the public, on various items where meterology interacted strongly with the public. There were statements on hurricanes, forecasting, and other items in addition to the one on weather modification. The first statement on weather modification was conservative in tenor. It did not directly refute Langmuir's claims, but only stated that although seeding could transform a supercooled cloud to ice crystals, it had not yet been proved that economically significant amounts of rain could be produced by seeding clouds. It further indicated that the large-scale effects proposed by Langmuir were not possible using present seeding methods. There was too much energy required for natural processes to be altered. Subsequently, through the years, new updated statements appeared. However, in the view of the weather modification community, the views they expressed, although not erroneous, did not place emphasis in the right areas and tended to be too conservative. They did, however, urge more research.

In 1968, the AMS took a first big step in providing a forum for weather modification. Under the auspices of its committee on weather modification, it held its First National Conference on Weather Modification, in April 1968, in Albany, New York. Prior to this, there had been meetings on cloud physics, a specialty important to weather modification but not embracing the full range of the field. The conference was held on the State University of New York campus (SUNY) at Albany. It was thought appropriate to hold the first weather modification conference there, since scientific weather modification had got its start in nearby Schenectady, and Dr. V. J. Schaefer was now a member of the staff at SUNY. In 1970, the Second National Conference on Weather Modification was

held at Santa Barbara, California, and North American Weather Consultants was afforded the honor of playing host.

The AMS also cosponsored with the Australian Academy of Science an International Conference on Weather Modification, held at Canberra in September, 1971.

The AMS has in recent years (from 1965 on) published many articles in its widely distributed Bulletin, on weather modification topics, including the various relevant bills introduced into Congress. This has brought the membership up to date on many events of long-term importance to meteorology in general and of immediate concern to weather modifiers.

In 1951, a group of weather modifiers and their clients formed the Weather Control Research Association. The real impetus for this developed when Stu Cundiff, an engineer for the California Electric Power Company, on the east side of the Sierra Nevada range, became concerned about the possible impact of the Southern California Edison Company seeding project on the Cal Electric project. The Edison Company project lay on the west slopes of the Sierra, just upwind of the Cal Electric project. Several meetings were arranged between North American Weather Consultants and John Battle of Precipitation Control Company, the two cloud seeders involved, and representatives from both of the utility companies. At one such meeting, Vincent Schaefer attended by invitation. It was decided to expand the basis and the result was the formation of the WCRA, an association for extending our understanding of weather modification.

Meetings were held twice yearly from then on, with the organization growing through the years, and changing its name to the Weather Modification Association in 1967. Besides weather modifiers, the membership includes representatives from a number of leading utilities, local, state, and federal government agencies, and also private individuals, some of whom were involved in weather modification projects, but most of whom were simply interested in keeping abreast of what was going on in the field. Representatives from the universities are also members, and this group became large and important after the establishment of the Bureau of Reclamation program during the 1960s. The membership has been predominantly from the western United States, but there are members from elsewhere in North America, and a few from overseas. At present the membership is 150.

Through the years the WMA has prepared several statements setting forth what the membership believed to be the proper mixture of private sector and government in weather modification and what the nature of regulation should be. A model state law was developed which it was hoped would provide guidance to state legislators considering new licensing and regulating laws, or updating old ones. The WMA instituted a certification program for weather modifiers, which required high standards of performance as exemplified by certain minima of experience in responsible charge of field projects, as well as standards of education. Thus far, 23 have qualified for certification under this program.

The American Society of Civil Engineers is another professional society which has taken an interest in weather modification. In 1960 I was invited to organize a symposium on weather modification which would afford the engineers of the Society's Irrigation and Drainage Division attending a division meeting in Denver, Colorado, an opportunity to learn about the current status of weather modification; this I did. Subsequently, another symposium was held at a specialty conference on research in Logan, Utah, in 1963. There were several

research panels at this meeting, each of which formulated the current status in its specialty, and listed research needs. Each panel drafted a report. The weather modification panel, of which I was Chairman, and members of which were outstanding weather modification researchers and cloud physicists, drafted some new concepts having to do with bringing the art of weather modification into wider use in the West. The economic value of additional water to the West was recognized.

This report made an impression on congressional leaders concerned about progress in weather modification and is reported to have smoothed the path toward involvement of the Bureau of Reclamation in weather modification.

Walter Gartska, an engineer with the Bureau of Reclamation who had followed cloud seeding with interest since its early days, and who had participated in the evaluation of the first Bonneville Power Authority cloud seeding project in the early 1950s, was quite active at this time in gaining support for expansion of the Bureau's rather limited activities in the weather modification field. Among other things, he conveyed to congressional leaders the latest information on cloud seeding, including the American Society of Civil Engineer's report. His activities along these lines were in many ways responsible for the emergence of the Bureau of Reclamation as a major supporter of research and development of usable weather modification methods for the amelioration of the perennial water shortage of western United States.

Subsequently, in the ASCE Irrigation and Drainage Division, there was formed a committee on weather modification. The first chairman was Finley Laverty, an engineer at that time with Los Angeles County Flood Control District, County of Los Angeles, an agency for whom NAWC worked in connection with their weather modification program. Laverty vigorously pursued the subject and got the ASCE to take positions on weather modification matters of national concern.

Several experienced members of the private sector have joined university staffs or government agencies as these groups have turned their attention to weather modification as a new field of research. Lewis O. Grant, now at Colorado State University, and others have already been mentioned. Eugene Bollay, cofounder of North American Weather Consultants, is now head of NOAA's weather modification effort, and Arnett Dennis, former vice president of Weather Modification Company, is now a staff member of the Institute of Atmospheric Sciences at the South Dakota School of Mines.

To many who are close to the scene, the solution to the various problems of weather modification seems to lie in the establishment of a consistent and well thought out national policy on weather modification. This would lead to the correct degree of control at the federal level. This seems logical since effects of seeding are interstate in nature.

On the other hand, a few feel that state laws, and interstate agreements, are adequate. State laws thus far have been the only mechanism for regulation. However, most existing state laws are considered inadequate because of lack of specification of qualifications required of the weather modifier. (A thorough treatment of the complexities of the local and regulatory situation is covered in Chapter 21.)

REFERENCES

1. Adderly, E. E., Rainfall increases down-wind from cloud seeding projects in Australia, *Proceedings of the First National Conference on Weather Modification, Albany, N.Y., April 28–May 1, 1968.*

2. Advisory Committee on Weather Control *Final Report*, Vols. I and II, Washington, D.C., Dec. 31, 1957.

3. Beckwith, W. B., Supercooled fog dispersal for airport operation, *Bull. Amr. Meteor., Soc.* **46,** 323–327, 1965.

4. Bergeron, T., On the physics of cloud and precipitation; *Proc.—Verbaux Assoc. Meteorol. Int. Union Geodesy Geophys., 5th General Assembly, Lisbon, 1933,* Paris, 1935, pp. 156–178.

5. Bergeron, T., The problem of artificial control of rainfall on the globe I, *Tellus VI* **1,** 32–43; II, *Tellus VI* **3,** 15–31, 1949.

6. Biswas, K. R., R. K. Kapoor, K. K. Kanuga, and Bh. V. Ramanamurty, Cloud seeding experiment using common salt; *J. Appl. Meteor.* **6,** 914–923, 1967.

7. Braham, R. R., Jr., The water and energy budgets of the thunderstorm and their relation to thunderstorm development, *J. Meteorol.* **9,** 227, 1952.

8. Braham, R. R., Jr., and J. A. Flueck, Some results of the Whitetop experiment, *Preprints, Second National Conference on Weather Modification, April 6–9, 1970, Santa Barbara, Cal.,* 1970, pp. 176–179.

9. Braham, R. R., Jr., and P. Speyers-Duran, Survival of cirrus in clear air, *J. Appl. Meteor.* **6,** 1053–1061, 1967.

10. Brown, K. J., and R. D. Elliott, Large scale dynamic effects of cloud seeding, *Proceedings of the First National Conference on Weather Modification, Albany, N.Y., April 28–May 1, 1968,* 16–35.

11. Church, J. F., and W. W. Vickers, Investigation of optimal design for supercooled cloud dispersal equipment and technique, *J. Appl. Meteor.* **5,** 105–118, 1966.

12. Dennis, A. S., and D. F. Kriege, Results of ten years of cloud seeding in Santa Clara County, California, *J. Appl. Meteor.* **5,** 684–691, 1966.

13. Donnan, J., D. Blair, W. Finnegan, and P. St. Amand, Nucleation efficiencies of AgI-NH₄I and AgI-NaI acetone solutions and pyrotechnic generators as a function of LWC and generator flame temperature, a preliminary report, *J. Weather Modif.* (Weather Modif. Assoc.) **2**(2), 155–164 1970.

14. Douglas, W. J., Toxic properties of materials used in weather modification, *Proceedings of the 1st National Conference on Weather Modification, Albany, N.Y., April 28–May 1, 1968,* American Meteorological Society, 1968, pp. 351–360.

15. Elliott, R. D., Cloud seeding area of effect numerical model, Report to Fresno State College Foundation, Aerometric Research Inc., Jan. 1969.

16. Elliott, R. D., P. St. Amand, and J. R. Thompson, Santa Barbara pyrotechnic cloud seeding test results 1967–70, *J. Appl. Meteor.*, **10**, 785–795, 1971.

17. Elliott, R. D., and K. J. Brown, The Santa Barbara II project—downwind effects, *Proceedings of the International Conference on Weather Modification, Canberra, Australia, Sept. 6–11, 1971*, pp. 179–184.

18. Elliott, R. D., K. J. Brown, and L. O. Grant, *Transactions of Seminar on Extended Area Effects of Cloud Seeding, Feb. 15–17, 1971, Santa Barbara, Cal.*

19. Elliott, R. D., and E. L. Hovind, On convection bands within Pacific Coast storms and their relation to storm structure. *J. Appl. Meteor.* **3**, 143–154, 1964.

20. Elliott, R. D., and E. L. Hovind; The water balance of orographic clouds, *J. Appl. Meteor.* **3**, 235–239, 1964.

21. Elliott, R. D., and E. L. Hovind, Heat, water, and vorticity balance in frontal zones, *J. Appl. Meteor.*, April 1965.

22. Elliott, R. D., and W. A. Lang, Weather modification in the Southern Sierras, *J. Irrigation Drainage Div., Proc. Am. Soc. Civil Engrs.* 5644 IR 4., 45–59, Dec. 1967.

23. Elliott, R. D., and R. F. Strickler, Analysis of results of a group of cloud seeding projects in Pacific slope watershed areas, *Am. Meteor. Soc. Bull.* **35**(4), 171–179, 1954.

24. Elliott, R. D., and J. T. Walser, Abrupt shifts in seasonal runoff relationships as related to cloud-seeding activity, *Proceedings of Western Snow Conference, 1963*, 1966, pp. 99–107.

25. Findeisen, W., Die Kolloidmeteorologischen Vorgänge bei der Niederschlagsbildung, *Meteorol. Z.*, **55**, 121–133, 1938.

26. Grant, L. O., C. F. Chappell, L. W. Crow, P. W. Mielke Jr., J. L. Rasmussen, W. E. Shobe, H. Stockwell, and R. A. Wykstra, An operational adaptation program of weather modification for the Colorado River Basin, *Interim Report to Bureau of Reclamation, Dept. Atmos. Sci., Colorado State University,* 1969.

27. Hall, W. F., T. J. Henderson, and S. A. Cundiff, Cloud seeding operations in the Bishop Creek, California watershed, *U.S. Weather Bur. Res. Paper* **36**, Jan. 1953. Also condensed version, Cloud seeding in the Sierra near Bishop, California, *Am. Meteor. Soc. Bull.* **34**(3), 111–116, March 1953.

28. Henderson, T. J., Tracing silver iodide under orographic influence, Paper at 24th National Meeting of the American Meteorological Society, Oct. 18–22, 1965.

29. Henderson, T. J., A ten year non-randomized cloud seeding program on the Kings River in California, *J. Appl. Meteorol.* **5**, 697–702, 1966.

30. Henderson, T. J., An operational hail suppression program near Kericho, Kenya, Africa, *Proceedings of the First National Conference on Weather Modification, April 28–May 1, 1968, Albany, New York*, pp. 474–483.

31. Henderson, T. J., Results from a two year operational hail suppression program in Kenya, East Africa, Preprints, *Second National Conference on Weather Modification, April 6–9, 1970, Santa Barbara, Cal.*, pp. 140–144.

32. Howell, W. E., Seeding of clouds in tropical climate, Paper No. 3354 *Transactions of Weather Modification Symposium, American Society of Civil Engineers,* 1962, pp. 349–371.

33. Howell, W. E., Cloud seeding and the law in the Blue Ridge area, *Bull. Am. Meteorol. Soc.* **46**, 328–332, 1965.

34. Howell, W. E., Twelve years of cloud seeding in the Andes of Northern Peru, *J. Appl. Meteor.* **4**, 693–700, 1965.

35. Howell, W. E., Cloud seeding against the 1964 drought in the northeast, *J. Appl. Meteorol.* **5**, 553–559, 1965.

36. Howell, W. E. and M. E. Lopez, Cloud seeding in southern Puerto Rico, April–July, 1965, *J. Appl. Meteorol.* **5**, 692–696, 1966.

37. Kraus, W. E. B., and P. Squires, Experiments on the stimulation of clouds to produce rain, *Nature* **159**, 489–491, 1947.

38. Langmuir, I., The production of rain by a chain reaction in cumulus clouds at temperatures above freezing, *J. Meteorol.* **5**, 175, 1948.

39. Langmuir, I., The control of precipitation from cumulus clouds by employment of various seeding techniques and II seeding increases probability of rainfall in New Mexico., *Bull. Am. Meteorol. Soc.* **5**, 169–170, 1948.

40. Langmuir, I., Analysis of the effects of periodic seeding of the atmosphere with silver iodide, *Final Report, Project Cirrus, Part II,*

Report No. RL-785, Contract No. DA-36-039-SC-15345, General Electric Laboratory, Schenectady, New York, 1953.

41. Lopez, M. E., and W. E. Howell, The campaign against windstorms in the banana plantations near Santa Marta, Columbia, 1956–57, *Bull. Am. Meteorol. Soc.* **42**, 265–276, 1960.

42. Lopez, M. E., and W. E. Howell, Cloud seeding at Medellin, Columbia during the 1962–64 dry seasons, *J. Appl. Meteor.* **4**, 54–60, 1965.

43. MacCready, P. B., Jr., and A. I. Weinstein, The Flagstaff cumulus studies and cloud modification program, program set-up and preliminary results, Met. Res. Inc., Altadena, Cal., 1965, 13 pp. + figs.

44. Mielke, P. W., Jr., and L. O. Grant, Cloud seeding experiment at Climax, Colorado, 1960–65, *Proceedings of the Fifth Berkeley Symposium on Mathematical Statistics and Probability*, Vol. 5, *Weather Modification*, University of California Press, Berkeley, 1967, pp. 115–131.

45. Mooney, M. L., and G. W. Lunn, The area of maximum effect resulting from the Lake Almanor randomized cloud seeding experiment, *J. Appl. Meteor.* **8**, 68–74, 1969.

46. Neyman, J., E. L. Scott, and J. A. Smith, Areal spread of the effect of cloud seeding at the Whitetop experiment, *Science* **163**(3374), March 28, 1969.

47. Neyman, J., E. L. Scott, and M. Vasilevskis, Statistical evaluation of the Santa Barbara randomized cloud seeding experiment, *Bull. Am. Meteor. Soc.* **41**, 531–547, 1960.

48. Orgill, M., J. E. Cermak, and L. O. Grant, Laboratory simulation and field estimaters of atmospheric transport dispersion over mountainous terrain, Ph.D. Thesis, Colorado State University, 1971.

49. Perez-Siliceo, E., A. Ahumuda A., and P. A. Mosino, Twelve years of cloud seeding in Necaxa watershed, Mexico, *J. Appl. Meteorol.* **2**, 311–323, 1963.

50. Rhea, J. O., P. Willis, and L. G. Davis, Park Range atmospheric water resources program, EG & G Inc., Boulder. Col., 1969, 385 pp.

51. St. Amand, P., L. A. Burkardt, W. G. Finnegan, J. A. Donnan, P. T. Jorgensen, L. Wilson, S. D. Elliott, R. F. Vetter, H. Sampson, M. H. Kaufman, F. K. Odencrantz; and W. Sand, *J. Weather Modif.* **2**(1) May, 1970: Pyrotechnic production of nucleants for cloud modification part I—general principles, pp. 25–32; part II—pyrotechnic compounds and delivery systems for freezing nucleants, pp. 33–52; part III—propellant compositions for generation of silver iodide, pp. 53–64; part IV—compositional effects on ice nuclei activity, pp. 65–97; part VI—Case study of apparent stimulation of convection by seeding a thunderstorm with "alecto" pyrotechnic devices, pp. 98–121.

52. St. Amand, P., W. G. Finnegan, and F. K. Odencrantz, Pyrotechnic production of nucleants for cloud modification—part VII—nucleation processes, *J. Weather Modif.* **3**(1), 1–30, April 1971.

53. Schickedanz, P. T., and F. A. Huff, An evaluation of downwind seeding effects from the Whitetop experiment. *Preprints, Second National Conference on Weather Modification,* April 6–9, 1970, Santa Barbara, Cal., pp. 180–185.

54. Simpson, J., and V. Wiggert, Models of precipitating cumulus towers, *Monthly Weather Rev.* **97**, 471–489, 1969.

55. Smith, E. J., Effects of cloud top temperature on results of cloud seeding with silver iodide in Australia, *J. Appl. Meteorol.* **9**, 800–804, 1970.

56. National Academy of Sciences–National Research Council, Weather and climate modification problems and prospects, *NAS-NRC Publ.* **1350**, 1966.

57. Weather and climate modification, National Science Foundation 66–3, Report of the special commision of weather modification; National Science Foundation, 1965.

58. Weather and climate modification, Weather Bureau, A Report to the Chief, United States Weather Bureau, U.S. Dept. of Commerce, July 10, 1965.

59. Water Resources Board, State of Calif., Modification Operations in California, *Weather Bulletin No. 16*, Sacramento, Cal., 1955.

60. Weinstein, A. I., A numerical model of cumulus dynamics and micro-physics, *J. Atmos. Sci.* **27** 246–255, 1970.

61. Weinstein, A. I., and L. G. Davis, A parameterized numerical model of cumulus convection, Report 11 to NSF (Grant GA-777), Dept. of Meteorol., Pennsylvania State Univ., 1968, pp. 43.

62. Weinstein, A. I., and P. B. MacCready, Jr., An isolated cumulus cloud modification project, *J. Appl. Meteor.* **8**, 936–947, 1969.

B

BACKGROUND

What are the meteorological bases for weather modification? They mostly deal with microphysics. The formation of water drops in clouds, the changing of water drops to ice crystals, and similar processes are discussed here for the benefit of the nonmeteorologist. This section of the book is intended to set the meteorological scene.

One of the most vexing problems faced by the experimentalist in this business is that he cannot easily tell what *is* going on. He frequently does not know the size distribution of water drops in the cloud he wants to modify—or what fractions are water drops or ice crystals—or how many natural ice nuclei are present in a region.

Chapter 4 gives a very good picture of measurement problems showing what can be done well, and what still can not be done. There are many gaps in our ability to measure all the important meteorological parameters.

One way in which many past weather modification experiments have failed has been by the lack of a valid statistical method of analyzing results to give firm answers. The use of randomized experiments, crossover designs, Wilcoxon tests, and other ways to analyze correctly are presented in Glenn Brier's chapter. This chapter is essential reading for anyone who wants to carry out a weather modification experiment and not be fooled.

B

BACKGROUND

3 | The Meteorological Background for Weather Modification

MORRIS NEIBURGER

HELMUT K. WEICKMANN

SCALES OF WEATHER PHENOMENA

Weather phenomena—rain, snow, hail, cold and warm spells, thunderstorms, gales, and floods—are the result of atmospheric processes that range in scale from the global circulation of air around the entire earth to the transfer of molecules of water at the surface of microscopic drops and ice crystals in clouds. Between these extremes are the large-scale disturbances of the order of thousands of kilometers, including the moving cyclones and anticyclones that are the major weather systems of temperate and high latitudes, the intermediate-scale phenomena of the order of hundreds of kilometers such as tropical storms (including hurricanes and typhoons), and relatively small circulations measuring tens of kilometers or less, which include land-and-sea breezes and the flow systems

Morris Neiburger, Professor of Meteorology, University of California, Los Angeles, Calif., wrote the sections on cloud condensation, droplet coalescence, and energetics of the atmosphere.

Helmut Weickmann, Director of Atmospheric Physics and Chemistry Laboratory, National Oceanic and Atmospheric Administration, Environmental Research Laboratories, Boulder, Colo., contributed the ice phase part in the section Formation of Precipitation and the section on the size distribution of raindrops.

associated with individual convective clouds, thunderstorms, and tornadoes.

These processes of various scales interact in a very complex fashion. The small-scale processes of condensation and formation of clouds are controlled by larger-scale processes which govern the occurrence of high moisture content, horizontal convergence, and upward motion. The energy for circulations at various scales, even the planetary general circulation, is largely released in the atmosphere as latent heat of condensation. In general, it may be considered that the larger-scale circulations are essentially horizontal, with the vertical motions that release energy representing small deviations, vertical velocities of at most a few centimeters per second occurring as slight departures from horizontal flow of the order of 10 m/sec. Embedded within the large-scale systems are the micro- and mesoscale circulations in which the vertical and horizontal speeds are of the same order, usually in the range of a few meters per second, but amounting to 30 m/sec or more in intense systems such as severe thunderstorms.

Although modification of the weather or climate on all these scales has been proposed, most actual attempts have been confined to the smaller scales. Not only do energetic considerations raise questions of feasibility regarding the larger scales, but the social and political implications constitute inhibiting factors. Although in this chapter we discuss to some extent the interaction of the large and small scales, the discussion is primarily concerned with the constraints which the large-scale circulations place on modifications of the small-scale systems, and only secondarily concerned with the influence of the smaller circulations in triggering changes in the large systems.

Proposals for modification of weather systems include alteration of the availability or utilization of solar energy by introduction of material (e.g., soot or ice crystals) into the air or changing the character of the surface on which it falls, adding heat to the atmosphere by artificial means, affecting the air motion directly by using fans or indirectly by altering the frictional resistance of the earth's surface, influencing the humidity by increasing or retarding evaporation, and changing the processes by which clouds form and give rise to precipitation by dispersing chemicals or plain water spray into the clouds, that is, by "seeding" these clouds. Of these proposed methods the last mentioned is the only category which is based on an intrinsic instability in the atmosphere and thus presents an innate likelihood to succeed. Most of the attempts at weather modification carried out so far have used cloud seeding, whether for the purpose of augmenting precipitation, dissipating fog, suppressing hail and lightning, stimulating the dynamics, or reducing the intensity of hurricanes. We concentrate, therefore, on the meteorological basis for the expectation of success in these attempts.

THE FORMATION OF CLOUDS

Broadly speaking, the processes of formation of clouds and precipitation may be divided into the *dynamic processes*, involving the motions of the air which give rise to the general conditions for their formation, and the *microphysical processes,* by which the individual drops or ice crystals form by condensation from the vapor and grow by interactions which include collision and coalescence. These two types of processes interact strongly. The upward motions determine the rate of cooling caused by expansion and thus control the rate at which condensation goes on. The release of latent heat in condensation and the drag of the particles formed affect the buoyant forces which accelerate the upward motions. The dynamic processes are prerequisite to the microphysical ones, but the microphysical processes are crucial to the occurrence of the potential instabilities which make it possible to produce large changes by relatively small artificial intervention.

Microphysical Processes

The microphysical study of the formation of clouds and precipitation concentrates on the behavior of individual cloud elements and the interaction between them. The initial formation and growth takes place by condensation on a nucleus, treated as an isolated solid or liquid particle in a field of diffusing water vapor. Only when the drops or crystals have grown to a considerable size or when drops and ice crystals are present at the same time does the presence of other particles have to be taken into account.

Condensation and Drop Formation. To explain the process of condensation, we turn to the concept of saturation. Consider a sealed tank partially filled with pure water, in the upper part of which there is dry air at the same temperature. Because of the thermal agitation of the water molecules some of them will escape from the water surface.

Those that escape will bounce around in the space above and some of them will bounce back to the water surface. As more evaporation goes on, the increased number of water molecules in the space above will lead to an increased number returning to the surface, and eventually the number leaving and returning will be the same. When this state is reached the vapor is said to be *saturated*. The vapor pressure in this equilibrium state is called the *saturation vapor pressure* e_s. It turns out that e_s depends only on the temperature T. The relationship between them is shown in Figure 3.1. At 0°C e_s has the value 6.1 mb, and it approximately doubles for every 10 deg Celsius. Thus at 20°C air "can hold" about four times as much water vapor as at 0°C.

The *relative humidity* U is defined as the ratio of the observed vapor pressure e

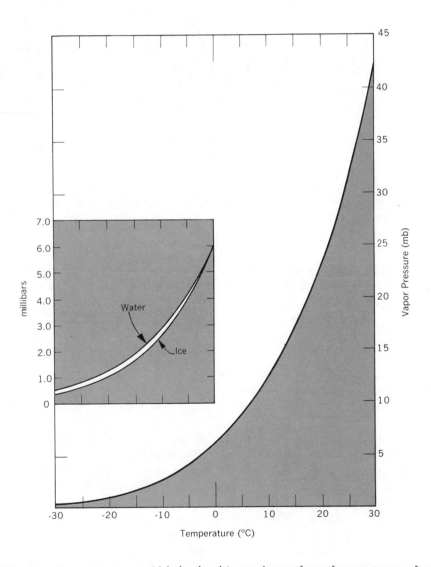

Figure 3.1 Saturation vapor pressure (right-hand scale) over plane surfaces of pure water, as a function of temperature. Insert: saturation vapor pressure (left-hand scale) over water and ice at temperatures below 0°C (reprinted by permission, from Byers[11]).

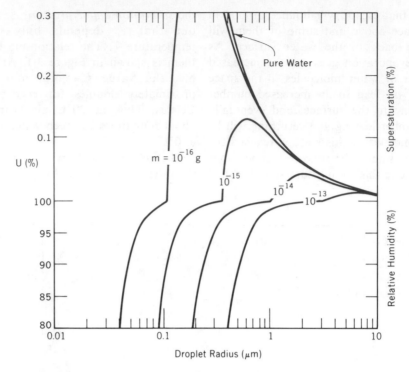

Figure 3.2 Relative humidity U (for values less than 100 percent) and supersaturation S (U − 100) in equilibrium with drops of salt solution containing various masses m of common salt (NaCl) (reprinted by permission, from Mason[46]).

to the saturation vapor pressure $e_s(T)$ for the observed temperature T. It is usually expressed in percent, so that

$$U = 100 \, e/e_s(T). \qquad (3.1)$$

If U exceeds 100 percent the air is said to be supersaturated, and the amount of excess is called the amount of supersaturation.

When a plane water surface is present, it is clear that the relative humidity cannot increase above 100 percent, since, as soon as it exceeds this value, more water condenses on the surface than evaporates, and the humidity returns to its equilibrium value. In the free air, however, the condensation must begin as tiny drops, and the equilibrium must be that with respect to a water surface of very small

radius of curvature. For surfaces with small radii of curvature, Kelvin showed that the equilibrium vapor pressure e_r is very large:

$$e_r = e_s \exp \frac{2\sigma}{r\rho_s R_v T}$$

$$\cong e_s \left(1 + \frac{2\sigma}{r\rho_s R_v T} + \ldots \right), \quad (3.2)$$

where σ is the surface tension and ρ_s the density of the liquid drops, e_s is the saturation vapor pressure over a plane water surface at the same temperature, and r is the radius of the drop. Thus the relative humidity in equilibrium with a drop exceeds 100 percent by an amount that varies inversely with the radius of the drop. For a very small drop it would be several hundred percent. The curve labeled "pure water" in Figure 3.2 shows the equilibrium relative humidity for drops larger than 0.4 μm.

If a wettable particle of a given size is present, condensation will occur on it when the humidity exceeds that in equilibrium with a drop of that size. Thus, as shown in Figure 3.2, a drop would form on a particle of radius 0.5 μm if the humidity exceeds 100.2 percent. For drop growth on wettable particles 0.1 μm in radius, the humidity would have to exceed 101 percent, and for 0.01 μm radius, 110 percent.

The equilibrium humidity over a droplet of solution is less than that for a pure water drop the same size. The equilibrium vapor pressure over a drop containing m grams of solute of molecular weight m_s is

$$e_r = e_s \exp\left(\frac{2\sigma}{r\rho_s R_v T} - \frac{3m_w \nu \phi m}{4m_s \pi \rho_s r^3}\right), \quad (3.3)$$

where m_w is the molecular weight of water, ϕ is the osmotic coefficient of the solute, and ν is the number of ions formed by dissociation per molecule of solute. Curves showing the variation of the equilibrium relative humidity U with r for drops containing various masses of sodium chloride are given in Figure 3.2. These curves can be used to indicate the behavior of particles containing a mass m of solute as U is increased. If U exceeds a critical value (78 percent for NaCl) the particle will pick up moisture and dissolve, becoming a droplet of the radius r for which U is the equilibrium value. If U is increased, r will increase to a new equilibrium value. For U = 100 percent and m = 10^{-15} g, for instance, r = .36 μm. As the humidity increases further, however, it reaches a critical value, slightly in excess of 100 percent, above which there is no longer an equilibrium size for the solution drop, and in fact as the drop grows larger it deviates farther from equilibrium. Thus condensation on

a solute particle—a hygroscopic nucleus—will take place at much lower supersaturation than on the same size nonsoluble nucleus.

In the troposphere there is normally a large number of hygroscopic nuclei, so that clouds form at very low supersaturations, effectively whenever U > 100 percent. The nuclei effective at the slight supersaturations that occur in natural clouds are called *cloud condensation nuclei* (CCN), in contrast to the smaller and nonhygroscopic particles, the Aitken nuclei, which are counted in instruments using high supersaturations. The concentration of CCN is usually in the range 25 to 1000/cm³, and this is consequently the usual number of drops in clouds and fog.

It is the large number of drops forming and competing for available water vapor that limits the size to which drops can grow by condensation. Because of this competition the modal radius of cloud drops is usually between 5 and 10 μm, and liquid drops rarely grow larger than 20 μm by condensation. These small drops fall relative to the air at rates that are smaller than the upward air speeds usually responsible for the production of the saturation leading to the condensation. Thus the drops cannot reach the ground as precipitation. The smallest precipitation drops have radii of 100 μm or more, and raindrops are usually greater than 1 mm in radius. Thus the mass of precipitation particles is 10^4 to 10^7 times the mass of the modal cloud drop. A crucial problem in cloud physics is the determination of the nature of the processes by which cloud drops grow to precipitation, and the circumstances under which they occur.

The occurrence of condensation involves the development of the necessary slight supersaturation. Unsaturated air can be brought to saturation either by adding water vapor—that is, by evap-

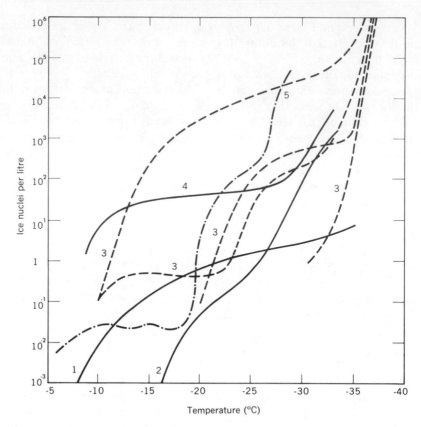

Figure 3.3 Examples of ice nucleus "spectra," that is, numbers of ice nuclei effective at various temperatures. (1) Findeisen-Schulz for 5 m/sec updraft; (2) Findeisen-Schulz for 20 m/sec updraft; (3) Smith and Heffernan; (4) Rau; (5) Murgatroyd and Garrod (reprinted by permission, from Fletcher[19]).

oration—to increase e to the existing value of e_s, or by cooling the air, thereby reducing e_s to the existing value of e. Producing condensation by evaporation sounds contradictory, but there are two situations in which it commonly occurs. One is the passage of cold air over a warm body of water. Rapid evaporation occurs because of the high equilibrium vapor pressure of the warm water. The vapor is rapidly cooled and condenses, forming *steam fog*. This phenomenon is seen in winter over unfrozen bodies of water at high latitudes. It is also frequent over heated open-air swimming pools at night. The second circumstance in which evaporation produces condensation is the falling of warm rain through colder air.

This sometimes occurs at a warm front. The evaporation of the warm drops may saturate the air below the frontal surface and contribute to the formation of warm-front type fog or stratus in the cold air ahead of the warm front.

The cooling of moist air to saturation may occur at constant pressure by radiation, conduction, or mixing, or it may be produced by adiabatic cooling as the pressure decreases, usually due to ascending motion. If the cooling is carried out at constant pressure the temperature at which saturation is reached is called the *dew point temperature* T_d. In adiabatic cooling, the condensation level, characterized by the condensation pressure p_c and condensation temperature

T_c, depends on the initial temperature, pressure, and mixing ratio. Clouds from which precipitation may occur almost always are produced by adiabatic cooling.

Ice Formation. Even when condensation occurs considerably below 0°C, it is usually in the form of liquid drops. It is an observational fact that clouds at temperatures down to −15°C, and lower, normally are composed of liquid drops rather than ice crystals. The reason for this is that the process of ice formation requires nucleation, just as formation of drops does. The nuclei of ice formation are much less plentiful than CCN. The number which are effective increases as the temperature decreases. Figure 3.3 shows some typical "spectra" of *ice nuclei* (IN), representing the number effective at various temperatures. The numbers range from less than $10/m^3$ at temperatures higher than −10°C to more than $1/cm^3$ below −30°C. At temperatures above −15°C, ice nuclei are scarce. At lower temperatures ice crystals are more likely to form, and once they are present in sufficient numbers the entire cloud tends to be transformed into ice crystals.

Atmospheric concentrations of ice nuclei vary considerably from location to location. Table 3.1 illustrates the difference in numbers of IN between rural and urban locations at −20°C. The concentrations vary over two orders of magnitude. Intensive anthropogenic sources in metropolitan areas include lead in automobile exhaust, and emissions from metal smelters. The concentration decreases with altitude in much the same way as other aerosol particles. An important threshold occurs at −40°C, where so-called homogeneous ice formation occurs; pure water without any type of nuclei crystallizes at this temperature.

The process of ice formation on ice nuclei is complex and not fully understood.

The ice nuclei are often crystalline particles whose surface initiates ice formation in bulk water or during condensation through the mechanism of epitaxy or oriented overgrowth. We know that ice may form by three different processes: (1) through direct deposition of the vapor on the nucleus in the humidity interval from ice saturation to water saturation; (2) through condensation with subsequent crystallization; and (3) through crystallization of the liquid phase by a nucleus floating in it or coming into contact with its surface, for instance, the surface of a drop. Organic or inorganic substances and gases can initiate the orientation of the H_2O molecules, which is necessary to form the ice lattice of the ice embryo. The nucleating substance cannot be soluble; it should, however, be wettable.

It is safe to say that ice formation in the atmosphere involves mostly a freezing process or a condensation-freezing process. Deposition of water vapor on a nucleus at ice saturation is not the rule. This is true also for the very low temperatures at which cirrus clouds form; they require water saturation; then condensation will occur, the product of which is the direct formation of an ice crystal without a visible water phase. At higher tempera-

Table 3.1 Comparative Values of Mean IN Concentrations

Location	IN l^{-1}, −20°C, Background	IN l^{-1}, −20°C, Urban
Barbados, 1969	1–2	
Colorado	0.5	20
Great Lakes	5	180
Great Lakes (above inversion)	<0.5	<0.5

Source: Paul Allee, NOAA (personal communication).

tures to almost -20 to $-25°C$ ice formation proceeds by both the freezing of a cloud droplet which has a freezing nucleus embedded and by the condensation-freezing process.

The microphysical conditions of ice formation have profound effects upon the development of precipitation in various cloud systems. The precipitation from convective clouds is mostly characterized by a small number of large drops, whereas the precipitation from deep layer cloud systems is characterized by a large number of small drops. This can be explained by the rapid upward motion in the former clouds in which the best nuclei respond first and find optimum conditions to grow—high liquid water content and a deep cloud system. Deep cyclonic layer cloud systems have entirely different conditions; they have updrafts of several cm/sec, relatively small amounts of precipitable water, but relatively large concentrations of ice crystals because they originate at the cirrus level, where most of the available aerosol particles act as ice nuclei. Conventionally, one postulates that one nucleus/liter is required to support sustained precipitation. This number has somewhat arbitrarily entered the field of weather modification when seeding amounts required to augment precipitation are computed. Through careful study of selected cases of snow precipitation it has been established that the concentrations furnished by nature amount to 5 to 40 crystals/liter.[66, 67]

Dynamic Processes of Clouds and Precipitation

We have seen that the supersaturation needed for the formation of cloud occurs principally by adiabatic cooling of rising air. The ascent of air can be associated with free convection, with the flow of air over mountains or fronts, or with regions of general convergence in the horizontal air flow.

Free convection occurs when the air is rendered hydrostatically unstable. The criterion for hydrostatic stability is expressed in terms of the *lapse rate*, which is defined as the rate of *decrease* of temperature with height,

$$\gamma = -\frac{\partial T}{\partial z}, \qquad (3.4)$$

where γ is the lapse rate, T is temperature, and z is height. If air rises without mixing or receiving or losing heat, it cools at the adiabatic rate γ.

$$\Gamma \equiv -\frac{dT}{dz} = \frac{g}{c_p} = 9.8 \times 10^{-3}°C/m,$$

where g is the acceleration of gravity and c_p is the specific heat of air. The adiabatic rate of cooling is approximately 1 deg/100 m. If $\gamma > \Gamma$, that is, if the temperature actually decreases more rapidly with height than the adiabatic rate of cooling, an unsaturated parcel of air displaced upward would find itself warmer than its surroundings and experience a buoyant force, tending to cause it to rise farther. The criteria for stability of unsaturated air are thus

$$\gamma > \Gamma \qquad \text{Unstable}$$

$$\gamma = \Gamma \qquad \text{Neutral}$$

$$\gamma < \Gamma \qquad \text{Stable.}$$

When condensation occurs in the rising air (or evaporation in descending cloudy air), the release (or absorption) of latent heat reduces the rate of temperature change below the unsaturated rate. We write

$$\left(\frac{dT}{dz}\right)_{\text{sat}} \equiv -\Gamma_s.$$

Unlike the unsaturated adiabatic rate Γ, which is constant, the saturation adiabatic rate Γ_s varies with temperature and pressure. At high temperatures it is less than $\Gamma/2$; at very low temperatures it approaches Γ. Its values for several temperatures and pressures are given in Table 3.2. In a fashion similar to the stability of unsaturated air, the criteria for stability of saturated air are

$\gamma > \Gamma_s$ Unstable

$\gamma = \Gamma_s$ Neutral

$\gamma < \Gamma_s$ Stable.

It is convenient to represent processes in the atmosphere on a thermodynamic diagram in which the coordinates are functions of the pressure and temperature. Figure 3.4 illustrates one diagram in common use, the skew T-log p diagram. The vertical coordinate is the logarithm of the pressure (really log

Table 3.2 Saturation Adiabatic Process Rate (°C/100 m)

Temperature (°C)	Pressure (mb)				
	1000	850	700	500	300
40	0.30	0.29	0.27		
20	0.43	0.40	0.37	0.32	
0	0.65	0.61	0.57	0.51	0.41
−20	0.86	0.84	0.81	0.76	0.68
−40	0.95	0.95	0.94	0.93	0.90

$1000/p$, so that pressure decreases upward). The horizontal coordinate is the temperature, but on a tilted scale, so that isotherms slope upward to the right. On this diagram the unsaturated adiabatic process is represented by curves sloping

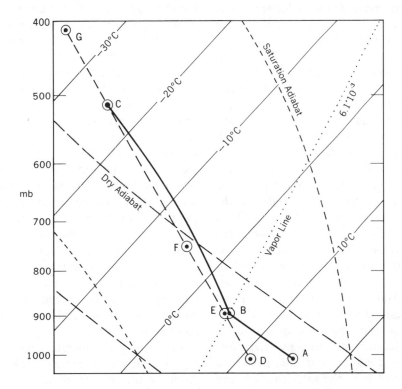

Figure 3.4 Thermodynamic diagram (skew T – log p diagram). Ordinate, pressure, labeled in mb. Solid lines sloping upward to right, isotherms, labeled in °C. Long dashed lines, unsaturated dry adiabatic process curves. Short dashed lines, saturation adiabatic process curves. Dotted lines, lines of constant saturation mixing ratio, labeled in parts per thousand. ABC, process curve which a parcel initially at 15°C, 1013 mb, and $w = 6.1 \cdot 10^{-3}$ would follow if lifted adiabatically. $DEFG$, typical sounding curve.

sharply upward to the left, and the saturation adiabatic process by curves sloping less rapidly to the left. The observed variation of temperature with decreasing pressure over a station can also be plotted on such a diagram. Curve $DEFG$ in Figure 3.4 is an example of such a sounding curve. By comparing the sounding curve with the adiabatic process curves, the stability of the air column represented can be determined. Thus we can see that below F $\gamma > \Gamma_s$ and above it, $\gamma < \Gamma_s$. If the air at any place in the portion of the air column represented by DF is saturated, it is unstable. The air at any position in the column represented by FG is stable, whether it is saturated or not.

Air can be rendered unstable by heating from below, as when the sun's radiation raises the ground temperature or when air moves from colder areas over warmer water or land, by advection of cold air at upper levels, or by vertical stretching of air columns. Once it is unstable the slightest disturbance will cause convection to occur. Rather than occurring randomly, the up-and-down motions are organized into patterns, cells in some instances, rolls in others. The upward-moving air is usually concentrated in a relatively small area in the center of the convective cell, the downward motion spread over a large area around it. If the rising air reaches its condensation level the upward flow is rendered visible in the form of cloud. The irregularities of the motion are shown in the bulging protuberances of the cumulus clouds. These pulsating irregularities have led some investigators to treat convection as a series of rising bubbles; others use a plume or jet to model the convective flow. In both treatments, air is mixed into the rising current from the sides and top, and this *entrainment* of environmental air reduces the buoyancy and liquid content of the rising air and limits the height to which it rises. Entrainment causes the rising cloudy air to cool faster than it would without mixing. Not only does this lead to smaller vertical velocities than would occur without mixing, but it also lowers the height to which convection extends, and thereby in some instances keeps the cloud from reaching temperatures low enough to activate the ice nuclei present and initiate the three-phase process of precipitation formation which will be described later.

When a cloud is transformed from liquid to ice, whether by the cloud attaining low-enough temperatures to activate the natural ice nuclei or by introduction of artificial nuclei, additional latent heat will warm the cloudy air and increase its buoyancy. Besides promoting the growth of precipitation-size crystals, cloud seeding with artificial ice nuclei may cause the cloud to grow to greater heights than it would naturally, increase the speed of convective currents, and stimulate the cloud replenishing process.

The adiabatic cooling which takes place when air flows over mountains or up frontal surfaces leads to widespread and continuous rain, in contrast to the spotty and intermittent clouds and showers produced by convection. However, if the air is *convectively unstable*, by which is meant that it is stable when unsaturated but becomes unstable by being lifted to saturation, showery precipitation, usually heavier than would occur otherwise, takes place in air ascending mountains or fronts. Except for the possibility of stimulating or augmenting convection, cloud seeding will not influence the dynamics of the flow up mountains or fronts.

In regions of general horizontal convergence ascent of air also takes place. Actually the motion of air at fronts is really an instance of upward flow caused by horizontal convergence. Cyclones—areas of low pressure around which the air flows counterclockwise in the northern hemisphere, clockwise in the southern hemisphere—have horizontal convergence associated with them, whether or not they have fronts in them. In contrast, anticyclones—high-pressure areas with the opposite sense of circulation—generally have horizontal divergence, particularly to the east of the center. Cyclones are thus characterized by widespread cloudiness and precipitation. If the air is unstable or subject to destabilizing influences, the rapid up-and-down motions of convective currents, with the associated cumuliform clouds, may be superposed on the widespread layer type cloudiness. The divergence and subsiding motion associated with anticyclones tend to inhibit convective activity and limit cloud growth to such small heights that precipitation cannot occur.

In tropical cyclones and hurricanes the horizontal convergence is concentrated at the eye wall and along spiral bands which radiate out from it. The highest clouds and heaviest precipitation occur there. Associated with the extremely rapid release of latent heat, the maximum winds occur in a ring around the eye wall. Attempts to reduce the severity of hurricane winds are based on the idea that if the area over which the latent heat is released can be broadened by massive cloud seeding the pressure gradient, and therefore the wind speed, will be reduced.

At the largest scale the flow of the atmosphere consists of a circumpolar vortex upon which disturbances—long and short waves, cyclones, and anticyclones—are superposed. These disturbances are initiated and grow as a consequence of instabilities in the basic flow pattern. The resulting vertical motions, clouds, and precipitation play a role in the release of the instability and growth of the disturbances. At present our understanding of the factors which control the stability of the large-scale circulation features is fragmentary. The first step

towards weather modification on larger scales must be the improvement of that understanding.

The new tools which have become available to meteorologists in the past two or three decades, particularly the high-speed digital computer and the instrumented satellite, have already expanded our understanding and give promise of enabling continued rapid strides. The computer has enabled the modeling of atmospheric processes far too complex to be dealt with otherwise. Included are models of convective clouds and precipitation processes. As these models become more complete and more realistic they will enable evaluation of the extent to which weather situations are subject to modification by available treatments. At present the treatments are being tried largely on an ad hoc basis.

THE FORMATION OF PRECIPITATION

It has been mentioned that drops formed by condensation are too small to fall as precipitation. They have radii mostly in the range 1 to 20 μm. Drops of this size fall with speeds between 0.01 and 5 cm/sec, and the upward motion of air that causes the adiabatic cooling that produces the cloud more than offsets the downward motion of the drops. If drops as small as this do fall out of the cloud into unsaturated air, they evaporate very quickly. Computations have shown that if $U = 90$ percent in the air below the cloud 10-μm drops would vanish within 1 m of the cloud base.

The smallest precipitation particles, drizzle drops, are about 100 μm in radius, and raindrops range from 0.5 to 3 mm in radius. The corresponding speeds of fall, 70 cm/sec to 9 m/sec, are large enough for the drops to move downward through the cloudy air faster than it is rising, and to pass through the unsaturated air below the cloud fast enough to reach the ground before being completely evaporated, even when the updrafts are fairly strong and the humidity below the cloud is low.

The key difference between cloud and precipitation is thus particle size. The change from a 10-μm cloud drop to a 1-mm raindrop involves increase in mass by a factor of one million. Growth by condensation on the CCN cannot produce this tremendous change in mass, because the CCN are so numerous. The two processes that appear to be able to do so are (1) collision and coalescence, and (2) the three-phase, or Bergeron-Findeisen, process. As the name implies, the collision and coalescence process consists of the collection of small drops to form a large one. The three-phase process consists of the transfer of water from numerous liquid drops to a few ice crystals by evaporation from the former and deposition directly in the form of ice on the latter. Once the ice crystals initiated by the three-phase process have grown sufficiently large to fall rapidly, they grow also by collecting cloud drops (riming), and their growth by collection may be more rapid than by deposition.

The Warm Rain Process

The collision and coalescence process is the only one which can cause precipitation to form in clouds which are entirely at temperatures above 0°C. For this reason it is called the *warm rain process*, even though it also acts in clouds with lower temperatures, in which the three-phase process may play the leading role in initiating the precipitation.

On the face of it the way the warm rain process acts appears obvious. Because the CCN have different sizes the drops forming on them also vary in size. Since the larger drops fall faster than the small

ones, they should overtake and collect those which are in their paths, becoming still larger, falling still faster, and sweeping up small droplets more rapidly. However, the process is not as simple as it seems. As the drops fall, the air ahead of them is pushed out of the way, and the small drops tend to be carried out of the path of the larger drops along with the air. Only a fraction of the drops in the path collide with the larger drops. This fraction is called the *collision efficiency* E_L. The drops which do collide may bounce off instead of coalescing. The fraction of those colliding which coalesce is called the *coalescence efficiency* E_S. The fraction of the drops in the path which is collected is called the *collection efficiency E*, and obviously $E = E_L \cdot E_S$.

The collision efficiency depends on the size of both the larger and the small drop. Figure 3.5 shows the results of recent theoretical computations[28] for different radii A of the large drops as a function of the ratio $p = a/A$ between the radius of the small drop and the radius of the large drop. It is seen that for $A \leq 15$ μm E_L is very small, 0.01 or less, and even for $A = 20$ μm $E_L < 0.05$. Earlier computations by Hocking[27] had indicated that $E_L = 0$ for $A < 19$ μm. The absence of this cutoff in collisions was first found by Davis and Sartor,[14] who obtained collision efficiency curves similar to the Hocking and Jonas curves in Figure 3.5.

The computation of E_L for these small values of A can be carried out fairly rigorously. For larger values of A, for which the Stokes linearization of the hydrodynamic equations cannot be used, computations of E_L require more approximations. Two procedures have been used recently, a superposition method and an application of the Oseen approximation. Because of the uncertainties of these approximations it is desirable to check experimentally the validity of the computational results. The experimental

evaluations, which give collection efficiencies rather than collision efficiencies,
are also difficult to carry out, and only a
few data are available. Figures 3.6a to e
show a comparison of the results of
various computations with the experimental results available. In Figure 3.6a
the results obtained for $A = 30$ μm by
Davis and Sartor (D-S)[14] using Stokesian
hydrodynamics and by Klett and Davis
(K-D)[33] using the Oseen approximation
are compared with experiments conducted by Picknett (P)[54] and Woods

and Mason (W-M),[65] the latter for values
of A averaging 33.5 μm. The Klett-Davis
computation appears to overestimate the
collision efficiency. In Figures 3.6b to e
computational results for larger A's obtained by Shafrir and Neiburger (S-N)
using a superposition technique[55] and by
Klett and Davis using the Oseen approximation are compared with experimental
results. The K-D curves are slightly
higher than the S-N curves and the experimental values, but the close
agreement of the two different computa-

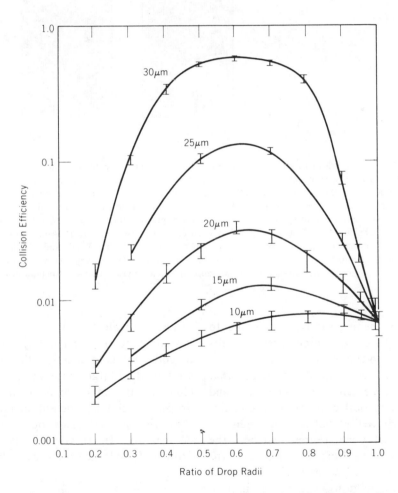

Figure 3.5 Collison efficiency E_L for several values (10 to 30 μm) of the radius A of the collector-drop as a
function of the ratio $p = a/A$ of the drop radii (reprinted by permission, from Hocking and Jonas[28]). The
comparisons are from a paper prepared by Neiburger for the International Conference on Cloud Physics,
London England, August 21–26, 1972.

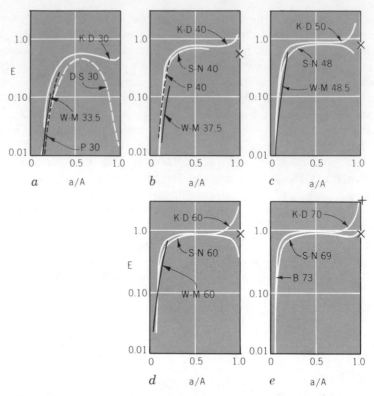

Figure 3.6 (a) Comparison of computed and experimental values of collision efficiency of drops with radius $A = 30~\mu$m. D-S, values computed by Davis and Sartor using Stokes hydrodynamics.[14] K-D, values computed by Klett and Davis using modified Oseen approximation.[35] P, results of experiments by Picknett.[54] W-M and experimental values by Woods and Mason.[65] (b–e) Comparison of computed and experimental values of collision efficiency of drops with radius A in μm given along curves. S-N, values computed by superposition technique. K-D, values computed by Klett and Davis using modified Oseen approximation. P, results of experiments by Picknett. W-M and ×, experimental values by Woods and Mason. B, experimental value by Beard.[2] +, experimental value by Telford et al.[58] These comparisons are from a paper prepared by Neiburger for the International Conference on Cloud Physics, London, England, August 21–26, 1972.

tions and the experiments lead one to conclude that the computational results are fairly good approximations.

A further conclusion from the fact that the computed collision efficiencies and the experimental collection efficiencies agree fairly well is that for the conditions of the experiments represented here the coalescence efficiency E_s is approximately unity. That E_s is smaller than 1 in conditions occurring in natural clouds has been indicated by experiments conducted at the University of Toronto[63] and at UCLA.[68]

The simplest way to treat the development of precipitation-size drops by collection is to deal with the growth of an individual large drop falling through a cloud composed of uniform-sized small drops, the cloud being sufficiently dense that the growth can be considered continuous. In this case the rate of growth of the large drop is given by

$$\frac{dA}{dt} = \frac{E(1 + p)^2 (V - v)L}{4\rho_s} = \frac{WL}{4\rho_s}, \quad (3.5)$$

where L is the liquid content of the cloud per unit volume, V and v are the fall

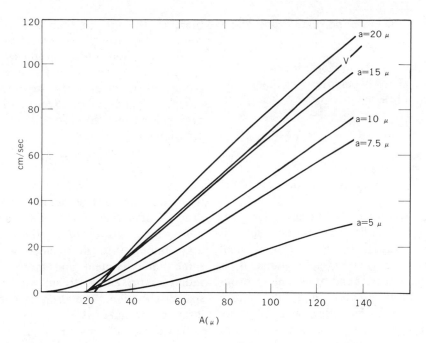

Figure 3.7 Collection coefficient W as a function of large drop radius A for clouds of droplets of radius a. The curve labeled V gives the terminal velocity of the large drops.

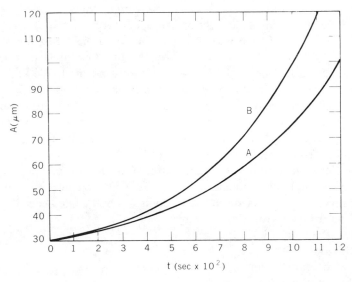

Figure 3.8 Growth of large drop, initial radius $A = 30\ \mu$m falling through cloud of smaller drops, liquid content $L = 1\ \text{g/m}^3$. Curve A, monodisperse cloud, drop radius $a = 10\ \mu$m; curve B, polydisperse cloud with K-M spectrum, average radius $a = 7.5\ \mu$m.

speeds of the large and small drops, and

$$W \equiv E(1 + p)^2(V - v). \qquad (3.6)$$

W is called the collection parameter.

If the cloud is polydisperse, the growth rate would be

$$\frac{dA}{dt} = \frac{\pi}{3} \int_0^A n(a)a^3 W(a)\, da, \qquad (3.7)$$

where $n(a)$ is the number of cloud drops per unit volume with radii between a and $a + da$.

The variation of W, the collection coefficient, with A for several values of a is shown in Figure 3.7. Since E is so small for $A < 20$ μm, it is seen that W is practically zero, and thus unless drops with radius greater than 20 μm are present growth by collection will not occur. It will be noted that the variation of W with A is approximately linear over a wide range. If we set $W = kA - B$, Equation 3.5 can be integrated, yielding

$$t_1 = \frac{4}{kL} \ln \frac{kA_1 - B}{kA_0 - B}$$

$$= \frac{9.2}{kL} (\log W_1 - \log W_0) \qquad (3.8)$$

for the time of growth from A_0 to A_1. For $a = 10$ μm $k = 6.8 \times 10^3$/cm and $B = 15.4$ cm/sec.

Application of Equation 3.8 shows that the growth by collection is slow while the drop is small and increases rapidly when the drop becomes larger. Curve A in Figure 3.8 gives the variation of radius with time for a drop initially 30 μm in radius falling through a cloud of 10 μm radius droplets with liquid content $L = 1$g/m³. It shows that it takes more than 400 sec for the drop to grow from 30 to 40 μm, but less than 100 sec for it to grow from 90 to 100 μm.

If the cloud contains drops of various sizes with the same mean volume radius and the same liquid content, the fact that some larger drops are present contributes more to the growth than the fact that some of the drops are smaller than the mean size, and the drop grows more rapidly. It has been found that the distribution of drop sizes in clouds of various types can frequently be represented by the formula given by Khrgian and Mazin (K-M)[35]:

$$n(a) = \frac{1.45La^2e^{-3a/\bar{a}}}{\bar{a}^6}, \qquad (3.9)$$

where $\bar{a}$ is the average radius of the cloud drops. Curve B in Figure 3.8 shows the more rapid growth of a 30-μm radius drop falling through a cloud with the same liquid content and mean volume radius as for curve A, but with a K-M size distribution. The drop is one-third larger at the end of 1000 sec.

Because clouds in actuality are polydisperse with size distributions typified by Equation 3.9, it is clear that rather than considering the growth of individual large drops falling in a cloud of small ones, we should be concerned with the interaction of drops of all sizes and the resulting change in size distribution. In a complete treatment of variation of drop size spectrum changes caused by condensation or evaporation, air motions, gravitational settling, and the formation of small drops by the breakup of large ones should be taken into account. We limit ourselves to considering the change due to hydrodynamic capture. In treating this problem it is convenient to consider the size spectrum in terms of drop mass instead of radius. We write $n(m)$ for the number of drops/cm³ having mass between m and $m + dm$. The change in $n(m)$ because of collisions and coalescence is

$$\frac{\partial n(m)}{\partial t} = \frac{1}{2} \int_0^m n(M)n(m - M)$$

$$\times K(M, m - M)\, dM$$

$$- n(m) \int_0^\infty n(M)K(m, M)dM, \qquad (3.10)$$

where $K(m,M)$, the rate of interaction of drops of mass m and M, is given by

$$K(m,M) = \pi A^2 W = \pi (A + a)^2 E(V - v).$$

(3.11)

The integration of Equation 3.10 for realistic drop size distributions cannot be carried out analytically. Numerical solutions for two K-M distributions are shown in Figures 3.9 and 3.10. Figure 3.9 shows the change with time caused by collection for a cloud with $\bar{a} = 4.5$ and $L = 1$ g/m^3. The initial distribution has practically no drops with radius greater than 30 μm; consequently the growth is very slow. Even after 3200 sec there is very little liquid in drops larger than 40-μm radius. The initial distribution (see Figure 3.10), in which total liquid is the same but $\bar{a} = 7.5$, has a number of drops larger than 30 μm. As a result there is

rapid growth of the larger drops and a second maximum in the spectrum develops, with more liquid in the drops greater than 100-μm radius than in the smaller ones by 2000 sec.

The computations of drop growth and change of drop spectra presented above used the collision efficiency E_L for E; that is, in them it was assumed that the coalescence efficiency E_S is unity. If E_S is smaller than 1 the growth by collection will be still slower, so that the requirement is even greater for drops with radius greater than 20 μm to be present in order that precipitation be initiated by the warm rain process. It is important to determine under what circumstances drops that large will occur in clouds.

Several possible factors that might cause their formation by condensation have been proposed. They include (a) the presence of an unusually small number of

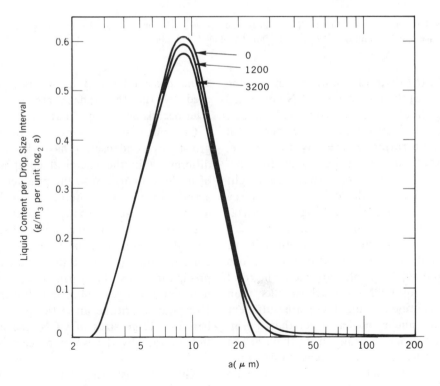

Figure 3.9 Evolution with time (given in seconds above each curve) of K-M drop size spectrum with initial average drop radius $\bar{a} = 4.5$ μm and liquid content $L = 1$ g/m^3.

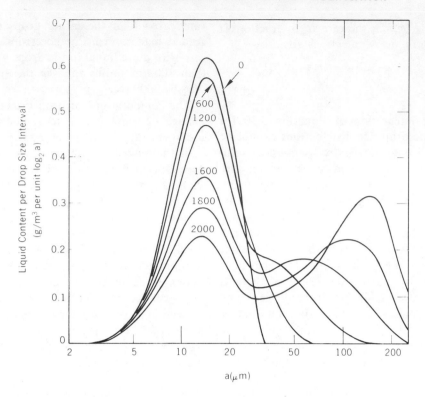

Figure 3.10 Evolution with time (given in seconds above each curve) of K-M drop size spectrum with initial average drop radius $\bar{a} = 7.5\ \mu$m and liquid content $L = 1$ g/m³.

CCN, (*b*) the presence of unusually large CCN, (*c*) the presence of CCN composed of particularly hygroscopic substances, and (*d*) fluctuations of the degree of supersaturation caused by turbulence. Another suggestion has been that accelerations due to turbulence might greatly increase the collection efficiency for small drops, making the initial presence of larger ones unnecessary.

It has been found that over the oceans, especially in tropical regions, conditions (*a*) and (*b*) frequently are met. The total number of CCN is much smaller than over continents, and there are sufficient "giant" nuclei, nuclei larger than 1 μm, for the drops to grow larger than 20 μm by condensation in a reasonable time. This is consistent with the fact that most of the confirmed instances of precipitation falling from clouds in which the temperature was above 0°C, so that there

is no possibility of it having been initiated by the three-phase process, were over oceans or in tropical areas.

Computations of models of drop growth by condensation have shown that differences in the chemical composition of nuclei do not lead to the production of significantly larger numbers of large drops. Turbulent mixing similarly has been shown to be incapable of rapid enough development of large drops by condensation to permit the initiation of precipitation by the warm process or to materially increase collection efficiencies. Except in maritime air with small numbers and large sizes of CCN, the initiation of rain by the warm process is unlikely.

This circumstance suggests the possibility of artificially initiating rain by the warm process, either by introducing water drops of sufficient size directly, say, by spraying drops larger than 30 μm into the cloud, or by introducing giant hygroscopic nuclei. For maximum yield the water drops or nuclei should enter the base of the cloud so that they can grow while being lifted in the updraft as well as during the period of fall after the drops have achieved terminal velocities larger than the speed of the rising air.

The Three-Phase Process

The three-phase or Bergeron-Findeisen process postulates that for the formation of precipitation a mixed cloud is necessary, consisting of water drops and ice crystals. Figure 3.1 shows that the vapor pressure over ice is less than that over water for a given temperature; therefore, if both phases are side by side in a water cloud, the ice crystals grow at the expense of the supercooled drops because of the vapor pressure gradient between the drops and the ice crystals, which causes the drops to evaporate and the ice crystals to grow by diffusion. This vapor pressure difference between the ice crystals and the water drops in the cloud environment is much greater than the vapor-pressure difference between the water drop and its environment (at a temperature of $-10°C$ by about a factor of 100); therefore the ice crystals take up by condensation so much more moisture that the droplets may start to evaporate. Consequently, they begin soon to fall faster than the remaining cloud droplets and begin to collect these through the riming process, all the way taking advantage of the vapor pressure difference between the water and ice. Bergeron postulated that this process is required for the production of rain drops and of a steady and healthy rate of precipitation. Recognizing the insufficiently small number of ice crystals forming in moderately supercooled clouds he postulated that for sustained precipitation to

occur a "seeder cloud" is as necessary as a "feeder cloud."[5] The seeder cloud was to be a cirrostratus or altostratus cloud, which both consist of plenty of ice crystals but have little precipitable water, while the feeder cloud was to be a low-level stratus cloud whose top temperature would be anywhere between, say, -5 and $-20°C$ while its base was usually near the freezing level. This cloud has a higher water content but a poor crystal production and in order to be effectively snowed out requires the ice crystals of a seeder cloud falling into it. The entire cloud system then is called "nimbostratus." Updraft velocities in these cloud systems are of the order of several centimeters per second, which is less than the fall velocities of ice and snow crystals, which are between $1/2$ and about 2 m/sec. In convective or orographic clouds with updrafts comparable to or higher than the fall velocities of snow crystals different precipitation mechanisms exist leading to graupel, soft hail, hail, and sometimes, giant wet snowflakes. In these clouds the process of riming dominates over diffusional growth, and the pure three-phase process plays a secondary role.

Diffusional Growth of Snow Crystals. The growth of snow crystals is considerably more complex than that of droplets mainly because of the more complex field of vapor diffusion around the various crystal habits.[34] Figures 3.11 and 3.12 convey an impression of some of the complex forms and crystal habits as they occur in natural clouds from the highest cirrus clouds to diamond dust and snow crystals near the surface. The rate of increase of crystal mass m_c due to condensation dm_c/dt follows the expression

$$\frac{dm_c}{dt} \simeq 4\pi CD(e_1 - e_0), \qquad (3.12)$$

where C is the electrostatic capacitance of a hypothesized conductor of shape and size identical to that of the ice crystal, D

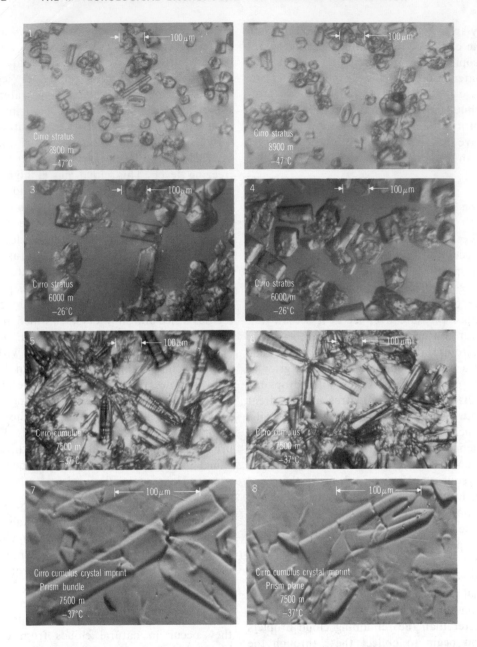

Figure 3.11 Typical ice crystal forms found in cirrus clouds. Photographs 1 to 4, cirrostratus crystals, single crystals in prism forms photographed in cloud top and base. Photographs 5 to 8 cirrocumulus crystals in form of prism bundles with hollow prism. Photographs 7 and 8 are imprints of crystals on lacquer which show the structure of the prism planes.

is the vapor diffusivity, e_1 is vapor density at infinity and water saturation, and e_0 is vapor density at the crystal surface. Values for C have been determined experimentally by J. E. McDonald[40] using brass scale models of ice crystals in a condenser field; he found that these values do not differ much from the theoretical values for the following shapes:

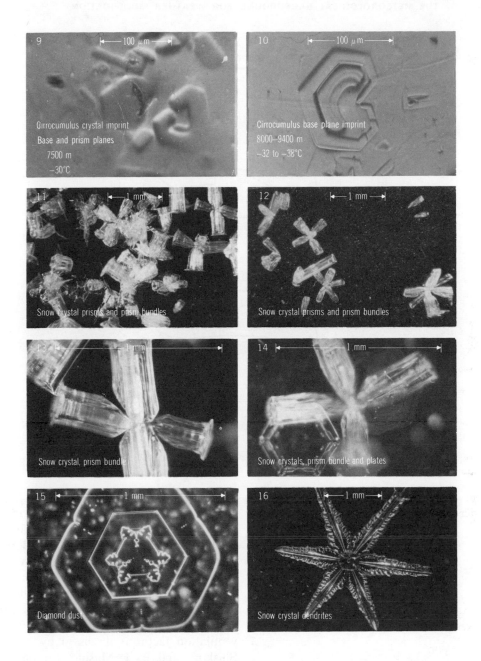

Figure 3.12 Photographs 9 and 10 are continuations of 5 to 8. They show the imprints of base planes of hollow crystals—note the tendency for a spiral characteristic. Closed hexagonal ring forms were also observed. Photographs 11 to 14 are snow crystals photographed at the ground surface. These crystals are typical of altostratus; they have formed at a low temperature in the cirrus level and have descended growing slowly at humidities near ice saturation. On falling through layers saturated with water plates or stars grow at their prism bases. Photograph 15, ice crystal as hexagonal plate indicating trigonal symmetry in the center. Photograph 16, large dendrite crystal.

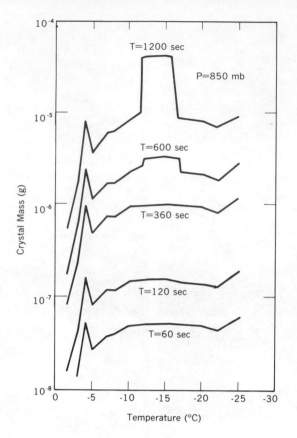

Figure 3.13 Mass of ice crystals grown by vapor deposition as a function of nucleation temperature and growth duration (see Cotton[13]).

Sphere of radius r:

$$C = 1$$

Circular disk of radius r:

$$C = 2r/\pi$$

Prolate spheroid, major and minor semi-axes a and b, respectively:

$$C = \frac{2a\epsilon}{\ln \left\{ (1 + \epsilon)/1 - \epsilon) \right\}}.$$

$$\epsilon = \left(\frac{1 - b^2}{a^2} \right)^{1/2}$$

The condensation of the molecules on the ice surface releases latent heat of condensation which affects e_0 and brings in the thermal conductivity K of the air.

The application of Equation 3.12 requires that the crystal is at rest with respect to its environment. In reality the crystal falls through the vapor field which causes a ventilation factor producing increased diffusion toward the crystal. This ventilation factor is discussed by N. S. Shiskin[56] and B. J. Mason[46] for an ice sphere for Reynolds numbers from 10 to 200 and found to be $1 + 0.23 \, \mathrm{Re}^{1/2}$; it is added to the equation as a factor for the growth rate. This ventilation factor may increase additionally through the fall attitude of the crystal which often rotates about a vertically pointing axis.

Another correction to the crystal

growth, difficult to assess, must be considered if the crystals fall through a water cloud and droplets enter the diffusion field and constitute local intensive sources of water vapor.[44, 51] It appears that this correction becomes significant in an environment of high liquid water content, that is, in thunderstorms and shower clouds. Dendrites often have air bubbles lined up along the mid rifts of the branches. This indicates that water in liquid form has frozen on these surfaces. These crystallization processes are related to the growth by riming, that is, by the collection of cloud droplets on branches, plates, columns, and spatial dendrites.

Following W. R. Cotton,[13] the final expression under consideration of $e_0/e_1 = S$ and inclusion of the heat released by riming $L_f\,[dm/dt]_R$ is

$$\frac{dm}{dt}\bigg]_v = \left\{ \frac{4\pi C(S-1)}{A_k + B_k} \right.$$

$$\left. - \frac{m_w L_s L_f \dfrac{dm}{dt}\bigg]_R}{R_a K_i T^2 (A_k + B_k)} \right\}$$

$$\times\,(1 + 0.23\,\mathrm{Re}^{1/2}), \quad (3.13)$$

Table 3.3 Crystal Habits

Temp. Range (°C)	Crystal Habit
0 to −3.0	Hexagonal plates
−3 to −5.0	Needles
−5 to −8.0	Prisms
−8 to −12.0	Hexagonal plates
−12 to −16.0	Hexagonal plates, dendrites
−16 to −25.0	Hexagonal plates
< −25	Prisms

where

$$A_k = \frac{m_w L_s^2}{K_i R_a T^2}$$

and

$$B_k = \frac{R_a}{m_w} \frac{T}{D_v e_s(T)}.$$

The following denotations are valid:

C	Crystal capacitance
S	Cloud saturation ratio
R_a	Gas constant of air
m_w	Liquid water content
L_s	Latent heat of sublimation
L_f	Latent heat of fusion
K_i	Molecular thermal conductivity
T	Cloud temperature (Kelvin)
Re	Reynolds number for ice crystal
D_v	Diffusivity of water vapor in air
$e_s(T)$	Saturation vapor pressure at temperature T.

This equation contains both the heat released by sublimation as well as by riming in the second term. The solution of this equation without the second term is shown in Figure 3.13 giving the mass of ice crystals grown by vapor deposition at water saturation as a function of nucleation temperature and growth duration. The growth conditions leading to this figure may modify in the presence of water droplets even though saturation humidity still prevails. Water droplets near the ice crystal surface may deform the diffusion field, steepen locally the gradient, and increase the rate of growth. It does contain, however, the influence of crystal habit through consideration of the form factor C for the crystal habits given in Table 3.3. Missing in this table, taken from an article by J. Hallett and B. J. Mason,[46] are the important spatial dendrites. Some of these crystals, which are the most efficient scavengers of the cloud water content in both the vapor and liquid form, originate as follows.

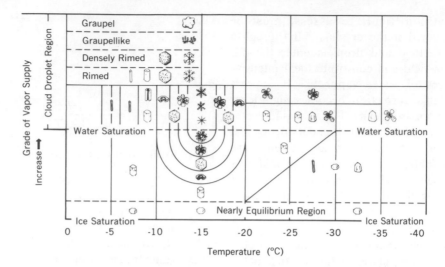

Figure 3.14   Temperature and humidity conditions for the growth of natural snow crystals of various types (see Magono and Lee[43]).

Cloud droplets that rise with the cloud may have embedded a freezing nucleus or come into contact with a freezing nucleus. When the threshold temperature of the nucleus is reached, the droplet freezes. The subsequently developing crystal habit is a function of the threshold temperature and the droplet size in a way that has been explored by C. Magono and H. Aburakawa,[42] and of the temperature and humidity of the atmospheric layers that it passes through. The temperature and droplet size determine whether it will crystallize into a single crystal or a polycrystal, and this determines whether a plane or spatial star will develop.[62]

Planar double dendrites and spatial dendrites form through freezing of cloud droplets that are carried in the water-saturated updraft to their threshold temperature between −15 and −25°C; the lower the temperature, the more likely the formation of a polycrystal for any droplet size. For normal cloud drop diameters of 10 to 20 μm, polycrystals begin to appear at about −15°C and determine the habit of the majority of crystals at −25°C; consequently it is between −15 and −25°C that planar and spatial dendrites form in a water cloud. There are at least two other processes by which these crystal habits may form: one is indicated by the snow crystals of Figure 3.12 (11–14); they indicate the tendency of growing dendrites at the end of the prism bundles from the base plane. The other is shown in a diagram from C. Magono and C. W. Lee[43] (Figure 3.14) in the temperature range between −20 and −25°C as a form in the cloud droplet region; it looks like a prism bundle whose prism planes have grown into side plates. We believe that this occurs when the prisms are entirely hollow; causing the base plane imprints to look like those in Figure 3.12 (9, 10). The figures show that they can be either closed or open because of spiral crystal growth. The great importance of these irregular types of snow crystals, which are found in a number of still different varieties caused by riming,[43] follows from the fact that they offer a large surface area for efficient growth through diffusion, as well as sharp edges causing high collection efficiencies for cloud droplets by riming. Figure 3.13 is qualitatively in agreement with Hallett's[23] laboratory studies in a diffusion cloud chamber, that is, also without ventilation and the proximity of water droplets.

Growth by Riming. The collection of cloud droplets by solid hydrometeors is called riming. Riming occurs in many forms and creates an infinity of new hydrometeors, from rimed snow crystals to conical graupel, and finally hailstones. The various forms are caused by numerous feedbacks between cloud microphysics and dynamics, the latter affecting updraft, cloud water content, and various environmental conditions, such as temperature and turbulence. The basic analytical expression for the rate of growth of hydrometeor mass caused by riming $\frac{dm}{dt}\big]_R$ is analogous to Equation 3.5:

$$\frac{dm}{dt}\bigg]_R = A_i E W (V_c - V_d). \quad (3.14)$$

Here A_i is the geometric captive cross-section, E the collection efficiency, w the liquid water content, and $(V_c - V_d)$ the difference in fall velocity of crystal and collected droplet, respectively. In most cases the droplet fall velocity can be neglected. This equation contains two parameters whose magnitudes are difficult to assess: one is the collection efficiency E and the other is fall velocity V_c. J. E. Jiusto[31] takes $E = 0.5$ for hexagonal plates and solid stellars and 1 for dendrites but with a reduction of the capture cross-section to $\pi r_c^2/2$. We believe that the crystals with the highest collection efficiency are the spatial dendrites for which we suggest a value of 1 with πr_c^2 for the capture cross-section. They have also the largest surface and are therefore more efficient than other crystals in growing by vapor diffusion. Since they develop in the temperature range in which nature's supply of ice nuclei becomes insufficient to snow out a potential precipitation system,

namely, nuclei concentrations below 5 to 40/1, this crystal habit is apparently nature's answer to this deficiency. The geometric capture cross-section becomes important for W. R. Cotton's[13] treatment of the graupel process. Given that a crystal always falls in a direction normal to the plane of its major dimension, $A_i = ac$ for needles and columns and $\pi a^2/4$ for plates, dendrites, and spherical ice particles. Here a stands for one of the secondary axes, while c stands for the primary axis. Geometric growth by riming is assumed to be in the direction of the a-axes for needles and columns and in the direction of the c-axis for plates and dendrites. Once a crystal has rimed to the extent that the a- and c-axes are of comparable size, the crystal is then considered to be a spherical graupel with corresponding geometric cross-section, fall velocity, and collection efficiency. Cotton accepts as the density for riming growth the U. Nakaya and T. Terada[53] value of 0.125 g/cm³. However, since this value has been derived from winterly graupel falls, we believe it to be representative for "snow pellets" rather than for graupel and that a more realistic value should lie between 0.5 and 0.7, in agreement with R. List.[37] Of course, Macklin and Macklin and Paine, quoted in Mason,[46] have shown that the density of accreted ice depends on the radius of the impacted droplet, the impact velocity, and the temperature and that densities over the entire range from 0.1 to 0.9 occur. Most certainly a value between 0.57 g/cm³ (randomly close-packed spheres of 0.9 g/cm³ density) and 0.67 g/cm³ (regularly packed spheres) would be more realistic than Nakaya's special value of 0.125 g/cm³. The efficiency of the riming process for dendrite crystals based upon Equation 3.13 under consideration of the release of latent heat of fusion is given in Figure 3.15. Comparison of Figure 3.13 with Figure 3.15 indicates that after

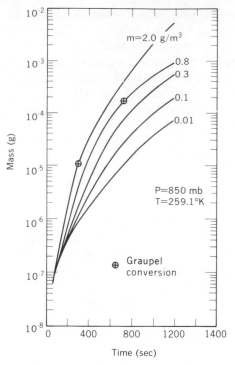

Figure 3.15 Total mass of dendrite grown by vapor deposition and riming as a function of growth duration (see Cotton[13]).

about 600 sec the riming process is more efficient for clouds with a liquid water content approaching 1 g/m³.

Snow Crystal Habit, a Three-Dimensional Interrelating Analysis Tool. Three basic precipitation systems can be clearly identified by the particle type which they discharge as well as by the typical cloud systems in which they originate. We wish to emphasize that the entire temperature range from −10 to −25°C is characterized by dendritic growth and complex shapes. The Magono and Lee[43] classification of snow crystal habits (Figure 3.14), which shows clearly the dependence of crystal habit on both temperature *and* humidity, will assist the reader in the following discussion. At −15°C, for instance, depending on humidity, all crystal habits from prisms (ice saturation) to dendrites (water cloud)

may occur. This modification of snow crystal habit is little appreciated in the literature, but it is important during seeding experiments where a water-saturated cloud may quickly change to ice saturation, causing a corresponding change in the habit. The crystal habit will not only change profoundly within the range of a few degrees Celsius but also within the range of a few percent of relative humidity. Thus crystals are extremely sensitive indicators of their environmental growth conditions, that is, of their cloud environment.

Modern computer methods so far have failed to model realistic snow-crystal forms, save for the most elementary habits; these habits may occur in seeded clouds but almost never in unseeded ones. Natural habits are complicated forms in which the six-sided star is the best known, but not the most frequently occurring one. Regularly formed stars have captured human imagination more than any other type since all ages; they were described by J. Kepler,[32] who stated that they are usually the first forms to fall when it starts to snow. Because of the double sensitivity of the snow crystal habit to temperature and humidity, we use the habits, in the subsequent description of snow precipitation systems, as miniature analogue computers which have solved and store the solution to the many subtle and intricate relationships between cloud dynamics and cloud microphysics. Equal habit comes from equal environment.

Description of Some Basic Snow Precipitation Systems. Considering the infinite variety of precipitation systems we acknowledge the fundamental significance of Bergeron's primitive yet genial parameterization of the precipitation mechanism into the existence of a "seeder" and a "feeder" cloud. The former, also called the "releaser" cloud, releases ice crystals at concentrations sufficient to snow out effectively the latter or "spender" cloud. This cloud, occurring at such high temperatures that little ice crystal formation takes place, is the real seat of the precipitable water. Bergeron's parameterization permits immediately the distinction of three precipitation systems, as follows.

Our System No. 1, overriding warm air over cold air on the back side of a large depression, caused the formation of a deep altostratus layer that continuously discharged very fine snow crystals. The overriding warm air is clearly shown on the radiosonde cross-section in Figure 3.16; in the shallow layer below the inversion, clouds but no ice crystals may form since the temperature is not suffi-

ciently low. From the homogeneously overcast sky fine snow fell whose crystals had the typical prismatic shapes of low-temperature origin which are typical of cirrus and altostratus [see Figures 3.11 (1–8) and, 3.12 (11–14)]. The ice character of the cloud follows from the dewpoint above 650 mbar, which represents ice saturation for the environmental temperature.

The total snow accumulation in 24 hr was not more than 2 to 3 in.; traffic problems were caused by icy roads and drifting snow. In Bergeron's terminology,[5] the snowfall was characterized by the existence of a "seeder cloud" but by the absence of a low-level "feeder cloud." As the crystals from these cloud systems usually do not aggregate their number

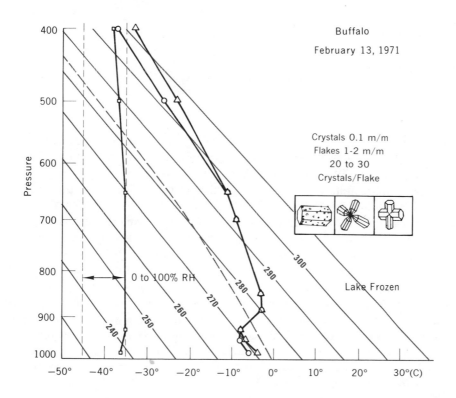

Figure 3.16 Atmospheric pressure (ordinate) vs. temperature (abscissa) diagram giving temperature and humidity profile. Curve with triangles is temperature, curve with circles dew point, curve with squares is relative humidity. Insert indicates crystal forms which originated on that day. They are prism and prism bundles as on Figure 3.11.

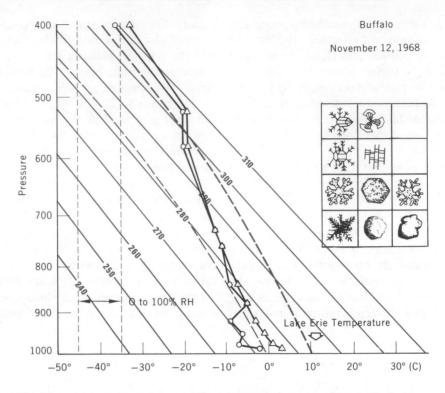

Figure 3.17 Atmospheric profile analogous to Fig. 3.16. Insert illustrates schematically the observed crystal forms which are divided into two main formation levels: (1) particles which originated as in Figure 3.16; (2) particles which originated in convective stratocumulus clouds over the warm lake. These are spatial dendrites, graupel, heavily rimed forms.

can be counted relatively easily; it is from such counts that we determined a crystal concentration of 5 to 40/1 as the natural concentration in precipitation.

A typical System No. 2 occurred on November 12, 1968. The synoptic situation on this day was similar to that described for System No. 1 except that low-level convection over warm Lake Erie contributed to cloud formation below the temperature inversion. A mixture of heavily rimed crystals of almost all possible forms, plane dendritic and sector stars, prisms and prism bundles with side plates, spatial dendrites, and graupel, were observed in the snowfall.

The forms are consistent with the radiosondes of this day (Figure 3.17, morning sonde). The radiosonde indicates the existence of a deep moist layer that

was ice saturated in its upper parts, which accounts for the formation of the prismatic crystal; the evening radiosonde indicated drying out from above but saturation up to −20°C, which allows for the formation of spatial dendrites. With the presence of a lower "feeder cloud" the precipitation efficiency was increased compared with the case of System No. 1, as was qualitatively evident from the low visibility and the heavier precipitation. The same habits of ice crystals occur with other intensive cyclonic disturbance.

The systems we shall call System No. 3 are shallow storms, as mentioned above. The one to be described has occurred over the Great Lakes Basin. The airflow was from the northwest to north over Lake Ontario, and the morning, as well as the evening, radiosonde of Buffalo, New

York, indicated the lapse rate was nearly moist adiabatic. The lake was still warm (+6°C), causing the formation of a convective stratocumulus layer whose top must have reached the 650 mbar level. A survey of the precipitation particles made along the south shore of Lake Ontario again showed a wide variety of crystal habit, with light to moderate riming and flaking. Figure 3.18 gives the temperature lapse rate of the Buffalo radiosonde for this day and the schematic drawings of the crystal forms in Nakaya's[50] classification.

The precipitation process that is active in this cloud system is no longer compatible with the Bergeron mechanism of a spender and releaser cloud. Apparently here the releaser cloud is missing, and the release of precipitaiton has to be accom-

plished by the spender cloud itself. While the precipitation intensity is always light when the releaser cloud is present and the spender cloud is missing, this is not necessarily true when only the spender cloud is available, since the rate of condensation in the convective cloud system is high. As long as the temperature is low enough, heavy precipitation may fall even from shallow systems.

An important mechanism that determines the efficiency of the precipitation mechanism is the self-cleaning process of clouds which free themselves from ice crystals generated during their active stage. These ice crystals, because of their small fall velocity, are taken with the updraft to the cloud top, leaving the cloud beneath

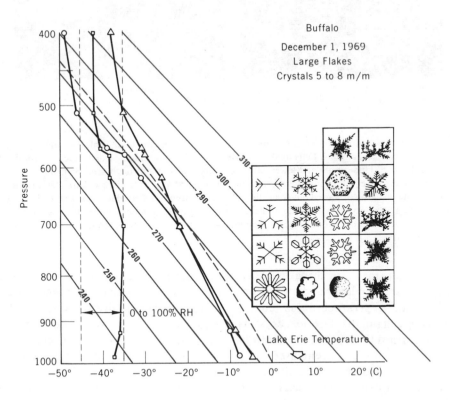

Figure 3.18   Atmospheric profile analogous to Fig. 3.16. Insert illustrates the observed crystal forms which have no high-level crystals but are entirely formed in convective stratocumulus.

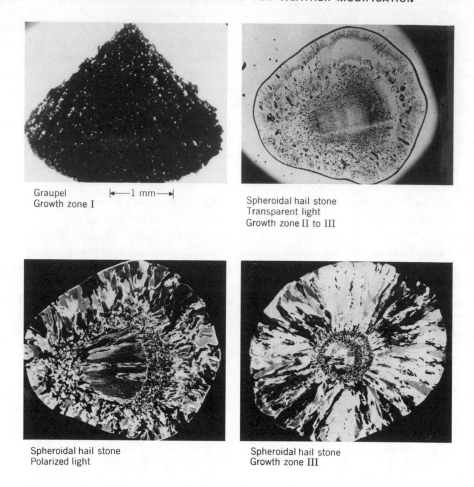

Figure 3.19 Graupel and hailstones. (For definition of growth zones see text, Table 3.4, and Figure 3.18.)
Upper left: typical conical graupel. Fall circulation base downward. Note a large cloud direction (drizzle) drop
as origin on the apex. Upper right and lower left: Cross-sectional views through same hailstone, upper right
seen in normal light, lower left in polarized light. Polarized light brings out the size and orientation of internal
crystallites. Note that growth started out conically and then changed to spheroidal. Cross-section lower left indicates the effects of aerodynamic molding in curved lines of crystallites (note particularly upper right part on
peripheral zone of hailstone). Lower right: spheroidal stone grown largely in Zone III (reprinted by permission, from List[37]).

icefree. On top they are dispersed horizontally in a shallow layer and away from
the cloud in a stratocumulus cumulogenitus mechanism. The ice crystals
that have formed throughout the depth of
the cloud are now concentrated in a
shallow layer in and around the cloud top,
and are ready to snow-out a neighboring
cloud whose updraft has ceased. If the
clouds are arranged like Bénard cells we

may have several neighbors whose spreading tops mix, perhaps concentrating the
crystal numbers over one cloud by a
factor of 5 or 6. We believe that this
process is an important mechanism for
the effective snow-out of cloud systems
whose releaser cloud is missing. When the
original cloud has become inactive also,
air no longer ascends through it and the
organized updraft decays into eddies and

turbulence. Ice crystals that fall into such clouds have excellent and prolonged conditions for growth, because their fall velocities are well within the velocities of the turbulent eddies.

Hydrometeor Formation in Convective Clouds

Much of what has been said about the warm rain process applies to the formation of solid hydrometeors in convective clouds. There are, however, important differences, the most important being that the process takes place in the supercooled cloud range, that the collecting particle may be a frozen droplet or an ice crystal, and that the product is in crystallized form. The formation of a solid kernel is by no means less efficient in a supercooled cloud than in a corresponding warm cloud, since, if the origin is an ice crystal, it has most likely a highly complex form caused by the great rate of condensation in a convective cloud. It is an efficient drop collector, therefore, and if it is a droplet, it is most likely a large drizzle droplet (see Figure 3.19).[37]

The riming process then proceeds analogously to the warm rain process; however, the collection efficiency in its dependence on size and fall velocity will change continuously and becomes a complicated function of the type of aerodynamic flow field around the collector particle R and its interaction with that around the collected droplet r. Irving Langmuir[36] found it to be dependent on a dimensionless parameter K

$$K = \frac{2\rho_1}{9\mu} \frac{r^2}{R} (V - v), \qquad (3.14)$$

where ρ_1 is the density of water, μ is the viscosity of air, and V and v are the velocities of the radius R and the radii r particles, respectively. We note that K is proportional to r, and to the ratio r/R, whereby one has to consider that for the case $r \equiv R$, $V \equiv v$ and therefore $K \equiv 0$. It is also proportional to $V - v$, and again it is necessary to remember that a large V will shorten the time period inside the cloud while a small V reduces E but increases the time period inside the cloud. Here the distinction between stratified and convective clouds becomes very important as in the latter the updraft will contribute to keeping the collecting precipitation particle inside the cloud for a longer time. Most important for the development of rimed precipitation particles is the cloud water content. The high water content of convective clouds will lead to compact forms like graupel, soft hail, and hail as the collected droplets have no time to grow to crystals before the next drop is collected. The heat of fusion liberated by the crystallization process can become so great that water remains unfrozen and begins to seep into the spaces in between the ice structure, thus creating spongy growth[37] or "slush" which can become molded by aerodynamic forces working on the hailstones. The ice structure of hailstones has been carefully studied with respect to its onion-ring character in order to try to understand their life history. Micrometer cuts have been made from hailstones and their structure has been studied in direct as well as polarized light with respect to air bubble structure, size and orientation of the crystallites, orientation of the c-axes of the crystallites, and impurities embedded in some of the rings. Although snow crystals lend themselves to use as aerological sondes so that equal crystal habit comes from equal cloud conditions, the same analysis has been so far essentially futile in the cases of hailstones. Table 3.4 gives characteristic properties of five growth zones of hailstones with respect to essentially three growth parameters—that is, air bubble structure,

Table 3.4 Characteristic Hailstone Properties

Zone	Bubbles	Crystallites	c-Axes	Growth Type
I	Opaque	Small	Random	Dry growth
II	Opaque to clear	Large	Radial	Dry growth
III	Clear	Large	Radial and/or random	Spongy growth
IV	Clear	Large	Tangential	Spongy growth
V	Clear	Large, radiating from center line	Tangential	Icicle growth

Special hailstones: (For definition of growth zones see Table 3.4.

Left picture: Cross section (polarized light) through world's largest hailstone, fallen at Coffeyville, Kansas, 3 Sept. 1970. Weight 766 gm, largest circumference 44 cm.

Lower left and right pictures: Cross shaped hailstone. Icicle-like protuberances probably formed as icicles by water running along their periphery probably due to rotation in the picture plane. Orientation of crystallites (in lower right picture) along protuberances is typical of icicle structure.

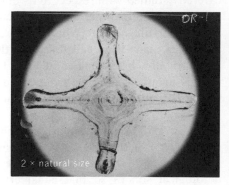

Figure 3.20 Special hailstones. (For definition of growth zones see text, Table 3.4, and Figure 3.21.) Top picture: Cross-section (polarized light) through world's largest hailstone, fallen at Coffeyville, Kansas, September 3, 1970. Weight 766 g, largest circumference 44 cm. Center and bottom picture: cross-shaped hailstone. Iciclelike protuberances probably formed as icicles do, by water running along their periphery probably due to rotation in the picture plane. Orientation of crystallites (in bottom picture) along protuberances is typical of icicle structure (photographs reproduced by kind permission of Charles and Nancy Knight, NCAR, Boulder, Colorado).

size of crystallites, and *c*-axis orientation.[61] Air, dissolved in water, is released during the freezing process. The bubbles become trapped in the ice if freezing proceeds quickly, presumably at low temperatures, but if freezing proceeds slowly at warmer temperatures, the air escapes and clear ice forms. Correspondingly, the size of crystallites is small if droplets are accreted at low temperatures, where they freeze immediately upon impact, but if this process happens at higher temperatures—either because of a warmer environment or an increased release of heat of fusion due to the collection of more cloud water—the newly accreted water becomes oriented upon freezing by the already existing ice surface and large crystallites form. Other stories are being told by the orientation of the *c*-axes of the crystallites, for instance, whether lobes have formed through accretion of small drops[9] or through an icicle-type process by liquid water running down along the lobe. In icicles the *c*-axes are oriented radially away from the center line of the icicle. Such an orientation is indicated in the strangely cross-shaped hailstone of Figure 3.20. For a description of this particular hailfall, see G. A. Briggs.[10]

A first attempt to assign the growth zones to atmospheric conditions in hailstorms is made in Figure 3.21. The figure is based on the List theory of hail growth which treats the heat transfer from the surface of the hailstone caused by convection and conduction, condensation, and evaporation, as well as the influence of accreted particles with their heat capacity and release of heat of fusion. The abscissa of the diagram is hailstone diameter in centimeters; the ordinate is air temperature. The parameters of the curves paralleling the growth zones are stone surface temperature, and where this would exceed 0°C, the ratio of ice-water content of the hailstones is given. A further parameter which List and J. Dussault[38] have incorporated in the figure is the liquid water content for Denver hailstorms[4] as computed for moist-adia-

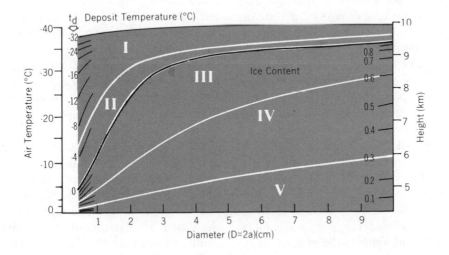

Figure 3.21 Suggested growth zones for various hailstone characteristics as explained in Table 3.4, based on List and Dussault, 1967.[38] Abscissa, hailstone diameter; ordinate left, air temperature; ordinate right, height above ground. Temperatures inside diagram on short dark lines reflect surface temperature of hailstone, 0°C temperature line is complete, numbers on short dark line on right side of diagram reflect the relative ice content of the hailstone at a surface temperature of 0°C. For further explanation, see text.

batic ascent, with a maximum content of 3 g/m³ at 8 km.

In higher elevations of the hail cloud, the cloud water content becomes increasingly mixed with ice crystals which form because of the low temperature or are entrained from the anvil. In such an environment the growth conditions should be enhanced as the water has to act simply as a binder and less heat of fusion must be transferred, although no growth characteristics are known that would be uniquely related to the growth in a mixed cloud. The simultaneous occurrence of clear ice coupled with small crystallites could be one such characteristic.

Another difficulty is related to our lack of knowledge of the microphysical conditions inside hail clouds. Soviet scientists point to the importance for the existence of an accumulation level of the liquid water just above the location of maximum updraft. If the maximum updraft is more than 9 to 10 m/sec even large water drops would not descend but remain suspended. However, accumulation of large amounts of liquid water act against the buoyancy and introduce for continuity reasons a strong component of horizontal divergence so that high liquid contents would at best be short-lived. A study of the List and Dussault theory of hail formation indicates that spongy hail with a high liquid water content (50 percent or more) will form already for a hail size of 2 cm diameter and a liquid water of 10 g/m³ over the entire layer from 0°C to $-20°C$ inside the updraft.[61] Since we have repeatedly observed the fall of soft, "slushy" hailstones of that size, one must conclude that moderately high accumulations of liquid water in the updraft do occur at times.

Figures 3.19 and 3.20 show a few typical hailstone habits and structures. Figure 3.19, upper left, is a conical graupel particle which has typically formed on a large frozen cloud or drizzle drop. Graupel fall with their bases forward. They form most of the hail embryos, and one frequently observes ellipsoidal stones like Figure 3.19, upper right and lower left, where the graupel particle grew in Zone II. Reaching a certain size, it began to tumble and to accumulate small crystallites, apparently the consequence of having been lifted with the updraft into Zone I. In descending, its growth continued again in Zone II as an ellipsoidal stone. The orientation of the crystallites in the upper right part of the stone suggests that they have been molded by aerodynamic forces and therefore makes it likely that the accumulation caused the formation of a soft spongy stone which later froze solid. An interesting story is told by the stone of the type shown at lower right in Figure 3.19. It is noted that almost its entire growth is characterized by a Zone II environment; this seems to indicate that the stone made little vertical movement during its lifetime inside the cloud and that therefore the updraft increased synchronously with the stone's increasing fall velocity. Figure 3.21 suggests that hailstones which display many alternating rings of clear and opaque ice may have acquired those particularly easily in higher elevations where Zones I, II, and III converge and can be crossed by relatively small vertical oscillations of the stone.

Lacking knowledge of many important dynamic properties of hailstorms, such as updraft velocities and their distribution, liquid water content, whether accumulated to high values in certain regions of the cloud or distributed moist-adiabatically throughout the cloud depth, the ratios of water to ice content in various cloud levels, and the influence of ice-water mixtures on hail growth, we have to admit that our understanding of the hail process is very limited. It is the more astonishing that we seem to be well on our way to suppressing hail formation operationally.

SIZE DISTRIBUTION OF RAINDROPS

The size distribution of raindrops would be of rather academic interest if it were not for the possibility of measuring the rate of rainfall by means of the radar echoes from raindrops. The great diffi-

culty of these measurements stems from the well-known fact that the radar return Z is proportional to ND^6, N being the concentration and D the drop diameter. This relationship introduces into the measurements a very great sensitivity toward the few largest drops of a distribution. On the basis of what we have said in the preceding sections about the formation of precipitation it is easily understandable that many factors enter into shaping the size distribution and af-

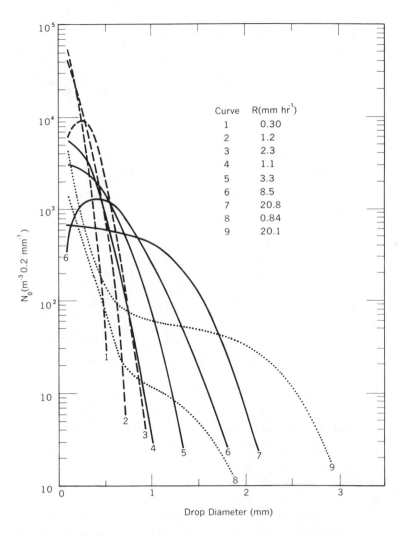

Curve	R(mm hr^{-1})
1	0.30
2	1.2
3	2.3
4	1.1
5	3.3
6	8.5
7	20.8
8	0.84
9	20.1

Figure 3.22 Raindrop distributions, as averaged from data of Blanchard.[7] Curves 1–3 are for measurements made at or near the dissipating edge of nonfreezing orographic clouds, while Curves 4–7 represent data taken at the cloud base. Curves 8 and 9 are for nonorographic raindrop distributions.

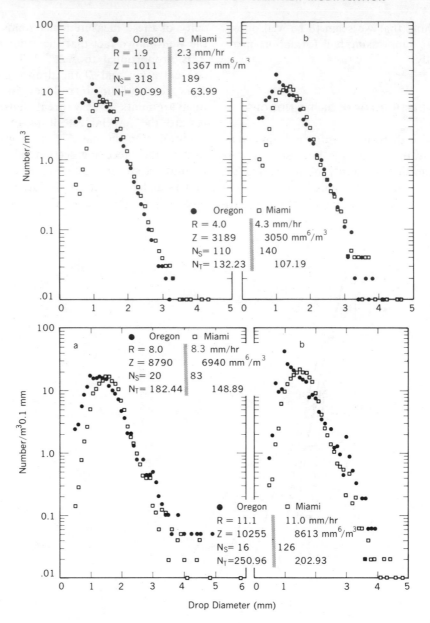

Figure 3.23 Raindrop spectra depending on rate of rainfall for Florida and Oregon (after Cataneo[12]). Denotations in figures signify: R; rate of rainfall in mm/hr; Z, radar return in mm^6/m^3; N_s, total number of cubic meters of rain sampled; N_T, total number of drops/m^3 for R.

fect the relationship between the median volume diameter and the rate of rainfall. Figure 3.22 shows examples of this dependency for orographic rains in Hawaii[7, 59] which are from the pure warm rain type, while Figure 3.23 shows the same relationship for rainfall of all types in Miami, Florida, and Corvallis, Oregon.[12] In the Hawaiian rains only one process contributes to the formation of raindrops, namely, the coalescence process. The raindrop spectra in Oregon and Florida are the result of coalescence, three-phase process, and the graupel process, whereby one might assume that in Oregon the three-phase-process and in warm Florida the coalescence process overshadows the others. Secondary effects in all three regions are caused by collisions of the raindrops below cloud base as well as by evaporation. Figure 3.22 is self explanatory showing the steady increase of drop size with the rate of rainfall; the denotations in Figure 3.23 are as follows: R is the mean rate of precipitation of all samples in mm/hr, Z is the radar return computed from the spectrum according to the relationship

$$Z = \sum_{D=0.5}^{D=7.9} n_D D^6, \qquad (3.15)$$

which takes into consideration all drops between 0.5 and 7.9 mm diameter D, and where n_D is the number of drops with diameter D. The number of rain samples taken is N_s (one sample being the number of raindrops/m³), and N_T is the average drop concentration/m³.

(Note that for the proper comparison of the figures the concentrations of the Hawaiian rains are listed in intervals of 0.2 mm, while the interval in the Florida and Oregon rain is 0.1 mm.) While small differences only appear to exist between the Florida and Oregon data, the differences become significant when the Z values are compared. For small rate of rainfall R, Z is larger for the Florida data, but as the rate increases the Z values shift and the Oregon values become greater. The relationship between Z and R is usually expressed in the form

$$z = AR^b. \qquad (3.16)$$

The famous Marshall-Palmer[45] raindrop size distribution that is being used in many theoretical studies involving the formation of precipitation obtains the constants $A = 220$ and $b = 1.6$. The Z-R relationship which can be derived for Oregon on the basis of the data in Figure 3.23 is

$$Z = 301R^{1.64}, \qquad (3.17)$$

and for Florida

$$Z = 286R^{1.43}. \qquad (3.18)$$

As radar would be a most convenient instrument for measurement of the rate of rainfall from cloud seeding, we like to call attention to the tenuous character of such measurements. Moreover, seeding may very well modify the raindrop spectrum in the direction of generating more and smaller drops, thus shifting the Z-R relationship to different and unknown factors A and b. We feel that the most intensive research into this matter has been performed by G. E. Stout and E. A. Mueller of the Illinois State Water Survey[57]; consequently we follow their results in the subsequent discussion. Two methods exist to determine the Z-R relationship for certain storms: (1) to measure the radar return and the rate of rainfall simultaneously, and (2) to measure the radar return and the raindrop size distribution and to compute the rate of rain. Both methods have very basic drawbacks because of the discrepancies of radar sampling area or volume and rain gauge sampling area or volume from which the drop spectrum is determined. Expressed in factors A and b for various conditions Table 3.5 gives results of the first approach from various authors, while Table 3.6 gives results obtained with the second approach by various authors.

For a detailed discussion of the physical significance of the tabulated data, the reader is referred to the original paper; here we like to impress the reader with the great variety that both factors A and b show, depending on rain type, location, method of determination, and experimenter. The reader may also be reminded that each relationship defined

Table 3.5   Radar Reflectivity and Rainfall Rate Relationships From Direct Measurements

Investigator	Geographical Location	Range of Applicability[a]	$Z = AR^b$		Comments
			A	b	
Doherty[17]	Ottawa, Canada	TRW	70	1.42	
		not TRW	38.4	1.63	
		$R < 10$	18.6	2.37	
		$R < 20$	25.9	2.02	
		$R < 40$	33.9	1.79	
		$R < 60$	38.2	1.69	
Berjuljew et al.[6]	Valday, USSR		340	1.5	The exponent is assumed equal to 1.5 and the coefficient determined from 2 years of rainfall
Wilson[64]	Norman, Okla.	TRW	45	1.43	Extremely low coefficient
		TRW	241	1.45	Extremely high coefficient
		TRW	183	1.18	Extremely low exponent
		TRW	141	1.72	Extremely high exponent
Aoyagi[1]	Tokyo		100	1.4	For diffuse radar echoes

Source: Stout and Mueller.[57]
[a]TRW, thunderstorm; R (mm/hr).

by A and b is the best fit to the variance of the data points around the regression line. The occurrence of a great variance is understandable due mainly to the great measuring difficulties; for instance, a sample of 44 m³ is required to be able to estimate the rainfall rate to within 10 percent with 95 percent confidence from spectrometer measurements.[48, 57] These authors then attempt to decrease the variance by various stratifications of the data. They investigate the influence of geographical differences, of different rain types, of the synoptic type of the existing weather situation, and of the degree and type of thermodynamic instability, among others. It is interesting that they find best agreement with the measured data and the least variance in the stratification by synoptic weather types. This interesting result may perhaps point to the im-

portance of the individual air mass type which is connected with certain synoptic situations. As the nuclei spectrum is considered to be a conservative air mass property and the nuclei spectrum will undoubtedly contribute to the formation of the raindrop spectrum, it is likely that the end result of the precipitation mechanism, the raindrop spectrum, reflects again the original cloud microphysics, namely, the nuclei spectrum.

ENERGETICS OF THE ATMOSPHERE

A major limiting factor in the feasibility of weather modification is the amount of energy required to produce any desired change. If the amount is large, the cost may exceed the value of the benefits to be expected from the change. Furthermore, the likelihood of undesirable side effects increases with the amount of artificial energy utilized in producing the change. It turns out that to produce any significant change in the atmosphere directly by supplying the entire necessary amount of energy is prohibitive except for very small volumes of air. The prevention of frost in orchards and the dissipation of fog in certain marginal instances has been achieved by direct addition of heat. Success in other modification efforts depends on initiating a redistribution of natural energy rather than supplying it arti-

Table 3.6   Radar Reflectivity and Rainfall Rate Relationships From Drop Size Spectra

Investigator	$Z = AR^b$		Comments
	A	b	
Marshall et al.[45]	220	1.6	Canada, widely accepted and used
Blanchard[7]	31	1.71	Orographic Hawaiian rain at cloud base
	16.6	1.55	Orographic Hawaiian rain within the cloud
Fujiwara[21]	80	1.38	Orographic Hawaiian rain
Hardy[25]	312	1.36	Arizona and Michigan rain with rates greater than 5 mm/hr
Imai[30]	700	1.6	One day of probably warm rain
	300	1.6	One day of continuous rain
	200	1.5	Air mass showers
	80	1.5	Prewarm front rain
Diem[16]	184	1.28	Overall average of different locations
	278	1.30	Entebbe, Uganda (tropical)
	240	1.30	Lwiro, Congo (tropical)
	176	1.18	Palma
	151	1.36	Barza Italy
	179	1.25	Karlsruhe, Germany, spring
	227	1.31	Karlsruhe, Germany, summer
	178	1.25	Karlsruhe, Germany, fall
	150	1.23	Karlsruhe, Germany, winter
	137	1.36	Axel Heiberg Land
Foote[20]	520	1.81	Tucson, Arizona
Dumoulin and Gogolombles[18]	730	1.55	France, highest coefficient
	255	1.45	Lowest coefficient
	426	1.5	Average of all observations, .95 correlation coefficient
Mueller and Sims[49]	286	1.43	Florida
	221	1.32	Marshall Islands
	301	1.64	Oregon
	311	1.44	Indonesia
	267	1.54	Alaska
	230	1.40	North Carolina
	372	1.47	Illinois
	593	1.61	Arizona
	256	1.41	New Jersey

Source: Stout and Mueller.[57]

ficially. To see this let us consider some aspects of atmospheric energetics.

The energy of atmospheric circulations comes ultimately from solar radiation. Of the radiation reaching the outer limits of the earth's atmosphere, which amounts to 338 W/m² when averaged over the earth's surface, only a small fraction (less than 20 percent) is absorbed directly by the atmosphere. Of the remainder about 35 percent is scattered and reflected back to space by the air, the clouds, and the earth's surface, and the rest (more than 45 percent) is absorbed at the ground, tending to raise its temperature. The average amount absorbed at the ground is about 150 W/m².

The main effect of solar radiation is to provide a source for heating the atmosphere from below. The energy added to the atmosphere from below shows up in five different forms: (1) kinetic energy of the winds, (2) turbulent energy, or kinetic energy of the smaller scale eddies, (3) potential energy, (4) internal or thermal, and (5) latent heat of evaporation or condensation. Various atmospheric processes change the energy from one to another of these forms, but ultimately they are transformed to thermal energy; thus, the dissipation of kinetic energy, which takes place principally in the kilometer next to the ground (the planetary boundary layer), converts the energy of the winds into heat. Estimates indicate that the rate of dissipation is about 2 W/m² in this layer and only a little more (3 W/m²) in the rest of the atmosphere. While total dissipation, 5 W/m², is only a little more than 3 percent of the solar energy absorbed at the ground, it is high enough that the atmosphere would be brought to rest in 3 days if the dissipation rate continued undiminished and the motion were not being restored by solar heating. The dissipation rate depends on the velocity, so as the atmosphere slowed

down the rate of dissipation would decrease, and computations show that rather than going to zero in 3 days it would take 6 days for the kinetic energy to be reduced to one-tenth of its original value. In actuality the loss of kinetic energy (KE) caused by dissipation is constantly being replenished by solar heating, but at the low efficiency indicated above—about 3 percent.

It is to be anticipated that if energy were added to the atmosphere artificially, it would be converted to motion with a similarly low efficiency. Thus it is reasonable to assume that to have a significant effect it would have to be supplied at a rate of the same order as the average rate at which solar energy is absorbed at the ground, 150 W/m². If this energy were introduced by burning fuel oil at 100 percent efficiency, the oil would be used at the rate of 100 barrel/km² hr. The area of a typical thunderstorm, 100 km², would require 10,000 barrel/hr of oil.

The energy released in a thunderstorm is actually much larger than this. The rate of precipitation from a thunderstorm is usually at least 1 cm/hr. The corresponding release of latent heat is $7 \cdot 10^3$ W/m², almost 50 times the above rate. Thunderstorms which produce 10 cm of precipitation over a 100 km² area are not unusual. In such a thunderstorm $2.5 \cdot 10^{16}$ joules are released, an amount of energy equivalent to the combustion of $4.8 \cdot 10^6$ barrels of oil. With such a tremendous release of energy it is not surprising that these large thunderstorms are accompanied by destructive winds.

Hurricanes also obtain their energy from the release of latent heat. The rate of precipitation is similar to that in severe thunderstorms, but the area over which it occurs is much larger, of the order of 10^5 km², and correspondingly the rate of release of energy is 1000 times or more as large as in an individual thunderstorm. This rate is equivalent to the explosion of 20 megaton bombs per minute.

That artificial release of large amounts of energy can produce local weather changes is shown by the fact that the atmospheric tests of nuclear bombs have been accompanied in some instances by showers, and that cumulus clouds, sometimes accompanied by small amounts of precipitation, have formed over large oil refinery fires,[26] forest fires, and with static tests of large space rockets.[47] There have been attempts to initiate showers in conditionally unstable air in southern France and in Africa by burning large amounts of oil rapidly,[15] but generally it has been accepted that the amount of energy required to produce any significant change by artificial heating is prohibitively large. Instead proposals for weather modification seek to use natural sources of energy. The fact that latent heat plays such a large role in atmospheric energetics, combined with the interest in precipitation augmentation as a primary objective, has led to emphasis on the precipitation process in weather modification efforts. Thus it is to be expected that cloud seeding will remain the principal method used in attempts to change the weather.

REFERENCES

1. Aoyagi, J., Areal rainfall amounts obtained by a 3.2-cm radar and a raingauge network, *Proceedings of the 11th Weather Radar Conference, Boulder, Colo.*, 1964, pp. 116–119.

2. Beard, K. V., A wind tunnel investigation of the terminal velocities collection kernels and ventilation coefficients of water drops freely falling in air, Ph.D. Thesis, University of California at Los Angeles, 1970.

3. Beard, K. V., and H. R. Pruppacher, An experimental test of theoretically calculated collision efficiencies of cloud drops, *J. Geophys. Res.* 73(20), 6407–6414, 1968.

4. Beckwith, W. B., Analysis of hailstorms in the Denver network, 1949–1958, *Physics of Precipitation, Geophys. Monogr.* (AGU) 5, 348–353, 1960.

5. Bergeron, Tor, Über den Mechanismus der augiebigen Niederschläge, *Ber. Deutsch. Wetterd.* 12, 225–232, 1950.

6. Berjuljev, G. P. et al., The results of radar measurements on areal rainfall in Valday, *Proceedings of the 12th Conference on Radar Meteorology, Norman, Okla.*, 1966, pp. 220–221.

7. Blanchard, D., Raindrop size distribution in Hawaiian rains, *J. Meteorol.* 10, 457–473, 1953.

8. Borovikov, A. M., Supercooling of water in the atmosphere and the phase of various types of clouds, *Proceedings of the International Conference on Cloud Physics, Aug. 26–30, 1968, Toronto, Canada*, pp. 290–294.

9. Browning, K. A., The lobe structure of giant hailstones, *Quart. J. Royal Meteorol. Soc.* 92, 1–14, 1966.

10. Briggs, G. A., Hailstones, starfish and daggers-spiked hail falls in Oak Ridge, Tennessee, *Monthly Weather Rev.* 96(10), 744, 1968.

11. Byers, H. R., *Elements of Cloud Physics*, University of Chicago Press, Chicago, 1965.

12. Cataneo, R., A comparison of raindrop size spectra between Miami, Florida and Corvallis, Oregon, *Trans. Illinois Acad. Sci.* 61(2), 165–170, 1968.

13. Cotton, W. R., A numerical simulation of precipitation development in supercooled cumuli, Report No. 17 to NSF, Dept. of Meteorol, Pennsylvania State Univ., University Park, Pa., 1970, 178 pp.

14. Davis, M. H., and J. D. Sartor, Theoretical collision efficiencies for small cloud droplets in Stokes flow, *Nature* 215, 1371–1372, 1967.

15. Dessens, H., A project for formation of cumulonimbus by artificial condensation, *Physics of Precipitation*, Helmut Weickmann, Ed., American Geophysical Union, Washington, D.C., 1960, pp. 396.

16. Diem, M., Rains in the arctic, temperate, and tropical zone, *Sci. Rept., Meteorol. Inst. Tech. Hochschule, Karlsruhe*, Contract DA-91-591-EUC-3634, 1966, 93 pp.

17. Doherty, L. H., The scattering coefficient of rain from forward scatter measurements, *Proceedings of the 10th Weather Radar Conference*, Washington, D. C., 1963, pp. 171–175.

18. Dumoulin, G., and A. Gogolombles, A comparison of radar values of precipitation intensities and rainfall rate from a raingauge, *Proceedings of the 12th Conference on Radar Meteorology, Norman, Okla.,* 1966, pp. 190–197.

19. Fletcher, N. H., *The Physics of Rainclouds,* Cambridge University Press, Cambridge, 1962, 381 pp.

20. Foote, G. B., A Z-R relation for mountain thunderstorms, *J. Appl. Meteorol.* **2,** 229–231, 1966.

21. Fujiwara, M., Preliminary report on Hawaii rain mechanism, *Tellus* **3,** 392–402, 1967.

22. Grunow, J., Snow crystal analysis as a method of indirect aerology. *Physics of Precipitation, Geophys. Monogr.* (AGU) **5,** 130–141, 1960.

23. Hallett, J., Field and laboratory observations of the crystal growth from the vapor, *J. Atmos. Sci.* **22,** 64–69, 1965.

24. Hallett, J., and B. J. Mason, The influence of temperature on the habit of ice crystals grown from vapour, *Proc. Roy. Soc. London Ser. A* **247,** 440, 1958.

25. Hardy, K. R., A study of raindrop size distributions and their variations with height, *Sci. Rept. No. 1,* Univ. Michigan, Contract AF 19(628)-281, 1962, 174 pp.

26. Hissong, J. E., Whirlwinds at oil-tank fire, San Luis Obispo, California, *Monthly Weather Rev.* **54,** 161–163, 1926.

27. Hocking, L. M., The collision efficiency of small drops, *Quart. J. Royal Meteorol. Soc,* **85,** 44–50, 1959.

28. Hocking, L. M., and P. R. Jonas, The collision efficiency of small drops, *Quart. J. Royal Meteorol. Soc.* **96,** 722–729, 1970.

29. Houghton, H. G., A preliminary quantitative estimate of precipitation mechanisms, *J. Meteorol.* **7,** 363–369, 1950.

30. Imai, I., Raindrop size distributions and Z-R, relationships, *Proceedings of the 8th Weather Radar Conference, San Francisco, Cal.,* 1960, pp. 211–218.

31. Jiusto, J. E., Crystal development and glaciation of a supercooled cloud. *J. Rech. Atmos.* **5,** 69–85, 1971.

32. Kepler, J., Strena vom Sechseckigen Schnee, translation by Fritz Rossman, *Dokumente zur Morphologie, Symbolik, und Geschichte,* W. Keiper, Berlin, 1943, 64 pp.

33. Klett, J. D., and M. H. Davis, Theoretical collision efficiencies cloud droplets at small Reynolds numbers, *J. Atmos. Sci.* **30,** 107–117, 1973.

34. Koenig, L. R., Numerical modeling of ice deposition, , *J. Atmos. Sci.* **28,** 226–237, 1971.

35. Khrigian, A. Kh., Ed. *Cloud Physics,* published for the U.S. Department of Commerce and the National Science Foundation, Washington, D.C., by Israel Program for Scientific Translations, 1963, pp. 68, 82 (392 pp.).

36. Langmuir, I., The production of rain by a chain-reaction in cumulus clouds at temperatures above freezing, *J. Meteorol.* **5,** 175, 1948.

37. List, R., Kennzeichen atmosphärischer Eispartikel. 1. Teil, *ZAMP* **9a,** 180–192, 1958.

38. List, R., and J. Dussault, Quasi steady state icing and melting conditions and heat and mass transfer of spherical and spheroidal hailstones, *J. Atmos. Sci.* **24,** 522–529, 1967.

39. Ludlam, F. H., The production of showers by the growth of ice particles, *Quart. J. Royal Meteorol. Soc.* **78,** 543–553, 1952.

40. McDonald, J. E., Use of the electrostatic analogy in studies of ice crystals growth, *ZAMP* **14,** 610–620, 1963 (Raymund Sänger Gedenkschft).

41. Magono, C., Structure of snowfall revealed by geographic distribution of snow crystals, *Physics of Precipitation,* H. Weickmann, Ed., *Geophys. Monogr.* (AGU) **5,** 142–151, 1960.

42. Magono, C., and H. Aburakawa, Experimental studies in snow crystals of plane type with spatial branches, *J. Fac. Sci. Hokkaido Univ.,* Ser. VII **3,** 85–97, 1968.

43. Magono, C., and C. W. Lee, Meteorological classification of natural snow crystals. *J. Fac. Sci. Hokkaido Univ. Ser. VII* **2,** 321–335, 1966.

44. Marshall, J. S., and M. P. Langleben, A theory of snow crystal habit and growth, *J. Meteorol.* **11**(104), 254–256, 1954.

45. Marshall, J. S., R. C. Langille, and W. McK. Palmer, Measurement of rainfall by radar, *J. Meteorol.* **4,** 186–192, 1947.

46. Mason, B. J., *The Physics of Clouds,* 2nd ed., Clarendon Press, Oxford, 1971, 277 pp.

47. Morris, David G., Initiation of convective clouds due to static firing of the Saturn V first stages, *Bull. Am. Meteorol. Soc.* **49**(11), 1054–1058, 1968.

48. Mueller, E. A., and A. L. Sims, The influence of sampling volume on raindrop size spectra, *Proceedings of the 12th Conference*

on Radar Meteorology, Norman, Okla., 1966, pp. 135–141.

49. Mueller, E. A., and A. L. Sims, Investigation of the quantitative determination of point and areal precipitation by radar echo measurements, *Illinois State Water Survey, Final Rept.,* Contract DA 28–043 AMC 00032(E), 1966, 110 pp.

50. Nakaya, U., *Snow Crystals,* Harvard University Press, 1954, 510 pp.

51. Nakaya, U., M. Hanajima, and J. Muguruma, Physical investigations on the growth of snow crystals, *J. Fac. Sci. Hokkaido Univ. Ser. II* **5,** 87 (265), 1958.

52. Nakaya, U., and K. Higuchi, Horizontal distribution of snow crystals during the snowfall, *Physics of Precipitation,* H. Weickmann, Ed., *Geophys. Monogr.* (AGU) **5,** 118–129, 1960.

53. Nakaya, U., and T. Terada, Jr., Simultaneous observations of the mass, fall velocity and form of individual snow crystals, *J. Fac. Sci. Hokkaido Univ.* **1,** 191, 1935.

54. Picknett, R. G., Collection efficiencies for water drops in air, *Aerodynamic Capture of Particles,* E. G. Richardson, Ed., Pergamon Press, 1960, pp. 160–167.

55. Shafrir U., and M. Neiburger, Collision efficiencies of two spheres falling in a viscous medium, *J. Geophys. Res.,* **68,** 4141, 1963.

56. Shiskin, N. S., On snow crystal growth in clouds, *Supplement to Proceedings of the International Conference On Cloud Physics, Tokyo and Sapporo,* 1965, pp. 136–146.

57. Stout, G. E., and E. A. Mueller, Survey of the relationship between rainfall rate and radar reflectivity in the measurement of precipitation, *J. Appl. Meteorol.* **7,** 465–474, 1968.

58. Telford, J. W., N. S. Thorndike, and E. G. Bowen, The coalescence between small water drops, *Quart. J. Royal Meteorol. Soc.* **81,** 241–250, 1955.

59. Weickmann, H. K., *Physics of Precipitation, Meteorol. Monogr.* **3**(19), 226–255, 1957.

60. Weickmann, H. K., The snow crystal as aerological sonde, *Artificial Stimulation of Rain, Proceedings of the First Conference, Physics of Cloud and Precipitation Particles, Woods Hole, Mass.,* 1967, 315–326.

61. Weickmann, H. K., The hailstorm, *Proceedings of the International Conference of Cloud Physics, Toronto, Canada,* 1968, pp. 400–410.

62. Weickmann, H. K., Snow crystal forms and their relationship to snowstorms, *J. Rec. Atmos.,* H. Dessens Memorial Issue **6,** 603–615, 1972.

63. Whelpdale, D. M., and R. List, The coalescence process in raindrop growth, *J. Geophys. Res.* **76,** 2836–2856, 1971.

64. Wilson, J. W., Storm-to-storm variability in the radar reflectivity rainfall rate relationship, *Proceedings of the 12th Conference on Radar Meteorology, Norman, Okla.,* 1963, pp 229–233.

65. Woods, J. D., and B. J. Mason, Experimental determination of collection efficiency for small water droplets in air, *Quart. J. Royal Meteorol. Soc.,* **90,** 373–381, 1964.

66. Gunn, K. L. S. The number flux of snow crystals at the ground, *Monthly Weather Rev.* **95,** 921–924, 1967.

67. Jiusto, J. E. and H. K. Weickmann, Types of snowfall, *Bull. Amer. Meteolog. Soc.* **54,** 1973 (to be published).

68. Levin, Z., M. Neiburger, and L. Rodriguez, Experimental evaluation of collection and coalescence efficiencies of cloud drops. *J. Atmos. Sci.* **30,** 944–946, 1973.

4 | Weather Modification Instruments and their use

R. E. RUSKIN

W. D. SCOTT

PERSPECTIVE

Early attempts at weather modification were characterized by "eyeball evaluation" and a minimum of instrumentation. Today many of our programs still rely on such estimates, as instrument development has been slow and erratic. Because of fiscal limitations, it appears that further development must be guided by our understanding of the physics of the phenomena. This means that numerical models should guide further laboratory and field experiments. Of course, these same models should utilize the experimental results to develop a new, sound physical framework. In this way two of man's tools, experimental measurements and numerical modeling, can be fully utilized with each supplementing the other.

The problem of measurement is aggravated by the wide range of time and size scales of the various processes to be measured. Weather modification pro-

grams involving cloud processes extend from active nucleus sizes of less than 0.1 μm (and still smaller gas-to-particle conversion processes forming these nuclei) to the size of the largest hurricane (1000 km); this is a dynamic range of 10^{13}. The general time and space scales of the various processes are shown diagrammatically in Figure 4.1. From this figure it becomes apparent that for every space scale there is a characteristic time scale.

With such a large dynamic range there are no instruments that can measure more than a small fraction of the total that we wish to know. Fortunately, in many cases detail with accuracies in the few percent range is not required. In some instances an order of magnitude estimate is sufficient. For instance, if we achieve even some accuracy in measuring ice nuclei, we are lucky since we do not yet understand all the mechanisms involved in the nucleation and freezing process, and a different ice nucleus count is expected for each type of processing by the instrument. On the other hand, the supersaturation present in a cloud or fog is generally so small that measurements of temperature and humidity must each be made with more accuracy than is

R. E. Ruskin is attached to the U.S. Naval Research Laboratory, Washington, D.C. W. D. Scott is with the National Hurricane Research Laboratory of the National Oceanic and Atmospheric Administration in Coral Gables, Florida.

possible with the present state of the art. It should be apparent that the accuracy needed for every type of measurement should be determined by calculating the expected effect of various measurement errors on the cloud process or on the complete cloud model. Results of this type of calculation may indicate that it is useless to attempt the measurement by the approaches available. Also, it is possible that numerical modeling can reduce the exacting instrumental requirements by combining the effects of several variables. The outcome of the calculation would specify the optimum quantities, positions, and times for verification of the physics. Indeed, this type of interplay between model and measured quantities

1. Atmospheric, Electric, and Magnetic Effects
2. Lightning
3. Ball Lightning
4. Tornado
5. Cu
6. Clear Air Turbulence
7. Cb
8. Stratus
9. Sea Breeze Front
10. Urban Effects
11. Hurricane
12. Cyclone
13. Planetary Wave
14. Monsoon
15. Drought
16. Climatic Change
17. Accretion
18. Rain
19. Snow & Hail
20. Drizzle
21. Break Up, Splintering, Precipitation Initiation

22. Condensation
23. Auto-Conversion
24. Nucleation
25. Gas-to-Aerosol Conversions
26. Coagulation
27. Removal from Tropo- and Stratosphere
28. Fast-Gaseous Reactions
29. Slow Gaseous Reactions
30. Increased CO_2 in Atmosphere

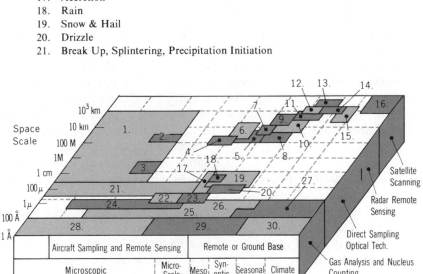

Figure 4.1   Time and space scales and applicable measurement for various storm processes.

presents the best possibility of understanding clouds and cloud modification.

Now let us narrow our perspective to instruments for the modification of clouds. In this case, instruments basically serve two purposes: first, they allow the assessment of the basic, natural state of the cloud as a baseline from which to observe and model the cloud processes. Second, they allow a measurement of the effect of our modification efforts on cloud processes. In serving both these purposes the instruments act as our eyes through which we can view the microphysical, dynamic, and other physical processes that make up the cloud.

In choosing instrumentation for weather modification, the least expensive and generally least productive is ground instrumentation. The most comprehensive instrumentation, area- and volumewise, are remote sensors, notably radars. However, the difficult field of cloud microphysical measurement is largely confined to aircraft instrumentation. These aircraft measurements are generally the most expensive and inefficient in manpower, but presently provide the main hope of improving our understanding; however, as will become apparent, instruments in several key areas have been lacking in appropriate development in terms of physical principles as well as operational ability.

PRIORITIES IN INSTRUMENTATION

Generally, modeling has been a combining of physical laws to produce a resultant mathematical model which in as many ways as possible shows characteristics akin to those observed in a natural cloud. At present, because of a general lack of well-qualified case studies, models are the principal means we have to determine the mechanisms of physical phenomena. So the following discussion relies heavily on modeling results. In viewing the results, the reader should be cautioned that "a model is no better than the measurements used to verify it."[13]

A list of measurements required to upgrade our knowledge is presented in Table 4.1. Of first importance is a description of the environment in which the cloud is situated in terms of an atmospheric sounding in near space and time. Modeling studies indicate that dramatic differences in cloud development can result from slight variations in the sounding. In the tropics, changes as small as $+0.2°C$ in the cloudy environment at cloud base or $-0.5°C$ in the cloud environment at the 400-mb level can increase the seedability (see Chapter 6) by more than a factor of ten.[69] Since present errors in the sounding can easily exceed $1°C$ from instrumental errors and lack of coherence in time and space, increased logistical and instrumental abilities for making more accurate soundings are top priority in our weather modification efforts.

Nearly all past modification efforts have been coupled with the development of ice in a cloud. At present, however, freezing and vapor deposition processes are still poorly understood. The main effort has been based upon the fact that liquid water is naturally present in clouds at temperatures colder than $0°C$. The dynamic mechanism of modification then consists of releasing this instability by freezing the water and increasing cloud buoyancy by release of the latent heat of fusion. To model the cloud development in a realistic way and describe its modification potential, we need to know the mechanisms by which ice crystals form. It appears that they can be formed by (1) vapor deposition on a nucleus, (2) freezing (either being nucleated by bulk or contact nucleation), or (3) condensing

Table 4.1   Measurement Interaction with Modeling

I. Measurements needed for developing and testing simplest cumulus models
   A. Complete thermodynamic sounding
      1. Near in space
      2. Near in time
   B. Cloud base height, updraft radius, and updraft velocity
   C. Cloud top height and tower radius as a function of time
   D. Radar echo RHI profiles as a function of time
II. Measurements needed for input to 2-D and 3-D dynamic models
   A. Basic data of set I (above)
   B. Surface temperature and moisture field data below cloud base
   C. Meso- or subsynoptic analysis of the kinematic properties of the wind field (flow, convergences, etc.)
   D. Cloud sizes, cloud number distributions, detailed cloud photographs
   E. Roughness characteristics for boundary layer parameterization
   F. Aircraft or Doppler wind field measurements
   G. PPI echo displays
   H. Dense rain gauge measurements
III. Detailed microphysical modeling requires even more detail
   A. Multilevel ice particle distributions and types
   B. Cloud base droplet concentrations and size distributions, or some indication of the shape of the curve, particularly in the large droplet end
   C. Detail of aerosol structure
      1. Cloud condensation nuclei (CCN)
      2. Chemical and physical characteristics
      3. Concentrations of ice nuclei effective through different processes
   D. Multilevel and time sequence of precipitation

Source: W. Cotton.[13]

on a combined cloud condensation nucleus (CCN) and ice nucleus, followed by freezing and growth by vapor deposition.

As a result of a lack of knowledge concerning when and where these mechanisms act, the number and kind of ice crystals formed in a cloud are unknown. In fact, so far there has been a 10^3 to 10^4 discrepancy between the number of ice crystals found in some clouds at temperatures slightly colder than 0°C[49] and the number of ice nuclei measured by present instrumentation methods. In modeling, however, completely different cloud development occurs when different numbers of ice crystals are formed in a cloud. Resolving this problem requires specific experimentation in the field and laboratory. These experiments should be capable of resolving sufficient detail so a reasonable one-to-one correspondence can be made between the aerosol's physical characteristics, chemical characteristics, and history, and the number and character of the ice crystals that are formed.

However, practically speaking, at the present time it appears that both the measurement of ice nuclei and perhaps the concept of ice nucleation need reassessment. Therefore, in lieu of ice nucleus measurements, direct measurements of numbers of ice crystals in appropriate conditions should be made. Indeed, crystal concentrations and their dependence upon atmospheric variables may some day lead us to the basic physical reasons for ice crystal production.

Instrument development is also required in the measurement of in-cloud temperatures and supersaturations. The problem, basically, is to measure the temperature and humidity of the air between the cloud hydrometeors without the device becoming wet and, in effect, measuring the wet bulb temperature. The temperature is used to estimate the density of the cloudy air and hence measure the buoyancy and heat flux in the cloud. Indeed, one of the poorer mathematical developments in our microphysical modeling is the calculation of cloud supersaturation or subsaturation above cloud base. Supersaturation influences the distribution of vaporous water on the spectrum of cloud condensation nuclei, cloud droplets, raindrops, and ice crystals, but cannot be measured with an accuracy better than about 5 percent in relative humidity (RH), whereas the total supersaturation seldom exceeds 1 percent (101 percent RH) in clouds.

Another measurement of concern in assessing the colloidal stability of clouds is a measure of the "tail" on the cloud droplet spectrum. This is the number of cloud droplets in the larger end of the droplet spectrum, the so-called "precipitation embryos." A measure of the number of cloud condensation nuclei active at different supersaturations can supply the basic data for models of precipitation development. However, our general lack of information regarding the partitioning of the condensed water does not allow us to predict the form of the "tail." Hence at this time, direct measurements of the cloud droplet size distribution with emphasis on the numbers of larger cloud droplets are important.

The importance of this "tail" in precipitation development is illustrated in Figure 4.2. Presented are numerical values of concentration of raindrops and hail which are produced at various times after the introduction of the cloud dro-

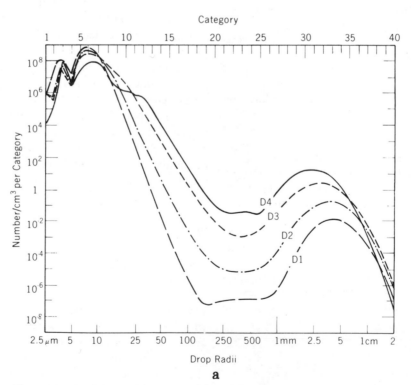

Figure 4.2a   Model calculations of precipitation development, curves (*a*).

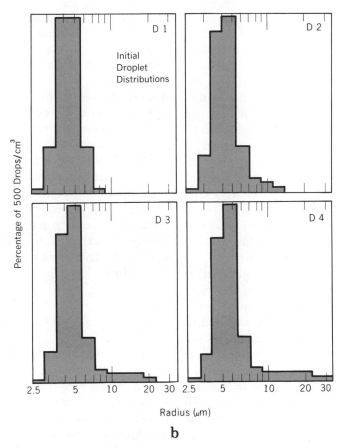

Figure 4.2b At various times after the four initial droplet distributions shown in (*b*) were introduced at cloud base (courtesy E. Danielson, NCAR, Boulder, Colorado).

plets into the model. The resulting distributions shown in the top of Figure 4.2*a* indicate how the spectra have developed from the various initial cloud droplet distributions shown in Figure 4.2*b*. Distribution D1 has evolved for 39 min; D2, 32 min; D3, 24 min; and D4, 22.5 min.

Note that the largest amount of precipitation is formed in the shortest time from the spectrum with the largest tail. Clearly our ability to assess the colloidal stability or the rainout of the cloud is dependent on our ability to determine the form of the cloud droplet distribution at cloud base.

In addition, we must include the remote measurements (in particular, the radar measurments) in a listing of priorities because of their past, present, and future capabilities. They alone can supply an overview of the cloud and should be a major contributor to any weather modification field program. Also measurements of turbulence would contribute to our knowledge of mixing processes, an area of great mystery at the present time. Finally, a list of measurement priorities would be wanting if it did not include the great need for real-time data processing.

Evaluations of Modification Results

What is to be measured, where it is to be measured, and when it is to be measured depend upon the specific goal of the program. For instance, if the production of water for ground use is of concern, rain gauges, stream height measurements, and/or radar reflectivities would be of prime concern. In the hurricane program (Project Stormfury), destruction is caused by the winds, storm surges (water), and rains. The destructive force of the winds varies with at least the second power of the wind speed, so the winds are the prime elements for measurement.

At this stage in the art of weather modification, measurements of the more important specific characteristics of the clouds and their changes are required to assure a correct evaluation of the program. In turn, the microphysical qualities of the seeding material must be assessed in the appropriate environmental conditions. Hence assessment of a full modification program could be divided into three portions, each an integral part of the whole: (1) monitoring of the total effect, (2) assessment of cloud alteration, and (3) testing of the effectiveness of the seeding material in the cloudy environment. The total monitoring (1) would generally be done on the ground after the event. For example, T. T. Fujita[20] has used a semiqualitative evaluation of the destructive character of the tornado and the magnitude of the winds based on actual destruction. Cloud assessment (2) must be done within the cloud's lifetime, preferably before the seeding and during at least the first 15 min after delivery of seeding agents. The cloud top heights of cumuli are usually measured to determine the effect of dynamic seeding.

The effectiveness of the seeding materials (3) should be tested in the same air parcel as that of the attempted modification by sampling directly (or remotely)

the parcel at an appropriate level within minutes of the seeding. For dynamic seeding the correct parcel should be located several thousand feet higher in altitude. This is an exceedingly difficult operation to accomplish by direct sampling and is best done by multiple air-craft penetrations. Even so, however, the chances of intersecting the correct parcel are small. Remote sensing, as discussed in a later section, appears to be the only hope for completing (3). Remember that if (3) is *executed properly* and there is no effect of the seeding materials, (2) and (1) cannot be expected to produce a con-clusive, positive result actually caused by the seeding.

INSTRUMENTATION PLATFORMS

Platforms for measurement in weather modification include ground observa-tories, mobile vans, aircraft, rockets, and sondes. Ground stations are particularly difficult to locate relative to storm or cloud systems unless there is some topographical preference for their development (i.e., wave clouds). Of course, radar and the remote sensors with large ranges permit more leeway in their location and are currently used with suc-cess to provide weather forecasters with large-area information.

The mobile vans give more freedom of movement and allow some adjustment to the natural phenomena. Ships lend similar adjustability to seagoing pro-grams. Generally speaking, though, our efforts require that either of these mobile platforms be operated in quite severe weather.

Aircraft often must be used, though the operating costs of large aircraft are perhaps 100 times that of a ground-based station, and it is considerably more dif-ficult to make measurements on a moving platform and still verify calibration of the instruments. The aircraft motion itself can disturb the cloud. To avoid this problem and the possibility of danger to the aircraft, sensors can be suspended on trailing wires. With slow-speed aircraft a trailing wire as long as about twice the turning radius of the plane permits placing an instrument nearly stationary below the center of a circle formed by the plane flying in a continuous tight turn.

A helicopter sonde developed at the Naval Research Laboratory (NRL) and now in use by the U.S. Navy is suspended, as shown in Figure 4.3, on a trailing cable 40 ft below the helicopter to permit measurements of the temperature, humidity, and pressure in the undisturbed air ahead of the wake of the rotor blades. This instrument is particularly advantageous for measurements in inversion layers, which are often quite important in weather modification.

Radio-controlled drones have been tested for possible use instead of manned aircraft to minimize personal danger in penetrating severe storms or other high-turbulence areas; however, drones can be flown only over restricted areas in controlled air space with no other aircraft. For safety they are designed to parachute to the ground whenever the engine fails or the radio-control signal becomes too weak. The slowest drones fly 400 knots and require precise control and timing, combined with accurate radar, transponder, and altitude tracking. Possible malfunction can cause loss of control and an uncontrolled power crash. Although this is highly unlikely, it is expensive in instrument and drone maintenance, if not in other property damage.

The use of lighter-than-air craft is also possible. In fact, at the present time the Air Force has a large meteorological blimp which operates out of Patrick Air Force Base, Florida.

Sailplanes may be used to advantage in determining the vertical motions of the air or clouds without the complexities required with large, powered aircraft. Because of the fidelity with which the sailplane follows the air motion, the only instrument required for measurement of

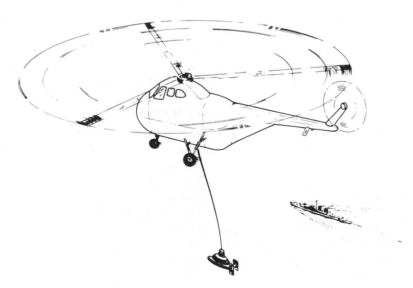

Figure 4.3 Helicopter sonde for recording fine-grain structure of temperature and humidity vs. altitude pressure for locating thin inversion layers (Courtesy NRL, Washington, D.C.).

vertical velocity is the variometer (discussed later). For instruments requiring low acoustical noise levels or low vibration the sailplane has an advantage; however, sailplanes have a small flying range and have extremely limited electrical power for instruments. Although limited to the availability of updrafts to keep airborne, some sailplanes have operated above 10 km. A new powered sailplane manufactured in Germany uses a small engine for launch and climb without a need for updrafts. However, most of the applications of sailplanes can be handled more conveniently by use of small conventional aircraft.

Kite-balloons, or "kytoons," have been used, both with radio telemetering and with wiresondes, to obtain data on thin inversions, winds, and electric fields up to about 1 km maximum altitude. Because of their kite action, in addition to buoyancy, they can be used in winds up to about 30 knots in the absence of turbulence from buildings or ship superstructure. When flown over land, FAA clearance must be obtained and "notams" to aircraft pilots must be requested a couple of months before use except in restricted areas.

Numerous special-purpose balloon-borne sondes have been developed. Notable among these is the frost point sonde with which Mastenbrook[45] has accumulated 8 years of monthly data through the stratosphere to serve as bench mark determination of water vapor and its natural variation prior to large-scale aircraft flights in the stratosphere, with their potential for inadvertent climate modification.

Rockets are used as delivery packages for seeding materials in the U.S.S.R. and have a limited capability for in situ measurements. In the United States they have been used to "trigger" lightning discharges and to collect "total air" samples at high altitudes.

Dropsondes have been invented with a wide variety of capabilities. Nelson[52] and Fujiwara[21] have developed raindrop sondes intended to measure raindrops in free fall. Similarly, C. Magono and co-workers at the University of Sapporo have developed snow crystal sondes based on the Formvar technique and electric charge sondes based on sensitive transmitting electrometers.

Weather telemetering buoys with recorders and transmitters are now frequently used in research programs. They have the unique ability to weather the storm and measure the sea state. Some are designed to be air dropped at appropriate sites in hurricanes for this purpose. Combinations of radiosondes and bathythermographs are now used to monitor the thermal structure of both the air and the ocean.

Satellites, of course, have come into operational use and are of most value in modification efforts when in stationary orbit. Besides their ability to describe the mesoscale in relation to global weather by observing cloud activity and motions, they are gaining the ability to make soundings throughout the atmosphere's depth using infrared and microwave remote sounding techniques.

STANDARD MEASUREMENTS

In weather modification programs, the standard measurements of temperature, pressure, and humidity are required to characterize the environment. These data are usually acquired by the use of a radiosonde. This device measures and transmits to a remote receiver information on the temperature, humidity, and wind structure of the atmosphere. These remote readings produce a profile of the atmospheric structure known as the "sounding." Meteorological stations throughout the world release radiosondes at approximately 10-hr intervals from specified locations.

For modification purposes, though, these standard soundings are not sufficient. The atmospheric structure shown by the sounding usually varies sufficiently in time and space (particularly in the lower levels) so that the temperature, humidity, and wind structure must be known in the immediate vicinity of the clouds. Since clouds generally do not form at specified locations, this means that mobile units and/or aircraft observations of the environmental structure or both are usually necessary.

Airborne Temperature Measurements

The measurement of free-air temperature and humidity aboard aircraft is complicated by the effects of changes in the temperature and pressure of the sample as it is accelerated to aircraft speeds from rest. At the skin of the aircraft or probe the air is heated and, at the leading edges, the pressure is increased. The heating of the air is caused by an increase in the air's total energy, including compression at the leading edge and friction over the horizontal surfaces. The net temperature rise at all aircraft surfaces is approximately the same and is equal to the square of the true air speed (mph) times 10^{-4} (temperatures in °C).

Similarly, the pressure rise at the leading edge may be calculated by considering the momentum transferred to the surface by an ideal fluid element. Indeed, the indicated air speed of the aircraft is usually measured using the dynamic or pitot-static pressure rise (the difference between the pitot pressure measured in a forward direction into the air stream and the static pressure measured perpendicular to it). The indicated air speed in mph is then simply the square root of the quotient: (the dynamic pressure rise in inches of water) $\div$ (4.92 $\times$ 10^{-4}. The measurement of true air speed is difficult because pitot pressure is sensitive to air density. True air speed is equal to the indicated air speed only if the aircraft is flying at standard sea level conditions. Otherwise, to obtain the true air speed, the indicated air speed must be corrected for air density by multiplying the square root of the ratio of air density at standard sea level to that at flight altitude.

It should be noted that the density at flight level can be determined accurately only if one already knows the true air temperature (which requires a knowledge of true air speed). Unless a flight computer is available, a series of approximations (iteration) is usually used to calculate the true air speed and true temperature simultaneously.

As an example of the corrections involved let us note that the true air speed is $\sqrt{2}$ times the indicated air speed when the air density is half of the standard sea level value, at about 20,000-ft altitude. At this altitude when flying at an indicated air speed of 180 knots (207 mph), the full dynamic temperature rise is 8.5°C and the skin temperature rise is 7.3°C. The temperature inside probes or pods is generally between these two temperatures. Similarly, the dynamic pressure rise at the leading edge would be approximately 0.8 psi. The pressure rearward at the trailing edge of a pod is roughly -0.8 psi (depending on the shape).

Aircraft observations are difficult at best and the standard measurements of temperature, pressure, and humidity are not exceptions. Originally, ordinary mercury-in-glass thermometers were used and corrections were made for dynamic heating and radiation effects. These thermometers, or thermometers using bimetallic elements, are still in use today and with appropriate corrections can be used

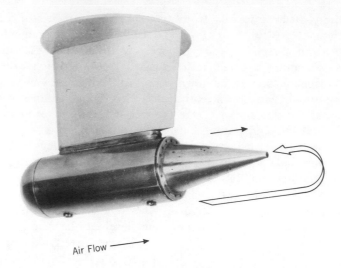

Air Flow ⟶

Figure 4.4 Reverse-flow thermometer to minimize droplets entering probe (Courtesy NRL, Washington, D.C.).

as checks on our electronic thermometers for flight in clear air. In clouds temperatures are usually measured using some technique to minimize the effect of liquid water. This is done by either a reverse flow housing (Figure 4.4) or by utilization of the vortex technique to spin off the water in the air (Figure 4.5). Two types of thermometers in common use on aircraft are the Rosemount probe and the axial-flow vortex probe. A third type used by AFCRL is the tangential-flow vortex shown in Figure 4.6. The Rosemount temperature probe* is detailed in cross-section in Figure 4.7. In its basic design it reduces the effect of droplets by inertial separation with the sensor recessed. The boundary layer is kept away from the

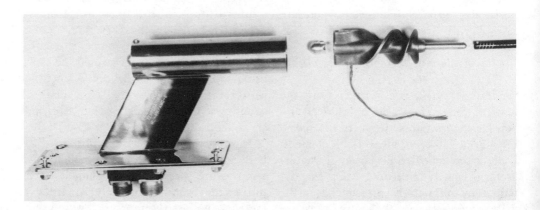

Figure 4.5 Exploded view of axial-flow vortex thermometer. Airflow is from left to right. At the right is a cylindrical carbon hygrometer element to install over thermometer element (Courtesy NRL, Washington, D.C.).

* Manufactured by Rosemount Engineering Corporation, Minneapolis, Minnesota.

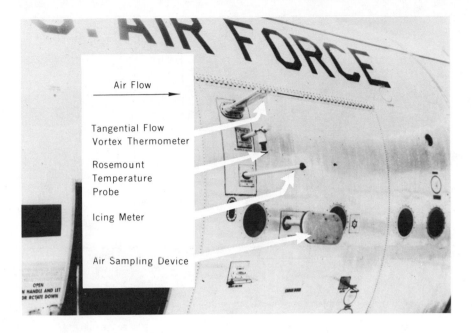

Figure 4.6 Cloud physics instrumentation on C-130 of R. Cunningham, AFCRL (photo courtesy R. Holle, EML, NOAA, Miami, Florida).

sensor by bleeding boundary-layer air out through the holes in the surface of the probe. The design also allows for deicing by the use of heat without an appreciable increase in the sensing error. The axial-flow vortex thermometer is an older,

widely accepted probe developed by NRL. A vortex is the only type of probe for temperature measurement from aircraft which does not require a dynamic heating calculation to correct the data for the true air speed of the plane. It is still in

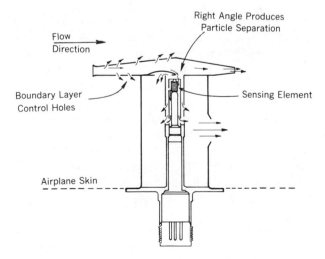

Figure 4.7 Internal configuration of Rosemount aircraft thermometer.

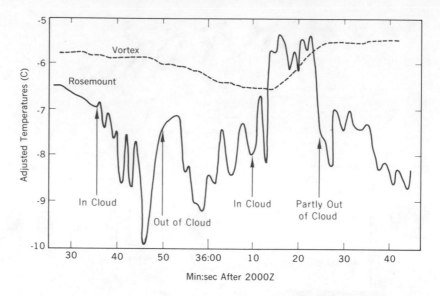

Figure 4.8 Comparison of temperatures measured with the axial-flow vortex and Rosemount thermometers.

use and has proved reliable, but it is no longer commercially available and is being replaced by the Rosemount probe.

Temperature measurements in flight might resemble the trace of Figure 4.8. Often the evaporation of liquid water lowers the temperature of the probe as the aircraft exits from the cloud (called "wet bulbing"). In this case, a similar effect caused by sublimation of ice may have caused the Rosemount temperature to lower shortly after emerging from the two clouds. In a similar way evaporation of water from a wet aircraft followed by sampling downstream can lower the temperature.

Icing, the freezing of supercooled water on aircraft surfaces, can cause a reverse effect. While in cloud, the freezing of supercooled water can release latent heat and result in a general warming; the warmer temperatures on the right side of Figure 4.8 may reflect this effect.

Note also in the figure the differences in the temperatures as measured with the two thermometers. First note that the absolute temperature, as measured with

the vortex thermometer, is about 1°C higher than the Rosemount value. This difference was corroborated during extended flights in clear air. Second, note the slow response of the vortex thermometer (>10 sec) and the time delay in the response. The Rosemount thermometer responds to fluctuations of less than 1 sec. Similar response can be obtained with a fast-response element in a vortex housing.

This exercise (Figure 4.8) illustrates the great need we have for redundancy in sensors to cover the space and time scale of interest. In this regard, we must remember that aircraft measurements in situ are valid only over the flight path. For a temperature measurement that is valid over a larger area, an infrared (IR) thermometer may be used. This instrument operates by measuring the thermal energy transmitted from a nearby source. In clear air the source is distributed over a long path; however, in clouds the source is the cloud droplets (which usually are at the wet-bulb temperature). Basically, the instrument collects the infrared light from the liquid, solid, or gaseous source, concentrates it with front-surfaced concave mirrors, and detects the output, received power. The temperature measured is the equivalent blackbody temperature. For precipitation-size drops falling at high terminal velocity from colder levels this temperature may be colder than even the wet-bulb temperature.

Airborne Humidity Measurements

When attempting to measure humidity of the air between droplets in clouds, it is impossible to isolate all the small droplets by centrifugal force without adding sufficient energy to evaporate some of them into the sample being measured. Most of the water can be excluded by the use of a rain deflecting entry designed at NRL. This entry is shown in Figure 4.9 as viewed from the bottom front (slipstream air flows from right to left in the figure). The air flows around the widest section at the front, leaving a droplet "shadow" at the hole near the rear (left). The screw head provides a pressure recovery to a pressure slightly above the free-stream static, so that air enters this hole. The hole in the round section near the front is at a position of maximum negative pressure (suction) to eject the sample. Changing the size of the screw head adjusts the resultant pressure of the sample at the dew point sensor attached to the entry inside the fuselage, so that no pressure correction to the data is needed. Normally any humidity instrument that responds to vapor density or relative humidity must be corrected for temperature rise in the housing and for the pressure of the sample. If the hygrometer responds to dew point, no temperature correction is needed.

The Cambridge Systems Division of E. G. and G. dew point hygrometer is perhaps one of the more reliable hygrometers in use at this time. Basically, this instrument is a thermoelectrically cooled mirror (Figure 4.10). A light is focused so that it reflects off the mirror onto a photodetector. The air sample is brought in contact with the mirror, and the mirror is cooled until dew forms on

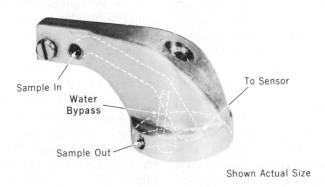

Sample In
Water Bypass
Sample Out
To Sensor

Shown Actual Size

Figure 4.9 Rain and droplet deflecting sample entry for dew point hygrometer mounted through aircraft skin, as viewed from front and below. Pressure at the entry hole is increased by screw head at left. Bottom hole has suction (courtesy NRL, Washington, D.C.).

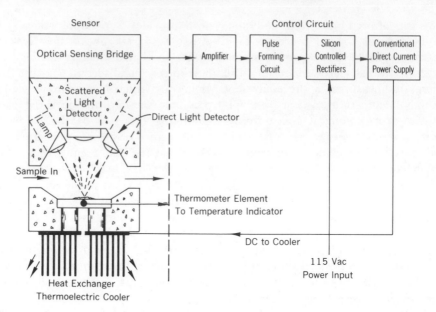

Figure 4.10 Schematic diagram of dew point hygrometer using thermoelectric (Peltier) cooling automatically controlled by photodetectors which sense dew or frost condensation on a cooled mirror (courtesy Cambridge Systems Div., E. G. and G., Cambridge, Massachusetts).

the surface, scattering the light beam. A minimum response time of about 1 sec can be obtained from this instrument, though the response time may approach 15 to 20 sec when operating near the limit of the cooling capability of the instrument. The primary maintenance required on these dew point hygrometers is an adjustment of the control point to allow for dirt or salt which gradually accumulates on the mirror. The sample entry for this hygrometer usually incorporates the water separator previously described (see Figure 4.9).

Another common technique for measuring humidity uses a material whose resistivity varies with the humidity. Though less reliable and less stable, the simplicity of the technique and operational success as an expendable element is attested to in the wide use of LiCl and carbon-coated humidity-sensitive sensors in radiosondes.

Another technique formerly in use operationally by NOAA's Research Flight Facility used the 1.37 μm spectral absorption band of water vapor to measure the water in the air by extinction of the beam in a cell with a 1-m path length. The density of water vapor is directly proportional to the square root of the fractional extinction of the light. The instrument has a sensitivity of better than 0.5 g/m^3.

It is possible to measure humidity using an absolute desiccant, either P_2O_5 (phosphorus pentoxide) or molecular sieves (substances with small molecular holes capable of retaining quantitative amounts of molecules smaller than the pore size). Both these techniques have been successful, but spurious irreversible absorption of water and other gases (and particulates) may give erroneous readings. Of course the absorbent (desiccant) must be weighed to determine the water content of the air and the mass of air sampled must be known.

Even the standard sling psychrometers have been adapted for use on aircraft and were, perhaps, one of the earliest airborne humidity sensors. The two mercury-in-glass thermometers are mounted either through the skin or outside the cabin window and read at intervals for a measure of wet- and dry-bulb temperatures. The technique works quite well in small aircraft at the lower altitudes and low aircraft speeds. The dynamic heating must be calculated; the ventilation rate must be about 10 to 20 knots; the wick must be kept clean and wet; and the pressure at the wick must be used in the psychrometric equation to calculate the vapor pressure and relative humidity or dew point.

One of the fastest response humidity instruments (a few milliseconds) is described by Ruskin[59] It is an instrument that uses spectral absorption in atomic hydrogen, Lyman-alpha line at the wavelength of 1215.6 Å. This spectral line is so highly absorbed by water vapor that the light is reduced to $1/e$ of its dry-air value in 0.5 cm of path length in saturated air at 0°C.[74] It can be used either for measurement of humidity or, as described under "Airborne Measurement of Liquid Water," with a sample preheater for the measurement of total water. An adaption of this instrument has been developed at NRL to measure the humidity of the air between particles of ground fog without interference from droplets or contamination and water on the windows. This measurement is achieved by using alternately a long and short path with the same optical components so that window effects are canceled. Although this is probably the first rigorous measurement of humidity of fog, it appears that at this time the same technique cannot be applied to aircraft in clouds without an inordinate amount of complications and aerodynamic development.

An alternate method of measuring humidity with a response time comparable to that of the Lyman-alpha hygrometer utilizes a microwave resonance cavity. The microwave refractometer uses a cavity flushed with the ambient air to measure the microwave refractive index. The cavity is mounted in such a manner as to eliminate the liquid water drops. An independent measurement of pressure and temperature is required in order to obtain humidity from the measured microwave refractive index.

Radiosondes and Dropsondes

For observations of vertical profiles of temperature, pressure, humidity, and winds, two types of sondes are in common use, the radiosonde and dropsonde. The radiosonde is released from the ground with a balloon and the dropsonde is dropped from an aircraft.

In the sondes, temperature is measured by use of a device (a metallic wire or a glass encapsulated thermistor) that has a known variation of electrical resistance with temperature. Pressure is measured using an aneroid (bellows) barometer or an hypsometer, a device which utilizes the boiling point of a liquid as a measure of pressure.

Humidity is measured using a carbon-coated or a lithium chloride hygrometer, devices which change their electrical resistance as a result of absorption of water vapor. In some European radiosondes, humidity is measured by the deflection of a goldbeater's skin* or the elongation of a human hair. Winds are measured using the drift of the balloon and transmitter, as measured by a tracking radio theodolite (rawinsonde).

Radiosondes have been in use for nearly half a century and certainly undergo more calibration checks and have undergone more development than

* An animal membrane which expands with increasing humidity.

Table 4.2   Dual Radiosonde Comparisons

Variable	Number of Flights	Number of Samples	Mean Difference	Random Difference
Temperature	46	1558	0.16 to 0.21	± .5°C
Pressure	46	1558	0.00 to 0.20	±1.9 mb
Dewpoint depression	46	767	−1.10 to −0.60	±3.6°C
Wind speed		3520	−0.1 to +0.1	± 3.1 m/sec
Wind direction		3520	5.7 to 6.7	±14.5°

Source: W. Hoehne, National Weather Test and Evaluation Laboratory, Sterling, Virginia.
Note: The random difference tabulated is the remainder of the spread in data after the mean difference has been removed from the data. The errors tabulated for wind speed and direction are due to the receivers.

perhaps any other of our measuring techniques. But errors exist and each of the quantities measured, for example, temperature, pressure, humidity, and winds, has a limited time and space scale of validity. Table 4.2 presents data acquired during dual radiosonde flights with two receiving stations. The mean difference reflects the fact that the lower radiosonde was exposed to warmer temperatures and higher pressures.

Dropsondes unfortunately have grown in notoriety because of data discrepancies since their design in the 1940s. Conceptually, they have the capability of giving a sounding close to the cloud in time and space (see Table 4.1) and should be capable of measuring the detailed structure of the actual environment at cloud base; however, they must be robust to withstand the large forces on entering the airstream. Also they must maintain resolution, particularly near the cloud base where the structure is quite variable, and they must fall fast enough to allow the aircraft to receive the data before it leaves the area. This latter requirement is a rigid one in operation where many cells may be seeded in a limited time, as with the hurricane seeding efforts.

A typical set of comparisons between dropsondes and radiosondes is shown in Table 4.3. The comparisons are relative to the Tampa and Key West radiosondes of midday. Note the relatively large mean differences (4 to 5°C) at the lower levels. This may be a real difference between a sounding over water and over land that results from an onshore breeze and the preferential heating of the ground, but it may also indicate instrumental problems.

The need for caution in accepting an atmospheric sounding is exemplified by recent measurements of humidity made prior to the Barbados Oceanographic and Meteorological Expedition (BOMEX). The results of comparisons between data taken from aircraft and radiosondes (from shipboard) show that the standard humidity measurements tend to be too low because of radiation effects which cause a general warming of the hygristor.

Airborne Updraft Velocities and Horizontal Winds

Airborne updraft velocities and horizontal winds are two of the most important measurements for studying cloud dynamics. The usual technique for measuring the strength of an updraft and its horizontal dimension (updraft radius), is to measure the rate of rise of an aircraft during penetration. The aircraft altitude is usually measured with a radar or laser altimeter, and corrections are applied to account for the attitude of the aircraft. In an equilibrium condition, the angle of attack, the pitch angle, the roll angle, and the yaw angle are used to describe the aircraft attitude and make it possible to calculate the updraft velocity.

In practice, it is usually assumed that roll and yaw effects are negligible; therefore the pitch angle (the fore-aft angle between the aircraft inclination and the horizontal) and the angle of attack (the angle between the relative wind and the aircraft inclination) specify the updraft velocity. The pitch angle is measured with a gyro located at the center of mass of the aircraft; the angle of attack is measured using a differential pressure probe or a wind vane either fixed or gimbaled. If this measurement is not available, the angle of attack can be estimated from the lifting equation for aircraft motion.[8] Even with no estimates of these angles, qualitative estimates of vertical motions can be made from the aircraft displacement. In any case, however, the accuracy of the technique is poor in small scales (< 0.1 km) with large aircraft owing to their inertia, and with any size aircraft for large-scale motions (> 10 km).

The instantaneous rate-of-climb meter,

Table 4.3 Dropsonde-Radiosonde Comparisons

Height	Mean Difference	Random Variation
Surface		
Pressure	− .1 mb	−4.2 to +2.7 mb
Temperature	−5.8°C	±1.6°C
Dewpoint depression	−2.9°C	−7.1 to +11.3°C
1000 mb		
Height	−5.8 m	−17.2 to +16.8 m
Temperature	−4.2°C	−1.3 to 1.1°C
Dewpoint depression	−9.0°C	−2.0 to +2.3°C
850 mb		
Height	−21.5 m	−25.5 to +18.5 m
Temperature	−1.4°C	−2.8 to 2.1°C
Dewpoint depression	−5.9°C	−2.7 to +2.9°C
700 mb		
Height	−28.5 m	−26.5 to +18.5 m
Temperature	−1.1°C	−1.0 to +1.4°C
Dewpoint depression	−10.8°C	−7.1 to +3.7°C

Source: Major L. A. Osburn, HQ Air Weather Service, Scott AFB, Illinois.
Note: Data were taken by WC-130 A/C #733, 741, and 493 on May 8, 1971, during the METCAL mission. Comparisons were made with both the Tampa and Key West radiosondes for 12Z.

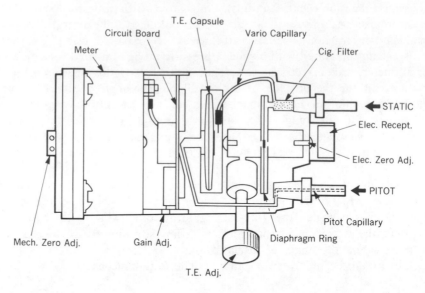

Schematic Sectional View of
Ball Variometer, Model 101-B

Figure 4.11   Schematic cross-section of variometer (courtesy of Ball Brothers Instruments, Boulder, Colorado).

or variometer, gives an estimate of vertical velocities with surprisingly good resolution. It definitely is superior to a "seat of the pants" estimate and can give a direct measure of the updraft radius, provided the pilot avoids making pitch corrections during the time of interest (difficult to avoid when in turbulence). The variometer measures the rate of static pressure changes, which are in turn related in a negative sense to changes in aircraft altitudes. Figure 4.11 shows a detailed sectional view of a Ball* variometer. The static pressure changes are sensed by a diaphragm coupling two chambers, each of which responds with different time constants to the pressure (because of the vario-capillary). The total energy capsule compensates for variations in aircraft velocity by sensing and correcting for dynamic pressure changes.

* Manufactured by Ball Brothers Instruments, Inc., Boulder, Colorado.

The most sophisticated technique, and perhaps the most accurate for cumulus scale motion, is the use of accelerometers on an inertial platform. This system consists of three sensitive accelerometers aligned in the three perpendicular directions by sensitive gyroscopes to maintain absolute orientation. The accelerometers are usually of the force-balance type, employing either strain gauges or an electromagnetic principle similar to that used in a galvanometer movement. A single integration of the acceleration gives the velocity, and a double integration gives the position. With the system it is possible to measure the two horizontal components and the vertical component of the wind.

A considerable amount of calculation must be completed to obtain the wind motions from the inertial platform and the wind vanes. Because of the large number of samples, a computer is necessary to process the data. The processing should be done in flight for navigational reasons, though it is possible with considerable sophistication to display the flight track and variables of interest on a cathode ray tube (CRT) display for use by the scientist on board.

However, it has been only since about 1965 that gyroscopes have been manufactured with the stability to discern detail in the draft scale (0.1 to 10 km). For longer-term stability and measurements of horizontal winds on larger scales, Doppler radar techniques are used. The technique measures the Doppler shift Δf_d in a radar return from the surface. The depression angle θ_d of the radar beam from the vertical is measured and the aircraft ground speed V_a is calculated from the equation (λ is the radar wavelength)

$$V_a = \frac{\Delta f_d \lambda}{2} \sec \theta_d. \qquad (4.1)$$

Included in every estimate of the winds is the aircraft's true air speed or the velocity of the aircraft relative to the air, U, in the equations.[58] The dynamic pressure is measured with the pitot tube and compared with a static pressure reading, made perpendicular to the airflow. The difference is the dynamic pressure difference, which is directly related to the square of the indicated air speed. To convert from indicated to true air speed, it is necessary to correct for the change of density with altitude in the manner described in the section, "Airborne Temperature Measurements."

Pitot tube systems have been in operation for nearly a century and are inherently reliable. They suffer from two basic flaws: they can be clogged with rime ice and their pressure readings undergo large fluctuations during heavy rain.

In summary, let us note that the validity of each of the techniques for wind measurements is limited. The inertial system measures acceleration directly so that velocity is obtained through a single integration and position is obtained

through a double integration. Hence errors in velocity and position tend to increase and become unbounded as time increases. The Doppler system and the variometer measure velocity directly (horizontal and vertical, respectively) so that position is obtained through a single integration and only position errors increase with time. The systems that measure position directly, such as the Omega system, the Loran system, and the other navigational aids that use radio-wave direction finders, have bounded errors which depend on the relative proximity of the transmitter. Using all three types of systems and weighting the wind and position values depending on time and location, it is possible to upgrade markedly our technical capability and, with feedback, even improve the outputs of each individual system. The AWRS system, being developed for the Air Force, is one system that employs this type of interaction. It should be capable of measuring winds to a few hundred centimeters per second and determining position to less than 1 mi. Such a system may be able to measure cloud drafts with sufficient detail to determine mass fluxes for verification of our numerical models.

Airborne Measurement of Liquid Water

Even excluding the problems associated with the presence of solid or mixed-phase (ice content), the measurement of liquid water content (LWC) in clouds is difficult because of the rapid spatial variations, as well as the spectral variations of droplet and drop sizes. As noted in Table 4.4 the liquid water exists in the cloud in the form of cloud droplets (1 to 50 μm diameter), raindrops (250 μm and larger), and intermediate sizes (50 to 250 μm).

One of the first widely used methods of measuring LWC in warm clouds was

Table 4.4   Sampling Requirements in Different Ranges

	Small	Intermediate	Large
Water drops	Cloud droplet	Drizzle drops	Raindrops
Ice crystals	—	Cloud ice	Snowflakes
Drop diameter or ice dimension	1–50 μm	50–250 μm	250 μm
Total expected concentration	1–10^3/cm^3	1–10^3/liter	1–10^3/m^3
Vol. sampled in 1 km of flight path	10 cm^3	10 liters	10 m^3
Sampling area	10^{-2}cm^2	1 cm^2	10^3 cm^2
Present instruments in common use	Slide impactor	Formvar replicator	Foil impactor

developed by Warner and Newnham in 1952.[80] This method employed a porous paper tape pretreated with a solution which was electrically conducting when wet, but not when dry. During flight the resistance of the tape as it moved over two fixed electrodes was a function of LWC and the aircraft speed, tape transport speed, and collection efficiency for the various sizes of droplets.

The most common aircraft instrument for measuring the small-droplet portion

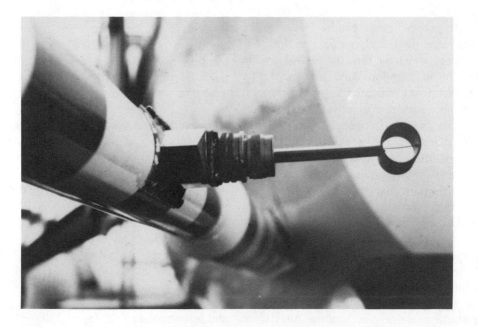

Figure 4.12   Johnson-Williams liquid water meter NOAA, probe mounted on the nose boom of a NOAA/RFF DC-8 (photo courtesy RFF, NOAA, Miami, Florida).

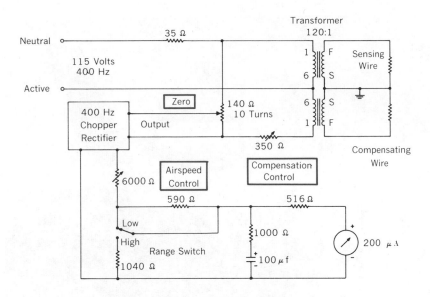

Figure 4.13 Electrical schematic diagram of Johnson-Williams hot wire liquid-water content meter described by Neel.[51]

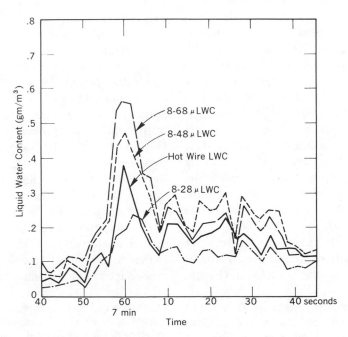

Figure 4.14 Comparison of liquid water content during a traverse through a small warm cumulus cloud. The Johnson-Williams hot-wire instrument response decreased for several droplet sizes larger than about 30 μm diameter as evidenced by these comparisons with the Knollenberg optical array measurements (courtesy R. Kollenberg, Particle Measuring Systems, Boulder, Colorado).

157

Figure 4.15 Hot-wire rainwater instrument of Levine, cone at front of boom; cloud water sensor wire strung on rectangular frame at side of boom on wing tip of NOAA/RFF DC-6 (photo courtesy RFF, Miami, Florida).

of the cloud water is the Johnson-Williams (J-W) hot wire instrument,[51] as shown in Figure 4.12.* The electric schematic can be seen in Figure 4.13. Basically, the instrument utilizes a hot nickel-iron wire with a known temperature coefficient of resistance. This wire is heated, and its resistance is measured using a bridge circuit. In dry air the wire maintains a steady-state temperature, a result of a balance between the electrical heat supplied and the heat removed by the dry airflow. However, if droplets are present in the air, they collide with the wire, cause cooling, and a temperature decrease which is directly related to the amount of liquid water in the airstream (Figure 4.14).

The instrument is primarily sensitive to cloud droplets below 20-μm radius. This means that for equivalent amounts of

cloud water and rainwater in a sample, only 1 percent of the total response will be due to the rainwater.

To create a liquid water system that would respond to both cloud water and rainwater Levine in 1965 designed an instrument which consisted of two portions, called the "cloud water instrument" and "rainwater instrument" (Figure 4.15)[39]. The "cloud water instrument" uses a plastic yoke strung with a heated wire similar in diameter to that of the J-W instrument, but with a greater length. The "rainwater instrument" is a threaded ceramic cone about 2 in. in diameter wound with a similar wire. In theory, the collection efficiency† of the cone-shaped sensor is such that a sorting

* Manufactured by Johnson-Williams, Inc., Mountain View, California.

† Collection efficiency is defined as the fraction of droplets in the air which collide with an object; the remaining droplets are diverted around the object by the air flow. Collection efficiency is generally higher for smaller obstructions, larger drops, higher air speed, and lower density air.

takes place that tends to make the rain-water instrument primarily sensitive to raindrops. Unfortunately, however, problems with electronic drift and atmospheric contamination (causing a partial electrical short on the surface of the cone) have so far made the outputs of the system difficult to interpret. In operation the two instruments have shown sufficient drift to mask the data during extended flights in clouds. Unlike the J-W instrument, this instrument does not have temperature compensation by a reference hot wire, but must be compensated manually.

A water-content instrument used in the Union of Soviet Socialist Republics is reported to be approximately equal in sensitivity for all sizes of drops. In this instrument a filter screen about 100 cm² is heated by high-frequency current and the resistance change with temperature provides a measure of the liquid water.

For liquid water content as high as 30 g/m³ a useful probe has been designed by T. Kyle of the National Center for Atmospheric Research (NCAR). This probe incorporates a heated grid whose temperature is servo-controlled. The amount of heating current required to maintain a constant temperature is a measure of the liquid or mixed-phase water being evaporated in the probe. The accuracy of this type of measurement is limited to the order of 1 g/m³ unless compensation can be provided for rapid fluctuations of the ambient temperature of the air and water entering the probe,

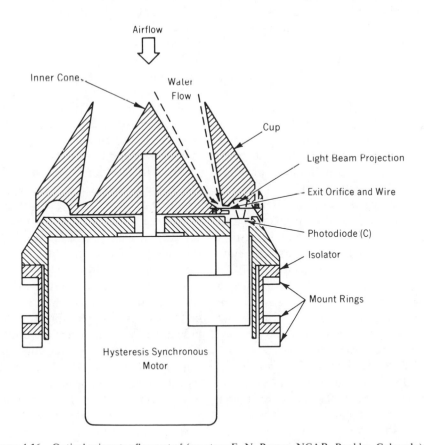

Figure 4.16   Optical rainwater flowmeter[5] (courtesy E. N. Brown, NCAR, Boulder, Colorado).

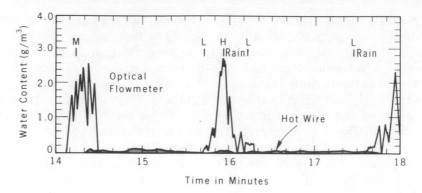

Figure 4.17 Comparison of Johnson-Williams hot wire cloud water content meter and Brown rainwater optical flowmeter in clear air showers (L = light shower, M = moderate, H = heavy by subjective estimate) (courtesy E. N. Brown, NCAR, Boulder, Colorado).

particularly when large drops are falling from colder altitudes.

Another instrument in development is a rotating cone designed by E. N. Brown, also of NCAR. This instrument is pictured in Figure 4.16.[5] Water, entering the annular space between the rotating double cone, is forced down into the conical base by centrifugal force. From there it is bled off through a small hole. In the hole is a small silver wire, which causes the formation of a web of water which is continuously spewed off the wire. A light source and a photodiode detector are used to measure the width of the web and the amount of liquid water is calculated using electronic analogue and digital processing. Figure 4.17 shows some typical results in a rain shaft below cloud base in Florida, September, 1970.

The instrument appears to give a reasonable measure of liquid water in the rainwater sizes. However, probably because of poor collection of smaller droplets, the instrument is relatively insensitive to cloud water. The instrument has undergone sufficient laboratory testing to guarantee reasonable reliability in operational programs. However, the design of the instrument suggests that it could be affected adversely by atmospheric contaminants. Deicing is not yet available for it.

The only aircraft instrument which measures total water, vapor, ice, and mixed phase in all sizes of precipitation and cloud drops is the NRL cloud evaporator instrument developed by Ruskin[59] and shown in Figure 4.18. The sample enters the 1-cm inlet at a speed about equal to flight speed so that no pressure increase is present to divert small drops (the flow is approximately isokinetic). In about 0.01 sec an open mesh heating element evaporates the liquid water. The resulting vapor is measured by the decrease in optical transmissivity of the air in the Lyman-alpha (1215.6 Å) line of the ultraviolet as discussed in the section, "Airborne Humidity Measurements." The instrument can resolve variations in millisecond times. The trace in Figure 4.19 shows evaporated "blobs" from individual particles of ice and liquid water. Unfortunately, the instrument is also sensitive to O_2 and CO_2 in the air, so that its calibration must be determined at each altitude and perhaps in every cloud pass. In the commercial instrument, a C.S.I. dew point measuring device is included

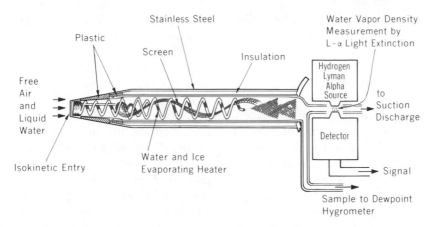

Figure 4.18 Schematic cross-section of cloud and rain total water (and ice and vapor) probe. Heater concentration at entry provides deicing; maze with heater and screens evaporate cloud water in 0.01 sec; Lyman-Alpha detector indicates changes in resultant vapor density in a few milliseconds (courtesy NRL, Washington, D.C.).

so that the calibration is essentially continuous.*

In closing this discussion, let us note that none of the above techniques can give a measure of liquid water that is better than about 20 percent. For raindrops this accuracy could be obtained in the absence of icing and contaminants on the surface with the old but simple technique of Vonnegut by collecting the water in a porous cup mounted in the air stream (see Vonnegut's capillary collector in Mason, 1971).[44] With supercooled water a suc-

cessful technique is the use of multicylinders, which utilizes the different collection efficiencies for cylinders of different sizes. The rime coat collected on the cylinders is a measure of the supercooled liquid water as well as the drop size distribution.[37] Icing measurements, however, are subject to errors from blowoff of liquid water.

A recently available alternative is to utilize our technological capability of measuring sizes and numbers of individual cloud drops and raindrops using

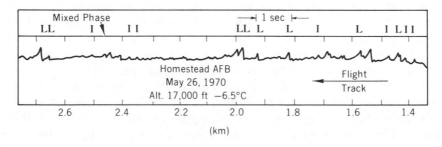

Figure 4.19 Lyman-alpha record of NRL total water instrument showing pulses from ice, symmetrical pulses (*I* on top line) and liquid drops (*L*).

* Manufactured by Cambridge Systems, a Division of E. G. and G., Cambridge, Massachusetts.

optical effects as discussed in the section, "Cloud Droplet Size Spectra."

SAMPLING IN CLOUDS

The basic problem in sampling is one of representation. The investigator must (1) decide on the time and space scales of interest, and (2) acquire samples over these scales as far as is practical. Then (3) each of the samples individually must contain a sufficient number of particles to be representative of the quantities on the small scale. That is, if requirement number (3) is met, the sample itself may be highly accurate, but at the same time may have no bearing whatever on the characteristics of that size scale of the cloud chosen under requirement number (1). In this regard, referring back to Figure 4.1, let us consider the time and space problems again. For example, an investigator may want to relate cloud elements over times of the order of seconds and distances of the order of 100 m. This means that samples taken over scales greater than these would be expected to smooth the data, whereas samples with very small scales could contain large random fluctuations. Models have been coded to resolve roughly these scales (of the order of seconds and 100 m); but instruments for direct sampling cannot physically acquire a sample of this size (approximately 10^6 m³). So, generally speaking, requirement (2) cannot be met for large drops with direct sampling.

Therefore we continue to make measurements in situ in the hope that variations in the measured quantity are not too great, awaiting help from technology. This help is likely to come from remote sensing.

Regarding requirement (3), the sample size itself must be appropriate to the portion of the size spectrum of interest. To illustrate this problem, let us divide the cloud particles into three size categories, extending over the size ranges listed in Table 4.4. In sampling raindrops, more than 1 m³ volume of air must be sampled over a cloud traverse (1 km) to assure any statistical validity. However, cloud droplets and the spectrum of cloud droplets can be measured with a sample volume of 10 cm³, but the extended "tail" on the droplet distribution (the intermediate-sized droplets) probably cannot properly be resolved in an air sample of 10 liters. These sampling requirements are summarized in Table 4.4, with the equivalent requirements for ice crystals. In the table, the term "cloud ice" has been used to represent ice particles between 50 and 250 μm. At present no adequate measurements of ice have been made, so the numbers shown could retain gross errors. In this regard, it should be remembered that in a water-saturated cloud ice grows at such a rate that the crystals are greater than 100 μm in about a minute.

Also for each raindrop there are millions of cloud droplets. Therefore in sampling raindrops the cloud droplets must be avoided either by physical exclusion from the sample or by electronic discrimination. Otherwise the cloud droplets can easily mask the raindrop sample.

The requirements for statistical validity can be briefly stated in the rule of thumb that 23 particles should be sampled in each desired size category.[12] Other important considerations in sampling are the collection efficiency of the sampling device and the effect of intervening or nearby aircraft surfaces. During collection by a sampling probe, the smaller particles tend to follow the airstream and so be deflected from impaction, whereas the momentum of the larger particles carries them onto the impaction surface despite the intervening airflow.[44] The effect is illustrated in Figure 4.20. Note that a 1-cm collecting surface will not collect an appreciable number of particles below 1 μm in diameter, and a 10-cm collecting surface will not collect many particles below about 4 μm. Of perhaps greater importance is the effect of the sampling support. If it is cylindrical and effectively about 10 cm in size (comparable to some models of the Formvar replicator described in the next section), one would expect relatively poor sampling of particles below 30 μm in diameter. (However, with airflow through the slit and strut the collection efficiency may be grossly altered.)

Other techniques to minimize sampling problems include the use of isokinetic flow. This technique draws the sample into a collector at a speed equal to the air speed outside the aircraft. For instance, an absolute filter system operating on an aircraft could collect even the smallest particles if the airflow through the sampling port were equal to, or greater than, the speed of the aircraft.

Yet another problem in sampling from aircraft is the effect of the aircraft surfaces themselves. Instruments located on probes projecting several feet upstream of the leading surface acquire the best samples; if attached to wing pods or other surfaces above or below the wings, they can acquire a reasonably representative sample. Instruments mounted on the fuselage or wing, however, usually result in some sacrifice in the quality of the sample. The fuselage, in particular, greatly disturbs the airflow. Sampling

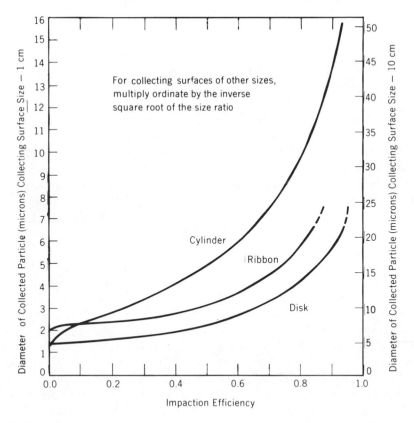

Figure 4.20 The effect of collecting surface shape and size on sampling (collection) efficiency calculated from Langmuir and Blodget[37] for an air speed of 100 m/sec for standard sea level and ideal flow conditions.

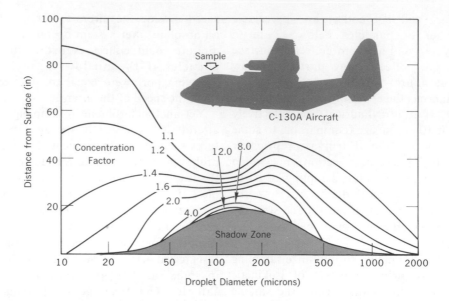

Figure 4.21 Concentration factors for particles resulting from airflow around a C-130 fuselage idealized as an ellipsoid of fineness ratio 2. Calculations based on radius of 85 in., velocity 300 ft/sec², and altitude 5000 ft (courtesy AFCRL, Hanscomb Field, Massachusetts).

from the fuselage should be done as far forward as possible and as far into the airstream as practical. In any case, the curvature of the fuselage is important and, for instance, the area near the windscreen is a particularly poor location because of high velocities and impaction that occur. In this regard, the sides of the fuselage are to be preferred because they are more gently curved.

Specifically, the effect is illustrated by Figure 4.21 for the C-130 aircraft. The data are theoretical calculations of the concentration factors expected at different distances from the aircraft. The location of the probe is on top of the aircraft, approximately where the fuselage becomes cylindrical. The calculations are such that the results are little more than qualitative, but certain features are apparent: (1) small particles (10 μm) that follow the streamlines are concentrated with a resulting error of no more than +50 percent, even close to the aircraft skin; (2) large (mm) particles, whose tra-

jectories are little affected by the aircraft, are accurately sampled near the aircraft skin; (3) particles that are collected by the aircraft or move relative to the airflow or both can be either eliminated or concentrated by a factor of 12 near the aircraft skin; (4) it is apparent that the probe should be located approximately 30 in. from the aircraft skin for a sampling error of about +50 percent; (5) in this case, the error seems to be relatively independent of particle size for sizes from 50 to 500 μm.

It is well to remember that the airflow about the aircraft is never as ideal as the theoretical curves would indicate due to turbulence, updrafts, and angle of attack effects. (Remember that the angle of attack changes as the gasoline is burned off in flight.) These effects are generally reduced by mounting on the side of the fuselage. Hence it may be necessary to locate the sampler as much as 48 in. from the fuselage. Also, the sampling characteristics presented are applicable only to spherical particles. Ice crystals with delicate dendritic structures might be expected to behave quite differently, perhaps having a greater tendency to follow the streamlines.

Apart from these problems are the effects upstream of the aircraft. These include dynamic heating of the air by adiabatic compression and air motion caused by the pressure disturbances. Indeed, these effects may be responsible for the distortions often seen in raindrop impressions on the aluminum foil impactor.

The gathering of water or rime ice on the exposed surfaces is another problem. Rime ice seems to clog up nearly every exposed hole. For humidity measurements, in particular, it is important that this water not enter the sampling port, and filter samples are easily ruined by liquid water. A technique which minimizes the effect "slices" the moist air away from the probe by a dividing plate. Clear air is then sampled a few inches from the skin. Another method of providing a "shadow" from droplets at a sampling entry is shown for the dewpoint sample in Figure 4.9. One way of minimizing the effect of icing is to mechanically disrupt the layer by a boot or external covering surface. Another way, used frequently, is to supply heat to the probe—a technique even effective with thermometers.

MEASUREMENTS OF CLOUD PARTICLES

The Replication Technique

The process of converting cloud and precipitation particles to plastic "skeletons" (replicas) was developed by Vincent Schaefer before 1941.[65] Since that time the technique has been used successfully for both ground-based and airborne sampling.[42] At this time assorted materials can be used to form the re-

plicas, including the common aerosol-spray lacquers and Eastman 910 adhesive. Each material requires a different experimental procedure.

The most widely accepted replicating material is Formvar. This is a plastic which is dissolved in chloroform or ethylene dichloride to make a solution of from 1 to 8 percent Formvar. In the simplest case, a glass microscope slide is simply dipped into that solution and held into the falling precipitation or the airstream. The particles impinge on the slide and are coated by the solution either by being submerged or by surface tension effects. The solvent then evaporates, leaving a plastic shell which may completely encapsulate the particle. With vaporized Eastman 910, the technique can resolve surface features of the particle to sizes as small as a few molecular dimensions. With Formvar dissolved in chloroform, the drying of the solution takes at least 10 sec, depending upon the temperature and the thickness used; ethylene dichloride tends to evaporate somewhat more slowly.*

After being replicated, the water or ice particle evaporates by diffusion through the rather porous plastic shell and only the plastic shell or "replica" of the original particle remains. Figure 4.22 shows several replicated snowflakes obtained in the Arctic using chloroform-Formvar solution on a handheld slide. (Remember that these crystals were chosen for their crystalline features. Usually crystalline forms are broken, aggregated, or melted and they are sometimes referred to as "junk" crystals.)

Commercial instruments are available which produce plastic replicas of the particles continuously on blank 16-mm movie film base. Problems with the instrument are many and can be classified

---

* Monsanto Chemical Company and Shawnigan Resins Corporation.

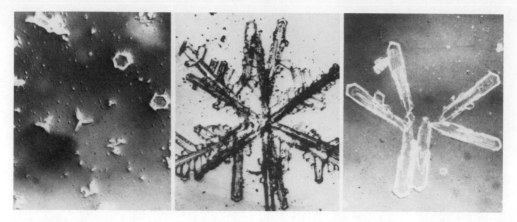

Figure 4.22 Replicas of ice crystals of millimeter size acquired on 1 × 2 glass slides coated with 2% Formvar chloroform solution: Slides were handheld in an Arctic snowfall by J. Pinnons, University of Washington, Seattle.

as (1) operational and (2) sampling. The operational problems include the manipulative difficulties of transporting the film to the sampling hole, coating the film with the plastic solution, and drying the plastic replicas. Hence the instrument contains a large number of sprockets and rollers, resembling a complex motion picture projector. It has complex plumbing, with pumps, tubing, and reservoirs. Also, the film is brought through a heated section containing a staggered train of rollers. As a result, the rollers and sprockets become gummed with plastic solution; the plastic solution clogs and reacts with the pump and tubing pieces; and the heater that drys the plastic tends to melt the ice crystals.

The sampling problems include breakup and melting of the particles during capture, the limited sampling range of the instrument, the difficulty of locating the sampling boom in the undisturbed air, and effects of airflow through the sampling slit. As a result, the film often is either blank or covered with particles, with many broken crystal pieces or splashed satellite droplets; and the fraction of the sample actually acquired is not known.

Viewing, interpreting, and retrieving the data from the film are no small tasks. Two general techniques are used, microscope viewing and projecting the image on a screen. The technique of projecting the images is operationally the most valuable because it allows several people to view the magnified crystals simultaneously. This tends to give a less biased assessment, an important consideration since all judgments are highly subjective, particularly if much of the ice was partially melted during the replication process. But it does not allow the replicas to be viewed with dark field illumination, a technique which best shows the intricate detail in the crystals. Viewing is also a particular problem because of the three-dimensional nature of the replicas, the distortion of images of small particles and detailed features by the optical system, and the interpretation of the mechanical distortions inherent in the rolling up of insufficiently dried film on the reel.

Improvements to the instrument include a system that allows the film to run continuously in a standby mode, thus minimizing clogging and obtaining a continuous record, the use of precoated film, the use of a particle decelerator, and numerous techniques to mark the film and identify the replicas. As mentioned, it appears that with experienced, professional personnel and by closely following the manufacturer's instructions on routine maintenance, with the various improvements, a reasonably operational performance can be achieved. It is obvious, however, that the capabilities of the instrument are extremely limited and results can be considered little more than qualitative, particularly with large liquid drops which may spatter and indicate spurious increases in small droplets.

Actual data from an NRL-developed instrument during flights at −3 to −5°C are shown in Figures 4.23 and 4.24. In these figures, notice the different types of signatures water drops can make. The frozen ones are identifiable by slight spurious crystal growth around them. Droplets as small as 1 μm can be distinguished, but 10^4 times as many are possibly missing because of the low collection efficiency for small droplets.

Ice crystals sampled in flight are shown in Figure 4.24. Here we see the regular features attributable to ice, but mostly fragments or unidentifiable chunks or both. Those on the left were probably from a single graupel pellet. The cluster of columns at the upper right was perhaps formed when a larger snowflake broke on impact, either with the sampling surface or the edge of the sampling slit. In any case, the number of ice particles in the air is greatly in doubt. With the instrument it appears that ice crystals above a certain size cannot be sampled intact because of

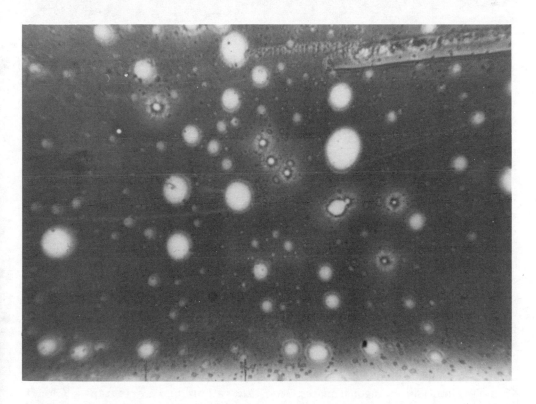

Figure 4.23 Replicas in red-dyed Formvar; frozen drops can be distinguished from supercooled liquid by spurious ice growth in clear area around the particles (courtesy NRL, Washington, D.C.).

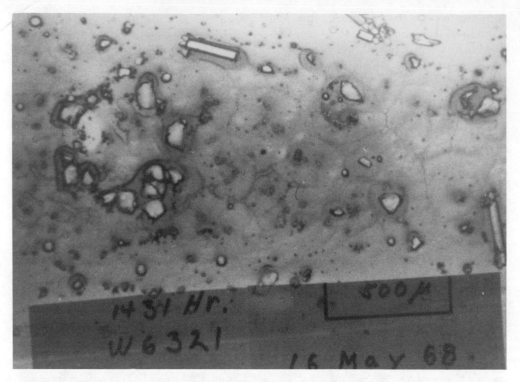

Figure 4.24 Shattered graupel on left; probably crushed snowflake upper right (courtesy NRL, Washington, D.C.).

shattering on impact with the film or because of the small sampling slit size. All in all, we see that the instrument can give no more than a rough estimate of the amount of ice.

A particle decelerator attached to the instrument can reduce the speed of the particles on impact by a factor of 5 and reduce the kinetic energy 25 times so that all except the most fragile snowflakes can be captured intact. From observations of crystal replicas with a 50-cm decelerating distance, it has been deduced that the maximum impact speeds in mph at which various types of crystals remain intact are as follows: thick plates with maximum dimension of 1.5 mm, 160 mph; 3-mm solid columns, 60 mph; 3-mm hollow columns, 40 mph; 1-mm thin hexagonal plates, 30 mph; 3-mm stellars, dendrites, and needles, 20 mph.[25] Maximum aircraft

speeds for each of these classes of crystals to be sufficiently decelerated are five times the crystal speeds listed.

In spite of its problems the replicator is so far the only instrument which can measure ice characteristics with any reliability, the possible exception being the aluminum foil impactor.[68] Other instruments are in development which may solve the ice discrimination problem, but the replicator is the only means of detecting ice in mixed-phase "mush" graupel at this time.

Visual Techniques

Let us not omit the value of visual observations and photographs. Clouds and their changes should be observed and photographed in every pose throughout the operation and, as well as possible, each should be documented. However, it must be kept in mind that the cloud is not an object with well-defined features. It looks different in transmitted light than in scattered light; it looks different in a light background than in a dark one; and it quite likely has a different apparent size at different angles of the sun. Also, its range must be known to establish any of its dimensions.

Parallax usually is quite a problem. Without range information from a radar echo, a range finder, or a calibrated distance, it can become quite difficult to say, for instance, whether the cloud base lowered or the cloud grew vertically.

The color of the cloud can indicate several things about the cloud. If the cloud is orangish and is not illuminated by orange light, it means that aerosol in the light path has removed some of the blue. Looking at a field of clouds, then, one would expect the orange clouds to be most distant from the observer. This distinction will not usually appear if the observer is not wearing polarized glasses because the light scattered by the background aerosol diminishes the contrast.

The optical effects in the cloud can quite often give definitive evidence on the constitution of the particles in the cloud. For instance, the presence of a true corona around the sun indicates the presence of droplets of a narrow size range; the angular distance between the colored rings decreases with increasing droplet size. The inverse effect, the glory, is formed about the shadow of the aircraft on a lower cloud having droplets with a narrow spectrum of sizes. Both effects are due to diffraction of the light rays by particles of size of the order of the wavelength of the light.

The rainbow itself gives a measure of raindrop sizes; as the colors become more distinct and red, the drops are more uniform and of larger sizes. Its counterpart,

the usually colorless cloud bow, gives an indication of the presence of incipient raindrops.

The appearance of the halo at 22° and 46° from the sun indicates the presence of ice with randomly oriented 120° and 90° crystal angles, respectively. The sun pillar is usually associated with the occurrence of needle crystals. Assorted other optical effects of ice crystals, including the beautifully displayed circumhorizontal arc indicate the presence of other crystal types.[26]

It is important to recognize that these observations can give answers about the presence of ice or water—and can be invaluable tools. For these observations, the only instrumental requirements are merely the presence of windows, open eyes, and informed minds.

The Measurement of Ice

The method most commonly used to measure the concentration and characteristics of ice crystals is the replication technique. The large data handling problem has greatly detracted from its use in operational programs.

Presently in the development stages are several optical instruments that attempt to supplant or supplement the Formvar replicators. These include the optical array instrument of Knollenberg.[34] The instrument uses the shadow cast by an individual particle to measure its size (Figure 4.25); the particle's size is measured by the number of optical fibers that are intercepted by the shadow. A digital logic system "watches" the shadow pass, measures the particles, tallies the measured values, and displays the output as a distribution of particle sizes on an oscilloscope.

Cannon[7] has installed a unique dual-image camera system on an NCAR sailplane. The system has the capability of

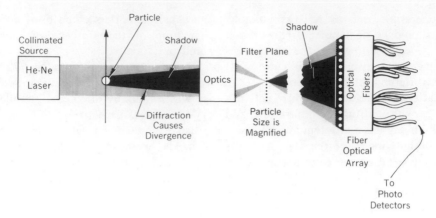

Figure 4.25 Schematic of optical array technique (full shadow area not shown) (courtesy R. Kollenberg, Particle Measuring Systems, Boulder, Colorado).

measuring both cloud droplets and precipitation-sized particles, and samples about 1 liter in 100 m of cloud traverse. As shown in Figure 4.26, a double light source is used; droplets larger than about 0.25 mm in the viewing volume act as lenses and produce two spots on the film. The distance between the spots is a measure of the droplet size. Ice crystals in the viewing volume act as opaque scatterers so that crystalline images appear on the film.

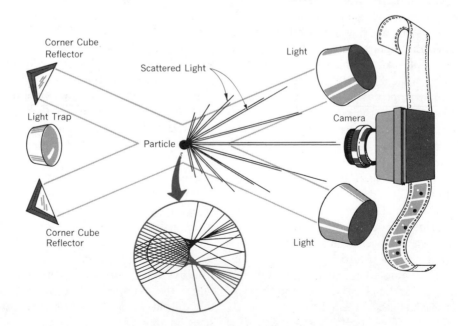

Figure 4.26 Schematic diagram of the dual-image cloud particle camera (courtesy T. Cannon, NCAR, Boulder, Colorado).

Ruskin[62] has devised a technique for measuring the net effect of a population of ice crystals by looking at the ratio of 40° to 100° scattering. This system applies the data of Huffman and Thursby[28] shown in Figure 4.27. The data indicate that an ice crystal population scatters about six times as much light at 40° as at 100°, whereas water drops scatter about 36 times more at 40° than at 100°. An optical system designed to look at this difference between the scattering at an angle of 100° and an angle of 40° should be able to distinguish between the presence of water and ice.

For a number of years, the subsun and halo phenomena have been used qualitatively to judge the presence of ice. Perhaps these phenomena can be made quantitative for measurement of ice

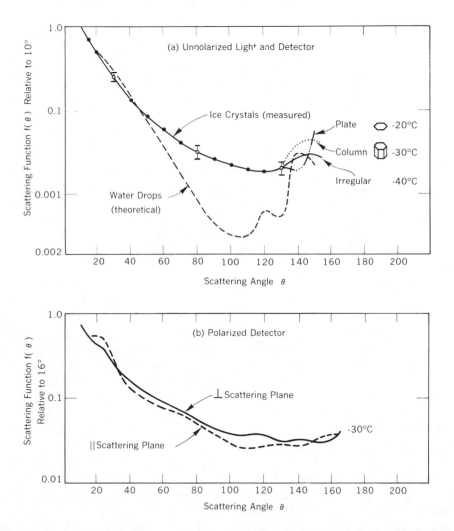

Figure 4.27   Measured intensity of scattered light from a population of ice crystals: (a) from Huffman et al.,[28] unpolarized source and detector, irregulars same size distribution as drops; (b) from Huffman,[27] polarized detector. Zero degrees is in the direction of light propagation (reprinted by permission, from Huffman[27]).

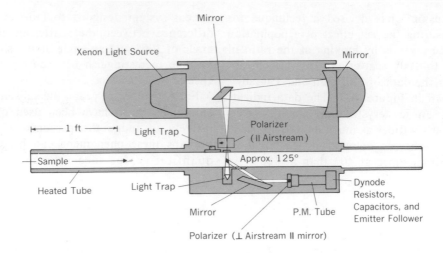

Figure 4.28   Schematic diagram of Mee Industries' ice-particle counter.

crystal numbers and types, as suggested by A. Frazer of Pennsylvania State University.

The Mee ice-particle counter* (Figure 4.28) attempts to capitalize on several optical characteristics of ice crystal.† The source light to the device is polarized and the detector (a P.M. tube) is covered by a polarizing filter oriented so that normally there is nearly complete extinction. Scattered light is measured at an angle of about 125° from the light source. The output is an electrical pulse corresponding to the presence of a particle. Some response is observed when water drops are present, and it appears that a pulse of equivalent height is produced by a 1-mm drop and an ice crystal 100 μm in size. A new model using IR and a 90° scattering angle has improved performance.

Other methods of detecting the presence of ice include the observation of symmetrical pulses of large size on the

Lyman-alpha total-water instrument (see Figure 4.19) as discussed under "Airborne Measurement of Liquid Water," and the use of electrical effects associated with ice.

Present-day electronics has the capability of detecting individual charges on precipitation particles. Mach and Hobbs[43] at the University of Washington and McTaggart-Cowan[46] at the State University of New York have attempted to use this idea for instruments to detect ice. Specifically, in both instruments the individual charges acquired by a wire stretched perpendicular to the airstream are measured and used as an indication of the presence of individual ice crystals. Both naturally acquired charges and charges produced by collison are measured depending upon the potential of the wire. At this time the technique shows promise but it is still overcome with electronic noise and data recording problems.

In retrospect, it appears that none of the above instrumental techniques presently can give a reliable measure of either the number of crystals or the type of crystals. Hence we should do the best

* Manufactured by Mee Industries, Altadena, California.
† T. Kyle of NCAR, Boulder, Colorado, and L. Radke[56] are presently developing similar instruments using the principle of extinction.

we can with visual indications of optical phenomena (halo, subsun, sun pillar, etc.), the presence of electrical effects, the bouncing of ice off the windscreen, or simply holding a "black glove" in the airstream. Historically, this refers to the days when a person held his hand, covered with a black glove, into a snowfall. Today, it refers to a black surface used for visual judgments regarding the characteristics of ice. A variation of this method suggested by Grant of Colorado State University is black velvet stretched across a hoop. This method has the advantage of capturing the particles and holding them for later viewing.

R. Cunningham of the Air Force Cambridge Research Laboratory (AFCRL) uses a 1-in.-diameter rod with a small, black, flattened surface covered with a calibrated white grid. The surface is viewed through a nearby window and an eyeball estimate is made regarding the sizes and numbers of crystals hitting the surface.

Only in such positive indications can we be sure of the qualities of the ice phase. However, our need for numbers of crystals requires that our judgments be quantitative as well. Let us hope technology fills this gap in the near future.

Cloud Droplet Size Spectra

The standard device for measuring distributions of cloud droplets has been the impactor slide. This is a small glass slide, covered with powder, oil, or other suitable material, that is held in the airstream. The small sampling volume requires that the sample be exposed for only a few milliseconds, so that ordinarily the slide is shot through an air gap with a gun, and the whole assembly has become known as the "slide gun." A similar but more quantitative instrument reported by Squires and Gillespie[71] uses soot-coated small round rods with a well-known collection efficiency, 100 rods being fired in rapid sequence.

Today, several instruments with real-time analyzing capabilities are being used operationally to measure the spectra of cloud droplets. These are the electrostatic probe[1, 32] and the optical disdrometers.[33, 63] A cross-section of the electrostatic probe is shown in Figure 4.29. Air containing the cloud droplets is sucked into the hole (250-μm diameter) in the right side at nearly sonic velocity. The droplets (or their fragments) then collide with the rounded rod directly behind the hole. A high potential on the rod (500 V) imparts a charge to the droplet fragments

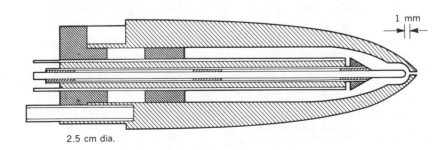

1 mm

2.5 cm dia.

Section of the Electrostatic Disdrometer Tip

Figure 4.29 Cross-section of the electrostatic disdrometer tip (courtesy J. Dye, NCAR, Boulder, Colorado).

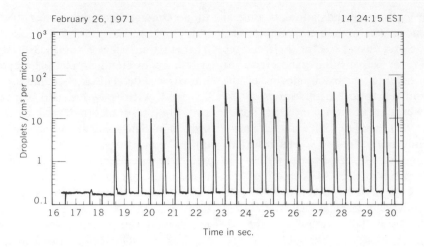

Figure 4.30 Changes in droplet size spectra from electrostatic disdrometer during one traverse of small cumulus in Florida. Size distributions for each 0.5-sec sample starts at 4 μm radius; each size category is 3 μm wide (courtesy J. Rye, NCAR, Boulder, Colorado).

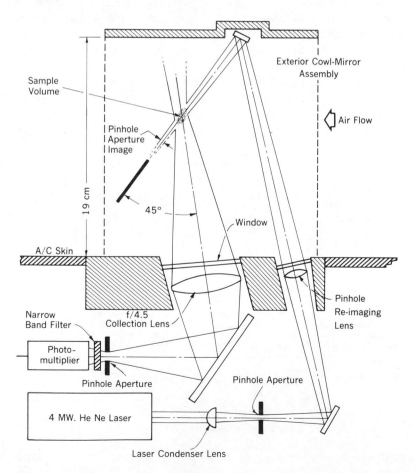

Figure 4.31 Optical diagram of the A.D. Little cloud particle spectrometer (courtesy R. Ryan, A. D. Little, Inc., Cambridge, Massachusetts).

as they leave, a charge that is related to the size of the original droplet. The output is analyzed for pulse size, converted to droplet size, and read out as a droplet distribution every 0.5 sec. A sample of the data during a pass through a small cumulus cloud in South Florida is shown in Figure 4.30. The data are analyzed in ten channels, with a 3-μm separation starting at 4 μm.

An instrument which measures the light scattered at 45° by single cloud particles has been developed by Ryan[63] and a schematic diagram of the device is shown in Figure 4.31. It is sensitive to cloud droplets between 1 and 30 μm and was used successfully during the early 1970's in flights aboard the NASA Convair 990. Typical particle distributions in various cloud forms are shown in

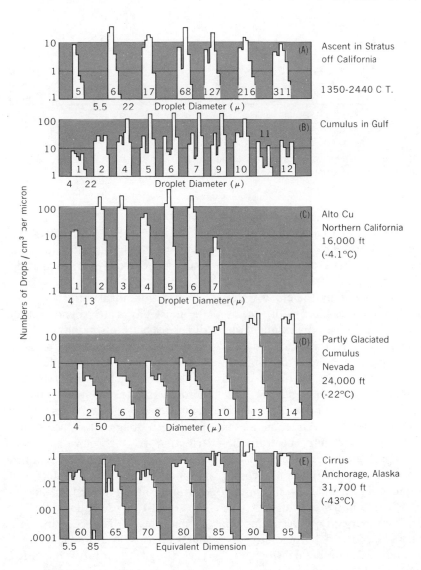

Figure 4.32 Cloud droplet size distributions at several times (seconds shown in each bar graph) during traverse across (A) stratus; (B) maritime cumulus; (C) continental altocumulus; (D) partially glaciated cumulus; (E) glaciated cirrus clouds. The size scale shown under one bar graph in each traverse is the same on all in that traverse (courtesy R. Ryan, A. D. Little, Inc., Cambridge, Massachusetts).

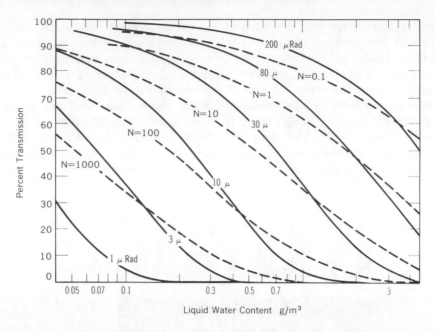

Figure 4.33 Curves to determine number of cloud droplets/cm³ and their average radius from optical transmission and LWC, calculated for 20-m path length (courtesy NRL, Washington, D.C.).

Figure 4.32. In the case of the strati-form cloud (Figure 4.32A) a passive microwave radiometer was present on board the aircraft which measured the total liquid water over a column from the thermal microwave emissions by the cloud water. The liquid water measured with the disdrometer agreed in an approximate way with the water content indicated by the radiometer. Note the broad distribution of large droplets in the maritime cloud (Figure 4.32B) and the narrow distribution of droplets in the continental cloud (Figure 4.32C). Also, as ice is formed in the cloud, the spectrum becomes diffuse (Figure 4.32D) and in a cirrus cloud layer the distribution becomes quite broad (Figure 4.32E). It appears that the instrument can detect the presence of ice indirectly by recording the appearance of a broad particle size distribution and small number density.

Optical scattering instruments for cloud droplets between 0.2 and 75 μm radius in three overlapping ranges are in use in the U.S.S.R.[35] These three instruments each utilize 90° dark-field scattering from the droplets inside a probe, using a method similar to that developed by Gucker and Rose in 1954.[23]

Extinction measurements are perhaps the most physically sound as they give a direct measure of the physical size of the particle from a measure of its shadow. Knollenberg[33, 34] has extended the method to the smaller cloud droplets (1 μm) and aerosol sizes using lasers and electronic light detectors. He tunes the laser to produce a doughnut-shaped intensity distribution in the viewing volume. This defines the sampling volume, discriminates against droplets which are partially out of the beam, and increases the signal-to-noise ratio by the use of the double pulses that occur in the detected light intensity as a droplet passes through the doughnut intensity profile. This unique feature coupled with improved optics allows the use of an extinction system to measure each cloud droplet as opposed to correcting for nonuniformity of the beam by statistical corrections to the size spectra.

If data from droplet spectra are to be used mainly to determine the number n, and the average droplet radius r, as a cloud develops precipitation, the less sophisticated techniques of Ruskin[62] can be used. The optical transmission over a 20-m path (a function of nr^2) is combined with a measurement of liquid water content (W) (a function of nr^3) (Figure 4.33). The technique used is rigorous for monodisperse droplets, but is quite satisfactory for the narrow spectra present in young clouds. In older clouds a qualitative measure of n can be obtained by using the cloud water as measured by the Johnson-Williams hot-wire meter, and applying the correlation only to the cloud water. The mean radius r can later be determined by combining n with a measure of total water (or ice) content as measured by the instrument discussed by Ruskin.[62]

The droplet size and number are calculated by combining the liquid water content, $W = \frac{4}{3}\pi r^3 n$, and the optical transmission I/I_o, where I is the light intensity at the detector (in cloud) and I_o is the light intensity at the source (equal to that at the detector in clear air). The equation relating the optical quantities is

$$I = I_0 \exp\left(-Q_{\text{ext}}\pi r^2 nL\right),$$

where L is the path length and Q_{ext} is the Mie extinction efficiency factor which approaches the value 2 for droplets with radius larger than about 4 times the light wavelength. The solutions of this equation with that above for W are shown graphically in Figure 4.33 for the case of $L = 20$ m.

Note that a continental cloud of 1000 droplets/cm³ and with 0.3 g/m³ of LWC has only 10 percent transmission, whereas the same LWC in a maritime cloud having 100 droplets/cm³ has 40 percent transmission. This figure emphasizes the extremely large range of transmissions that is experienced in clouds. A shorter path length is marginal (too little extinction) in the maritime clouds, and a longer path is marginal in continental ones (nearly total extinction).

Raindrop Measurements

In-cloud measurements of raindrops serve three purposes: (1) they may act as an integrated measurement of the processes involved in their creation. For example, their size distribution may indicate whether they are formed by warm- or cold-cloud processes; (2) they provide a measure of the precipitation developed which may or may not reach the ground; and (3) they may be used to calibrate airborne and, perhaps, ground-based radars.

One of the older techniques uses a metal foil backed by a screen or grid.[3, 11] The drops impact upon the foil and make impressions that give a direct measure of the sizes of the drops. Dead soft lead foil backed by a wire mesh was used in 1958 in a design by Brown.[4] Because each sampling surface had to be hand prepared, the technique was good only for single samples, though later it was adapted to care for multiple samples automatically.

The raindrop samplers in use today[16, 66] use a continuous strip of aluminum foil, 0.001 in. thick, 3 in. wide, and about 300 ft. long. A picture of the MRI continuous hydrometeor (foil) sampler* is shown in Figure 4.34. The sampler can be run at several speeds from about 0.5 to 2.0 in./sec. The backing surface, instead of a grid, is a series of parallel ridges, 250 μm apart. Its resolution is about 200 μm, about twice that of the dead soft lead.

---

* Manufactured by Meteorological Research, Inc., Altadena, California.

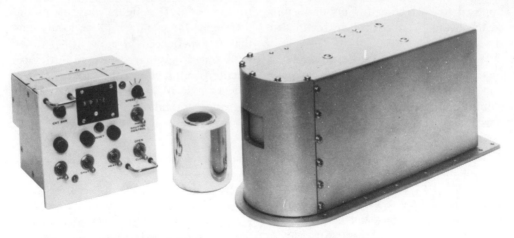

Figure 4.34   The MRI aluminum foil hydrometeor sampler and its control panel.

The operational reliability of the sampler is good. The instrument has a sample area about 1.5 by 1.5 in., captures about 4 m³ in traversing 1 km of cloud, and provides a meaningful sample. Its greatest problems are a tearing of the foil by the airstream, and corrosion or bending of the foil during storage. The procedure usually used to reduce the data involves photographing specularly reflected light off the back side of the foil and measuring projected images, a tedious and time-consuming task. A major source of error in the analysis lies in interpreting the shadowy images at the edge of the impressions. This leads to an uncertainty of about 10 percent in the drop size, which may be eliminated if calibrations and sampling are done systematically by the same individual. Automatic counting systems are now available to expedite data reduction.

The foil sampler technique can also distinguish ice characteristics, provided the number of lines is greater than 3 (750 μm). Data from the foil sampler which show remarkable resolution of detail in snow crystals are shown in Figure 4.35.

Several automatic techniques for the measurement of raindrops are under development. One is the impact transducer, developed by Sutherland and Booker[73] of Weather Sciences, Inc., Norman, Oklahoma. A drop impacting on the sensor head imparts a force to a piezoelectric crystal which in turn generates a pulse whose voltage is a direct measure of the momentum of the drop. Electronic analogue discriminators sort the pulses according to pulse height and electronic counters tally the pulses to produce a continuous raindrop distribution. In principle, the instrument is sound; but in operation, noise from acoustic and electronic sources has so far nearly defeated its purpose.

For the past several years R. Cunningham of AFCRL, Hanscomb-Field, Massachusetts, has continued to improve an optical scattering instrument for the measurement of individual raindrops. It presently is mounted in a wing pod of their C-130A aircraft (Figure 4.36). Through an optical system it illuminates the sample volume (viewing volume) with a plane of light, approximately 100 cm² in cross-section and about 5-mm deep. A photomultiplier tube views this volume at a scattering angle of 90° and produces electrical pulses corresponding to indi-

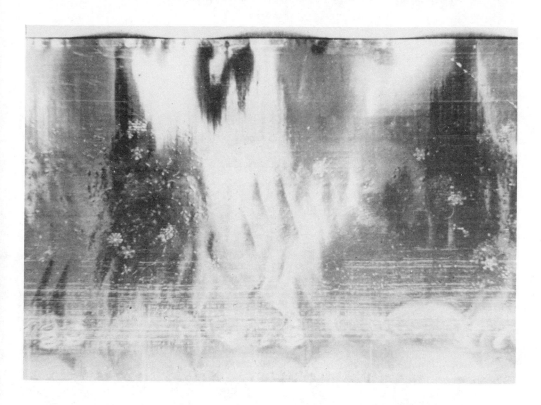

Figure 4.35 Impressions of stellar ice crystals on the aluminum foil impactor. Analysis of these data permitted an estimate of the fraction of ice in tropical storms (Scott and Dossett[69]) (photo courtesy NHRL, Miami, Florida).

Figure 4.36 Optical disdrometer with 100 cm² sampling area, mounted on AFRCL C-130 of R. Cunningham (photo courtesy of R. Holle, EML, NOAA, Miami, Florida).

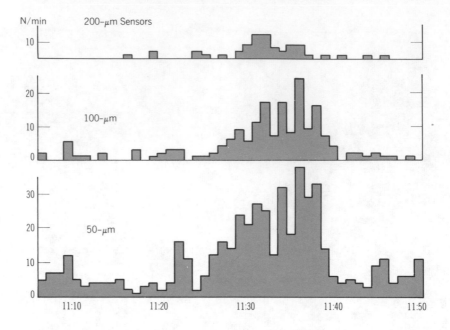

Figure 4.37 Data using the multispaced coils of Sasyo; these histograms show, from bottom to top, the time variation of number of drops counted by 50, 100, and 200 μm sensors for 1 min (reprinted by permission, from Sasyo[64]).

vidual particles entering the light. Electronic counters then sort the pulses. At present, it appears that the instrument will reliably detect raindrops above 0.5 mm in diameter. One of the major problems has been background light created by cloud droplets or pollutants. The effect is to raise the scattered light level and reduce the sensitivity of the instrument to raindrops. The great advantage of the instrument is its large sampling area (100 cm²), which is more than an order of magnitude greater than the sampling area of the foil impactor. Should it prove operational, it should be considered for use in operational programs.

Another raindrop-measuring device has been developed in Japan by a group at the Meteorological Research Institute of Tokyo.[64] Unfortunately, the instrument has only been tested on the ground at this time. In operation, it utilizes the finite electrical conductivity of raindrops. Several ceramic spools are bifilar wound with individual electrically separated wires. Each spool has a specific spacing between the turns. When drops above this limiting size impinge upon the spool, electrical contact is made between the two separate wires. A simple counting of the number of times that the wires are "shorted" together gives a direct measure of the number of drops above the limiting sizes of the various spools. The composite measurement from all the spools is the entire cumulative raindrop distribution. Data from the instrument are shown in Figure 4.37. The simplicity of the technique, the relative reliability of the system, and the sampling statistics attainable speak for its use in our operational programs.

THE CHARACTERISTICS OF AEROSOLS

Physical Properties

Aerosols and their precursors encompass scales all the way from molecular sizes (see Figure 4.1) to around 100 μm. They are small particles which can be characterized in part by the terms:

Background aerosol
Tropospheric aerosol
Stratospheric aerosol
Maritime aerosol
Continental aerosol

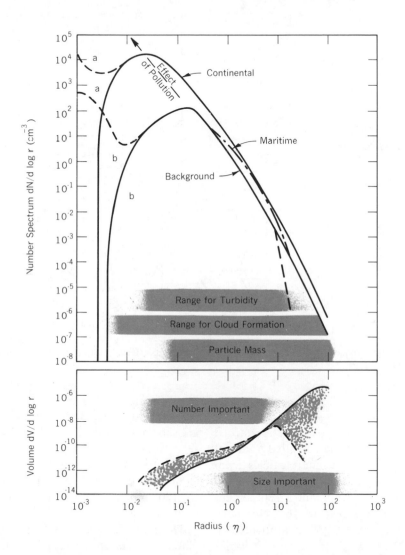

Figure 4.38 Typical comprehensive size distribution for the principal trospheric regimes and the size ranges important for turbidity, cloud formation, and mass concentration of particles. The lower curve presents the same data in terms of particle volumes, indicating the relative importance of particle number and size. The dashed curves a are distributions resulting from continuous formation of small particles; solid curves b without. (Adapted from SMIC/Inadvertent Climate Modification by permission of the MIT Press, Cambridge, Ma. Copyright 1971 by MIT).

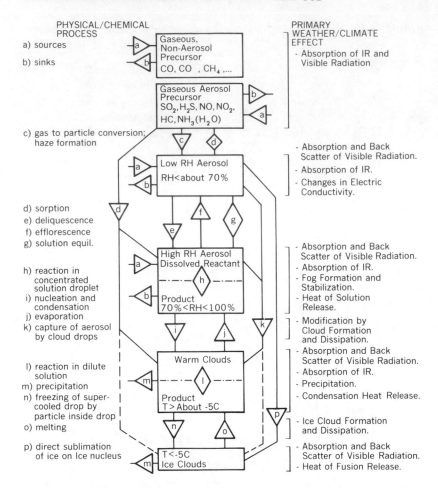

Figure 4.39 Tropospheric air chemistry and weather/climate effects (reprinted by permission, from Butcher and Charlson[6]).

The background aerosol is the general aerosol level in a relatively unperturbed environment. It is the so-called "natural aerosol" but may include particles from anthropogenic sources. In the past 50 years there has been about a 100 percent increase in its level in the Northern Hemisphere, and in 1972 it amounted to about 300 to 500 particles/cm³. It varies considerably near its sources but beyond 5 km or so above the ground is remarkably steady. Surprisingly, scavenging by raindrops or cloud droplets has little effect on the aerosol except in local areas. The overall spectra of these particles (number and volume spectra) are shown in Figure 4.38. The other terms refer to aerosols specific to certain levels or locations. Generally speaking, the tropospheric aerosol loading is about 10 to 100 times as large as that of the stratosphere.

The roles aerosols play in weather and climate effects are summarized in Figure 4.39. The aerosols are formed by one or more of the following processes: (1) photochemical reactions between trace gases, (2) absorption of gases in particles or droplets with subsequent chemical reaction, (3) aeolian (wind) effects on dust, surf, and so on, (4) homogeneous reactions, or (5) through interactions during scavenging processes. The particles can be inorganic (e.g., NaCl, $(NH_4)_2SO_4$, volcanic ash, or clay material) or they can be organic (soot, bacteria, or oil globules). But present measurements indicate that, though the background aerosol has a multivaried character, sulfates in combination with ammonia are the most common. As indicated in the figure, the relative humidity (RH) provides one of the controls on the form of the aerosol. Above 50 percent RH most aerosol particles contain some water on their surfaces due to their hygroscopic nature. At 80 to 90 percent RH the wetted particle sizes begin to affect visibility, and haze or smog layers form. Of course, a sufficiently large quantity of pollutants alone will also reduce the visibility. At 100 percent RH (or a little above) fogs and clouds form.

Cloud Condensation Nuclei (CCN)

The number of droplets that form in a cloud is determined largely by the number of cloud condensation nuclei (CCN) in the aerosol. These CCN are the special particles that serve as condensation sites on which each fog or cloud droplet forms at the relatively low supersaturations (generally < 1 percent) present in clouds or even lower supersaturations (< 0.1 percent) in fog. Figure 4.40 shows the effect of supersaturation on the measured number density of CCN in typical maritime and continental air masses.[76]

The importance of aerosols in weather and climate effects is evident in several situations, most import of which is the effect on the colloidal stability of clouds. Maritime clouds with 50 to 100 CCN/cm³ active at the cloud supersaturation (about 1 percent) often develop

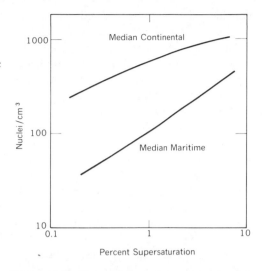

Figure 4.40 Comparison of continental and maritime CCN which were activated at various supersaturations typical of clouds (courtesy NRL, Washington, D.C.).

enough precipitation to show on 10-cm radar in 10 to 15 min after start of vigorous growth; usually the clouds produce rain before even reaching the freezing level. On the other hand, clouds in continental air masses having 1000 to 2000 CCN/cm³ active at even lower supersaturations (about 0.5 percent) frequently require an hour for the smaller droplets to grow to precipitation sizes. The large difference in number density of CCN between maritime and continental air is caused primarily by differences in the natural sources of the CCN. However, in localized areas pollution effects on CCN can also be important. For instance, if the number of CCN over an area were increased, the reflectivity or albedo of the clouds would increase. If the albedo of clouds were increased by 50 percent, this could cause approximately a 1 percent increase in the cooling rate of the earth, an immediate effect equivalent to the heating which might be contributed by the atmospheric buildup of CO_2 at the year 2000. This calculation doesn't consider the increased cloud cover and thickness that might accompany such a change in CCN, or on the other hand, the possible decline in clouds and rainfall because of reduced surface evaporation at the resultant colder temperatures. On a smaller scale there is already evidence of increased cloud cover in polluted areas, sometimes with decreased, but more often with increased rainfall, probably because of the increased CCN reducing the tendency for rainout from small clouds and allowing them to grow over the heat island of the city into cumulonimbus (Cb) that are effective rain producers.

In the weather modification field CCN play important roles in that (a) their addition to clouds and fog (usually as "giant" nuclei $> 1 \mu$m) is the primary method so far employed for modification at temperatures warmer than 0°C, and (b) there is considerable evidence that the potential for precipitation augmentation by AgI seeding is greater when the number density of the CCN (hence cloud droplets) is large. This effect results from several factors: (1) because of the greater colloidal stability of those clouds which develop with a large number density of CCN more time is available for seeding before the cloud is raining naturally; (2) the droplets are closer together and so have an increased probability of contact nucleation; that is, the ice nuclei reach the droplets more easily, which may result in a more rapid release of the latent heat of fusion; (3) the larger surface area of the droplets present with numerous, small droplets increases the rate of evaporation of the liquid phase, leading to a speed-up in the sublimation to ice by the Bergeron-Findeisen process, thereby also speeding the release of the latent heat of fusion; and (4) if secondary ice embryos are formed naturally by a mechanism involving large hydrometeors, this process would be slower in the continental type cloud in which the large hydrometeors

form more slowly; therefore, it is less likely that these clouds will be seeded by nature before the AgI can be introduced. Field measurements to prove the validity of all these mechanisms are difficult and are still in progress, but if the major mechanisms for changes are validated, they should improve our capability for modification by AgI seeding.

The measurement of the number of CCN may be done using a thermal gradient diffusion cloud chamber of the general design of Twomey based on earlier instruments by Langsdorf and Wieland.[38, 75, 81] In the chamber there are two parallel plates, separated by a relatively small spacing (1 to 1.5 cm). The top plate is heated or the bottom one cooled so that there is a temperature difference between them, and a strip of damp blotting paper covers each of the two opposing surfaces. This arrangement establishes a linear profile of temperature and water vapor pressure through the effects of diffusion of heat and mass. Owing to the nonlinear dependence of the equilibrium vapor pressure of water on temperature, however, the vapor becomes supersaturated approximately in the center of the space between plates (or a little above center). Then, when a sample is introduced between the plates, a cloud of droplets forms. The number of droplets formed is counted on a photomicrograph of the sample volume and is equal to the number of CCN active at the supersaturation present in the diffusion chamber.

A more automatic instrument* is the one developed by Radke (Figure 4.41).[17, 56, 57] This instrument uses for detection a modification of the integrating nephelometer of Charlson[9] arranged to look at the droplets formed in a diffusion cloud chamber. The number of droplets is infer-

* Available commercially form MRI, Altadena, California.

red by assuming that the cloud droplets grow to known size at specific times after sample introduction, based on a measurement of the total amount of light scattered in the sampling volume. The operation of the instrument is automatic with sequential sampling, equilibration, measurement of scattered light, and recording of the inferred CCN number density; however, the instrument has two basic failings: (1) it is affected adversely by the presence of a large amount of background aerosol, and (2) the mean droplet size at the instant of measuring the scattered light is affected by the humidity of the sampled air and by the composition of the CCN, both of which influence the equilibrium time. Continuous-flow types of CCN diffusion chambers with automatic readout are being developed at Desert Research Institute, Reno, and at NRL.

Another instrument, widely used since the turn of the century before the advent of thermal gradient diffusion cloud chambers is the "Aitken particle counter," "total condensation nucleus counter," or simply "condensation nucleus (CN) counter" (not to be confused with CCN counter). Some models of this instrument use a detection technique not unlike that of the integrating nephelometer; others use optical extinction. However, this instrument operates at high supersaturations, about 300 percent, so that it counts the total number of aerosol particles. One model of the instrument presently available is hand operated and the output from a photocell is read on a dial which, when calibrated, gives a

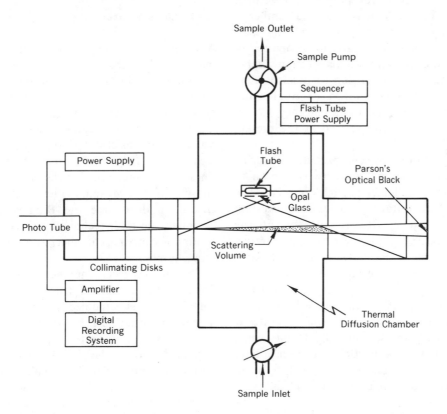

Figure 4.41 Schematic diagram of Radke diffusion cloud chamber with integrating nephelemeter used to measure number density of CCN.[56]

reading of numbers of CN. The technique has been automated for rapid sequential sampling by General Electric Company and by Environment, One, Inc., both of Schenectady, New York, and is quite usable in some aircraft operations. At the smaller concentrations (below about 300/cm³), however, the device becomes inaccurate. With such concentrations a large chamber is required and the results must be interpreted by reference to an instrument such as that developed by the University of Missouri at Rolla using a photomicrograph to count the particles.

Remember that Aitken nucleus counters, even when used at low expansion ratios, do not provide a measure of nuclei (CCN) active at the low supersaturations found in nature and should not be used for this purpose.[60, 61, 75]

At this time, for purposes of weather modification, a measure of the number of CCN (with a diffusion cloud chamber) and the integrated scattering properties of the aerosol (with an integrating nephelometer) should suffice to describe adequately the aerosol. (Ice nucleation abilities are important, but they cannot be properly characterized at the present time.)

If a more advanced particle measuring system is desired or there is a need to consider the larger particles in relation to the "precipitation nuclei," the system developed by R. T. Whitby at the University of Minnesota is attractive. This system basically consists of three counters with overlapping ranges. Particles larger than 0.3 μm are measured using one of a number of counters available commercially which operate by measuring the scattered light that occurs when individual particles intercept a light beam. In the range 80 Å to 0.5 μm an electrostatic particle counter* is used.

* Whitby Aerosol Analyzer is manufactured by Thermo-Systems, Inc., Minneapolis, Minnesota.

The device charges the particles negatively and collects a portion of them on a positive electrode. Those particles with mobilities less than a preselected critical value (larger particles) remain in the air and are collected downstream on an absolute filter. The negative current from the filter is measured by a sensitive electrometer. Varying the voltage on the capturing electrode gives a measure of the different mobilities (and therefore particle size). If the charging of all particles is uniform, the instrument is directly calibratable in terms of size and has been tuned by Whitby's group to give excellent results. A third instrument, an automatic Aitken particle counter, helps to establish the calibration of the electrostatic counter.

Ice Nucleus Measurements

As mentioned earlier, recent information indicates that the concept of ice nucleation is so ill defined that no single instrument simulates all the types of ice nucleation. However, the importance of ice nucleation to weather modification necessitates a brief description of present techniques for ice nucleus measurements. These techniques are, for the most part, considered by Mason,[44] Fletcher,[19] and Grant[22] and include the Bigg-Warner chamber, the filter technique, the NCAR continuous mixing chamber, and the settling chamber.

The Bigg-Warner chamber is a cylindrical chamber cooled below 0°C and containing a supercooled sugar solution in the bottom.[79] In some models the sample is taken under pressure and expanded, cooling the air and forming ice crystals on the nuclei. The small crystals fall into the supercooled sugar solution, grow to visible size, and are counted. Variations of the technique include expansion of the air sample into a vacuum and the use of a sausage skin filled with warm water to form a cloud in the chamber without expansion.

A promising modification is the technique adopted by the Japanese and discussed by Ohtake.[53] He uses a warm sponge in the top of the chamber to supply vapor for a cloud which settles through the cooled air sample, an experimental configuration called a settling chamber. The chamber holds approximately 10 liters of air. Large chambers of 1-m³ volumes are also commonly used as settling or expansion chambers to measure the nuclei produced by artificial seeding materials. The output is the number of nuclei produced at a given temperature per gram of seeding material—the so-called nucleation efficiency.

The filter technique is perhaps the most widely used today.[22, 48, 72] The aerosols are caught on the surface of a filter and later processed by exposing the filter to an atmosphere which is supersaturated with respect to ice. The technique is relatively free from apparatus error but does suffer from migration of nuclei into the pores and from the masking effect, the covering of active particles by inactive ones or the depletion of vapor by hygroscopic substances. In developing the filter, care must be taken that the supersaturation is controlled with no wet areas warmer than intended. One instrument that achieves good control was developed at the Hebrew University, Jerusalem.*

The automatic counting technique is the most attractive from the viewpoint of field operations. The NCAR counter supplies this capability using a continuous-flow mixing cloud chamber.[36] The sample air is cooled in the cloud chamber to form

a cloud where the particles freeze and fall out. The particles then are sucked down through a small capillary at the bottom of the chamber. The capillary size is such that the airflow becomes sonic with an associated sonic front. This front is disrupted when the particle passes, producing a sonic pulse. This pulse is detected by a microphone upstream and counted as a measurement of an ice nucleus. The mixing cloud chamber suffers from an inability to maintain a given supersaturation; therefore this instrument can activate, deactivate, or otherwise eliminate nuclei. Similarly, the expansion chambers tend to produce unrealistic supersaturations and remove smaller aerosol particles either by dissolution or capture by the wall.

None of the techniques can measure the nuclei that are active by all three mechanisms. Spraying droplets on the filters has been used as a technique for simulating contact nucleation, with only partial success. Freezing nuclei are measured by counting the number of drops (formed from precipitation) which freeze at a given temperature[78] But this technique also does not properly simulate the history of the nuclei. It appears that an appropriate measurement of ice nuclei which cause freezing by the different mechanisms is presently beyond the scope of our weather modification efforts; hence at this time we should put our main emphasis on measurements of ice crystal concentrations actually present in a cloud.

Chemical Properties of Aerosols

Further understanding of the way in which cloud condensation nuclei (CCN) and ice nuclei (IN) act requires an understanding of the chemical composition of the particles. Today the chemical composition of some types of particles can be

* Thermal Diffusion Ice Nucleus Counter available from Maimahd Scientific Instrumentation, Rishon Le-Zion, Israel.

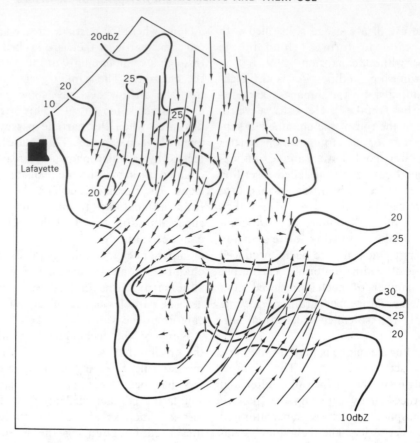

Figure 4.42 Particle horizontal motion field inside a convective storm observed at a mean altitude of 400 m by a dual Doppler radar system. Although there is a more extensive region which is covered by one of the radars, the dual Doppler observations are limited to regions covered by both radars. The echo intensity is also shown by contour lines. The contour labeled 30 refers to precipitation intensity of 15 mm/hr. Note the strong convergence between the two radar cells and vorticity in the southwest part (courtesy R. Lhermitte, University of Miami, Florida).

measured; noteworthy among techniques are the electron microprobe, the flame spectrometer, and the use of sample pretreatment. The electron microprobe measures the chemical composition of individual particles > 0.5 μm by irradiation with an electron beam. The wavelengths of the emitted X-rays are a measure of the atomic composition of the aerosol particles.

The use of flame emission spectrometry has been proved operational for sodium-containing particles $\gtrsim 0.06$ μm.[57] It uses an acetylene flame which draws in

ambient air by aspiration. The mixture burns the air with the particles, and each sodium particle emits a pulse of orange, sodium-d light. A photomultiplier tube detects the light pulses, which are sorted and counted and converted to equivalent particle sizes.

The pretreatment technique deduces chemical composition by sample processing. For example, Twomey measured CCN remaining in samples after preheating.[77] Ordinary salt (NaCl) is quite heat resistant, whereas a common constituent of aerosols, ammonium sulfate, $(NH_4)_2SO_4$, decomposes below 400°C. The results of flight experiments indicate that most CCN are volatile even over the ocean, and they are most likely ammonium sulfate.[15]

REMOTE MEASUREMENTS

As the field of weather modification progresses, there will be an increasing need for more detailed control and monitoring; but even so, our successes often rely on subtle changes. Hence we see the continuing need for more detail in our measurements with increased statistical validity. In situ measurements are limited physically in location and time as well as sample volume. Therefore, remote measurements often offer the most meaningful solution. Every day we see the value of the weather radar; in any operation radar PPI and RHI data can provide the primary semiquantitative indicators of storm activity and changes from modification. Further quantification

of the data using calibrated radars with isoecho contours should be an essential goal in every modification project.

In addition, ultrasensitive radars may be used and can be made sensitive to backscattering originating in the turbulent atmosphere itself. Dual Doppler radars as used by R. Lhermitte at the University of Miami can sense the three-dimensional wind field (Figure 4.42), small-scale variability, and its time development whenever precipitation particles are present to provide a radar target.[40, 41] In Figure 4.43 is shown a vertical profile of precipitation-particle fall velocities observed by a vertically pointing Doppler radar. In this figure the trace brightness is a measure of the amount of radar signal intensity in the Doppler frequencies corresponding to the various hydrometeor fall velocities shown in the ordinate scale. When the ice or snow is falling from the higher altitudes (right-to-left in the figure) the fall speed becomes sufficiently well organized to give a usable signal at about 10 km. At that

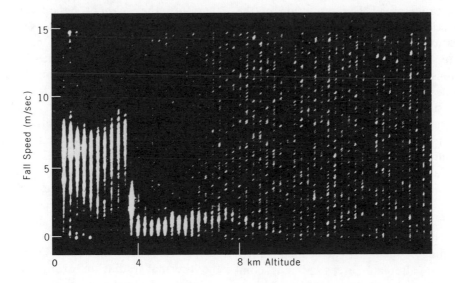

Figure 4.43   Range-velocity display of vertically pointing Doppler radar. Ice crystals (shown falling from right to left, 300 m/sweep) produce a detectable return signal at about 10 km with 1 m/sec velocity, increasing to 2 at 8 km. Below the freezing level, 4 km, the fall speed increase to 9 m/sec, indicating the presence of raindrops (courtesy R. Lhermitte, University of Miami, Florida).

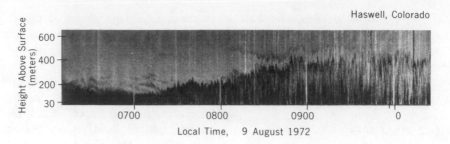

Figure 4.44a An example of the breaking up of the temperature inversion as detected by the acoustic sounding technique (courtesy G. Little, Wave Propagation Laboratory, NOAA, Boulder, Colorado).

altitude the speed is about 1.5 m/sec, increasing to 2.5 m/sec at 8 km. At the melting level, about 4 km, the fall speed quickly increases to 9 m/sec.

An acoustic sounding system has also proved itself in sensing the temperature and velocity structure of the atmospheric boundary layer. When the acoustic Doppler measurements were compared with measurements made from a tethered ky-toon (called a Boundary Layer Profiler (BLP) and operated by NCAR), the resulting comparison indicated the BLP values contained a slightly positive systematic error.[2] The resolution of the acoustic technique to air temperature variations is shown in Figure 4.44*a*. Its value in mapping wind profiles is demonstrated in Figure 4.44*b* showing profiles obtained from three separate monostatic systems (single units in which the same reflector serves both the transmitter and receiver). The temperature profile, showing multiple inversion layers, was obtained with a vertically pointing monostatic system.

The temperature profile can also be obtained by a combined radar and acoustic technique, called the radioacoustic sounding system.[54] The sound waves produce alternate areas of air compression and expansion, which in turn alter the electromagnetic index of refraction of the air. An optimum effect is obtained when the radar wavelength is twice the acoustic wavelength. The big advantage of the combined use of acoustic and microwave active sensing is its lack of dependence on the existence of either a precipitation target or a pronounced variation of temperature or wind.

Operations of both the microwave (radar) and the acoustic techniques are possible in cloud or fog. Strong surface winds or heavy precipitation tend to make the present designs of the acoustic systems unusable. High ambient noise temporarily disables acoustic systems; therefore they cannot be operated on board aircraft. The acoustic sounding system has not as yet been proved effective above an altitude of a few hundred meters, but the modification used by Marshall has been shown to operate consistently to an altitude of 1.5 km.[54]

Microwave sensing can also be used in the passive mode. In this case the microwave brightness temperature is observed along a line integral, or through an air column. Utilizing, for instance, emissions of oxygen in the 60 GHz intense band of oxygen, the integrated temperature of the column can be detected because the temperature determines the intensity of the thermal radiation of the oxygen. Scanning at a series of angles from the horizontal to the vertical, it is possible to obtain a full temperature profile (sounding). The technique is cumbersome mathematically and, even so, does not yield a truly unique solution. Nevertheless, the technique will give a profile of temperature with a root-mean-square error of about 1°C up to about 3 km in clear weather.

The presence of liquid water alters the measured brightness temperature due to the significant energy absorption by water molecules at 60 GHz. Indeed, the liquid water drop emissions (with Mie scattering) in the microwave region have been used as a measure of liquid water[14] and show promise of presenting a measure of total liquid water integrated over a column of indefinite length. Fortunately, the liquid water measurement is nearly independent of the particle size spectrum, since scattering is not important at the wavelengths used. Unfortunately, the technique hasn't yet been able to detect the difference between water and ice. Also, the emission saturates in very thick clouds because of attenuation in the storm front, so that one measures the blackbody (physical temperature) of the cloud instead of the liquid water content.

The use of multiple frequencies may, in part, alleviate these difficulties.

Optical (lidar) probing techniques have also been used quite successfully to detect the presence of layers in the atmosphere, and can be used, for instance, to detect the height or slant range of cloud base. Lidar probing techniques are particularly sensitive to particulate matter density. Lidars using Raman scattering have the unique capability of identifying atmospheric gases and slant visual range through aerosols or light fog, but this technique is limited by background noise from scattered sunlight.

Of these techniques, optical and acoustic probing are of limited use in dense cloud because of absorption and poor penetrating ability; however, acoustic sounding can sometimes be used to locate the top of fog bands, because of the temperature discontinuities. Radar techniques have the unique capability of observations through dense storms and, indeed, are useless without the presence of precipitation particles.

Infrared remote sensing techniques

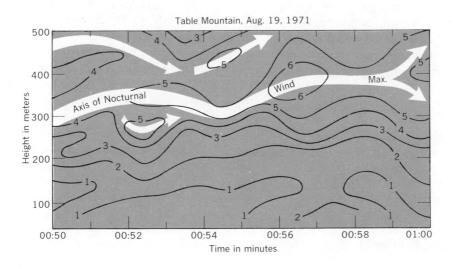

Figure 4.44b Time section of isotachs for horizontal wind (m/sec). Winds were derived solely from acoustic Doppler measurements using three separate monostatic systems (courtesy G. Little, Wave Propagation Laboratory, NOAA, Boulder, Colorado).

may also be of great value in profiling the temperature of the cloud surface and even obtaining atmospheric soundings by the use of multiple-wavelength spectrometers operated in a manner similar to that described for the passive microwave techniques. It is possible, perhaps, that a temperature change accompanying a modification effort might be sensed by viewing the seeded portions of the cloud from above with an IR remote sensor. This procedure is complicated by cooling at the cloud edges and, when the modified, more buoyant parcel rises, cooling by expansion.

In summary, we see that remote sensors offer hope for the future in the field of weather modification, though most operational instruments may yet be many years in coming. The remote measurements can be made over an indefinite column or can be averaged over a given line or path, or given as a profile over the entire path. A measurement can encompass both the large scale and the fine structure, or can measure the spectrum of the fluctuating components or the three-dimensional flux.

Whether velocity, temperature, humidity, aerosol concentration, or hydrometeors are measured, the techniques offer the unique ability to scan and measure over an extensive atmospheric region. Wherever practical we should work toward substituting remote in place of in situ measurements. As the techniques develop and achieve the range and resolution required for our efforts in weather modification, remote sensing techniques should place our research on a whole new frontier.

THE MEASUREMENT OF ELECTRICAL PARAMETERS

Several atmospheric electrical quantities can be utilized as detectors of effects of weather modification. In a program to reduce lightning, measurements of electric field and local field changes produced by lightning are essential. Also, polarity reversals in the electric field and intensification may indicate the presence of ice. Since small ions migrate to the larger aerosol particles, a decrease in the electrical conductivity of the air caused by loss of the small ions may be used to indicate the presence of aerosol or early stages of fog formation. For further information on the measurement of electrical parameters, the reader is referred to Israel in 1970[29] and Fitzgerald in 1965.[18]

The standard techniques for measuring electric field are the field mill and the radioactive probe. The field mill consists of either a rotating cylindrical shutter or a set of flat vanes. These vanes are conducting surfaces which are insulated from local ground except through a resistance-capacitance (RC) network, and are alternately shielded and exposed to the ambient field by the shutters. The induced charge traveling to and from the surfaces as they are shielded and exposed is detected, amplified, and rectified in a phase-sensitive circuit to produce an electrical output which is proportional to the ambient electric field. A set of six flat-vane type field mills mounted symmetrically on an aircraft allows measurements of the three components of the electric field. Alternatively, the difference signal between two mills mounted axially on each wing tip gives the lateral component of the horizontal field. If the horizontal field is zero, the vertical component is determined by flying in a turn at a calibrated bank angle. The vertical and lateral components may also be measured by a single cylindrical mill on the nose of the aircraft.

The radioactive probe electric field meter uses a radioactive source to ionize the air and bring the probe to the potential of the air. It is quite reliable but lacks the dynamic range of the field mill; it is often used for airborne and ground-based measurements because of its low cost. On aircraft, two radioactive sources are required to measure one component of the electric field; the spacing between the sources should be at least 0.5 m to minimize the effect of changes of the effective volume of the ionized cloud around the sources with changes in wind.

A crude indication of electric field is possible by measuring point discharge current from sharp points aligned in the direction of the field component of interest. In still wind conditions, the voltage between the points is proportional to the square root of the point discharge current. With wind present, the point potential is roughly proportional to the ratio of the current to the wind speed.

The electrical conductivity of the air due to the number density and mobility of small ions can be measured with a Gerdien ion meter. This is a device which collects ions on the center electrode of a cylindrical capacitor. Across this capacitor is imposed a radial field which is maintained at a level small enough so that the current from ions migrating across the space is proportional to the voltage. Air is drawn through the radial field and, depending on the sign of the potential placed between the electrodes, ions of the opposite sign are collected and the small resultant current is measured.

Space charge density is another item of interest. It may be determined by measuring the electric field inside a screened Faraday cage and applying Poisson's formula. A second method is to measure the charge collected on an aerosol filter. However, the only practical method of measuring space charge on aircraft in clouds is by measuring the gradients of the electrical field and calculating the space charge density using Poisson's formula. For sources of the field located at a considerable distance from the aircraft this technique may result in gross errors.

AIRCRAFT SYSTEMS FOR DELIVERY OF SEEDING AGENTS

While delivery systems are not a part of weather modification instruments, the same personnel are frequently responsible for instrumentation and delivery systems. Therefore a brief summary of seeding methods is presented.

Three types of airborne silver iodide (AgI) delivery systems are in common use: burners, flares, and hoppers. Burners are used primarily for horizontal seeding (usually at cloud base in updrafts). The same burners can also be used to distribute tracer materials, either radioactive or with distinctive chemical characteristics.

The Lohse burner, the MRI Skyfire burner,* and the Patten burner † are used commercially.[55] Most burners are run at a single temperature, but the Patten burner has provisions for adjusting its temperature from 700° to 1200°C. Burners of all these types can operate with sodium iodide-silver iodide-acetone or ammonium iodide-silver iodide-acetone. The sodium iodide mixture can be burned over a wide temperature range without change in ice nuclei activity and produces complex hygroscopic crystals; the ammonium iodide mixture produces simple crystals of silver iodide. Though the measurements of ice nuclei are certainly not absolute and can be used only for relative, qualitative judgments, indications are that the ammonium iodide is more effective at warmer temperatures (close to 0°C) and that the sodium iodide mixture loses its effectiveness if in-

* In use by the U.S. Forest Service.
† Developed under the auspices of the Atmospheric Physics and Chemistry Laboratory, Boulder, Colorado, and Research Flight Facility, Miami, Florida.

jected into the cloud at temperatures above 0°C.*

Pyrotechnic silver iodide flares are manufactured by a number of companies: Olin Weather Systems, East Alton, Illinois; Colspan Environmental Systems, Boulder, Colorado; Atlantic Research Company, Alexandria, Virginia; and Lee Wilson Engineering, Holister, California. Flares are also produced by the Naval Ammunition Depot, Crane, Indiana. The Olin 1055 flare is used in the Florida cloud seeding project conducted by EML, Miami, Florida.[70] Project Stormfury uses a variation of the LW-83, high silver iodate pyrotechnic flare† manufactured by the Navy's Crane, Indiana facility. All producers can furnish two types of flares, the dropping flares and the trailing flares (end burning). The dropping flares are used for vertical drops from aircraft and are similar to a shotgun shell or a flare-pistol cartridge. The end-burning flares are similar to warning flares (fusees). Flares contain silver iodate with or without an auxiliary oxidizer such as potassium nitrate, together with varying amounts of aluminum, magnesium, and synthetic resin binder. The sequence of dropping flare burn starts with an electrical activation which ignites a match-mix, which ignites an expulsion charge, which in turn ejects the cartridge and ignites the silver iodate charge. Dropping flares are intended to be dropped directly into the growing updrafts and to seed the cloud over a kilometer or more of vertical depth. Burner seeding is intended to be gradual and more controlled.

Hoppers are used to dispense materials in a solid form; the dry ice dispenser normally used is similar to the ice chopper used to crush ice from a home refrigerator. For modification of warm fog and clouds, dry salt or urea is commonly dispensed from a hopper with an auger feed, though it may be pumped in a solution (e.g., urea-ammonium nitrate) to nozzles in the wings. (Distribution from the wing tips can use the wing-tip vortices to help mix the material into the air.)

Other methods of ice phase seeding also under development include an expansion nozzle with a high expansion ratio to create ice by homogeneous nucleation at very low temperatures. Dispersion of liquid propane for supercooled fog modification is used operationally at Orly Airport, Paris, France, and at several U.S. Air Force installations. In Colorado, experiments are underway by the Denver Research Institute to seed with metaldehyde using steam. Metaldehyde is one of the organic materials which is reported to have a high temperature threshold of nucleation. Also, Colspan Environmental Systems presently manufactures small, light rockets to be launched from ground or aircraft. The rockets are intended to burn at a preselected area in the supercooled cloud and release the nucleating material. Rockets are presently in operation in the United Soviet Socialist Republic and are claimed to be an effective delivery system for use in hail suppression. In the United States, hazards to aircraft and the possibility of rockets falling back to earth have precluded their general use.

In the last few years work on the development of new fuels and delivering systems for cold- and warm-cloud seeding and tracer studies has increased. Generally, cost comparisons show that the cost of using burners is presently about one-fourth the cost of equivalent flares. The unique seeding characteristics of droppable pyrotechnics, however, warrants the increased cost for specific applications. However, each delivery system

* Ice nucleus measurements are not cross-comparable. That is, comparisons of data from different instruments or under different conditions are likely to contain gross errors.
† Developed by the Naval Weapons Center, China Lake, California.

supplies unique operational capabilities and the cost of seeding agents is usually secondary.

SYSTEMS DESIGN

Lacking in almost all of our programs is the concept of systems design, the design of an integral system to make a specific measurement. This concept is closely related to the concept of representativeness of the sample. Every portion of the measuring chain—the sensor, the amplifiers, the data processors, the recorders—must retain the integrity of the sample in spite of electronic noise, acoustic noise, and severe environments of high temperatures, humidity, wind, icing, and salt spray. Many systems (e.g., pressure sensors, thermometers) without electronic output have operated successfully for decades and because of their simplicity and inherent reliability should never be abandoned. However, the great need for real-time output and the present capabilities to achieve it with electronic digital and analogue techniques indicate a great importance of electronics in our future work.

In these systems all-pervading electronic noise is a problem and requires consideration on an aircraft whenever an alteration in instrument configuration is considered; a power supply is relocated; or the instrument is located close to such noise producers as radios, radars, motors, solenoids, etc. This is because electrical noise is generated by near fields and far fields (radio waves). Radio (plane) waves radiated by intentional and unintentional sources are transmitted, reflected, reradiated, and absorbed as unwanted noise. Fortunately, however, most transmitted radio waves are of a high frequency (commonly called radio frequency interference, RFI) and the penetrating power decreases with increasing frequency so that good copper shielding and the use of good grounding practice can virtually eliminate RFI noise.

Noise generated by near fields is not so easily eliminated. Such noise is basically of two types: (1) high-frequency (>10 kHz) electrostatic noise, and (2) low-frequency (< 1 kHz) magnetic noise. The electrostatic noise is relatively easy to eliminate and can usually be removed by shielding with copper or aluminum, but the magnetic noise is eliminated only by ferromagnetic materials of high permittivity (two materials are NETIC and CO-NETIC, manufactured by the Perfection Mica Company). Fortunately, the magnetic noise is greatly attenuated by distance, and diminishes as the inverse third power of the distance, compared with an inverse second power attenuation of electrostatic noise.

Practically speaking, the best solution to the electrostatic noise problem is the use of shielded wires, but magnetic noise is best eliminated by physical removal of the noise generator. Electrostatic noise is generally high impedance, being typically around 10^7 MΩ or greater. This means that live pickup can be minimized by termination in as low an impedance as possible. Magnetic noise, on the other hand, is low impedance and, for instance, is a particular problem with thermocouples. In our systems, high impedances are usually associated with high voltages and lower frequencies, whereas low impedances are associated with low voltages and high frequencies. Typically a high impedance instrumentation amplifier would have an impedance of 10^5 to 10^{13} Ω and be responsive above about 10 mV, whereas a thermocouple would supply a total of 100 μV at 10 Ω. In all design, size is an important consideration. Noise-collecting surfaces can be eliminated, for instance, by building an amplifier, power supply, and all-associated hookup within a well-shielded sensing head—with only the output, at low impedance and high voltage levels, connected to the recording and processing equipment. Close coupling is particularly important for charge detectors and thermocouples, where the slightest charge (caused, for instance, by triboelectric effects within the cables due to twisting vibration) can greatly reduce the signal-to-noise ratio of the system. Here, installation of a high impedance field effect transistor (FET) or one of the microamplifiers can produce an output at a low impedance and a high level.

Power supplies also are not always simply a matter of commercial acquisition of a power converter of high stability. Regulated power supplies compensate for low aircraft voltage by increasing their input current, thereby increasing the load on the plane's system momentarily so that its voltage drops further until its regulator brings it back up, then the reverse effect occurs. This sequence can occur rapidly enough to appear as noise pulses throughout the instrumentation. When the repetition frequency of electrical noise pulses gets into the proper relationship with the chopper frequency of an instrument amplifier, the data may become highly erroneous. Often an inexpensive dry battery can give a lower noise circuit, but they have inherent drift and cannot be used in applications needing appreciable current drain. The recording system also requires consideration because few instruments have compatible outputs without some minimal signal processing and averaging. Basically, three types of signals are found in meteorology*: (1) those signals of slow variation and little electronic noise that can be recorded on a slow recording system (i.e., temperatures, winds, pressures; (2) those signals of high variation that have appreciable changes in less

than 1 sec; and (3) those varying signals with high variation which require recording rates of 1 msec or less.

Slowly varying signals are averaged in time in such a way that the average is compatible with the physical interests of the researcher. The signals are recorded at a time interval at least twice as fast as the required averaging time. The averaging is necessary to smooth out any signal fluctuations due to noise and supply a truly representative signal over the time interval. In this regard, for instance, when the dew point T_d of the air is measured, the average water vapor density is proportional to the average of $\exp(-T_d)$ since the vapor density is an exponential function of the temperature. Generally, however, a large number of measurements of the dew point are made and it is assumed that an average dew point can be used to calculate the vapor density. Considering the exponential, we see that, with electronic noise added to the signal, averaging in this way can bring gross errors into the measurement. It is far better to calculate the vapor density as the measurements are made, taking into account the type of noise expected and such things as the time constants of the instruments by using analogue techniques or the modern digital computers.

Signals of the second type are even more subject to errors introduced by these procedures and generally cannot be recorded properly on our airborne systems even with sophisticated recording techniques. Direct recording of such rapidly fluctuating signals during a flight of several hours can produce dozens of magnetic tapes which are likely to be overwhelming. Measured quantities with these relatively stringent response requirements include turbulence spectra, eddy fluxes, and liquid water measurements.

Signals of the third type, if recorded

---

* This classification obviously is scale dependent.

as raw data, would require an entirely excessive storage capacity. It is more practical to proceed with on-board data processing before recording. The measurement of individual electronic pulses from cloud particles is an example. In this case, it is of first importance that decisions be made regarding whether a given signal is noise or real, utilizing such characteristics as pulse rise time, fall time, and pulse height. Afterward, a tally of real pulses is made, and finally a spectrum (histogram) of pulse sizes is recorded.

Most important is that every portion of the process from sensing to recording of atmospheric information be considered, including effects of the fuselage in disturbing the sample before being sensed, the amount of sample acquired at the sensor, the response of the sensor to the variable or sizes of hydrometeors, the lead wires and impedances, as well as the interfacing with signal conditioning, processing, monitoring, and recording devices.

An example of instrument system integration is the use of a minicomputer for real-time data processing on board aircraft. Sierra Research, Inc., Boulder, Colorado, operated a Data General Nova minicomputer on board a Cessna aircraft to process and display the radar isopleths, aircraft track, and wind vectors in real time for the purpose of directing the operation. The radar signals are digitally averaged over 20 to 100 pulses to increase the signal-to-noise ratio; then they are processed into 100 to 1000 range gates (corresponding to isoecho contours).The computer stores the complete data in a set of arrays for 360° of azimuth, converts the data from polar coordinates, and, allowing for variation of signal with range, corrects the data to true Z values. Also, the computer continuously monitors the performance of the radar and

adjusts the receiver power to allow for changes in calibration. Finally, the computer generates the selectable isopleths of the radar return and displays them on a cathode ray tube.

Following this trend in the use of minicomputers, it may be possible in the near future to feed the computer aircraft data acquired near the cloud to upgrade the sounding data and actually run a model calculation in flight to guide the operation. This type of approach may be of particular importance in the case of warm-cloud seeding or wherever predictions of the updraft profile are necessary to optimize the seeding operation.

Instrument Calibration

It is contrary to the basic philosophy of science to use an instrument of unknown capabilities to measure an unknown quantity, yet we in the atmospheric sciences commit this type of heresy without a second thought. Before data can be said to have any quality they must be calibrated with the known. In particle measurements, this means supplying the instruments with particles of known size and character at a known velocity. The velocities are aircraft velocities, and the sizes cover a wide range so that usually size calibration and appropriate velocity cannot be obtained simultaneously. A compromise procedure is to supply the instrument with a stagnant sample in the laboratory and check the response. Then the instrument is moved to a natural environment and tested, say, on a mountaintop with real hydrometeors. Finally, if the instrument shows the desired qualities, it is tested on board aircraft in real clouds.

Complete testing of most instruments in a simulated environment is nearly impossible with our present facilities. For

Figure 4.45 Whirling arm facility. Aircraft engine is used only for speeds generally above 500 knots top speed. Only the automotive engine (foreground, below rotor housing) normally drives arm up to the speeds of most cloud physics aircraft (courtesy NRL, Washington, D.C.).

instance, Cornell Aeronautical Laboratories, Buffalo, New York, or the Fog Facility at the State University of New York, can supply a stagnant fog with known characteristics, but they cannot supply a calibrated fog at aircraft velocities. Natick Laboratories in Massachusetts can supply temperatures down to −70°C and air velocities to 40 mph, but they cannot supply a fog of known character and their chamber is too small to accommodate crystals or raindrops with appreciable fall velocity. The Air Force Arnold Test Center in Tennessee has impressive wind tunnel capabilities. It can supply fog, rain, and snow at speeds up to sonic velocities in a chamber as large as 16 ft. across; however, the facility is intended for the testing of aircraft engines, is in frequent use, and costs $1,000/hr. An additional problem with any wind tunnel is that changes of fog characteristics with speed are as difficult to calibrate for the tunnel as for the instruments being calibrated. The fog droplets evaporate or condense because of aerodynamic temperature changes.

There is a real need for a national facility for the testing of instruments. No single facility now can incorporate all requirements, as it would have to have a size comparable to that of a small cloud to accommodate the residence times of the larger particles and simulate the turbulence and air motion. An adequate facility would have to contain a high-speed vehicle to move the instrument through the cloud. Further, it should have the capability of providing cold and warm environmental conditions to simulate the thermal conditions of the cloud. These conditions might partially be met by building a climate-controlled structure over the whirling arm shown in Figure 4.45. This facility provides the only means now available to achieve continuous aircraft speeds within centimeters of calibrated, stationary thermometers while, at the same time, maintaining a stream of known hydrometeors in a free fall.

In terms of size, the best approximation to a whole cloud simulator facility is the climate hangar at Eglin Air Force

Base, Florida. It has room for several air-craft and will accommodate the C-5A. Ice and fog can be supplied with some airflow produced by fans. Still, all these facilities fall short of providing a realistic cloud environment; it appears that any final tests will have to be made using the clouds themselves as test environments. Also, this necessitates that we be content with instrument comparisons rather than absolute calibrations.

A practical solution to the instrument testing might be to test instruments operated on the ground, together with instruments of a second type that can operate in the air. Then the latter instruments could form secondary standards, and further comparisons could be made on board aircraft. Using this "bootstrap" technique some estimation of accuracy could be placed on the data from each instrument.

GROUND-BASED INSTRUMENTATION

Several of the aircraft and remote sensing instruments can be at least cross-compared by reference to ground-based instruments for "ground truth." For example, radar determinations of the rainfall rate R are dependent on an accurate "Z-R relationship" (Z is the radar return signal), a relationship which varies with variations of drop spectrum, since Z is approximately proportional to the sixth power of the radius. A large network of closely spaced standard rain gauges has been used for the calibration. In the network raindrop size spectra were determined by drop cameras. [31, 50] Today easier data reduction may be achieved by use of a momentum disdrometer.[30]

The Z-R relationship is even more reliable in estimating snowfall. Snowfall measurements are largely confined to a determination of total accumulation and its change with time using "snow pillows"

and snow depth gauges. A system developed for the Atomic Energy Commission by the Idaho Falls Reactor Test Station utilizes the nuclear radiation absorbed by the snow cover. Several of these sensors provide an automatic readout of snow cover when interrogated by telecommunication links.

The other important standard ground-based instruments are covered in several texts and are not covered here.[24, 47]

FUTURE OF INSTRUMENTATION

Basically, our instruments fit into three classes. The first class consists of instruments with qualities as absolute standards that, perhaps, require tedious data analysis and have little or no ability to supply real-time measurements (e.g., photographic instruments). The second class consists of instruments capable of high resolution and response in nearly real time (e.g., instruments with electronic readout). The third class of instruments consists of those instruments that give an overview of the cloud and, in a way, a partial analysis (radar data).

Of these three classes of instruments, the older instruments, the replicators, foil impactor, etc., fit into the first class because they have qualities (hopefully) as standards, require tedius data analysis, and have little or no ability for real-time analysis. Several instruments presently in development for measurements in situ generally have real-time capabilities and some have on-board data analysis capabilities; they fit into the second class. The remote measuring techniques give an overall view and fit into the third class.

It is to be hoped that in the future the role of present instruments of class one will be largely taken over by more automatic ones. New instruments will be added for applying such techniques as

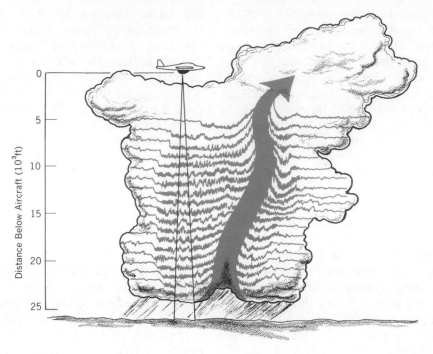

Figure 4.46 Concept of the future potential of airborne Doppler radar to provide detailed multilevel measurement and map of velocities in clouds (courtesy R. Lhermitte, University of Miami, Florida).

scanning electron microscopy, lidar spectroscopy, infrared radiometry, and holography to supply basic calibration points for other instruments. Data from instruments of class two will be systematically collected and used to supplement and quantify data from instruments of class three. Instruments of class three (i.e., calibrated PPI, RHI, and Doppler radars, and IR thermometers, all with real-time analyzing computers) will be the workhorses of weather modification. For example, the adaptation of Doppler radar to aircraft use should open a whole new area in the analysis of cloud motions. The technique is illustrated in Figure 4.46. In a single pass above a cloud, the vertical motion structure over the entire cloud depth can be obtained. In our modification programs, this technique may make it plausible to locate specific portions of the cloud which would be susceptible to modification in terms of increased dynamic activity as well as ice content (see Figure 4.43).

We can look forward to the evolution of new measuring concepts. These will include the development of deterministic (not statistical) correlations between variables closely allied in time and space (i.e., the relationship between aerosol particle properties and numbers of ice crystals which form in the same air parcel). Also, the concept of dynamic variables may emerge. These are such things as trajectories of raindrops or the number of collisions per second between ice crystals and hail particles in a given storm. These measurements are contrasted with the ordinary passive measurements of temperature, pressure, particle concentration, etc. Scott in 1970 developed an instrument that was built completely within a simulated hailstone to measure some of these quantities.[67] The simulated hailstone was made of copper and contained a detector and a transmitter to convey the information to a remote receiver.

Also, new concepts of measuring such quantities as temperature are required. The large effect of a very small change in the virtual temperature at cloud base makes it obvious that we need measurements that are accurate and are averaged over the appropriate, large areas so that accuracies of a few tenths of a degree relative to the virtual temperature of the nearby environment will have meaning. In this regard, NRL has found Caribbean clouds in which no temperature excess was apparent under cloud base, but humidity increases under cloud base were sufficient to give several tenths of a degree of virtual temperature excess at cloud base, accounting for the cloud buoyancy.

If an instrument were available to measure air density, the cloud buoyancy could be ascertained more directly. Toward this end National Hurricane Research Laboratory (NHRL) and the Atomic Energy Commission (AEC) initiated the development (with technical coordination by NRL) of an instrument intended to measure the air density directly using the extinction of a beam of beta particles. So far, nuclear instrumentation problems have frustrated this development, but it is hoped that the future may permit a solution.

Perhaps the new laser techniques will give remote measurements of density appropriate for relatively large areas. In any case, our attempts at understanding and recording phenomena in support of weather modification are certain to require the best that technology can provide in coming years.

ACKNOWLEDGMENTS

Special thanks go to B. Patten, Research Flight Facility (RFF), NOAA, for major contributions to the section on Delivery Systems; to W. Finnegan, Naval Weapons Center, China Lake, California, for reviewing that section; and to H. Friedman, RFF, for compiling the list of Aircraft Research Platforms. The authors are also indebted to H. Mason and G. Conrad, RFF; A. Miller and J. Simpson, Experimental Meteorology Laboratory, NOAA; G. Little, Wave Propagation Laboratory, NOAA; R. Cunningham, AFCRL; and R. Lhermitte, University of Miami for reading the manuscript. E. M. Trent, NRL, and Mr. and Mrs. J. Pinnons provided technical editing on their own time; B. Julian, NRL, provided technical assistance; and Mrs. W. Scott typed two drafts on her own time. The entire staff of EML, in particular, W. Cotton and J. Levine, participated in exhilarating discussions which contributed to the development of this chapter. Many recent developments discussed here were reported to the Symposium on the Measurement of Cloud Elements held in June, 1971 at the University of Chicago, by the following contributors: R. R. Braham, Jr., and R. G. Knollenberg, University of Chicago; W. R. Cotton, EML; D. E. Culnan, Bureau of Reclamation; B. A. Kunkel and R. M. Cunningham, AFCRL; E. F. Danielsen, T. G. Kyle, and J. E. Dye, NCAR; D. M. Takeuchi, MRI; H. A. Friedman, RFF; U. Katz, Cornell Aeronautical Laboratory; W. M. Ketcham, University of Utah; L. D. Nelson, State University of New York; R. E. Ruskin, NRL; W. D. Scott, NHRL; R. T. Ryan, A. D. Little, Inc.; R. L. Smith, Eglin AFB; and J. T. Sutherland, Weather Sciences, Inc.

REFERENCES

1. Abbott, C. E., J. E. Dye, J. D. Sarton, An electrostatic cloud droplet probe, *J. Appl. Meteorol.* **11**, 1092–1100, 1972.

2. Beran, D. W. and S. F. Clifford, Acoustic Doppler measurements of the total wind vector, *Preprints of the Second Symposium on Meteorological Observation and Instrumentation*, 1972.

3. Bigg, F. J., M. McNaughton, and T. J. Methven; The measurement of rain from aircraft in flight, *Tech. Note No. 1*, Mech. Eng. 223, Royal Aircraft Establishment, Farnbough, England, 1956, 14 pp.

4. Brown, E. N., The technique for measuring precipitation particles from aircraft, *J. Meteorol.* **15**, 462–466, 1958.

5. Brown, E. N., A prototype optical flowmeter for the measurement of liquid content, *NCAR Technical Note TN/EDD-61*.

6. Butcher, S. S., and R. J. Charlson, *An Introduction to Air Chemistry*, Academic Press, 1972.

7. Cannon, T. W., A camera for photographing airborne atmospheric particles, *Image Technology*, April/May 1970.

8. Carlson, T. N., and R. C. Sheets, Comparison of draft scale turbulence velocities computed from gust probe and conventional data collected by a DC-6 aircraft, *NOAA Tech. Memo ERL NHRL-91*, 1971.

9. Charlson, R. J., H. Howath, and R. F. Pueschel, The direct measurement of atmospheric light scattering coefficient for studies of visibility and air pollution, *Atmos. Environ.* **1**, 469–478, 1967.

10. Cobb, W. E., and H. J. Wells, The electrical conductivity of oceanic air and its correlation to global atmospheric pollution, *J. Atmos. Sci.* **27**, 814–819, 1970.

11. Cornford, S. G., A note on some measurements from aircraft of precipitation within frontal clouds, *Quart. J. Royal Meteorol. Soc.* **92**, 105–113, 1966.

12. Cornford, S. G., Sampling errors in measurements of particle size distributions, *Meteorol. Mag.* **97**, 12–16, 1968.

13. Cotton, W. R., The need for broad spectrum, multiphase, cloud particle samples for verification of numerical models, *Proceedings of the Symposium on the Measurement of Cloud Elements, Chicago, 1971*, W. D. Scott, Ed., 1974.

14. Decker, M. T., and E. J. Dutton, Radiometric observations of liquid water in thunderstorm cells, *J. Atmos. Sci.* **9**, 785–790, 1970.

15. Dinger, J. E., H. B. Howell, and T. A. Wojciechowski, On the source and composition of cloud nuclei in a subsident air mass over the North Atlantic, *J. Atmos. Sci.* **27**, 791–797, 1970.

16. Duncan, A. D., The measurement of shower rainfall using an airborne foil impactor, *J. Appl. Meteorol.* **5**, 198–204, 1966.

17. Engelman, R., and W. G. N. Slinn, et al., *Precipitation Scavenging, (1970), AEC Div. Tech. Info. Symp. Ser.* **22**, 500 pp., 1970.

18. Fitzgerald, D. R., Measurement techniques in clouds, *Problems of Atmospheric and Space Electricity*, S. C. Coroniti, Ed., Elsevier, New York, 1965.

19. Fletcher, N. H., *The Physics of Rainclouds*, Cambridge University Press, 1962, 390 pp.

20. Fujita, T. T., Proposed characterization of tornadoes and hurricanes by area and intensity, *Satellite and Mesometeorology Research Project (SMRP), Res. Paper No. 91*, Dept. of Geophys. Sci., Univ. of Chicago, 1971.

21. Fujiwara, M. An improved raindrop radiosonde with filter paper, *Tellus* **19**, 403–404, 1967.

22. Grant, L. O., Ed., *Proceedings of the Second International Workshop on Condensation and Ice Nuclei, Atmos. Sci. Paper No. 172*, Colorado State Univ., Ft. Collins, 1971, 160 pp.

23. Gucker, F. T., and D. A. Rose, A photoelectronic instrument for counting and sizing aerosol particles, *Brit. J. Appl. Phys.* **5**, Suppl. 3, 1954.

24. *Handbook of Meteorological Instruments*, Parts I, II, and III, Her Majesty's Stationary Office, London, 1966.

25. Hobbs, P. V., L. F. Radke, A. B. Fraser, J. D. Locatelli, C. E. Robertson, D. G. Atkinson, R. J. Farber, R. R. Weiss, and R. C. Eastern, Studies of winter cyclonic storms over the Cascade Mountains (1970–1971), Contribution from the Cloud Physics Lab., University of Washington, *Research Report* **6**, 1971.

26. Humphreys, W. J., *Physics of the Air*, McGraw-Hill, New York, 1940, pp. 501–536.

27. Huffman, P. J., Polarization of light scattered by ice crystals, *J. Atmos. Sci.* **27**, 1207–1208, 1970.

28. Huffman, P. J., and W. R. Thursby, Jr., Light scattering by ice crystals, *J. Atmos. Sci.* **26**, 1073–1077, 1969.

29. Israel, H., *Atmospheric Electricity*, Israel Program for Scientific Translations, Jerusalem, IPST Cat. No. 1995, 1970.

30. Joss, J., and A. Waldvogel, Raindrop size distributions and Doppler velocities, *Preprints of 14th Radar Meteorology Conference, Tucson*, 1970, pp. 153–156.

31. Jones, D. M. A., and L. A. Dean, A raindrop camera, *Research Report No. 3*, U.S. Army

Contract No. DA-36-039, SC-42446, Ill. State Water Survey, Urbana, Ill., 1953.

32. Keily, D. P., and S. G. Millen, An airborne cloud-drop-size distribution meter, *J. Meteorol.* **17**, 349–356, 1960.

33. Knollenberg, R. G., Particle size-measurements from aircraft using electro-optical techniques, *Proceedings of the Electro-optical Systems Design Conference, 18–20 May, 1971, West Anaheim, Calif.*, 1971, pp. 218–233.

34. Knollenberg, R. G., Comparative liquid water content measurements of conventional instruments with an optical array spectrometer, *J. Appl. Meteorol.* **11**, 501–508, 1972.

35. Laktionov, A. G., N. K. Nikiforova, V. V. Smirnov, G. I. Scelchkov, and O. A. Volkovitsky, New automatic equipment for the investigation of drop and crystal microstructure in clouds, *Volume of Abstracts International Cloud Physics Conference*, Royal Meteorological Society, London, 1972, p. 9.

36. Langer, G., J. Rosinski, and C. P. Edwards, A continuous ice nucleus counter and its application to tracking in the atmosphere, *J. Appl. Meteorol.* **6**, 114–125, 1967.

37. Langmuir, I., and K. B. Blodgett, Mathematical investigation of water droplet trajectories. Final Report under Army Contract W-33-038-ac-9151, Report No. RL-225 from the General Electric Company, 1949, 46 pp.

38. Langsdorf, A., A continuously sensitive cloud chamber, *Phys. Rev.,* **49**, 422, 1936.

39. Levine, J., The dynamics of cumulus convection in the trades, a combined observation and theoretical study, Ph.D. Thesis, Dept. of Meteorology, Woods Hole Oceanography Institute, Mass., 1965.

40. Lhermitte, R. M., Study of wind field in a convective storm by dual Doppler radar, *Proceedings of the Cloud Physics Conference, Ft. Collins, Colo.*, 1970, pp. 163–164.

41. Lhermitte, R. M., Dual Doppler radar observation of convective storm circulation, *Preprints of 14th Radar Meteorology Conference, Tucson*, 1970, pp. 139–144.

42. MacCready, P. B., Jr., and C. J. Todd, Continuous particle sampler, *J. Appl. Meteorol.* **3**, 450–460, 1964.

43. Mach, W. H., and P. V. Hobbs, The electrical particle counter, Contributions from the Cloud Physics Laboratory; Research Report III, Results of Aircraft

Measurements in Winter Storms (1968–69) over the Cascade Mountains, Univ. of Washington, 1969, pp. 80–95.

44. Mason, B. J., *The Physics of Clouds*, Oxford Press, 1971, 675 pp.

45. Mastenbrook, H. J., The variability of water vapor in the stratosphere, *J. Atmos. Sci.* **28**, 1495–1501, 1971.

46. McTaggart-Cowan, J. D., G. G. Lala, and B. Vonnegut, The design, construction, and use of an ice crystal counter for ice crystal cloud studies by aircraft, *J. Appl. Meteorol.* **9**, 294–299, 1970.

47. Middleton, W. E. K., and A. F. Spilhaus, *Meteorological Instruments*, 3rd Ed., University of Toronto Press, 1953.

48. Mossop, S. C., N. S. C. Thorndike, R. T. Meade, and K. J. Heffernan, The use of membrane filters in measurements of ice nucleus concentration, *J. Appl. Meteorol.* **5**, 703–709, 1966.

49. Mossop, S. C., R. E. Ruskin, and K. J. Heffernan, Glaciation of a cumulus at approximately $-4°C$, *J. Atmos. Sci.* **25**, 889–899, 1968.

50. Mueller, E. A., Study on intensity of surface precipitation using radar instrumentation, *Quart. Tech. Report No. 10*, U.S. Army Contract DA-36-039 SC-75055, Ill. State Water Survey, Urbana, Ill., 1960.

51. Neel, C. B., A heated-wire liquid-water-content instrument and results on initial flight tests in icing conditions, *NACA Res. Memo RMA54123*, 1955, 33 pp.

52. Nelson, R. T., The design, construction, and testing of a raindrop dropsonde, *Proceedings of the Symposium on the Measurement of Cloud Elements, Chicago, 1971*, W. D. Scott, Ed., *NOAA Tech. Memo.*, 1974.

53. Ohtake, T., Cloud settling chamber for ice nuclei count, *Second International Workshop on Condensation and Ice Nuclei, Atmos. Sci. Paper No. 172*, Colorado State University, Ft. Collins, 1971, pp. 92–96.

54. Parry, H. D., The role of ground-based remote sensors in the improvement of local weather forecasts, *Preprints of Second Symposium on Meteorological Observation and Instrumentation, San Diego*, 1972, pp. 44–51.

55. Patten, B. T., J. D. McFadden, and H. A. Friedman, Airborne cloud seeding systems developed and utilized by the Research Flight Facility, *Proceedings of the International Conference on Weather Modification, Australia, Sept. 6–11, 1971*, pp. 358–360.

56. Radke, L. F., W. D. Scott, and C. E.

Robertson, Interaction of cloud condensation nuclei and ice nuclei with cloud and precipitation elements, *Precipitation Scavenging, 1970*, R. Englemann and W. G. N. Slinn, Eds., *AEC Div. Tech. Info. Symp. Ser.* **22**, 37, 1970.

57. Radke, L. F., and P. V. Hobbs, Measurement of cloud condensation nuclei, light scattering coefficient, sodium containing particles, and Aitken nuclei in the Olympic Mountains of Washington, *J. Atmos. Sci.* **26**, 281–288, 1969.

58. Ross, I., Inertial navigation and gust measurement from meteorological research flight aircraft, *Meteorol. Mag.* **95**, 370–376, 1966.

59. Ruskin, R. E., Measurements of water-ice budget changes at −5°C in AgI-seeded tropical cumulus, *J. Appl. Meteorol.* **6**, 72–81, 1967.

60. Ruskin, R. E., Considerations in the measurement of pollution effects on the number concentration of cloud condensation nuclei, *J. Appl. Meteorol.* **10**, 994–1001, 1971.

61. Ruskin, R. E., and W. C. Kocmond, Summary of condensation nucleus investigations at the 1970 International Workshop on Condensation and Ice Nuclei, *Second International Workshop on Condensation and Ice Nuclei, Atmos. Sci. Paper No. 172*, Colorado State Univ., Ft. Collins, 1971, pp. 92–96.

62. Ruskin, R. E., The measurement of cloud elements: The use of transmission and scattering techniques to measure ice and water, *Proceedings of the Symposium on the Measurement of Cloud Elements, Chicago, 1971*, W. D. Scott, Ed., *NOAA Tech. Memo*, 1974.

63. Ryan, R. T., H. Blau, Jr., P. C. Von Thuna, and M. L. Cohen, Cloud microstructure as determined by an optical cloud particle spectrometer, *J. Appl. Meteorol.* **11**, 149–156, 1972.

64. Sasyo, Y., Studies and developments of the meteorological instruments for cloud physics and micrometeorology (II): New methods of measurement for larger cloud drops, *Papers in Meteorology and Geophysics, XX*, Meteorological Research Institute, Tokyo, Japan, 1969, pp. 27–40.

65. Schaefer, V. J., A method of making snowflake replicas, *Science* **93**, 239, 1941.

66. Schecter, R. M., and R. G. Russ The relationship between imprint size and drop diameter for an airborne drop sampler, *J. Appl. Meteorol.* **9**, 123–126, 1970.

67. Scott, W. D., An instrumented "Hailstone" for cloud physics research, *Preprints of Conference on Cloud Physics, Ft. Collins*, 1970, pp. 109–110.

68. Scott, W. D., and C. K. Dossett, An estimate of the fraction of ice in tropical storms, Project STORMFURY Annual Report, 1970, National Hurricane Research Laboratory, Miami, Fla., 1971.

69. Scott, W. D., and C. K. Dossett, Modeling the seeding effect in Hurricane Ginger, Project STORMFURY Annual Report, 1971, National Hurricane Research Laboratory, Miami, Fla., 1972.

70. Simpson, J., W. L. Woodley, H. A. Friedman, T. W. Slisher, R. S. Sheffie, and R. L. Steele, An airborne pyrotechnic cloud seeding system and its use, *J. Appl. Meteorol.* **9**, 109–122, 1970.

71. Squires, P., and C. A. Gillespie, A cloud-droplet sampler for use on aircraft, *Quart. J. Royal Meteorol. Soc.* **78**, 387–393, 1952.

72. Stevenson, C. M. An improved millipore filter technique for measuring the concentrations to freezing nuclei in the atmosphere, *Quart. J. Royal Meteorol. Soc.* **94**, 35–43, 1968.

73. Sutherland, J. L., and D. R. Booker, An airborne momentum sensing raindrop spectrometer, *Preprints of Conference on Cloud Physics, Ft. Collins*, 1970, pp. 101–102.

74. Tillman, J. E., Water vapor density measurements utilizing the absorption of vacuum ultraviolet and infrared radiation, Chap. 43 in *Humidity and Moisture Measurements and Control in Science and Industry*, Vol. 1, *Principals and Methods of Measuring Humidity in Gases*, Reinhold, New York, 1965, 704 pp.

75. Twomey, S., Measurements of natural cloud nuclei, *J. Rech. Atmos.* **1**, 101–105, 1963.

76. Twomey, S., and T. A. Wojciechowski, Observations of the geographical variation of cloud nuclei, *J. Atmos. Sci.* **26**, 684–688, 1969.

77. Twomey, S. The evaporation of submicron aerosol particles, *J. Rech. Atmos.* **5**, 93–100, 1971.

78. Vali, G. D. Knowlton, An automated drop freezer system for determining the freezing nucleus content of water, *Second International Workshop on Condensation and Ice Nuclei, Atmos. Sci. Paper No. 172*, Colorado State Univ., Ft. Collins, 1971, pp. 75–79.

79. Warner, J., An instrument for the measurement of freezing nucleus concentration, *Bull. Obs. Puy de Dome* **2**, 33–46, 1957.

80. Warner, J., and T. D. Newnham, A new method of measurement of cloud-water content, *Quart. J. Royal Meteorol. Soc.* **78**, 46, 1952.

81. Weiland, W., Condensation of water vapor on natural aerosol at slight supersaturation, *Z. Angew. Math. Phys.* **7**, 428–460, 1956.

5 | Design and Evaluation of Weather Modification Experiments

GLENN W. BRIER

"The most important questions of life, are, for the most part, really only problems of probability."

Laplace, *Théorie Analytique des Probabilitiés.*

Much of the controversy and confusion that followed Schaefer's discovery of the dry ice seeding technique revolved around attempts to answer what may appear to be a relatively simple or elementary question. This question was illustrated in a timely cartoon about rainmakers, which showed two clergymen gazing through a church window at the falling raindrops and asking each other, "Is it theirs or ours?" As sometimes happens, the cartoonist, with his penetrating insights and touch of humor, put his finger on a basic scientific question—here, that of evaluating weather modification activities— "Has the attempt to alter the weather process produced any result different from what would have occurred in the absence of the modification effort?" In principle, it might appear that it is simple to answer this question, since all

that is necessary is to compare the observation following seeding with the prediction of what would have happened if nature had taken its normal course. Unfortunately, in most practical cases, meteorological theory and knowledge of weather processes do not permit us to make a sufficiently accurate prediction for a meaningful comparison, and it has been necessary to resort to more sophisticated concepts and methods to answer whether it is "theirs or ours." A considerable amount of confusion or misunderstanding resulted when these concepts or methods were introduced in the early 1950s, at a time when the scientific community and general public were being showered with fantastic and controversial claims and counterclaims of success and disputes among amateur and commercial rainmakers, involved scientists, and public servants. In this chapter we attempt to give a brief description of some of the techniques of design and evaluation and the concepts upon which they are based,

Glenn W. Brier is a professor in the Department of Atmospheric Science, Colorado State University, Fort Collins, Colorado.

206

to indicate why controversy has arisen in some areas, and hopefully, to point out some of the possible future developments.

DESIGN

The Roles of Randomization and Replication

The initial experiments with dry ice could be interpreted without any ambiguity since they dealt with systems that are very stable or persistent in time or space. Supercooled water in a laboratory cold box or in a large stratus cloud deck can exist for long periods of time without any visible change. The introduction of a seeding agent produces obvious changes within seconds or minutes, and evaluation is mainly a matter of observation. There is the implicit prediction that the conditions existing before the experiment would continue to persist, but there is no need to postulate a formal framework, and one might say that "there is no need for statistics," since variability is so low. Furthermore, the experiment can be repeated easily at a different time or location with consistent results. The situation is quite different, however, when we deal with phenomena such as storms, hurricanes, or cumulus clouds. In Chapter 6 it is pointed out that a cumulus cloud may go through its complete life cycle in an hour or so, and that the response of a cloud depends upon when in its life cycle the treatment is administered, as well as upon the general environmental conditions, which may differ widely from day to day or from one area to another. The weather scientist recognizes the large natural variability of rainfall and cloud characteristics in space and time, and sees the need for appropriate statistical methods to cope with the problems of uncertainties, for, as expressed by F. Mosteller and J. W. Tukey in 1968, "One

hallmark of the statistically conscious investigator is his firm belief that however the survey, experiment, or observational program actually turned out, it could have turned out somewhat differently."[25]

Statistical methods designed to handle these problems were developed by R. A. Fisher[13] in connection with the design and analysis of comparative experiments in biological and agricultural research, where large and only partly controllable variability is present. The basic ideas involve (1) *replication*, from which a quantitative estimate can be made of the experimental "error" or the variability of the response to a treatment, and (2) *randomization*, a process of allocating treatments to the experimental material by tossing an unbiased coin (or equivalent procedure), which may make it possible for the experimenter to attribute whatever effects he observes to the treatment and to the treatment only. Together, these two principles enable one to make a valid assessment of the uncertainty of the result in terms of a probability statement or by setting confidence limits.

Unfortunately, the term "randomization," which had a definite meaning that was well understood in the field of agricultural experimentation, was interpreted by some to mean "non-sensical" or "haphazardly," and the erroneous conclusion was drawn by some that statisticians were recommending that field experiments be conducted without proper consideration of physical processes and without regard to logistic and operational problems. There was a tendency to contrast "physical" and "statistical" evaluation, although today this tendency has largely disappeared, partly as a result of conferences and discussions such as those reported in the Skyline Conference in 1959, which stated, "The degree of success so far achieved by various programs in weather modification is, in large measure, due to detailed and skilled analysis of data which combines sound statistical techniques and enlightened meteorological insight."[35] The statements on statistical design in the report of this conference still remain valid, although progress in meteorological and statistical theory along with field experimentation has led to refinements not visualized at that time. Today we can think of a big "physical experiment" as one where the overall process spanned by the test can be partitioned into a chain of subprocesses, in each of which the physics is understood and can be measured or observed. Quantitative statistical checks can then be used for each process to improve the test decisiveness and act as a check on our theory or preconceived ideas. Although this role of statistics in hypothesis testing is an important one, it is not the only one, as we shall see in later sections.

Some Typical Designs

Many types of designs have been used in cloud seeding experiments and operations. Usually it has been desirable to tailor the design according to the needs of each particular investigation. Numerous examples are given in Chapters 6, 7, 12, and 13 and the reader is referred to the NAS report written in 1973, "Weather and Climate Modification, National Policies and Programs," for a more complete discussion on this topic and for further references.[40] An excellent review article by P. A. P. Moran[24] on the methodology of rainmaking experiments is very helpful. In this section we are primarily concerned with a few basic designs and some of the problems that arise when they are used. Additional problems are discussed in later sections.

Target-Only Design. This design involves a single area (or cloud) and is often used in commercial rainmaking opera-

tions where the emphasis is on augmentation of the water resources with little or no effort made to obtain rigorous evaluation or scientific information. The results of such operations may sometimes be compared with a "historical" period when no seeding took place, but the value and validity of such comparisons is usually open to question for a number of reasons. When randomization is used, leaving some of the experimental periods unseeded to act as a "prediction" of what should happen naturally, it becomes possible to make a scientific evaluation of the seeding effect. However, such experiments are not likely to be very sensitive in detecting small or moderate seeding effects since the signal may be masked by the noise of large natural variability in precipitation, and months or years might be required before definitive conclusions could be drawn. In the case of the randomized winter orographic projects reported by L. D. Grant and P. W. Mielke, Jr., this design was the only practicable one to use.[16] The experiment (Climax I) ran for 6 years and gave evidence for different seeding effects according to various meteorological stratifications. The experiment was repeated (Climax II) with an additional 5 years of operation, confirming the earlier results and supporting the physical model that had been formulated.

Another circumstance where the single area or single cloud design is used is in preliminary or exploratory work where the scientist is checking out his equipment or making provisional tests of his hunches or intuitive beliefs. Randomization procedures have been criticized on the basis that an inadequate number of experimental units are available to leave some untreated, or that rigorous statistical controls inhibit the scientist from making full use of his intuitive and subjective beliefs. In this regard, C. R. Blyth states that, "The use of subjective beliefs has the advantage of getting us ahead very rapidly if our beliefs happen to be correct, and the disadvantage of getting us ahead just as rapidly—up a blind alley—if our beliefs happen to be wrong. On balance, it is very clear to me that maximal use of subjective beliefs is desirable, especially in the early stages of a scientific problem, for the simple reason that without using them I wouldn't even get started. However, in the end, all subjectivity has got to be removed.... [In] the testing of a model against reality ... total and uncompromising objectivity is absolutely essential."[3] It is at this latter stage that the careful atmospheric scientist has insisted upon proper statistical controls, and it is the failure to distinguish between this testing state and the earlier exploratory phases, where hypotheses are being formulated, that has led to some of the confusion and needless controversy.

Target-Control Design. This design involves a single area that is seeded on a randomized basis and a nearby control area that is never seeded, and (presumably) not affected by the seeding. In some cases, two or more control areas have been used, but this generalization does not change the basic concept. This design is often better than the target-only design, since the precision of the experiment may be increased by utilizing a control area that is a good predictor variable or covariate, making it possible to detect a real effect of seeding in a shorter period of time; however, problems of contamination may arise with the use of this design, since the control area is generally chosen close enough to the target area to ensure a reasonably good correlation between target and control area precipitation. This contamination can be of two types. The first is the spread of the seeding material into the control area, especially when ground generators are used and there is little con-

trol or knowledge as to where the seeding material goes. It should be emphasized that this contamination does not invalidate the test of significance; it only makes it more difficult to detect real effects. In the case where seeding has no effect, the treated and untreated events will have the same expected value of precipitation and any observed differences between the two groups will be random and due to natural variations. On the other hand, if the seeding material has an effect (presumably in the same direction in both areas), there is less opportunity for the target and control areas to respond differently. In the Israeli experiment reported in Chapter 13, it was felt that the control areas were contaminated on some occasions and that the real effects of seeding were greater than those reported.

The second type of possible contamination is the spread of the effects of the seeding material beyond the intended target through "dynamic contamination," as discussed in Chapter 6. If such contamination exists, then the test situation might become quite complicated, especially if the effect in the control area is opposite to that in the target area. Although very little is known about this, it is clear that the design of any weather modification program requires that consideration be given to the possibility of effects outside the nominal target.

The target-control design has also been used to evaluate nonrandomized commercial seeding projects. In these projects the evaluation procedure usually consists of comparing the observed target precipitation in the seeded periods with a "prediction" or estimate obtained from a target-control regression equation based on a historical period. One difficulty with this method is the questionable assumption that the relationship between the target and the control area is the same during the experimental period as it was during the historical period. Brier and Enger[8] found that the relationship between the rainfall of a target and a control area in Arizona was not stable with time. Neyman, Scott, and Wells[30] and others have emphasized this point and presented additional supporting evidence. Other deficiencies in this procedure have been described by J. Neyman, E. L. Scott, and M. Vailevskis[29] and K. A. Brownlee[9] and summarized by L. D. Calvin.[10]

Crossover Design. This design has considerable appeal from a purely statistical point of view since, in the absence of pretest or historical data, it is much more efficient than the target-only and target-control designs, as pointed out by Moran.[23] S. Wu, J. S. Williams, and P. W. Mielke[41] have discussed the efficiency of the design with respect to alternative designs when historical data are available. The crossover design involves two areas, only one of which is seeded at a time, with the area for seeding selected randomly for each time period. The Israeli experiments described in Chapter 13 made use of this design.

The same difficulty arises with the crossover design as with the target-control design with respect to the assumption that the effects of seeding are limited to the target area. An additional factor has to be considered when the crossover design is used. This is the effect of seeding in an area after the seeding has been terminated, known as the "persistence effect." Evidence presented by Grant[15] suggested that freezing nuclei counts remain high for many days after seeding, and E. G. Bowen[4] presented evidence indicating a general decrease in the positive effect of seeding in Australia over a period of years which was consistent with the hypothesis of a carry-over effect from one time period to another. One explanation suggested was that the

seeding nuclei were trapped on the vegetation in the area and were blown into the air at a later time. The question has remained controversial since others have found no evidence of such effects; however, it would appear to be desirable to take the possibility of such effects into consideration when designing experiments, and E. M. Smith[36] has made several suggestions related to detecting and investigating the persistence effect in crossover design.

Other Designs.   A variety of designs has been used for the numerous experiments and operations in weather modification that have been conducted during the past 25 years. Some of these designs involved a moving or floating target, perhaps with a moving control area as well. In the Whitetop cloud seeding experiment,[14] a varying target area was used which was called the plume and defined as that portion of the research area where the winds carried the seeding material. The remaining portion of the research area was called the nonplume and was used as one of the controls. In other experiments the moving target was a cloud or a group of clouds in a band, where special equipment such as radar was used to estimate the growth of the cloud tops or the precipitation falling. In all these cases the complexity was introduced in order to incorporate more realistic meteorology into the design, which often greatly increased the problems of statistical analysis and data processing.

ANALYSIS

Significance Tests

The purpose of designing and carrying out a particular experiment is to collect a sample of data from which the investigator can draw valid inferences. In the presence of considerable unreliability or uncontrolled variation, any inference from the particular to the general must be attended with some degree of uncertainty, but by means of probability theory and statistics the nature and degree of the uncertainty may themselves be capable of rigorous expression. In appraising a difference, such as the amount of precipitation with and without seeding, a statement can be made that the true difference lies between certain limits, plus a probability that the statement is correct. In testing whether average rainfall can be increased over an area, for example, a test of significance is used—a rule for deciding, from examination of the data, whether or not to reject the hypothesis of no effect. Some of the most commonly used tests are discussed here.

The t-Tests. These classical tests are used to compare sample means of two groups A and B, and have a wide variety of forms. Their use and theoretical background are described in practically every elementary text in statistics and are not discussed in detail here. Basically, the test consists of the ratio

$$t = \frac{\text{Sample mean } A - \text{sample mean } B}{\text{estimated standard deviation of numerator}},$$

and is a test of the "null hypothesis" (H_0) that there is no difference between A and B. Tables are available for the distribution of t showing the probability P of securing an extreme or a more extreme sample result than the one observed under the assumption that the null hypothesis (H_0) is true. In using the tables, care must be taken as to whether they refer to one-tail or two-tail P values. In a rain augmentation project where a test is being made to produce an *increase* in precipitation, the one-tail value of P is appropriate, which is one-half of the two-tail P value; however, in actual practice one might want to use an additional one-

sided tail for a *decrease* in precipitation, for this possibility should not be overlooked since it appears to have happened in the past.

A common error in reporting a P value is to use phrases such as "there is only one chance in twenty that the null hypothesis is true." The computed probability P is used to scale our judgement of whether the deviation in question can be explained on the basis of chance or not. The "null hypothesis" is never *proved*, but if P is sufficiently small the "chance" explanation becomes unreasonable, and we *discard* the hypothesis and conclude that on the contrary there was some real *cause* back of the event. We can still be wrong in a particular instance, but if we choose a P value low enough we can be confident we will not be wrong very often.

The t tests are appropriate when the measurements being treated are independent and are described adequately by the normal distribution whose probability density function is given by

$$f(x;\mu,\sigma) = \frac{1}{\sqrt{2\pi}\sigma} \exp\left[-\frac{1}{2}\frac{(x-\mu)^2}{\sigma^2} \right]$$

$$(-\infty < x < \infty),$$

where μ and σ are the population mean and standard deviation, respectively. With precipitation data, normality is not usually found, so methods must be devised to alleviate this difficulty if normal distribution theory is to be used in evaluating cloud seeding experiments. Attempts have been made to normalize precipitation distributions by transforming the observed variable. Also, it has been found that precipitation data can be represented adequately by a gamma distribution with a density function

$$f(x;k,\mu) = \frac{1}{\Gamma(x)} \mu^k x^{k-1} e^{-\mu x}$$

$$(x \geqslant 0, K > 0, \mu > 0),$$

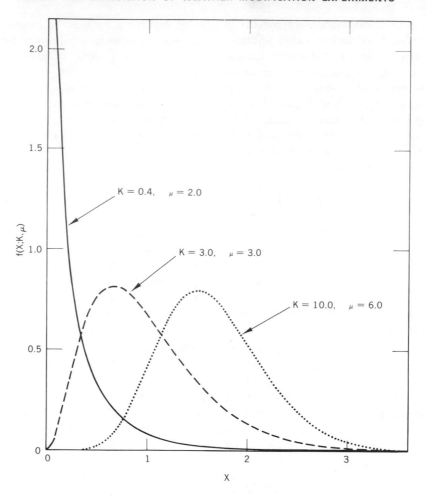

Figure 5.1 Graphs of several gamma distributions.

where k is the shape parameter, μ is the scale parameter, and Γ denotes the complete gamma function. The graphs of several gamma distributions are illustrated in Figure 5.1. G. L. Barger and H. C. S. Thom[2] were the first to point out the usefulness of gamma distributions in fitting rainfall observations and subsequently numerous other investigators have confirmed their results. Simpson[34] used the distribution in evaluating rainfall from seeded cumuli. Moran[24] has extended the ideas to include a five-parameter bivariate gamma distribution for dealing with target-control area designs. The whole topic is discussed in considerable detail in "Weather and Climate Modification: National Policies and Programs."[40]

The C (α) Tests. When the data are normally distributed and a test is to be made whether seeding produces a constant increment (the additive hypothesis) on the target area, the t-test and standard regression methods are applicable and give an optimum test of the null hypothesis. There is, however, considerable evidence suggesting that the real effect of seeding, if it exists, is multiplicative rather than additive. The difficulty now arises of handling a nonstandard hypothesis. Partially motivated by the problem of finding a satisfactory test for the situation frequent in weather research, J. Neyman in 1959 developed the "$C(\alpha)$ tests."[26] These tests, which are further described by Neyman and Scott[28] and other investigators, have a number of desirable properties and can be shown to be optimal under certain circumstances. In answer to the question as to what to use in practice, Moran[24] states, "undoubtedly there would be some preference from the meteorological side for the $C(\alpha)$ test but no real certainty exists," and suggests the use of Monte Carlo methods (simulations) to study the performance of this (and other) tests since a few hours of simulations on a high-speed computer is inexpensive compared with the cost of a rainmaking experiment.

Nonparametric Tests. There are a number of these tests that are used to avoid assumptions about the shape of distribution from which the sample is drawn. Siegel[32] describes many of these tests but only a few of the more popular ones are discussed here.

The *sign test* is used if one has a series of n differences obtained, for example, by subtracting the value for treatment A from treatment B. One counts the number of positive differences if one is testing whether treatment B produces *larger* values than treatment A. If the number of positive differences is r, the well-known binomial theorem is used to determine the probability P of observing r or more successes in a sample of n trials, on the assumption that positive and negative differences are equally likely. (If differences are exactly zero they are excluded from the sample.) For example, if $n = 8$ and $r = 7$, we find $P = 9/256 = 0.03$ and reject the null hypothesis at the 5 percent level. On the other hand, if $n = 8$ and $r = 5$, then $P = 93/256 = 0.36$, and

we accept the hypothesis that positive and negative differences are equally likely and conclude that there is no significant difference between treatments A and B. It should be noted that this test ignores the absolute magnitude of the difference and thus does not make use of all the information available in the sample.

The *Wilcoxon signed rank test* takes some account of the size of the differences and is applied to matched or paired observations. For example, in a cloud modification experiment, two similar clouds are chosen as experimental units and a random selection is made to seed one of them. The intensity of the radar echo is measured for each of them, and the difference between the treated and control cloud in each pair is tabulated and used as a test variate. To apply the test, the differences are ranked according to their absolute size and then the original sign is reattached to each difference. The values are then summed, and appropriate tables are used for deciding whether to accept or reject the null hypothesis.

The *Wilcoxon-Mann-Whitney test* is a popular test that often goes by other names and is used to test the difference between the location parameters of two distributions, neglecting their shape and supposing the two samples independent (unlike the Wilcoxon signed rank test). To perform the test, the measurements of both samples are pooled and ranked from least to greatest. The sum of the ranks for either one of the samples is used as a test statistic. It is compared with published tabulations to determine whether it is unreasonably large or small compared to a random selection of ranks.

Other nonparametric tests, such as the *squared rank tests* of A. M. Mood and M. Taha, are described by P. W. Mielke.[21, 22, 38] Duran and Mielke[12] have studied the performance of the Wilcoxon-Mann-Whitney and Taha tests when used

Table 5.1   Relations Between Mean Rainfall $\bar{R}$ and the Mean of the Transformed Values $\bar{T}$

	$\bar{R}$	$\bar{T}$	Inverse Transform = R^*	Bias $R^* - \bar{R}$
Original distribution	6.13	1.409	2.94	−3.19
5% incresae	6.44	1.419	3.05	−3.39
10% increase	6.74	1.429	3.17	−3.57
15% increase	7.04	1.437	3.26	−3.78

on distributions likely to be found in weather modification. Provisions have been made for the problem of tied measurements, so it now appears that a number of useful tests are available for application.

Use of Transformations

As discussed earlier, the problem of skewness and other lack of normality in precipitation distributions has been attacked by using the gamma distribution to fit the observations. Another approach is by using transformations to aid in obtaining normality and in meeting the underlying assumptions of equality of variances. Commonly used transformations are (1) the square-root transformation; (2) the cube-root transformation; and (3) the logarithmic transformation. With appropriate transformations, the t-test and some of the nonparametric tests such as the Wilcoxon-Mann-Whitney test are known to have considerable power—the probability that the test detects an effect when there actually is one. This is an important consideration in choosing a test and designing an experiment, especially because of the controversial nature of rainmaking where great emphasis has been placed in the past in testing whether

any effect of cloud seeding exists at all. But hypothesis testing should not be the only goal of weather modification experimentation, for we are interested also in the economic value of a seeding procedure. For this purpose, we need to know the expected increase in the actual precipitation and the distribution of the possible increase. But if we have used a transformation, a problem arises, for it is the increase in the expectation of the transformed value which can be estimated, and a direct estimate of the increase in the expected value of the untransformed variate may be difficult. For example, if a square-root transformation has been applied to the original observations, x, in a sample and the sample mean $\bar{T}$ of the transformed values determined, then the retransformed mean ($\bar{T}^2$) is biased and does not give an unbiased estimate of the mean ($\bar{x}$) of the original variable x. This was recognized by Neyman and Scott,[27] and methods of correcting for the bias have been devised by them and others.

An illustration of this bias is shown in the following example, taken from a simulation study using actual daily precipitation observations from a network of 18 stations in a target area for a period of 36 hypothetical operational days of seeding. The 648 values of precipitation, recorded in millimeters, ranged from 0 to 243 mm. The transformation used was

$$T = (1.0 + R)^{1/4}$$

where R was the precipitation measured in mm. This transformation has been found useful when 24-hr amounts for individual stations are analyzed. The mean rainfall $\bar{R}$ for the 648 values was 6.13 mm, and after the transform was made on each of the 648 values, the mean value $\bar{T}$ was 1.409. The inverse transform

$$R^* = \bar{T}^4 - 1.0$$

was then computed, giving a value of 2.94 or a bias $(R^* - \bar{R})$ of -3.19.

To show the effect of possible changes due to a modification experiment, each of the original values of rainfall was then increased by 5 percent and the transform made on each and new values of $\bar{R}$, $R,^*$ and $\bar{T}$ computed. These values are shown in Table 5.1 along with the corresponding

values for 10 percent and 15 percent increases in the original rainfall values. Figure 5.2 shows the results in graphical form. We note that in this instance, the relationship is nearly linear, but that a 10 percent increase in rainfall, for example, does not result in a 10 percent increase in either $\bar{T}$ or its inverse transform; however, this graph could be used for correcting for the bias. Studies like this can be helpful in designing or evaluating an experiment and are easily performed on a high-speed computer.

Use of Covariates

The power or sensitivity of an experiment can be increased greatly if predictive devices are used. In fact, the design and analysis of weather modification experiments is intimately related to the meteorological prediction problem. As

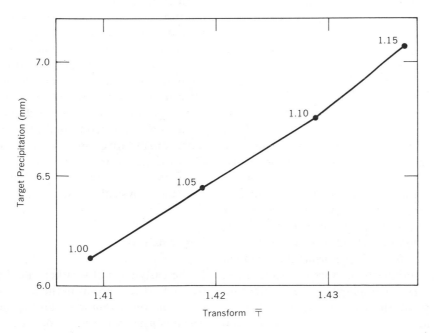

Figure 5.2 Relation between mean target precipitation $\bar{R}$ and the transform $\bar{T}$ for hypothetical increases of 5, 10, and 15 percent in precipitation, where $T = (1.0 + R)^{1/4}$.

pointed out earlier, the evaluation of any attempt to modify the atmosphere depends upon a comparison between some weather parameter y and an estimate $\hat{y}$ of what would have happened naturally. The estimate $\hat{y}$ can be considered to be a forecast, whether it is explicit or implied by the experimental design. For example, in the randomized target-only design, the mean precipitation for the control cases is essentially a prediction of what should have happened naturally. The difficulty is that under ordinary circumstances the prediction error

$$e = \hat{y} - y$$

may be as large as (or larger than) the magnitude of the effect we are trying to detect. Obviously we must make all possible efforts to improve our estimate $\hat{y}$. Every effort should be made to find effective predictors so that the number of periods required for each seeding experiment can be reduced to a minimum.

The prediction $\hat{y}$ might be primarily statistical, or based on a physical model, or some combination of the two. In statistical experiments it is common to use a linear equation

$$\hat{y} = b_0 + b_1 x_1 + b_2 x_2 + \cdots b_p x_p \qquad (5.1)$$

where $x_1, x_2, \cdots x_p$ are known as the covariates or predictor variables and the b's are regression coefficients chosen to fit some set of data or theory. Usually, the fit is not a perfect one and is made in the sense of a least-squares approximation to the "truth." The variable y might be some measure of precipitation, radar intensity, cloud growth, or hail occurrence, for example, while the x's might be measurements of cloud moisture, lapse rate, pressure gradient, or some other indicator of the current state of the atmosphere; however, there is no general requirement that the relation between y and x's be statistically determined, for knowledge of the physics can be used to

give increased statistical control. Statistical control is not inimical to physical control, for together they can increase the power and validity of the experiment by providing a better prediction of $\hat{y}$ and a quantitative assessment of its degree of uncertainty. From the statistical point of view, a good physical theory can offer a much better set of covariates or predictor variables combined in a more useful form than the standard linear function of Equation 5.1.

The importance of physically based covariates has been demonstrated convincingly in the Florida cumulus cloud experiments (Chapter 6). The use of the "seedability" concept as a covariate is of particular interest in experimental design, since it involves not only a prediction of what should happen naturally but in addition provides a quantitative estimate of the expected response of the cloud to the treatment. However, in predicting rainfall from individual clouds, it appears that the most useful predictor variable is the amount of rain that has fallen immediately preceding the beginning of the experimental period. It seems likely that the greatest improvement in test decisiveness will result from research oriented toward providing better stratification techniques for classifying possible seeding situations, and from making possible more quantitative and accurate predictions of the natural behavior of weather systems which would be based on combined physical-statistical models.

Data Analysis

In the early days of weather modification research, the emphasis in the use of statistics was on hypothesis testing, a role which is usually not very popular, at least when compared with the more exciting and creative exploratory research where the emphasis is on the formulation of models of reality and gaining an understanding of the processes. Today there is more emphasis on using the techniques of statistics to provide indications of what the data are trying to tell us in regard to the physical relationships, to help us discover or expose relevant factors. New methods of data analysis have been developed, such as those that appear in papers by F. Mosteller and J. W. Tukey.[25, 39] It is not only a question of new or old methods but the changing attitude which makes data analysis a no-holds-barred attempt to wrest all the information possible from the data. The statistical summary of the Whitetop data by Flueck[14] is a good example of a report presented from the point of view of data analysis.

The methods of data analysis include multiple regression procedures, graphical analysis, new visual display mechanisms, auxiliary measurements and designs, and the use of partitioning that is post hoc stratification. Examples of strata or factors that have been found useful for understanding physical processes and designing new experiments are cloud temperature, storm types, and wind direction. Numerous specific examples are given in other chapters of this book.

The advent of the high-speed computer has been a great help in the development and use of data analysis procedures, especially in regard to simulation experiments. It is now considered highly desirable that no field program should be initiated until simulation analyses have been computed, using observations of the same kind as those to be obtained in the experiments. Such analyses can be used as a basis for decisions on the size of the target area, the length of the experimental unit (day, storm, or month) and an appropriate statistical technique to be used. The theoretical power of a proposed test can be compared with its actual power, and various tests can be compared

for robustness—the lack of susceptibility to the effects of nonnormality. The effects of transformations can be examined and bias corrections made as illustrated, or by the more general methods discussed in considerable detail.[40] All these procedures involve additional time and money, but weather modification experiments are generally expensive and a reasonable effort to make use of the numerous methods of data analysis should result in benefits that easily outweigh the costs.

SOME PROBLEM AREAS

Sources of Variability

High Natural Variability. We have seen earlier that the large variations in time and space of atmospheric phenomena affect the atmospheric prediction problem, thus making evaluation more difficult. This variability presents problems in the measurement of important variables which are needed to specify the current state of the atmosphere. Questions are raised as to the representativeness of a measurement or set of measurements. What is the sampling error in the measurement of mean precipitation over an area, and how does it depend on the density of the observational network? What kind of measurement should be used to obtain an estimate of some quantity? For example, in estimating the total rainfall over an area a few rain gauges might give an unbiased estimate of the mean rainfall but with a large error or variance. On the other hand, a modern radar which covers the spaces between the gauges and integrates over the whole area might give estimates with a lower variance but with some bias caused by the calibration factor. In the statistical literature there is often stress on the "best unbiased estimate," but in actual practice we are often more concerned with mean-square-error (MSE), to which both bias and variance make a contribution, as indicated by the formula

$$\text{MSE} = b^2 + \sigma_e^2,$$

where b is the bias (mean error) and σ_e^2 is the variance of the errors. We should not take it for granted that the "best unbiased estimate" is always the one to use. This is just one of many problems that arise. On some of these progress has been made and on others much more research is needed.

The Treatment Effect Model. There is increasing evidence that seeding effects may be either positive or negative, depending upon initial conditions that are not well understood, and a random effects model has been proposed. Unfortunately, in most cases, present physical knowledge does not permit a precise formulation of a seeding effect model. To design an experiment with appropriate tests, consideration must be given to the form of the effect, and the closer the model of the treatment is to the actual physical process, the more is learned from the experiment. In developing a significance test for the null hypothesis, this is especially important, for the calculation of its power will depend upon the alternative hypothesis. In the early days of rain-making, many of the hypotheses were vague or indefinite, and this made effective testing difficult. (Bertrand Russell once said that it is more important that a hypothesis be exact than that it be true!) As described earlier, attempts have been made to devise tests for nonstandard hypotheses, but it is easy to agree with the statement by Moran that " . . . a much more important problem than testing of the theory that seeding both increases and decreases precipitation is that of finding what meteorological situations cause the effect to be positive in some cases and negative in others."[24]

Uncontrolled Background Effects. Randomization permits valid sampling distributions and although it controls bias *on the average*, it should not be assumed that it solves all the problems connected with sampling from variable populations. For example, there is no assurance that randomization will provide the same number of treatment and control cases, unless special designs are used. Furthermore, and more importantly, there is no assurance that for any particular experiment randomization will balance out the uncontrolled variation and provide equivalent clouds for the two (or more) test groups. The failure to obtain equivalent groups for experimentation has been referred to as "uncontrolled background effects" (see Chapter 6), and analyses of the Whitetop experiment by Flueck[14] and Lovasich, Neyman, Scott, and Wells[20] indicate that there was a background effect not balanced out by the randomization. This is another reason for making efforts to find significant covariates, preferably before the experiment begins, so that they can be incorporated into the design and analysis.

Instrumental and Measurement Errors. This type of problem is common to most experimental work and need not be discussed here in any detail. In weather modification research we are especially concerned with the density of the rain gauge network and the errors in the measurement of rain, snow, or hail intensity, particularly in remote areas where automatic recorders must be used. In studying the physics of precipitation processes, we are interested in measuring (or counting) natural and artificial nuclei, cloud droplets, etc. The difficult problems associated with these measurements constitute another source of variance in the already complicated task of evaluation.

Data Processing. Modern methods of meteorological and statistical analysis rely upon large and complex data systems to supply the necessary information. As these systems have become more complex, they have tended to become unwieldy and difficult to keep in order. Any individual analyzing data from weather modification projects realizes that the data are often incomplete and subject to errors and wild measurements. The problem of "cleaning and drying" data becomes a difficult and time-consuming task, although a necessary one, for no analysis is more reliable than its component parts, one of which is the reduction of data. Various methods have been devised for the editing and verification of data, and work in other fields has contributed to the solution of some of the problems, but much more work needs to be done. A related question has to do with the preservation and exchange of data and rights to access to it. Many weather modification projects or experiments are sponsored with public funds, and, even with privately financed projects, the public interest is involved when attempts are made to modify the environment. Suggestions have been made for the establishment of data centers for the preservation and exchange of data from various weather modification projects, and no doubt this proposal will receive increasing attention if weather modification becomes recognized as an important national program.

Sources of Possible Bias

Selection of Experimental Units. In a nonrandomized experiment or commercial-type operation, it might be possible for the experimenter to select clouds that could be expected to precipitate naturally or to seed only those storms that could be expected to have a higher target-to-control precipitation ratio than normal, and then take credit for the increased precipitation. We have

seen that one purpose of randomization is to eliminate this type of bias, but it appears that there are sometimes subtle types of bias that can still creep into well-designed experiments. One of these is the selection bias, discussed by Stigler,[37] that *may* be introduced when suitable experimental subjects arrive sequentially. For example, suppose in a cloud seeding experiment that randomization is by pairs and that the second cloud is selected by the experimenter after the treatment (or nontreatment) of the first. If, after the experiment on the first cloud had been performed, the experimenter knew what the treatment was, a deliberate or subconscious bias could affect the selection of the second cloud. Even though the instructions on the first cloud were not known to the experimenter, he might try to guess what the instruction was on the basis of visual observation of the first cloud (the questionable value of such "eyeball" evaluation is discussed in Chapter 6.)

Brier[7] discussed this type of situation and suggested methods for testing for and estimating this bias. Briefly, these methods involve using a covariate and comparing the seeding effect of the first clouds of the randomized pairs with that of the second clouds. No evidence of such bias was found in the Caribbean and Florida series of cloud seeding experiments reported in Chapter 6.

Statistics of Time Series. The validity of significance tests and estimates of standard errors depends upon the assumption of independence of the observations. In meteorological data where considerable spatial and temporal correlation is likely to exist, these assumptions may not always be valid. Another question about meteorological time series is whether the characteristics of the series specified by quantities such as means, variances, and regression coefficients are stable in time. This has been discussed earlier in the section on using regression equations based on a historical period to evaluate nonrandomized commercial seeding projects. This problem of the stability of distributions is relevant to the question of the applicability of the Bayesian approach to evaluation and to the question of whether randomization can be dispensed with when physical modeling reaches a point where a nearly perfect forecast of some phenomena, say the growth of a cumulus cloud, can be made. Suppose that a number of such predictions have been verified, and at some later time a seeding operation is performed, with no control cases. Furthermore, let us assume that, unknown to us, the background of effective ice nuclei in the atmosphere has changed, due to a man-made cause or a natural cause such as a volcanic explosion, a change in solar radiation, or an influx of meteor dust. An unusual apparent response of the cloud might erroneously be ascribed to the seeding agent used, rather than to the changed environmental conditions which could be responsible. The point is that it is doubtful whether any careful scientist would ever have absolute confidence in both his physical understanding and the detailed knowledge of the initial environmental conditions to dispense entirely with randomization. Some control cases would be desirable, if only for psychological reasons.

Partitioning, or post hoc Stratification. The methods of data analysis have been found useful in adding new dimensions to past experiments and in designing new experiments. Among these is the procedure where the investigator stratifies the data according to factors or variables that were not originally controlled in the investigation, but now appear meaningful to him. When the results of such investigations are presented in terms of significance levels, Calvin[10, 11] and others have pointed out the dangers of interpreting these as probability levels and suggest that they be considered as no more than indices. This is the problem of multiplicity and selection which exists when many tests are applied to the same data. This does not mean that such methods of analysis should not be used, but that scientists must carefully distinguish between "indications" and "conclusions" and insist upon independent data to test hypotheses formulated from post hoc analyses.

Optional Stopping and Other Causes. This problem arises if an experiment or operation to augment rainfall is stopped when one has enough precipitation in the target area, something like always dropping out of a poker game when one is ahead. Harris[17] studied this question in connection with an evaluation of commercial rainmaking projects using the historical regression method as control and tentatively concluded that a 5 to 8 percent bias was possible. Brier, Carpenter, and Kline[6] made a small study on similar commercial operations to investigate the effects of starting during a dry period. These results suggested that the bias, if any, was negative; that is, any real positive effect of seeding would be underestimated, making it more difficult to detect such effects. In any case, it appears that more quantitative investigations are needed on both these questions as well as on the more general topic of sequential experimentation related to weather modification.

PROPOSED DIRECTION OF RESEARCH

Areas for Improvement and Advancement

Predicting the future direction of progress in design and evaluation methods could be as difficult as forecasting the weather or the response of a cloud to a seeding treatment; however, it is possible to recognize some of the present problem areas and where and what the needs are, and the trends in current technology that might lead to some solutions.

Predictions from Physical Models. No doubt the continued development of more realistic models of atmospheric behavior will make major contributions to test decisiveness by providing better predictors or covariates and clarifying the proper treatment effect model to be used under varying conditions. The development of the physical theory and increased understanding of the processes should give us better guidance by telling us what to test, how and what to measure, and what auxiliary measurements are most useful for hypothesis testing and data analysis. Should we be paying more attention to duration or intensity of precipitation, or to its frequency? Can we decrease the frequency of hail or the intensity of hurricane winds without affecting the beneficial rains? These are just some of the questions that we would hope to answer in the next decade.

Measurement Techniques and Instruments. The rapid development and improvement of measuring devices during the last 25 years can be expected to continue, probably with emphasis on remote sensing devices, such as those mounted on satellites or aircraft. Within the next decade we can expect to have accurate and reliable automatic radar information on the integrated rainfall over an area. Instrument specialists could point out numerous other possibilities.

Design and Analysis Procedures. Advances in statistical theory and practice should provide some improvements in experimental design, testing and estimating procedures, and data processing and analysis. No new or revolutionary concepts are expected, but some of the older tools can be sharpened up or modified. Simulations on high-speed computers can be used to compare the relative merits of various tests, by calculations of power and robustness, before field experiments are started. Improved systems for the storage, retrieval, and processing of data should permit data analysis to be kept up to date. In the past the complete detailed analysis of the experimental data was not finished (or even started) until months or years later. If data analysis can be kept current during the course of an experiment, then there is the opportunity to formulate new or modified hypotheses based on the findings, while there is still time for verification or testing during the remainder of the experiment. This is not meant to imply that basic changes should be made in the original design, but that auxiliary designs can be incorporated along with appropriate and valid tests. This seems to be an area where the significant contributions could be made to scientific advances.

Bayesian Analysis

In the past few years there has been an increasing interest in the Bayesian approach to problems of inference and decision in many different areas. In the area of weather modification Simpson and Pezier[33] have outlined a Bayesian approach to the evaluation of multiple cloud seeding experiments and Boyd[5] has examined the problem of hurricane modification within the framework of decision analysis. Bayesian inference combines prior information or probabilities, whether objective or subjective, with the information provided by the data to produce a posterior distribution which yields a Bayesian "significance" test; that is, the calculation of an odds ratio in favor of or against a null hypothesis. Bayesian inference has been a controversial topic, as discussed in Mosteller and Tukey[25] and elsewhere. In many practical cases classical and Bayesian methods give essentially the same results. But in cases where considerable earlier information is available, they may differ since classical methods do not as a rule incorporate such information explicitly into the analysis, unless it is incorportated into the design of the investigation. Bayesian statisticians would argue that their methods can be applied successfully to problems where the orthodox methods fail completely or can be used only with great difficulty. Opponents will argue that the method is weak because of the need to include prior information in a quantitative form. For example, in a weather modification experiment intended to alter the frequency distribution of natural precipitation falling from a cumulus cloud, one is concerned with more detail than merely the distribution function chosen to represent the past observations. There is also the question of the constancy or stability of this function with time (e.g., variations in storm types, climatic trends, or man's influence on the environment). In any case, more application of the Bayesian approach to evaluation in weather modification can be expected and this might help to clarify the situation.

The Bayesian approach to decision, when combined with techniques such as decision analysis, offers an attractive approach to handling some difficult problems. Weather modification is more than a scientific topic; it involves economic, social, and political consequences, as well as others. It is not difficult to find situations where an increase of 0.5 percent of rainfall would more than pay for the cost of a weather modification operation, including the necessary scientific research. Since at this time there is no reasonable chance of detecting such a small amount, it is clear that decisions on weather modification are not going to be made on the grounds of physical effects alone. Lindley,[19] a Bayesian methodist, argues that the only reasonable way to reach a sensible decision is to describe the three major steps he considers important. These involve (1) the quantification of the uncertainties in terms of values called probabilities, (2) description of the various consequences of courses of action in terms of utilities, and (3) on the basis of calculated probabilities, taking the decision that has the greatest expected utility. Agreement with these principles does not imply that the task is an easy one. The first step often may involve problems of subjectivity and personal opinions. Important contributions to this problem have been made recently in a basic paper by Savage,[31] who has discussed an approach to methods of eliciting personal probabilities and has shown the relationship to statistical decision theory. Difficult problems arise in the application of the second step. Consider, for example, the possibility of modifying a hurricane for the purpose of changing its course. In assigning utilities, how many drownings in Key West balance x dollars of property damage in Miami? How do you compare a politician's villa and a number of more modest dwellings of less influential people? These are some of the types of questions that could arise if scientific progress ultimately gives us considerable ability to control the weather. (These questions arise whether or not you attempt to quantify them in a Bayesian framework.) Additional problems related to the decision to seed hurricanes are discussed in the recent article by Howard, Matheson, and North.[18] It would be an interesting turn of events, if upon attaining sufficient scientific knowledge and the technical means of controlling the weather, we found that we didn't want to do so because of inability or unwillingness to make the hard decisions of a political and social nature that would be required.

SUMMARY AND APPRAISAL

In these few pages I have tried to give a brief description of some of the techniques used to evaluate weather modification experiments and projects and to indicate how the problem of evaluation and the methods used have added fuel to the flames of controversy that already existed among atmospheric scientists, between the enthusiastic believers at one extreme and skeptical scoffers at the other. We have seen how the conflict was intensified by the initial introduction of statistical concepts of design that were not well understood, partly due to problems of semantics, such as the use of the term "randomization." Disputes among members of the statistical community, exemplified by the arguments over the statistics in the ACWC report (see Chapter 1), did not help matters. Fortunately, today most of the questions have been resolved. There is general agreement about the need for adequate experimental controls, and increased understanding of atmospheric processes along with physical modeling has provided more specific hypotheses to be tested, as opposed to some of the vague claims made earlier. Increasing use of the methods of data analysis where the emphasis is on discovering or exposing relevant factors—on formulating rather than testing hypotheses—has helped to give statistics a more positive image as a constructive tool, but there is no need to put undue emphasis on these points or to think that these problems are unique to weather modification. Such conflicts are likely to happen in any young science with a developing technology, especially one that involves the day-to-day affairs of many people, for scientific conclusions based on experimental evidence are always subject to possible attack on statistical grounds, as elegantly phrased and put in proper perspective by R. A. Fisher in the opening pages of his classic book on experimental design.[13]

. . . Since the interpretation of any considerable body of data is likely to involve computations, it is natural enough that questions involving the logical implications of the results of the arithmetical processes employed, should be relegated to the statistician. At least I make no complaint of this convention. The statistician cannot evade the responsibility for understanding the processes he applies or recommends. My immediate point is that the questions involved can be dissociated from all that is strictly technical in the statistician's craft, and, when so detached, are questions only of the right use of human reasoning powers, with which all intelligent people, who hope to be intelligible, are equally concerned, and on which the statistician, as such, speaks with no special authority. The statistician cannot excuse himself from the duty of getting his head clear on the principles of scientific inference, but equally no other thinking man can avoid a like obligation.

REFERENCES

1. Advisory Committee on Weather Control, *Final Report,* Vol. 2, 1957, 422 pp.

2. Barger, G. L. and H. C. S. Thom, Evaluation of drought hazard, *Agron. J.* **41** (11), 519–526, 1949.

3. Blyth, C. R., Comment, another look at Bode's law, *J. Am. Stat. Assoc.* **66,** 566–567, 1971.

4. Bowen, E. G., The effect of persistence in cloud-seeding experiments, *J. Appl. Meteorol.* **5,** 156–159, 1966.

5. Boyd, D. W., R. A. Howard, J. E. Matheson, and D. W. North, Decision analysis of hurricane modification, Stanford Research Institute, Menlo Park, Calif. (NOAA, National Weather Service, Systems Development Office, Techniques Development Laboratory, Silver Springs Md., Contract No. 0-35172, SRI Project 8503, Final Report, 1971, 206 pp.

6. Brier, G. W., T. H. Carpenter, and D. B. Kline, Some problems in evaluating cloud seeding effects over extensive areas, *PFBMSP* (Weather Modification Experiments). University of California, Berkeley Press, Vol. 5, 1967, pp. 209–221.

7. Brier, G. W., G. F., Cotton, J. Simpson, and W. L. Woodley, Cloud seeding experiments: lack of bias in Florida series, *Science* **176,** 163–164, 1972.

8. Brier, G. W. and I. Enger, An analysis of the results of the 1951 cloud seeding operations in central Arizona, *Bull. Am. Meterol. Soc.* **33,** 208–210, 1952.

9. Brownlee, K. A. Statistical evaluation of cloud seeding operations, *J. Am. Stat. Assoc.* **55** (291), 446–453, 1960.

10. Calvin, L. D., Statistical aspects of weather modification, American Association for the Advancement of Science, Washington, D. C., Dec. 30, 1967.

11. Calvin, L. D., Book review of proceedings of the fifth Berkeley symposium on mathematical statistics and probability, Vol. 5, Weather modification, *J. Am. Stat. Assoc.* **64,** 1085–1087, 1969.

12. Duran, B. S., and P. W. Mielke, Robustness of sum of squared rank test, *J. Am. Stat. Assoc.* **63,** 338–344, 1968.

13. Fisher, R. A., *The Design of experiments,* 7th edition, Hafner, New York, 1960.

14. Flueck, J. A., Final report of project Whitetop, Part 5, Statistical analysis of the ground level precipitation data, University of Chicago, 1971.

15. Grant, L. O., Indications of residual effects from silver iodide released into the atmosphere, *Proceedings of Western Snow Conference,* 1963, pp. 109–115.

16. Grant, L. O. and P. W. Mielke, Jr., A randomized cloud seeding experiment at Climax, Colorado, *PFBSMP* (Weather Modification Experiments), University of California Press, Berkeley, Vol. 5, l967, pp. 115–131.

17. Harris, T. E., Possible effects of optional stopping in regression experiments, *Weather and Climate Modification, Problems and Prospects, NAS-NRC Pub.* **2**(1350), 174–183, 1966.

18. Howard. R. A., J. E. Matheson, and D. W. North, The decision to seed hurricanes, *Science* **176,** 1191–1202, 1972.

19. Lindley, D. V., *Making Decisions,* Wiley-Inter-science, New York, 1971.

20. Lovasich, J. L., J. Neyman, E. L. Scott, and M. A. Wells, Hypothetical explanations of the negative apparent effects of cloud seeding in the Whitetop experiment, *Proc. Natl. Acad. Sci.* **68**(11), 2643–2646, 1971.

21. Mielke, P. W. Note on some squared rank tests with existing ties, *Technometrics* **9,** 312–314, 1967.

22. Mood, A. M., On the asymptotic efficiency of certain nonparametric two-sample tests, *Ann. Math. Stat.* **25,** 514–522, 1954.

23. Moran, P. A. P., The power of a cross-over test for the artificial stimulation of rainfall, *Australian J. Stat.,* **1959,** 47–52.

24. Moran, P. A. P., The methodology of rain-making experiments, *Rev. Int. Stat. Inst.* **38,**-105–119, 1969.

25. Mosteller, F. and J. W. Tukey, Data analysis, including statistics, *The Handbook of Social Psychology,* 2nd ed., G. Lindzig and E. Aronson, Eds., Addison-Wesley, Reading, Mass., 1968, Vol. 2, 819 pp.

26. Neyman, J., Optimal asymptotic tests of composite hypothesis, *Probability and Statistics,* U. Grenander, Ed., Almqvist and Wiksell, Stockholm, 1959, pp. 416–444.

27. Neyman, J. and E. L. Scott, Correction for bias introduced by a transformation of variables, *Ann. Math. Stat.,* **31,** 643–655.

28. Neyman, J. and E. L. Scott, Asymptotically optimal tests for composite hypotheses for randomized experiments with non-controlled prediction variables, *J. Am. Stat. Assoc.* 699–721, 1965.

29. Neyman, J. E. L. Scott, and M. Vasilevskis, Statistical evaluation of the Santa Barbara randomized cloud-seeding experiment, *Bull. Am. Meteorol. Soc.* **41**(10), 531–547, 1960.

30. Neyman, J., E. L. Scott, and M. A. Wells, Statistics in meterology, *Rev. Int. Stat. Inst.* **37**(2), 119–148, 1969.

31. Savage, L. J., Elicitation of personal probabilities and expectations, *J. Am. Stat. Assoc.* **66,** 783, 1971.

32. Siegel, S., *Nonparametric Statistics for the Behavioral Sciences,* McGraw-Hill, New York, 1956.

33. Simpson, J. and J. Pezier, Outline of a Bayesian approach to the EML multiple cloud seeding experiments, Office of the Director, NOAA, Environmental Research Laboratories, Boulder, Colo., *NOAA Tech. Memo. ERL-OD-8,* 1971, 43 pp.

34. Simpson, J., Use of the gamma distribution in single-cloud rainfall analysis, *Monthly Weather Rev.* **100,** 309–312, 1972.

35. Skyline Conference on the Design and Conduct of Experiments in Weather Modification Proceedings (1959), National Academy of Science, National Research Council, 24 pp.

36. Smith, E. M. (1967), Cloud seeding experiments in Australia, *PFBSMS* (Weather Modification Experiments), University of California Press, Berkeley, 1967, Vol. 5, pp. 161–176.

37. Stigler, S. M. Cloud seeding experiments: possible bias, *Science* **173,** 850, 1971.

38. Taha, M. A. H., Rank test for scale parameter for asymmetrical one-sided distributions, *Publ. Inst. Stat. Univ. Paris* **13,** 169–180, 1964.

39. Tukey, J. W., *Exploratory data analysis,* Addison-Wesley, Reading, Mass., 1970.

40. National Academy of Science, Weather and climate modification: national policies and programs, Report to committee on atmospheric sciences, 1973.

41. Wu, S., J. S. Williams, and P. W. Mielke, Jr., Some designs and analyses for temporally independent experiments involving correlated bivariate responses, *Biometrics* **28,** 1043–1066, 1972.

C
PRECIPITATION
MANAGEMENT

Can we make rain? Would it have rained anyway? Did we decrease the rain downwind of the target? Did we in fact decrease the total rain? Substantial differences of opinion have arisen on how much man can change the weather. These were quite heated at about 1950 and have not disappeared yet. Probably the biggest area of disagreement is in rainmaking.

Nobel laureate Dr. Irving Langmuir once said:

The analysis of the rainfall data shows that a concentration of 1 milligram of silver iodide per cubic mile of air is enough, under the synoptic conditions that prevailed in New Mexico, to give a 1-in-3 chance that heavy rain will occur during any one day at any

place. Assuming the atmosphere to be 5 miles thick, one thus finds that to get a 30 percent chance of rain per day within a given area in New Mexico, the cost of the silver iodide is only $1.00 for 4,000 square miles.

If similar conditions prevailed over the whole United States, the cost per day to double the rainfall would be only of the order of a couple of hundred dollars.

Dr. Mason, Chief of the British Meteorological Office, said recently:

Having studied developments in this field very closely for about twenty years, but without having been directly involved either personally or officially, I cannot avoid the conclusion that, with some notable exceptions, such operations have generally failed to conform to the accepted principles and standards of scientific experiment and analysis, and are therefore incapable of providing objective answers to such questions as to whether, in what circumstances, and to what extent, it is possible to modify precipitation by artificial seeding. I believe that the majority of responsible meteorologists share my concern that, in several countries, politicians and entrepreneurs, ignorant or impatient of the scientific facts and problems, are initiating and conducting major weather modification projects without the benefit of proper scientific direction, advice and criticism, and that this may have serious repercussions on the reputation of meteorology as a science and a profession. It is no secret that such operations are being promoted, under such euphemisms as "weather engineering" and "weather management" projects, on the premises that the basic assumptions and techniques are already proven and that the remaining problems are largely of an engineering or logistic nature. But no smokescreen of management or decision-making jargon can hide the fact that the protagonists of this approach are continuing to use the same inadequate concepts and techniques that have failed to provide convincing answers during the past twenty years.

Which is true, that we understand how to do almost everything in this business or almost nothing? Reality lies somewhere in between, but probably nearer the nothing end of the spectrum than the everything.

This section of the book considers, fairly thoroughly, rainmaking and its variations, snowmaking and fog removal.

Two of the best experiments carried out in recent years in precipitation management are (1) Florida cumulus cloud modification by Dr. Joanne Simpson, discussed in Chapter 6; and (2) Rocky Mountain snowpack enhancement —the Climax experiment by Professor Lewis Grant, reviewed in Chapter 7. These experiments have both been carried out in a controlled manner and so have yielded measured answers. Both have shown increases in precipitation under certain conditions. Both are being continued to expand the area of understanding. These two experiments show how work in this field should be done. Many other experiments have not been carried out so well.

6 | Cumulus Clouds and Their Modification

JOANNE SIMPSON
ARNETT S. DENNIS

Cumulus clouds are of immense importance to man's livelihood and to his life. First, they produce more than three-fourths of the rain that waters our planet, dominating in the thirsty tropical and subtropical areas. Second, giant cumuli constitute the firebox of all severe storms, such as the hurricane, thunderstorm, hailstorm, tornado, and squall line. Third, cumulus clouds are a vital part of the machinery driving the planetary wind systems; and fourth, they act as a valve regulating the income and outgo of radiation from the earth.

Although the atmosphere is fueled by solar heating, the sun's energy is fed into it only indirectly, first being absorbed at the earth's surface and then supplied to the air from below, mainly in the form of latent heat in gaseous water vapor. Cumulus clouds are the "converters" whereby the gaseous water vapor is condensed into liquid water droplets, turning "latent" heat into "sensible" heat

Joanne Simpson is Director of Experimental Meteorology Laboratory NOAA, and Adjunct Professor at the University of Miami, Miami, Florida.
Arnett S. Dennis is at the Institute of Atmospheric Sciences, South Dakota School of Mines and Technology, Rapid City, South Dakota.

which can be sensed by a thermometer and used to drive circulations. On the average, equatorward of about 38° latitude the planet earth receives more radiant energy than it loses, but poleward of about 38° more energy is lost to space than that received from the sun. Since the 1930s, meteorologists have learned how the tropical driving circulation operates by means of thermally direct (Hadley) cells, with ascent concentrated in a narrow "equatorial trough" zone and mean sinking motion elsewhere. At high levels, the tropical cells export energy and momentum poleward to make up the radiation deficits and drive the circulations in temperate and polar regions.

Cumulus clouds form vital links in nearly every part of this engine. The water vapor fuel is mainly provided by the tropical oceans. Above these warm seas, growing day and night, trade cumuli (Figure 6.1) act as fuel pumps which carry the energy upward, releasing about 20 percent through local rainfall to drive the trade wind systems and making the rest available for these vast bands of easterlies to carry equatorward. In the "equatorial trough" the firebox function of the atmosphere is performed by only 1500 to 5000 giant cumulonimbus towers

Figure 6.1 Typical trade-wind cumulus clouds. (*a*) Oceanic clouds in "streets" lined up with the vertical wind shear, (*b*) Foreground: cumuli over a tropical island. Background: large cumulonimbus clouds (courtesy C. True, NHRL, NOAA, Miami, Florida).

Figure 6.2 Tropical oceanic cumulonimbus clouds. (*a*) Cumulus clouds. Tropical oceanic "hot towers" in the equatorial trough zone. (*b*) Cumulonimbus clouds. Clouds of a typical weak oceanic trade-wind disturbance, showing "orphan anvils" (courtesy C. True, NHRL, NOAA, Miami, Florida).

Figure 6.2   Tropical oceanic cumulonimbus clouds. (*c*) Large, multicellular cumulonimbus cloud east of Rapid City, South Dakota, Antenna of 10-cm radar of National Center for Atmospheric Research in foreground. (Project Hailswath photo courtesy G. G. Goyer).[16]

(Figure 6.2) concentrated in about 0.1 percent of the area.

These huge clouds are the cylinders of the heat engine, as well as a crucial part of its heat pump. Clustered into wavelike and vortical deformations of the flow, they raise and convert energy at a rate nearly one million times that of all human power consumption in the world. The atmosphere, however, runs an inefficient engine. Only about 2 percent of this vast energy is actually utilized in driving winds; most of it is lost to space by radiation. The role of cumuli in global energy budgets is documented in more detail by J. S. Malkus.[30]

In temperate latitudes, cyclones drawing on the energy stored in air mass contrasts are a main driving mechanism in the planetary westerlies. At high levels, these winds snake around the globe in a wave pattern. With their instabilities, these westerlies govern the fluctuating weather in which so many of us live. Over the midlatitude oceans, cumulus clouds have been found to be a key in the occasional explosive deepening of winter storms, which often bear winds of near

hurricane force. Their rapid deepening can trigger a major readjustment in wind pattern around a whole hemisphere.[43]

It is clear from the foregoing that systematic modification of cumulus clouds holds enormous potential impact upon man and his environment. Significantly enough, cumulus clouds are one of the few geophysical phenomena upon which controlled experiments have already led to definitive results, as we see later in this chapter. If we could indeed alter large numbers of these clouds controllably, a powerful tool would be supplied for managing water resources and for mitigating the disastrous destruction from severe storms. Reaching farther, cumulus cloud modification could conceivably trigger important changes in planetary circulations and, if sustained, possibly might even bring about modification of regional climates.

What portions of these possibilities are near realization today, how much lies in the foreseeable future, and what must be postponed to a later generation or put aside as sheer speculation? It is our purpose not just to provide tentative answers to these questions, but to build sufficient scientific foundations so that the reader can evaluate their soundness and prospects. To achieve this goal, it is essential to understand what is known and what is not known about cumulus clouds, and how modification experiments are carried on and evaluated.

THE DYNAMICS AND PHYSICS OF INDIVIDUAL CUMULI

Cloud science began in the 1930s and has made great strides since World War II. In its infancy, the microphysics and dy- namics of clouds were mainly approached as separate disciplines and worked on by two different groups of researchers. Today they are progressively combining into one discipline, with the same workers contributing advances in both areas. Cloud physics or microphysics concerns the particles in a cloud, namely, the condensation and freezing nuclei, the water drops and ice crystals, their origin, growth, and behavior. Cloud dynamics deals with the relation between forces and motions in clouds, with the purpose of predicting and understanding the structure and life cycles of the updrafts and downdrafts.

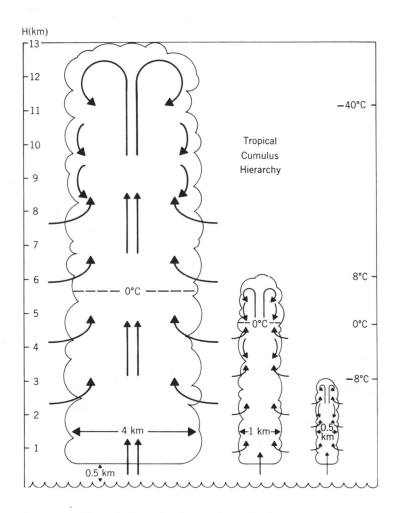

Figure 6.3   Schematic hierarchy of tropical cumulus clouds.

In Figure 6.3 we see a hierarchy of three important sizes of tropical cumuli. Naturally they come in all sizes, but we have just picked three here to clarify the explanation. Aircraft studies have suggested that cumulus towers sometimes resemble a buoyant bubble or "thermal" or when more vigorous, a growing plume or jet. Both forms have a vortexlike internal circulation near the top. The energy driving the motions comes from the condensation of water vapor into liquid cloud drops. The cloud drops grow to 5 to 20 μm radius by the condensation process; the growth in radius slows down in direct proportion to the radius increase. For condensation growth a small supersaturation must be maintained by expansional cooling of the air in the updraft, which serves to sustain it in the face of removal of vapor by the growing droplets. In the early stages there are normally between 20 and 2000 drops/cm³, depending on the effective *cloud condensation nuclei* (CCN) present and the updraft. Since the condensation process by itself would require more than the lifetime of a cumulus to produce a drop even as large as 40 μm, clearly cumulus rain requires additional mechanisms. The best candidate processes are (a) the collision and coalescence mechanism of creating big water drops, and (b) the Bergeron ice crystal growth mechanism as discussed in Chapter 3.

Most cumulus clouds evaporate without ever producing precipitation at the earth's surface. The fact that many convective clouds begin to dissipate at about the time rain emerges from the base led to the idea that the clouds are destroyed by the precipitation forming within them; the "precipitation brake" is not fully understood, but some aspects of it are discussed shortly.

A useful concept in studying the behavior of shower clouds is the "precipitation efficiency," defined here as the ratio of the mass of precipitation reaching the ground to the total mass of water vapor passing upward through the cloud base. The precipitation efficiency (PE) is obviously zero in small non-precipitating cumulus clouds. It increases with cloud size, up to a point, other conditions being equal. A PE of 10 percent has been quoted for small thunderstorms in the central United States, and it exceeds 50 percent for some large thunderstorms.[7] However, there is some evidence that clouds with updrafts in excess of, say, 25 m/sec are inefficient. The precipitation efficiency of an isolated hailstorm in South Dakota with a water vapor flux of 5 kton/sec* has been calculated at only 3 percent.[13] Considerable moisture is lost by evaporation at the edges of such clouds in their downdrafts and often much of the condensed water is pumped away through a large anvil of mostly frozen particles streaming downwind from the cloud top (Figure 6.2b). The icy anvil evaporates much more slowly than the liquid portion of the cloud body.

We said that the cumulus engine is created and maintained by the release of condensation heating. To build up the science of cumulus dynamics, we must find out just how the heat release drives the motions. Briefly, when the clouds are formed, the release of latent heat (of condensation) makes the cloudy air warmer than its surroundings and thus less dense, creating buoyancy. The buoyancy establishes and supports the updraft, maintaining the slight supersaturation required to keep the condensation process going.

Like people, cumulus clouds have a life cycle. They are born, they grow up, and eventually age and die, but unlike people, the fatter they are, the longer and more

---

* One metric ton = 10^3 kg. Therefore, one kton = 10^9 g. An acre-ft of water weighs 1.23 kton.

vigorously they live and the taller they grow. Small trade cumuli (Figure 6.3) usually enjoy an active lifetime of only 5 to 10 min, and the medium ones thrive on the order of 30 min at most. The total amount of water vapor passing upward through the base of a small cumulus cloud may be 1 kton or less. A giant cumulonimbus in a hurricane or squall line can be active from one to several hours. It may process 10 kton/sec or more than 50,000 kton during its lifetime, while producing heavy rain, lightning, and possibly hail. But at all times, natural existence is a desperate struggle for a cumulus cloud; its life is a precarious balance between the forces of growth and those of destruction. We have just described the buoyant growth forces. What are the destructive forces and how do they work?

The science of cumulus dynamics began in 1946 with the discovery of resistive forces and their documentation by measurement. Before 1946 meteorologists had been largely unaware of interactions between clouds and their surroundings. Ignoring them, they had derived very beautiful stability criteria (called the "parcel theory") for cloud growth; these oversimplified criteria depend only upon the upward rate of temperature decrease (lapse rate) in the surroundings.[17] The increased interest in tropical meteorology stimulated by World War II revealed that there was something drastically wrong with predictions that were made using "parcel" concepts. According to the parcel idea virtually all tropical clouds should penetrate into the stratosphere, topping at the level of about 100 mb or 54,000 ft.

But in real life, only the fat giant clouds actually penetrate to the tropopause. An even more obvious deficiency of parcel theory is that most tropical clouds top out at heights between 3 and 6 km, at the very levels where the parcel buoyancy is greatest. This observation suggests some kind of drag or friction force in near balance with the buoyancy.

Just after World War II, Henry Stommel[50] postulated that cumuli were "entraining" or mixing into themselves air from their drier surroundings. Using the newly available aircraft measurements of temperature and humidity made inside and directly outside of clouds, Stommel devised a mathematical and graphical method of computing the rate of entrainment[27] which is expressed mathematically as $(1/M) \, dM/dz$ where M is the mass in the cloud element and z is the vertical coordinate. In small trade cumulus clouds (Figure 6.1), calculations from observations show that $(1/M) \, dM/dz$ is about 10^{-5}/cm. This means that the mass entrained is 1 in 10^5 cm or 1 km—or that in a rise of 1 km, the cloud entrains into itself just about as much environment air as it originally contained. This large dilution with dry air drastically reduces the moisture lifeblood of the cloud, and thereby cuts down its latent heat release and buoyancy.

Following the discovery of entrainment, the Woods Hole group made many advances in understanding entraining cumuli in a shearing wind field. By a "shearing wind field" we mean that the horizontal wind vector is varying with height in the atmosphere. The simplest case, common in the tropics, is that of a speed change only, with the wind strength increasing or decreasing upward while the direction remains unchanged. This situation is illustrated in Figure 6.4.[25, 26] When the wind increases upward, the cumulus moves more slowly than the air surrounding it, because it is bringing up slower-momentum air from below. When the wind decreases upward cumuli will, conversely, move downstream faster than the wind. We were thereby able to explain why small clouds are so often *not* moving with the wind speed. In the case of large cumuli, there are additional, more complicated reasons why clouds move sometimes with a considerably different speed and direction from that of the wind.[38]

We were also able to show with both theory and aircraft observations that the cloud entrains mainly on its upshear side (tail of white arrows), and it sheds moist air or "detrains" on its downshear side (head of white arrows). Thus the cloud imparts its heat, moisture, and momentum to its surroundings, a vitally important exchange which had been virtually ignored before 1946. A cumulus is not an inert entity, but rather a dynamic balance between rapid growth and equally rapid destruction. A component of its motion is often caused by growth on one side and dissipation on the other.

After the discovery of entrainment, one of the next major problems was to specify what cloud and environmental factors control it. This problem is by no means completely or satisfactorily solved today. Great progress on it was contributed by laboratory experiments in the 1950s (see Figure 6.5). Upside-down laboratory "clouds" were made in water tanks using miscible fluids very slightly denser than water, released by overturning a semicir-

cular cup at the top of the tank.[41] These experiments showed that the entrainment rate decreases in direct proportion to the increase in horizontal size of the buoyant element.[40, 51]

In the case of the real atmosphere, observations leave little doubt that the giant cumulonimbi of Figure 6.2 are diluting at a slower rate than are the tiny cumuli in Figure 6.1, so the inverse dependence of entrainment on horizontal size is verified at least to first order.[29] Now we can understand much better the cloud hierarchy of Figure 6.3. Since the largest cloud shown has a diameter eight times that of the smallest, it will require about 8 km instead of 1 km to be diluted approximately 50:50 with the surrounding air. Its central core is protected from the inroads of dry air by a much greater volume per unit surface area, and hence the tower can bulge upward through most of the whole depth of the troposphere before losing buoyancy. The "hot towers" performing the firebox function of the hurricane and equatorial trough zone are 3 to 6 km across[35] and so can pump the sea-warmed surface air virtually undiluted to the tropopause and sometimes

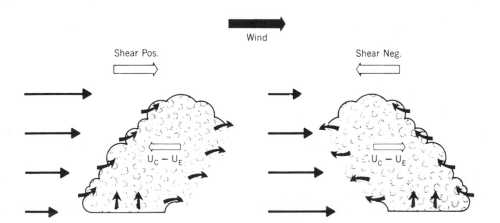

Figure 6.4 Illustration of how cumulus clouds lean with the vertical wind shear; on the left, the cloud velocity U_C is less than the wind speed U_E, while the reverse is true of the cloud on the right.

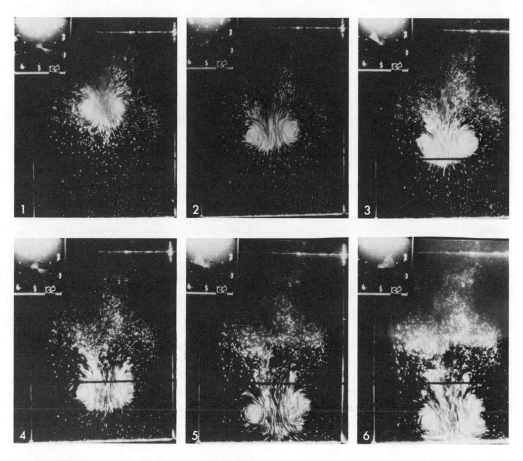

Figure 6.5 Development of laboratory tank "cloud" as a function of time. A blob of salt solution is released into a tank of pure water. The "cloud," being denser than its surroundings, moves downward. Its vortexlike circulation is made visible by neutral-density, white painted particles (courtesy Dr. P. M. Saunders).[40]

even into the stratosphere. The factors controlling the diameters of cumulus towers are still not completely documented, except that we know that the big ones are found in those locations where the large-scale airflow is convergent at low levels. In addition to entrainment, we will see that additional destructive forces are provided by the weight of suspended water substance and by downdrafts which may act to compensate the rising motions within the clouds. Cumuli may also be destroyed if their source of moist, low-level air is exhausted or cut off.

CUMULUS MODELS

The power and the credibility of any weather modification effort is enhanced by many factors when a model exists prescribing quantitative relationships for the modified versus unmodified system. When we have a model, we are saying that we understand the important physics of the system and the causality of how the modification works. Furthermore, when a working model exists, generally it can predict the effect of the modification upon more than one measurable quantity, thus increasing our power to test the validity of the modification hypothesis.

Cumulus processes can be modeled by analogy in the laboratory or mathematically by solving a set of differential equations describing their behavior, so that the solutions predict the cloud properties as a function of space and time, given specified initial and boundary conditions. The equations, their physical background, and their solutions are what we commonly refer to when we speak of "cumulus models." A word of caution is in order here regarding all models of atmospheric processes. First, we cannot yet formulate the exactly correct equations governing them. Second, even if we could, no foreseeable computers would be large enough or fast enough to solve them, and third, even if the foregoing were realized, it would not be profitable to try to solve the exact equations exactly because our measurements can never hope to be accurate enough to specify the "initial conditions" with commensurate precision. Hence all meteorological models are, and will remain, hierarchies of approximations and simplifications. Their adequacy must be judged by the degree to which they predict the phenomenon in question; the predictions must always be checked by measurements, which in their turn are never perfect.

In several important areas of cumulus modification it has been hypothesized that mainly the microphysics or particle structure is changed, and that alterations in the dynamics or motion field are sufficiently small and/or unimportant so that ignoring them will not lead to serious errors in model predictions.[4, 20] If this is justified, then the modeler's task is much less difficult than it is in situations where dynamic changes are involved or when dynamical-physical feedbacks strongly influence particle growth. Today, many meteorologists are coming to believe that we can advance little further, even in microphysical modification activities, without better understanding and modeling of the complex dynamical-physical interactions.

Concerning cumulus dynamics, models were virtually prohibitive before the introduction of the high-speed computers in the early 1950s because key processes are highly nonlinear. The accelerations driving the motions are created by horizontal density differences, but in turn the densities are altered by the motions of the air and of the particles, which thus change the accelerations in a constantly operating "feedback loop."

The reason that computers permit, for the first time, meaningful approaches to this type of problem follows: the nonlinear differential equations describing cumulus motions cannot be solved "analytically" by formal mathematics, but we can obtain "numerical" solutions in specific cases by feeding the equations into a computer, which then solves them by successive approximations or iterations, completing work in a few seconds which might take months to reproduce by hand.

Two classes of numerical models are under development to simulate cumulus dynamics. The more sophisticated and at first glance more rigorous approach is to take the complete hydrodynamic equations of motion, energy, mass, etc., and set them up in finite difference form, that is, divide up the space occupied by the cloud and its surroundings into a grid of discrete points separated by distances much smaller than the cloud. Then the equations are solved by the computer in a series of finite time steps beginning with a prescribed "initial condition," which is usually no motion but a diffuse blob of slightly less dense air at the center of the grid. The density field then initiates motions and the computer prints out results for the "model cloud" every few minutes. The output at each grid point consists of velocity, temperature, water content, etc. These are called "field-of-motion" models.[37, 39]

The second type of dynamic models we call "entity" models. In these a cumulus is likened to a jet or a plume, a buoyant bubble, or some other physical entity that can be seen and recognized. The forms have been suggested by time-lapse motion pictures of real clouds, by laboratory experiments (Figure 6.5) and by the results of numerical experiments.[1]

A major simplification offered by entity models is the reduction of the number of differential equations; the primary one to be solved is usually for the rise rate of the entity or the updraft at its core. Key features of the models are semi-empirical laws derived from measurements or theories concerning the entities; these permit the complete specification of every "parameter" in the equations except for the cloud properties we are trying to predict. For example, many current entity models are based on the inverse radius entrainment law. Although plainly oversimplified, these models have been surprisingly successful. Their virtues and shortcomings are described later in the framework of the modification experiments on cumulus dynamics which stimulated their development.

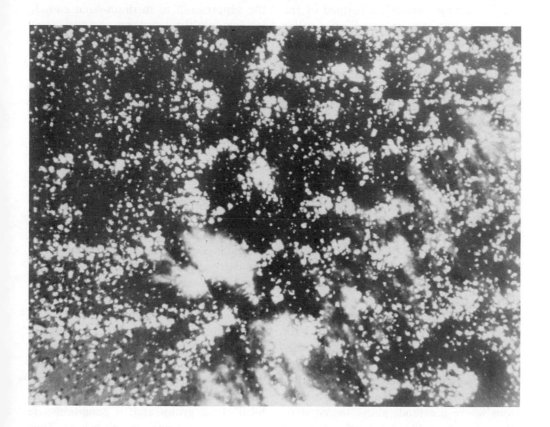

Figure 6.6 Photograph of polygonal pattern in tropical Atlantic cumuli (just east of Barbados) taken by the Apollo 10 astronauts in May, 1969. The picture covers a width of about 200 km (courtesy Dr. J. Kuettner, ERL, NOAA, Boulder, Colorado).

CUMULUS GROUPS AND INTERACTIONS

A cumulus cloud rarely occurs in isolation. Nearly always, cumuli form in groups, lines, patterns, and clusters (Figures 6.1a, 6.6). Some of the most important effects of cumuli, such as heavy rains and severe weather, occur when two or more cumulonimbus clouds join forces to become a merged complex, or "merger." For example, in Florida, a single isolated cumulonimbus thunderstorm may produce in its lifetime 100 to 2000 acre-ft of rainfall, while a merger of two or more such clouds often produces 5000 to 50,000 acre-ft.[56]

Moreover, some of us believe that nearly all large cumuli are formed by the merger of adjacent or successive smaller ones. Hence it is vital to study and experiment on cumulus interactions. Unfortunately, in 1973 we are no farther ahead in this task than we were in 1945 in treating individual cumuli. And the problem is at least 10 times more difficult because it involves the mesoscale (10 to 100 km) of atmospheric phenomena, virtually unmeasured and unknown. Numerical models of cumulus groups and interactions are in their beginning stages.

Some study, with aircraft and satellites, has been devoted to the formation and patterning of oceanic cumulus groups.[33] Over the tropical oceans, cumulus bases are found at roughly 600 m. The "subcloud" layer is neutrally stable and homogeneous; that is, water vapor is quite evenly distributed both in the vertical and horizontal. The vertical thickness of the homogeneous layer is often deformed upward, sometimes in a wavelike manner, and in those regions where it extends above the condensation level* of the subcloud air, small cloudlets break out, most commonly

* See Chapter 3 for explanation of "condensation level."

lined up with the wind shear (Figure 6.1a). Causes of vertical deformations of the homogeneous layer are not all well-known or modeled; they are sometimes caused by localized warm spots in the ocean[28] and perhaps more commonly by patterns of convergence set up by uniform heating, surface friction, and wavelike phenomena.

The most spectacular displays of oceanic cumuli are found in disturbances, both of the tropical variety (whose extreme form is the hurricane) and the extra-tropical frontal cyclone. In these regions, the cumuli are associated with lines or bands of strong convergence in the low-level wind flow. Whether the clouds are the hen or the egg has not yet been resolved. A sign of a disturbance is the suppression of medium-sized cumuli, leaving only the very large and very small ones in the sky. *The modification potential of cumuli almost surely varies between fine weather and disturbed conditions,* an hypothesis of the utmost importance to be borne in mind throughout this and succeeding chapters. Time-lapse motion pictures suggest that larger oceanic cumuli are built up by aggregation and conjunction of small cloudlets. At least until they reach the precipitating stage, individual oceanic cumuli do not have "roots" in updrafts or downdrafts extending below their bases. The mixed layer appears no different under cloud groups than in the intervening clear spaces. However, a major change comes about when a cloud reaches the shower stage and falling precipitation induces a strong cold downdraft which penetrates all the way to the sea surface. These downdrafts are of vital importance in accelerating sea-air exchange in tropical disturbances.[14] They can both kill off the local cloud group and, if conditions are right, initiate a new one elsewhere. They are possibly relevant in man's potential ability to change the atmosphere via cumulus modification, as we see later.

DESIGN AND EVALUATION OF CUMULUS MODIFICATION EXPERIMENTS

Two immense obstacles have opposed man's attempts to modify the weather. The first is the enormous amount of energy expended in natural atmospheric processes. A large thunderstorm releases as much energy as the fusion energy of a hydrogen superbomb while even a moderate-strength hurricane converts, through its cloud systems, 400 bombs worth in a single day. Hence puny man must seek an Achilles heel in the system, which when struck, can instigate a sizable reaction from an energetically small trigger. Fortunately, we have seen that lives of cumulus clouds are a precarious seesaw between growth and destructive forces; recognition of this, in fact, led to the discovery of some of the triggers we describe.

A more serious obstacle, one which stands in the way of our sound evaluation and judgment of modification efforts, is the enormous natural variability in atmospheric phenomena. The most striking lesson these writers have learned from 30 years of cloud study is that a cumulus cloud can do virtually anything all by itself, without any interference by man. In a field of identical-looking cumuli, one or several can explode to thundering cumulonimbus, while the rest humbly die. Given two apparently identical convective storm systems, one can rain pitchforks, while the other, with indistinguishably similar looking clouds, remains dry.

High natural variability has at least two unfortunate consequences. The first is that the average layman grossly overestimates what has been and can be accomplished in weather modification, particularly since the news media tend, for example, to play up those cases where floods follow cloud seeding. The second, and more crucial consequence as far as experiment design is concerned, is that we just cannot conduct a meaningful modification experiment without sound and rigidly enforced statistical controls.

"Eyeball" evaluation is of little value. A great deal of the unproductive controversy surrounding the modification efforts of the past two decades has arisen because people have kept making assertions like the following: "I just know the seeding was a success: the seeded clouds behaved very differently from the neighboring unseeded clouds—in fact, I have just *never* seen any clouds behave the way those seeded clouds behaved." The cloud expert will reply, "Just stick around for awhile, chum, and you'll see natural clouds behaving that way."

Seriously, what is the solution to this discouraging dilemma? In the current state of our knowledge, randomization is a most satisfactory procedure. That is, unknown to the experimenters, an honest coin must be tossed (or a sealed instruction opened) which says "treatment" or "no treatment" for the given experimental situation, be it single cloud, target area, or day. By this procedure we can obtain, with minimum bias, a treated and untreated sample of cases which can be compared by the methods described in Chapter 5 to arrive at not only a quantitative assessment of the treatment effect, but equally importantly, an estimate of how confident we are that the measured differences between treated and untreated populations are due to the treatment and not due to chance fluctuations. Our models must, in fact, still be tested by the results of the experiments and so we cannot yet have full confidence, without randomization, that the model does indeed confirm any modification hypothesis.

Productive modification experiments are more likely if backed by a model. An example of this situation is the single cloud experiments discussed in the section, "Dynamic Seeding." In cases where we do not yet have a model, as in multiple cloud or area seeding experiments, we may hope that randomization and judicious statistical analysis lead us not only to evaluation of seeding effects, but to a start at modeling them.

The design strategy of a cumulus modification experiment depends mainly upon what effect we desire, and then upon many other factors such as technique availability, possible side effects, and expense.

Nearly all cumulus experiments that have been actually and deliberately conducted so far have involved "seeding" the clouds with some kind of small particles. In some cases, the particles are dispersed from the ground, counting on air currents to get them into the clouds. In most of the others, aircraft are used to dispense the seeding materials. Aircraft seeding can be executed by flying with dispensing systems or generators upwind of the target clouds, by circling below cloud base and relying on updrafts, or by dropping materials from above directly into cloud top or by flying horizontally through the cloud. Seeding from ground generators has been going out of fashion in the past decade primarily because of the difficulty of reliably introducing enough material into designated clouds in designated areas, and secondarily because of possible decay, contamination, or undesired voyaging of the seeding materials. Aircraft seeding has been taking over because it affords accurate targetting and opportunity for measurement and observation. Aircraft are, however, rather costly for prolonged usage. For years, scientists in the Soviet Union have successfully seeded cumuli with artillery shells and rockets (Chapters 10 and 11). Radar is used to determine the parts of the clouds to be seeded. This economic and, quite possibly, more effective technique might be useful in other nations in

those experimental situations where air traffic problems could be circumvented.

Rain augmentation in cumuli has been attempted in two ways, by increasing or accelerating the coalescence process and by initiating or increasing ice particle growth in the presence of supercooled water. In cumuli whose tops extend above the freezing level (roughly 4 to 4.5 km above sea level in tropical or summer air masses), these processes are virtually inseparable, although we must try, for explanation purposes, to separate them. Historically, ice-phase seeding was discovered first and has been practiced more extensively; the proper historical perspective is developed in Chapter 1. Here we begin with coalescence seeding for the sake of clarity and logical development.

CUMULUS MODIFICATION TO ENHANCE COALESCENCE

The growth of cloud drops (~ 5 to $20\ \mu$m) to precipitation size (> 0.5 mm or thereabouts) by coalescence in cumuli is related to the updraft speed and dimension, water content, cloud lifetime, and the initial drop-size distribution, which in turn is controlled by the size spectrum of the condensation nuclei (CCN). Most favorable to growth by coalescence is the presence of large drops among much smaller ones—the large ones descend relative to the small ones and hence are able to collide with and "collect" them. Over the oceans, giant sea salt nuclei (1 to 10 μm in radius) may be a major factor in setting off coalescence, contributing to the observation that marine cumuli rain more readily than their counterparts over land.

Beginning with one of the great patriarchs of weather modification, Nobel laureate Irving Langmuir, coalescence models attained by the late 1960s a high degree of sophistication with two reservations. The first is that some factors controlling "collection efficiency" (see Chapter 3) still must be determined empirically; second and more important, dynamic interactions and entrainment were left out. Early coalescence models were based on "continuous" collection or accretion, where a single large drop falls through a uniform distribution of water (the small drops). Later models[2] treat the much more difficult subject of "stochastic collection" where the large drops are also permitted to collect each other.*

An exciting aspect of coalescence with possibly enormous impact upon modification is the so-called "Langmuir chain reaction."[22] As raindrops grow large, they become unstable and start breaking up. The size at which this breakup begins is not well-known, but by the time drops have reached 6-mm diameter, the breakup probability is very high. Each fragment can in turn grow to breakup size and repeat the process. If the cumulus updraft is strong enough to sustain the near-breaking drops (nearly 10 m/sec is the terminal velocity for a 6-mm raindrop), a rapid acceleration of rain growth would be possible. Both further modeling and further measurements are required to ascertain the frequency and conditions for the natural occurrence of chain reactions and to assess man's chances of instigating one.

It seems plausible to hypothesize that the introduction of artificial precipitation embryos could shorten the time required for the appearance of rain and could sometimes increase the precipitation efficiency in short-lived clouds. Methods attempted for introducing artificial embryos have included the use of water

spray[8] and hygroscopic powders and solutions.*

An alleged difficulty in the water spray approach to coalescence modification is that large quantities of water must be injected into clouds to produce significant effects. A way to mitigate this operational problem is to use hygroscopic materials as seeding agents. With these, calculations suggest that introduction of 1 kg of material into a cloud could result in the collection of 5 to 10 kg of cloud water into artificial raindrop embryos.

One of the main materials used in hygroscopic experiments is fine powdered salt (sodium chloride). It is difficult to grind this material to a desired and controlled size spectrum and to prevent the particles from clumping before delivery. Introduction of many too-small condensation nuclei into the cloud could have an adverse effect on coalescence growth. Another method is to use sprays of hygroscopic materials such as ammonium nitrate and urea. This technique is alleged to have the advantage that a particle released may be small enough, say 10 μm in diameter, to be spread by small-scale turbulence and carried readily up to cloud base by thermals and yet grow sufficiently as the relative humidity increases to be able to capture other cloud droplets when it reaches cloud base. However, it is not known what the best particle size is, nor do we have the ready technology to control the spray output to conform to a specified optimum size. There still are unresolved questions about the efficacy of ground-based generators in getting adequate amounts of seeding materials into cumuli.

Numerous salt-seeding experiments

* These models are just now being adapted for incorporation into dynamic models.

* Hygroscopic substances are those which attract water or encourage condensation of water vapor into liquid water upon themselves, at relative humidities that are less than 100 percent for pure water. See Chapter 3 for fuller discussion.

were carried out during the 1950s, principally in the tropics, but also in Great Britain, France, and the United States. Most were started by individual scientists or commercial firms and were short-lived and inconclusive.

The dynamic effects associated with currently employed hygroscopic seeding agents are almost surely much smaller than those that can be produced by artificial ice nucleants introduced in some supercooled clouds.

Computer studies to examine cloud particle growth in the framework of both the entity and field of motion types of dynamic model are now being done.[3] Although this surely is a vast improvement over assuming fixed, nonentraining updrafts, some meteorologists question whether either the microphysical or the dynamical simulations are even now sufficiently close to nature to be reliable in coalescence modification field work. With this reservation, the current models have confirmed the results of the earlier, more simplified ones. They suggest that artificial embryos of 20 to 100 μm diameter introduced into the lower part of a continental cumulus cloud can lead to the formation of rain 10 to 15 min before natural rain could form. They also show that large quantities of seeding agents, up to or exceeding 100 kg per cloud, are required unless a chain reaction can be initiated. This result suggests the desirability of experimenting with those cumulus clouds having updrafts strong enough to support a Langmuir chain reaction and of specifying the conditions favoring chain reaction growth.

Consequently, in South Dakota, hygroscopic seeding experiments have turned to vigorous convective clouds, using radar as an important tool in evaluation. Up to the end of 1971 radar data had been collected from 41 salt seeded and 38 unseeded test cases, each consisting of a cluster of towering cumulus or cumulonimbus clouds. Preliminary examination of the radar data, which is recorded on magnetic tape for ease in data reduction, has led to two important results. First, it is quite definite that first radar echoes in new cloud towers appear closer to cloud base in salt seeded than in unseeded clouds, the average height of first echoes above cloud base being roughly 1.5 and 3 km for salt seeded and unseeded clouds, respectively.[11] Second, rainfall from clouds of moderate depth, say, 5 km, can be increased by salt seeding but the apparent effect decreases with increasing cloud size (Figure 6.7). It is difficult to make quantitative estimates because of uncertainties regarding radar reflectivity-rainfall rate relationships, but the results to date suggest that increases of several hundred acre-feet can be produced in small cumulonimbus clouds where the natural precipitation processes are inefficient.

A tropical salt seeding program was conducted in 1968 and 1969 on St. Croix, V.I., by a group from Pennsylvania State University. Interpretation of results has been hampered by incomplete data and the limitations of the cloud physics simulation in the current one-dimensional models. Many other programs on salt seeding have been undertaken recently, but remain inconclusive owing to lack of either statistical controls or physical measurements and understanding. The factors determining the optimum amounts and sizes of seeding particles, and the place and time to introduce them into cumuli are not yet firmly known. Some cumulus experts believe that improved modeling together with improved knowledge of cloud physics-dynamics are necessary prerequisites before large-scale randomized hygroscopic seeding programs will be justified. Others believe in going ahead with the programs,

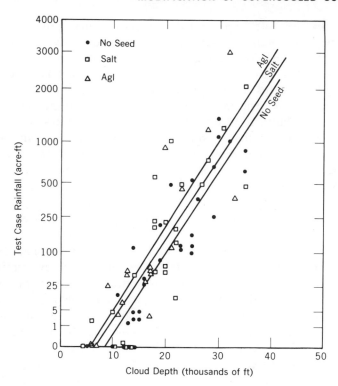

Figure 6.7 Scatter diagram showing test case rainfall (plotted with a cube root transformation) as a function of cloud depth for Project Cloud Catcher cases of 1969–1970. Rainfall amounts are estimates based on X-band radar data. Covariance analysis shows about a 10 percent chance that salt and no-seed regression lines are the same, but less than 5 percent chance that AgI and no-seed regression lines are the same.

designing them in such a way as to increase basic knowledge regardless of the practical success.

MODIFICATION OF SUPERCOOLED CUMULI BY ARTIFICIAL GLACIATION— METHODS AND BASIC PRINCIPLES

When cumulus clouds tower above the freezing level (about 4 to 4.5 km), much of their liquid drop content usually remains supercooled. Laboratory evidence suggests, however, that below temperatures of about −40°C, liquid drops will freeze "homogeneously," that is, all by themselves without any foreign substances. So if the cloudy air is suddenly cooled enough, the cloud should glaciate.

This was the great discovery in 1946 by Vincent Schaefer[42] when he originally introduced dry ice (solid CO_2) into clouds, first in the laboratory and then in the real atmosphere (see Chapter 1). Dry ice has a temperature of about −78°C. When small pellets of it are introduced into a supercooled cloud, the adjacent air becomes highly supersaturated and is also cooled below the critical threshold of −40°C. Myriads of liquid droplets are formed, and they as well as the preexisting supercooled drops are transformed into ice crystals. The latest laboratory measurements indicate that 1 g of dry ice can produce nearly 10^{12} (one million million) ice crystals in a supercooled cloud in the temperature range −2 to −12°C.

To deliver dry ice into supercooled cumulus towers, it has been necessary to dump it into their tops from aircraft, which is expensive. It is also necessary to have grinding apparatus to produce pellets of controlled size. The major advantage of dry ice relative to other artificial glaciation methods is that it works at warmer temperatures and evaporates above the ground, leaving no residue which possibly could harm either people or the environment, or which could persist to accidentally seed clouds other than those intended.

The other main way to induce glaciation in supercooled clouds is by means of the introduction of artificial freezing nuclei. The first discovered and most widely used method is with silver iodide (AgI). Silver iodide was originally used by Vonnegut because its crystal structure most closely resembles that of ice. The crystal structure of the much cheaper substance, lead iodide, is only slightly less close to that of ice. Lead iodide has had widespread usage in the famous hail suppression experiments in the Soviet Union (see Chapter 11).

There is a common, oversimplified explanation of how silver iodide glaciates a supercooled cloud. It says that, because of the resemblance in crystal structure, the droplets in the cloud react as if the silver iodide particles were ice crystals, so that ice forms and grows on the AgI particles while the water drops evaporate, except those frozen by direct contact with silver iodide. Although it is reasonable to expect that the resemblance in crystal structure is vital, ice nucleation by silver iodide has been found to be extremely complex and has by no means been fully understood or optimized.

Artificial freezing nuclei may have two classes of important modification effects on supercooled cumulus clouds. The first is to increase precipitation efficiency. The ice particles, acting as "precipitation embryos," may grow at the expense of the water droplets. Under favorable conditions these embryos may grow big enough to fall out as snow or rain. This is because of the lower saturation vapor pressure prevailing over ice at supercooled temperatures, as explained earlier. For simplicity, we shall call the seeding concept based on this precipitation efficiency improvement alone the "static approach." The second class of possible effects is dynamic. In glaciating a water cloud, heat is released—the so-called "latent heat of fusion." Since the lifeblood of a cumulus updraft is buoyancy, or density deficiency relative to the surroundings, conceivably warming the cloudy air could decrease its density enough to affect its motions, that is, alter its dynamics. Seeding experiments using this concept are said to use the "dynamic approach." Even though effects due to the precipitation embryos produced are in nature often inextricably combined with the dynamic effects, for clarity we have separated experiments involving artificial glaciation of supercooled clouds into the two categories, depending upon which effects the investigators were seeking to produce.

SEEDING SUPERCOOLED CUMULI—STATIC APPROACH

Until the 1960s, virtually all operational supercooled seeding efforts and most randomized scientific experiments were focused on increasing the precipitation efficiency; that is, they were based on the static approach. In those key experiments described in this section, it has been believed that dynamic effects were small and/or negligible.

Since a cumulus contains roughly 100 tiny droplets/cm^3, one million of them would occupy about 10 liters of cloudy air. Their combined water content is required to build one raindrop.[36] Therefore, we would need something like one artificial ice nucleus per ten liters of cloud air, or perhaps 1 nucleus per liter if we allow the rain embryos reasonable further growth by coalescence as they fall. Consider a seeding material that yields 10^{12} active nuclei per gram, such as dry ice or some silver iodide generators at supercooled temperatures of -6 to $-19°C$. Since the supercooled volumes of typical cumuli are in the range of 1 to 10 km^3 or 10^{12} to 10^{13} liters, about 1 to 10 g of seeding material would be required for an individual cumulus.

In a review of the worldwide application of the static approach to seeding cumuliform clouds to produce rain increases, there appears to be only one type of situation where positive significant results have been obtained. The successful trials have been confined to seeding convective storms in California (Chapter 2), Israel (Chapter 13), Australia (Chapter 12), Switzerland, and possibly elsewhere. These weather systems are quite different from fields of towering white cumuli superimposed on a blue sky. Furthermore, in these stormy situations, dynamic changes and important cloud interactions cannot be precluded. Dynamic effects are next examined, first in their least complicated context.

DYNAMIC SEEDING OF ISOLATED SUPERCOOLED CUMULI

So far, we have discussed cumulus modification in terms of altering just the cloud particles, namely just microphysics. Nearly always these efforts have been attempts to increase precipitation efficiency. That is, from the given amount of water vapor entering the cloud by ascent through its base, a greater fraction would be (hopefully) caused to fall to the ground as rain, over that which would have fallen naturally. In the 1950s, however, while scientific and practical applications of static seeding were in progress, many meteorologists increasingly contended that the strength, size, and duration of the vertical air currents, namely dynamics, has a far stronger control on cumulus precipitation than does the microphysics.

To illustrate the importance of dy-

Table 6.1   Rain from Cumulus Clouds as Function of Precipitation Efficiency[a]

	Big Cloud	Middle-Sized Cloud	Little Cloud
Air density			
ρ (g/cm^3)	$\sim 10^{-3}$	$\sim 10^{-3}$	$\sim 10^{-3}$
$q(g/g)$	18×10^{-3}	18×10^{-3}	18×10^{-3}
A(cm^2)	12.6×10^{10}	0.78×10^{10}	0.20×10^{10}
w(cm/sec)	200	100	50
Δt (sec)	3600	1800	600
R(100% PE) (kton)	1633	25.3	1.1
R (50% PE) (kton)	816	12.6	0.5
R (10% PE) (kton)	163	2.5	0.1

[a] See Figure 6.3.

namic control on cumulus precipitation, let us return to Figure 6.3 and compare the precipitation from each of the three typical clouds. The upward flux of water vapor through cloud base may be calculated from the formula:

$$\text{Flux (vapor)} = \rho q A w \Delta t, \qquad (6.1)$$

where ρ is the air density in grams per cubic centimeter, q is the specific humidity of water vapor in grams per gram of moist air, A is the cloud base area, w is the average updraft velocity through cloud base, and Δt is the lifetime of the updraft. Typical values of these properties and the resulting fluxes and rainfall, with varying precipitation efficiencies, are shown in Table 6.1.

Table 6.1 shows forcefully that a giant cumulonimbus with even 1 percent precipitation efficiency brings down more rain than does an ordinary warm cumulus with 100 percent precipitation efficiency. Could we deliberately make giant cumuli from small ones? If so, why did meteorologists not follow this avenue from the outset of seeding, rather than devote decades attempting to manipulate just the precipitation efficiency? The answer to this is a paradox. It is partly historical, illustrating the principle that for a feasible weather modification experiment, theory (and/or model) and technology must meet, together with motivation and economics.

Invigorating cumulus updrafts by artificial glaciation was originally suggested by Langmuir and was almost certainly produced in some of the pioneering Australian experiments in which large amounts of dry ice (10 to 150 lbs) were dumped into individual cumuli (see Chapters 1 and 12). Notwithstanding, in 1958 the prospect of artificially altering cumulus dynamics was dismissed as impossible by the famous cloud physicist J. E. McDonald.[36] The concept arose again in 1963 in a context which was, except

for technology, completely apart from weather modification.

Since the end of World War II, the first author of this chapter and colleagues had been attempting to model the dynamics of a single cumulus, with the dream of someday incorporating the interactions of the droplets with the updrafts. To test and improve the crude early beginnings, an elderly amphibious aircraft was instrumented and flown into hundreds of tropical oceanic cumuli,[27] and artificial clouds were created and measured in laboratory tanks (Figure 6.5). By the late 1950s, access to a primitive electronic computer permitted the first nonlinear "field of motion" cumulus model which used some of these observations to grow a simulated cloud from an initial density perturbation in a resting fluid.[34]

Based on all these results together came the breakthrough by J. Levine,[23] namely, the framework of a relatively tractable one-dimensional "entity" model. This model simulated the rising phase of an individual cumulus tower. The achievement of this breakthrough was twofold: first, the rate of rise of the tower was expressed by a relatively simple ordinary differential equation, whose components could be specified (after many assumptions and parameterizations) from knowledge of cloud base conditions and an environmental sounding of temperature and humidity. If necessary, the equation could be solved by hand integration. Second, when the equation was solved for the height of cloud tops (defined as the level where the rise rate goes to zero) and internal tower properties, the results were sufficiently realistic to compare favorably with aircraft and photographic observations.

Using evidence from the laboratory and computer, as well as time-lapse pictures of real clouds, Levine hypothesized that the internal motions in a cumulus tower resemble those of a spherical vortex. This hypothesis permitted him to adapt the classical rate-of-rise equation for a buoyant vortex to a cloud tower, namely,

Vertical Acceleration
= Buoyancy − Drag

$$\frac{dw}{dt} = w\frac{dw}{dz} = \frac{qB}{1 + \gamma} - \frac{1}{M}\frac{dM}{dz}\,w^2, \quad (6.2)$$

where w is the tower ascent rate, t is time, z is height, g is the acceleration of gravity, B is tower buoyancy, γ is the so-called "virtual mass" coefficient, and $(1/M)\,dM/dz$ is the entrainment rate, or the dilution rate of the cloud's mass with the outside air as it rises.

Buoyancy is reduced and drag is created by entrainment. Buoyancy dilution is maximum at low levels in warm air masses (where the specific humidity difference is maximum between saturated cloud air and unsaturated surroundings). In tropical cumuli this eating away of cloud fuel is the predominant effect of entrainment. Once the entrainment rate is specified, we are well on the way to solving Equation 6.2 for the rate of rise w (zero at maximum level achieved by the tower) and other cloud properties as functions of height z. The solution applies just to the rounded cap as it rises, as if the observer were riding with the tower; later one-dimensional models treated "steady-state" profiles and time-dependent plumes.

The postulate regarding entrainment dependence is crucial, because it permits solution of the equation. Most existing one-dimensional models are based on the inverse-radius entrainment law, namely,

$$\frac{1}{M}\frac{dM}{dz} = \frac{2\alpha}{R}, \quad (6.3)$$

where R is tower radius, and α is the coefficient of proportionality, derived empirically from measurements in the labo-

ratory and, very roughly, on real clouds. When entrainment is better understood, more accurate relationships may be substituted for Equation 6.3 in the models.

Cloud buoyancy is cut down not only by entrainment, but also by the weight of the liquid or solid particles, that is, the water drops, ice crystals, hailstones, etc., that it carries. A typical modest water content of 1 g/m³ can subtract more than the equivalent of 0.5°C from the buoyancy, which, in small cumuli and near cloud tops, is often as large as the entire buoyancy. Thus if any precipitation falls out of the tower as it rises, we must be able to specify the amount that leaves in each height step in order to calculate the remaining weight of water substance. This requires a precipitation growth and fallout scheme in the model, an example of which will be outlined shortly.

One remaining parameter needs specification before solution of Equation 6.2 is possible, namely, the "virtual mass" coefficient γ. In rising laboratory plumes, this fictitious buoyancy reduction occurs because the ascending tower must push aside the surrounding fluid as it rises. No direct verification of this effect has been possible with real clouds. Steady-state models, therefore, assume $\gamma = 0$ and the Experimental Meteorology Laboratory (EML) series of NOAA discussed here uses the laboratory value for the spherical vortex, namely, 0.5.

The calculation is best understood if we make the entrainment calculation first, directly after specifying the cloud radius R. With an environment sounding, entrainment can then be computed either graphically or by machine and its output is the cloud temperature and humidity at each level (assuming in-cloud saturation) and the amount of water vapor condensed into liquid. After applying the precipitation fallout scheme, we then compute cloud buoyancy and integrate Equation 6.2 upward in a "marching" scheme, as-

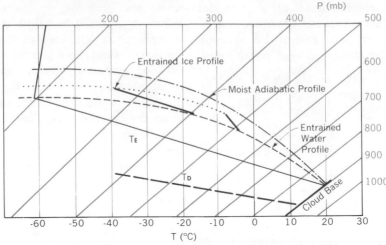

Figure 6.8 Tephigram illustrating the principle upon which dynamic cumulus seeding is based. The environment actual and dew point temperatures are given by the solid (T_E) and dashed (T_D) lines, respectively. The warming due to freezing is shown by the slanted solid lines; the lower right-hand one is postulated to be the effect of seeding. The upper part of the curve for the seeded cloud is thus to the right of and hence warmer than that of the natural cloud, whose heat of fusion is realized more slowly and at a higher level. The abscissa is temperature in °C; the slanted ticks are pressures in mb. (courtesy J. McCarthy).[24].

suming a small value of w (say, one-half to one m/sec) at cloud base. When w goes to zero, the tower has achieved its maximum height.

In early trials of the model on natural tropical clouds, we either assumed no precipitation or assumed that a fixed fraction of the condensed drops fell out in each height interval; that fraction was adjusted to give agreement with the very fragmentary cloud water measurements then available. Despite these oversimplifications, valuable insight into relationships between cloud top heights, dimensions, buoyancy, temperature, and water content was obtained. It became clear, for example, that if the middle cloud in Figure 6.3 had its buoyancy increased by 50 to 100 percent, it could (under many tropical conditions) grow as tall as the big cloud on the left.

In 1961, the invention of pyrotechnic generators of silver iodide at the Naval Weapons Center enabled massive seeding from aircraft, and hence modification experiments on hurricanes could be undertaken seriously (Chapter 14). We saw that this exciting invention could also readily supply a cumulus with one ice nucleus for every one of its cloud drops, and rapidly release the latent heat of freezing.

In Figure 6.3 the middle cumulus has about 8×10^{16} supercooled cloud drops* (1 ice particle per liter is still only one for each 100,000 drops). From a 1-kg pyrotechnic (efficiency about 10^{13} ice nuclei/g at $-10°$C) we get 10^{16} ice nuclei, or more than enough to rapidly glaciate the cloud. Freezing 1.5 to 3.0 g water per kg of cloudy air, a typical liquid water content, would raise the cloud temperature by 0.5 to 1.0°C caused by the heat of fusion and about 0.8 to 0.9°C more due to the vapor deposition occurring when we proceed from water to ice saturation.

Our modeling experience led us to seize

---

* Assuming 100 drops/cm³ and a supercooled volume $V = A \, \Delta Z = \Pi R^2 \Delta Z \simeq 0.785 \times 10^{15}$ cm³ where $\Delta Z = 1$ km vertical thickness above 0°C.

upon the pyrotechnics as a tool to make a real-life experiment in cumulus dynamics and as a marvelous opportunity to test our model. The idea behind the experiment is illustrated in Figure 6.8. When the latent heat is released rapidly by AgI seeding at or just above $-4°C$ level, cloud temperature excess and buoyancy is increased. In the natural cloud, the figure shows the latent heat of fusion being released much more slowly at a higher level; in real clouds most of it may not be available to invigorate the updraft. If the real clouds top between about 14,000 and 18,000 ft, they may not be high enough to draw on natural freezing, but on the other hand, sudden artificial freezing could give their dwindling buoyancy a life-saving "shot in the arm."

Florida Experiments

In August, 1963, 4 days of cumulus experimentation were "bootlegged" as a part of the practice for Project Stormfury, underway over the ocean south of Puerto Rico; thus began the first deliberate attempts at the modification of cumulus dynamics by seeding or dynamic seeding.

Pyrotechnics were dropped into the tops of six medium-sized (tops at about 20,000 to 24,000 ft) supercooled cumuli. Four of these exploded spectacularly (Figure 6.9), first about doubling their height and then expanding horizontally into a giant long-lived cumulonimbus. There were four initially similar looking cumuli selected as controls for comparison; all of these failed to grow and dis-

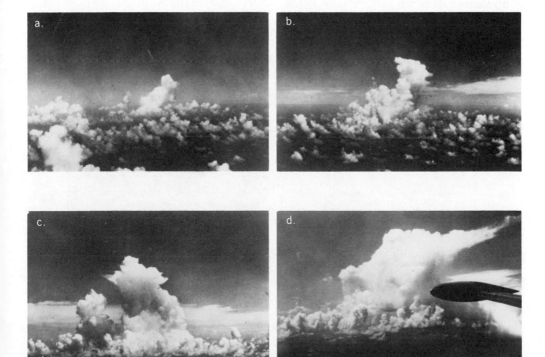

Figure 6.9 "Explosion" of a tropical cumulus following seeding. (*a*) Cloud at seeding time, top about 7.5 km. (*b*). Cloud 9 min later. (*c*) Cloud 19 min after seeding. (*d*) Cloud 38 min after seeding, with top at about 12 km. It is now a full-blown cumulonimbus.

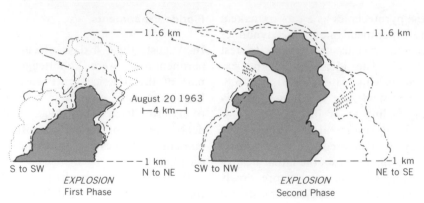

Figure 6.10 Growth of the cloud of Figure 6.9 following seeding. Constructed from a series of photographs made at known distances and directions from the cloud.

sipated in the normal 15 to 20 min life-time.

Following the field experiment, a "seeding subroutine" was introduced into the model which froze the cloud water and released its latent heat linearly in the height interval −4 to −8°C (roughly 5.3 to 6.2 km). These calculations showed that, with seeding, the test clouds indeed could have grown as observed, while unseeded clouds would have failed to grow much above about 20,000 ft.[46] The horizontal expansion of many of the seeded clouds following their vertical

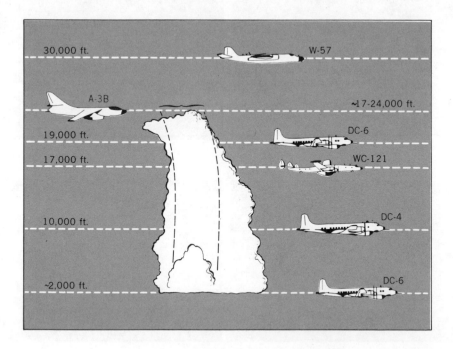

Figure 6.11 Field design of single cumulus airborne pyrotechnic seeding program. Pyrotechnic AgI generators are dropped at 100-m intervals in bracketed zone. The vertically stacked aircraft are heavily instrumented; they make one cloud penetration prior to seeding and several afterward. The seeding decision is not known to the scientists so that seeded and control clouds are studied identically.

growth (Figure 6.10) was not predicted and is still not fully explained. In these cases, the whole updraft system is invigorated. Factors involved may be the protection afforded against entrainment by the falling ice showers at cloud edges and a possible destabilization of the environment by cooling at the melting level, and perhaps enhanced convergence due to pressure field alterations.

Many members of the scientific community doubted that the enormous growth of the seeded clouds was caused by seeding, on the grounds of high natural variability and possible bias by the experimenters in their cloud selection. Consequently, a carefully randomized, much more extensive dynamic seeding experiment was executed in the same lo-

cation in 1965.[45] Six aircraft were used as shown in Figure 6.11 to provide detailed "before and after" depictions of the selected clouds. A seventh (not shown) directed the rest by radar and conducted precise photography.

Sealed instructions were opened secretly on the seeder aircraft so that project scientists did not know whether a "GO" cloud they selected was seeded or a control.

Analysis was made combining field measurements, model results, and statistics. Seeded clouds grew an average of 1.6 km higher than controls (significance better than 1 percent), seeded and control populations were distinctly separate, and the model showed considerable skill in predicting top heights and internal

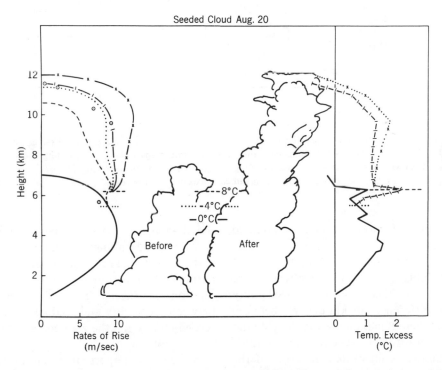

Figure 6.12 Some of the one-dimensional model results applied to the growing tower of the seeded cloud illustrated in Figures 6.9 and 6.10. Curves show the rise rate (left) and temperature excess (right) of the tower as it rises through a given level H, the left scale, in kilometers. Unseeded properties are shown by solid lines. Various seeding subroutines (see Table 6.3) are being tested with the upper curves. Circles are the photographically measured rise rates of the actual cloud tower. Note the increase in temperature excess caused by the seeding, which leads to the taller growth of the seeded model cloud.

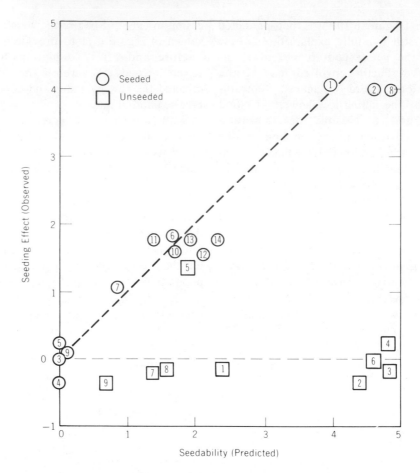

Figure 6.13 Seedability versus seeding effect for the 14 seeded (circles) and 9 control (squares) clouds studied in 1965. Note that seeded clouds lie mainly along a straight line with slope 1 (seeding effect is close to seedability) while control clouds lie mainly along straight horizontal line (showing little or no seeding effect regardless of magnitude of seedability). Units of each axis are in kilometers.

properties of modified and unmodified clouds.

The important concepts of "seedability" and "seeding effect" were defined with this experiment. Seedability is the difference (in kilometers) between the predicted maximum top height and the predicted unseeded top height. Seedability is illustrated in terms of the model in Figure 6.12. Figure 6.13 shows the relationship between seedability and seeding effect for the 1965 seeded and control clouds. If model and data were perfect, all seeded clouds should lie along

the diagonal line with slope 1, and all control clouds along the horizontal line with slope 0. The success of the experiment is illustrated by the 0.97 correlation coefficient ($p < 0.5$ percent) between seedability and seeding effect for seeded clouds and the zero correlation for the controls.

An equally important result of these experiments was that *different growth regimes followed seeding, depending on the initial conditions of the cloud-environment system*. In addition to explosive growth, seeded clouds could undergo two other growth regimes, illustrated in Figures 6.14 and 6.15. We can now diagnose which regime will predominate by examining the soundings, as explained by Figure 6.16. On rainy, humid days with a deep unstable layer (Figure 6.16*a*), all cloud radii will grow naturally, so seedability is small. Figure 6.16*b* illustrates the small stable dry layer in midtroposphere (20,000 to 26,000 ft) that limits natural growth and offers maximum seedability. When the air is too dry near natural cloud tops, the rising seeded towers will cut off (Figure 6.14);

when adequate moisture is present with these lapse rates, spectacular explosive growth can follow dynamic seeding. Under dry, stable, or drought conditions (Figure 6.16*c*), the inversion may be too strong and dry for any growth to follow seeding, or in the extreme, cumulus tops may be suppressed below the $-4°C$ level, eliminating AgI seeding potential altogether.

That the rainfall should be increased when a seeded cloud grows explosively seems logical to expect, but had to be demonstrated. Therefore, in 1968 the dy-

a Cloud at seeding time.   *b* Ten minutes later.

c Eighteen minutes after seeding, when tower has reached 11 km and cut off.

Figure 6.14 Photographs illustrating "cutoff tower" regime which often follows dynamic seeding of a single cumulus (see Figure 6.16*b*). (*a*) Cloud at seeding time. (*b*) 10 min later. (*c*) 18 min after seeding, when tower has reached 11 km and cutoff.

Figure 6.15 Typical "no growth" regime. At 12 min after seeding cloud looks unchanged except that the top has glaciated.

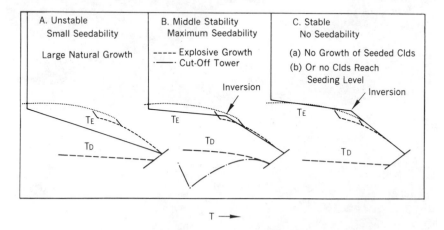

A. Unstable Small Seedability Large Natural Growth	B. Middle Stability Maximum Seedability ----- Explosive Growth —·— Cut-Off Tower	C. Stable No Seedability (a) No Growth of Seeded Clds (b) Or no Clds Reach Seeding Level

$T \longrightarrow$

Figure 6.16 Soundings illustrating the characteristic environments for the four main growth regimes of single cumuli following dynamic seeding (Courtesy J. McCarthy).[24] (a) Unstable. All cumuli grow whether seeded or not. (b) Inversion in mid-levels. Unseeded cloud tops there; increased buoyancy enables seeded tower to grow past inversion and reach unstable region above. T_D dashed, enough moisture for explosive growth. T_D dash-dotted, dry mid-layer, cutoff tower regime probable. (c) Dry, inversion conditions characteristic of tropical droughts. Either seeded clouds fail to grow following seeding or tower does not reach the $-4°C$ level.

namic seeding experiment was moved to south Florida where the rainfall could be measured by a calibrated 10-cm radar.[54, 55] We first had to develop self-consuming pyrotechnics[47] and to simulate precipitation growth and fallout in the model.

A precipitation scheme evolved for use in the EML-series models is outlined in Table 6.2.

Each of the steps in Table 6.2 involves an equation. Autoconversion is the process by which raindrops first form from cloud drops, or on giant salt nuclei.[21] Its rate depends on the droplet spectrum at cloud base and will be faster for the more disperse spectrum and larger drops in maritime clouds. In most models, collection of cloud drops by raindrops is described by a simple "continuous collection" equation where the rate of growth of rainwater depends on the mass of rainwater, the mass of cloud water, and a collection efficiency near unity; sophisticated "stochastic

Table 6.2 Precipitation Scheme—EML Cumulus Models (One-Dimensional)

1. Autoconversion from cloud drops to precipitation.
2. Collection of cloud drops by precipitation drops.
3. Precipitation has Marshall-Palmer spectrum.
4. Compute volume median drop diameter of precipitation.
5. For volume median drop, compute terminal fall speed from diameter.
6. Compute fraction fallout in each height step Δz from the formula

$$\text{Fraction Fallout} = \frac{\text{Time for tower to rise } \Delta z}{\text{Time for median drop to fall through distance } R}$$

7. Subtract fallout from preexisting water content.

Table 6.3 Seeding Subroutine—EML Cumulus Models (One-Dimensional)

Define "slush" region as occurring between -4 and $-8°C$ in seeded cloud. All listed changes proceed linearly in this interval.

1. Fusion heat release of 60% total liquid water content at $-4°C$.
2. Proceed from water to ice saturation.
3. Proceed from water to ice hydrometeor spectrum (still Marshall-Palmer with different slope for same mass of water substance).
4. Ice collection efficiency remains one.
5. Reduce terminal velocity of ice precipitation to 70% of that of droplets of same mass.

collection" equations are being put in the more advanced models.

The Marshall-Palmer precipitation spectrum simply says that the distribution function of raindrop size is linear on semilogarithmic paper, with drop number in each size category diminishing with size; the slope of the line is determined by the rainwater content. The spectrum assumption has been verified well by foil impactor samples in active towers. Once the spectrum and water content are specified, the volume median diameter follows; from this the terminal velocity is computed from an empirical equation. It should be emphasized that this model calculates only the precipitation growth in and fallout from a single cumulus tower; no existing cumulus models can yet adequately calculate the total rain reaching the ground from an entire cloud or cloud system.

A precipitation scheme requires a more sophisticated "seeding subroutine." The current EML version is shown in Table 6.3.

Fortunately, cloud top heights are insensitive to the assumptions in Table 6.3,since several of the latter are not

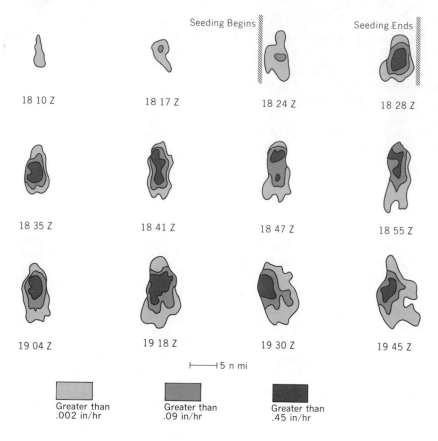

Figure 6.17 Contoured cloud base echoes showing (below) equivalence of contours to rainfall rate obtained from radar calibration. Area within each contour is measured and summed over time to give rain volume falling from cloud base.

founded on adequate observations. For example, we know little about ice spectra, collection efficiencies, and terminal velocities in tropical clouds; these depend, for example, upon whether the ice particles are snowflakes, hail embryos, junk ice fragments, or some mixture. The icing levels in large cumuli pose severe hazards to aircraft, and further, the instrumentation problems raise extremely formidable obstacles to recording the needed data (see Chapter 4).

In 1968 and 1970, randomized dynamic seeding experiments on single clouds in south Florida reconfirmed the 1965 results on cloud growth. More important,

they showed conclusively that the rainfall from the seeded clouds greatly exceeded that from the controls. The rainfall increase was proportional to seedability.[48, 54]

In this experimental series, the model was run in real time in advance of launching the experimental aircraft. On days of poor seedability, namely, wet conditions with excessive natural growth (Figure 6.16*a*) and dry suppressed conditions with no seeded growth (Figure 6.16*c*), missions were not usually launched, avoiding the "dud" cases of 1965 and saving expensive aircraft time.

Altogether, 52 "GO" clouds were obtained, 26 seeded and 26 controls on 19 operating days. The total rainfall from each cloud was evaluated by radar. The radar calibration was checked against rain gauge records for each of the 2 years. The cloud base echoes were planimetered (Figure 6.17) to get the rain in each 10-min period in each contour interval and then summed. The average rainfall difference between seeded and control clouds was 271 acre-ft, significant at better than the 5 percent level. The single-cloud seeding effect on precipitation was calculated to exceed a factor of 3. A diagram similar to Figure 6.13 was constructed for these experiments. In addition to excellent height predictions, model predictions of seeded minus control tower rain fallout correlated with observed seeded minus control rain to above 0.9 ($p < 0.5$ percent). However, the model predicted smaller amounts owing to reasons cited earlier.

A particularly important result of the experiments was that the rainfall increase from seeded clouds was very large and positive on "fair" days, but it was small and negative on "rainy" or "disturbed" days. The fair day increases were highly significant; there were too few rainy days flown for the decrease to be significant. The key results are summarized in Table 6.4. A "rainy" day is defined as one in which the radar echo coverage exceeded

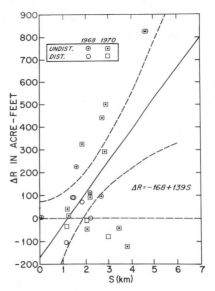

Figure 6.18 Empirical relation between rainfall increase ΔR (acre-feet) and seedability as tentatively established for south Florida. Note the rainfall decreases apparently associated with seedabilities (S, in kilometers) of about 1 km or less. Different empirical $S - \Delta R$ relationships may be applicable in other areas and/or other weather conditions. The 95% confidence band for the regression line is shown by the dashed curves.

13 percent at 1800 GMT (12 p.m. local daylight time). In south Florida, about 50 percent of the total rain falls on "rainy" days.

As we go from a fair to a rainy day, control cloud rainfall increases greatly, as expected, but seeded rainfall also decreases. Careful case studies showed smaller horizontal expansion and shorter lifetimes—that is, weaker "explosions" of seeded clouds on rainy "socked in" days. The reason is not yet clearly understood.

Figure 6.18 shows an empirical relation between seedability and seeded rainfall increase in south Florida. The seedability threshold for rain increase is at about 1 km. Owing to a small sample, this graph should not be taken too seriously; it shows, however, what can perhaps

Table 6.4   1968 and 1970 Key Results, Florida Single Cumulus Dynamic Seeding

Stratification	Rainfall R (acre-ft)		
	Seeded	Unseeded	Average Difference
	n $\bar{R}_s$	n $\bar{R}_{ns}$	$(\bar{R}_s - \bar{R}_{ns})$
Fair	22 459	20 89	370
Rainy	4 297	6 411	−114
All	26 434	26 163	271

eventually be done to learn about rain augmentation potential from dynamic seeding for a given area. Maps of seedability as a function of location and season have been made with the one-dimensional models and the library of radiosonde observations for each station.[52] Needed to accompany these maps are radar population studies showing the time distribution of seedable clouds, as done by EML for south Florida.[19]

South Dakota Experiments

The silver iodide seeding experiments conducted in the Northern Great Plains by the South Dakota School of Mines and Technology for the U.S. Bureau of Reclamation[9] have followed a different pattern from those of the Experimental Meteorology Laboratory, but the results of the two series of experiments are consistent if allowance is made for the differences in air mass and cloud characteristics in South Dakota as compared with Florida.

The one-dimensional steady-state cloud model developed at Pennsylvania State University has been adapted for use in the Northern Great Plains experiments by Mr. John Hirsch. Hirsch has modified the handling of the water substances to include solid precipitation (graupel). Precipitation fallout is treated after the methods developed at the EML. Additional changes concern the value of the entrainment parameter α in equation 6.3 which has been adjusted 25 percent downward to 0.15. With this adjustment to α, reasonable agreement was achieved among observations of updraft radius R_o, updraft profile, and cloud top height. This study, based on aircraft and radar data, reemphasized the previously observed fact that updraft radii in the Northern Plains can be as small as one-tenth the radii of the visible clouds. Because up-

draft radii are difficult to measure, the present practice is to use observed shower heights to determine initial updraft radii, rather than vice versa. This approach yields estimates of updraft speeds which compare favorably with aircraft measurements.

The atmosphere on unstable summer days over the Northern Plains does not feature the weak inversions around 6 or 8 km sometimes observed in Florida. Originally, it was thought that the absence of such inversions would make it difficult to increase rainfall by dynamic seeding, as it precludes the spectacular explosive growth situations sometimes encountered in the tropics. However, as modeling and radar studies continued, it was realized that increases in cloud height of a few hundred meters are possible and might have significant impact upon total rainfall. It is an interesting point that such increases in cloud height are almost impossible to document. The natural clouds on a given day over western South Dakota may have tops ranging all the way from 5 to 15 km above sea level, and the detailed observations which would be necessary to predict a cloud top to within 500 m appear prohibitively complex and expensive. Nevertheless, the associated potential rainfall increase is worth pursuing.

The best data on dynamic seeding in the Northern Plains come from a comparison of the silver iodide seed and no-seed cases of Project Cloud Catcher in 1969–1970.[11] This was a three-way randomized experiment (no-seed, silver iodide, salt) using "floating" target areas and has already been mentioned in connection with hygroscopic seeding. The hypothesis governing the silver iodide seeding was that it could stimulate cloud development by latent heat release and thus increase rainfall through intensification and/or enlargement of updrafts feeding the cloud. However, silver iodide seeding was conducted at moderate rates of only a few hundred grams per 1-hr test case to minimize possible reduction of precipitation efficiency by cloud glaciation at warm temperatures. The objective was to move the region of cloud glaciation, which normally occurs in cumuli of western South Dakota in a nonlinear fashion between about -20 and $-40°C$, down to the -5 to $-25°C$ region. This "light" seeding treatment has been simulated in the Hirsch model and shown to have significant dynamic effects, although not as great as those produced by massive seeding. Most clouds in the area with natural tops at 7 to 10 km above sea level show some additional growth (in the model) as a result of light AgI seeding, with a typical increase in cloud height (seedability) being 500 to 1000 m. As radar estimates of total rainfall versus maximum observed echo height for shower complexes in the area have shown that the rainfall roughly doubles for every 2-km increase in height, the predicted increases in cloud height would produce substantial increases in rainfall, provided the natural height to rainfall relationship was maintained. This would require maintaining precipitation efficiency and relationships such as height versus cloud diameter and height versus cloud lifetime. The precipitation processes within the model are being refined in the hope that they will ultimately provide reliable estimates of precipitation efficiency in both seeded and natural clouds.

Analysis of the taped radar data suggests that the silver iodide seed cases of Project Cloud Catcher so far have yielded roughly 40 percent more rainfall per test case than the no-seed cases, with the increases contributed entirely by cases where cloud depth did not exceed 8 km (Figure 6.7). While the 40 percent apparent increase in rainfall is of marginal statistical significance, if only the radar rainfall estimates are considered, confidence in the correctness of the seeding treatment is increased by a study which shows that first echoes in clouds seeded with silver iodide tend to appear near the $-8°C$ level and significantly closer to cloud base than in unseeded clouds, although not as close to cloud base as in salt seeded clouds.[11]

It is tempting to ascribe the apparent rainfall increases to increases in cloud height. While the best indication of the 40 percent increase comes from a covariance analysis utilizing maximum observed radar echo height as a predictor, which would suggest that the rainfall increase is related to an increase in precipitation efficiency rather than cloud height, most of the test cases consist of a succession of cloud towers, and the succeeding treatment is aimed at introducing silver iodide into new towers as they appear around existing showers. It may be that the dynamic effects are occurring in these cloud towers, which increase in size and total rainfall production without necessarily increasing the height of the tallest cloud occuring during the 1-hr test case. In this connection, the existence of multiple cells within most South Dakota thunderstorms is relevant. However, there are alternative explanations for the apparent rainfall increases so that conclusions regarding the mechanisms responsible for them are only tentative.

Dynamic seeding experiments on single cumuli have been executed in Pennsylvania, Arizona, Australia, and Africa with similar results to those in Florida and South Dakota. These experiments in cumulus modification have been both unusual and satisfying because it has been possible to achieve significant positive results on two variables (vertical growth and precipitation) with a relatively small sample of cases. The experimentation has also been productive of advances in understanding and modeling cumulus processes. But what of the practical usefulness of this work? Can we hope, for example, to apply it to water management economically? Can it be applied in other cases of weather modification than water management?

In considering the rain augmentation problem, suppose that by flying two aircraft in tandem we could seed 10 cumulus clouds in a daily operation (a reasonable goal). If the figures in Table 6.4 can be applied, we might expect to gain 2700 acre-ft of water. At the rate of $50 per acre-ft* this would be a benefit of $135,000. If the aircraft were well instrumented and flew 10 hr all told at $600 per hr and if 100 AgI pyrotechnics were expended at $14 each, the benefit-to-cost ratio would exceed 18. Could we reasonably expect to find 10 isolated seedable cumuli in a day over a needy watershed? What of possible side effects? And most important of all, would there be cloud interactions; would these help or hurt us, or does the answer depend on both the context and the experimental approach?

CUMULONIMBUS MERGERS, DYNAMIC SEEDING OF MANY CUMULI OVER A TARGET AREA, AND APPLICATIONS

From Project Cloud Catcher in South Dakota and from natural cloud studies in Florida, it was discovered that merged systems or "mergers" produced much more rainfall, by an order of magnitude, than the sum of their component clouds (Figures 6.19 and 6.20). Radar measurements showed that a single merger may produce 5,000 to 50,000 acre-ft of water, compared to 100 to 2000 from the vigorous isolated thunderstorm.

Can mergers be induced by dynamic seeding? If so, can multiple cloud dy-

* The approximate cost of municipal water in south Florida, clearly only a rough approximation to a realistic cost assessment.

namic seeding lead to rain increases over sizable areas without adverse side effects? In taking the step from single to multiple cloud experiments over whole areas, the scientific problems become more difficult by at least a factor of ten. Natural rain variations are so great that seeding effects must be large to be distinguishable by classical statistics even with 100 to 200 experimental cases. Cloud interactions and mesoscale motions (10 to 100 km in size) may allow greater natural variations and may also permit seeding effects to propagate in space and time.

Cloud group modeling is in its early infancy. At EML, a three-dimensional model with the shape of south Florida has been devised which predicts areas favorable for convection arising from interaction of the heated boundary layer over the peninsula with the larger-scale flow. While the initial steps toward introducing moisture were taken in 1972, it will be many years before area seeding experiments can be simulated on multiple cloud models.

In the field, multiple dynamic cumulus seeding began in Florida in 1970 and 1971[48] in a 4000 nmi^2 target area (Figure 6.21). The experiment design is outlined in Table 6.5. The "daily suitability criterion" requires some explanation. It is intended to select for experimentation fair days with high seedability S (in kilometers). N_e is the number of hours with echoes in the target area between 10 a.m. and noon local time. This parameter is introduced for the purpose of screening out naturally rainy days. It has a maximum value of 3 if echoes are present in the area on each of the 3 hours.

The execution of the south Florida randomized area experiment has encountered serious obstacles, some of them sociological. The western half of the target area is heavily agricultural; its predominant crop is tomatoes, which are

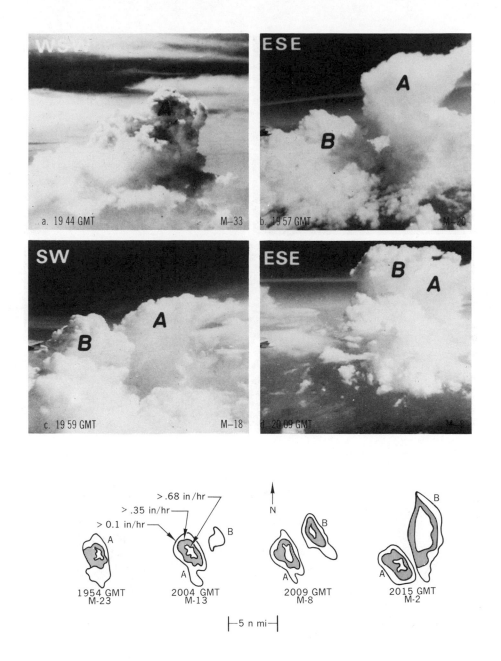

Figure 6.19a Photographic and radar documentation of a merger. Cloud *A* is seeded and *B* is not. Top, photographs before merger. Camera direction in upper left. Below, radar echoes before merger. *M* is time of merger. Times in minutes relative to merger time.

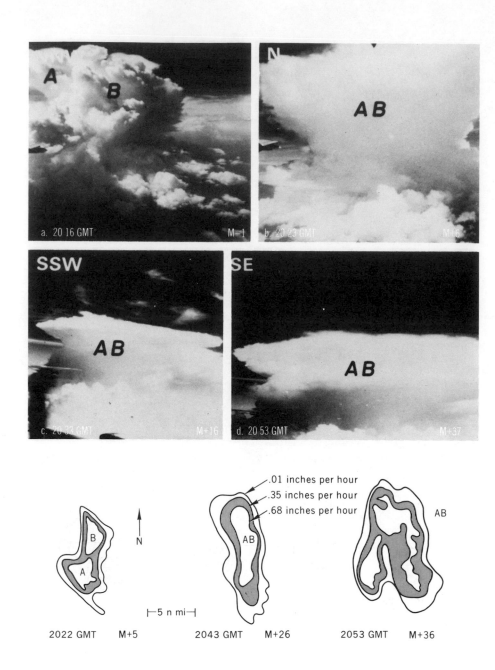

Figure 6.19b Top, photographs after merger. Note giant system at $M + 37$. Below, radar echoes after merger.

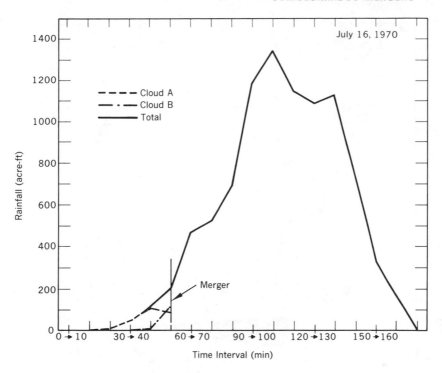

Figure 6.20 Precipitation history of cumulonimbus merger shown in Figure 6.19. This system produced about 9000 acre-ft of rain in its lifetime. Rainfall is plotted against time (10-min intervals relative to seeding Cloud *A*).

harvested during April and May. Rain on tomatoes during harvest ruins the fruit, so that the farmers are understandably bitterly opposed to any seeding programs over that area during that period, which is unfortunately also the optimum season for cumulus clouds. The second best period for cumulus, namely, July and August, is also hurricane season, during which the NOAA seeding aircraft have been involved on a priority basis with the hurricane modification experiments (see Chapter 14).

Owing to these difficulties, only 13 "GO" days have been obtained in the randomized area experiment in 1970 and 1971, seven seed cases and six controls. Since with this limited sample, there is no immediate hope of evaluating the seeding effect over the whole target area, an inter-

mediate goal has been established, namely, first evaluating the seeding effect on "floating targets." The floating targets consist of all cloud echoes that have undergone seeding penetrations* and all those echoes merging with them. The total target rainfall thus consists of that from the floating targets and that from the unmodified clouds that also happen to be in the target area.

Results of the first two experimental years are shown in Table 6.6.

The average floating target rainfall for the seven seeded cases is 2.6×10^4 acre-ft, while it is 0.67×10^4 acre-ft for the five fair control cases, which is smaller by

---

* Whether it was actually a seeded or control day was unknown to the scientists and aircraft flights were conducted identically (and the seeding button pushed) on both types of occasions.

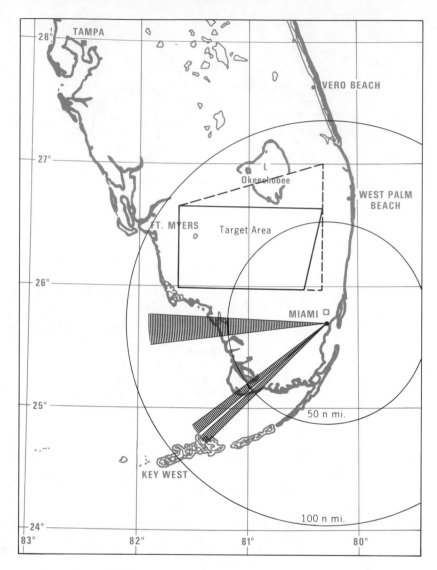

Figure 6.21   EML area in south Florida for randomized multiple cumulus seeding experiments. 1970 area, solid outline; 1971 area, dashed outline. Shaded regions are "blind cones" of the University of Miami 10-cm calibrated radar.

nearly a factor of four. Is this difference significant? With classical statistics (Mann-Whitney-Wilcoxon test; see Chapter 5) the difference is significant at the 10 percent level using a one-tailed test; this result says that we do not have a large enough sample to conclude safely that the seeded-control difference is not due to chance. Using Bayesian statistics and some unproven assumptions, we can estimate the floating target seeding effect and its significance. The tentative (1972) best estimate of the floating target seeding effect on rainfall is about a factor of 3, with almost no probability that the effect is negative and a slight probability that it could exceed 5. The dangerous assumptions lie in asserting that we have accurately specified the unmodified floating target rainfall distribution using just the existing control cases and some other knowledge gained from single cloud studies. The next step in this experiment must be an accurate determination of these natural fluctuations. Then, unless our present assumptions are grossly in error, about 12 more seeded cases should be adequate to evaluate the "floating" target seeding effect. The last column in Table 6.6 offers further encouragement, in that the correlation between number of mergers experienced by the floating targets and their rain production is 0.9. The demonstration of this relation supports the physical hypothesis behind the experiment, namely, that dynamic seeding can promote merger and thereby can enhance cumulus rainfall.

Regarding total target rainfall, we encounter a more sticky ball of wax. The

specter should be clearly raised that we could find large positive effects in floating targets and simultaneously find the effects in a larger fixed target area incorporating the floating ones to be undetectable or even negative.

The area experiments of the South Dakota School of Mines and Technology have been underway for several years. The Rapid Project[12] from 1966 to 1968 utilized a randomized crossover design with north-seed and south-seed days in target areas of 700 mi² each. The days were stratified in advance in accordance with the synoptic situation, with the principal distinction being between shower days and storm days. Shower days were characterized by a moist unstable atmosphere, but no "trigger." Storm days were characterized by similar conditions plus a trigger (positive vorticity advection at 500 mb), which generally led to

Table 6.5   Design of Florida Multiple Cloud Seeding Experiment

Fixed Target Area—Randomization by Day
↓
Continuous UM/10-cm Radar Surveillance of Target
↓
Daily Suitability Criterion } $S - N_e \geq 1.00$? No → No Operations
Yes
↓
Target Suitability } Suitable Clouds in Target?   No → Return to Base
Yes
↓
Randomizer Determines Seeding Decision
↓
Carry out Seeding Instruction
↓
Acceptability Criterion } 6 Seeded Clouds or Expenditure of 60 Flares?   If No → No Area Analysis
Yes
↓
Do Area Analysis

Table 6.6 Dynamic Cumulus Seeding in Florida—Randomized Area Experiments. 1970–1971 Floating Target Analysis

Date	Action	FT Rainfall (acre-ft $\times 10^4$)	Rank	FT Depth (in.)	FT Mergers
1970					
6/29	S	0.16	12	1.29	3
6/30[a]	NS	3.08	—	2.72	9
7/2	S	1.11	5	1.28	6
7/7	NS	0.78	8	1.04	6
7/8	S	8.96	1	2.49	18
7/8	S	4.56	2	4.39	12
1971[b]					
6/16	S	0.23	11	1.66	1
7/1	NS	0.26	10	1.02	6
7/12	NS	0.35	9	1.31	4
7/13	S	1.58	4	2.49	11
7/14	S	1.66	3	1.23	7
7/15	NS	0.96	7	1.68	8
7/16	NS	1.00	6	2.07	8

[a] Rainy day, discarded from sample.

[b] Adjusted values of rainfall. Radar-raingage comparisons made in both years indicated that in 1971 the radar rainfall was underestimated by 1.75 relative to 1970. Ranks are not changed by the adjustment.

widespread convective activity, sometimes organized in squall lines.

Seeding was conducted principally from aircraft with generators releasing roughly 300 g AgI per hr. The project design differed from the Arizona and Whitetop experiments in that seeding was conducted only when supercooled clouds were present and then in updraft areas below cloud base.

The Rapid Project is apparently the only one to date on convective clouds yielding evidence of increases in average rainfall under any predetermined set of conditions throughout a designated target area. Quite good evidence for such increases was obtained from the shower days, with suggestions that the rainfall may have been increased by a factor of 2 or more on such days (Table 6.7). Rainfall was quite scarce on shower days and calculations show that the apparent increase was equivalent to the production of a single shower of a few hundred acre-ft over the seeded target area on each shower day. At some stations the statistical difference between north-seed and south-seed days, as evaluated by a rank test, proved significant at the 5 percent level.

The results on storm days were mixed and pose interesting questions. On storm days with northwesterly winds aloft the data suggested a decrease in total rainfall (Table 6.7). On storm days with southwesterly winds the data suggested rainfall decreases in the areas where the seeding was accomplished and possibly even upwind of such areas, but substantial increases in rainfall about 20 mi downwind of the silver iodide release areas. There was also evidence that hail was suppressed, but throughout the surrounding region rather than in the seeded target itself. Concerning these paradoxes the experimenters reason, "Effects in the unseeded target would not require physical transport of the seeding agent from one target area to the other. A decrease could arise, for example, from widespread subsidence related to the intensification of convection in the seeded target area." Dennis and Koscielski[10] have coined the term "dynamic contamination" to cover these and other possibly even more subtle effects of seeding upon the surrounding regions not necessarily reached by the seeding agent.

A further 2-year randomized experiment over a larger target area of 5300 mi² in northwestern South Dakota has not resolved the various possibilities. In

Table 6.7 Average Rainfall per Gauge per Test Case (in mm) and Seed/No-Seed Ratios in Target Area for All Days of Given Type on the Rapid Project

	SW Flow	NW Flow
Shower Days		
North Area		
Seed N	0.99	0.92
Seed S	0.69	0.64
Seed/No-seed ratio	1.4	1.4
South Area		
Seed S	1.8	1.1
Seed N	0.30	0.10
Seed/No-seed ratio	6.0	11
Storm Days		
North Area		
Seed N	2.2	0.89
Seed S	2.8	4.3
Seed/No-seed ratio	0.78	0.21
South Area		
Seed S	2.4	1.4
Seed N	3.0	3.2
Seed/No-seed ratio	0.80	0.44

that project the target area and two smaller control areas to the south and west were instrumented with about 100 rain gauges. Randomization was imposed by reserving 2 days of every 8-day time block during the two summers as no-seed days. The radar and rainfall data collected were compatible with any or all of the following postulated effects:

1. An increase in cloud heights in the target area.
2. A shift toward lighter rains in the target area achieved either by initiation of light showers or by suppression of heavy rainfalls, or both.
3. A decrease in rainfall in the control areas, due mainly to a decrease in the number of rainfall events.

Additional analyses utilizing hourly rainfall observations at 130 stations in South Dakota, North Dakota, and eastern Montana have shown that the area of rainfall deficiency on seed days in the target and control areas was actually part of a large area of deficiency extending over much of the three states and, furthermore, that the rainfall deficiency on seed days existed from 0200 local time onward, or for 6 to 10 hr before any seeding was undertaken. This is another example of an uncontrolled background effect dominating the outcome of a randomized experiment. However, the failure of the evidence for a rainfall suppression effect to stand up under close scrutiny in this case does not mean that the possibility of such an effect can henceforth be ignored.

This and a second type of dynamic contamination have been observed by EML scientists in Florida. The second type occurs when a huge cumulonimbus anvil, from a seeded or natural cloud, extends over the target area. A single anvil may cover thousands of square miles and its shadow completely wipes out the cumulus activity below by cutting off solar radiation.

The possibility of "dynamic contamination" is of utmost importance in weather modification. First, it renders the randomized crossover target procedure suspect and possibly untenable, since even clouds over a control area upwind of a seeded area could be dynamically or thermally influenced by the seeding. Second, it raises the specter of extensive downwind, large-scale, and persistent effects of seeding.

With these vital questions unanswered, disastrous and damaging droughts have forced dynamic seeding into attempted practical applications prior to the establishment of the required sound scientific foundations, posing serious moral dilemmas to the scientists involved.

In the winter and spring of 1971, south Florida experienced a record drought. By April 1, the water deficit exceeded 3 million acre-ft. Drinking supplies for the coastal region were threatened by saltwater intrusion, while raging grass and muck fires in the Everglades compounded the already severe damage to plant and animal wildlife. At the request of the state government, NOAA/EML undertook a dynamic seeding program from April 1 to May 31 in two target areas, including the largest possible fraction of the important watersheds consistent with avoiding most of the tomato harvest.[49]

Additional precautions were undertaken to avoid potential severe weather side effects of cumulus invigoration and merger. In Florida, virtually all severe weather such as squalls, lightning, hail, and tornadoes are associated with cumulonimbus mergers. Severe weather is increasingly likely with increasing instability and vertical wind shear. Criteria were evolved which required removing seeding areas farther upwind of populated areas as these parameters increased, halting seeding altogether if actual warnings were likely to be issued by the Weather Services.

Important scientific and practical benefits were apparently gained from the program, although the latter cannot be confirmed without resolution of the seeding effect in the randomized area experiments. First, within the 61-day operational period during a prevailing severe drought, 21 days presented seedable clouds in the target area. Owing to aircraft limitations, only on 14 days were clouds actually seeded; this, however, exceeded the anticipated number. Second, *frontal* cumuli associated with dissipated fronts were found to be highly seedable. One such merger alone produced in excess of 50,000 acre-ft, or more than 25 percent of the seeded cloud rain for the whole experiment. Since most cumuli in Flordia's dry periods and seasons are associated with old cold fronts, this result is very encouraging, with the reservation that these situations are also those most prone to severe weather.

Rainfall from all seeded clouds in the program was assessed (from radar and rain gauge records) to be more than 180,-000 acre-ft. Using floating target, single cloud results, modeling, and other criteria, a conservative estimate of the rain added by seeding was about 100,000 acre ft. Although this amount is a "drop in the bucket" compared to the several million acre-ft shortage, it is likely that the $165,000 program had a benefit-to-cost ratio considerably exceeding 10. The factor would be 32 if $50 per acre-ft (municipal water price) is used for the value of the water produced, which is admittedly a poor measure.

Even if much of the seeded rain evaporates without direct usage by man or the ecology, any rain at all can be of immeasurable value in breaking the vicious self-aggravating cycle of a tropical drought. Rain may wash some of the excessive number of cloud condensation nuclei out of the air and dampen the ground so that fewer particles remain aloft to divide the precipitable water into such a large number of small drops that none can fall as rain. EML studies show that burned Everglades vegetation particles serve as very active CCN's.[18] Cooling the ground and wetting the lowest air works against another major drought-maintaining factor, namely, the 3000 to 5000 ft rise in cloud base from coastal to inland Florida which sets in as the normal swamps become dry. Model calculations show that raising the cumulus base is a most effective way of reducing their rain potential. Thus even evaporating seeded rain can result in increased rain production from later natural clouds. Finally, in 1971 rain from seeded clouds extinguished more than 26 Everglades fires, on the one occasion when a seeded merger delivered 50,000 acre-ft.

Despite the uncertainties related to dynamic contamination and other unknowns, the state of South Dakota has launched an operational program of weather modification to increase rainfall and suppress hail from summer convective clouds. Impetus for the program has come from the research results obtained at the South Dakota School of Mines and Technology and from a target-control analysis of about 25 project seasons of commercial seeding programs in the state. Most of these projects were conducted with aircraft releasing silver iodide in updraft regions below cloud base. The target-control analysis has suggested an overall positive effect of roughly 10 percent of the expected rainfall over the 25 project seasons. Results for individual seasons vary considerably and results for the individual months range from "decreases" of perhaps 50 percent to "increases" of as much as 100 percent. The results are not surprising in view of the large natural variability of rainfall in the region.

The South Dakota state progam was operating in only two or three project areas in 1972, but was expected to spread to cover most of the state within 1 or 2 years. Seeding is conducted from aircraft burning pyrotechnic devices and acetone generators charged with silver iodide-ammonium iodide solution and guided from radar equipped field offices. Weather forecasts and cloud model predictions to guide seeding attempts are provided from the School of Mines in Rapid City, which has access to a time-share computer network operated by the U.S. Bureau of Reclamation as part of Project Skywater. Associated economic and sociological studies will be undertaken by university groups throughout the Northern Great Plains region, while experimental seeding programs under Project Skywater are expected to continue in the Rapid City area and in western North Dakota.

To combat droughts in the Philippine Islands, Okinawa, and Texas, airborne dynamic seeding programs have been undertaken. The experimenters claim success mainly based on visual observations. In Africa, several water management oriented dynamic seeding programs are underway. Some of these are beginning by constructing the optimum scientific foundations, namely, randomized single cloud experiments, closely based on radar measurements and modeling.

ARE THERE OTHER WAYS THAN SEEDING TO MODIFY CUMULI?

So far all cumulus modification approaches discussed in this chapter have involved "seeding" with some sort of small particles. All these modification methods require preexisting cumuli. We want now to inquire, whether there may be cumulus modification methods involving techniques other than seeding. In particular, it would be valuable to know whether "warm" cumuli might be dynamically invigorated, whether existing cumuli could be artificially dissipated, and whether and how man might create cumuli out of a clear sky, and if he could do this, whether he could induce the formation of precipitating cumuli.

The most advanced of these methods involves altering the radiative properties of cumuli; so far carbon black has been the substance used. Some theoretical and experimental work was done developing this approach in the 1950s, which was dropped in a promising stage. The carbon particles were dispersed from an aircraft into its exhaust wake. Typically, about 5 lb of material, of particle size about 1 to 0.1 μm in diameter, if totally deagglomerated gives a concentration of 100 to 100,000 particles/m^3 in 2.2 km^3 of air. Calculations based on the radiative properties of the material, the sun, the cloud, and the air mass indicate that the cloudy air would be warmed 0.2 to 0.4°C/min while losing 0.03 to 0.1 g/m^3 of liquid water in the same time interval with an average 1 percent solar absorption/m^3.

Carbon black experiments were conducted both to dissipate cumuli and to invigorate them. In the dissipation experiments, the carbon black is dispersed on the edges and outer surface of the cloud. An exploratory dissipation experiment was run on eight warm cumuli (maximum top height 12,000 ft) in Georgia in 1958 by a group at the Naval Research Laboratory (NRL). All eight clouds were reported to have dissipated in 5 to 24 min after treatment with 1½ to 6 lb of carbon black. The experimenters brought out the following points:

1. All except one of the treated clouds were vigorously growing when treated.
2. There was some indication that increasing the amount of carbon increases the dissipation rate.

3. Flights through some similar cumuli without dropping carbon produced no noticeable results; the mechanical action and exhaust wake of the aircraft above had no apparent effect.
4. There was some indication of grayish virga forming below the treated clouds.

Clearly, cumuli this size usually dissipate naturally in 5 to 24 min, so that no firm conclusions could be drawn without a randomized experiment. A sequel nonrandomized dissipation experiment in New England and the tropics (by another group) gave less apparently successful results; many of those clouds treated were stratocumulus.

As part of the NRL experimental series described above, carbon black was dispersed near the natural condensation level in clear air, with the purpose of inducing formation. On four out of five occasions using 1½ lb of dry powder, a small cumulus formed. On the one unsuccessful try, the carbon black package may not have opened. A sixth attempt used 6 lb of carbon black in 5 gal of water. In this case, a 1-mi line of cumuli formed along the track, 3000 ft in thickness. Although critics concluded that the 5 gal of water itself, finely dispersed in saturated air, would result in a visible cloud, it seems unlikely that this factor alone could build the cloud thickness to 3000 ft.

Small cumulus clouds have also been produced successfully in clear air by the introduction of salt particles.[53] The physics behind this approach is that, due to its hygroscopicity, water vapor condenses on salt particles at relative humidities of only 80 to 90 percent. An artificial cloud made by the release of dry particles of sodium chloride near Hawaii is shown in Figure 6.22.

While initiation of small cumuli by salt and/or carbon black can be of great value scientifically in studying convective processes, these cloudlets are most likely to begin and end their lives as small, "single-bubble" puffs, of extremely limited vertical development, unless the environment is in such marginal equilibrium that one warm puff could trigger a sustained convective plume.

Nature herself often performs cloud modification experiments which man can, if he is sufficiently ingenious, sometimes model, comprehend, and simulate under conditions in which the results might usefully be applied. Among the most fascinating natural cumulus experiments are those performed by heated islands surrounded by ocean (Figure 6.1b). "Streets" of cumuli, often precipitating, frequently extend downwind of small (50 to 100 km²) tropical islands for 100 km or more.[29, 30, 31, 32]

Black[6] suggested simulating a heated island by a 5 × 20 km asphalt coating on the ground to induce rain-producing cumuli in needy tropical and subtropical areas. Asphalt coatings commonly show midday temperature excesses above 11°C over grassy surroundings, and maintain a sizable temperature excess for about 20 hr out of the 24. Thermal "mountains" of 1000 m or more were calculated to correspond to asphalt coatings of reasonable lengths (about 20 km). Black estimated the potential benefits from the excess rain associated with actual tropical mountains of similar heights to his heat "mountains" to be as much as 50 cm rainfall per year. If the asphalt coating is assumed to require renewal each 5 years, the excess precipitation would be induced at $9 per acre-ft, with a benefit-to-cost ratio exceeding 5, the value depending on how water cost is assessed. Obstacles have so far prevented actual execution of this experiment.

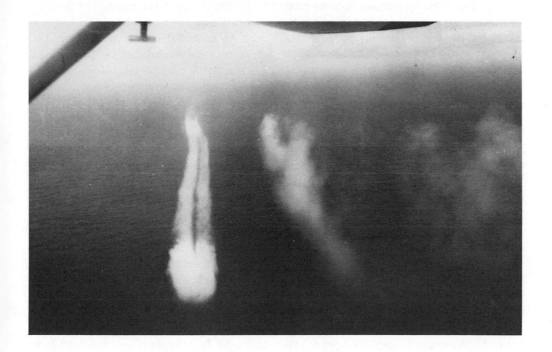

Figure 6.22 Small cloudlets produced by dumping powdered salt from aircraft near cloud base level over the ocean off Hawaii (Courtesy A. H. Woodcock).[53]

EXTENDED AREA AND PERSISTENCE EFFECTS. POTENTIAL IMPACTS OF CUMULUS MODIFICATION

Whether or not cumulus precipitation can be artifically increased over target areas of a few hundred or a few thousand square miles is still unresolved. It is even more unresolved whether attempts at cumulus modification within such target areas have extensive effects *outside* the target areas and possibly persisting for hours, days, or even months beyond the cessation of the treatment.

Some meteorologists have argued, without concrete evidence, that when we increase the rainfall from a cloud or cloud group that we must be "robbing Peter to pay Paul," namely, that rain must be decreased elsewhere. Others have argued, with no better evidence, that man can willfully increase the total amount of rain falling on the earth's surface. Surely this is one of the most important unanswered questions in precipitation management. In the realm of cumulus modification, it is a question without even tentative answers as yet; what follows in this section is therefore highly speculative.

In the silver iodide seeding of winter orographic and convective storms, however, in the mountainous western portion of the United States, persuasive evidence does exist for sizable downwind effects (see Chapters 2 and 7). Apparent 50 to 100 percent increases in rainfall on seeded occasions have been found 80 to 150 mi downwind of seeding sites. In Project Whitetop, on the other hand, there were apparent extended area and persistent rainfall decreases of as much as 90 percent reaching 120 mi downwind of the seeding and persisting 5 hr after its cessation, particularly on the south wind days. In Australia (Chapters 1 and 12), modification experts have claimed that positive effects of seeding on rainfall have persisted even from one year to the next!

Persistence was deduced to explain the decrease with time of the ratio of seeded-to-control rainfall in some single targets, and in some pairs of targets where the randomized crossover design was used. Still other scientists have argued for seeding effects on rainfall upwind of seeded areas.

Extended space and time effects of cloud modification need not be confined to rainfall, but could affect radiation and energy budgets, momentum transports, boundary layer processes, severe weather manifestations, wind circulation patterns, etc. If any of these prove real or even plausible, they constitute simultaneously an exciting, in some instances useful, but also potentially hair-raising, aspect of weather modification, involving unbelievable avenues for legal and sociological complications, as well as scientific controversy. For example, conceive of the interstate fracas which could break out if it were believed that a seeding project in Colorado changed the rainfall in parts of Nebraska, or of the international fracas if such an extended effect were alleged to cross national boundaries in the troubled Middle East!

In cumulus experiments, what could be the possible causes of extended effects in space and/or time? These might include but need not be confined to the list in Table 6.8.

Let us consider these possibilities in the order listed in Table 6.8.

Attempts to trace silver iodide crystals or specially introduced tracer compounds such as indium trichloride through convective cloud complexes have yielded some unexpected results. Experiments in Illinois thunderstorms and Alberta hailstorms have found tracer compounds precipitated to the ground in regions not anticipated from available wind observations. So far, the distances involved have not exceeded 100 km.

The importance of the transport of silver iodide to regions downstream of the target area and its persistence could be reduced by the decrease in AgI effectiveness as an ice nucleant, at the suggested rate of about one order of magnitude per hour. Decay is likely to be more rapid than this in direct sunlight and sometimes slower under cloud conditions or at night. At night, one could anticipate ice-nucleating activity from silver iodide crystals released 100 to 200 km upwind from an observation site, particularly with dynamic area-seeding experiments in-which tens of kilograms of AgI may be released in a day's work.

Ice crystals produced by silver iodide (or dry ice) could conceivably be carried

Table 6.8 Possible Causes of Extended Space and/or Time Effects

1. Physical transport of the seeding agent.
2. Physical transport of ice crystals produced by a seeding agent.
3. Changes in radiation and thermal balance, as for example, from cloud shadows or wetting of the ground.
4. Evaporation of water produced.
5. Changes in the air-earth boundary, such as vegetation changes over land or changes in the structure of the ocean boundary layer following cloud modification.
6. Dynamic effects:
   a. Intensified subsidence surrounding the seeded clouds, compensating for invigorated updrafts.
   b. Advection or propagation of intensified cloud systems which subsequently interact with orography or natural circulations.
   c. Cold thunderstorm downdrafts, either killing local convection or setting off new convection cells elsewhere.
   d. Extended space-time consequences of enhancement or suppression of severe weather owing to cumulus modification.
   e. Alteration, via altered convection, of wind circulation patterns and/or their transports which could interact with other circulations, perhaps at great distances.

for very long distances by upper winds before falling into supercooled clouds to act as nucleating agents. In both tropics and temperate latitudes "orphan anvils" from natural cumulonimbus clouds are found several hundred miles and many hours from their site of origin. Figure 6.23 shows an extensive anvil streaming out from an exploding seeded cumulus in Florida, a not uncommon event with a strong jet stream over the seeding area.[44]

In addition to their nucleating potential, "orphan anvils" could have important radiative impacts. Where solar radiation striking the ground directly maintains convection, as over Florida in summer, the shade of a single anvil often wipes out cumuli over a sizable fraction of the southern peninsula extending outward in any direction from the target area, depending on winds aloft. Over tropical oceans, convection is maintained around the clock by the warm ocean, so that shutting off incoming solar radiation does not suppress cumuli. There the anvils could be most important in reducing the outflux to space of infrared radiation from the surface.

Evaporation of water and/or wetting of the ground produced by seeding could have many different, possibly interacting, effects depending upon the locale of the seeding *and the initial conditions of the system.* In droughts, we have suggested how cloud base could be lowered and CCN production suppressed by these mechanisms. Over a heated island, cooling of the ground by precipitation was found to destroy the island effect. Evaporation of falling rain is a major factor in starting and maintaining the thunderstorm downdraft, while wetting of the ground can change the soil and vegetation, etc. The latter effect is postulated by some Australian scientists as a possible cause of year-to-year persistence in rainfall increases from seeding.

While extended space-time dynamic

Figure 6.23 A Florida seeded cloud with a long anvil, May 27, 1968.[23] (*a*) Seeded cloud 6 min after seeding. (*b*) Same cloud 1 hr and 44 min after seeding. Note long anvil extending more than 80 mi to eastward.

effects of cumulus modification are potentially complex, widespread, and subtle, the writers believe they have observed many of those listed, in their own dynamic cumulus seeding experiments in Florida and South Dakota. The first one, compensating subsidence, is believed to be a main way in which we may be robbing Peter to pay Paul. However, no one yet knows how to specify where the compensating subsidence will occur; its location and local strength almost surely depend on the meso- and larger-scale wind patterns and their convergence. Even where we have observed cleared-out areas surrounding seeded complexes, we cannot be sure whether Peter has been robbed of more or less than Paul has been paid. Preliminary results in Florida do not indicate so far that there are compensating rain decreases surrounding seeded systems.

In Florida, we have clearly documented the propagation of seeded cumulonimbus complexes out of the target area and their interaction with the coastal sea breeze, to produce heavy rainfall at least 50 mi away from the seeding site up to 5 hr after seeding had ceased. If the seeding did indeed cause the observed explosive growth of the original seeded clouds, the

extended space and time effects follow. More explicitly, we may alternatively postulate that the seeding was a necessary, but not sufficient, condition to cause the observed chain of events, or better perhaps, that the extended effects could be triggered by the seeding because of the special initial conditions of the cloud-environment systems.

The effects of cold cumulonimbus downdrafts to suppress or propagate convection have not yet been explicitly studied in connection with modified cumuli. However, they are becoming documented in connection with natural cumulonimbi.[57] Since in the tropics no observable differences have been detected between cumulonimbi whose stature has been brought about by dynamic seeding and those created by nature, cold downdrafts could quite probably play a role in extending the effects of cumulus modification experiments.

As described in other chapters, cumulus modification has been attempted to suppress lightning and hail. It has also been alleged that under some conditions, inducing cumulonimbus growth by dynamic seeding could result in severe weather phenomena that might not have occurred had increased cloud growth and merger not been induced. The alteration, either way, of hail or lightning, could quite possibly give rise to extended effects. Hail swathes can be as much as 20°C colder than their surroundings and persist for 2 to 24 hr. Lightning sets off extensive forest fires and perhaps more subtle electric effects in the air, which could affect future condensation, coalescence, and even cumulus dynamics.

The items from Table 6.8 discussed so far are mainly extended space-time effects on the mesoscale. Many of them have been deduced from statistical analyses and rain gauge records only, with an unfortunate lack of direct evidence regarding cloud processes. New tools such as tracer techniques, infrared photography, radars, and satellites must be adapted to document the large-scale effects of modification. At EML, considerable promise is found from calibrating enhanced satellite photographs with radar to measure precipitation changes over extended areas, including oceans. Mesoscale modeling is in its infancy and must be pushed hard to the point where it can be used to supplement the most sophisticated statistics and observations.

Next we must proceed to the last topic in Table 6.8, the vitally important possibility of cumulus modification upon synoptic and larger-sized circulations in the atmosphere.[44] Both types of current hurricane modification experiments (see Chapter 14) are based on the postulate that we can alter the storm scale of motion by massively seeding hurricane clouds. This postulate is based on nearly two decades of hurricane research in which the role of cumuli in driving and maintaining the hurricane motions has been well documented. Results of full-scale field experimentation to date warrant some optimism that a storm-sized system may be markedly changed by means of cumulus modification.

Another promising but untried area lies in the "explosive" deepening of middle-latitude oceanic cyclones. Very sudden intensification of winter storms often occurs in the Gulf of Alaska and off continental east coasts. An important factor in the deepening process appears to be sea-air heat flux, coupled with increasingly tall cumulus convection. In fact, when the cumuli remain stunted, satellite studies indicate that deepening fails to take place.

If the seedability in some potential deepening situations were to prove adequate, the explosive development of the storm could possibly be induced artificially. But why would anyone wish to consider such an experiment? The reason is that explosive deepening of marine cyclones, particularly in the Gulf of Alaska, has been found to trigger major changes in the entire wind circulation pattern over the whole Northern Hemisphere. Other "teleconnections" between local triggers and hemispheric circulation adjustments are becoming subjects of serious investigation.

In pursuit of the possible effects of cumulus modification upon really large, namely, planetary, scale circulations, we use the vast progress of the last decade that has been made in documenting the role of cumuli as both combustion cylinders and fuel pumps in many large-scale circulation branches, particularly in the tropics.

The primary newly gained knowledge that renders questioning the role of cumulus modification here meaningful, rather than ridiculous, is the concentration of the cumulus function. Studies of the atmosphere's firebox, namely, the equatorial trough zone[30] have shown that its entire vertical transport and heat release function is carried out by a few thousand (1500–5000) cumulonimbus hot towers active at one time around the globe, mainly clustered in vortical disturbances and/or over continents. In tropical storms and disturbances, the firebox function is conducted by 100 to 400 active hot towers occupying only 2 to 4 percent of the storm area. Identification of these towers has now been made possible from satellite pictures using simple enhancement techniques. The role of cumuli in large-scale flows is fourfold:

1. Latent heat release by precipitation, which provides (by warming) the pressure gradients that drive many wind systems.
2. Radiational (discussed earlier).
3. Vertical transports of heat, moisture, vorticity, and momentum.
4. Possible creation of relative vorticity and/or its reorientation from a vertical to horizontal plane.

It is therefore clear that a successful outcome of area seeding to increase cumulus precipitation is not necessarily required, except for item 1, for cumulus modification to affect large-scale flows.

Last, we should consider another direct dynamic effect of cumuli on planetary circulations not yet included in most large-scale numerical models. Gray [15] has emphasized the role of these towering clouds in the vertical flux of horizontal momentum. Other researchers have shown that vertical momentum fluxes in the tropics are important in driving the circulation of the Hadley Cell, which in turn supplies midlatitudes, partly through direct infusion by the subtropical jet stream. It is thus not inconceivable that massive modification of equatorial cumuli might one day change the weather in middle America. More important than this speculation is that we are gaining the means of testing it.

In reaching this point, we have in a sense brought cloud modification in a full circle, from the pioneering days of Schaefer and Langmuir.[22, 42] A quarter of a century ago, Langmuir postulated that seeding clouds in New Mexico affected the subsequent weather in Boston. His suggestion met with violent reactions from many meteorologists ranging from bitter hostility to angry ridicule. Today, enough about the atmosphere, the role of cumuli, and modification potential has been learned so that the wise meteorologist neither asserts nor ridicules anything. On the contrary, he inquires how adequate understanding and documentation can be gained to formulate and test hypotheses meaningfully.

OUTLOOK

After 25 years of effort, the promise and excitement of cumulus modification are great, while the concrete accomplishments are modest. Definitive sound results are confined to single cloud experiments.

In water spray seeding these definitive results involve the frequency and early initiation of rain echoes within clouds but treatment has not yet been related to volume of rain falling out. Seeding with hygroscopic agents has been shown to hasten the initiation of precipitation in convective clouds and to increase rainfall from some clouds of moderate depth, say 3 to 8 km. Seeding supercooled clouds with dry ice or silver iodide can also initiate precipitation formation, and silver iodide seeding for dynamic effects has given definitive results for both cloud growth and increased volume of rain falling from cloud base and reaching the ground.

Major advances in cumulus experimentation and in confidence in their results have derived from the development and use of numerical models. Despite their imperfections and the criticisms incurred, one-dimensional cumulus models originated the concept and exploitation of dynamic "seedability." They have provided virtually priceless criteria for launching and evaluating seeding missions; they provide unique means for quantitatively examining physical-dynamical interactions; they have provided pioneering classifications of seeding climatologies; and last but not least, they serve both as springboards and test beds for more advanced and sophisticated modeling efforts.

Concerning area experiments on cumuli, *none* have yet attained definitive positive results, although one (Whitetop), involving the static approach, appears to have obtained negative results (Chapter 1). However, the apparent rainfall increases at four to five Rapid Project stations (on shower days) were significant at the 5 percent level and may indicate some real area effect, even if only a redistribution between the two targets.[5]

Extended space and time effects of cumulus modification are likely to exist, but their nature, magnitude, extent, and functional dependence await documentation, which will require modeling advances and detailed measurements combined with judicious applications of statistics.

Our most important conclusion is that cumulus modification depends on not only the amount, nature, and method of the treatment, *but also on the initial conditions of the cloud-environment system.* An identical treatment may lead to entirely different results even in the same target area, depending on the initial cloud circumstances and the prevailing weather. Data stratification and probably detailed documentation of cloud behavior will be necessary to unravel causality. The returns from blind statistics applied to masses of averaged rain gauge data have probably diminished to zero or lower. As a corollary, one of the most futile questioning methods in weather modification seeks "yes" or "no" answers, such as "does seeding cause rain increase?"

It is clear from this chapter that most of the major advances in cumulus modification have originated from scientists primarily involved in basic research. Many of the breakthroughs, or their necessary

backgrounds, were not at first intended to apply to modification at all but had the purpose of advancing man's knowledge of cumulus processes.

We believe that the optimal way to reach useful cumulus modification is to devote increased resources to a judicious balance between basic research and controlled experimentation in the field, in the laboratory, and on the computer. The experimentation is likely to reach economic application faster and more firmly if it is mainly devoted to establishing sound causality, understanding, and prediction.

ACKNOWLEDGMENTS

The first author wishes to thank her colleagues Roscoe Braham, William Cotton, Alan Miller, Robert Ruskin, William Woodley, and Helmut Weickmann for continual help, advice and criticism in preparing this chapter, and her husband, R. H. Simpson, for patient and enthusiastic encouragement. Connie Arnhols and Robert Powell did their usual fine work on manuscript and figure preparation.

We also wish to thank Eugene Bollay for a fine job of editing and Fridel Settle for making the numerous revisions of the manuscript.

REFERENCES

1. Arneson, G., R. S. Greenfield, and E. A. Newburg, A numerical experiment in dry and moist convection including the rain stage, *J. Atmos. Sci.* **25**, 404–415, 1968.
2. Berry, E. X., Cloud droplet growth by collection, *J. Atmos. Sci.* **24**, 688–701, 1967.
3. Biswas, K. R., and A. S. Dennis, Calculations related to formation of a rain shower by salt seeding, *J. Appl. Meteor.* **11**, 755–760, 1972.
4. Bowen, E. G., The formation of rain by coalescence, *Australian J. Sci. Res.* **A3**, 193–213, 1950.
5. Boyd, E. I., South Dakota cloud seeding

evaluation, 1965–71, *WMA: J. Weather Modif.* **4,** 172–194, 1972.

6. Black, J. F., Weather control: Use of asphalt-coatings to tap solar energy, *Science* **139,** 226–227, 1963.

7. Braham, R. R., The water and energy budgets of the thunderstorm and their relation to thunderstorm development, *J. Meteorol.* **9,** 227–242, 1952.

8. Braham, R. R., L. J. Battan and H. R. Byers, Artificial nucleation of cumulus clouds, *Meteorol. Monogr.* **2** 11, 47–85, 1957.

9. Dennis, A. S., Modifying precipitation by cloud seeding, *J. Soil Water Conserv.* **25,** 88–92, 1970.

10. Dennis, A. S., and A. Koscielski, Results of a randomized cloud seeding experiment in South Dakota, *J. Appl. Meteorol.* **8,** 556–565, 1969.

11. Dennis, A. S., and A. Koscielski, Height and temperature of first echoes in unseeded and seeded convective clouds in South Dakota, *J. Appl. Meteorol.* **11,** 994–1000, 1972.

12. Dennis, A. S., and M. R. Schock, Evidence of dynamic effects in cloud seeding experiments in South Dakota, *J. Appl. Meteorol.* **10,** 1180–1184, 1971.

13. Dennis, A. S., C. A. Schock, and A. Koscielski, Characteristics of hailstorms of western South Dakota, *J. Appl. Meteorol.* **9,** 127–135, 1970.

14. Garstang, M., Sensible and latent heat exchange in low-latitude synoptic scale systems, *Tellus* **19,** 492–508, 1967.

15. Gray, W. M., The mutual variation of wind, shear and baroclinicity in the cumulus convective atmosphere of the hurricane, *Monthly Weather Rev.* **95,** 55–73, 1967.

16. Goyer, G. G., W. E. Howell, V. J. Schaefer, R. A. Schleusener and P. Squires, Project Hailswath, *Bull. Am. Meteorol. Soc.* **47,** 805–809, 1966.

17. Hess, S. L. *Introduction to Theoretical Meteorology,* Henry Holt, New York, 1959, 362 pp.

18. Holle, R. L., Effects of cloud condensation nuclei due to fires and surface sources during south Florida droughts, *J. Appl. Meteorol.* **10,** 62–69, 1971.

19. Holle, R. L., Populations of parameters related to dynamic cumulus seeding over Florida, *J. Appl. Meteorol.* **13,** April, 1974.

20. Howell, W. E., The growth of cloud drops in uniformly cooled air, *J. Meteorol.* **6,** 134–149, 1949.

21. Kessler, E., On the distribution and continuity of water substance in atmospheric circulations, *Meteor. Monogr.* **10** (32), 1–84, 1969.

22. Langmuir, I., The production of rain by chain-reaction in cumulus clouds at temperatures above freezing, *J. Meteorol.* **5,** 175–192, 1948.

23. Levine, J., Spherical vortex theory of bubble-like motion in cumulus clouds, *J. Meteorol.* **16,** 653–662, 1959.

24. McCarthy, J., Computer model determination of convective cloud seeded growth using Project Whitetop data, *J. Appl. Meteorl.* **5,** 818–822, 1972.

25. Malkus, J. S., Effects of wind shear on some aspects of convection, *Trans. Amer. Geophys. Union* **30,** 19–25, 1949.

26. Malkus, J. S., Recent advances in the study of convective clouds and their interaction with the environment, *Tellus* **4,** 71–87, 1952.

27. Malkus, J. S., Some results of a trade-cumulus cloud investigation, *J. Meteorol.* **11,** 220–237, 1954.

28. Malkus, J. S., Trade cumulus cloud groups: Some observations suggesting a mechanism of their origin, *Tellus* **9,** 33–44, 1957.

29. Malkus, J. S., Recent developments in the study of penetrative convection and an application to hurricane cumulonimbus towers, *Cumulus Dynamics,* Pergamon Press, London, 1960, pp. 65–84.

30. Malkus, J. S., Large-scale interactions, Chap. 4, *The Sea,* Interscience, New York, Vol. l, l962, pp. 88–294.

31. Malkus, J. S., Tropical rain induced by a small natural heat source, *J. Appl. Meteorol.* **5,** 547–556, l963.

32. Malkus, J. S., and A. F. Bunker, Observational studies of the air flow over Nantucket Island during the summer of l950, *Papers Phys. Oceanog. Meteorol.* (Mass. Inst. Tech. Woods Hole Ocean Inst.) **13**(2), 1–47, 1952.

33. Malkus, J. S., and H. Riehl, *Cloud Structure and Distributions over the Tropical Pacific Ocean,* University of California Press, Berkeley, l964, 229 pp.

34. Malkus, J. S., and G. Witt, The evolution of a convective element, A numerical experiment, *The Atmosphere and Sea in Motion,* Oxford University Press, 1959, pp. 425–439.

35. Malkus, J. S., C. Ronne, and M. Chaffee, Cloud patterns in Hurricane Daisy, 1958, *Tellus* **13,** 8–30, 1961.

36. McDonald, J. E., The physics of cloud modification, *Advances in Geophysics,* Vol. 5,

Academic Press, New York, 1958, pp. 223–303.

37. Murray, F. W. and L. R. Koenig, Numerical experiment on the relation between microphysics and dynamics in cumulus convection, *Monthly Weather Rev.* **100,** 717–732, 1972.

38. Newton, C. W., Dynamics of severe convective storms, *Meteorol. Monogr.* **51**(27), 33–58, 1963.

39. Orville, H. D., and L. J. Sloan, A numerical simulation of the life history of a rainstorm, *J. Atmos. Sci.* **27,** ll48–ll59, 1970.

40. Saunders, P. M., Penetrative convection in stably stratified fluids, *Tellus* **14,** 177–194, 1962.

41. Scorer, R. S., and C. Ronne, Experiments with convection bubbles, *Weather* **11,** 151–154, 1956.

42. Schaefer, V. J., The production of ice crystals in a cloud of supercooled water droplets, *Science* **104,** 457–459, 1946.

43. Simpson, J., Some aspects of sea-air interaction in mid-latitudes, *Deep Sea Res., Suppl.* **16,** 233–261, 1969.

44. Simpson, J., Cumulus cloud modification: Progress and prospects, *A Century of Weather Progress,* American Meteorological Society, l970, pp. 143–155.

45. Simpson, J., G. W. Brier, and R. H. Simpson, Stormfury cumulus seeding experiments 1965: Statistical analysis and main results, *J. Atmos. Sci.* **24,** 508–521, 1967.

46. Simpson, J., R. H. Simpson, D. A. Andrews, and M. A. Eaton, Experimental cumulus dynamics, *Rev. Geophys.* **3,** 387–431, 1965.

47. Simpson, J., W. L. Woodley, H. A. Friedman, T. W. Slusher, R. S. Scheffee, and R. L. Steele, An airborne pyrotechnic cloud seeding

system and its use, *J. Appl. Meteorol.* **9,** 109–122, 1970.

48. Simpson, J., and W. L. Woodley, Seeding cumulus in Florida: new 1970 results, *Science* **172,** 117–126, 1971.

49. Simpson, J., W. L. Woodley, and R. M. White, Joint federal-state cumulus seeding program for mitigation of 1971 south Florida drought, *Bull. Am. Meteorol. Soc.* **53,** 334–344, 1972.

50. Stommel, H., Entrainment of air into a cumulus cloud, *J. Meteorol.* **4,** 91–94, 1947.

51. Turner, J. S., The "starting plume" in neutral surroundings, *J. Fluid Mech.* **13,** 356–368, 1962.

52. Weinstein, A. I., Ice-phase seeding potential for cumulus cloud modification in the western United States, *J. Appl. Meteorol.* **11,** 202–210, 1972.

53. Woodcock, A. H., and A. T. Spencer, Latent heat released experimentally by adding sodium chloride particles to the atmosphere, *J. Appl. Meteorol.* **6,** 95–101, 1967.

54. Woodley, W. L., Precipitation results from a pyrotechnic cumulus seeding experiment, *J. Appl. Meteorol.* **9,** 242–257, 1970.

55. Woodley, W. L., and A. Herndon, A raingage evaluation of the Miami reflectivity-rainfall rate relation, *J. Appl. Meteorol.* **9,** 258–264, 1970.

56. Woodley, W. L., B. Sancho, and J. Norwood, Some precipitation aspects of Florida showers and thunderstorms, *Weatherwise* **24,** 106–119, 1971.

57. Zipser, E. J., The role of organized unsaturated convective downdrafts in the structure and rapid decay of an equatorial disturbance, *J. Appl. Meteorol.* **8,** 799–814, 1969.

7 | Weather Modification for Augmenting Orographic Precipitation

LEWIS O. GRANT

ARCHIE M. KAHAN

THE REQUIREMENT AND POTENTIAL

Shortly after the cloud seeding experiments of the General Electric group, headed by Irving Langmuir and Vincent Schaefer (General Electric Research Laboratories, 1951), Tor Bergeron[2] made "an inventory of actual tropospheric clouds and cloud systems" to evaluate their potential for weather modification. He concluded that the "main possibility for causing considerable artificial rainfall might be found within certain kinds of orographic cloud systems." Twenty-three years of subsequent research has shown that orographic clouds do in fact provide one of the most productive and manageable sources for beneficial weather modification.

Efforts to increase orographic precipitation were started shortly after the initial discoveries of Schaefer in 1946.[44, 45] The greatest concentration of effort was in the western United States. This region is, of course, highly dependent on irrigation water supplies and virtually all this water comes from the mountains that are interspersed throughout the region. Except for the coastal plains of the West Coast and the highest elevations of the mountainous areas, evapotranspiration greatly exceeds precipitation throughout this region. Average annual runoff is produced by generally less, and in many areas considerably less, than 1 in./year of the precipitation that falls. In some mountainous areas, however, annual runoff to as much as 20 to 40 in./year occurs at higher elevations. Such mountainous areas, however, constitute less than 5 percent of the total land area. It is this water that provides the basic supply for domestic use and for irrigated agriculture, industry, and recreational uses throughout the western United States. Water has been in extremely short supply during past droughts (partly from lack of storage development and effective use)

Lewis O. Grant is a Professor in Department of Atmospheric Sciences at Colorado State University, Fort Collins, Colorado.

Archie M. Kahan is Chief, Division of Atmospheric Water Resources Management, Department of Interior, Bureau of Reclamation, Denver Federal Center, Denver, Colorado.

and the supply is being increasingly strained even during normal years. Storage facilities are now reaching a capacity nearly equal to the potential natural supply. The atmosphere is the source for these natural water supplies and any real increase, in contrast to improved usage, must also come from the atmosphere. Political, economic, engineering, and ecological considerations limit the potential during the foreseeable future for substantial surface water importation from areas with more abundant supplies.

Since natural water supplies originate in the mountains of the west and since physical considerations indicate that these areas offer the greatest potential for weather modification, early and continued efforts of weather modification have been concentrated on precipitation augmentation from orographic clouds. Commercial groups, such as power companies and irrigation groups, took the early initiative. Analyses of these seeding programs served as the primary basis for the conclusion of the 1957 President's Advisory Panel on Weather Modification[15] that "the most probable effect of cloud seeding operations in the mountains of the west, was a 10–15 percent increase in precipitation." The indications were that seeding in a number of separate projects produced an overall positive effect. For some areas, the indicated potential was not nearly so clear. The conclusions of the Advisory Committee were, however, sufficiently encouraging to stimulate increased support for research. This increased research effort, primarily supported by the Federal Government, has produced a continually improving technology which can systematically and predictably utilize the potential for obtaining additional water supplies from seeding over the mountains. Physical models of orographic cloud processes under natural and seeded modes can be

used to provide seedability criteria, and, for differing criteria, to define the amount of the precipitation increase which can be achieved. These models, along with information on the climatological variations of their important parameters, can be used to determine the potential water resources that can be realized from weather modification in various geographical areas.

THE NATURE OF OROGRAPHIC CLOUDS

Airflow patterns over a mountain barrier and the thermodynamic and microphysical (cloud particle) scale characteristics of resulting clouds impose important controls on the requirement and on the methodology for artificial seeding to augment precipitation. The characteristics of these clouds must thus be considered on each of these scales.

Characteristics of Orographic Clouds

Orographic clouds are formed as moist air is lifted over the terrain. The lifted air with water vapor present, but not yet saturated, expands and cools at a rate of about 5.4° per 100 ft of lifting. With sufficient lifting and a continuation of this rate of cooling, air parcels frequently reach saturation since at cooler temperatures less water vapor is required for air parcels to become saturated. Once saturation is reached, small liquid cloud droplets are formed by condensation. The visible location of the edge of the cloud on the upwind side of the mountain barrier is close to where this condensation takes place. Once condensation of liquid droplets has occurred, these droplets move with essentially the speed of the wind as the air progresses over the mountain. As the air parcel subsequently descends in the lee of the mountain, its temperature and, consequently, the water holding capacity of the air again increase. Surviving liquid droplets or ice crystals are then evaporated to satisfy the increased water holding capacity of the air at the now higher temperature. This zone in which the evaporation takes place delineates the lee, or downwind edge, of the mountain cloud. In the case of a water cloud, the evaporation takes place rapidly and the cloud boundary is sharply defined. When ice crystals are present, sublimation of these particles takes longer because of the greater size of ice crystals and the lower saturation pressure of ice. In this case the cloud boundary is less distinct.

The upper boundary of mountain clouds is determined by the depth and stability of the moist air mass approaching the mountain barrier. Blanket clouds, those in the lowest levels which envelop the barrier, usually extend to an altitude of 12,000 to 15,000 ft msl along the coastal range of the western United States and to 15,000 to 20,000 ft msl over the Rocky Mountains. This gives a range of orographic cloud tops from just above the mountain peaks to elevations some 6000 to 8000 ft above the general crest of the barrier. In lateral extent, orographic clouds can extend along a mountain barrier for hundreds of miles. Most frequently they extend along the mountains of the western United States for some 100 to 200 mi at any specific time. This is determined by the zone, or jet, of strong moist airflow preceding or following a general storm disturbance passing through the region. The storm disturbance serves to increase the velocity of flow against the barrier and, consequently, the rate of lifting and cooling of the air mass. It also serves to advect large quantities of moist air to the mountain barrier.

The blanket orographic clouds described are of primary interest from the standpoint of weather modification. There are a number of other types of clouds which form over mountains. Spectacular wave clouds at high elevations, commonly at altitudes of 20,000 to 50,000 ft msl, can frequently be observed. These have virtually no potential for increasing water supplies, although they present interesting opportunities for certain cloud physics research. Certain of these wave clouds stand stationary with condensate forming on their upwind side and then evaporating shortly thereafter on the lee side of the wave. This process in some cases can go on for many hours and produce a persistent cloud for cloud physics or seeding experiments.

A still different type of mountain cloud is one which indirectly results from the mountain lifting but is primarily convective in character. Through a combination of forced lifting and surface heating effects, an air parcel is lifted to saturation. When the thermal structure of the atmosphere is sufficiently unstable, the heat released in the formation of liquid condensate produces an air parcel warmer than the surrounding air. The parcel is then buoyant and can continue to rise in a manner not directly dependent on the mountain barrier. The formation and life cycle of these mountain induced convective cloud systems are complex and tightly controlled by the characteristics of the mountain barrier. These clouds can play an important role in determining the precipitation climatology of the mountains, or in some cases, in the creation of mesoscale cloud systems that can move from the mountains and affect a much larger area. Unstable, moist air masses are required for the formation of mountain induced clouds of this type. Specific mountain characteristics play a major role in initial formation and subsequent development of such convective cloud systems. Critical orographic factors include (1) mountain barrier orientation, (2) extent and size of mountain, and (3) height of mountain relative to the convective cloud base. For example, the orientation of a mountain controls how the variable heating of the mountain slope takes place. This is of major importance for the north-south oriented Rocky Mountains from Mexico to Canada. Their orientation allows for the solar heating to progressively shift from the eastward facing slopes in the morning to the westward facing slopes in the afternoon. Further, this barrier frequently blocks the westward extension of the moist air from the Gulf of Mexico. This blocked moist air, overlying the eastward facing mountains, heats during the morning hours. The convective cloud systems which form frequently are so extensive and intense that they control the mesoscale circulation over the High Plains and even the Midwest as they move eastward later in the day. The lateral extent of the mountains also provides a vital control on the mountain induced convective cloud systems. T. Fujita[17] has shown, for example, the close parallel between precipitation amounts and the elevation of the mountain slopes for the isolated San Francisco mountains in northern Arizona. He has also shown that the height of the convective cloud base in relation to the height of the mountain causes significant variation in the rainfall patterns around the mountains. In general, precipitation will tend to be heavier (1) on the windward slopes when cloud bases are below the ridge top and (2) on the lee side when cloud bases are higher. This effect is readily observed in the Colorado Rockies, where precipitation is generally heavier on the western slopes during fall, winter, and early spring, the seasons when cloud bases are near or below the peaks. During late spring and summer, when convective type clouds are formed, cloud bases are several thousand feet above the peaks (frequently 16,000 to 18,000 ft msl) and most precipitation occurs on the eastern slopes, and on the plains, as the mountain induced cloud systems produce a dynamic control on the eastward moving mesoscale cloud systems.

Further discussion in this chapter will concentrate on precipitation augmentation from blanket type orographic cloud systems that form primarily during the colder season. Weather modification of the mountain convective cloud systems can more appropriately be considered with the seeding of convective clouds. Justification for this limitation of the scope of this chapter includes (1) the importance of the blanket, orographic cloud on the water supplies of the west (they provide more than 90 percent of the annual runoff in most sections) and (2) the more advanced state of the art for seeding clouds of this type.

Microphysical Characteristics of Orographic Clouds

There are several factors that tend to produce uniformity in the microphysical characteristics of wintertime orographic clouds. This results from the location and manner in which these clouds form and from the time scale during which they exist. Wintertime orographic clouds of the United States, and other continental areas at similar latitudes north or south of the equator, occur in the strong belt of westerly winds that circle the earth. This zone of westerlies is characterized by moderate to strong winds even over mountain ranges that rise only a few thousand feet. Wind velocities at mountaintop levels are commonly in the range of 20 to 40 mph and are occasionally stronger. Although notable exceptions do occur, most mountain barriers have a lifting section generally less than 20 to 40 mi wide over which air is lifted to condensation and transported over the barrier. Many of the primary mountain ridges of the intermountain basin of the west and of the Rocky Mountains produce the primary lift for a distance as short as 10 mi or less. With such short lifting distances and the existence of moderately strong mountaintop winds, it is apparent that the microphysical properties are primarily established in a short time interval of consistently less than 1 hr, and frequently less than ½ hr. This restricted time for cloud particle growth strongly affects and limits the size to which liquid droplets can grow, either from vapor diffusion or from droplet interaction and coalescence. In some areas, particularly along the west coast ranges of the Cascades and Sierras, the upwind lifting section of the ranges can be much wider and, at least with some meteorological situations, cloud droplet growth and broadening processes can proceed for several hours. These cases might provide sufficient time for the development of significant droplet interaction and the formation of coalescence type precipitation. The opportunities for this process are frequently increased by the presence of low-level clouds over the coastal and valley areas upwind of the mountains. These low clouds on occasion exist for many hours so that some large droplets already exist before the lifting over the mountain barrier is initiated. The presence of these conditions is not well documented in the literature and the frequency of their occurrence for the coastal ranges of the west coast is not clear. Cloud droplet size spectrum broadening from droplet coalescence over the interior mountains is clearly limited in most wintertime clouds. Some process other than droplet coalescence is required to disrupt the colloidal stability of these clouds and produce precipitation particles that can grow large enough to fall out of the cloud system in the time interval available. Although conditions in these clouds are generally unfavorable for the development of coalescence type

precipitation, their temperature structure is generally suitable for the development of precipitation by a process which requires the coexistence of subcooled liquid water and ice crystals. As is discussed later in this chapter, the ratio of the mass of subcooled liquid water to the concentration of ice crystals is critical to the cloud precipitating efficiency.

In addition to the short time for particle growth, several other factors play a major role in establishing the microphysical structure of wintertime orographic clouds. These include the activation characteristics of both the condensation and ice nuclei entering the cloud system, ice crystal multiplication processes, the characteristics of overlying higher clouds, and the temperature structure of the blanket clouds as controlled by their elevation and geographical location.

Concentrations of cloud condensation nuclei, and consequently cloud droplet concentrations, are generally low along the coastal areas. This tends to favor coalescence growth processes and possibly reduce the requirement for seeding to promote an ice formation process. In contrast, concentrations of condensation nuclei in the interior areas increase with the continentality of the site and tend to increase the requirement for ice phase seeding to disrupt the colloidal stability of the clouds. Concentrations of ice nuclei do not correspond well with concentrations of condensation nuclei and, in the mean, are found to be relatively uniform for differing locations and elevations. Large concentrations of ice nuclei, however, may be found in areas downwind from some pollution sources or downwind of weather modification projects.

The concentrations of ice crystals in some clouds may be controlled by "ice multiplication" processes rather than by "primary" ice nuclei as measured in contemporary ice nuclei counters. The problem of "ice crystal multiplication" needs considerable additional research and can be considered only in a general way in describing the microstructure of orographic clouds and in seeding models. Mossop[36] has concluded that ice multiplication processes should be less rapid in (1) clouds with bases colder than 0°C, (2) clouds with higher droplet concentrations, and (3) clouds with relatively narrow cloud droplet spectra. Orographic clouds over the higher mountains of continental areas thus might be expected to have a minimal ice multiplication effect in comparison to more maritime clouds at lower elevations. Hindman[26] and Grant[19] have reported that the average ice crystal concentrations, in clouds over the Colorado Rockies, are not very different from observed concentrations of primary ice nuclei activated at corresponding temperatures. Vardiman and Grant[50, 51] show, however, that concentrations of ice crystals considerably greater than expected from primary ice nuclei observations are frequently associated with convective elements embedded in the orographic cloud. This suggests an important additional role that the thermal stability, which affects the development of convective elements, may play on the orographic cloud microphysical characteristics.

The characteristics of overlying cloud systems in some cases probably drastically affect the microstructure of mountain clouds and their precipitating efficiency. In some geographical areas, and clearly many of those along the west coast of the United States, upper cloud systems, above the orographic blanket clouds, precede advancing storm systems. Ice crystals formed at the much colder temperatures of the upper clouds may settle into the upper portions of the orographic clouds. This can greatly increase the ice particle concentrations above the natural concentration expected for the orographic cloud lying at a lower elevation and warmer temperature. In most cases, this effect would tend to increase the natural precipitating efficiency of these clouds and reduce the requirement for seeding.

The elevation and location of the barrier over which clouds form can also play a vital role on the microstructure of the overlying blanket clouds. Many orographic clouds along the west coast of the United States typically have bases a few thousand feet above sea level and as a rule have several thousand feet of cloud thickness at temperatures greater than freezing. These orographic clouds frequently have tops at elevations from 10,000 to 15,000 ft msl, such that about half of the cloud is at temperatures greater than about −6°C and all of the cloud is warmer than −20°C. This is in contrast to areas of the Rockies where cloud bases are typically at elevations around 10,000 to 12,000 ft msl (temperature at the base is frequently 0 to −10°C) and tops around 19,000 to 20,000 ft msl (temperature generally less than −20°C). These clouds, in contrast to the lower coastal orographic clouds, generally have small portions of the cloud warmer than −6°C. Cloud temperatures as controlled by the location and elevation of a mountain barrier thus affect the microstructure and consequently the precipitating efficiency of the respective clouds. This factor, if not offset by other processes, would more frequently produce a requirement for ice phase seeding in the lower-elevation, warmer coastal orographic clouds than in those formed over the higher mountain ranges farther inland in the western United States. The cloud elevation and location also exert an important control on the amount of available condensate within the respective clouds. For similar updraft speeds, lower-elevation, warmer orographic clouds can be expected to have substantially greater amounts of liquid water than orographic clouds over the higher inland mountain ranges. Typically, major portions of the clouds over the higher mountains are at such cold temperatures that little liquid water exists and natural ice nucleation can produce clouds completely composed of ice particles. Even in each of these extremely different examples, however, an important percentage of all clouds are typically in a temperature region -6 to $-20°C$ where subcooled liquid water and ice crystals coexist.

THE PHYSICAL BASIS FOR SEEDING OROGRAPHIC CLOUDS

Shortly after the initial demonstration by Schaefer that the microstructure of super-cooled clouds could be artificially modified, Bergeron[2] considered the potential for effective modification of various cloud types. He emphasized a natural precipitation process whose optimal use of cloud water would depend on the cloud temperature, the size and concentration of cloud droplets, and the ratio of cloud droplet concentrations to ice crystal concentrations. Using this approach he concluded that substantial potential for precipitation augmentation should occur in clouds produced by persistent and substantial upward motions (as present in orographic clouds) which take place in clouds colder than freezing and warmer than $-10°C$. Ludlam,[31] followed a somewhat different approach and evaluated orographic cloud precipitation efficiency by comparing the removal of cloud condensate by the vapor growth of ice crystals with the supply of vapor to the cloud. Ludlam also showed that any artificially induced increase in precipitation depends on the ice crystal fall trajectory, in relation to the cloud system, as well as on the precipitation ef-

ficiency within the cloud itself. The transfer of the water from small subcooled liquid droplets to ice crystals must produce particles large enough to settle to the mountain slopes before sublimation takes place below cloud base or to the lee of the mountain barrier. He concluded that the potential for precipitation increase over the mountains of central Sweden might be as high as 100 percent. His treatment did not limit the potential to only clouds slightly colder than $0°C$ but instead to clouds with deficient concentrations of snow crystals. The natural concentration of active ice nuclei is, of course, highly dependent on temperature, resulting in specific temperature requirements for having optimal ice nuclei concentrations.

Ludlam's general approach served as a framework for a systematic analysis of the potential for increasing snowfall from orographic clouds in investigations at Colorado State University during the 1960s. Major portions of the following material are based on these studies. The Colorado studies had the additional objective of developing a technology that could serve as a basis for operational programs for increasing water supplies if the reality of adequate potential was established.

Cloud Water Availability

The limiting conditions for the production of precipitation from an orographic cloud is the mass of water condensate formed from vapor as the air mass is lifted over the topographic barrier (this neglects a small additional contribution to the precipitation by the vapor growth of ice crystals at humidities between water and ice saturation). The mass of condensed water is determined by the temperature and humidity of the air approaching the clocking mountain barrier and by the rate at which it is carried over the barrier. A detailed description of the rate at which orographic cloud condensate is formed is quite complex since the vertical motion field is generally the result of storm scale and local convective components of vertical airflow in addition to the orographic lifting. In most mountain areas, however, the orographic component dominates and accounts for major portions of the mountain precipitation. A general estimate of the relative contribution of the dynamic, or storm, component on the precipitation can be made by reference to nearby, upwind precipitation stations at sites not favored by orography. Along the west coast of the United States precipitation at such upwind stations is commonly in the range of ¼ to ½ of that experienced in the mountains. This would indicate that ½ to ¾ of the mountain precipitation is related to the orography. The substantial contribution of the orographic effect over the Colorado Rockies can be noted in

Figure 7.1 which shows the mean west to east normalized precipitation profile across Colorado during the winter months. This shows that around 80 percent of the precipitation over the central Colorado Rockies is related to orography. Commonly, precipitation rates and amounts of precipitation at mountain sites during passage of frontal or convergence zones are not greater, and frequently less than in nearby nonmountainous areas. The storm, however, serves to intensify the circulation. This brings moist air into the area and increases the convective stability in the lower layers following storm passage. The orographic influence usually is most apparent immediately preceding and following these storm systems.

Since an orographic cloud is generally stationary and approaches a steady-state condition for an extended period of time, a mean rate at which condensate forms is usually adequate to describe the mass of water condensed from vapor for most

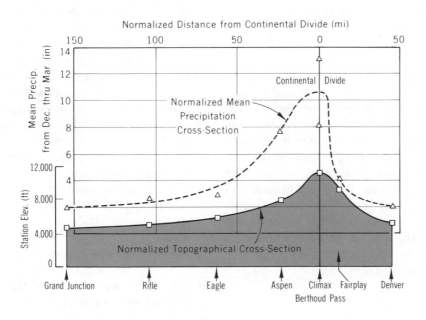

Figure 7.1 Normalized east-west cross-sections of topography and mean precipitation for the central Colorado Rockies. Mean precipitation computed for period December through March (1960–1964).

weather modification considerations. One of the most attractive aspects of orographic clouds for weather modification is, of course, the persistent cloud that can be seeded for many hours. The assumption of a mean rate for the formation of condensate is thus consistent with the mode for seeding operations which can be employed.

The mean rate of condensate production per unit volume can be expressed in terms of (1) the mean vertical velocity caused by orographic lifting, and (2) the difference in the saturation specific humidity for the layers in which the orographic cloud is formed. Mean vertical motions in orographic clouds over large mountain barriers are commonly in the range of 10 to 40 cm/sec. The saturation specific humidity is a function of cloud temperature which in turn is related to the elevation of the orographic barrier and its associated cloud.

Consumption of Cloud Water to Form Precipitation

An optimum utilization of cloud water for the production of precipitation occurs when all cloud water is consumed to form precipitation. An inefficient use of cloud water for the formation of precipitation occurs when all or a portion of the cloud water reevaporates to the lee of the barrier. The basic process by which cloud water is utilized for the formation of precipitation in "cold" (subcooled) orographic clouds is by the vapor growth of ice. The rate of growth of single ice crystals by vapor deposition is usually expressed as

$$\frac{dm}{dt} = 4\pi S_i C G' f_1 f_2,$$

where m = mass of crystal, S_i = supersaturation relative to a plane ice surface,

C = electrostatic capacity factory (this is a function of crystal shape), G' = a thermodynamic function, f_1 = ventilation factor of airflow past the ice crystal, and f_2 = correction factor for vapor field to that of supercooled cloud.

The rate of consumption of cloud water per unit volume by the vapor deposition can be determined by considering the combined growth of all ice crystals per unit volume within the cloud. For maximum utilization of cloud water, the growth rate of all ice crystals in a unit volume must proceed at the same rate as cloud water is supplied to the same volume by the condensation of water vapor.

The collection of subcooled water droplets by ice crystals, a form of accretional growth, can be expected to remove additional cloud water when the rate of formation of condensate exceeds the rate at which it is consumed by vapor growth of ice crystals. While the accretional consumption of water is an important backup mechanism for utilization of orographic cloud condensate, it usually does not have the potential for fully utilizing cloud water not consumed for ice crystal growth by vapor deposition. Use of cloud water by accretion to produce rimed particles with greater fall velocities can, however, be a vital process in certain clouds where increased fall velocities are required for particles to reach the ground in short time intervals. Such rimed particles can have fall velocities in the range of 1 to 2 m/sec in contrast to 20 to 70 cm/sec for unrimed ice crystals.

Efficiency of Natural Clouds

It can be shown, based on the above considerations, that under certain cloud temperature conditions the rate of water consumption for ice crystal growth is less than the rate at which condensate becomes available to the cloud. Considerable losses of cloud water to reevaporation on the lee side of the mountain barrier result under these conditions. In other cases, ice growth by vapor deposition proceeds at a rate sufficient to utilize cloud water as it condenses. In still other situations, the supply of ice crystals can be so great that particle growth is restricted by the supply of cloud water available and fall trajectories of the ice crystals relative to the ground may be adversely affected. Figure 7.2[23] as prepared from Figures 11 and 13 of a 1970 article by C. F. Chappell[7] shows, as a function of cloud

temperature, the relative rates for the (1) formation of cloud condensate from vapor (Curve B), (2) consumption of cloud water for growth of ice crystals (Curve A), and (3) observed natural precipitation (Curve C). Average values are shown and data have been normalized to show water equivalent per precipitation day so all data can be compared readily. The precipitation data used in Figure 7.2 are for the High Altitude Observatory (HAO) near Climax, Colorado. The rates of precipitation (Curve C) are determined from the precipitation observations for 131 nonseeded cases of 251 randomized seeding experiments

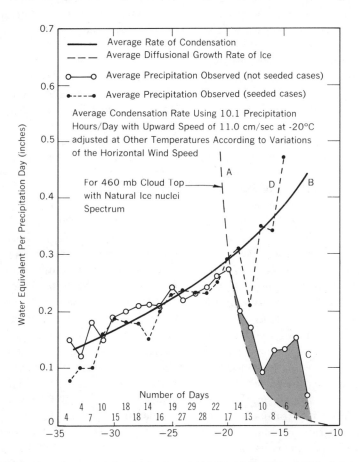

Figure 7.2 Distribution of nonseeded precipitation at HAO as a function of 500-mbar temperature compared to a theoretical distribution computed using the mean diffusional model. Precipitation data are from Climax I sample (251) and are a running mean over a two-deg temperature interval (Grant et al.[23]).

(Climax I) carried out near Climax, Colorado, by Colorado State University during the period 1960–1965. Since cloud top measurements are not available for each case, 500-mb temperature has been used to represent the cloud top temperatures. R. W. Furman[18] using radar and other visual and aircraft observations, shows that the 500-mb level is close to the mean elevation of cloud tops near Climax. This 500-mb level undoubtedly could not be used to represent cloud top temperatures over many other mountain barriers.

Curve B in Figure 7.2, the potential condensate curve, shows that, other factors being equal, more condensate is available in the warmer clouds. Curve A shows that the consumption rate of cloud water for the growth of ice crystals increases very rapidly at progressively colder temperatures. The change in curvature at around −15°C reflects the maximum growth rate of individual ice crystals which occurs at about that temperature. The controlling parameter in converting cloud water to ice by depositional growth (Curve A) is the concentration of ice crystals available for growth. Since the natural production of primary ice crystals is an exponential function of decreasing temperature, the concentration of ice crystals formed near cloud top, the coldest portion of the cloud, can control the mean ice crystal concentration throughout the orographic cloud. Effective primary ice nuclei concentrations in natural clouds generally increase by a factor of 10 for each decrease of 4°C in temperature. Thus the coldest 4°C layer of subcooled cloud can be expected to produce roughly 90 percent of the ice crystals in the cloud which are formed from primary ice nuclei. In most orographic clouds the settling rate of ice crystals is greater than the mean upward airflow, thus permitting the redistribution, through settling, of the

many ice crystals formed at the colder temperatures from the upper to the lower (warmer) portions of the cloud. This is in contrast to convective clouds which generally have upward vertical velocities greater than the fall velocity of individual ice crystals and for which other particle formation and cloud mixing processes are required.

It is readily apparent that the potential condensate curve (Curve B) and the water consumption curve for ice growth (Curve A) cross sharply at around −20°C. It can be observed clearly that at temperatures colder than this intersection the potential for utilizing cloud water to grow ice crystals is at least equal to, and generally considerably greater than, the rate at which condensate is formed. Natural nucleation processes thus supply adequate ice forming nuclei at these colder cloud temperatures, and artificial seeding that would provide additional ice nuclei is not required. On the other hand, at cloud temperatures warmer than the intersection of these two curves, condensate becomes available at a rate increasingly greater than the rate at which it can be consumed by the growing ice crystals. Thus a natural precipitation process dependent upon ice crystal growth by vapor deposition on primary ice nuclei becomes progressively less efficient at warmer temperatures. Natural processes in these warmer orographic clouds do not supply sufficient ice nuclei to utilize all available condensate, and artificial seeding to provide additional ice nuclei in these warmer orographic clouds should improve the utilization of cloud water for precipitation. The crossover temperature of −20°C at Climax applies to the 500-mb level. The best estimate of actual average cloud tops for this data sample is at 460 mb, a level where temperatures would be 4 to 5°C colder than at 500 mb. Thus the critical cloud top temperature at the crossover point from these calculations should be about $-24°C$.

Curve C (Figure 7.2) shows the actual precipitation measured near Climax on the randomly selected unseeded days of the sample. It can be noted that the actual precipitation rate corresponds closely to the rate at which condensate is provided when temperatures were colder than at the intersection of the potential condensate and ice growth curves. As would be expected, the natural precipitation efficiency of these colder cloud systems was very high since virtually all condensate was used to produce precipitation. The actual precipitation from clouds with 500-mb temperatures warmer than $-20°C$, on the other hand, was less than at colder temperatures despite the greater supply of potential condensate. The difference between actual precipitation and potential condensate, in general, becomes greater at increasingly warmer temperatures. These variations of actual precipitation, as a function of cloud temperature, agree very well with changes to be expected from physical considerations. The warmer clouds had a low precipitating efficiency which should be increased by a supplemental supply of ice particles to improve the utilization of cloud condensate.

Despite the good agreement between precipitation expected and the actual precipitation as shown in Figure 7.2, one important discrepancy can be noted. The drop-off in observed precipitation starts near the intersection of the potential condensate (B) and ice growth (A) curves as expected. At warmer temperatures, however, precipitation remains above Curve A and appears to have an important peak near $-15°C$. The differences between Curves C and A probably represent the extraction of cloud

water by processes other than through the growth of ice crystals by vapor deposition. At least a portion of this additional use of cloud water must come from the removal of cloud droplets by collection on falling ice particles (riming). Ice multiplication related to the dendritic ice crystal growth at this temperature or greater ice growth by vapor deposition than assumed in the calculations could also contribute to this difference. The difference, by one or more of these processes, does constitute an additional utilization of cloud condensate. As can be seen from Figure 7.2, however, the potential condensate is still considerably greater than the net precipitation resulting from all processes. The potential for weather modification is therefore represented by the difference between the potential condensate (Curve B) and the actual precipitation (Curve C). It should be expected that additional processes by which cloud water is used, other than by growth of ice crystals by vapor deposition, would vary with geographical area, barrier elevation, and storm type. A comparison of the differences between C and B type curves for other specific areas can serve as a basis for estimating the potential for weather modification for the respective mountain watersheds.

The dependence of precipitation efficiency on cloud microphysics, in contrast to storm dynamics, can be verified indirectly by looking at the success in forecasting precipitation for the Colorado Rockies using only synoptic weather considerations. This was done by considering the forecasts prepared for the Climax area by the Denver office of the U. S. Weather Bureau. Cloud microphysical considerations were not used in preparing the forecasts. The declaration of experimental days for the Climax randomized experiment was based on a special forecast prepared each

Table 7.1   Distribution of Nonseeded Experimental Days, Experimental Precipitation Days and Experimental Zero Precipitation Days with the Concurrent 500-mb Temperature for the Total Climax (623) Sample

Temperature Category (°C)	No. of Experimental Days	No. of Precipitation Days	No. of Zero Days	Percentage of Precipitation Days to Experimental Days (Correct Forecast)
−40 to −36	2	2	0	100.0
−35 to −31	17	16	1	94.2
−30 to −26	64	60	4	93.8
−25 to −21	129	103	26	79.8
−20 to −16	88	50	38	56.8
−15 to −11	21	7	14	33.3
Total or average	(321)	(238)	(83)	(73.9)

Source: Chappell.[7]

morning. The criteria for declaring experimental days was a forecast of measurable precipitation at Leadville, Colorado, which lies at the southern end of the Climax experimental area. Randomization was made subsequent to the declaration of an experimental day. Six hundred twenty-three days have been considered in this analysis. Table 7.1[7] shows the percentage of the experimental cases properly forecast on nonseeded days as a function of 500-mb temperature. It can be noted that the percentage of correct forecasts of measurable precipitation was extremely high, greater than 90 percent when 500-mb temperatures were −26°C or colder. These represent cases when an efficient cloud precipitation process would be expected and failure to consider microphysical processes would not effect forecast accuracy. At progressively warmer temperatures, however, the influence of microphysical processes on precipitation occurrence and efficiency should play an increasingly important role. As would be expected, since microphysical processes were not considered in preparing the forecasts, forecasting success dropped off rapidly at progressively warmer temperatures and was less than 50 percent in the warmest category.

RESULTS OF FIELD EXPERIMENTS

Encouraging results, though not conclusive, were associated with the early experiments to increase water supplies by cloud seeding in the western mountains of the United States. These activities were evaluated and reported by a number of investigators using various analysis techniques (many of these are listed in the Final Report of the Advisory Committee on Weather Control, Appendix D, 1957). Perhaps the most important interpretation of these efforts is that of the President's Advisory Committee.[15] This analysis was based on the post hoc analysis of commercial seeding efforts. It concludes that the most probable result of seeding of orographic clouds for precipitation augmentation was an increase of 10 to 15 percent. This analysis employed statistical techniques for evaluating field seeding programs and considered the results of a large number of programs. Although it was generally accepted as the most complete interpretation available, it was criticized on several points. The seeding programs that were analyzed had not been conducted on a random basis and concern was expressed that effects other than weather modification could account for the observed changes in precipitation. Further, these analyses produced little insight into criteria for defining which weather situations or meteorological criteria were producing desired effects.

A research program to (1) evaluate weather modification potential, (2) determine seedability criteria, and (3) develop a technology for seeding orographic clouds in the central Colorado Rockies, was initiated in 1960 by Colorado State University. The research has emphasized an investigation of the physical processes taking place in mountain clouds during seeding and a field program of randomized seeding to serve as a basis for definitive statistical analyses. The primary support for the research was provided by the National Science Foundation. The field seeding program was carried out in the vicinity of Climax, Colorado. The initial seeding experiment (Climax I) took place from 1960 to 1965. The basic aspects of this experiment were repeated during the 1965–1970 period as an independent replication (Climax II) of the earlier experiment. The nature of the results expected during the Climax II experiment were declared in advance, based on results described in the earlier experiment. Analyses have been made for all Climax II data in a manner similar to that used for Climax I. Analyses have also been made for a subsample (designated IIB) of the Climax II data.

This sample eliminates cases when upwind seeding by other groups probably contaminated the Climax experiment. It has been prepared by eliminating days from both the seeded and nonseeded cases when the airflow was from the directions that would permit contamination. The use of this subsample does not substantially alter the overall results and probably provides a more meaningful sample.[34] The basic design of these experiments provided for the following:

1.   Randomization in obtaining the seeded and nonseeded samples. The randomization was restricted to the extent that large blocks of experimental units (20 to 40) had the same number of seeded and nonseeded cases.

2.   An experimental unit with a 24-hr interval of time.

3.   Observations of meterological variables for all stages of the precipitation process to be made as extensively as feasible for both seeded and nonseeded cases.

4.   The use of Standard Weather Bureau stations to the southwest, west and northwest of the experimental site as control stations.

5.   An experimental day based on a forecast of at least 0.01 in. of precipitation at Leadville, Colorado, during the 24-hr experimental period. This forecast was prepared by the U. S. Weather Bureau in Denver. The forecasters had no access to the subsequent seeding decision which was made on a random basis after the declaration of an experimental day.

6.   Generators to be turned on $\frac{1}{2}$ to 1 hr prior to the beginning of the experimental day, and shut off $\frac{1}{2}$ to 1 hr prior to the end of the day, depending upon their respective distances from the prime target area of the experiment. The prime target area was considered to be the summit vicinity of Fremont Pass.

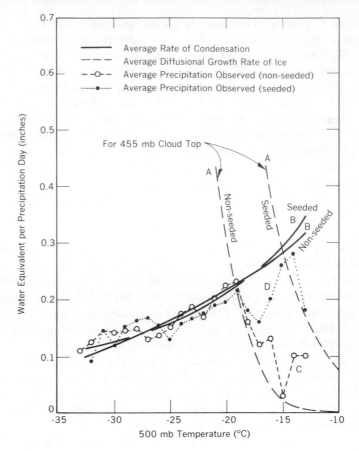

Figure 7.3   Distribution of seeded and nonseeded precipitation at HAO as a function of 500-mbar temperature. Precipitation data are from Climax II (372) sample and values are a running mean over a two-deg temperature interval (courtesy C. Chappell, ERh, NOAA, Boulder, Colorado[7]).

7. The use of Colorado State University's modified "Sky-fire" needle type ground generator for seeding.

The Climax Experimental Results

The average rate of formation of condensate (Curve B), the average rate of cloud water consumption by ice growth (Curve A), and the average rate of precipitation for nonseeded days (Curve C) for the Climax I randomized experiment were shown in Figure 7.2. Curve D, not previously discussed, shows the average rate of precipitation on the 120 randomly selected seeded days of the ex-

periment. While the precipitation rate on unseeded days followed the potential condensate curve to the intersection with the ice growth curve and then dropped off rapidly, precipitation on seeded days continued to increase at warmer temperatures and, in general, continued to follow the potential condensate curve very closely even at warmer temperatures. This constitutes good agreement between results of field experiments and expectations from physical considerations.

The basis for both the physical treatment used and for the reality of the seeding effects are strengthened even further by a similar analysis of the independently replicated Climax II experiment. These are shown in Figure 7.3. Again, on nonseeded days, there is good agreement between precipitation and the potential condensate curves only up to the intersection with the ice growth curve. This is again in contrast to the precipitation curve for the seeded days, which continues to follow the potential condensate curve even at warmer temperatures. Thus in both the Climax I and Climax II experiments, the actual amount of precipitation on seeded days corresponds closely to the amount of cloud condensate available for the production of precipitation. This is clearly not the case for nonseeded days in either experiment for cloud temperatures warmer than the intersection of the condensate and ice growth curves. Physical considerations show that requirements for additional ice nuclei existed at the warmer cloud top temperatures. When these additional requirements for ice-forming nuclei were artificially met on the seeded days with warmer cloud top temperatures, the precipitation efficiency was increased substantially over that for nonseeded days.

Two additional randomized orographic experiments have been carried out by Colorado State University. These were conducted in the Wolf Creek Pass area of southern Colorado and in the Monarch Pass area in the central part of the state. These programs were randomized on an annual basis (for 6 years) to make possible an analysis of stream flow which, in the Colorado Rockies, has an annual cycle. During these experiments all events were seeded in one year and none in the next. In addition to the stream flow analyses, other evaluations are being made that treat precipitation during seeded and nonseeded events in a manner similar to that used for the analysis of the Climax precipitation data. In doing this, one makes the questionable assumption that the distribution of daily precipitation amounts are the same from one year to the next. The computations of (1) the available condensate, (2) the water consumption for ice growth, and (3) the precipitation curves on nonseeded days, however, is independent of seeding and the randomization scheme. These curves can be used to evaluate the potential for seeding. Comparisons of precipitation amounts on seeded days have only been made to show the consistency with the physical expectations and with the results of Climax. Differences in precipitation between the seeded and nonseeded days cannot be considered with the same confidence as those in the analysis of the Climax data where daily randomization was used throughout the experiment.

The preliminary analyses incorporating the first 5 of the 6 experimental years (3 seeded, 2 not seeded) of the Wolf Creek Pass experiment show (Figure 7.4) that, again as at Climax, the rate of natural precipitation (Curve C) agrees well with the rate of formation of available condensate (Curve B) for temperatures colder than the intersection of the condensate and ice growth (Curve A) curves. In contrast, it can be noted that the natural precipitation rate (Curve C) at warmer temperatures is considerably greater than would be expected from ice growth by vapor deposition on primary ice nuclei (Curve A). It is still, however, considerably less than the potential amount of condensate (Curve B) available, and the drop in utilization of all condensate occurs near the intersection of the condensate and ice growth curves. This suggests that other precipitation growth mechanisms are much more significant in the lower, warmer clouds at Wolf Creek than in those at Climax. This demonstrates the value of these physical considerations in providing an "identifier" or "signature" for the relative roles of vapor growth of ice, and the other precipitation growth

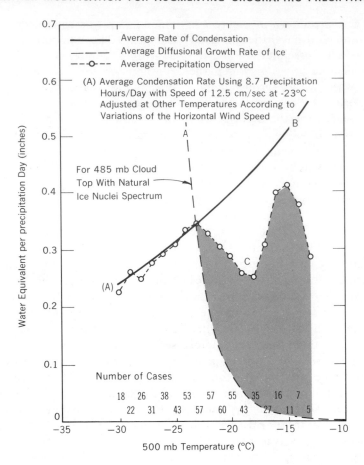

Figure 7.4 Distribution of nonseeded precipitation at Wolf Creek Pass as a function of 500-mbar temperature compared to a theoretical distribution computed using the weather modification potential model. Precipitation data are from Wolf Creek (441) sample and values are a running mean over a four-deg temperature interval (courtesy C. Chappell, ERh, NOAA, Boulder, Colorado[7]).

processes, in a specific area. Again, as in the Climax experiment, the precipitation on seeded days was comparable to the amount of condensate available even at the warmer temperatures.

The above interpretations of seeding effects on orographic clouds are based on physical considerations. In addition, the statistical significance of the differences between the precipitation on seeded and nonseeded days has been determined for the Climax experiments. The precipitation for all seeded cases was greater than all of the unseeded cases by

9, 13, and 39 percent, respectively, for Climax I, Climax II, and Climax IIB data samples. Differences of these magnitudes could have reasonably been expected by chance. (The Climax IIB sample is significant at near the .05 probability level but reflects the use of more cases with airflow from directions that give higher cloud temperatures.) The above results are to be expected since a comparison using the total experimental sample includes individual days when physical considerations show that increases, no change, or decreases should

Table 7.2   Ratios of Seeded to Nonseeded Mean Precipitation Amounts (Wilcoxon Statistic One-sided p-Values): A and B for Increases, C for a Decrease for 500-mb., Temperature Partitions (Climax I—251 Experimental Units, Climax IIB—296 Experimental Units, Climax I and Climax IIB Combined—547 Experimental Units)

	A −20 to −11°C	B −26 to −21°C	C −39 to −27°C
Climax I	1.85 (.206)	0.97 (.436)	0.78 (.041)
Climax II B	1.74 (.034)	1.28 (.230)	1.18 (.732)
Total Sample	1.75 (.045)	1.10 (.350)	.89 (.150)

occur. This emphasizes one of the difficulties that has been encountered during the past 20 years in evaluating weather modification efforts. Programs that have shown significant results, either increases or decreases, are likely to have been those that experienced cloud characteristics that provided a preponderance of events in the categories favoring increases or decreases in precipitation. The design of the Climax experiments provided for meteorological stratification to test the significance of differences in the precipitation for groupings of events that could be expected to have a similar response to seeding.

Tables 7.2 through 7.5 show for the Climax experiments the magnitude and significance of differences in precipitation between the seeded and nonseeded days. Precipitation data are for observations taken from the most representative grouping of gauging stations near the center of the specified target. These tables present comparisons in precipitation for meteorological stratifications based on 500 mb temperature, 700 mb equivalent potential temperature, and 700 mb wind direction and velocity. The 700 mb equivalent potential temperature combines information on both cloud thermal structure and moisture availability on experimental days. It can be seen from Tables 7.2 and 7.3 that large positive differences in precipitation, which are generally statistically significant (well beyond the 1 percent level for the 308 to 327° K equivalent potential temperature

Table 7.3   Ratios of Seeded to Nonseeded Mean Precipitation Amounts (Wilcoxon Statistic One-Sided p-Values): A and B for Increases, C for a Decrease for 700-mb Equivalent Potential Temperature Partitions (Climax I—251 Experimental Units, Climax IIB—296 Experimental Units, Climax I and Climax IIB Combined—547 Experimental Units)

	A 308 to 327°K	B 295 to 307°K	C 281 to 294°K
Climax I	1.77 (.131)	0.96 (.496)	0.66 (.037)
Climax II B	1.76 (.002)	1.16 (.764)	1.28 (.716)
Total sample	1.70 (.003)	1.05 (.750)	.78 (.13)

Table 7.4   Ratios of Seeded to Nonseeded Mean Precipitation Amounts (Wilcoxon Statistic One-Sided p-Values): A, B, C for Increases, D for Decreases for 700-mb Wind Velocity Partition (Climax I—251 Experimental Units, Climax IIB—296 Experimental Units, Climax I and Climax IIB Combined—547 Experimental Units)

	A 0 to 8 m/sec	B 9 to 11 m/sec	C 12 to 14 m/sec	D 15 to 28 m/sec
Climax I	1.08 (.460)	1.24 (.236)	1.44 (.043)	0.73 (.093)
Climax II B	0.99 (.568)	1.17 (.367)	3.02 (<.001)	0.87 (.278)
Total sample	1.03 (.600)	1.23 (.300)	1.90 (<.001)	.70 (.13)

category), occurred on the randomly selected seeded days for the warmer temperature categories. The excess precipitation on the seeded days for the total sample is 75 percent for the -11 to $-20°$C, for 500-mb category and 70 percent for the 308 to 327°K equivalent potential temperature category. These warmer temperature categories correspond to the temperature regimes where the rate of formation of condensation under natural conditions exceeds the amount of water consumed by diffusional growth of ice. Thus they should be temperature categories in which precipitation increases should occur. In the colder categories these advantages should not and do not appear. In the coldest category, seeded days generally received less precipitation. This suggests an abundance of natural nuclei already available.

Table 7.4 shows the dependence of seeding potential on the rate at which condensate becomes available in the orographic cloud. This is presented in terms of the wind velocity near cloud base (approximately 700 mb) at Climax. It can be noted that the ratio of precipitation for seeded to nonseeded days increases with wind velocity (and consequently with the rate of formation of condensate) to a peak for the 12 to 14 m/sec category. The advantage in the precipitation for seeded days is very large (+90 percent for the

total sample) with a high statistical significance ($P < .001$). The indicated decrease at still higher velocities suggests that seeding caused a reduction in size and fall speed of ice crystals to the extent that in some cases precipitation that would have reached the mountain under natural conditions did not reach the surface network before being carried rapidly over the mountain barrier. This could be affected by the specific location of the seeding generators used during the experiment.

The importance of wind direction can be noted in Table 7.5. Airflow patterns from the southwest and from the northwest are more nearly perpendicular to the complex mountain barrier near Climax and consequently produce the most marked orographic lifting. Large positive advantages in precipitation for seeded cases with a high statistical significance is noted for these directions.

An estimate of the overall change in precipitation that could be expected in the Climax area from weather modification can be made by considering the changes in precipitation to be expected for each of the cloud temperature categorizations and from the frequency of occurrence of those categories. This has been done by using changes in precipitation to be expected within the 308 to 327°K, 295 to 307°K, and the 281 to 294°K equivalent potential temperature categories for the

Table 7.5   Rations of Seeded to Nonseeded Mean Precipitation Amounts (Wilcoxon Statistic One-Sided p-Values): A, B, C for Increases, D for Decreases for 700-mber Wind Direction Partitions Climax I—251 Experimental Units, Climax IIB—296 Experimental Units, Climax I and Climax IIB Combined—547 Experimental Units)

	A 190 to 250°	B 260 to 300°	C 310 to 360°	D 10 to 180°
Climax I	1.94 (.031)	0.85 (.863)	1.41 (.095)	0.72 (.218)
Climax II B	2.27 (0.13)	1.06 (.508)	1.84 (0.28)	0.83 (.520)
Total sample	2.08 (.004)	.92 (.800)	1.54 (.024)	.77 (.360)

combined Climax I and Climax IIB experimental units. The frequency of occurrence of the respective categories has been determined from the total experimental sample of 623 events. Table 7.6 shows the estimated change in precipitation at Climax for a seeded winter with normal precipitation.

It can be seen that if all cloud events are seeded then approximately 11 percent more wintertime precipitation should occur. When seedability criteria are considered, and seeding is conducted only when favorable events occur, the increase in precipitation from seeding should amount to about +17 percent. This is probably representative of the potential for precipitation increase in most of the northern and central portions of the Colorado River Basin. Nearly as much return, around +15 percent, could be achieved by seeding only the events within the warmest temperature category. This would require seeding only during the

Table 7.6   Estimated Changes in Precipitation at Climax for a Seeded Winter Season With Normal Precipitation (~14.00 in.)

700-mb Equivalent Potential Temperature (°K)	Seasons Precipitation (%)	Natural Precipitation (in.)	Percentage Change (%)	Seeded Event Precipitation (in.)
A. All events are seeded				
308–327	21	2.94	+70	5.00
295–307	49	6.86	+5	7.20
281–294	30	4.20	−22	3.28
	100	14.00		15.48
Total increase = 1.48 in. = ±11%				
B. Only favorable events are seeded				
308–327	21	2.94	+70	5.00
295–307	49	6.86	+5	7.20
281–294	30	4.20	—	4.20
	100	14.00		16.40
Total increase = 2.40 in. = +17%				

periods when about 21 percent of the natural precipitation is accumulating. The change expected from seeding in other areas could be more or less than this, depending on the relative frequency of the respective temperatures and on the precipitation efficiency of the natural clouds. These parameters can be reasonably evaluated for sites for which (1) upper air data is available for estimating the rate of formation of condensate and cloud temperature and (2) precipitation data are available for estimating the rate of natural precipitation as a function of cloud temperature.

Statistical analyses have also been used to evaluate the relative contribution of changes in precipitation duration and in precipitation intensity to the changes in total precipitation observed on the seeded days.[9] These analyses show that the total change in observed precipitation is primarily controlled by a change in the duration of precipitation events, rather than by a change in intensity during times of natural precipitation. The main effect of seeding for the warmer clouds appears to have been an earlier initiation, or an extension of the precipitation episode, so that the period of precipitation was substantially longer. For the coldest clouds the period of precipitation with seeding is observed to be considerably shorter. These results are consistent with the concepts of natural cloud microstability at the warmer temperature and cloud overseeding at the coldest temperatures.

Detailed descriptions of the statistical tests applied and the results for various groupings of precipitation data on and near the target area are available in the open literature.[7, 9, 20, 23, 25, 32-34] The results presented above are descriptive of the overall findings.

The cloud water balance and statistical interpretations discussed above are supported by laboratory and field observations. These observations show that appropriate concentrations of ice nuclei were produced by the seeding equipment, and that the seeding materials did arrive in the target clouds. The seeding was carried out using two or three generators during each seeded event. Generators were located in upwind valleys with airflow feeding directly over the Rocky Mountain barrier. These generators were tested in the Colorado State University isothermal cloud chamber. They produce about 4×10^{15} active ice nuclei/g of AgI effective at $-20°C$, 7×10^{14} effective at $-15°C$, and 4×10^{13} effective at $-10°C$.[25] These generators are inefficient at temperatures warmer than about $-10°C$. Clouds with their tops warmer than $-10°C$, however, are extremely rare in the central Colorado Rockies during winter.

Ice nuclei were continuously monitored in the target area during the course of the experiment. Substantial increases (frequently by a factor from 10 to 10^3) in ice nuclei were routinely observed following seeding.[20, 25, 42] While the routine ice nuclei observations were from a mountain station, they do not provide a description of the vertical distribution of ice nuclei within the mountain cloud. This has been studied using tethered kites and aircraft probes for a limited number of cases. Observations over a wind tunnel model of the Climax area have also been made in an attempt to better describe the diffusion and transport characteristics under a broader range of environmental conditions. Laboratory observations and the limited field observations for appropriate temperature conditions have been found to be in good agreement.[6, 22, 39] The wind tunnel studies have shown that valley filling and subsequent transport of seeding materials over the mountain barrier can be expected under many flow and stability conditions when orographic clouds occur.

Evaluation of stream flow changes associated with precipitation changes are not feasible for the Climax experiment since randomization was on a daily basis and winter precipitation accumulates as snow that melts and runs off the following summer. Seeding carried out on an annual basis during the randomized experiment in the Wolf Creek Pass area does provide a basis for making estimates of stream flow changes associated with seeding. Figure 7.5[23] shows a preliminary comparison of seeded area stream flow with that for a control area for nonseeded (historical and randomly selected) years. (The confidence intervals shown as dashed lines in this figure should be slightly curved with a broadening of the intervals near the extremes.) It can be noted that the historical correlation is very good, $r = .97$, and the coefficient of variation is low, ≈ 0.1. The likelihood of receiving the actual stream flow observed in two of the randomly selected seeded years, 1965 and 1967, is low, with their individual p values less than .05. The third year, although having a somewhat higher expectancy, is still low. The probability of the combined stream flow for the three seeded years equaling or exceeding the observed value by chance is very low, $p = .005$. The observed precipitation excess during seeded years of 18.9 percent (228,000 acre-ft) is in reasonable agreement with the overall change in

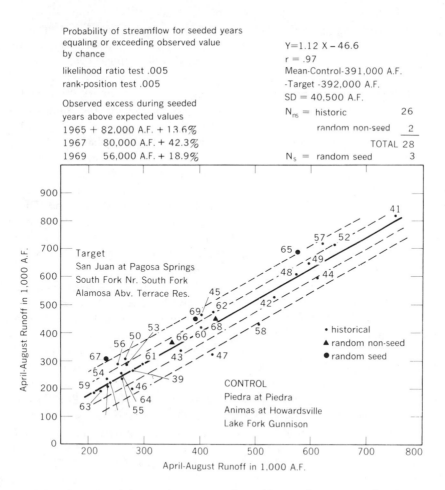

Figure 7.5 Comparison of seeded area (target) stream flow with control area stream flow from nonseeded (historical and randomly selected) and seeded years (Grant et al.[23]).

Table 7.7   Research Experiments for Seeding of Orographic Clouds

Group	Area for Field Studies
Bureau of Reclamation Pilot Project	San Juan Mountains of southern Colorado
CSIRO	Tasmania
Desert Research Institute	Sierra Nevada Mountains
EG and G, Inc.	Park Range in northwest Colorado
Fresno State University	Sierra Nevada Mountains
Montana State University	Bridger Range in Montana
New Mexico State University	Jemez Mountains of New Mexico
North American Weather Consultants	Mountains near Santa Barbara, California
Utah State University	Wasatch Mountains in northern Utah
University of Washington	Cascade Mountains in Washington
University of Wyoming	Elk Mountain and Wind River Mountains in Wyoming

precipitation observed for the seeded cases.

Orographic Cloud Research Experiments in Other Geographical Areas

A number of good weather modification research experiments for evaluating the effects of seeding in orographic clouds have been completed recently or are in progress. These are primarily in the western United States and are largely supported by the Bureau of Reclamation (the exception is Commonwealth Scientific and Industrial Research Organization(CSIRO)). Table 7.7 lists these experiments. All these experiments involve randomized seeding except for the University of Washington and the University of Wyoming programs.

The following is a brief summary of the results of statistical analyses of randomized field seeding experiments that have been reported. A summary of the physical studies carried out along with many of these randomized experiments, and particularly in connection with certain of the other experiments, is beyond the scope of this chapter.

CSIRO Australian Experiments. Randomized seeding experiments in Australia carried out before 1965 gave differing results.[3] The Snowy Mountains and New England experiments showed positive results for which the significance was marginal at or near the 5 percent level. The South Australia and Warragamba experiments were slightly negative and not statistically significant. The CSIRO investigators[1, 3, 4] have emphasized the trend in the results with time. The general trend has been for more favorable results in the early years followed by deteriorating results as the experiments continued.

The Tasmania experiment was consequently designed to escape possible persistence effects from the seeding by operating only in alternate years. Smith[48] concludes from this experiment that the "...cloud seeding caused an increase in rainfall of about 15 to 20 percent in the autumn and winter." For these seasons the precipitation for seeded events was substantially greater for the total target and for both halves of the target in every operational year 1964–1970 (1964, 1966, 1968, 1970). Although analyses according to meteorological stratification have not yet been emphasized, orographic influences with clouds in the temperature range for most favorable opportunities for seeding might be expected during the fall and winter months. The larger indicated increases and greater statistical significance for the fall season may reflect warmer cloud temperatures and a greater requirement for artificial nuclei during that period.

EG & G Experiment. An interesting randomized seeding experiment was carried out by EG & G[43] in the Park Range of northwest Colorado during the 1968–1969 winter season. This involved seeding by aircraft for 1 hr of a 6-hr block during which intensive data collection was emphasized. Precipitation for a 3-hr interval during which the first hour had been seeded (seeded sample) was compared with a 3-hr unseeded block. Once an experimental unit of 6 hr had been established, randomization was used to establish whether the first or last 3-hr interval should be the seeded block. Seeding efforts were carried out using the output from a physical model to delineate seedability. Analyses, when temperature stratification was adjusted for differences in cloud depths between the Park Range and Climax, gave results similar to those obtained at Climax. An increase of > 100 percent for seeded cases was indicated at Rabbit Ears Pass for the warmest category of indicated cloud top temperatures ($\geq -20°C$), and the difference was significant at the 3 percent level. A decrease of 24 percent was noted for cloud tops $< -20°C$ but with less statistical significance. This statistical analysis was supported by physical observations, including silver analysis in snowfall, ice crystal sampling, radar observations, etc. These respective studies gave consistent indication of a seeding effect on the cloud.

Montana State University Experiment. The analyses of randomized seeding in the Bridger Range east of Bozeman, Montana, are now in a preliminary stage. Mitchell, Super, and Yaw[35] conclude that "meteorological stratifications are an important key in the analysis. The specific stratification used in the Climax studies will, in general, not apply for the Bridger Range clouds which are much lower in elevation and normally affected by stronger winds." The authors believe that "some of the meteorological parameters used at Climax will be used in the stratification of the Montana State University experiment data, but with different groupings." They report that "several of the stratifications tried thus far suggest that seeding in the northern Rockies does influence precipitation. This is especially true of the lowest group of 700 mb equivalent potential temperature."

New Mexico State University Experiment. The analyses of randomized seeding over the Jemez mountains of northern central New Mexico are now in a preliminary stage.[30] The investigators report that, "the results to date include an indication of overall apparent positive effects of seeding and of the probability that this effect can be maximized by seeding only under select favorable conditions." This experiment provides opportunities for investigating the effect of seeding on winter orographic clouds under substantially different conditions from those experienced in the other orographic experiments in the western United States. Preliminary analyses show greater precipitation on the seeded days for the warmer temperature ranges of 500-mb temperature and 700-mb equivalent potential temperature. The range of temperatures at which this is noted correspond closely with the range of temperatures at which positive seeding effects are indicated for the Climax experiment. In contrast to the model considerations and the results of Climax, however, a second peak is observed at very cold temperatures. This second peak has not been explained. The sample size producing this peak is small. Positive seed/no seed ratios are obtained when the airflow is from directions between 220° and 310°. Winds from this sector would experience sharp orographic lifting as they rise over the Jemez mountains.

North American Weather Consultants. An important randomized seeding program was carried out in the Santa Barbara, California area (Santa Barbara I) in the period 1957–1959. Early published analyses of this project were inconclusive. Reanalyses,[12, 24] however, have subsequently been made to categorize the events into cloud temperature groupings indicated as being important to seeding success. The basis for making these specific categorizations of events was not available at the time of the original study. When analyzed in this manner, large advantages in precipitation following seeding (> 100 percent with cloud top temperatures of > −20.5°C) are apparent for the warm cases where a requirement for seeding would exist. Reduced precipitation occurred for temperatures −25°C and colder, as would be expected from physical consideration. A second randomized experiment has been carried out in the Santa Barbara area (Santa Barbara II) in the period 1967–1970. Pyrotechnic flares were used for the seeding in this experiment. As in the Santa Barbara I experiment, large and statistically significant greater precipitation amounts were observed for the seeded cases in the warmer temperature categories. Although not statistically significant, increases in precipitation were also observed at colder temperatures. The difference in the nature of the results at the colder temperatures are believed by Elliott[24] to be related to: (1) an important effect of convection in the

convection bands which were the target for seeding, and (2) the use of pyrotechnic flares, which produce effective ice nuclei at warmer temperatures, during the Santa Barbara II experiment in contrast to the use of an AgI solution complex during the earlier experiment.

Utah State Experiment. An aircraft randomized seeding program, generally similar to that employed by EG & G in the Park Range, has been carried out by the Utah Water Research Laboratory of Utah State University over a large mountain area of north central Utah.[8] Four hour randomized blocks were used during an 8-hr operational period. The first 2 hours of the "seed" block were actually seeded; the remaining hours were not. The results, although not yet statistically significant, due to the still small sample sizes, are consistent with the physical basis for seeding cold orographic clouds. That is, the precipitation over the higher terrain is substantially greater during the seeded intervals with cloud top temperatures in the range −13 to −23°C. With temperatures in the still warmer −6 to −12°C range, the large positive values are displaced further downwind. When cloud top temperatures were −24°C or colder the precipitation, as would be expected was apparently less than during the nonseeded period over most of the target area.

Summary of Results of Field Experiments

A number of well conducted and carefully analyzed experiments of weather modification to augment orographic precipitation are giving results consistent with expectations from physical considerations of "cold" orographic cloud processes. It seems clear that for many orographic clouds with cloud top temperatures at least in the range from about −10 to around $-20°C$, a substantial weather modification potential exists and that this potential is being realized from artificial seeding. The basic results from the field experiments also confirm that most orographic clouds with temperatures colder than around $-24°C$ have naturally efficient ice phase cloud water utilization processes and thus offer little or no potential for precipitation augmentation. These experiments show that again, as would be expected from physical considerations, clouds with cold temperatures, less than around $-28°C$, generally experience precipitation reductions from seeding. The smaller ice crystals present in these cases may drift downwind and, when a cloud of supercooled water is present in that area, continue to grow and augment the precipitation at that location.

The simple parameterization of cloud top temperature is so crucial that it alone delineates in a general way the weather modification potential from many orographic clouds. In some orograhic clouds, the coldest levels to which the seeding materials are transported rather than the cloud top temperature would constitute the critical cloud temperature. Cloud top temperature still plays an important role in determining seedability even for these clouds, since it largely controls the natural supply of ice crystals from higher regions of the cloud.

Modeling of cold orographic cloud and cloud seeding processes shows that consideration of additional factors such as rate of moisture flow, ice crystal trajectories, etc., are vital to optimal seeding and targeting of snowfall. Incorporation of the results of continuing cloud physics research, such as improved information on ice crystal multiplication, can readily be incorporated into existing average or time-dependent models. Considerable additional model development is required for orographic clouds with cloud top temperatures warmer than about $-10°C$, clouds in which droplet coalescence processes dominate, and for orographic cloud systems in which induced convection plays an important role. Present models of orographic clouds and seeding processes, however, can now serve to define the technology required for realizing a substantial portion of the existing potential.

THE TECHNOLOGY FOR SEEDING "COLD" OROGRAPHIC CLOUDS

Technology Assessment

The technology for seeding orographic clouds was established in the late 1940s and early 1950s by the commercial weather modification operators. This technology was developed on the general assumption that additional water supplies are required, and that specific users should pay the program costs. The technique developed involved releasing silver iodide smokes, generally from ground sites, along the upwind slopes of the mountain barrier to be seeded when clouds were present. This basic approach still appears sound for the seeding of orographic clouds over many mountain barriers. Substantial improvements in all aspects of the operating programs have, however, resulted from continuing research and developmental programs.

The greatest deficiencies of the past operational programs, in probable order of importance, were the following:

1. The lack of criteria for recognizing the seedability of specific clouds.
2. The lack of specific information as to where the seeding materials would go once they were released.
3. The lack of specific information as to downwind or broader social and ecological effects from the operations.
4. The lack of detailed information

on the efficiency of seeding generators and material being used for seeding clouds with differing cloud temperatures.

Important advances have been made in correcting all these deficiencies. Criteria for recognizing seedability and knowledge of seeding generators and seeding material characteristics are now adequate for incorporation into a "good" technology. Additional research is needed in the other areas. Specific problem areas needing emphasis in continuing research include (1) the dispersal and transport of the seeding materials, (2) "ice multiplication" and its incorporation into models, and (3) human dimension problems including political, legal, social, economic, and ecological aspects. Overall, a refined technology now exists for supplementing water supplies from many mountainous areas. The Bureau of Reclamation is using a pilot project concept to bridge the gap between the research experiments and fully operational projects. Pilot projects are envisioned as large-scale experiments designed to furnish technological and environmental data and to test, on large watershed scale, the operational systems, procedures, and techniques that have evolved through research. The next section describes how the first of these pilot projects, the Colorado River Basin Pilot Project, was conceived and is being carried out.

The Bureau of Reclamation Pilot Project

Background. The Bureau of Reclamation was commissioned by Congress in 1961 to advance "... research on increasing rainfall by cloud seeding. . . ."[46] The Atmospheric Water Resources Management Program, often termed "Project Skywater," which began with this commission, is oriented toward early development of practical and feasible techniques to increase water supplies by weather modification. With this orientation, application of existing methods and findings and use of existing expertise have been sought from the beginning. Nearly all program effort is conducted through contracts with universities, private firms, and other government agencies.

By spring, 1966 the Bureau of Reclamation had eight winter orographic weather modification experiments underway throughout the mountainous areas of the Western States. These areas were (1) the Park Range in northern Colorado, (2) Elk Mountain in southern Wyoming, (3) the Jemez Mountains in northern New Mexico, (4) the Wasatch Mountains in northern Utah, (5) the Cascade Range in Washington, (6) the central Sierra Mountains in California, (7) the Bridger Range in Montana, and (8) the Lake Tahoe area in Nevada and California. The purposes of the deliberate spread of projects were to study cloud and precipitation modification under various climatological and terrain conditions and to begin development and adaptation of a cloud seeding technology in these more critical water supply areas. Four of the eight experiments were in or immediately adjacent to the Upper Colorado River Basin.

In 1966 the Bureau introduced the pilot project concept as a necessary step between research experiments and fully operational projects (Bureau of Reclamation, 1966). Pilot projects were envisioned as the main focus of the entire Skywater program and were to be large-scale field experiments designed to furnish technical and environmental data and tests of the operational systems, procedures, and techniques evolved through research. Adapting operating systems to the particular region and resolving the environmental and benefit-cost effectiveness questions were to be key aspects of any forthcoming pilot projects. Authority to ". . . initiate pilot weather modification projects . . ." was given the Bureau in 1968.[47]

Actual initiation decisions for the Colorado River Basin Pilot Project rested largely on two events: (1) the positive evaluations of the Climax experiment by Colorado State University as reported at the Fifth Berkeley Statistical Symposium[25] and the preliminary evaluations of additional Climax data and data from the state of Colorado sponsored cloud seeding experiment at Wolf Creek Pass, Colorado; and (2) the passage of the Colorado River Basin Project Act in September, 1968. Besides authorizing the Central Arizona Project, Title II of the act directs the Secretary of the Interior to ". . . conduct full and complete reconnaissance investigations for the purpose of developing a general plan to meet the future water needs of the Western United States." The investigations are to consider means of water augmentation, including weather modification. This final report to Congress on the reconnaissance investigation is due June 30, 1977.

Project Objectives and Goals. The main objective of the pilot project is to provide sound scientific and engineering evaluations of precipitation increases over a large area by an operational-type application of cloud seeding techniques employed and criteria developed through the Climax, Colorado, experiment. Because of its planned size and operational orientation, the project will afford the first major opportunity and meaningful climate for assessing social and environmental problems associated with weather modification and for appraising technical performance factors. These objectives are oriented toward learning definite answers on the technological factors and the environmental and economic feasibility considerations involved in producing large quantities of additional stream flow in the Colorado River Basin.

The project goal is to give the Bureau of Reclamation a supportable basis for recommending or not recommending an operational cloud seeding program in the required Title II report of the Colorado River Basin Project Act.

The present and projected water supply deficiencies in the Colorado River Basin have often been reported and discussed. Concern and controversy over the availability and quality of the limited stream flow from the mountains of this basin are as serious as any of the nation. Preliminary studies indicate that some 2 million acre-ft of stream flow could be added annually to the Colorado River by cloud seeding.[11, 27] Operational costs for producing this new water supply are estimated at $1.00 to 1.50/acre-ft, although "hidden costs" such as snow removal and environmental protection would increase overall costs. Benefits from such additional water could range up to $25 million annually and higher. The recent analyses of seeding results and other studies confirm the early availability of practical techniques, the preliminary augmentation potential estimates, and the expected costs.

For a critical water area such as the Colorado River Basin, producing some 2 million acre-ft annually of usable additional water supply would be a substantial contribution and would ease many problems, particularly the international water quality question. The possible highly favorable benefit cost ratios and the flexibility of use, largely through present water and power facilities, point to weather modification as a practical tool worth major investigation and test in the Colorado River Basin.

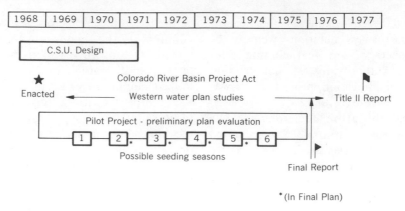

Figure 7.6   Schedule for Western Water Plan Studies.

Project Plan. A contract was awarded Colorado State University in April 1968 for a design of an operational adaptation (pilot project) in the Colorado River Basin. Economic and social considerations and seeding suspension criteria were included in the comprehensive plan in addition to statistical and meteorological design aspects. Development of a general seeding model from their experiences was included.

Passage of the Colorado River Basin Project Act had an important effect on plans for weather modification development in the Colorado River Basin. First, it gave a specific directive for sound and complete answers on weather modification to be incorporated in a major, comprehensive water plan. Second, it set an early deadline for results and a report on cloud seeding feasibility. Third, this deadline necessitated an acceleration of planning and initiation of the project. In order to plan and install instrumentation and seeding equipment, conduct cloud seeding for four winter seasons, thoroughly evaluate the data, and prepare a full report to be incorporated in the required 1977 Title II report, an early start was dictated. Figure 7.6 depicts the planned schedule for the project.

Experience in analyzing cloud seeding experiments and commercial operations indicate that at least 4 or 5 years are usually necessary to give sufficient data for sound evaluations with statistical confidence above 90 percent when precipitation increases are in the 10 to 20 percent range.

Deciding on the evaluation was a key aspect in the design.[21] To give the results the desired water resource orientation, a strong cloud physics monitoring and analysis approach was rejected in favor of quantitative surface precipitation and hydrologic measurements and evaluation. Although cloud and ice crystal changes with seeding as measured by radar and airborne sensors would be meaningful and significant to cloud physicists and other scientists, these changes would be a step removed from the desired results— more snow on the ground and water in the streams.

There are two basic schemes in evaluating weather modification effects, the target-control comparison and the seed/no-seed analysis for the same area. A target-control comparison was rejected as a prime evaluation scheme since coefficients of correlation for either precipitation or stream flow amounts between proposed target areas and available controls were generally less than 0.70.[38] This low correlation occurs because large areas covering whole mountain massifs were to be target areas. Suitable control areas would of necessity be in widely separated mountain massifs affected differently by the same storms, thus giving varying storm and seasonal precipitation amounts and resulting runoff. With expected precipitation increases in the 10 to 30 percent range, evaluations with statistical confidence of 0.90 or better could not be expected using target-control methods within 4 years. If seasonal stream flow evaluation would be used, only four affected data points and about 35 historic, unaffected points could be used.

There was concern that not using target-control analysis methods, which are straightforward and which could include stream flow as a major evaluation factor, would be less acceptable to water resources engineers and planners. The question of the variability in seasonal precipitation-runoff relationships would be bypassed if direct stream flow increases could be shown with confidence. The conclusion from people who would be appraising the results was that this group would be satisfied without a direct stream flow evaluation if a strong precipitation evaluation could be made. If a solid statement could be made on the amount of snowpack and snowfall increase, they felt its relation to stream flow could be handled. Design emphasis was placed on sound and dependable precipitation evaluation, using seed/no-seed analysis, with stream-flow evaluation a secondary consideration. The precipitation data and additional stream flow data, along with advances in mathematical watershed modeling, would give improved precipitation-runoff relations from mountainous areas for use in the

eventual comprehensive appraisal of the project.

The initial plan proposed was for a two-area project, a northern area in the mountains around Granby in northern Colorado and a southern area in the San Juan Mountains of southwestern Colorado.[29] Seeding of suitable storms (warmer cloud tops) would be randomized on a yearly basis. Number of suitable storm days for typical winter season and the project areas were estimated as follows:

	Northern Area (Granby)	Southern Area (San Juans)
1 season	20 days	40 days
4 seasons	80 days	160 days

At least 40 days each of seeded and non-seeded events are needed for stable statistical analysis. Thus four typical winter seasons, two seeded and two unseeded in each area, would give sufficient data for evaluations.

By the end of January, 1969, the planning, design, and field reconnaissance for the project had advanced to a stage where a full scientific and engineering review was warranted. To afford an opportunity and mechanism for a comprehensive and frank review, a special conference was held.[41] Although the overall design and scientific basis for the project were accepted without major reservations, the yearly randomization was questioned.

Subsequent studies led to abandonment of yearly randomization in favor of a daily randomization where part of the suitable storms would be seeded each year. The basic concern was that a particular type of storm or broad weather pattern could be prevalent throughout a winter season. With a yearly randomization, if all suitable storms were seeded in years under a particular pattern and none

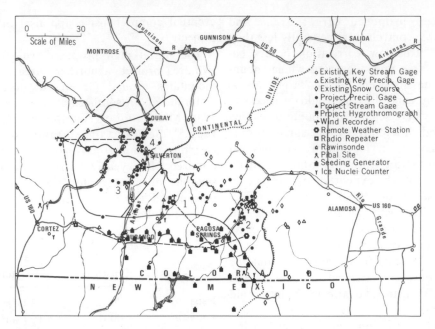

Figure 7.7 Colorado River Basin Pilot Project.

during a different pattern, an unbalanced evaluation could result. This consideration, along with less emphasis on hydrologic evaluation, led to selection of the following experimental design:

1. A 24-hour experimental day for operations and seeding.
2. A 50–50 daily randomization of seed and no-seed.
3. The 24-hour snowfall amounts will be the prime evaluation data.

Withdrawal from the yearly randomization meant paired project areas were not needed for efficient use of men and seeding equipment. Because of anticipated budgetary problems, it was felt advisable to initially plan a project for only one area. The southern project area of 3300 mi² in the San Juans was selected. In general, the San Juans offer the highest seeding potential in the Colorado River Basin and best evaluation possibilities. Problems there would be typical of what must be faced eventually if cloud

seeding is to be developed into a practical means of water resources management. One of the purposes of the project was to define and resolve such problems.

A map shows the project area and the equipment and instrumentation placement (Figure 7.7). The project area was partitioned into four subareas designated by number according to highest priority for seeding. Subarea 1, generally between the Animas River and Wolf Creek Pass, was considered best for seeding because there are few people in the area; it is directly exposed to the major southwestern storms; it contains Lemon and Vallecito Reservoirs, two Bureau of Reclamation reservoirs where additional water is needed; although it contains the San Juan Wilderness Area, access for data collection is less difficult; and most important, studies showed stream flow and snow courses correlated well within the area, confirming the relatively regular precipitation patterns and uniform watershed characteristics. Subarea 4 was determined less opportune for seeding and access for evaluation data collection is more difficult.

Only Subareas 1 and 2 are included for the project seeding. These two areas furnish a target area of about 1300 mi².

Operational Design. Of particular importance to the project design is the diurnal minimum of precipitation during the late morning hours. This diurnal effect led to definition of the 24-hr experimental day as being from 11 a.m. to 11 a.m. mst. Seeding carryover effects or measurement errors would tend to be minimized between experimental units if the division was during a minimum precipitation period. The 11 a.m. start of an experimental day also was near optimum for forecasting and seeding operation considerations. Forecasts could be made from analyzed 1200 *Z* (5 a.m.) observations and operational decisions made and implemented prior to 11 a.m.

Experimental days are established based on the morning forecasts if the following criteria are met:

1. Precipitation of 0.01 in. or greater is forecast.

2. Cloud-top temperature or in absence of this datum, 500-mb temperatures of −23°C or warmer are forecast to exist over the project area for at least 3 hr.

3. Mean wind field from the surface to 700 mb is forecast to be toward the mountain slopes of the project target area (approximately 170° through 300°).

4. There are no seeding restrictions in effect.

Only after the decision to call an experimental day is the randomized seed/no-seed decision made. A randomized schedule of 400 days has been prepared by the Statistical Laboratory at Colorado State University to give the 50–50 split in intermixed blocks of 10, 20, and 40 days.

Experimental days randomly determined for seeding are seeded using a variable seeding rate as based on the Climax model. Effective seeding rates to be used are determined by cloud top, or 500-mb forecast temperatures, as monitored by rawinsonde and mountaintop telemetered weather stations throughout the day as follows:

1. 140 g/hr for temperatures −14°C and warmer.

2. 20 g/hr for temperatures −15 to −23°C.

The nuclei production efficiency requirements for the generators are at least 10^{15} effective particles/g at −20°C and at least 10^{14} effective particles/g at −15°C.

Seeding is accomplished by 33 ground-based generators using butane with a fuel mixture of acetone and silver iodide with sodium iodide as a mixing agent. Thirteen of the generators are radio-controlled and 20 are operated by local farmers and ranchers with telephoned instructions. The location of these generators is shown on the project map.

Four mountain communities are located in the northern portion of the area—Ouray, Telluride, Silverton, and Lake City, Colorado. Objections to the project, based largely on snow removal and avalanche hazards, were voiced from this area where silver mining is the main economic occupation during the winter. The avalanche hazards on Red Mountain Pass, U.S. Highway 550 between Ouray and Silverton, are among the most serious in the nation.

Because of these objections, an important part of the project plan is the suspension of seeding during situations which could possibly cause pronounced effects on the public and the environment in and near the project area.[29] Seeding is suspended when:

1. The snowpack exceeds 150 percent of long-term averages.

2. Severe avalanche potentials exist or are forecast.

3. Severe storm conditions exist or are forecast.

4. Various other critical situations exist as appraised by other Federal and state agencies.

Chances are that about 1 year out of 5 could have suspensions in effect during most of the season. After two winter seasons of seeding, 1970–1971 and 1971–1972, there have been two suspensions called, one during big game hunting season when heavy early snows occurred and the other over a New Year's holiday season when heavy traffic over the mountain passes and snow conditions were causing snow removal problems.

Expected Results. About 90 snowfall days during a typical winter season in the San Juan Mountains furnish a mean snowfall of 24 in. (water equivalent) as averaged over the entire area. Only 40 of these days have the warmer cloud tops suitable for seeding, but about 14 in. of the mean seasonal precipitation occur on these days. Results of the Climax and Wolf Creek experiments indicate seeding should cause an average increase of 55 percent in precipitation from suitable clouds in the San Juans.

With a 50-50 randomization, an average 3.8-in. precipitation addition or a seasonal 16 percent increase is anticipated.

10 in. from 50 unsuitable days (no seeding)	10.0 in.
7 in. from 20 suitable days (no seeding)	7.0
7 in. from 20 suitable days (seeded with 55 percent increase)	10.8 ——— 27.8 in.

The additional 3.8 in. of precipitation caused in the initial target areas during the project to test seeding techniques should yield about 250,000 acre-ft annually of additional stream flow from rivers originating in the area. This additional water will be a "bonus" benefiting downstream water users.

Social and Environmental Considerations. Although much active support and acceptance of the project have been evidenced in Colorado, strong objections have also been raised. Farmers, ranchers, and those with water resources interests generally support the project because of the additional water and possibilities for future operations. Objections are generally raised by those with naturalist interests and those living in and near the project whose livelihood is not dependent on water and where snow is a real hindrance or danger.

Objections voiced were centered in the four towns within the proposed project area, particularly Ouray. Although misconceptions over the magnitude of cloud seeding effects and definite anti-Federal Government attitudes were prevalent, there were valid concerns over avalanche hazards and snow removal problems in the mountain communities and mining areas. The possible effects on deer and elk, the timber industry, and the summer tourist trade also worried many in the area. A series of public meetings were held in each of the communities to hear these concerns and inform people of the project plans. These meetings had three principal results:

1. Subareas 3 and 4 were deleted from the seeding target area.

2. An intensive avalanche research program was undertaken in the Red Mountain Pass area.

3. The seeding suspension criteria were made more restrictive.

Petitions have been circulated requesting a court injunction against the project but no definite legal activity has yet occurred. Damage suits against the Bureau of Reclamation and its contractors have also been threatened.

A full public information program is underway, including weekly news release of all seeding dates and times, circulation of a newsletter on significant project events, and preparation of a technical movie entitled "Mountain Skywater" for wide distribution. This information, plus the fact that the first two winters had much less snowfall than normal, has—at least temporarily—lessened social pressures against the project.

Many ecological questions have been raised on possible effects of increasing snowpack. A comprehensive ecological research program is being made to monitor and study the possible effects. Wildlife, fishery, and plant ecology (especially in the alpine areas) are among the various environmental factors being investigated. This work is conducted by Colorado State University, the University of Colorado, and Fort Lewis College.

Conclusions. Because of below-average frequency of storms and seeding suspensions there have been only 44 experimental days with snowfall during the first two seasons. Very preliminary looks at these limited data indicate increases are being achieved both in the target area and in the Rio Grande Basin immediately downwind. The preliminary analysis of the first season indicated about 40 percent more precipitation fell on 17 seeded days than on 17 unseeded but similar seedable days.[29] This increase matches expectations and confirms the model and seeding technique.

The objections and public concerns at the beginning of the project show that local groups must be informed and involved early in the development of a cloud seeding project. Information on cloud seeding and meetings to discuss the project can ease many of the apparent fears. Acceptable design compromises and safeguard criteria can be established through open and sincere efforts to satisfy specific problems. Because a severe winter has yet to be experienced by the project, perhaps the real test of the public opinion efforts has not risen.

Experiences from the pilot project indicate possible decisions for an enlarged operational program must be backed by much public discussion, well defined objectives and justification, and a thorough project design. Ownership of the additional waters produced by cloud seeding and their regulation within the existing water right compact structure is a major legal problem to be resolved. The first half of the experiences of the Bureau's project in the San Juans are meeting the planned objectives of scientific answers and social-environmental problem definition.

REFERENCES

1. Adderley, E. E., Rainfall increases downwind from cloud seeding projects in Australia, *Proceedings of First National Conference on Weather Modification, Albany,* 1968, pp. 42–46.

2. Bergeron, T., The problem of an artificial control of rainfall on the globe; general effects of ice-nuclei in clouds, *Tellus* 1, 32–50, 1959.

3. Bowen, E. G., Lessons learned from long-term cloud-seeding experiments, *Proceedings of International Conference on Cloud Physics, Tokyo and Sapporo,* 1965, pp. 429–433.

4. Bowen, E. G., Review of current Australian cloud-seeding activities, *Proceedings of First National Conference on Weather Modification,* AMS, 1968, pp. 1–7

5. Brown, K. J. and R. D. Elliott, Mesoscale changes in the atmosphere due to convective band seeding, *Preprint of Third Conference on Weather Modification,* AMS, 1972, pp. 313–320.

6. Cermak, J. E., M. M. Orgill, and L. O. Grant, Laboratory simulation of atmospheric motion and transport over complex topography as related to cloud seeding operations, *Proceedings of Second National Conference on Weather Modification,* AMS, 1970, pp. 59–65.

7. Chappell, C. F., Modification of Cold Orographic Clouds, *Atmos. Sci. Paper No. 173,* Department of Atmospheric Science, Colorado State University, 1970, 196 pp.

8. Chappell, C. F., Airborne seeding of wintertime Wasatch Mountain clouds during Project Snowman, *Preprint of Third Conference on Weather Modification,* AMS, 1972, pp. 129–132.

9. Chappell, C. F., L. O. Grant, and P. W. Mielke, Jr., Cloud seeding effects on precipitation intensity and duration of wintertime orographic clouds. *J. Appl. Meteorol.,* **10**(5), 1006–1010, 1971.

10. Colorado River Basin Project Act, *PL 90-537,* 90th Congress, 2nd Session, 1968 (Note: House of Representatives Report No. 1861, Confidence Report—S.1004, September 4, 1968).

11. Crow, Loren W., Report on major sub-basin target areas for weather modification in upper Colorado River Basin, *Report 53,* Denver, Colo., 1967.

12. Elliott, R. D., Cloud seeding area of effect numerical model, Aerometric Research, Inc., Report, Santa Barbara, Calif., 1969.

13. Elliott, R. D. and K. J. Brown, The Santa Barabara II Project—downwind effects, *Proceedings of the International Conference on Weather Modification, Canberra, Australia,* 1971, pp. 179–184.

14. Elliott, R. D. and B. N. Charles, Summary of recent development in weather modification and implications for space technological developments, unpublished Report, North American Weather Consultants, 1969.

15. *Final Report of the Advisory Committee on Weather Control,* 1957, Advisory Committee on Weather Control, U.S. Government Printing Office, 1958.

16. Final Report Project Cirrus, General Electric Research Laboratory, Report under Signal Corps Contract No. W-36-039-sc-38141, 1951.

17. Fujita, T., Mesoscale aspects of orographic influences on flow and precipitation patterns, *Patterns of the Symposium on Mountain Meteorology, Atmos. Sci. Paper No. 122,* Atmospheric Science Department, Colorado State University, 1967, pp. 131–146.

18. Furman, R. W., Radar characteristics of wintertime storms in the Colorado Rockies, *Atmos. Sci. Paper No. 112,* Department of Atmospheric Science, Colorado State University, 1967, 53 pp.

19. Grant, L. O., The role of ice nuclei in the formation of precipitation, *Proceedings of the International Conference on Cloud Physics, Toronto,* Canada, 1968, pp. 305–310.

20. Grant, L. O., J. E. Cermak, and M. M. Orgill, Delivery of nucleating materials to cloud systems from individual ground generators, *Proceedings of Third Sky Water Conference,* Bureau of Reclamation, 1968.

21. Grant, L. O., C. Chappell, L. Crow, P. Mielke, Jr., J. Rasmussen, W. E. Shobe, H. Stockwell, and R. A. Wykstra, An operational adaptation program for the Colorado River Basin, *Interim Report,* Contract No. 140-06-D-6467, Colorado State University, Fort Collins, Colorado, 1969.

22. Grant, L. O., C. F. Chappell, and P. W. Mielke, Jr., The recognition of cloud seeding opportunity, *Proceedings of the First National Conference on Weather Modification,* AMS, 1968, pp. 372–385.

23. Grant, L. O., C. F. Chappell and P. W. Mielke, Jr., The Climax experiment for seeding cold orographic clouds, *Proceedings of International Conference on Weather Modification, Canberra, Australia,* 1971, pp. 78–84.

24. Grant, L. O. and R. D. Elliott, The cloud seeding temperature window, paper being prepared for submission to *J. Am. Meteorol.,* 1972.

25. Grant, L. O. and P. W. Mielke, Jr., A randomized cloud seeding experiment at Climax, Colorado 1960–1965, *Proceedings of the Fifth Berkeley Symposium in Mathematical Statistics and Probability,* Vol. 5, University of California, 1967.

26. Hindman, E., Snow crystal and ice nuclei concentrations in orographic snowfall, *Atmos. Sci. Paper No. 109,* Department of Atmospheric Science, Colorado State University, 1967, 83 pp.

27. Hurley, P. A., Augmenting Colorado River by weather modification, *J. Irrigation and Drainage Div. Am. Soc. Civil Engrs.* **94**(IR4), 363–380, Proceedings Paper 6271, 1968.

28. Hurley, P. A., Colorado Pilot Project: Design hydrometeorology, *J. Hydraulics Div. Am. Soc. Civil Engrs,* **98**(HY5), 811–4826, Proceedings Paper 8901, 1972.

29. Kahan, A. M., Status of cloud seeding—1972, presentation, *American Water Resources Assocation Symposium—Watersheds in Transition,* Fort Collins, Colo., 1972.

30. Keys, C. G., Jr., J. V. Lunsford, F. D. Stover, R. D. Wilkins, and M. Lentner, Jemez atmospheric water resources research project, *Interim Progress Report No. 6*, January 1, 1972 through June 30, 1972, Contract No. 14-06-D-6803, August 1972, 91 pp.

31. Ludlan, F. H., Artificial snowfall from mountain clouds, *Tellus* **7**, 277–290, 1955.

32. Mielke, P. W., Note on some squared rank tests with existing ties, *Technometrics,* **9**, 312–312, 1967.

33. Mielke, P. W., L. O. Grant, and C. F. Chappell, Elevation and spatial variation effects of wintertime orographic cloud seeding, *J. Appl. Meteorol.*, **9**, 476–488, 1970; Corrigendum, **10**, 842.

34. Mielke, P. W., L. O. Grant, and C. F. Chappell, An independent replication of the Climax wintertime orographic cloud seeding experiments. *J. Appl. Meteorol.*, **10**(6) 1198–1212, 1971.

35. Mitchell, V. L., A. B. Super, and R. H. Yaw, Preliminary results of a randomized winter orographic cloud seeding experiment in the northern Rocky Mountains, *Preprint of Third Conference on Weather Modification*, AMS, 1972, pp. 125–128.

36. Mossop, S. C., Concentrations of ice crystal in clouds, *Bull. Am. Meteorol. Soc.* **51**, 474–479, 1970.

37. Mossop, S. C., The multiplication of ice crystals in clouds, *Proceedings of International Conference on Weather Modification, Canberra, Australia*, 1971, pp. 1–4.

38. Nakamichi, H. and H. J. Morel-Seytoux, Suitability of the Upper Colorado River Basin for precipitation management, *Univ. Hydrology Paper No. 36*, Colorado State University, Fort Collins, Colo., 1969.

39. Orgill, M. M., Laboratory simulation and field estimates of transport-dispersion in mountainous terrain, Ph.D. Thesis, Civil Engineering Department, Colorado State University, 294 pp.

40. *Plan to Develop Technology for Increasing Water Yield from Atmospheric Sources*. Bureau of Reclamation, U.S. Department of the Interior, Washington, D.C., 1966.

41. Skywater Conference V, *Proceedings U.S. Department of the Interior*, Bureau of Reclamation, Denver, Colo., 1969.

42. Reinking, R. and L. O. Grant, The advection of artificial ice nuclei to mountain clouds from ground-based generators, *Proceedings of the First National Conference on Weather Modification*, AMS, 1968, pp. 433–445.

43. Rhea, J. O., P. Willis, and L. G. Davis, *Park Range Atmospheric Water Resources Program, Final Report to Bureau of Reclamation*, EG & G Inc., Boulder, Colo., 1969,

44. Schaefer, V. J., The production of ice crystals in a cloud of supercooled water droplets, *Science* **104**, 457–459, 1946.

45. Schaefer, V. J. and I. Langmuir, The Modification Produced in Clouds of Supercooled Water Droplets by the Introduction of Ice Nuclei, paper Presented to the AMS Annual Meeting, 1946.

46. Senate Report No. 1097, Committee on Appropriations, 87th Congress, 1st Session, September 1961.

47. Senate Report No. 1405, Committee on Appropriations, 90th Congress, 2nd Session, July, 1968.

48. Smith, E. J., E. E. Adderley, L. Veitch, and E. Turton, A cloud seeding experiment in Tasmania, *Proceedings of International Conference on Weather Modification, Canberra, Australia*, 1971.

49. Super, A. B., B. H. Yaw, C. C. Grainger, and G. Langer, Development of a practical system for direct detection of cloud seeding effects, *Preprint of Third Conference on Weather Modification*, AMS, 1972, 119–124.

50. Vardiman, L. and L. Grant, A case study of ice crystal multiplication by mechanical fracturing, *Volume of Abstracts, International Cloud Physics Conference*, London, 1972, pp. 22–23.

51. Vardiman, L. and L. Grant, A study of ice crystal concentrations in convective elements of winter orographic clouds, *Preprint of Third Conference on Weather Modification*, AMS, 1972, pp. 113–118.

8 | The Mitigation of Great Lakes Storms

HELMUT K. WEICKMANN

STRUCTURE AND PROPERTIES OF GREAT LAKES CLOUD SYSTEMS

The Great Lakes Basin encloses an area of more than 750,000 km² and has a water surface of 250,000 km². Thirty-two million people live within the basin, 28 million of them taking their water requirements directly from the basin hydrological system. The present daily total water consumption for industrial and private use is 15,600 million gal. Between November 4, 1966 and January 25, 1969, the average amount of snowfall from a typical Lake Erie "lake storm" was nearly 250,000 million gal., equivalent to a 16-day supply of water.[10, 13] The 23 snowstorms that occurred during the period mentioned above produced a 1 year supply of water. As the hydrology of all lakes is delicately balanced, it is obvious that the artificial manipulation of the atmospheric source of lake water may have very significant influences important to human activities. In the following discussion we deal primarily with the mechanism of Lake Erie snowstorms and their mitigation. It is unavoidable, however, to call attention to more far-reaching applications of weather modification in that area.

Helmut K. Weickmann is Director of Atmospheric Physics and Chemistry Laboratory, Environmental Research Laboratories, National Oceanic and Atmospheric Administration, Boulder, Colorado.

In early winter specific kinds of intense snowstorms form over the Great Lakes Basin. These have their origin in and owe their severity to the combination of continental polar air outbreaks and the warm lake surfaces over which the cold air flows. The continental cold air, already in neutral equilibrium through mechanical mixing on its way across the Canadian plains, is destabilized when it contacts the warm lake surfaces and forms intensively precipitating convective cloud systems, which nevertheless are quite shallow, seldom reaching above the 4000 m level. The more intensive storms are often associated with the passage of an upper level trough.[11, 20]

The long-term climatology of precipitation for Buffalo, New York, located as it is at the downwind eastern end of Lake Erie, indicates periods—singularities—when snowfall is particularly frequent. Figure 8.1 shows the cumulative daily snowfall for November and December of the past 97 years for Buffalo, New York. Three periods stand out clearly: mid-November, the end of November, and mid-December. Typical precipitation patterns of Erie and Ontario lake storms are shown in Figures 8.2 and 8.3.[11, 13] The highest yield of precipitation is connected with storms caused by a long fetch of the air over the water. Both the

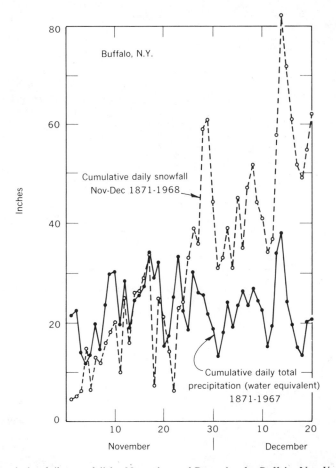

Figure 8.1 Cumulative daily snowfall for November and December for Buffalo, New York, 1871–1968.

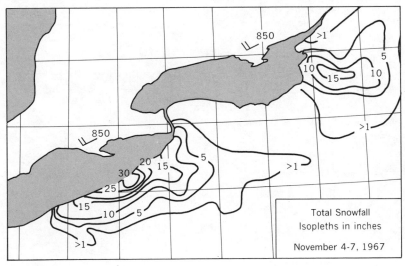

Figure 8.2   Snowfall in intense winter storm of November 4–7, 1967 (after Jiusto[13]); 850-mb flow west to southwest.

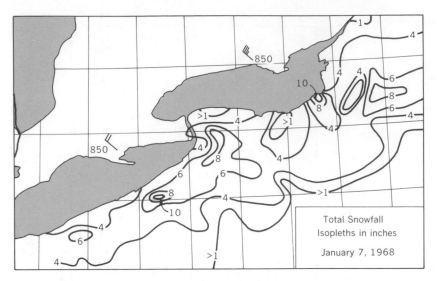

Figure 8.3 Snowfall in weak winter storm of January 7, 1968 (after Jiusto[13]); 850-mb flow northwesterly.

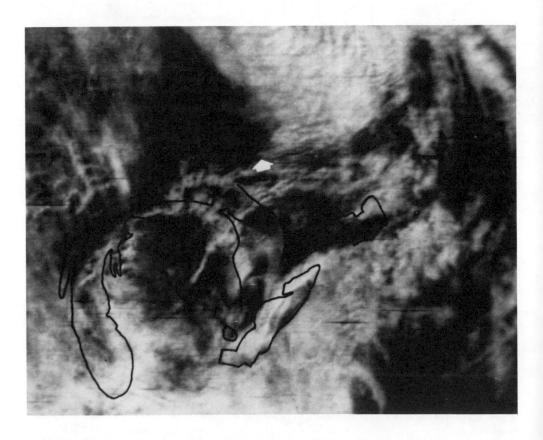

Figure 8.4 Satellite picture of snowstorm of November 15–16, 1969.

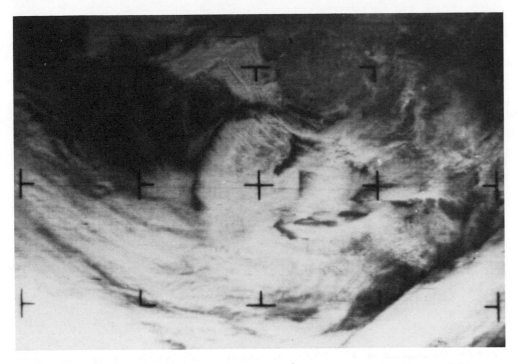

Figure 8.5   Satellite picture of storm of December 2, 1966. Note cloudfree west coast of Lake Michigan.

moisture source and the convective instability of the lake storms originate from the lakes, and since the maximum intensity occurs over the downwind shore areas, the heaviest precipitation falls along

Table 8.1   Time Periods New York State Thruway Closed Due to Lake Storms

Date	Closing Period (hrs)[a]
December 23, 1960	5.4
December 10, 1962	20.1
December 19, 1963	5.1
November 21, 1964	22.7
January 30, 1966	36.0
December 2, 1966	13.3
January 1, 1969	27.1
January 10, 1969	13.1
January 9, 1970	19.5

[a] Data were supplied by Mr. A. Fisch, Director of Operations, New York Thruway Authority.

the shore belt. Important industrial complexes and traffic arteries are close to the shore and are subject to operational delays and damage from the severe snowfall. Table 8.1 shows periods during the past 10 years when the New York State Thruway was closed between Buffalo, New York, and the Pennsylvania state border along the Erie lake shores.

Satellite pictures of typical storms are given in Figures 8.4 and 8.5. Figure 8.4 shows the storm of November 15 and 16, 1969; Figure 8.5 shows the storm of December 2, 1966,[21] with pronounced bands over both Lake Erie and Lake Ontario. A few interesting properties of the November 15–16 storm system can be noted, for instance, the pronounced storm line originating over Saginaw Bay. Apparently the wind direction developed exactly along the orientation of Saginaw Bay with resulting acceleration over the reduced drag narrow water surface and convergence from the sides. The ensuing cloud

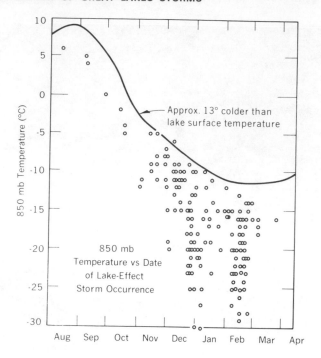

Figure 8.6   Difference between 850-mb temperature and lake surface temperature during lake storms (reprinted by permission from Jiusto and Holroyd[10]).

line crosses Lake Huron and can be recognized for at least 150 mi downstream. One notes also that, contrary to the storm of December 2, 1966, downstream of the Great Lakes a void in the cloud formation appears, perhaps due to snow-out from the lake storms.

Important information concerning the lake storms is conveyed by Figure 8.6.[8, 10] Here the 850-mb temperature has been plotted against date of occurrence of lake storms. In addition a temperature line is shown which is 13° cooler than the lake surface temperature. Such a temperature difference would exist under a dry-adiabatic lapse rate between the lake surface and 850 mb. The figure shows that this is a limiting curve for normally much greater differences of temperature. The heat transfer from the water to the air is not 100% efficient, and the lake surface can be considerably warmer than would be required for a dry adiabatic lapse rate.

The figure confirms that lake storms are typified by steep lapse rates (super adiabatic over at least shallow surface layers) and consequently by unstable convective conditions. This instability forms the basis for the explanation of the mesoscale dynamics and structure of these storms.

The Mesoscale Numerical Model

A realistic mesoscale numerical model of the typical winter lake storm has been developed by Lavoie.[15] The key assumption for the model is that the atmospheric layer in which the storm clouds develop is well mixed and can be represented by a constant potential temperature. This layer is topped by an inversion with stable lapse rate above. The mixing as the air moves across Canada is essentially mechanical; over the lake additional mixing occurs due to convection. The thermal structure of the model is given in Figure 8.7. The heat transfer from the surface is parameterized using a drag coefficient that depends only on the surface roughness, assuming that the lapse rate is nearly adiabatic. Other assumptions are as follows:

1. Vertical gradients of potential temperature and vector wind are negligible within the mixed layer, except insofar as they contribute to fluxes of heat and momentum.

2. Hydrostatic equilibrium holds.

3. The air is incompressible along a streamline.

4. Moisture effects can be neglected as a first approximation.

5. Well above the mixed layer a surface can be identified where the induced disturbance in the pressure and temperature fields become negligible.

6. The upper boundary of the mixed layer acts as a barrier to momentum and heat flux.

7. Momentum and heat flux through the lower boundary of the mixed layer can be parameterized through use of a drag coefficient that is independent of lapse rate.

The model utilizes the equations of (1) motion; (2) hydrostatic equilibrium; (3) energy; (4) state; and (5) continuity. The equations of motion can be expressed in the form:

$$\frac{\partial u}{\partial t} = - u \frac{\partial u}{\partial x} - v \frac{\partial u}{\partial y} + fv$$

$$+ \frac{g}{\theta_h} \left[\theta - \frac{\theta_h + \theta_H}{2} \right] \frac{\partial h}{\partial x}$$

$$+ g \frac{(h - z_S)}{2\theta} \frac{\partial \theta}{\partial x} + (H - h)F_v$$

$$- F_2 + g \frac{(H - h)}{2\theta_h} \frac{\partial \theta_h}{\partial x}$$

$$- C_D \frac{\sqrt{u^2 + v^2}}{h - z_s} u,$$

where

$$F_v \equiv \frac{g}{2\theta_H} \frac{\partial \theta_H}{\partial x} \quad \text{and} \quad F_2 \equiv \frac{\theta}{\theta_h} \left(\frac{1}{\rho} \frac{\partial p}{\partial x} \right)_H$$

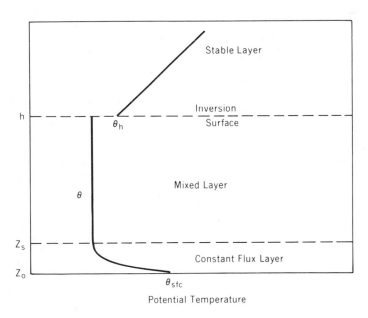

Figure 8.7 Thermal structure of Lavoie mesoscale model.

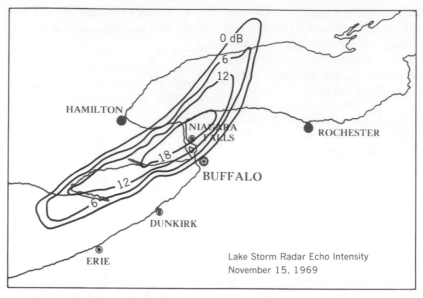

Figure 8.8a   Radar contour lines of storm on November 15–16, 1969.

with a corresponding equation obtained for $\partial v/\partial t$. These expressions are useful in that they offer insight into the mechanics of the flow. The term containing $\partial h/\partial x$ defines a positive acceleration if the slope of the inversion, $\partial h/\partial x$, is negative (sloping downward in direction of the flow) as long as the average potential temperature above the inversion is greater than the potential temperature in the mixed layer. The term containing $\partial\theta/\partial x$ defines a positive acceleration when θ increases as the cold air moves over the warm lake surface. The acceleration term containing the drag coefficient C_D is always negative but diminishes in magnitude over the water.

The numerically computed steady-state solution of the resulting equations defines a deformation of the inversion surface which shows that the inversion surface height decreases as the air moves from the land out over the water and increases as the air moves from water to land. The greater the temperature difference

between land and water, the greater is the acceleration and divergence on the upwind side of the lake and the deceleration and convergence on the downwind side.

The solution of the partial differential equations of the model employ finite difference equations on a 40×50 point rectangular grid covering an area of approximately 1200×1200 km centered on Lake Erie. The axes of the grid system are oriented parallel and perpendicular to the long axis of the lake. Grid spacing is 12 km parallel to the long axis and 6 km to the cross-lake direction. Beyond the boundaries of Lake Erie, the grid spacing is increased exponentially to minimize the propagation of deleterious effects from the boundaries.

For verification of a mesoscale model in the case of November 15–16, 1969, Figure 8.4 gives a satellite picture of the intensive lake storm that developed on November 15, 1969. During the long duration of the storm, the axis of the band moved slowly back and forth between the northern and southern lake shores. Such motions appear to be a typical property of lake storms and are probably the consequence of variations of the upstream flow conditions.[22] At the time of the satellite picture, the radar return on the PPI scope of the National Weather Service 10-cm radar was recorded as illustrated in Figure 8.8*a*, where the isolines are lines of various attenuation levels. The heaviest precipitation at that time was located within the 18-dB attenuation contour. Upper air soundings showed the inversion height was about 1500 m upwind of the lake, whereas at Buffalo the height was 3000 m. Given the observed upstream rawinsonde potential temperature, and inversion height, and a lake temperature of 10°C, the model computed the deformation of the inversion (thickness of the mixed layer), as shown on Figure 8.8*b*. The isolines indicate an increase of the inversion level over the original 1500 m. Note that the maximum height of the inversion is now at about 3000 m. The levels, as

shown in Figure 8.8*b*, include the effects of the release of latent heat of condensation.[4]

The deformation of the inversion at the downwind lake end is of greatest importance for the development of precipitation. The clouds not only have a much greater depth and contain therefore more water to snow out, but also the fact that their tops reach into colder strata makes the precipitation mechanism more efficient.

One can fit a cumulus model under the inversion and can compute the precipitation that it yields. An attempt to predict the total amount of snowfall by this method has been made by R. L. Lavoie.[16]

The Seeding Hypothesis

It follows from the observed mesoscale structure of these storms, as well as from the numerical model, that the heaviest

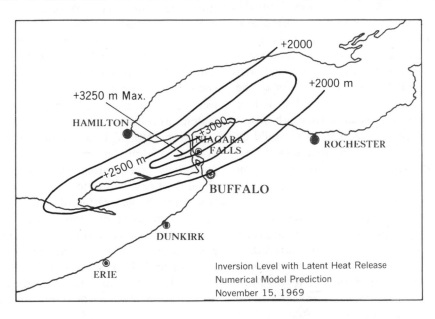

Figure 8.8b Deformation of inversion surface computed from mesoscale model with inclusion of heat of condensation.

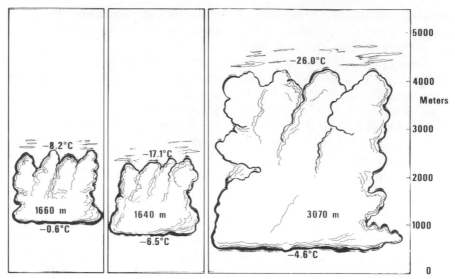

Figure 8.9 Schematic drawing of three typical lake storm cloud systems.

precipitation falls near the shore and decreases in intensity farther downwind. This is partly because of the diminishing intensity of the convective cloud systems as they move inland and away from the warm convective moisture source and partly because of the fact that the more intensive convective clouds produce heavy precipitation particles like graupel and rimed snowflakes which fall close to the shore. If one were able to produce lighter precipitation particles through over-seeding of the clouds, the snow would be advected over a larger area inland. We postulate therefore the following seeding hypothesis:

The precipitation process of severe winter lake snow storms can be modified through seeding in such a way that, instead of graupel and rimed snowflakes, unrimed flakes form, whose crystals have grown through diffusion. The modified flakes fall more slowly and are carried farther by the wind, thus shifting the bulk of the precipitation away from the coast and farther inland.

The Cloud and Precipitation Systems of Lake Storms

The Cloud Systems. Figure 8.9 illustrates three typical phases of the convective cloud systems over the Great Lakes. The physical parameters by which the three systems are defined, such as standard deviation or the range of the average conditions, cannot as yet be given. We were able to distinguish between three systems as follows.

System No. 1 is a nonprecipitating system which can be classified as potentially precipitating, inasmuch as the cloud thickness and water content are sufficient for the generation of snow or rain. The concentration of freezing nuclei is too low, and at the same time the cloud droplet concentration is high enough to minimize the probability of random freezing of large droplets so that an effective precipitation mechanism is not initiated. The system is colloidally stable. This system is most promising as a potential source for atmospheric water management. It is discussed below more thoroughly.

System No. 2 precipitates naturally. The cloud depth is comparable to System 1; however, the temperatures are colder with the cloud top temperature sufficiently cold to provide appreciable ice crystal nucleation. We have modified this system through seeding and its effects became apparent through aircraft observations of the cloud and surface observations of the snow crystal habits. We have been unable to detect the effects on the rate of precipitation with a snow gauge network. Very careful snow gauge measurements using a dense network may shed light on this problem; however, it is well known that precise measurements of the rate of snowfall are difficult to accomplish. To make things even more complex, areal effects of seeding are involved, resulting in the intensification of some snowstorms and the partial suppression of others.

System No. 3 is typical of severe winter lake storms. It is about twice as deep as the other two systems and is more extreme in its base and top temperature range. This system is the primary target for our seeding effort because it is the heaviest snow producer. To date, no No. 3 systems have been seeded.

The basic ingredient of all systems is a convective stratocumulus layer, which, in order to produce precipitation, must be thicker than 1000 m and must exist within a certain temperature range before natural freezing nuclei become effective. In many cases these layers are topped by one or sometimes several stratified stratocumulus layers 100 to 300 m thick. The uppermost of these layers is located at the inversion that is typical of these Great Lakes cloud systems as described above.

The origin of these upper stratocumulus layers is not well understood. Several possible mechanisms exist, such as (1) formation as stratocumulus cumu-

logenitus; (2) formation due to momentum transfer in a process similar to that involved in the formation of pileus clouds which develop when a cumulus penetrates an inversion, (3) the lifting of the inversion due to the mesoscale flow conditions such as have been modeled by Lavoie, and (4) radiative cooling and mixing.[9] It is probable that all four processes contribute in varying degrees. Process 1 may not directly result in the formation of clouds, but it may contribute to the transport of moisture into the layer just below the inversion, which then can reach saturation through the combined action of processes 3 and 4.

Most difficult to assess is process 2. We believe that in Great Lakes cloud systems, this process may play a role when the inversion is sufficiently high to permit the formation of a deep convective layer, that is, when the inversion is somewhat above the 2000-m level. In the more prevalent lake cloud systems that average 1600-m in depth, momentum transfer most likely plays a secondary role.

Mesoscale Structure of Lake Storm Cloud Systems. Infrared remote probing methods were used for the mesoscale survey of the lake storm cloud systems. An aircraft equipped with a downward looking radiometer sensitive in the spectral range of 10 to 12 μm flew above the cloud system and measured the cloud top temperature. Figure 8.10 shows such a plot for the weak lake storm of December 7, 1968. The cross-section was obtained on a northerly heading from Jamestown, New York, crossing Lake Erie to the Canadian shore. The abscissa gives time and location, the dotted line from 1720 to 1728 Universal Time (UT) represents the elevation of the underlying terrain; Lake Erie was crossed between 1729 and 1740 UT. The general direction of the airflow was from the west,

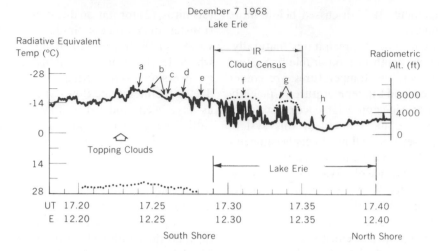

Figure 8.10   Infrared radiometer remote cloud census of lake storm on December 7, 1968. Abscissa given in Universal Time (UT) and local time (Eastern Standard Time).

that is, emerging from the plane of the drawing, but with an onshore component towards the south shore. The most northern line of cumulus clouds marked "g" appears at about lake center. Nearer the south shore, another line of cumulus clouds is evident with higher reaching tops, marked by "f." South of the southern lake shore the cloud system becomes a stratocumulus layer that still has considerable structure as the letters "e" to "a" indicate. There is an indication that the cloud tops rise with the underlying terrain features. In this area the clouds reach into temperatures colder than $-15°C$ which provided that natural precipitation was released.

The individual building blocks of these convective cloud systems are cumuli of several scales, from fair weather cumulus and cumulus mediocris, to cumulus congestus, and even to small cumulonimbus in which lightning and thunder occur. We have observed that cumulonimbus occur when the convective clouds of the lake storm reach into an upper altostratus canopy of a passing upper-level trough. Strictly speaking, the clouds are therefore not typical Cb, but Cu

congestus reaching into an upper layer of ice clouds. The convective energy stems from the temperature difference between the cold continental air and the warm water surface, enhanced by the normally steep lapse rate in the cold air.

Cases No. 1 and 2 (Figure 8.9) were the types more frequently experienced so far. They are typical of weak or moderate lake storms and are of such depth that cumulus mediocris form their building elements. The life history of cumulus mediocris has been described elsewhere.[28] Their life cycle consists of two major phases: a convectively organized phase followed by a convectively unorganized or turbulent phase. The formative or "bubble" phase of the cloud is characterized by the assent of a short-lived (5 to 10 min) vortex element. This is followed by a second or "coasting" phase, in which the potential energy generated in the form of liquid water is used up. The depth of the clouds characterized by this life history is most often 1500 to 2000 m, and their diameter 1000 to 5000 m, that is, the same characteristic dimensions as individual cells of lake storm clouds. The unequal number of observations of both phases support the suggestion that the lifetime of the coasting phase is 5 to 6 times the lifetime of the bubble phase.

McVehil[22] studied the wind field of these storm types and found the divergence to be about 10^{-4}/sec. The wind field divergence can be expressed as

$$\text{div } \mathbf{v} = \frac{\partial v_x}{\partial x} + \frac{\partial v_y}{\partial y} + \frac{\partial v_z}{\partial z} = -\frac{1}{\rho}\frac{d\rho}{dt}.$$

Since the right side of this equation is nearly zero, the updraft velocity can be computed by

$$v_z = -\int_{z_8}^{h}\left\{\left(\frac{\partial v_x}{\partial x}\right) + \left(\frac{\partial v_y}{\partial y}\right)\right\} dt.$$

The vertical speed at an elevation of 1000 m for a divergence of 10^{-4} becomes 10 cm/sec.

Suppose we consider a lake storm such as the December 3, 1969 storm which was observed by means of the Buffalo WSR-57 radar. After seeding, the storm band visibly disintegrated in the PPI presentation and became 8 to 10 individual showers with an average diameter of about 6 km. In agreement with the previous discussion of the life cycle of cumulus mediocris, one to two out of the eight shower clouds making up the band may be expected to be in their active stage. If these caused the mesoscale divergence of 10^{-4} over the area of the lake storm, they would have an updraft velocity of about 1 m/sec at 1000 m elevation. The remainder of the cumulus mediocris clouds would be coasting and dissipating, consuming their potential energy and liquid water content as the result of growth of the snow crystals which descend from the cloud tops. The presence of downdrafts would call for additional updraft velocities. Even though there are cloudfree areas, the relative humidity is likely to be everywhere above ice saturation, and crystals will grow wherever they fall.

Development of Precipitation. Of paramount importance for the formation of precipitation in a convective cloud system is, of course, the top temperature of the clouds. For temperatures warmer than about -14 to $-15°C$, the concentration of natural freezing nuclei is insufficient to produce precipitation other than a few graupel. These may fall mixed with heavily rimed snow crystals. Of course, the number of freezing nuclei that become active at a certain temperature depends also on the amount of water which is made available at this temperature. For this reason the flux of water (liquid water content times the updraft velocity) has an influence on the number of ice crystals developed. The concentration of freezing nuclei increases by about a factor of 10 for each $4°C$ temperature decrease. The natural threshold for the development of efficient precipitation appears to be set at temperatures between -15 and $-20°C$. Downwind of cities and industrial areas, anthropogenic nuclei exist which initiate precipitation at higher temperatures. We have observed such artificial snow showers when the top temperature was as high as $-10.8°C$. This case occurred on November 25, 1968, on which day a convective cloud layer existed over Lakes Erie and Ontario with the base at 900 m and $-2°C$ and the tops of 1500 m and $-10.8°C$. The radar echoes of the showers that developed downwind of Buffalo, Toronto, and Oshawa are shown in Figure 8.11. Also shown is the flight path of the aircraft below cloud base; the numbers along the flight path signify freezing nuclei concentrations per liter as measured with an NCAR (National Center for Atmospheric Research) counter.

Nuclei and Ice Crystal Concentration. Table 8.2 shows average concentrations of freezing nuclei as measured in aircraft below 400 m msl in three areas: over the

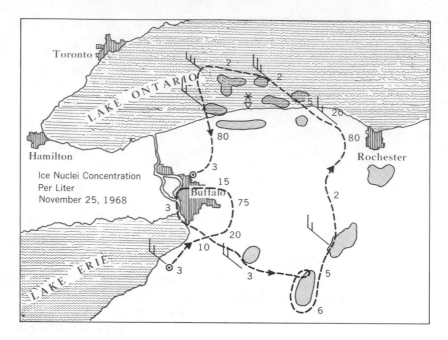

Figure 8.11 Anthropogenic snow showers downwind of urban and industrial pollution centers. November 25, 1968 freezing nuclei counts taken at −18°C.

Great Lakes region, over Colorado, and over the tropical ocean near Barbados, West Indies. The concentrations were obtained with a carefully calibrated NCAR freezing nuclei counter for a temperature of −20°C. Concentrations above the boundary layer are less than 1/liter while in the boundary layer the concentrations extend over 3 orders of magnitude when we compare the background value in an uncontaminated area with the values in industrial urban centers of the Great Lakes Basin. The significance of these variable concentrations is discussed more thoroughly in subsequent paragraphs.

The threshold temperature of −15°C given above for an efficient precipitation mechanism is, of course, no magic temperature, because it depends on the natural background of freezing nuclei concentrations. In uncontaminated air the temperature may be as low as −20°C. Only 1 freezing nucleus/150 liters was measured on February 12, 1971, over

western Lake Erie at 3000 m altitude. On the same day we were unable to detect an undersun in cloud tops as cold as −22°C, while on December 4 in eastern Lake Erie glaciation of stratocumulus was complete at −19.5°C, as observed from a bright undersun.

One unsolved problem is to determine the concentration of single crystals which nature provides to sustain an intensive snowfall. Conventionally, 1 crystal/liter is assumed to be required. Our observations indicate that this number is at least one order of magnitude too small. The correct concentration can be observed under conditions when snow crystals fall from a deep altostratus without being deformed by riming, diffusional growth in a water cloud, or aggregation. In a number of cases we have observed that in such deep cyclonic precipitation systems, about 5 to 20 single crystals fall on 1 cm² in 10 sec. with a fall velocity between ½ and 1 m/sec. This amounts to a range of concentration

Table 8.2 Average Measured Concentrations of Freezing Nuclei/Liter Effective at −20°C, Below 4000 m msl

	Background	Urban
Barbados, 1969	1–2	
Colorado	0.5	20
Great Lakes	5	180
Great Lakes (above inversion)	<0.5	<0.5

of 5 to 40 crystals/liter. For the concentration of spatial dendrites which form in a certain temperature range around −20°C we have found from one set of observations about 125 crystals/m³. Since their number not only depends on a temperature threshold but also on the volume of air rising past this threshold they may easily vary by a factor of 10 before the updraft becomes fast enough to take them along and discharge them somewhere else. Similar observations have been reported for orographic snowfall over the Continental Divide in the Rocky Mountains, when the 500-mb temperature of −20°C was related to the crystal concentration and updraft.

The Modification Potential of Nonprecipitating Cloud Systems in the Great Lakes Basin

Cloud System No. 1 occurs within a cloud temperature range in which nature does not develop precipitation, but in which seeding may produce intensive rain or snow showers. We have conducted a preliminary study of the past 10 years' climatology to determine, within the Great Lakes Basin, the frequency of occurrences of moist layers with top temperatures between −5 and −14°C. Figure 8.12 shows the result for Buffalo, New York: the abscissa is time, and the ordinate is the thickness of the moist layer. The columns thus illustrate occur-

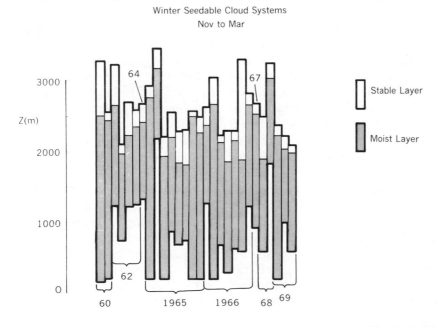

Figure 8.12   Frequency of seedable winter cloud systems 1960–1969, November to March, Buffalo, New York.

Table 8.3   Artificial Precipitation Potential—Winter, Great Lakes Basin 1960–1969 (Radiosonde Data)

	Number of Releases with Precipitation at Time of Radiosonde	Number of Occurrences of Artificial Precipitation potential at Time of Radiosonde	Mean Thickness from Base of Moist Layer to Base of Stable Layer (m)	Mean Lapse Rate (°C/100 m) of the Moist Layer to the Base of the Stable Layer
Flint	58	23	1598	0.42
Dayton	50	20	1767	0.45
Green Bay	53	35	1729	0.36
Buffalo	113	28	1766	0.47
Sault Ste. Marie	61	18	1704	0.38

rence and thickness of seedable cloud systems that are without release of precipitation.

Table 8.3 also gives the average layer thickness and lapse rate. We have, however, no data concerning the areal extent of the nonprecipitating cloud systems.

Visible Seeding Effects in Stratocumulus Layers. Previous seeding experiments aimed at dissipation of supercooled stratus and stratocumulus decks revealed that the simple Bénard cell model does not always apply.[1] When thin stable stratocumulus layers were seeded with dry ice from above, the seeded line was observed to spread homogeneously to both sides, usually at speeds from about 1 to 3 m/sec. The homogeneity of the spreading indicated that the in-cloud diffusion mechanism was a kind of Bénard cell circulation that distributed the seeding agent and the ice crystals formed by the seeding agent homogeneously throughout the cloud layer. When the cloud deck was convective (steep lapse rate) and thicker (> 600 m) the seeded line no longer spread homogeneously, but produced holes ir-

regularly lined up left and right along the seeded path. In these cases the cloud deck must have consisted of individual, though, "interlocked," clouds whose updraft velocities have exceeded the fall velocities of the artificial crystals; therefore, a cloud in its formative stage did not dissipate after seeding but carried the crystals aloft and ejected them laterally at its top. From here they could be deposited into a neighboring cloud whose updraft had ceased and which could readily be dissipated by the injected ice crystals. In the next section we return to this "self-cleaning" mechanism of the growing cloud, which is important for the formation of natural precipitation, and which causes a concentration of ice crystals in the top layer of the cloud system.

Figure 8.13 shows the updrafts and downdrafts measured with a vertically pointing Doppler radar during a lake effect storm on January 4, 1968, while heavy snowfall occurred at the radar site.[22] The observed Doppler velocities were adjusted to remove the downward velocity component of the precipitation particles. The section indicates a fairly well-organized pattern of alternating up and downdrafts. From the transit time across the radar beam, cell diameters of 1 to 2 km are obtained. Vertical velocities of 3 to 4 m/sec were frequently observed with maxima of up to 6 m/sec. The cloud depth in this storm was more than 2000 m, and the clouds appear in the cross-section of Figure 8.13 typically like cumuli that spread out along the inversion. The suggested three-dimensional cross-section through such a convective layer is shown in Figure 8.14.

This cloud pattern is consistent with an analysis of the divergence-convergence conditions for the lake storm of November 3–4, 1966.[21] Maximum values of convergence were 3×10^{-4}/sec. with the greatest convergence occurring below 1700 m, and the maximum divergence between 1700 and 3300 m.

It is noteworthy that an identical band structure has been analyzed for snow bands originating over the Japan Sea west of Ishikara Bay of the Island of Hokkaido by Z. Yanagisawa and M. Fujiwara.[30]

The reaction of a convective layer to seeding depends not only on its structure but also on the seeding method. Figures 8.15 and 8.16 show the areal spreading of two seeding patterns in stratocumulus on December 6, 1968. In Figure 8.15, seeding was conducted at the cloud base with 2100 g of silver iodide dispersed by means of pyrotechnic flares. In Figure 8.16 seeding was conducted at the cloud tops with dry ice at the rate of 10 lb/mi. The thickness

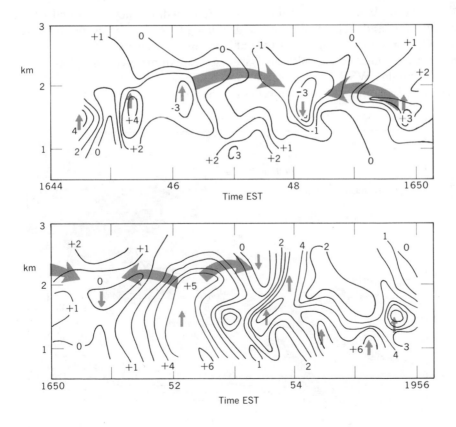

Figure 8.13 Up- and downdrafts in lake storm clouds measured in meters per second with a vertically pointing Doppler radar.

Figure 8.14 Schematic cross-section through convective stratiform cloud system.

of the cloud deck was 1650 m; temperatures at base and top were −6.5 and −14.5°C. The data of Figures 8.15 and 8.16 are based on a very careful analysis of the Buffalo WSR-57 radar which showed the development of strong echoes after seeding; they withstood an attenuation of 36 dB and were connected with intensive showers. The radar echo areas were planimetered and plotted in Figure 8.17 against time after the appearance of the first echo.

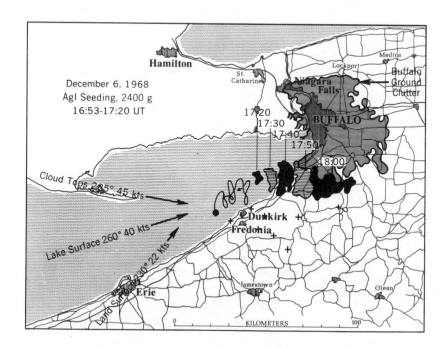

Figure 8.15 Radar echoes following AgI seeding of December 6, 1968.

Seeding in the cloud tops with dry ice causes the formation of a vertical "curtain" of artificial ice crystals resulting in a linear area spreading of the seeding effect. Seeding with AgI at cloud base requires about 20 min before the areal rate of spreading parallels that of the dry ice seeding. Consequently, in the case of AgI, much time is lost in the beginning with the upward diffusion of the seeding agent and its distribution throughout the entire cloud layer. Once this is accomplished the rate of areal spreading equals that of dry ice seeding. The ice crystal concentration in these cases 20 min after seeding was computed as 60/liter for the dry ice seeding and 10/liter for the AgI seeding.

Corresponding results on the rate of lateral spreading were obtained in another experiment when dry ice and AgI were introduced about 500 ft below cloud tops. The rate of spreading was measured by making repeated passes over the seeded areas.

Wind shear is an important factor in the propagation of the seeding effect throughout a cloud deck. This is shown in Figures 8.18 and 8.19. Figure 8.18 gives the effect of shear in velocity only, as observed on November 21, 1969. Here the cloud deck was 2000 m thick, and the base and top temperatures were -9 and $-22°C$, respectively. Wind directions at the top and bottom of the cloud system were identical, while the speeds differed as follows from Figure 8.18. The cloud deck was seeded at the base with 2400 g AgI from pyrotechnic flares during a 22-min period. The seeding path is shown in Figure 8.18. Shown also is a flight track flown about 1 hr later well above the cloud tops to observe the location of the intensive undersun which marked the glaciated area caused by seeding. We then transposed geometrically the seeding path downwind using the upper and lower cloud level wind speeds and obtained the hatched areas positioned within the flight path

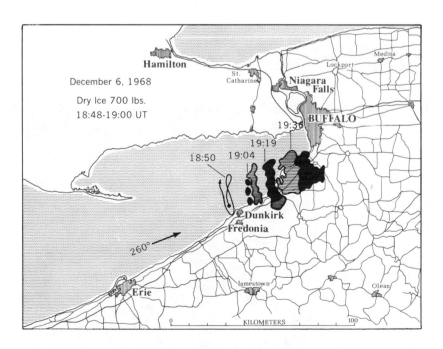

Figure 8.16   Radar echoes following dry ice seeding of December 6, 1968.

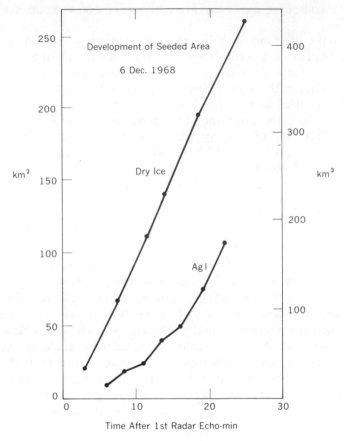

Figure 8.17 Development of seeded areas after seeding with dry ice and AgI on December 6, 1968.

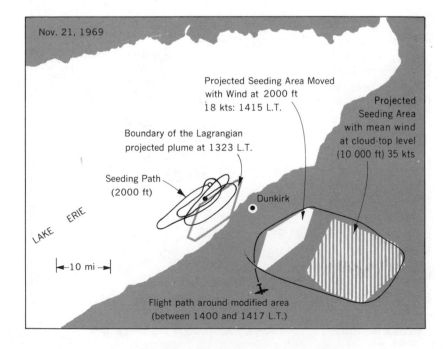

Figure 8.18 Seeding information for experiment on November 21, 1968.

which encircled the modified area. It is apparent that the boundaries of the modified volume were determined by both wind fields. The strong effect visible in the top layer showed that the seeding result diffused effectively upward where it traveled with the higher wind speed, releasing ice crystals and seeding the lower clouds.

A similar result was analyzed for the more complex case of shear in speed and direction. Figure 8.19 shows the wind field on December 3, 1969 at cloud base and cloud tops as well as the aircraft flight path. Base temperature was −4°C, top temperature was −18°C, and in-cloud seeding was accomplished at 4000 ft (1300 m) in two repeated racetrack patterns (see figure 8.20) using the acetone AgI

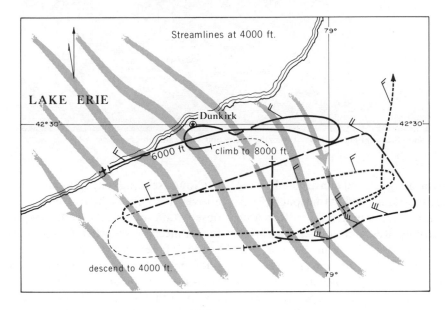

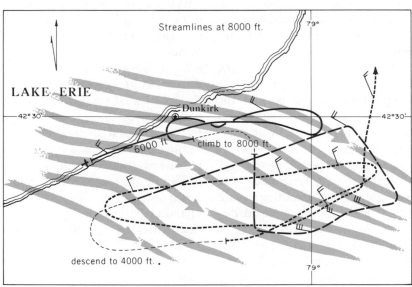

Figure 8.19 Seeding and analysis flight path on December 3, 1969. Shaded lines give low- and upper-level wind fields. Note the longitudinal scale has been expanded for clarity.

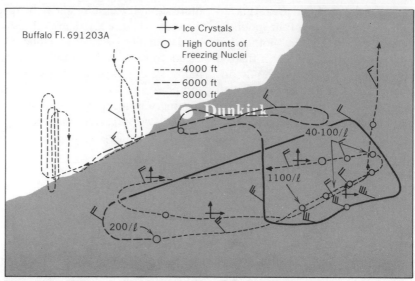

Figure 8.20   Flight analysis of freezing nuclei and ice crystal observations.

generator designed and constructed by the National Oceanic and Atmospheric Administration/Environmental Research Laboratories Research Flight Facility.[26] Cloud depth was 1850 m. Figure 8.20 shows observations of ice crystals and high freezing nuclei concentration along the E-W direction flight loops which were made while descending from 8000 to 4000 ft (2700 to 1300 m). It is particularly noteworthy that the high freezing nuclei concentrations were measured at the lower level in a region where, at that time, the seeding material could have arrived only by traveling an appreciable part of the distance at the speed of the upper-level wind. The air was likewise filled with "diamond dust," ice crystals, at all levels— without doubt a consequence of the AgI seeding. Locally spectacular results were observed almost 2 hr after seeding began, showing that the seeding agent had diffused into a volume of about 70 × 80 × 1.85 km, or 10,400 km³. It may be noted (Figure 8.20) that the true areal nuclei distribution showed considerable variation; this variation was also apparent in the surface snow crystal observations.[29]

Cloud seeding experiments in such convective layers have permitted the observation of lateral nonisotropic diffusion mechanisms when flying above the seeded layer and observing optical phenomena, such as undersuns, which show the location of the ice crystals. It was observed that streams of crystals diffused laterally away from a convective cloud top in the seeding line and moved in between the surrounding convective cloud elements.

Targeting of Seeding Effect.   The influence of wind shear on the seeding area becomes important if one attempts to affect a certain target area. Usually when a seeding experiment was carried out, an attempt was made to locate the seed in such a way that the seeded area drifted over the target area. An average wind for the depth of the cloud layer was determined from rawinsonde data; 20 to 30 min was assessed for the seeding to become effective in the production of precipitating snow crystals, and on this basis a seeding location was determined which was the center of a racetrack pattern with the long side 10 mi in length and perpendicular to the wind direction. Strangely enough, more often than not the seeded area drifted past the target area. While this discussion on the effect of wind shear explains part of the difficulty, it appeared to us that the seeding technique also was not without effect on the drift of the seeded area.

A case that indicates the different response of a cloud deck to dry ice or AgI seeding, with respect to the drift and development of the generated snow showers, was the experiment of December 6, 1968. The cloud base over the lake was at about the 300 m level; the solid convective layer reached to about 1500 m at $-10.3°C$ with tops of cumulus clouds above and below that level. A Sc layer reached from 1800 to 2100 m, at a constant temperature of $-11.7°C$. Figure 8.15 shows the echo development as observed with the Buffalo Weather Bureau WSR-57 radar when the cloud base was seeded along the flight path with 2100 g of AgI pyrotechnics from 1153 to 1220 local time.

From 1348 to 1400 local time the clouds were seeded with 700 lb of dry ice as we flew at the tops of the convective layer (1800 m) in and out of dense cumulus tops (Figure 8.16). In both cases the development of echo areas were carefully plotted using the WSR-57 records. The comparison of the echo areas indicate that the dry ice area drifts in the direction of the lower winds, while the AgI seeded area drifts rather with the upper winds. The drift of the seeding agent with the upper wind had also been observed on November 21 and December 3, cases which have been discussed above.

These circumstances must depend on the different dispersal methods for these agents: dry ice is dispersed just above the cloud tops in the form of pellets which, by falling through the cloud, affect the whole layer almost instantly like a seeding "cur-

tain"; AgI is dispersed along a line at cloud base, from where it has to be carried by diffusion and in-cloud updraft to the cloud tops where it is most effective. Lateral diffusion in the cloud top level is caused by either the Bénard cell circulation or a stratiform Sc layer or the general divergence of airflow near the cloud tops. The bulk of the ice crytals are generated in the atmospheric layer near the cloud tops. From here they are released to descend and snow out the convective Sc layer. Consequently, the seeding pulse will travel with the upper-level wind rather than with an average wind. When seeding is done with dry ice, each pellet will leave behind a high-density trail of ice crystals which will grow by diffusion to snow crystals and quickly aggregate to form flakes in agreement with the arguments of aggregational growth. These will be diffused throughout the entire layer and therefore travel with an average wind direction. This simplified mechanism will depend on other factors as well, such as the temperature levels in the cloud layer, the depth of the cloud layer, the degree of "convectivity," and the water content, in short, all the cloud system properties which affect the formation of precipitation.

The Microphysical Precipitation Model

As mentioned before, the natural production of precipitation elements in convective stratiform cloud systems is practically nonexistent as long as the top temperatures of the clouds are warmer than $-15°C$. Since the number of natural freezing nuclei increases by about a factor of 10 for every $4°C$ temperature decrease, the efficiency of the production of natural snow crystals increases rapidly as the temperature of the cloud lowers. At $-16°C$ the large vapor pressure difference between the ice crystal and the surrounding water cloud causes the ice crystal to grow rapidly and to develop plane dendritic branches. These branches are efficient collectors of cloud droplets and, depending on the rate of condensation and the liquid water content of the cloud, large, more or less heavily rimed crystals will develop. Under proper conditions riming alone can determine the growth process and graupel or soft hail will form. At convective cloud top temperatures of -20 and $-25°C$, the crystals formed are spatial dendrites, and their concentration has again increased by a factor of 10 or 100. Because of both the large surface area and the spatial arrangment of the branches, these forms are most efficient in scavenging from a cloud its vapor and liquid water content. The increased efficiency for removing water compensates for the increased rate of condensation and water storage which often accompanies deeper cloud systems.

If such clouds are seeded it is most likely that their rate of precipitation will not increase, but it should be possible to eliminate riming and to have the crystals grown only by diffusion. This will change other properties, particularly their fall velocity, and this in turn can lead to redistribution of the precipitation at the surface. Also, since the clouds are convective, seeding may release additional heat of fusion sufficient to modify the rate of condensation and the water storage in the cloud as well, thus initiating an increased rate of precipitation.

Interaction Between Cloud Dynamics and Microphysics.   The interaction between cloud dynamics and microphysics has been illuminated from various points of view. Lavoie et al.[16] fitted a Davis-Weinstein cumulus model under the inversion of the Lavoie mesoscale lake storm model and studied the development of precipitation in a natural and seeded cloud under consideration of entrainment and of the

evolution of snow crystal precipitation. The cloud model was integrated vertically in 100-m steps until the updraft velocity vanished. In contrast to deep cumulonimbus and maritime convective clouds, where a large portion of the precipitation is generated during the growth stage, lake effect cumuli develop only the "kernels" of precipitation during the growth stage.

To obtain an estimate of the precipitation potential of the cloud in this numerical model, the updraft velocity is set equal to zero throughout the cloud structure after the development of the cloud is complete. Each nucleated class of ice crystals that has been carried to the top of the cloud is then integrated downward (in order of the fastest falling to the slowest) through the cloud profile. The slower falling ice crystals encounter cloud liquid water contents that have been depleted by those crystal classes that have preceded it. The precipitation rates (melted) and crystal mass and geometry of each class are then predicted at the ground.

The most noticeable result of this numerical treatment was the remarkable difference that appeared between the natural precipitation and the seeded cases in the analysis of the conditions that existed on December 14, 1968: the natural rate of precipitation with particle concentrations below 50/liter turned out to be immeasurably small, while in the seeded case with particle concentrations between 50 and 100/liter, the rate became more than 25 m/m hr. Computer Z-values for the radar return differed by several orders of magnitude. Interestingly, this was at variance with the observations. The cloud line which was seeded on that day was under surveillance by an M-33 radar stationed nearby and in their report the authors state explicitly that "the time lapse movies of the radar showed no important effect from the seeding."[16] This indicates in our opinion that the theory seriously underpredicts the natural rate of precipitation. On this day Jiusto and Holroyd[10] reported that natural precipitation consisted of *heavy* graupel whose density must have been larger than the value of 0.12 g/cm³ used in the theoretical derivation. Furthermore, we believe that the rates of precipitation for the seeded case can be more realistically modeled than that for the natural case because the simple crystal habits which form under seeding conditions are numerically tractable while the complex natural habits are not.

Additional Microphysical Consideration. A slightly different point of view which invokes the need for complete glaciation of the cloud if the seeding hypothesis is to be verified has been studied by Jiusto[12] who postulated that a concentration N_C of crystals must exist which will cause complete glaciation (for snowfall redistribution). The condition is that the vapor extraction from the cloud equals the

vapor supply from the rate of co. sation:

$$N_C = \frac{dm/dt \ \text{(supply)}}{dm/dt \ \text{(extraction)}}$$

The time derivative of the ice supersaturation S_i in a mixed water-ice cloud is expressed by

$$\frac{dS_i}{dt} = \zeta_1 u - \zeta_2 N_C \left(\frac{dm}{dt}\right)_c - \zeta_3 N_d \left(\frac{dm}{dt}\right)_d$$

where u is the updraft velocity. The first term on the right side of the equation is the rate of condensation, the second term contains the water and vapor extraction through diffusional and riming growth of the ice crystals, and the third term involves the evaporation of cloud droplets in the presence of a mixed cloud. The cloud model is based on a steady state mesoscale updraft rather than a cumulus model. This may limit its applicability but on the other

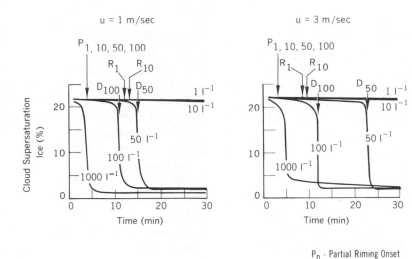

Figure 8.21 Overseeding conditions for various crystal concentrations and updrafts (Reprinted by permission, from Jiusto and Holroyd[10]).

...ure (°C)	Frequency (%)
..0 to −5.0	83
−5.1 to −10.0	9
Colder than to −10.0	8

hand the mesoscale interaction of a family of cumulus clouds embedded in a convective Sc layer has not been modeled.

Jiusto postulates that a seeding rate exists which converts a water cloud into an ice cloud. This seeding rate he defines as the "overseeding" rate. Practically, overseeding has occurred when the water vapor pressure of the cloud has decreased from water saturation to ice saturation. These conditions are shown in Figure 8.21. The initial properties of the cloud were as follows: $T = -20°C$; $P = 800$ mb; $S_i = 21.5$ percent; $N_D = 200/m^3$; $r_D = 7$ μm; $w = 0.43$ g/m^3. The figure shows the decrease of water saturation to ice saturation in the cloud as a function of updraft velocity (rate of condensation), u, and crystal concentration, n. Also indicated is the occurrence of certain growth conditions, such as P for partial riming, R for heavy riming, and D for diffusional growth. It is seen that with 1000 crystals/liter, the cloud as specified above will glaciate in about 5 min, whereas little difference exists between the 1 and 3 m/sec updraft conditions.

During the field experiments, we have succeeded in attaining crystal concentrations of 1000/liter by seeding, as postulated by the Jiusto theory. It should consquently be possible to redistribute the snowfall.

Observations during the field phases of the project have revealed a mechanism that was not foreseen as a complexity to the experiment at the inception of the project. This is the ability of nature to form snowflakes. Jiusto and Holroyd[10, 12] on the basis of surface observations of snow crystals and snowflakes precipitating from seeded clouds, postulated that snowflakes may form in both unseeded and seeded clouds, but that the flakes formed in the latter may fall faster because they consist of smaller crystals that will be more closely packed, that is, will have a greater density, than natural snowflakes.

The oldest reliable source of statistics on snowflake formation as a function of temperature is Dobrowolski's report of snow crystal observations in the Antarctic during the First Polar Year, 1897–1899.[3] He found a marked temperature dependence of snowflake occurrence as given in Table 8.4. The table conveys a strong dependence on the temperature which is confirmed by Magono,[19] who states that snowflakes form at temperatures warmer than −10°C.

Only a few laboratory investigations deal with the formation of snowflakes. As pointed out by R. E. Hallgren and C. L. Hosler[7] and by N. H. Fletcher[5] the formation of snowflakes is not only a matter of collision but also a matter of adhesion or, in my opinion, of interlocking of crystals of suitable shape. The process of adhesion depends on the temperature and the humidity of the environment. According to Hallgren and Hosler, the adhesion is strongly diminished if the crystals are brought into contact in a dry environment. Fletcher[6] assumes that within 10°C from the melting point, the surface is covered by a thin quasi-liquid film whose thickness at 5°C is calculated to be about 40 Å.

The interlocking of snow crystals of the dendritic and spatial dendritic types is supported by their fall attitude, which is mostly a rotating mode, and by the presence of a supercooled water cloud, that is, high humidity. On the basis of these considerations, one may expect that flaking becomes important in the temperature range of dendrites and spatial dendrites. This, however, does not necessarily apply to other crystals. Observations of prism bundles which form in a cirrostratus or altostratus cloud and have consequently a great fall distance and much opportunity to collide have a poor tendency to form flakes. Also, the humidity being at ice saturation may not support the adhesion process.

After we found through direct observation and theoretical analysis that aggregation of snow crystals into snowflakes is an important process and may prevent the occurrence of indivdual crystals due to seeding at least in the temperature interval from 0 to $-10°C$ and for certain crystal habits from 0 to $-20°C$, it was necessary to know how aggregation will influence the fall velocity of snowflakes, because the reduced fall velocity of the crystals after seeding was one of the prime assumptions of our snowfall redistribution concept.

It appears that nature has an almost limitless choice of combinations when it comes to the snowflake properties that influence their fall velocity: combinations of various crystal habits, of rimed and unrimed crystals, size, shape, and density of the flake. It is therefore not surprising that a clear velocity size dependency does not exist, that usually a range of fall velocities is given for a certain size, and that even inverse relationships exist with the fall velocity decreasing with size.

The experience of the observations of Japanese scientists has been condensed by Magono,[18] who gives analytical expressions for the fall velocities of rimed and nonrimed flakes, taking into account the aerodynamic form drag of the flakes as well as the drag caused by the air going through the open structure of the flakes. If

r is the radius in cm and the fall velocity u in cm/sec, then

$$u = 132 \left(\frac{r}{0.40 + 0.63r} \right)^{1/2} \quad \text{nonrimed flakes}$$

$$u = 194 \left(\frac{r}{0.45 + 0.60r} \right)^{1/2} \quad \text{rimed flakes}$$

M. P. Langleben[14] studied fall velocities from a smaller sample of observations, but he distinguishes flakes which are built from crystals having different habits. He expresses the fall velocity v (cm/sec) versus the melted diameter D of the flake (cm), and he finds, among others, the

$v = 218 \, D^{0.274}$ for flakes formed from plates and columns

$v = 178 \, D^{0.372}$ for flakes formed from dendrites, surface temperature 28 to 30°F

For flakes formed from rimed dendrites, surface temperature 25°F, he obtains $v = 210 \, D^{0.283}$. If these three equations are solved for $D = 0.2$ cm, that is, for equal mass of the flakes, one obtains in the order listed 140, 98, and 133 cm/sec, suggesting from this limited sample that flakes formed from plates and columns have a higher fall velocity than flakes from dendrites. Jiusto concludes from this observation, and the fact that seeding produced columnar snow crystals, that seeding may produce snowflakes with higher fall velocities than that of the natural flakes.

Suppose now the cloud is seeded and the large spatial dendrites have changed to many small plates which form a flake. Their density can be estimated from the experimental data of Hallgren and Hosler.[7] These authors find that a collecting ice sphere and crystals from 7 to 18 μ diameter acquire a flake density of about 0.02 g/cm³ at $-20°C$ and 0.054 g/cm³ at $-6°C$. This is in the same order de as the density of natural nowflakes and, consequently, ze difference will determine all faster. Apparently there is ertainty in our knowledge of nships that only actual quantitative measurements of the rate of snowfall before and after seeding will decide whether the original seeding hypothesis is valid.

SUMMARY OF SEEDING EXPERIMENTS

Seeding experiments were conducted during the winters of 1968, 1969, 1971, and 1972. Intensive lake storms (System No. 3) have not as yet been seeded. Seeding agents were dry ice, pyrotechnic flares burning an AgI compound, and an acetone mixture of AgI and NaI, released from the NOAA Research Flight Facility airborne generator or from the Cornell Aeronautical Lab aircraft. A network of surface snow gauges supplemented by mobile snow observation stations was available in 1969 in the southeast Lake Erie shore regions. Other facilities were a vertically pointing Doppler radar at Dunkirk, New York, the National Weather Service WSR–57 10-cm radar at Buffalo, New York, rawinsondes, an instrumented NOAA DC–6 aircraft, and a two-engine cloud physics and seeding aircraft belonging to the Cornell Aeronautical Laboratory. In 1968 Pennsylvania State University participated with an M-33 radar located at Jamestown, New York, and with a cloud physics research aircraft. The State University of New York participated with mobile field units for snow crystal research and with a field station at the Fredonia (New York) Lakeside Laboratory. The general target area for the project was the area southeast of Dunkirk, New York, and along the lake shores.

A complete discussion of these experiments and their results is given in Weickmann,[29] which summarizes in more detail the 4 years of field work conducted.

Even though it has been possible in a number of cases to trace and follow the seeding agent from the moment of release on its way through the cloud system to its final discharge and appearance in snow crystals, partly as the center nucleus and partly as AgI crystals scavenged along the branches of a dendrite, we did not obtain a quantitative measurement of precipitation released or redistributed from the cloud systems. This is mainly because of the great difficulties of measuring quantitatively the rate of snowfall under field conditions. The results obtained from these field experiments are pertinent, however, to the overall problem of weather modification and shall therefore be summarized.

Seeding Techniques

Whatever seeding method has been used, it was always made sure that the seeding agent was released at the cloud tops or in the cloud base in a way that without any doubt it became effective as planned. We observed and measured the diffusion of the seeding agent and of the artificial ice crystals throughout the cloud deck, using radar as shown in Figures 8.15 and 8.16, or airborne measurements of freezing nuclei, or simply eye observation of the seeding effect, when flying above the cloud deck. All these observations kept us convinced that the seeding agent was released properly and was diffused effectively throughout the cloud layer. This was also true for the case where the seeding agent was released from the surface and, drifting over the warm lake, ascended convectively into the cloud layer over the lake. Only in cases where multiple cloud layers occurred which were separated by inversions did silver iodide released at the cloud base fail to affect the entire cloud system and could become trapped underneath an inversion.

Results

Evaporation from the lakes of the Great Lakes Basin, particularly during the early and mid-winter, forms an abundant source of water vapor. This explains the frequent formation of supercooled stratocumulus layers over that area. These layers occur in all kinds of convective regimes, which appear to reach from a homogeneous Bénard cell type organization to a much looser organization between the individual cloud elements approaching interlocked, yet individual, cumulus clouds. Updrafts in the Bénard cell regime can be assumed to be consistent with the linear spreading velocity of a seeded line which has previously been found to be between 0.67 and 1.25 m/sec.[1] Updrafts in the more convective Sc layers have been found by Doppler radar measurements to be as great as several meters per second. Under these conditions even heavy seeding rates no longer lead to homogeneous spreading of a seeded line because the ice crystals are blown up to and are discharged sideways at the cloud top. If, as often occur, the convective layer is topped by a nonconvective thin Sc layer, this layer may become overseeded and will distribute the ice crystals over large areas of the convective layer. If this layer is missing, the crystals are "handed over" from individual cloud to individual cloud in an irregular fashion.

With respect to the response of stratocumulus layers of the Great Lakes Basin area to seeding, we may divide them into the following categories:

I. Convective layers, nonprecipitating
   A. Thickness less than 1000 m
   B. Thickness more than 1000 m
      1. Seeding rate about 10 nuclei/liter
      2. Seeding rate about 100 to 1000 nuclei/liter
II. Convective layers, precipitating (line storms)
III. Stable layers

The response of the category IA type of cloud system has been studied and described previously.[1] Within about 35 min of seeding (10 to 20 lb of dry ice had been used per mile) the cloud layer dissipates. The type of weather pattern which followed dissipation of about 170 km² on December 12, 1953, over Thunder Bay in Michigan is shown in Figure 8.22. The figure shows Thunder Bay, Michigan, taken from below cloud base after the overcast from the Sc layer had dissipated. It illustrates the man-made clear sky above; in the background one sees the last man-made snow showers (snow not reaching the surface) while sun shines in the foreground, initiating the formation of new cumulus clouds with bases far below the upper cloud deck. Most likely the moisture for these clouds comes from the downward transport of moisture due to snow-out of the upper clouddeck following seeding. Such man-made Cu clouds have frequently been observed.

Category IB is characterized by a thickness greater than 1000 m, by a steep lapse rate, and by its occurrence in all temperature ranges. This category occurs with or without a thin Sc layer on top and in a large convectivity range. It appears to respond differently to moderate and heavy seeding rates. If the rate is moderate and causes the formation of 10 to 50 ice nuclei, then moderate to heavy precipitation develops. In this case the crystal concentration agrees with that provided by

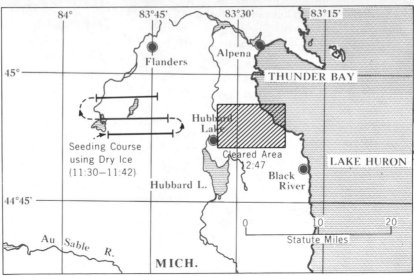

Figure 8.22 Thunder Bay, Lake Michigan, man-made weather panorama through artificial dissipation of stratocumulus deck, 750-m thick.

Figure 8.23 Artificial snow shower near Dunkirk, New York, over south shore of Lake Erie. Depth of Sc deck 1500 m, November 25, 1968.

Figure 8.24 Alternate view of snow shower on November 25, 1968. Sunshine behind the shower testifies to efficient snow-out.

Table 8.5 Initiation of Precipitation Table

Date	Convective Layer		Stratiform Layer		Seeding	
	Base	Top	Base	Top	Type	Result[a]
	m °C	m °C	m °C	m °C		
13 Nov 68	900 −7	2000 −9			650 lb dry ice, racetrack	$\overset{*}{\triangledown}$ 35 min A.S.
25 Nov 68	900 −2	1800 −9.5	2100 −10	2400 −10.8	250 lb dry ice, circle	$\overset{*}{\triangledown}$ 1 26 min A.S.
29 Nov 68	700 −2	1800 −10.8			450 lb dry ice, 3 legs	$\overset{*}{\triangledown}$ 1 1/3 in./hr
2 Dec 68	300 ±1	2400 −11			550 lb dry ice	$\overset{\cdot}{\triangledown}$ Over lake
3 Dec 68	1000 +1.5	2600 −8	2700 −11.3	2900 −12.3	2400 g AgI, fusees	$\overset{\cdot}{\triangledown}$ Over lake
6 Dec 68	300 −6.5	1500 −12.0	1800	2100 −14.5	2100 g AgI, fusees	$\overset{*}{\triangledown}$ 1−2 Over lake
					700 lb dry ice	$\overset{*}{\triangledown}$ 1−2 36 min A.S.

[a] Legend: $\overset{*}{\triangledown}$ means snow shower; $\overset{\cdot}{\triangledown}$ means rain shower; A.S. means after seeding; exponent 1 or 2 indicates intensity—1, moderate and 2, heavy.

nature in the Bergeron-Findeisen[2] process and it may be this circumstance which so effectively snows out the precipitation potential of these clouds. The man-made weather looks like Figure 8.23 or Figure 8.24. Both pictures are taken along the southern shore of Lake Erie looking east toward the rear of a man-made snow shower. Figure 8.24 indicates the effective snow-out as sunshine occurs "behind the front." The photographs were taken on November 25, 1968. The conditions for this day and alternate cases are listed in Table 8.5. Particularly well documented are the cases of November 25, 1968 and December 6, 1968. For both cases we have

the crystal concentration after seeding: on November 25 they were 20/liter, while for the December 6 case they are given on page 335). Both cases led to heavy snow showers; they were particularly conspicuous on December 6, on which day we penetrated with the aircraft the heavy snow squalls that had developed over Lake Erie, according to Figures 8.15 and 8.16.

Figure 8.25 illustrates category IB2. The picture was taken 1¼ hr after seeding ended, 64 km downwind of the seeding location (racetrack pattern 22 km long, 5 ½ km wide), with the size of the seeded area roughly 2 t × 21 km. The cloud deck was 1.2 km thick, and the total modified volume at that time was roughly 500 km³. With a total of 600 lb of dry ice seeded and a measured efficiency of 7.5×10^{10} crystals for the temperature range -8.5 to $-10.8°C/g$ of CO_2, the modified cloud volume at that time still had a concentration of 45 crystals/liter. Visual observation of the cloud formations in the area indicated that both cloud forming and cloud dissipating mechanisms had been at work and lines of intensively growing clouds formed about parallel to the long axis of the seeded racetrack pattern. The natural irregular convectivity of the cloud deck had become organized into bands whose clouds were formed by more powerful cumulus clouds than those forming naturally in the environment. This areal effect was, in our opinion, caused by two circumstances: a high seeding rate and the peculiar seeding pattern that evolves in the rapidly moving (28 knots) cloud deck. In this pattern, caused by the cloud motion and due to the aircraft flying in Eulerian coordinates and turning on one end of the racetrack with the wind and on the other against the wind, there are always two legs on top of each other, separated by an 8 to 10 km distance. Our experience with this kind of seeding pattern leads us to believe that in the heavily seeded curtains downdrafts form due to descending ice crystals, whereas cloud development is possible in the in-between spaces; cloud formation was even dynamically enhanced through the entrained ice crystals. This caused the intensive cloud lines to form.

Category II shows that when precipitation was already in progress, the radar echoes were usually arranged in the form of more or less well defined lines.

Figure 8.25   Area seeding effect on November 22, 1971. See text.

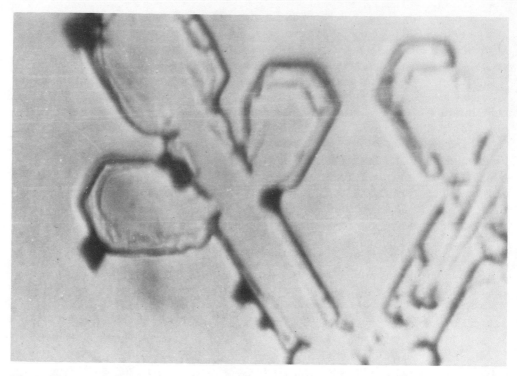

Figure 8.26   Individual AgI analysis on snow crystals collected during seeding experiment on December 3, 1969.

These line storms occurred in any intensity and with any cross-lake or along-the-lake wind direction. The moderate variety of such storms was seeded on December 7 and 14, 1968 and November 21 and December 3, 1969. On December 7 and 14, 1968, the radar echo would not indicate any modification in its pattern after seeding, even though on December 7 a large glaciated area could be observed from above the clouds, as well as special crystal habits on the surface as described by Jiusto.[12] A heavy snow shower was observed underneath the glaciated area which consisted of a great number of small snow crystals, typically caused by seeding, which were aggregated in flakes.

On December 14 the seeding effect could be traced across several stations of the snow gauge network through silver analysis in the precipitated snow,[27] but the radar again showed no effect from

seeding. On November 21 and December 3, 1969, the line storms formed on a cross-lake wind direction. Seeding effects were observed from above the cloud deck due to the glaciation of a large area, and from surface observers due to the change of snow crystal habit in heavy showers underneath the glaciated area, while the radar echo indicated decay of the storm lines through the transformation of a solid echo into several individual showers and a re-formation of the line storm to the west of the seeded one. There can be no doubt, however, that within the area affected by the seeding, snow showers occurred that were both particularly intensive and characterized by an especially high number of small snow crystals partially assembled in snow flakes. Investigations of the Formvar replica of the small crystals for AgI, following the Parungo-Rhea[25] method, indicated that they had formed on an AgI nucleus. Also, scavenging of AgI nuclei by naturally formed dendrites occurred (Figure 8.26). There can be no doubt that redistribution of precipitation was effectively initiated in these cases, but it is regrettable that quantitative precipitation measurements could not be obtained. We are convinced that statistical design of the experiment and randomization into seeded and nonseeded storms is unfeasible, since the definition of an experimental unit is not possible. The difference from storm to storm due to varying wind directions and shear within the cloud layer is so great and at the same time affects so profoundly the outcome of the seeding that we would not dare to treat them as one unit. We wonder to what degree this may be true in other randomized projects in weather modification as well.

In such cases as Category III, a cloud layer could consist of a number of individual layers which are separated by inversions. In these cases it makes a difference whether seeding is carried out from above with dry ice or from below with silver iodide; in the former case, the entire cloud deck becomes affected while in the latter case the transport of the seeding material from below is restricted. For dry ice seeding the seeding effect could be observed as being a short intensive snow shower after which no further modification takes place.

CONCLUSION

So far, we have not seeded a severe lake storm and have not tested the seeding hypothesis. What can be said is that we are well able to affect a storm in the direction postulated in the seeding hypothesis, but that it is not clear what kind of response the storm will give. It is possible that the crystals aggregate and that little changes compared to nonseeded storms, but the aggregation of snow crystals is not such a certain process, that the seeding hypothesis should be abandoned. We have repeatedly observed the downwind redistribution of snowfall, a fact which should be established for severe lake storms before a final verdict can be spoken, It seems to us that by our open-minded experimentation we have made another important observation: the significance of the shallow precipitation systems for winter precipitation in general and as a potential source of artificial precipitation enhancement specifically.

This research constitutes one of the few studies available which have shown unambiguously without post-experimental data manipulation that artificial precipitation can be generated and that the cases in which "nature misses her chance" do occur often enough to warrant their continued exploration. The need for two future studies is clearly outstanding: (1) to establish the frequency of occurrence and the areal extent of the shallow precipitation systems as a potential atmospheric water resource and (2) to determine quantitatively the amount of artificial precipitation potential of such systems.

These precipitation systems belong to the phenomena of the planetary boundary layer and are therefore usually less than 2000 m thick. Their artificial precipitation potential can be estimated from the rate of condensation and the assumption that the efficiency of the artificial precipitation is 50 percent. The potential rate of precipitation is given in Table 8.6.

As mentioned in the body of the paper the shallow cloud systems are a common phenomenon in regions where air moves over warm bodies of water, where it moves over large upslope distances and reaches the condensation level, or where it is embedded in a large cyclonic system. The Great Lakes Basin is a particularly attractive region because of its large bodies

Table 8.6   Estimate of Induced Rate of Precipitation

Geographic Area	Stratus Cloud Thickness (m)	Temperature Range (°C)	Updraft Velocity (cm/sec)	Induced Precipitation (mm/hr)
	3000	0 to −22	10	1.73
Great Lakes	3000	0 to −22	50	8.05
	2000	0 to −15	50	4.50
Colorado	2000	0 to −10	10	0.46
(Front range)	2000	0 to −10	50	2.30

of warm water in early winter, and because of the high value of freshwater supplies to its cities. These supplies are currently limited because of the tightly controlled hydrology of the interconnected lake bodies. The city of Chicago, for instance, is allotted an amount of 1700 ft^3/sec; the present cost to the consumer is $2.06/1000 ft^3. Consider Lake Michigan only. If 1 in. of rain could be produced in a 20-km wide area running the length of the lake (500 km), we would produce a 30-day supply of water for Chicago which at present costs the consumer about $10 million! Such an artificial precipitation enhancement effort would do little more than recycle the water which evaporates along the upwind side of the lake back into the lake along its downwind side. Exploration of this potential appears to be well worth the effort.

Lake storm cloud systems normally constitute the spender clouds identified in Tor Bergeron's[2] classification of precipitation systems. Efficient rain-out and snow-out of a cloud system require a releaser cloud which releases the precipitation particles in the form of ice crystals generated in a cirrostratus or altostratus cloud. The concentration of these particles, according to our observations, is within the range of 5 to 40/liter. In the Great Lakes region around Buffalo such concentrations can easily be reached for cloud temperatures of −20°C, since the peculiar life history of these convective systems causes a concentration of ice crystals in the top layer. Since the air is cleaner and the top temperature higher in the High Plains region it is quite conceivable that the available crystal concentration becomes insufficient, so that artificial seeding will materially increase the rate of precipitation. This is certainly the case where the top temperature increases to temperatures warmer than −15°C. We have further made the point that within the range of temperatures from −15 to −25°C, nature produces the most efficient crystal habit for snow-out which may well compensate for large deficiencies of crystal concentrations. It is safe to say that snow-out processes are generally more efficient than rain-out processes. This makes warm convective clouds more attractive for artificial rain production than cold convective clouds. In between lie the shallow cold precipitation systems that constitute a largely unexplored source of artificial precipitation. What makes this source attractive is the scale over which it can be applied: it is not limited to the one suitable convective cloud in a population of many but it can be applied over mesoscale areas which are covered by the proper shallow cloud system.

We feel we have opened a door into a new potential of precipitation enhancement. We know that a much greater effort than we were able to muster is necessary if we want to exploit the space beyond.

REFERENCES

1. aufm Kampe, H. J., J. J. Kelly, and H. K. Weickmann, Seeding experiments in subcooled stratus clouds, *Meteorol. Monogr.* **2**(11), 86–111, 1957.

2. Bergeron, Tor, uber den Mechanismusder ausgiebigen Niederschläge, Berichte, Deutsch. Wetterd., U.S.-zone (Bad Kissengen) pp. 225–232.

3. Dobrowolski, A., Résultats du voyage du S. Y. "Belgica," *Meteorologie,* La neige et le givre, Anvers, Belgium, 1903.

4. Eadie, W. J., R. J. Pilié, and W. C. Kocmond, Modification and modeling of lake-effect weather, Final Report, Contract E22–39–70(N), NOAA, Cornell Aeronautical Lab, Inc., Buffalo, N.Y., 1971, 101 pp.

5. Fletcher, N. H., *The Physics of Rainclouds,* Cambridge University Press, 1962, 386 pp.

6. Fletcher, N. H., *The Chemical Physics of Ice,* Cambridge University Press, 1970, 271 pp.

7. Hallgren, R. E., and C. L. Hosler, Preliminary results on the aggregation of ice crystals, *Geophys. Monogr.* (AGU) **5**, *Physics of Precipitation,* H. Weickmann, Ed., AGU Publ. No. 746, Waverly Press, Inc., Baltimore, Md., 1960, 435 pp.

8. Holroyd, E. W. III (1971), Lake-effect cloud bands as seen from weather satellites, *J. Atmos. Sci.* **28**, 1165–1170, 1971.

9. James, D. G., Observations from aircraft of temperatures and humidities near stratocumulus clouds, *Quart. J. Royal Meteorol. Soc.* **85**, 120–130, 1959.

10. Jiusto, J. E., and E. W. Holroyd III, Great Lakes snowstorms, Part I. Cloud physics aspects, Atmospheric Sciences Research Center, State University of New York, Albany, N.Y., Grant No. E22–49–70(G), ESSA-APCL, 1970, 142 pp.

11. Jiusto, J. E., D. A. Paine, and M. L. Kaplan, Great Lakes snowstorms, Part II. Synoptic and climatological aspects, Atmospheric Sciences Research Center, State University of New York, Albany, N.Y., Grant No. E22–49–70(G), ESSA-APCL, 1970, 58 pp.

12. Jiusto, J. E., Crystal development and glaciation of a supercooled cloud, *J. Res. Atmos.* **5**, 69–85, 1971.

13. Jiusto J. E., and M. L. Kaplan, Snowfall from lake-effect storms, *Monthly Weather Rev.* **100**, 62–66, 1972.

14. Langleben, M. P., The terminal velocity of snowflakes, *Quart. J. Royal Meteorol. Soc.* **80**, 174, 1954.

15. Lavoie, R. L., A numerical model of the atmosphere on the mesoscale with application to lake effect storms, Ph.D. Thesis, The Pennsylvania State University, University Park, Pa., 1968, 102 pp.

16. Lavoie, R. L., W. R. Cotton, and J. B. Hovermale, Investigations of lake effect storms, Dept. of Meteorology, Pennsylvania State University Final Report to ESSA-APCL, Boulder, Colo., Contract No. E22–103–68(N), 1970, 127 pp.

17. Lavoie, R. L., A mesoscale numerical model of lake-effect storms, *J. Atmos. Sci.* **29**, 1025–1040, 1972.

18. Magono, C. (1953), On the growth of snowflakes and graupel, *Sci. Rept., Yokohama Nat. Univ. Sect. I,* 18–40, 1953.

19. Magono, C. (1960), Discussion remark, *Geophys. Monogr.* AGU **5**, *Physics of Precipitation,* H. Weickmann, Ed. AGU Publ. No. 746, Waverly Press, Baltimore, Md., 1960, p. 263 (435 pp.).

20. McVehil, G. E., and R. L. Peace, Jr., Project Lake Effect—A study of interactions between the Great Lakes and the atmosphere, Report No. 1, Contract No. DA–28–043–AMC 00306(E), Cornell Aeronautical Lab, Inc., Buffalo, N.Y., 1965, 65 pp.

21. McVehil, G. E., J. E. Jiusto, R. A. Brown, and R. L. Peace, Jr., Project Lake Effect—A study of lake effect snowstorms, Final Report No. E22–49–67(N), ESSA, Cornell Aeronautical Lab, Inc., Buffalo, N.Y., 1967, 80 pp.

22. McVehil, G. E., C. W. C. Rogers, and W. J. Eadie, The structure and dynamics of lake-effect snowstorms, Final Report No. E22–89–68(N), ESSA, Cornell Aeronautical Lab, Inc., Buffalo, N.Y., 1968, 49 pp.

23. Mielke, P. W. Jr., L. O. Grant, and C. F. Chappell, Randomized orographic cloud seeding results for eight wintertime seasons at Climax, Colorado, *Second National Conference on Weather Modification, April 6–9, 1970, Santa Barbara, Cal.,* 1970, pp. 66–69.

24. Paine, D. A., and J. E. Jiusto, Synoptic vs microscale influence on Great Lakes snowstorms, *Conference on Cloud Physics, Aug. 24–27, 1970, Ft. Collins, Colo.,* 1970, pp. 191–192.

25. Parungo, F., and J. O. Rhea, Field use of a simple technique for identifying silver iodide particles as snow crystal nuclei, *J. Appl. Meteorol.* **9**(4), 651–656, 1970.

26. Patten, B. T., J. D. McFadden, and H. A. Friedman, NOAA Research Flight Facility

capabilities for weather modification research. Part III. Airborne cloud seeding systems developed and utilized by the Research Flight Facility, *Proceedings, International Conference on Weather Modication, Canberra, Australia, Sept. 6–11, 1971,* pp. 358–360.

27. Warburton, J. A., and M. S. Owens, Silver analyses in lake effect studies—Lake Erie experiments, 1968, Contract No. E22–9–69(N), APCL-ESSA, Desert Research Institute, University of Nevada, Reno, Nev., 1969, 58 pp.

28. Weickmann, H. K., A. R. Tebo, and F. R. Jones, Temperature and humidity conditions in cumulus mediocris, *Conference on Cloud Physics, Aug. 24–27, 1970, Ft. Collins, Colo.,* 1970, pp. 177–178.

29. Weickmann, H. K. The Modification of Great Lakes winter storms, *NOAA Tech Report,* ERL 265–APCL 26, 1973, 103 pp.

30. Yanagisawa, Z., and M. Fujiwara, Structure of weak snowband in relation with propagation mechanisms as revealed by radar, *Conference on Cloud Physics, Aug. 24–27, 1970, Ft. Collins, Colo.,* 1970, 193–194.

9 | Fog

BERNARD A. SILVERMAN

ALAN I. WEINSTEIN

The fog comes on little cat feet. It sits looking over harbor and city on silent haunches and then moves on.

Carl Sandburg

Fog, a source of inspiration to the poet, water to the farmer, and frost protection to the orchard grower, is also a major source of inconvenience to the traveler. Despite the technological advances made in electronic aids, fog continues to be the most serious hazard to navigation in the air, on land, and at sea (Figure 9.1). Modern high-speed transportation systems are periodically brought to a complete standstill by sieges of dense fog, dramatizing the fact that these sophisticated systems are no match for this insidious weather phenomenon. More important than the inconvenience it causes, fog is responsible for a large percentage of the transportation industry's losses in life, property, and revenues.

The advent of the jet age has made the fog problem of particular concern to aviation. The ever-increasing density of air traffic, especially jet aircraft with limited endurance, is placing increasing demands on the acceptance rates of airports. Military aviation, in particular, demands the assurance of uninterrupted operations. In recent years, the Department of Transportation has found it necessary to limit the number of flights at the nation's busiest airports during periods of low ceiling and visibility. The loss of revenue by one fog at a major airport, because of aircraft diversions, delays, and cancellations, is estimated to be $100,000. The cost of one fog occurrence in the era of jumbo jets is expected to rise to $500,000. Militarily, the cost is frequently measured in lives rather than money.

The oceangoing vessel is probably the oldest victim of fog. Modern ocean liners and freighters are as vulnerable to the perils of fogbound harbors, canals, and sea lanes as were the Spanish galleons of old. During the winter months, as much as 2 hr/day of shipping through the Panama Canal are lost due to fog. When sea transportation does proceed in fog, some of the most catastrophic shipwrecks can occur. In 1956, for example, the luxury liner Andrea Doria collided on its maiden voyage with the Stockholm off New York and sank. The toll was 51 human lives and millions of dollars in property.

The automobile is the primary land victim of fog. Some of the most spectacular multiple car collisions, involving up to 100 vehicles, have been caused by fog. Costs in excess of $300 million/year are incurred by fog-associated accidents

Bernard A. Silverman and Alan I. Weinstein are attached to the Air Force Cambridge Research Laboratories, Bedford, Massachusetts.

Figure 9.1 A blanket of fog rolls past the Golden Gate Bridge into San Francisco Bay. The thick, murky cloud ties up transportation, slowing down automobile, ship, and airplane traffic.

on the nation's highways. This figure is expected to rise as the number of cars and the cost of repair continue to increase.

Considering the dependence of our society on its transportation systems, it is no wonder that fog has historically been the focus of attempts to modify it. Fog was, in fact, the subject of the first scientifically designed weather modification effort of any kind. In a milestone program that was disclosed to the public in 1938, Houghton and Radford[5] presented several reasonably feasible methods of dissipating warm fog over airports. Some successful experiments were conducted but not followed up, because it was believed that instrument landing systems would make fog dissipation unnecessary.

The seriousness of the fog problem during World War II prompted the English to develop a thermal dissipation method called "FIDO." Further developments of the FIDO system in England and in the United States were abandoned after 1953, however, mainly because it was considered too expensive to warrant routine use by commercial aviation. With the advent of the jet age, the fog problem became acute again, and activity in warm fog dispersal research was intensified. Although the basic concepts of warm fog dispersal that are being pursued are not new, modern engineering technology is being employed in an effort to make these techniques practicable.

Another landmark event in the history of fog dispersal and, indeed, the science of weather modification in general occurred in 1946 when Dr. V. J. Schaefer and his co-workers at the General Electric Company demonstrated that snow and subsequently clearings could be produced in supercooled stratus clouds by seeding with dry ice pellets.[12] The theory underlying their results has formed the basis for much of the weather modification efforts that have been pursued since that time. The only reliable operational application of this theory achieved thus far is supercooled stratus clouds and fog dissipation.

Sound weather modification practice requires a thorough understanding of the natural weather phenomenon. It is, therefore, essential to review, at the outset, the state of knowledge on the morphology of fog. The physical basis and requirements of fog modification are considered against this background. The various methods of fog modification are then discussed in some detail, the chapter concluding with an outlook on the future prospects of fog modification technology.

MORPHOLOGY OF FOG

According to accepted definition, the name *fog* is given to any cloud that envelops the observer and restricts his horizontal visibility to 1000 m or less. It is composed of numerous minute water droplets or ice crystals in colloidally stable equilibrium with their environment. Unlike most other cloud forms, fog can persist from a few hours to several days dissipating naturally under the influence of strong vertical mixing or solar heating.

Fog Types

Fogs are traditionally classified according to the cause of their formation. Saturation of the air in the presence of sufficient condensation nuclei near the earth's surface is necessary for fog formation. Since condensation nuclei are abundantly available in the atmospheric boundary layer, condensation commences and fog forms when the air is brought to its saturation point either by cooling to its dew point or by increasing its moisture content. There are numerous moisture

transport and cooling mechanisms in the atmosphere which give rise to fog, each process being associated with a generic fog type. Since the classification of fogs by causal mechanism is of more importance to the science of fog forecasting than fog modification, only a brief description of these fog types is presented here. The reader is referred to the treatise by Byers,[1] from which the following summary has been extracted, for a comprehensive description of the traditional classification of fogs. A categorization of fogs that is more germane to fog dispersal technology will be presented thereafter.

Fogs formed primarily by ambient air cooling are called air mass fogs. Such fogs are broken down further according to cooling mechanism into advection, radiation, and upslope fogs. More frequently than not, more than one cooling mechanism is operative, with one process being dominant.

Advection-type fogs are formed when there is a transport of air between regions of contrasting temperature. Such fogs are, by their very nature, coastal and open water phenomena. Warm, moist air moving at gentle to moderate velocities over a water surface that becomes progressively or suddenly colder downwind, provides the conditions that are favorable for its formation. Such conditions occur when (1) air from the summer-heated land is transported over cooler coastal waters on the east coast of continents or over large inland bodies of water; (2) warm ocean air is transported over a cold ocean current; (3) warm tropical air is transported poleward over the ocean. Steam fog, on the other hand, occurs when cold air from a chilled land mass in winter is transported over the warmer water surfaces of oceans, lakes, and rivers.

Radiation fogs are formed when stagnant moist air near the ground becomes progressively cooler during a cloudless night because of an excess of outgoing radiation. Valleys are particularly subject to radiation fog. Air that is cooled at higher elevations drains into the valleys where it accumulates, resulting in dense fog as radiation cooling lowers the temperature of the air further.

Upslope fog is the air mass fog which results when stable air is adiabatically cooled to its saturation point as it is gradually orographically uplifted.

Fogs formed by bringing air to its saturation point by the addition of water vapor are called frontal fogs. When warm rain falls through cold air at a frontal surface, the air becomes saturated by the evaporation of the rain. If the air through which the warm rain falls is initially unsaturated, it will be cooled to its dew point by the evaporation process.

From the standpoint of dispersal, fog can be classified into three general types according to its constitution and temperature, that is, ice fog, supercooled fog, and warm fog. This classification system is more relevant to fog dispersal than that based on causal mechanism, since the method of modifying each type is quite different. The differences in a given fog type due to the various ways that it forms may only require some variation in the application of the modification method.

Ice fog is a suspension of ice particles that occur at very low temperatures during clear, calm conditions. Ice fog is rare at temperatures warmer than $-20°F$ ($-28.9°C$) and increases in frequency and density with decreasing temperature until it is nearly always present to some degree at air temperatures of $-50°F$ ($-45.5°C$) in the vicinity of water vapor sources. Such sources are the open water areas of streams and rivers, herds of animals, but especially the multiple sources of moisture associated with man-made activities. Because of the very cold temperatures, the addition of only very small amounts of water vapor are sufficient to bring air to its saturation point and form ice fog.

Supercooled fog is composed of water droplets that exist at below-freezing temperatures. Although bulk water freezes at $32°F$ ($0°C$), water droplets have been observed to remain in a liquid state at temperatures as cold as $-40°F$ ($-40°C$), the temperature at which pure water droplets freeze. Ice crystal formation is inhibited by the lack of suitable ice embryos in the atmosphere. It has been found that more than 80 percent of the clouds warmer than $15°F$ ($-9.4°C$) contained liquid, but nearly half of them were mixed liquid and ice. By $-5°F$ ($-20.5°C$), only 10 percent were liquid clouds, although 30 percent contained both supercooled drops and ice crystals.

Warm fogs consist of water droplets at above-freezing temperatures. They are both colloidally and thermodynamically stable. They are the most common type of fog and the most difficult to artificially disperse.

Climatology

Fog is largely a localized weather phenomenon. The moisture and cooling that are required for its formation are greatly influenced by local geographical and meteorological conditions. The spatial distribution of fog frequency is, therefore, highly discontinuous, with great changes in fog frequency occurring over small distances. A reasonably true account of fog occurrence can be attained only through observation. Because of a pilot's and ship captain's need to see in order to navigate, fog occurrence is best documented at airports, lighthouses, and harbors, and in ships' logs.

Figure 9.2, taken from recently published data by Guttman,[4] shows the worldwide frequency of fog days. A fog day is here defined as occurring whenever the visibility falls below ⅝ mi some time during the day. It can be seen that fog is primarily a coastal phenomenon. With the notable exception of western Europe, the regions of the world where fog occurs most frequently are almost all coastal. They form mainly as a result of cooling of the warm, moist ocean air as it passes over cold ocean currents. The two foggiest regions of the world, the west coasts of South America and Africa, are dominated by the cold Humboldt and Benguela currents, respectively. To a lesser degree, the Canaries and West Australian currents create fog conditions off the northwest coast of Africa and the western coast of Australia, respectively. All these regions experience the same frequency of fog in each season. The California current is responsible for the high fog frequency on the California coast in the summer months. In all cases, the fog occurs almost exclusively as warm fog.

The prevalent summer sea fogs in the Grand Banks area of Newfoundland are caused by the passage of air from the warm Gulf Stream waters to the cold ocean currents in the vicinity of the Banks. A similar juxtaposition of the warm Japan current and the cold currents from the Bering Sea produce an area of high fog frequency in the sea between Japan and Korea during the summer months. Similar conditions are also found off the southeastern coast of South America where the warm Brazil current meets the colder waters from the west. In all these cases, the fog is also predominantly warm.

Tropical air fogs occur over western Europe in winter as the warm marine air is cooled as it passes over the cold continent. Radiation fogs also occur in the valleys of western Europe during winter. Both types of fog are occasionally supercooled. Other areas where fog occurs frequently and the temperatures are below freezing are the northwestern United States, Alaska, and Greenland.

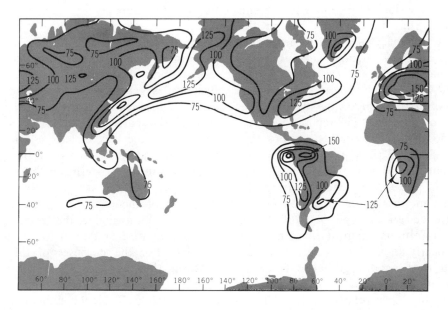

Figure 9.2   Worldwide fog climatology. Isopleths represent the number of days per year on which the visibility falls to ≤ 5/8 mi some time during the day.

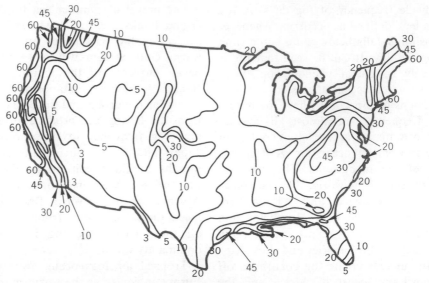

Figure 9.3   United States fog climatology. Isopleths represent the number of days per year on which the visibility falls to ≤ 1/4 mi (reprinted by permission, from Court and Gerston[2]).

Some of the fogs in the far northern latitudes are undoubtedly ice fogs.

Figure 9.3 shows the frequency of occurrence of days with fog in the United States. In this figure fog is defined as restricting visibility to ¼ mi or less. It can be seen that the regions having the highest frequency of fog are the Pacific Coast, the New England Coast, the Appalachian valleys, and the Pacific Coast valleys, all of which experience more than 60 days of fog/year on the average. The mean fog frequency per reporting station for 256 first-order weather stations, approximately 50 percent of which are air terminals, is 27 days/year. The 20 airports in the United States having the greatest number of air carrier operations experience an average of 2700 hr of below-minimum visibility weather per year, nearly all of which is caused by fog.

Fog that occurs at below-freezing temperatures accounts for only 5 percent of the fog occurrences in the United States. These fogs are restricted to the

northern latitudes in winter. Warm fog accounts for the other 95 percent of the cases.

Physical Structure

Despite the research interest in fog phenomena during the past three decades, the amount of consistent data on fog structure is generally meager. Measurements of the vertical structure of fog are particularly sparse. The deficiency stems partly from the lack of adequate instrumentation and partly because of the great variability in fog properties with fog type and age. While statistically valid measures of the physical characteristics of fog are not available, the specification of representative or typical fog values is possible. For the purposes of this discussion, the fogs are categorized according to their constitution; that is, ice and water fogs. Table 9.1 summarizes the physical properties of these fogs.

Most of the published observations on ice fog were obtained at various locations in Alaska.[10] The ice fog was concluded to result almost entirely from water vapor discharged into the atmosphere by human activity, that is, heating plants, power plants, moist air vents, vehicle exhausts, and the exhausts of oil and coal space heaters. The concentrations, size distributions, and solid water content of the ice-fog particles varied from place to place, depending on the local temperature, humidity, and moisture supply rate. The ice particles that formed were of three principal types: hexagonal plates, prisms, and droxtals. A droxtal is a tiny spherical ice particle, about 3 to 10 μ in diameter, that is formed by the direct freezing of supercooled water droplets at temperatures colder than $-22°F$ ($-30°C$). The concentration of ice crystals increased from about a few particles/cm^3 at $-22°F$ ($-30°C$) to 700 particles/cm^3 at $-49°F$ ($-45°C$). The range of equivalent water content of ice fogs was found to be 0.01 to 0.18 g/m^3 with a mean value of 0.1.

When the air temperatures are in the upper range for ice formation, that is, $-4°F$ ($-20°C$) to $-22°F$ ($-30°C$), the water vapor condenses rapidly as relatively large ice crystals which soon fall out. The concentration of the ice crystals is usually too low to reduce the visibility significantly. At temperatures lower than $-22°F$ ($-30°C$), small ice crystals and droxtals are formed which remain suspended for longer periods. The relative number of droxtals increases rapidly with decreasing temperature and are responsible for the low visibilities associated with ice fog.

There is no distinction between the physical properties of supercooled and warm fogs in the atmosphere. The physical properties of a water droplet fog,[6] whether it be warm or supercooled, are related to the method of formation, however. Radiation fogs that are characteristic of inland valleys tend to have a high concentration of small droplets. Advection fogs that are typical of coastal and oceanic regions have, on the other hand, lower concentrations of relatively large droplets. It is not uncommon to find drizzle falling out of advection fogs that have existed for several hours. These differences between advection and radiation fogs are consistent with the nature and concentration of condensation nuclei in marine and continental environments. The thickness of a fog layer can range from a few tens of meters for a radiation fog to several hundreds of meters for advective type fogs. The winds are

Table 9.1   Fog Characteristics

Fog Parameter	Ice Fogs	Water Fogs	
		Radiation	Advection
Average particle diameter (μ)	8	10	20
Typical particle size range (μ)	2–30	5–35	7–65
Equivalent water content (g/m^3)	0.10	0.11	0.17
Particle concentration (cm^{-3})	150	200	40
Horizontal visibility (m)	200	100	300

generally lighter and the visibility lower in radiation fogs. There is also usually more heterogeneity in space and time in radiation fog, with conditions changing significantly during a 10-min period.

FOG MODIFICATION CONCEPTS

The objective of fog dispersal is visibility improvement. Equation 9.1 defines the meteorological visibility, V, as the distance at which a black target can be just detected against a horizon sky in daytime with a contrast threshold of 2 percent, that is,

$$V = \frac{3.912}{\pi \sum_{i=1}^{M} (K_i N_i r_i^2)} \qquad (9.1)$$

where N is the number concentration of fog droplets or ice crystals, r is the fog particle radius, K is the scattering efficiency of a fog particle, and the summation, Σ, is taken over all fog particles. For visible light and spherical fog droplets, the scattering efficiency has a constant value of approximately two.

It can be seen from Equation 9.1 that the visibility can be improved by either decreasing the number concentration of fog particles, decreasing their radius, or both. Because of the inverse square relationship between visibility and radius, a decrease in radius by, for example, a factor of 3 results in ninefold increase in visibility. A similar decrease in the number concentration, on the other hand, results in only a threefold increase in visibility.

Methods designed to decrease the radii of the fog particles do so by evaporation. Modification techniques that are aimed at decreasing the number concentration of fog particles involve their physical removal.

Physical Removal Methods

A number of ways to accomplish the physical removal of fog particles have been proposed. In general these methods can be divided into those in which fog droplets or ice crystals are removed from suspension in air by precipitating them onto suitable surfaces; those in which fog particles are caused to fall out by gravity after agglomeration among themselves or after collection by larger particles that have been introduced for this purpose; and those in which the volume of fog-laden air is totally replaced by a similar volume of clear air. Electrostatic precipitation of the fog particles and filtration by screens, baffles, and forests are examples of the first type of physical removal method. Coalescence of the fog particles induced by ultrasonic sound waves and scavenging of the fog particles by electrically charged or neutral seeding particles, or by organic chemicals known as polyelectrolytes, are examples of the second type. Use of large fans on patchy ground fog is an example of the last type of physical removal method.

All the air to be cleared of fog by the physical removal methods must, of necessity, be acted upon directly. Because of the low number concentration of fog particles, the scavenging and agglomeration methods are inefficient. The size of the apparatus required for the application of the filtration and replacement methods are so large as to render them impractical. For these reasons, none of the physical removal methods has thus far proved to be operationally feasible.

Evaporation Methods

Evaporation methods are among the most promising approaches to fog dispersal. Since fog is a suspension of water droplets or ice crystals that are in equilibrium with the water vapor in the air, the fog particles can be induced to evaporate either by removing some of the water vapor from the air or by increasing the capacity of the air to retain additional water in vapor form. Removal of water vapor can be achieved by condensation on hygroscopic materials, that is, materials that have a particular affinity for water vapor in subsaturated atmospheres. Chemical desiccants such as calcium chloride, sodium chloride, and urea are examples of hygroscopic materials that have been proposed for this purpose. Raising the temperature of the fog environment by the application of heat is the most obvious way of increasing the water vapor capacity of the foggy air. Mixing of the fog with drier air of natural or man-made origin is another way of achieving the subsaturated environment that is required to promote the evaporation of the fog particles.

It is important to note that the water vapor content of saturated air decreases with decreasing temperature. Table 9.2 illustrates the impact that this physical fact has on the magnitude of the reduction in relative humidity that must be achieved by the two evaporative methods, as a function of temperature, to accommodate a fog having a water content of 0.17 g/m³. At a temperature of 68°F (20°C), the water vapor content of saturated air is 100 times larger than the fog water content, and therefore the relative humidity of the air need be reduced only slightly to accommodate the evaporating fog particles. At 32°F (0°C), the ratio of saturation vapor content to fog water content is only 5 and the required relative humidity reduction is consequently higher. At −40°F (−40°C), the saturation vapor content is comparable to the fog water content, causing the required reduction in relative humidity to be extremely large.

At temperatures below 5°F (−15°C), the required relative humidity falls below 90 percent and the energy and logistical requirements of the evaporative methods become prohibitively large. These methods are, therefore, only feasible for warm and some supercooled fog conditions.

Table 9.2   The Relative Humidity as a Function of Temperature[a]

| Temperature | | Relative Humidity (%) | |
°F	°C	By Water Vapor Removal	By Heating
68	20	99.0	99.0
50	10	98.2	98.2
32	0	96.5	96.6
14	−10	92.8	93.3
−4	−20	84.2	86.3
−22	−30	62.5	72.7
−40	−40	3.2	50.8

[a] That must be achieved to accommodate a fog water content of 0.17g/m³ by the water vapor removal and increased vapor capacity (air heating) evaporation methods.

Prevention Methods

All the modification techniques described thus far involve dissipation of existing fog. Another appealing fog modification concept is fog prevention. The prevention of fog involves control of the moisture, nuclei, and/or cooling that are required for its formation. The use of evaporation inhibiting chemicals on open water surfaces and the reduction of moisture pollution from man-made sources are examples of possible moisture control methods. It has been suggested that fog could be prevented by treating the natural nuclei with condensation-retardant chemicals. It has been shown, however, that this method delays fog formation by only approximately 10 min and that when it does eventually form, it is considerably more dense than it would have been otherwise. It has also been suggested that the nocturnal cooling required for the formation of radiation fog could be suppressed by the artificial creation of a cloud layer to greatly reduce the outgoing radiation.

Fog prevention operations must be conducted on a scale that is generally one to two orders of magnitude larger than those required by the physical removal and evaporation methods which attempt to clear only a portion of the fog. Fog prevention methods are, therefore, only feasible for application in those situations where the dispersal methods are not feasible and/or the sources of moisture and nuclei are limited in size and number.

GENERAL REQUIREMENTS OF FOG DISPERSAL

Since the majority of all fog modification efforts are aimed at improving visibility at airports, the rest of this section addresses this problem. Similar factors are important for other applications such as visibility improvements on highways or shipping lanes. The only differences in most cases are the dimensions of the clearing and the level to which the visibility must be improved. In the latter respect aircraft operations, because of the high speeds involved, have the most stringent requirements.

The size and shape of the region in which visibility must be improved, and the level of improvement necessary to permit landing operations, depend upon the level of sophistication of the electronic landing aids present at the airport. The least sophisticated systems, called Category I landing systems, require a minimum visibility of 2400 ft and a decision height of 200 ft. The more elaborate Category II systems require visibilities of ≥ 1200 ft and a decision height of 100 ft. The most sophisticated systems, Category III, are further subdivided into three categories, reflecting the difficulty in achieving the goal of very low visibility systems. Category IIIa, b, and c systems have no decision height specification but require minimum visibilities of 700, 150, and 0 ft, respectively. As of January 1, 1971, all major airports in the United States had Category I landing systems, 14 civilian airports in the continental United States had Category II systems, but no airports had any Category III systems, although one or two test facilities were being planned.

Aircraft generally land on a 3° glide slope. The distance from the decision

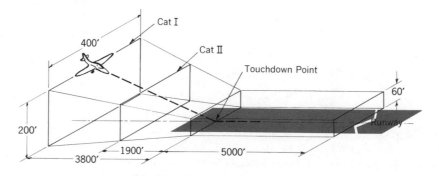

Figure 9.4   Regions to be cleared by fog dispersal techniques at airports. Dimensions are in feet.

height to the touchdown point for Category I and II landing systems are 3800 and 1900 ft, respectively. The general dimensions of the volume that must be cleared for Category I and II landing systems are shown in Figure 9.4. The volume of fog to be cleared for Category I and II landing systems are approximately 280 million and 170 million ft³, respectively.

It should be noted that a pilot need not see his touchdown point upon reaching his decision height; rather he must be able to identify the approach lights to decide if he is properly aligned for a safe landing. During rollout after touchdown the pilot must have sufficient visibility to maintain alignment on the runway. Since the aircraft is rapidly decelerating during rollout, the visibility need not be as good as that required on approach and, of course, it need only exist up to the height of the cockpit (less than about 50 ft for most aircraft). Under most conditions aircraft can taxi to the terminal in unmodified fog under the guidance of radar or follow-me vehicles.

In consideration of fog dispersal methods the wind condition, particularly the variability in wind speed and direction, is a very important factor. The wind velocity not only determines the volumetric dissipation rate required to create and maintain a clearing over an airport, but it also dictates the quantity and configuration of the necessary apparatus. Fog is never found in absolutely calm air, the wind velocity usually being from 1 to 20 mph. Thus, even for fogs having only relatively little mean motion with respect to the runway, any modification method would have to be applied either continuously to all air entering the volume of interest or in a single operation throughout a considerably larger volume. To keep an airport clear of fog in the presence of, for example, a 5 mph cross-runway wind, it is necessary to renew the

cleared zone once every 40 sec as new fog is continuously being introduced on the upwind side. The energy and logistic requirements of a fog dispersal system generally increase with increasing wind velocity.

The extent and duration of a clearing produced by some fog dispersal methods depend to a large extent on the intensity of turbulence in the fog layer. If a fog dispersal method is to be operationally effective, the clearing must be produced in substantially less time than it takes the fog to refill the cleared volume by turbulent diffusion. The relative rates of the opposing processes determine the usable lifetime of the clearing. Evaporation methods are, in general, superior to physical removal methods with respect to this factor because the limiting effects of turbulence are partially reduced by the lowered relative humidity of the cleared air. Since the characteristic diffusion time is directly proportional to the square of the width of the cleared zone, the influence of this factor becomes less important as the size of the clearing is increased.

SUPERCOOLED FOG DISPERSAL

The simplest of all weather phenomena to modify is supercooled fog or stratus clouds. The reason for this relative simplicity lies in the fact that supercooled fog, although colloidally stable, is in a thermodynamically metastable state. The supercooled water droplets exist at an energy level above that of the more stable ice phase. The addition of only a small amount of energy is required to induce their transition to the more stable ice phase, the final result of which leads to the dispersal of the fog.

The artificial dissipation process is based on the physical fact that the equilibrium vapor pressure over ice is less than that over water at the same temperature. When ice crystals are introduced into a supercooled, water saturated fog, they grow by vapor deposition and thereby cause the water droplets to evaporate. As the ice crystals grow, they are disseminated throughout the fog under the action of turbulent mixing which is intensified locally by the latent heat of fusion that is released as the affected volume is transformed from water to ice. The ice crystals spread laterally at a rate of about 1 m/sec, stopping only when the ice crystal concentration is significantly reduced by fallout. After 10 to 20 min, crystals start reaching the ground. Fallout continues up to 1 hr or more after the initial introduction of the ice crystals. Usable clearings form within 30 to 60 min, depending on fog thickness, temperature, and wind conditions. A single line of ice crystals generally produces a clearing that is 1½ to 2 mi wide.

Seeding Technology

Two techniques can be used to create the ice crystals necessary to initiate the artificial dissipation process. The first technique is based on seeding with tiny particles about 1 μ in diameter which have a crystal structure that is very similar to that of ice. These particles serve as embryos on which ice can grow. The particles are called ice or freezing nuclei and the initiation of the ice growth process is known as heteorogeneous nucleation. Silver iodide is the most commonly used artificial ice nucleus. Lead iodide and some organic materials have also been found to be effective nucleating agents. They are active at temperatures that are colder than 23°F (-5°C).

The second technique by which ice crystals are introduced into a supercooled fog involves homogeneous nucleation.

The ice crystals are formed by the local cooling of the air to below -40°F (-40°C), the critical temperature at which nucleation of ice occurs spontaneously without the aid of special nuclei. The cooling necessary to initiate homogeneous nucleation is produced by seeding with dry ice, that is, solid carbon dioxide that exists at temperatures as low as -112°F (-80°C), or by the instantaneous vaporization and expansion of refrigerants, such as liquid propane, that are sprayed into the fog. This technique is effective in producing ice crystals at temperatures as warm as 30°F (-1°C).

The reader is referred to C. A. Knight[7] for a review of the physics of the heteorogeneous and homogeneous nucleation processes.

Operational Programs

The technology required to carry out operational supercooled fog dissipation programs has been available since the early to mid 1950s. Operational programs at airports, however, did not come into practice until the early 1960s. At present, they are being conducted in the United States, France, and the Soviet Union. Both airborne dry ice seeding and ground-based propane systems are being used. Because of the high frequency of occurrence of supercooled fog at temperatures warmer than 23°F (-5°C), the airborne silver iodide seeding technique is not routinely used on an operational basis.

Dry Ice Seeding.   In a typical operation, the seeding of the fog is accomplished with an aircraft that has been fitted with a dry ice crusher/dispenser (Figure 9.5). Dry ice cakes are stored on board the aircraft and are loaded into the crusher, as required, during the seeding operation. The usual seeding pattern consists of 5 to

30 parallel lines, 5 to 6 mi long and ½ to 1½ mi apart, flown just above the fog at a distance between 45 and 60 min upwind of the airport to be cleared. The dispensing rate is generally about 15 lb/mi. Since about half the crushed dry ice is in the form of power that probably vaporizes before reaching the fog deck, the effective seeding rate is about 7 lb/mi.

The airborne dry ice seeding technique is employed in airport fog dispersal operations conducted by the Air Weather Service (AWS) at Air Force bases in Alaska and Western Europe, by the Air Transport Association at commercial airports in the United States, and by the appropriate Soviet authorities at airports in Russia. During the years 1968–1972, AWS fog dispersal operations assisted or expedited 736 aircraft departures and 686 arrivals at Elmendorf Air Force Base, Alaska. They were also responsible for the successful completion of 256 aircraft departures and 172 arrivals at nine operating locations in Europe during the years 1970–1972. The clearing effects produced by one of the dry ice seeding operations at Elmendorf AFB is shown in Figure 9.6.

Dry ice seeding operations have been used to disperse supercooled fog at 13 airports in the north-central and northwestern sections of the United States. A considerable reduction in revenue losses due to fog has been achieved ever since the initiation of the fog dispersal operations in 1963. During the winter 1969–1970, for example, fog dispersal operations costing approximately $80,000 resulted in a saving of over $900,000 in airline operating expenses.

Supercooled fog dispersal operations are conducted at approximately 15 of the largest airports in Russia. They reported

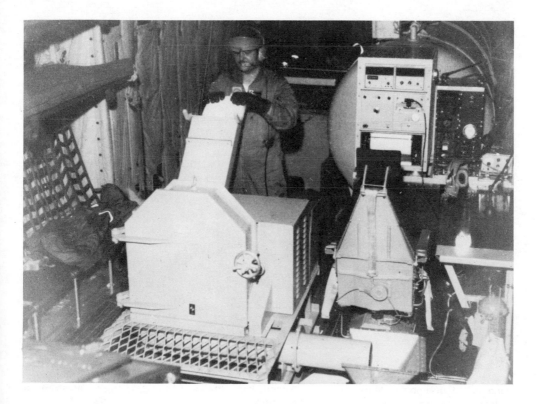

Figure 9.5   Dry ice crusher/dispenser installed in an air weather service WC-130 aircraft.

Before   After

Figure 9.6 Supercooled fog dispersal operation using airborne dry ice seeding as carried out by the air weather service at Elmendorf AFB, Alaska, on December 20, 1968.

Figure 9.7 Liquid propane dispenser at Orly Airport, Paris, France. This unit is part of a ground-based supercooled fog dispersal system operated by the Paris Airport Authority.

that 80 percent of their operations at the Moscow airport were successful between 1964 and 1967. Fog dispersal operations were credited with permitting 284 takeoffs and 143 landings that would not have otherwise been possible. An even higher success rate was achieved in later years.

Propane. Supercooled fog dispersal operations using a ground-based liquid propane system are conducted by the Air Weather Service at Fairchild Air Force Base, Washington, and by the Paris Airport Authority at Orly Airport, Paris, France. The AWS system consists of 21 dispensers, four near the airport for calm-wind situations and the rest placed at various distances upwind along the most common wind directions. Each unit is capable of dispensing 10 gal of propane/hr. Since its installation in 1969, fog dispersal operations have been responsible for 97 aircraft departures and 49 arrivals that would not have been possible otherwise.

The French system has been operating at Orly Airport since 1964. A centrally controlled ground installation consisting of 60 dispensers of the type shown in Figure 9.7 is used at an operating cost of $20/hr. During the winter of 1970–1971, 340 aircraft arrivals and 284 departures were made possible by the operation of the Orly Airport system.

WARM FOG DISSIPATION

Warm fogs are among the most stable cloud systems in the atmosphere. In contrast to supercooled fog there is no latent phase instability in warm fog that can be exploited to promote the artificial dissipation process. Warm fog dispersal methods are necessarily "brute force" in character. Whatever energy is required to dissipate the fog must be supplied by the dispersal method. Careful engineering is required to make any warm fog dispersal technique reliable, cost-effective, and free of detrimental side effects.

The development of warm fog dispersal methodology in the past few years has been based on an interactive program of computer model simulation of the artificial dissipation processes and field experimentation. The acceleration of progress in the engineering of warm fog dispersal techniques can be attributed to this approach. Three techniques, all designed to promote the evaporation of the water droplets, have been found to be effective in improving the visibility in warm fog: (1) mechanical mixing of the fog with drier, warmer air from above, (2) drying of the air with hygroscopic chemicals, and (3) heating of the air.

Helicopter Downwash Mixing

The physical basis of warm fog clearing with helicopters rests primarily on the principle of downwash mixing. The helicopter, during the clearing operation, either hovers or moves slowly forward in the clear air above the fog layer. The downwash action of the helicopter forces this relatively dry, clear air downward into the fog. The wake air, on descending, entrains and mixes with the fog. If the relative humidity of the air above the fog is approximately 90 percent or less, the resulting air mixture becomes subsaturated and the fog droplets are thereby caused to evaporate. The dimensions of the clearings created by the helicopter wake are much larger than the helicopter itself, usually by a factor of 10 to 20. In shallow fogs the wake of large helicopters may be strong enough to physically push the foggy air aside and replace it with the clear air from above.

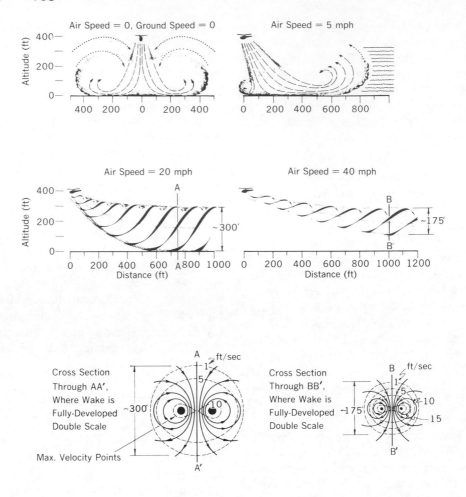

Figure 9.8 Schematic representation of a helicopter rotor wake at different forward air speeds (after Plank et al.[11]).

The dimensions of the rotor downwash wake of a helicopter and, therefore, its fog clearing capability depends primarily on its weight, its forward speed, and the thermal stability of the fog layer. The wake of a medium-size helicopter weighing approximately 25,000 lb such as a CH-3E helicopter, in the hover and forward motion flight modes has the general appearance indicated in Figure 9.8. The dimensions of the wake, particularly its penetration depth, increases with the weight of the helicopter. Under otherwise analogous conditions, the wake is deep but narrow when the thermal structure of the fog layer is relatively unstable and shallow but wide when the layer is relatively stable. Fog layers of thickness greater than the penetration distance of the wake cannot be effectively cleared by helicopters. If, however, the wake penetration distance exceeds the fog thickness, the dimensions of the clearing will be enhanced because the bottom part of the wake impinges on the ground and spreads laterally.

The typical appearance of a hover-produced clearing in a fog is shown in Figure 9.9. When the helicopter is flown in forward motion above the fog at an air speed of 15 mph, a cleared trail, as shown in Figure 9.10, is produced behind the helicopter. The clearings usually appear within 30 to 45 sec and persist for 3 to 8 min after the helicopter clearing operation is terminated.

The most extensive, quantitative tests of the helicopter downwash mixing technique of fog dispersal were carried out jointly by the United States Air Force and United States Army during September 1969 at Lewisburg, West Virginia. The results of these tests and subsequent theoretical analyses have demonstrated that (1) cleared zones large enough to permit helicopter landing operations can be created by the downwash of single, medium-size helicopters in most fog situations where the fog depth is less than 300 ft, and (2) single or multiple helicopters can create and maintain continuous clearing of fog over airports in situations where the fog is 200 ft in depth, the refilling diffusion velocity is less than 1 ft/sec, and the cross-runway component of the wind is less than $2\frac{1}{2}$ mph. The reader is referred to the report on the West Virginia experiments by Plank, Spatola, and Hicks[11] for a more complete discussion of the fog clearing capability of helicopters.

Fog clearing with helicopters has, to date, been used in military operations only in Southeast Asia. The evacuation of 29 wounded personnel was made possible by helicopter fog clearing operations. It was also instrumental in permitting the resupply of two outposts and the landing of 42 aircraft.

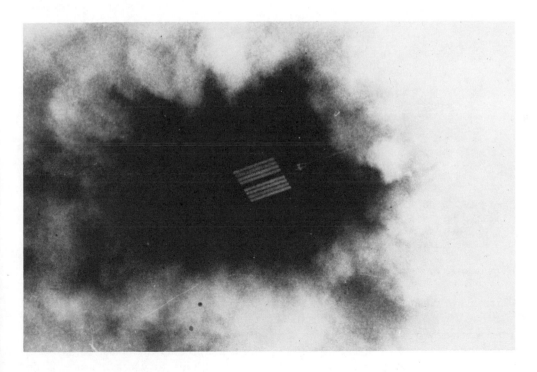

Figure 9.9   Fog clearing created by a CH-47a helicopter operating in the hover mode (after Plank et al.[11]).

First Clearing Pass

08 23:50   08 24:55

Third Clearing Pass

08 29:30   08 32:25

Fifth Clearing Pass

08 37:12   08 37:42

Figure 9.10 Cleared trail created in fog by a CH-47a helicopter at a forward air speed of 15 mph (after Plank et al.[11]).

Hygroscopic Particle Seeding

When hygroscopic substances in the form of either dry particles of solution droplets are released within a fog they absorb water vapor, and the air, in drying, causes the fog droplets to evaporate. Figure 9.11 illustrates the four phases that characterize the artificial clearing process. During the seeding phase, carefully sized hygroscopic particles are introduced into the fog above and upwind of the intended target zone, usually an airport runway. If the seeding particles are introduced from the ground, they must be blown up to the altitude to which clearing is desired. Because of their great affinity for water vapor, the hygroscopic particles grow rapidly by condensation as they fall under the action of gravity. Most of the hygroscopic particles grow about three times larger than their initial size before falling out of the fog layer in approximately 5 min. The visibility improves as the fog droplets evaporate in response to the vapor deficit that is created. The clearing first appears at the seeding level and spreads, in time, to the ground. The maximum visibility improvement on the ground occurs approximately 10 min after seeding. The clearing is advected past the target zone by the wind field and eventually refills under the action of turbulent mixing. If further clearing of the target zone is required, additional applications of seeding material are required.

The quantity of seeding material that is required to produce an operationally useful clearing by the process described above is dependent upon the size and chemical nature of the hygroscopic material, the fog water content, and the environmental wind field. The influence of these parameters on the clearing process has been evaluated by means of interactive computer simulation studies and field experimentation. The reader is referred to the works of Kocmond,[8] Kunkel and Silverman[9] and Silverman[13] for a more complete discussion of the results of these investigations.

The clearing effectiveness of the hygroscopic treatment is critically dependent on the size of the seeding particles. As the seeding particle size decreases, the quantity of hygroscopic material to attain a given visibility improvement decreases but the residence time of the particles in the fog increases. If the hygroscopic particles are too small, however, they will remain in suspension and contribute to lower visibility. Excessively large particles, on the other hand, fall out too rapidly for water to condense on them efficiently. The optimum size particle is one that can grow large enough to fall out of the fog and produce the required visibility improvement in reasonable times using as small a quantity of hygroscopic material as possible. Figure 9.12 shows the maximum visibility produced by seeding a fog, having an initial visibility of 85 m, with various sizes and concentrations of urea particles. It can be seen that the optimum size seeding particle in this case is 30 μ in diameter. Hygroscopic particles approximately 30 μ in diameter are, in general, most efficient.

It should be pointed out that the urea concentrations shown in Figure 9.12 are based on seeding with uniform-size particles. The generation of monodispersed size distributions of hygroscopic particles is not practicable, however. The seeding concentrations required to produce opera-

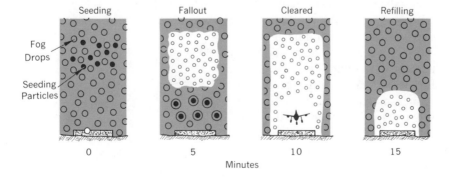

Figure 9.11 Schematic representation of warm fog modification by airborne hygroscopic particle seeding.

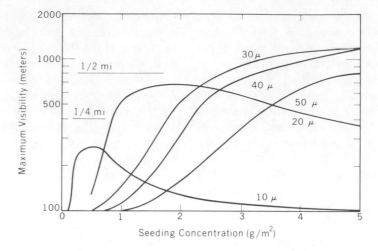

Figure 9.12 Computed maximum visibility at the ground as a function of urea seeding concentration and particle diameter. The initial visibility was 85 m (after Weinstein and Silverman[15]).

tionally significant improvements in visibility with realistic, commercially available size distributions of hygroscopic particles are two-to-three times greater than those required for idealized uniform-size distributions having the same modal diameter.

The most important property of an effective seeding agent is that it maintain its affinity for water vapor as the material concentration gets progressively more dilute as the particle grows by condensation. Since the clearing of the fog near the ground is of greatest practical value, the seeding agent must be able to continue to remove water vapor after it has fallen through some depth of fog and becomes quite dilute. Some materials such as calcium chloride take on water at a very low relative humidity, that is, 45 percent, but rapidly lose their affinity for water vapor after becoming dilute. Other materials such as lithium hydroxide require a higher relative humidity, that is, 83 percent, before they take on water, but maintain a relatively high growth rate as they become dilute. Table 9.3 lists some of the more effective hygroscopic materials and their effi-

ciency. Fog clearing efficiency is herein defined as the ratio of the maximum vertical visibility from the ground produced by a given mass of hygroscopic material to that produced by sodium chloride. Sodium chloride is taken as the standard because it has been the most

Table 9.3 Some of the More Effective Hygroscopic Chemicals and Their Fog Clearing Efficiency

Hygroscopic Chemical	Fog Clearing Efficiency
Lithium hydroxide	4.10
Ammonium chloride	2.06
Lithium chloride	1.88
Sodium hydroxide	1.85
Potassium hydroxide	1.31
Ammonium nitrate-urea-water	1.17
Acetamide	1.02
Sodium chloride	1.00
Urea	.93
Disodium phosphate	.85
Ammonium nitrate	.84
Calcium chloride	.62

widely used in fog clearing experiments. With the exception of acetamide, which is prohibitively expensive, all the hygroscopic materials that are more effective than sodium chloride are also highly toxic to plants and animals or corrosive to metal surfaces or both. As is the case with sodium chloride, these materials cannot be used operationally at airports and other populated areas. Of these materials, only ammonium nitrate-urea-water has been actively pursued for limited operational use where corrosion and/or ecological factors are not critical. The primary technical problem with this material is that it is disseminated in liquid form through high-pressure nozzles. Generation of the required seeding concentrations in the correct particle size distribution is beyond the current state-of-the-art of spray technology.

Urea is the best, safe hygroscopic seeding material. Raw urea cannot be used, however, to seed warm fog because it has a soft, friable crystalline structure which fragments easily during handling, producing large numbers of submicron particles that contribute to a degradation rather than an improvement in visibility. Conventional sizing methods such as mechanical milling and sorting are, therefore, not suitable for urea. Microencapsulation technology, whereby single crystals are chemically packaged inside thin, harmless shells, has been exploited to provide for the sizing and stabilization of the urea particles, thereby optimizing its efficiency as a warm fog seeding agent. The microencapsulation process produces a narrow size distribution that is completely devoid of the very small particles. Microencapsulation technology is widely

Figure 9.13 Fog clearing at McClellan AFB, California created by airborne seeding with microencapsulated urea particles.

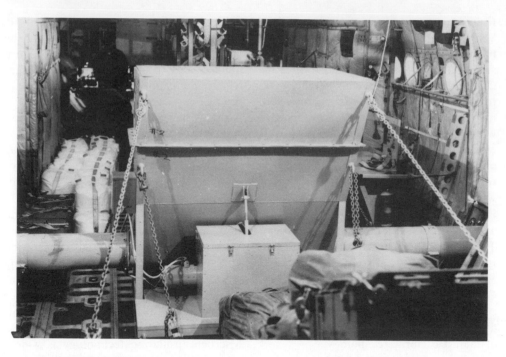

Figure 9.14 Hygroscopic particle dispenser installed in a U.S. Air Force WC-130 aircraft.

used in industry to produce such consumer products as timed release cold capsules and carbonless multiple copy paper.

An example of a clearing in warm fog that was produced by seeding with microencapsulated urea particles is shown in Figure 9.13. A fog over McClellan Air Force Base, California that was 100 ft in depth was seeded with 1200 lb of microencapsulated urea particles over a distance of 1800 ft. The seeding operation was carried out by a C-130 aircraft that was fitted with a specially constructed motor-driven auger-feed dispenser (Figure 9.14). Seven minutes after seeding the 700-ft wide by 1500-ft long clearing appeared over the approach end of the runway. By 36 min after seeding the hole was completely refilled by turbulent mixing.

This experiment demonstrates the inhibiting effects of one meteorological variable, turbulence, on fog clearing effectiveness. Another variable that influences the effectiveness of seeding is fog water content. The quantity of seeding material required to produce a given percentage increase in visibility increases linearly with increasing fog water content. Because visibility is inversely related to the water content, that is, fogs of high liquid water content have very low initial visibility, the quantity of seeding material required to raise the visibility to an operationally useful level increases almost as the square of the increase in liquid water content.

As a result of the counteracting influences of turbulence and vertical wind shear, and the great difficulty in targeting the clearing, single line hygroscopic particle seeding is not operationally feasible. Application of the seeding material over a wide area is required to produce operationally useful clearings. The size of the clearing and, therefore, the quantity of seeding material to produce it increases as the speed of the wind and its variability in direction increases. For the normal range of wind conditions in fog, a seeding area of 1 to 10 mi² is required to ensure proper targeting of the clearing. Approximately 8 to 10 passes by one or more seeding aircraft must be made to cover areas of this size. Wide-area seeding patterns can be executed under the guidance of standard airport aircraft control radar provided that the wind at the seeding level and its expected variability during the 15 min period it takes for the clearing to develop can be determined.

Table 9.4 gives model computations of the seeding material and cost requirements of operational warm fog dissipation for average radiation and advection fog conditions that are associated with a 3-knot cross-runway wind.[15] The seeding concentrations and costs shown in Table 9.4 should be multiplied by approximately 0.5 and 2.0 to account for the range of fog conditions likely to be encountered. The cost figures are based upon a cost of $0.50/lb for microencapsulated urea, the supplier's projected cost of the material if purchased in large quantities. The costs per aircraft landing are based on assisting one aircraft operation during a 5-min seeding period and are, therefore, considered to be conservative. Operational warm fog dissipation by airborne hygroscopic particle seeding, while technically feasible, is nevertheless costly.

Ground-Based Heating

One of the oldest and most successful methods of dissipating warm fog is by ground-based heating through the combustion of hydrocarbon fuels. Sufficient thermal energy must be provided to evaporate the fog droplets and to raise the temperature of the air sufficiently to accommodate the additional water. This quantity of heat is easily calculated from the basic laws of physics. Figure 9.15 shows how the required amount of heat varies as a function of air temperature and fog liquid water content.

The energy requirements given in Figure 9.15 are theoretical minimum

Table 9.4 Requirements for Operational Warm Fog Modification by Airborne Seeding with Microencapsulated Urea

	Category I	Category II		Category IIIa	
Visibility goal					
(mi)	$\frac{1}{2}$	$\frac{1}{4}$		$\frac{1}{8}$	
(m)	800	400		200	
Fog type	Both	Radiation	Advection	Radiation	Advection
Seeding Concentration					
(g/m²)	2.25	1.0	1.6	0.6	0.4
(lb/mi²)	12.8K	5.7K	9.2K	3.4K	2.3K
Total Material					
(lb/hr)	80K	36K	58K	22K	14K
(lb/landing)	6.7K	3.0K	4.8K	1.8K	1.2K
Cost					
($/hr)	40K	18K	29K	11K	7K
($/landing)	3.3K	1.5K	2.4K	0.9K	0.6K

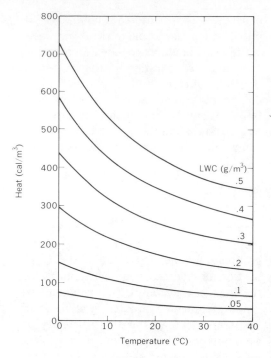

Figure 9.15 Heat required to dissipate fog as a function of air temperature and fog water content.

values needed to completely dissipate the fog. In practice the values have to be increased to account for the need for rapid, but not necessarily complete, evaporation of the drops, the additional water vapor introduced into the air by the combustion of the fuel, and the uneven spatial distribution of heat. The first two factors can readily be incorporated into the calculations used to produce Figure 9.15. The effect of the last factor, the uneven heat distribution, however, cannot be evaluated as easily because the physics of heat plume systems is not completely understood. In order to ensure adequate heat to account for all the uncertainties described above, the values given in Figure 9.15 are generally multiplied by a factor of two or more, in designing operational warm fog dispersal systems.

The rate at which the heat must be applied is primarily a function of the wind speed. Except for the near calm condition, the heating rate increases with increasing wind speed. Large heating rates are required, however, during very light wind conditions to compensate for the vertical heat losses caused by buoyancy.

Based on extensive tests conducted since 1936, thermal fog dissipation systems were installed at 15 airfields in England during World War II to ensure the safe arrival of RAF aircraft following offensive sorties in Europe. The thermal fog dissipation systems, called FIDO (Fog Intensive Dispersal Of), consisted of pipelines along the runways through which aviation fuel was pumped under low pressure and burned as it escaped through small holes in the pipes. Despite problems of smoke and ignition failure, the FIDO systems were highly successful. The first operational landing by an aircraft returning from a mission was made in November, 1943. By the end of the war, over 2500 aircraft containing approximately 10,000 airmen had landed with the aid of FIDO. On one occasion 85 aircraft were landed with the aid of FIDO in an 8-hr period. The reader is referred to the report of Walker and Fox[14] for the most comprehensive review of the FIDO work.

After World War II, further development of FIDO was pursued by the United States Navy at the Landing Aids Experimental Station (LAES), Arcata, California. Advances were made in the mechanical, electrical, and combustion design of FIDO. A high-pressure fuel flow system was used to avoid the preheating required by the old low-pressure system and to permit the use of cheaper diesel fuel or fuel oil. Special burner heads were used to reduce the unwanted production of smoke, and heat-coil igniters were used to minimize ignition failure. The improved FIDO system was tested at the Arcata Airport during the period

1946–1950 as a component of the other integrated landing aids which were also being evaluated. In 94 percent of the test cases, minimum visibility and ceiling conditions were produced by the FIDO system in dense advection fog that occurred in association with a moderate onshore wind.

As a result of the successful fog dispersal operations at Arcata, a FIDO system was isnstalled at Los Angeles International Airport (LAX) in 1949 to serve commercial aviation. The LAX FIDO system, which cost about $1,325,000 to install, was improved further by providing for automatic ignition and monitoring of visibility and wind from a central location. Operation of the system was, however, still plagued with problems of fuel leakage, ignition, dirt, and smoky burners. Fog dispersal operations were abandoned in December, 1953 when it was realized that the FIDO system, as installed, was not capable of clearing fog conditions more severe than $\frac{1}{8}$ mi visibility. This limitation was largely caused by inadequate maximum heat outputs and gaps in the burner lines at cross-runways and at the ends of the runway. It was generally concluded that an effective FIDO system was too expensive to warrant routine use by commercial aviation. Advances in electronic landing aids made at that time permitted the lowering of visibility and ceiling minima to such an extent as to considerably reduce the amount of air traffic disrupted by fog.

In 1947, the United States Air Force undertook the development of a thermal fog dissipation system that would overcome the main disadvantages of FIDO in military operations, that is, the permanence of the installation and the inherent objection to mixing fire and aircraft. A mobile or transportable system which employed the exhaust heat from jet engines to dissipate the fog was developed.

In response to a request by Berlin Airlift Operations, a system consisting of seven J-33 jet engines combined with afterburning in an aspirator was constructed in 1949. Preliminary tests of a single engine at Arcata were favorable. With the closing of the Arcata test facility after the 1949 fog season, the full jet engine system was sent to Alaska for testing in ice fog during 1950 and 1951. The system failed to dissipate the ice fog. The combustion of the fuel produced a liquid water content of 0.02 g/m³/°F rise in temperature which at −40°F (−40°C) actually intensified the fog. It was not realized that this was the true reason for the failure of the J-33 tests, so the discouraging results on ice fog were generalized and all activity on the development of the jet engine system was discontinued.

Interest in the jet engine exhaust heat approach was revived in 1958 when it was realized that this technique should be successful in dissipating fog at above-freezing temperatures. Development efforts were initiated in both the United States and France. After conducting tests in 1958 to determine the fog clearing capability of a jet engine, United States Air Force scientists recommended that the jet engine fog dispersal technique be used as an emergency capability at all airbases where jet aircraft are available. Use of jet engines in a fixed underground installation was not recommmended at that time because it was not considered to be cost-effective. In the early 1960s the French conducted experiments using jet aircraft parked alongside a runway which were successful in clearing the fog. In 1968 the Air Weather Service conducted similar tests using four C-141 four-engine aircraft and again demonstrated the feasibility of this technique.

The increased losses in revenue with the advent of the jet age made fixed installation heating systems economically attractive for at least those airports

Figure 9.16 Turboclair fog dispersal installation at Orly Airport, Paris, France. The installation was constructed by Bertin et Cie for use by the Paris Airport Authority in dissipating warm fog.

having a high volume of air traffic. Further development of the jet engine technique by the French led to the installation of a sophisticated thermal fog dissipation system, called Turboclair, at Orly Airport, Paris, in 1970. The initial Turboclair installation consisted of eight jet engines in underground chambers alongside the upwind edge of the runway, as shown in Figure 9.16. Each unit was equipped with a remotely controlled, directional outlet grid to distribute the exhaust heat over the desired part of the runway. The installation was capable of clearing fog over a distance of 300 m in the approach zone and 300 m along the runway. In 1972 another four engines were installed to clear an additional 600 m in the touchdown and rollout zones of the runway. The complete installation cost about $3 million to install and costs about $3400/hr to operate. Tests of the Turboclair system showed that it was capable of improving the visibility in the approach and touchdown zones of the runway from below minimum to at least Category II landing conditions. The turbulence generated by the jet engines did not create any problems for the landing of the test aircraft, even with automatic pilot on during their approach. Based on the results of these tests, in 1972 the French Ministry of Transportation authorized French air carriers to use Turboclair to assist in landings in fog in accordance with established airport minima criteria. The reader is referred to the report of Fabre[3] for a review of the history of the development of the Turboclair system.

In 1970, development of fixed installation thermal fog dissipation systems in the United States was also resumed. By the application of modern heat and meteorological engineering technology, the United States Air Force is developing a thermal fog dissipation system that promises to be cost-effective, meet air quality standards, and be safe for aircraft operations. The Federal Aviation Administration is considering the development of a similar system for application at civilian airports.

ICE FOG ELIMINATION

At present, there is no practical method of dispersing ice fog once it has formed. Although the methods of warm fog dispersal are, in principle, applicable to ice fog, the extremely large energy and logistic requirements to make them effective at very cold temperatures make them economically unsound. Ice fog can, however, be prevented by controlling the man-made nuclei and moisture sources which lead to its formation. Various suppressive and corrective measures have been devised to alleviate the ice fog problem at Eielson Air Force Base, Alaska. Many of the unnecessary moisture sources have been eliminated. Emission of moisture from many of the remaining sources has been greatly reduced by use of condensate return systems. A modern central power plant was constructed which permitted the shutting down of three diesel- or oil-fired power substations on the base. Combustion studies resulted in revised operating procedures for the coal burners used in the laundry, mess halls, motor pools, and power station which increased the coal burning efficiency of its boilers and decreased the amount of moisture by about 10 to 15 percent. Less successful engineering attempts have been made to eliminate the large amount of moisture from the stack effluents of the central power plant, its cooling pond, and the exhaust of vehicles.

Despite the engineering attempts to eliminate major moisture sources at Eielson AFB, it is probable that there will always be some ice fog to hamper base activities. Even the small moisture inputs that result from the operation of clothes dryers, exhaust fans, vehicles, aircraft, etc., will continue to produce local pockets of ice fog on the base. Ice fog has not been eliminated completely but its severity has been lessened.

Conditions favoring the formation of ice fog could largely be avoided at new air bases. The runways should be located as far as possible from the remainder of the base and should be placed upwind of water vapor sources. The only source of moisture would then be the aircraft. Aircrews report that the visibility reduction due to departing aircraft is drastic, lowering the visibility to near zero, but in 5 to 10 min the visibility is restored to its original value.

OUTLOOK

Limited operational systems to dissipate supercooled fog at airports are being successfully employed in the United States and Europe. Large airports are increasingly favoring fixed, ground-based systems using propane over airborne dry ice seeding systems. The initial investment in ground-based installations is larger than that for airborne systems, but the relative ease of operation makes it worthwhile. Smaller airports and locations where supercooled fog is an infrequent but costly phenomenon will tend to favor the airborne dry ice system. The airborne dry ice system will also be used to increase the amount of winter sunshine over cities that are plagued with persistent supercooled stratus clouds. In any case, the investment necessary to implement either system is unquestionably less than the losses suffered by users due to fog.

Inasmuch as warm fog is the most prevalent and, therefore, the most troublesome fog type, considerable effort has been invested in developing methods for its dispersal. This investment is rapidly approaching the point where returns will be forthcoming. Recognizing that the clearance of large areas of warm fog with the expenditure of small amounts of energy is incompatible with physical reality, current efforts are primarily concerned with the application of modern technology to the engineering of proven "brute force" techniques. Each of the three techniques seriously considered for operational implementation appears to have its applications. Helicopter downwash mixing is simple and inexpensive but restricted to shallow radiation fog. It is most dependable for application in situations where relatively small clearings are needed, such as those required to facilitate helicopter landing or rescue operations, or small but congested sections of roadways. Wide area hygroscopic particle seeding is applicable to deeper, but not all, warm fogs. The seeding equipment is inexpensive but the seeding material is costly and the seeding operation is relatively difficult to execute. It is most appropriate for use in applications where the fog must be dissipated and mobility is essential. The thermal technique is effective in all warm fog situations. Although relatively simple and inexpensive to operate and maintain, it is extremely costly to install. Its use is therefore economically justifiable only for urgent military purposes or in cases, such as Los Angeles International Airport, where its frequency of operation will rapidly amortize the cost of the initial installation.

Ice fog can be prevented by controlling the man-made sources of moisture and nuclei that lead to its formation. It is, however, not within the capabilities of the present state of the art to dissipate ice fog once it has formed. A promising new approach to ice fog dissipation that is presently being investigated is based on the hypothesis that a widespread stratus cloud can oftentimes be artificially created in an arctic atmosphere and such cloud decks will radiationally induce sufficient surface level warming to evaporate the ice fog. Continued research on ice fog and new dispersal concepts such as this one may, in time, result in the development of an effective dissipation technique.

There is no doubt that fog can be eliminated by artificial means. Further experimentation with existing techniques to establish criteria that can be used to make the techniques more successful as well as to decrease the cost and the hazards of operation is desirable. There is still room for new ideas in both methods and equipment. Empirical testing will in time lead to standardization of fog dispersal techniques.

REFERENCES

1. Byers, H. R., *General Meterology*, McGraw-Hill, New York, 1959, pp. 540.

2. Court, A. and R. D. Gerston, Fog frequency in the United States, *Geograph. Rev.* **56** (4), 543–550, 1966.

3. Fabre, R., Aeroport D'Orly—Installation De Denebulation Turboclair, Report by the Aeroport De Paris and Societe Bertin et Cie, Jan. 1971, p. 39.

4. Guttman, N. B., Study of worldwide occurrence of fog, thunderstorms, supercooled low clouds, and freezing temperatures, Report for the Commander, Naval Weather Service Command, NAVAIR 50-1C-60, by the NOAA Environmental Data Service, Asheville, N.C., 1971, p. 131.

5. Houghton, H. G. and W. H. Radford, On the local dissipation of natural fog. I. On the possibilities of fog dissipation, *Papers Phys. Oceanogr. Meteorol.* **6** (3), 13–26, 1938.

6. Jiusto, J. E., Investigation of warm fog properties and fog modification concepts, First Annual Summary Report, NASA Contract No. NASr-156, NASA Report CR-

72, CAL Report No. RM-1788-P-4, Cornell Aeronautical Laboratory Inc., Buffalo, New York, 1964, p. 8.

7. Knight, C. A. (1967), *The Freezing of Supercooled Liquids*, D. Van Nostrand, Princeton, N.J., 1967, p. l45.

8. Kocmond, W. C., R. J. Pilie, W. J. Eadie, E. J. Mack, and R. P. Leonard, Investigation of warm fog properties and fog modification concepts, Sixth Annual Summary Report, NASA Contract No. NASW-1933, CAL Report No. RM-2864-P-1, Cornell Aeronautical Laboratory Inc., Buffalo, N.Y., 1970, pp. 3–38.

9. Kunkel, B. A. and B. A. Silverman, A comparison of the warm fog clearing capabilities of some hygroscopic materials, *J. Appl. Meteorol.* **9**, 634–638, 1970.

10. Ohtake, T., Studies on ice fog, Final Report on National Center for Air Pollution Control, Public Health Service Contract No. AP-00449, University of Alaska Report UAG R-211, Geophysical Institute, University of Alaska, Fairbanks, Alaska, 1970, p. 177.

11. Plank, V. G., A. A. Spatola, and J. R. Hicks, Fog modification by use of helicopters, *Environmental Research Papers No. 335*, AFCRL-70-0593, 28 October 1970, Air Force Cambridge Research Laboratories, Bedford, Mass., 1970, p. 154.

12. Schaefer, V. J., The production of ice crystals in a cloud of supercooled water droplets, *Science*, **104**, 457–459, 1946.

13. Silverman, B. A., Warm fog modification by airborne hygroscopic particle seeding, Ph.D *Thesis*, Division of Physical Sciences, The University of Chicago, Chicago, Ill., 1972, p. 144.

14. Walker, E. G. and D. A. Fox, The dispersal of fog from airfield runways, A Record of the Work of Technical Branch F of the Petroleum Warfare Department 1942–1946, Ministry of Supply, London, 1946, p. 321.

15. Weinstein, A. I. and B. A. Silverman, A numerical analysis of some practical aspects of airborne urea seeding for warm fog dispersal at airports, *J. Appl. Meteorol.*, **12**, 1973.

D
PROGRAMS IN
OTHER
COUNTRIES

The United States has been the chief player in the weather modification game, at least since the Schaefer discovery of dry ice seedability of supercooled clouds. Many other countries are conducting programs on a continuing basis also, with the Soviet Union having the largest of these programs. Academician Fedorov, the Chief of the Hydrometeorological Service, has 75,000 people who work for him on all problems of weather and oceanography. By contrast, our U.S. National Weather Service has about 6000 employees. Only a small fraction of these two organizations actually work on weather modification. The Soviet program in weather modification, described in Academician Fedorov's Chapter 10, includes fog dispersal, precipitation augmentation, and hail abatement and is clearly considered an important national effort. In hail abatement, the Soviet Union leads the United States in its program.

Dr. Sulakvelidze's Chapter 11 describes the progress of hail suppression work in the Soviet Union. They routinely protect 8 million acres of rich agricultural land and they claim a 70 percent reduction of crop damage by hail by this protection. There are some in the United States who view this claim of damage reduction with skepticism, but a team of U.S. meteorologists who visited the Soviet Union projects a few years ago returned with a general feeling that the Soviets might indeed be achieving considerable damage reduction.

It is probably not surprising that Australia and Israel, both water poor and both advanced technologically, should attempt to develop means for rain augmentation. Their programs, discussed in Chapters 12 and 13, represent major efforts in weather modification research. The Israeli program has been a well designed and carefully controlled experiment, yielding well documented positive answers—clearly one of the best experiments carried out anywhere. The Australian program has also been well conceived and well executed, but has given a puzzling answer that there seems to be a persistent effect of AgI seeding, maybe lasting years—how can this be? This is one of the unsolved mysteries of this business.

Academician Fedorov, in the summary of his chapter, makes a very important statement that, "Man is becoming master of the earth. . . . There is nothing absurd in dreams of a 'struggle for meteorological mastery.' Soviet scientists harbor a deep concern about cessation of the arms race, disarmament, and the maintenance of peace. . . ."

I'm sure that all scientists everywhere and especially those in the United States share Dr. Fedorov's feelings and hope strongly that this technology that we are developing will be used in achieving those important objectives.

10 | Modification of Meteorological Processes

YE. K. FEDOROV

Active intervention in meteorological processes is but one of the many issues that have developed as human society interacts with its natural environment. An entire spectrum of problems related to this interaction has now taken on an unparalleled urgency. Progressive depletion of natural resources, pollution of the natural environment, and disturbance of ecological equilibrium are by now of concern not only to specialists, but also to growing segments of the general public in all countries.

As a result of such problems, some experts[11, 15] arrive at uncomfortable conclusions in estimating the outlook for the future development of society. They foresee a serious crisis in man's interaction with his environment within the next 50 to 100 years if the fundamental trends in the characteristics of man's development continue as they have in recent decades.

There are two basic causes for concern: the question of whether the world's natural resources will continue to be able to satisfy the increasing requirements of our society, and the problem of whether

man will have a debilitating impact on his natural environment. At one time the alarm expressed was principally over the adequacy of reserves of nonreplaceable natural resources, such as coal, oil, and other fossil fuels. Many economists predicted that the reserves of several vital mineral resources would be exhausted before the end of the 20th century. However, in recent years the increase in known deposits of many of these resources is easily outdistancing their utilization. Also man has rapidly expanded his ability to obtain the substances he needs from many different kinds of natural raw materials, and to derive energy from new sources. Thus, for the time being, the exhaustion of reserves of nonreplaceable natural resources would not seem to present a very immediate threat to mankind.

On the other hand, when it comes to the use of such natural resources as fresh water, timber, soil, ocean biota, and others—resources that are constantly being replenished by the earth's established natural processes—we may in the not too distant future approach total utilization of the "income" part of such budgets. For example, at the present time we are using only about 10 percent of the

Ye. K. Fedorov is Director, Hydrometeorological Service, Moscow, USSR.

freshwater available in the runoff of rivers; however, according to the calculations of many hydrologists, within 50 to 100 years we will be using almost the entire runoff of earth's rivers to meet the needs of industry, irrigation, and public power consumption if patterns of human water use do not change. Similarly, at the present time we are using about 50 percent of the timber growth; within 100 years, obviously, we will be using the entire output. About 70 percent of the total increase of commercial fish in the oceans is now being recovered; within less than 20 years the catch will approach 100 percent of that increase.

What, then, will we do when we are using up vital replaceable natural resources as fast as they are being replaced? Many experts[11,15] believe that by that time mankind's consumption of any given natural resources must be brought into balance with natural recuperative capabilities, and not that further industrial development (and consequently increases in consumption as well) must be halted. However, the human race learned long ago not merely to harvest the resources it found in the natural environment, but also to accelerate their replacement. Many thousands of years ago man began to set apart from the wilderness around him a field or a herd of animals, and to invest enormous effort in tending it, in an attempt to increase its yield. Man's deliberate restoration of the natural environment and cultivation of its resources are going forward on an ever-increasing scale. I believe that at the present time there are many kinds of measures needed to foster and increase the rate of replacement of crucial natural resources of our planet. Examples are the increase of freshwater supplies through acceleration of the water cycle, and the development of an increased food supply by aquaculture in the worldwide ocean basin.

An especially disquieting and un-foreseen effect upon the natural environment which has become particularly important in the last 20 to 50 years is pollution of the natural environment. For the most part pollution has proceeded with extreme rapidity. In technologically developed countries with large concentrations of industry and population, pollution has reached such proportions that all elements of the public are alarmed by it. This has led certain scientists[11] to call for a halt to further industrial development, and in particular to the industrialization of the developing nations. Otherwise, in their opinion, the stress on the natural environment of the planet will increase to such a point that natural processes will not be able to cope with it. But pollution is not an inevitable consequence of industrial activity: means have long since been developed not only for cleaning up industrial effluents to keep harmful substances from entering the atmosphere and the water environment but also for restructuring technological processes so as to prevent almost completely discharges into the surrounding environment.

Even if we could eliminate the significant pollution of the natural environment causing manifest harm to man, we still could not help but exert an increasing influence upon the environment in a multitide of other ways. Thus all human activity is accompanied by the evolution of heat. At the present time this heat totals only about 0.01 percent of the heat which reaches us from the sun. However, if mankind continues to increase his use of energy sources at today's tempo, within a few decades the proportion of the heat generated by man will increase to 1 to 2 percent of the solar input.[3] Such a large release of energy might have serious environmental consequences.

Nor can we avoid a certain number of changes in the composition of the surrounding environment to "enrich" or, in industrial processes and recycling of materials, a certain portion of the substances used in industry will escape into the surrounding environment to "enrich" or, in other cases, to poison it. Consequently, we are faced with the unavoidable necessity of disturbing the equilibrium of processes established in nature and the occurrence of both deliberate and inadvertent change in the structure and composition of the environment. We are faced with new sources of energy and with associated changes in natural thermal balances. We must learn to recycle environmental constituents, especially water. We must learn to prevent the harmful consequences of such change.

From the time that people ceased to shelter themselves in trees or caves and began to build dwellings, they began to acquire skills (and later a whole body of knowledge) applicable to building: a knowledge of the strength of materials, the behavior of various structures under stress, and other such fields. It is this that has enabled us to design and build all kinds of structures. Similarly, at the present time, we need to develop skills enabling us to calculate, just as carefully and exactly, all reactions occurring in nature which are affected by our activities: an understanding of the effects on our environment as a result of substances added to it, changes in the thermal balance, alteration of the water cycle, or others.

The situation is complicated by the fact that the natural environment is not a frozen structure, but rather a complex web of processes interlinked in all directions. These processes frequently occur under unstable conditions, for which comparatively slight changes can result in consequences on a much greater scale. This can be both good and bad. It can be good in that theoretically it enables us to attempt to produce major changes by means of comparatively slight expenditures of energy and materials. On the other hand, the fragility of natural processes constrains us to behave with caution because of the possibility of unforeseen effects.[6,8,9]

Active intervention in meteorological processes is in my opinion a primary objective, and an extremely important and interesting one, in the overall problem of man's effect on the natural environment. In this article, I wish to set forth my own views on matters of weather and climate modification and to give a brief outline of the work being done in this field in the Soviet Union. I also believe it would be well to present a little more detail about the first modification efforts of Soviet scientists in the 1930s, which are not too well known abroad. An annotated bibliography of basic Soviet research in weather and climate modification, carried out in recent years, separate from my own reference list of citations, is included as an appendix to this chapter. A separate chapter is devoted to one of the most interesting and most practical applications achieved in the USSR—hail suppression.

EARLY SOVIET WEATHER MODIFICATION EFFORTS

The main difficulty in weather modification has always been considered to be the enormous rate of production of energy involved in meteorological processes. Actually, the basic source of this power—a stream of radiant energy from the sun to the earth—can be estimated at 10^{13} kw. The rate of energy production of an individual process in the atmosphere can be estimated at magnitudes from 10^6 kw (an ordinary cloud) to 10^9 to 10^{10} kw (processes governing weather over a significant area).

However, the enormous energies involved are only one aspect of the matter.

Another aspect is the fact that atmospheric phenomena involve complex interactions and conflicts between opposing influences. On one hand, for example, an air mass retains its physical characteristics over a significant period of time. Opposing this conservative tendency, however, is the ability of an air mass, under certain circumstances, to change its physical properties abruptly. Thus, for example, when air contains water vapor in a concentration close to saturation, the slightest cooling forms a fog or cloud, which radically changes the optical properties of this volume of atmosphere, and creates another condition for its heat exchange with the ground.

The atmosphere is extremely sensitive to the presence of certain impurities, in amounts which are quite insignificant compared to the mass of the air containing them. The formation of a cloud requires not only a vapor concentration which is sufficient under the particular conditions, but condensation nuclei as well. The physical properties of the nuclei govern the degree of supersaturation needed for the formation of droplets, the rate of growth of droplets, and the kinetics of the further development of the cloud. Despite this, the total mass of condensation nuclei needed for the formation of a cloud system with a volume of tens of thousands of cubic kilometers is 1 kg or less.

The contradictory nature of the development of meteorological processes, and the instability and abrupt shifts in their development in response to small stimuli, make it possible to seek methods of control. However, the atmosphere lacks the permanent channels of control which are characteristic of machines or living organisms. Consequently, we must seek, within a complex tapestry of atmospheric processes occurring over a period of time, the chains of interrelated phenomena which can be manipulated as channels of control. An example of the development of methods for controlling weather processes is the technique, now rapidly being perfected, of cloud and rainfall modification. Experiments in cloud modification have a long history.

For many decades such experiments were carried out by independent amateurs, without sufficient scientific foundation. An example is the well-known practice of shooting at hail clouds with a mortar. In the 1930s the Soviet scientist V. N. Obolenskiy and his colleagues first posed the problems of weather modification and rainmaking as serious scientific topics, despite the fact that at that time all leading meteorologists considered them impossible of solution.

By 1932 the USSR Hydrometeorological Service had already assigned these scientific research problems to the Institute of Artificial Rain,[13] later reorganized into the Institute of Experimental Meteorology. V. N. Obolenskiy, B. P. Veynberg, N. A. Bulgakov, M. A. Aganin, and many other scientists carried out detailed research into the natural processes of development of clouds and precipitation, and based on this, on means of cloud modification as well.[1, 4, 17, 18, 23] They seeded a supercooled cloud with crushed ice in order to stimulate crystallization of drops. They also employed hygroscopic substances and electrically charged particles, exposed a cloud to ionized radiation, and in 1937–1939[21, 22] reported the first successful instances of cloud modification under field conditions. These studies were of substantial theoretical interest. However, once the war began, they were forced to curtail their work.

Until the end of the 1930s western scientists were predominantly skeptical about the feasibility of modification. This position was expressed, for example, by Professor Marvin, then head of the U.S. Meteorological Service, in 1931.[20] The first scientifically grounded ideas about the feasibility of modification of meteorological processes to be expressed in western countries were voiced by T. Bergeron[2] and V. Findeisen[10] at end of the 1930s. Much later, in 1945 and 1946, attempts to modify clouds and obtain additional rainfall were begun in the United States.

In these and following years, Schaefer, Langmuir, Vonnegut, and other scientists turned their attention to the very effective method of modifying clouds by crystallizing supercooled water drops. Instead of using ice as others had, they employed a very efficient means—dry ice. They obtained the first successful and important field results.[14,19,24] As a result of excessive publicity, however, these experiments raised exaggerated expectations as to the feasibility of modification, to the point that there was even speculation about the international "struggle for meteorological mastery."[1] Gradually, however, this uproar died down, and interesting and important results were published by many serious scientists in the United States. The reader is, of course, acquainted with numerous additional articles from other countries which have greatly advanced our understanding of the structure of clouds and the laws of their natural development.*

RECENT SOVIET WORK

In recent years Soviet scientists have compiled voluminous data about the phase state, microstructure, water content, and macrostructure of clouds of stratiform and cumuliform varieties obtained from the observation of several

* The author makes no further citation of references by Soviet authors, assuming that the reader will familiarize himself with the annotated bibliography appended to this chapter.

thousand clouds in various stages of development and under varying conditions. The data so obtained are used in theoretical computations and in solving various practical problems.

Methods have been worked out for using data from radar observations to estimate the reserves of water in stratiform and convective clouds. These results seem promising for the development of modification techniques designed to stimulate precipitation. Means have also been developed for estimating the possible effectiveness of using this sort of modification in several regions of the country.

A theory of the formation of clouds, and in particular of hail, has been developed. Many methods and devices have been worked out for determining various characteristics of clouds and precipitation, both from an airplane and remotely, principally by means of radar. One of them, which must be mentioned, is an apparatus for measuring condensation nuclei; another is a two-beam radar system used in hail suppression, which makes it possible to estimate the size of large cloud particles.

The initial phenomenon leading to the formation of a cloud is the cooling of a certain volume of air which contains water vapor. The cooling commonly begins when a volume of air ascends into a colder layer of the atmosphere. Numerous articles have been devoted to the study of convective currents, jets and bubbles, and air-mass cumulus clouds. Several researchers have developed methods for stimulating updrafts in order to induce convection or facilitate breakup of inversion layers impeding the development of convection. They developed a different type of heater, the so-called "meteotron," originated by Dessens, which creates a rising air current. Others have attempted to stimulate updrafts by changing the albedo of the underlying surface, as for example by darkening the ground.

Soviet scientists have also been active in this field in recent years, performing experiments with a strongly heated, high-velocity jet obtained by using several powerful turbojet engines at once. This made it possible to study the movement and dissipation of jets to altitudes of 2 to 3 km, under various conditions, as well as the formation and development of the clouds they created. It was shown in these experiments that it was possible to obtain rainfall in flat country a few hours earlier than would have occurred naturally. Along with this, various means of suppressing updrafts were considered and field-tested for the first time (vertical takeoff of jet aircraft, dumping of powder, and other techniques). The feasibility was shown, both theoretically and experimentally, of breaking up powerful developing cumulus clouds by creating downwardly directed air currents in their upper portions. For the time being, however, despite the interest of these results, attempts to modify clouds by suppressing or stimulating updrafts have not developed beyond the stage of isolated experiments.

We now turn to a different class of studies, based on the elementary physical and chemical processes which made possible the development of a cloud. The processes to be considered are the condensation, aggregation, and crystallization of cloud particles—with a view to explaining the mechanism of their natural operation and determining possible routes of stimulation or suppression.

Many scientists have paid particular attention to condensation nuclei. However, for a long time all they studied were particles acting as nuclei at considerable supersaturations, which are never seen in the real atmosphere and which play practically no role in the formation of clouds. A cloud forms at supersaturations measured in fractions of a percent. Only in recent times have serious technological difficulties been overcome so that condensation nuclei operating under such conditions could be studied. The first devices for measuring the concentration of nuclei of varying activity in the real atmosphere were developed in the Soviet Union. These devices made it possible to study the relationships between the concentration and microstructure of nuclei and the amount of supersaturation required for condensation. In addition, data were obtained on the typical distribution of condensation nuclei in the atmosphere under a variety of conditions.

The process of formation of precipitation particles is to a significant degree determined by the manner in which conversion of the condensation nuclei into cloud droplets occurs. The dynamics of this process are determined, on the one hand, by the properties and concentration of the particles from which the droplets form and, on the other hand, the rate of change of external parameters, primarily air temperature. The first particles formed from the most active condensation nuclei begin to absorb vapor intensively and thus hinder growth on less active particles. This self-regulating process, the influence of various factors which govern the extent to which supersaturation occurs, and the concentration of the drops formed have been studied in detail by Soviet scientists in recent years, and we can now speak with confidence of the creation of a theory of the initial stage of condensation.

It is well known that condensation growth of cloud droplets does not lead to the narrow size distribution of particles prescribed by classical condensation theory. The search for mechanisms which would offer a satisfactory explanation of the experimental data led to the creation of a theory of stochastic condensation, which takes into account fluctuations of vapor in the system due to turbulence, nonuniform distribution of droplets, and a number of other factors. Priority in and development of these lines of investigation belong the Soviet scientists. As a consequence of this work, certain Soviet researchers are working the problem of stimulating or suppressing the activity of nuclei.

The phenomenon of aggregation is responsible for the growth of droplets and the formation of precipitation in warm clouds. Growth of droplets due solely to condensation proceeds extremely slowly, and can scarcely lead to the formation of precipitation. Studies have shown that gravitational aggregation, occurring as drops fall at different rates, can form precipitation out of warm clouds. However, it begins only after a sufficient number of drops reach sizes significantly larger than those of the remaining drops. Many aspects of the initial stages of aggregation are still unresolved, particularly in regard to aggregation of small droplets. In recent years, therefore, Soviet scientists have devoted increasing attention to developing a theory of the initial stages of aggregation, and to experiments along these lines in large chambers where conditions approximating those existing in a real cloud can be created. Thus numerous theoretical studies on the effect of turbulence on the rate of aggregation of droplets have been carried out. At present, work is proceeding on a more exact definition of the size of the coefficient of capture for small droplets, based on a more correct allowance for the strength of hydrodynamic interaction.

Numerous agents have been found to date for stimulating aggregation; however, none is effective enough. Examples are comparatively large water drops sprayed into the cloud to induce gravitational aggregation or the scattering of electrically charged bubbles of light plastic ribbons. Some effect is obtained, but the quantity of water or other materials which must be introduced into the cloud is still too great for the reaction to be of practical use. Many countries are devoting considerable attention to the possible use of hydroscopic solid particles, or drops of solutions. However, sufficiently effective results have not as yet been obtained.

The most realistic means of cloud modification, and the only one of any practical use at the present time, is the stimulation of the crystallization of supercooled drops. This method is related to the fact that although an atmospheric aerosol contains many particles capable of becoming condensation nuclei (as a rule such particles are found in abundance in the atmosphere), it contains hardly any particles which can serve as crystallization nuclei. The line of research based on this fact was the one that produced successful and effective field results and, naturally, was the one that attracted the interest of many researchers. Methods of stimulating crystallization of supercooled droplets are being applied successfully at the present time. There are a number of substances which are used. Some of them—dry ice, liquid propane, and certain others—are effective because they drop the temperature sharply. Fine particles of certain other substances (AgI, PbI_2) are used as crystallization nuclei.

Amounts of 100 to 200 g of solidified acid or several grams of iodine compounds will suffice to crystallize supercooled droplets in a cubic kilometer of cloud. Thus the ideal of control, the use of a small perturbation to produce a substantial result, is completely realized. The crystallization of supercooled water clouds is considered in a great number of references, although certain features of the reactions described are still not completely clear at the present time. In recent years our scientists have also studied the activity of various substances which stimulate crystallization and have found the optimal methods and norms for seeding clouds with them.

In the USSR, crystallization of supercooled droplets is used for three purposes to seed low-lying clouds and fogs, to suppress the development of hail, and to investigate the possibility of stimulating precipitation. Seeding of stratus and stratocumulus clouds, particularly low-lying ones which impede the takeoff and landing of aircraft, is done with very high reliability. This is also true of fogs of supercooled water droplets, at temperatures of $-4°C$ and less.

Various techniques and devices, airborne and ground-based, are at present used to seed clouds with dry ice, AgI, PbI_2, and other reagents. Thus we have developed several types of automatically operating airborne CO_2 seeders, ground-based seeders which shoot CO_2 in a vertical ascending stream at low altitude, pyrotechnical shells and small rockets which scatter iodine compounds during combustion, and large rockets and antiaircraft projectiles which deliver these same compounds into clouds at an altitude of several kilometers.

When clouds are seeded, the energy which was previously reflected from their upper surface is incident upon the earth. The lower layer of the atmosphere is heated by the earth, and a distinct change in the natural meteorological process may result. A small number of Soviet experiments in seeding clouds over a territory of several thousands of square kilometers did actually change the state of the lower layer of the atmosphere. The daytime temperature was increased by 7 to 8°C. The area cleared of overcast was maintained for a significant period of time, and it must be assumed that a phenomenon generally reminiscent of a

weak anticyclone was caused. Changes in the amount of thermal energy entering the lower layer of the atmosphere over the cleared territory reached a total of 10^{14} cal in these experiments. It can be assumed that the development of a cleared area of tens of thousands of square kilometers, which is completely practicable to create by modification techniques at the present time, can serve for a while as a useful means of influencing larger-scale meteorological processes in certain situations.

From a practical point of view, the use of the crystallization reaction for hail suppression is of the utmost interest. A large task force of Soviet scientists is working on this problem at the present time. Despite certain discrepancies in present hypotheses of the formation and development of hail clouds, certain fairly simple and general conclusions can be expressed. Thus it can be assumed that large hail develops when ice crystals are initially formed in a zone of the cloud where there are comparatively large supercooled drops. Freezing on the initial crystals when they collide with these comparatively large drops leads to rapid growth of the hail, and its subsequent fall to earth. Thus one way of preventing the growth of large particles is to crystallize supercooled drops in the zone where they would otherwise encounter initial ice crystals. Mass crystallization could result in stopping the rapid growth of hailstones. Graupel or fine hailstones will form and as they fall will become drops of rain. Soviet scientists are presently performing numerous hail suppression experiments to protect agricultural crops. In 1971 the area protected reached 4 million hectares. Five or six years of experimental work over large areas attest to the fact that the amount of hail damage is four times less in the protected zone than in the unprotected regions.

Precipitation Augmentation

An enormous amount of research all around the world is being devoted to ways of stimulating precipitation. Significant numbers of such experiments were performed over a test area set up in 1958 by the Ukrainian Scientific-Research Hydrometeorological Institute in the Dnieper-Petrovskiy region of the Ukrainian Soviet Socialist Republic. The test area consisted of two sections, target and control, of equal size (50 × 75 km), furnished with a dense network of rain gauges. The air space over the target section was scanned by several radars to monitor air flight patterns, modification operations, and observations of cloud and precipitation conditions. Similar experiments were done in the winter on another test area in the Obninsk region (Moscow and Kaluzhskiy regions). The results were evaluated by the use of specially designed snow surveys.

Several series of experiments to modify clouds in order to induce precipitation were carried out in the Northern Kazakh steppe regions, in the mountainous Uzbek regions, and in other regions of the country. As a result of these studies it was found that during systematic modification, precipitation on the target section was increased by approximately 10 to 15 percent in comparison with other regions. To estimate the possible increase of precipitation during cumulus cloud modification, randomized experiments in seeding cumulus clouds were carried out, and the quantity and other characteristics of precipitation from each seeded and unseeded cloud were measured from airplanes. It was established that seeding powerful cumulus clouds increases the probability of incidence of precipitation from them. The amount of precipitation from a seeded cloud is approximately 1.5 to 2 times greater than from an unseeded one. At the same time, no effect on the

vertical size of the cloud was noted after seeding. Experiments on stimulating precipitation were also carried out on stratus clouds in mountainous regions.

From our point of view, it is important to pin down whether, during cloud modification, augmentation or redistribution of precipitation from the cloud occurs. With this in mind, studies were made for several seasons of test areas whose control section was extended 80 to 100 km beyond the modification (target) region along the path of motion of the seeded cloud system. It is not as yet possible to speak with any certainty about the results of these experiments. Data from one region seem to indicate that the primary effect is the redistribution of precipitation, not the generation of additional precipitation out of the cloud. Another test area gave the opposite result.

Experiments to augment precipitation carried out in the USSR up to the present time, as well as those in other countries, have indicated that approximately 10 to 15 percent more water can be obtained by seeding. Such a result cannot be considered effective for purposes of supplementary irrigation. Nevertheless, stimulation of precipitation from cumulus clouds can in certain instances be used in an applied manner. Thus in recent years in the USSR we have developed a technique for extinguishing forest fires in regions where there is a high probability of convective cloudiness arising. In 1971 such methods were used to put out 35 fires which started in various regions of Siberia, taking in a total area of around 10^5 hectares. It may be assumed that in years to come the use of artificially induced precipitation for this purpose will become more widespread.

It is curious that one figure for precipitation augmentation—about 10 percent—recurs practically unchanged in an enormous number of experiments carried out by different means in widely different regions of the globe. A 10 percent increase of precipitation implies extraction of about 50 to 70 percent of the water contained in the treated clouds occurring over the experimental territory which would not ordinarily produce precipitation by natural means. It must be borne in mind that in natural precipitation the cloud, as the experiments of Soviet researchers have shown, produces approximately 10 to 20 times more water than is contained in it at any given moment; that is, over a period of time it acts like a peculiar kind of water generator, forming the water vapor found in the air into water droplets or crystals, which then fall to earth. The real object, if we are to obtain a significant augmentation of precipitation, is to find some way of stimulating a nonprecipitating cloud to behave in this manner. If we could do this, we could depend on obtaining approximately the same amount of precipitation from seeding as is obtained from cloud systems by natural processes. If we do not succeed in doing this, then we shall never obtain an increase in precipitation which is more than a certain fraction of the water existing in and around the cloud.

We do not as yet sufficiently understand the laws governing the natural process of formation of precipitation in a cloud. It is known that every cloud begins to develop under the influence of factors external to it, such as, for example, the ascent of air in convective currents or by movement along the surface of a frontal boundary. However, at a certain stage in the cloud's development, the energy released in the process of condensation becomes sufficient to draw into the cloud an equal or a yet larger amount of vapor. From this moment on the reaction becomes self-sustaining. Is there some means by which we can move a cloud forward to this stage of a self-sustaining reaction? It is obvious that this is the crux

of the whole question of stimulating the water-generating properties of a cloud.

Underestimating the interaction of a cloud with the surrounding environment has in its time led many researchers to an overly optimistic view of the possible results of modification. In exactly the same way, underestimating the influence of internal factors in the development of a cloud fosters a certain pessimism about the results of modification. Studies of the interaction of the cloud as a whole with the surrounding environment, a study of equilibrium and dynamics of water in clouds, and a study of the water reserves of cloud systems are obviously the key, at the present time, to a correct evaluation of and choice of means for modification aimed at a substantial increase of precipitation.

At the present time we in the Soviet Union have embarked upon a series of broad field experiments whose object it is to look into the practical possibilities of obtaining economically profitable additional precipitation from systematic modification of all suitable clouds. One location for this is Sevan Lake, where a systematic increase in precipitation by an average of 10 of 15 percent over a comparatively small basin would have substantial impact on the cycle of the lake.

Hail Control

In studying the possibilities of and developing the means for modification of meteorological phenomena, we encounter the need, at a certain point in the studies, to make the transition from theoretical work, laboratory experiments, and preliminary field experiments, to a broad field experiment. It is obvious that we cannot get a reliable estimate of the result of a future field experiment on the basis of theoretical results or laboratory experiments, as is usually done in the development of new technological processes in industry. Meteorological experiments, moreover, are extremely complicated and very expensive. Thus great difficulties and high risks are peculiar to problems of atmospheric modification. The difficulties and risks are usually aggravated by unavoidable inadequacy in our ability to predict the natural development of the process to be modified. To obtain a reliable estimate of the method of modification to be tested, we must rely either on very large numbers of instances of its application in a limited region, which requires an extended period of time, or on a sharp expansion of the areal extent of the experiments.

Thus in developing hail suppression methods, for example, our scientists protected a territory of several tens of thousands of hectares over the course of two seasons. Completely successful protection was achieved, but everyone understood that no serious conclusions could be drawn from this, since the probability of hail falling on precisely that small area was very slight. Only after we had protected an area of up to 10^6 hectares, and worked on it for two or three seasons, could we estimate the effect of the modification with any degree of confidence, and persuade ourselves that immense sums of money would not be spent in vain.

We strongly believe that at the present time it is essential to set up sufficiently large-scale precipitation stimulation experiments, even though it is quite likely that results which are positive from the economic point of view will be obtained only in certain regions and only under very specific and, it may be, uncommon conditions.

In connection with organizing a broad effort to avert hail damage, a special analysis group was created whose responsibilities included considering the results of intervention in hail processes, making preliminary estimates of the effectiveness of the methods developed, and generating recommendations and suggestions for creating a single objective technique for analyzing effectiveness. An analysis of the results of modification of hail processes has shown that a sound estimate of effectiveness can be obtained by a simultaneous consideration of changes in the areal coverage, intensity, and duration of all cases of hail incidence occurring during modification. This group concluded that at the present time the most reliable and objective estimate is one based on a change in the area damaged by hail. For an average of 5 to 8 years of hail suppression operations in various regions of the USSR, the area of hail damage was 3 to 10 times less compared to the average annual value in the years before protection, with the statistical reliability of the results averaging 90 to 95 percent. More precise estimates of the results of hail suppression are impossible because of the great space-time variability of areas of hail damage. The coefficient of variation of annual areas of hail damage is approximately 50 to 100 percent, and the coefficient of correlation between protected and control areas in the absence of modification is equal to .3 to .5. Errors in determining the effect of modification over a year are in the tens of percents. This is the order of magnitude of error in provisionally determining the savings realized. Although the problem of devising more reliable methods for assessing the results of modification is very important, the fact that the magnitude of the effect achieved significantly exceeds the possible errors obviates any doubt as to the high econimic effectiveness of the hail suppression work.

CHANGING THE CLIMATE

Even given the concrete progress being made, we are still carrying out cloud modification only within a very narrow range of conditions. Nevertheless, experimental modification of local features of the weather enables us to pose questions about the possibility of modifying processes on a significantly greater scale—regional, and perhaps even global. The basic questions in estimating the feasibility of modification of large-scale meteorological phenomena, it would seem to me, are questions about the overall unity of climate and the stability of the processes producing climate.[7, 15] Climate is the result of a unique temporary equilibrium of the very complex set of hydrometeorological processes which exert their influence over the entire surface of the globe. We can conceptualize the interaction of these processes in a general, qualitative way; however, it is not as yet possible to calculate the quantitative relationships existing among them.

It is well-known that in an earlier period of the earth's history, the climate was radically different from what it is today; that is as a whole, hydrometeorological processes on the earth were in states of temporary equilibrium different from the present one. What explains the transition from one state to another? No generally accepted theory of climatological changes presently exists. Various hypotheses are put forth about the reasons for and factors in climatic change. Some of them explain changes in the climate by fluctuations in the sun's activity; others concentrate on variation in volcanic activity, which chokes the atmosphere with combustion products and thus alters the earth's thermal balance. Changes in the tilt of the earth's axis and, consequently, the latitudinal positions of various regions of the globe, as well as shifts in the positions of continents and oceans when the tilt of the earth's axis remains constant, in and of themselves necessarily involve changes in climate.

However, there are also hypotheses which explain changes in climate by the transition of the fabric of hydrometeorological processes from one state to another without any essential changes in the structure of the earth, the tilt of its axis of rotation, or the distribution of continents, oceans, and mountain chains, and also without any substantial change in the amount of energy coming from the sun. These hypotheses undoubtedly deserve serious consideration, since otherwise it is difficult to explain the substantial, although less dramatic, climate changes which have indisputably occurred over the last 10 to 20 thousand years of our planet's history.

If several states of temporary equilibrium of the sum total of hydrometeorological processes are possible without there being any change in the basic characteristics of the earth's structure, then theoretically this opens up the prospect of large-scale climate modification. This, naturally, could be a change of the climate as a whole over the entire globe, or it could be of one hemisphere. Estimates and calculations of the different climates which are possible without changing the basic features of our planet's structure are of very great significance, to our way of thinking. Soviet scientists are studying this question.

What means exist for modifying climate? Obviously, one such means is to change the nature of the underlying surface over large enough areas. We are well aware of the great role played by the underlying surface in the initiation and development of atmospheric processes. At the present time it is difficult to ascertain the exact quantitative characteristics of the relationship between properties of the underlying surface, and possible climatological changes. Nevertheless, it can apparently be assumed that in addition to direct relationships, self-generating reactions and various kinds of "feedback" relationships will at times occur.

As one example, we might recall the hypothesis, supported by a number of studies, that someday the ice sheet in the Arctic Ocean will again disappear, because the changing pattern of circulation of the earth's atmosphere will favor the maintenance of a relatively warm, open Arctic Ocean. In a similar manner, a trend toward transformation of the atmosphere circulation could at some future period change the temperature of the surface layer of water over broad expanses of the ocean. Or temperature could obviously change as a result of some effect on the ocean currents, possibly from large hydrotechnological installations. As is known, in at least one case the sudden shift in the course of the El Niño off the coast of South America caused quite a substantial deviation from normal weather patterns for a period of several months, creating major problems for inhabitants of a large seacoast region.[16]

A number of hypotheses for changes of climate have been suggested which relate to the alteration of major ocean currents, or the creation of new currents: for example, changes which might result from the construction of dams or pumping stations in the Bering Strait. It should be noted that the attention attracted to these suggestions is not merited, since the most important factor in the argument—an estimate of the impact on climate—is not, unfortunately, scientifically well-founded. Snow cover can be an objective for modification. Acceleration of thawing or, on the other hand, delay of thawing or artificial formation of snow cover over significant areas, would seem to affect the heat exchange between the atmosphere and the surface of the earth. And that in turn should cause certain changes in the meteorological processes. At the present time thawing of snow cover and the ice surface is being accelerated over areas of tens of square kilometers by scattering soot, coal, dust, or the like. Stratus clouds are seeded in the winter to create a thin snow cover on territories of significant size.

A major feature of the underlying surface is its relief. It is well-known that large relief features have an appreciable impact on atmospheric circulation. However, even minor details of relief, such as forest belts, as many researchers are showing, can exert a distinct effect on meteorlogical processes. Thus, for example, it is obvious that minor relief features—slight prominences—can explain the well-known unevenness of precipitation. It is possible that small but specially designed and constructed installations could exert some sort of modification effect on the state of the lower layer of the atmosphere; for example, they could stimulate turbulent motions.

Reclamation alters evaporation substantially: prospects are that it will be carried out over significant areas. Although the main role in the water cycle is played by it planetary component—the transport of water from the ocean over the continents—it is still most important to ascertain whether or not comparatively slight changes in local links in the chain of events in the water cycle affect rainfall in one region or another of a continent.

And, finally, one possible means of modifying the phenomena which produce climate is to intervene in processes as they evolve in the stratosphere. The small amount of matter in the upper layers of the atmosphere and the slight energy transferred from the stratosphere to the troposphere as a result of phenomena related to solar activity, it seems to us, raise the possibility of substantial artificial changes in the properties of the upper layers of the atmosphere, changes so great as to change the nature of the activity of the lower layers. Many of these possible means of modifying the climate have been considered repeatedly in the writings of Soviet scientists.

We have mentioned above some phenomena which could serve as their own sort of initial reactions, as triggering mechanisms for climate modification. However, actual changes in the climate occur only when these reactions set off a chain of events, unfolding in a determined direction, until the desired change develops in the general circulation pattern of the atmosphere. This change, to be effective, must prove to be persistent, and it should be predicted without error. Herein, and not in matters of technology or expenditures of energy, it seems to me, lies the fundamental and staggering difficulty of the problem.

It is scarcely possible to achieve such a change in large-scale atmospheric circulation on an applied basis in the coming two to three decades. Yet it is nonetheless essential to get to work on it, since we are already exerting a rapidly escalating impact on the atmosphere on the water environment.

The rapidly increasing input of additional heat into the lower layer of the atmosphere as a result of industrial operations and human activity in general; the introduction of new compounds into the atmosphere, that is, combustion products and industrial pollutants; the alteration of properties of the earth's surface (roughness, albedo, and others); the change in the water cycle caused by reclamation—all these are inevitably reflected, in one way or another, in the sum total of the hydrometeorological processes governing climate. We need not list here all the potential ways in which man's activities affect the climate, since the variety of such effects are recognized,

and this question was thoroughly considered in 1971 at a specially convened international seminar, where leading specialists in climatology, including representatives of Soviet science, participated.[12] It is our view that we must take a broader look at the relevant disciplines in order to understand in as short as possible a period the way in which changing factors in our environment can affect climates, in order to be able to prevent chains of reactions which could affect the climate in possibly undesirable ways.

CONCLUSION

It is scarcely possible to avoid the increasing impact of man's activity on the climate; an example is the heat produced by man. In the not too distant future, therefore, we shall be faced with the problem, if not of global change, then perhaps, rather, of stabilizing the present climate, to which the economies of all the nations of the earth are geared, in one way or another. We must also seek to put into operation the channels of control which will compensate for inadvertent modification, or channel it in the direction we desire. For example, theoretically, limiting the additional heat generated by human activity is not the sole means of avoiding undesirable changes in the heat budget of the planet; it is also possible to increase the other parts of the budget, such as radiation or reflection.

It is hardly necessary to argue the reality of this problem. Even if it is going to require several decades to implement its solution on an applied basis, the time has now arrived for broad organization of the requisite studies. It is easy to see that the problem of global or regional climate modification is, of its very essence, international and requires the unified efforts and coordinated operations of all nations.

Returning to the view stated at the beginning of this article, it must be stressed that mankind is ever more rapidly approaching that stage in its interaction with nature in which it will require practically all the natural resources of the earth and will need the capability to cope with elemental phenomena on a large scale over the entire globe. In other words, man is becoming master of the earth. It is evidently no accident that our entry into space coincides with this development.

There is hardly any need to argue that under these conditions all mankind must present a united front to the world around it. There is no other alternative. There is nothing absurd in dreams of a "struggle for meteorological mastery." Soviet scientists harbor a deep concern about cessation of the arms race, disarmament, and the maintenance of peace and believe that the guarantee of freedom, independence, and progress for all peoples must be our common concern and responsibility.

REFERENCES

1. Aganin, M. A., Effect of impurities in atmospheric water upon processes of coalescence of drops, *Izv. Akad. Nauk SSSR, ser. Geofiz.,* **1940** (3).

2. Bergeron, T., *Mém. de 5me Ass. Gen. De l'Union Géophys. et Géograph.,* 1935, p. 156.

3. Budyko, M. I., Vliyaniye cheloveka na klimat [Man's influence on climate], GIMIZ, 1972.

4. Bulgakov, N. A., Some considerations with regard to artificial rainmaking, *Vestn. Elektrotekh.* **3,** 104, 1930.

5. *Bull. Am. Meteorol. Soc.,* N6, 1953.

6. Fedorov, Ye. K., Oxnovnyye problemy Gidrometeorologicheskoy sluzhby [General problems of the Hydrometeorological Service], General Assembly, AN SSSR, Moscow, 1946.

7. Fedorov, Ye. K., Status of and prospects for solution of problem of modification of weather and climate, *Trudy Vsesoyuznogo Nauchnogo Met. Soveschchaniya* [Proceedings of the All-Union Scientific Meteorlogical Conference] **1,** 48, 1962.

8. Fedorov, Ye. K., A threat which must be averted, *Priroda* **9,** 63, 1970.

9. Fedorov, Ye. K., The interaction of man and environment, *Bull. At. Scie.,* 5, Feb. 1971.

10. Findeisen, W., *Meteorol. Z.* **55,** 121, 1938.

11. Forrester, J., *World dynamics,* 1971.

12. *Inadvertent Climate Modification,* The Massachusetts Institute of Technology, 1971.

13. Kramaley, Scientific research problems of the Institute of Artificial Rain, *Vestnik Yedinoy Gidrometeorologicheskoy sluzhby* [Bulletin of the United Hydrometeorological Service] **2,** 4, 1932.

14. Langmuir, I., *J. Meteorol.* **5,** 175, 1948.

15. Meadows, D. H., D. L. Meadows, J. Randers, and W. Behrens III, *The Limits of Growth,* 1972.

16. Murphy, R. C., Oceanic and climatic phenomena along the west coast of South America during 1925, *Geograph. Rev.,* 16, 1926.

17. Obolenskiy, V. N., Conditions of cloud and fog stability, and conditions for extracting precipitation from them, *Meteorologicheskiy Vestnik* [Meteorological Bulletin] No. 244, 1931.

18. Obolenskiy, V. N., Role of ions, neutral and charged dust particles, and chemically active nuclei in the formation of clouds and fogs, *Zh. Geofiz.* ["Geophysics" Journal] **IV** (1), 91, 1934.

19. Schaefer, V., *Science* **104,** 457, 1946.

20. The Rainmakers, *Sci. Am.* Aug. 1931.

21. *Tr. Lening. Inst. Eksp. Meteorol.* [Proceedings of the Leningrad Institute of Experimental Meteorology], Issue 1, 1937.

22. *Tr. Nauchn. Issled. Inst. Gidrometeorol.* [Proceedings of the Scientific Research Institutes of the Hydrometeorological Service], Ser. 1, Issue 1, 174, 1941.

23. Veynberg, B. P., Toward a theory of inducing precipitation from clouds by seeding them with electrified sand, *Zh. Russ. Fizikokhim. obshchest.* [Journal of the Russian Physics and Chemistry Society], Issue 6, 1924.

24. Vonnegut, B., *J. Appl. Phys.* **19,** 959, 1948.

Appendix

MICRO- AND MACROSTRUCTURE OF CLOUDS

Studies of the Micro- and Macrostructure, Water Content, and Phase State of Clouds

Bichiashvili, A. D., R. I. Doreuli, A. I. Kartsivadze, V. A. Lapinskas, and T. B. Salukvadze, Certain characteristics of radar reflections from rainfall and hail, *Trudy Vesesoyuznogo nauchnogo soveshchaniya po aktivnym vozdeystviyam na grodovyye protsessy* [Proceedings of the All-Union Scientific Conference on Modification of Hail Processes], Tbilisi, 1964.

Borovikov, A. M., V. V. Kostarev and A. B. Shypyatskiy, Some results of radar observations of the evolution of powerful-cumulus and cumulonimbus clouds and the results of modification, *Proceedings of the All-Union Scientific Conference on Modification of Hail Processes*, Tbilisi, 1964.

Borovikov, A. M. and B. I. Demidova, The question of the phase states of clouds of various forms, *Trudy TsAO.* [Proceedings of the Central Aerological Observatory—Tsentral'naya aerologicheskaya observatoriya], No. 64, 1965.

Gavrilenko, N. M. and Z. M. Yansovskaya, Water content and capacity (power?) of convective clouds during various synoptic processes, *Trudy UkrNIGMI* [Ukrainian Hydrometeorological Institute—Ukrainskiy nauchno-issledovatel'skiy gidrometeorologicheskiy institut], No. 61, 1966.

Leonov, M. P. and G. I. Perelet, *Aktivnyye vozdeystviya na oblaka v kholodnoye polugodiye* Cloud modification during the cold part of the year], Gidrometeoizdat [Hydrometeorology Publishing House], Leningrad, 1967.

Minervin, V. Ye., Seasonal and geographical distribution of water content of clouds, *Trudy TsAO*, No. 55, 1964.

Minervin, V. Ye, Water content in crystalline clouds, *Trudy TsAO*, No. 64, 1965.

Minervin, V. Ye., Fluctuations of water content in stratiform clouds, *Trudy TsAO*, No. 71, 1966.

Nevzorov, A. N., Distribution of largedrops in liquid-drop stratiform clouds, *Trudy TsAO*, No. 79, 1967.

Skatskiy, V. I., *Issledovaniya vodnosti kuchevykh oblakov* [Studies of the water content of cumulus clouds], Gidrometeoizdat, Leningrad, 1968.

Prikhot'ko, G. F., *Iskusstvennyye osadki iz konvektivnykh oblakov* [Artificial precipitation from convective clouds], Gidrometeoizdat, Leningrad, 1968.

Experiments on Convection:

Imyanitov, I. M. and T. V. Lobodin, On zones of non-uniformity in thunderclouds, *Trudy GGO* [Proceedings of the Main Geophysical Observatory—Trudy Glavnoy Geofizicheskoy Observatorii], No. 157, 1964.

Mastushkov, R. S. and S. M. Shmeter, The effect of vertical movements in powerful convective clouds upon wind field, *Izv, AN SSSR, FAiO* 4(3), 1968.

Shmeter, S. M., Interaction of cumulonimbus clouds with wind field in surrounding atmosphere, *Izv. AN SSSR, FAiO* 2(10), 1966.

Shmeter, S. M. Structure of field of meteorological elements in zone of cumulonimbus clouds, *Trudy TsAO*, No. 88, 1969.

Shmeter, S. M. and V. I. Silaeva, Vertical currents within cumulonimbus clouds, *Meteorologiya i gidrologiya* [Meteorology and hydrology], No. 10, 1966.

RADAR STUDIES

The Use of Polarization Methods of Radar Study of Clouds and the Results of Their Modification:

Minervin, V. Ye. and A. B. Shupyatskiy, Radar study of phase state of clouds, *Izv. AN SSSR, FAiO* II(9), 1966.

Morgunov S. P. and A. B. Shupyatskiy, Estimating effectiveness of modification by the polarization characteristics of an echo-signal, *Trudy TsAO*, No. 57, 1964.

Minervin, V. Ye., S. P. Morgunov, and A. B. Shupyatskiy, Polarization studies of the structure of cumulonimbus clouds, Proceedings of III All-Union Conference on Radar Meteorology, Gidrometeoizdat, Moscow, 1968.

Pavlov, N. F. and V. F. Stepanenko, Technique for and results of experimental study of polarization properties of meteorological objectives, *Proceedings III All-Union Conference on Radar Meteorology*Gidrometeoizdat, Moscow, 1968.

Minervin, V. Ye., S. P. Morgunov, and A. B. Shupyatskiy, Measuring polarization characteristics of signal from cloud of artificial particles, *Trudy TsAO*, No. 95, 1971.

Radar Methods of Measuring Microstructure of Clouds and Precipitation:

Gorelik, A. G., I. V. Gryts'kin, L. A. Penyaz' and V. V. Tsykunov, Some results of combined radar and ground measurements of micro-structure of precipitation, *Proceedings of III All-Union Conference on Radar Meterology*, Gidrometeoizdat, Moscow, 1968.

Abshaev, M. T., Radar methods for measuring micro-structure characteristics of clouds, *Proceedings III All-Union Conference on Radar Meteorology*, Gidrometeoizdat, Moscow, 1968.

Abshaev, M. T. and Yu. A. Dadali, Possibilities of micro-structural studies of clouds and precipitation by radar methods, *Trudy VGI* [Proceedings of the High-Altitude Geophysical Institute—Vysoko-gornyy geofizicheskiy institut), No. 5, 1966.

Abshaev, M. T., Radar characteristics and microstructure of convective clouds, *Trudy VGI*, No. 5, 1966.

Minervin, V. Ye., S. P. Morgunov, and A. B. Shupyatskiy, One-time returning of cumulus and cumulonimbus clouds according to radar observations, *Trudy TsAO*, No. 5, 1971.

Orderly and Turbulent Motions in Clouds:

Zhupakhin, K. S. and V. T. Lenshin, Correlation links of statistical parameters of echo-signals from cumulonimbus clouds with rates of updrafts in them, *Trudy GGO*, No. 224, 1968.

Zhupakhin, K. S., Technique of operational estimate of turbulent motions in clouds, *Trudy GGO*, No. 224, 1968.

Zhupakhin, K. S., Operational statistical analysis of echo-signals from clouds, showers and thunderstorms, *Trudy GGO*, No. 224, 1968.

Gashina, S. B., I. M. Imyanitov, I. I. Kamaldina, Ye. M. Sal'man, and Ye. V. Chubarina, Relation of radar characteristics of clouds to their turbulent and electrical states, *Trudy GGO*, No. 173, 1965.

Mel'nichuk, Yu. V., Structure of horizontal pulsations of wind in precipitation according to radar data, *Proceedings of III All-Union Conference on Radar Meterology*, Gidrometeoizdat, Moscow, 1968.

Gorelik, A. G. and V. A. Patsaeva, Study of structure of turbulent current by means of radar in near-earth layer of subfrontal clouds and precipitation, *Proceedings of All-Union Conference on Radar Meteorology*, Gidrometeoizdat, Moscow, 1968.

Gorelik, A. G. and A. A. Chernikov, Some results of radar study of structure of wind field at altitudes of 50–70 m, *Trudy TsAO*, No. 57, 1964.

Gorelik, A. G., V. V. Kostarev, and A. A. Chernikov, Coordinate-Doppler Wind observation method, *Trudy TsAO*, No. 57, 1964.

Gorelik, A. G. and V. A. Patsaeva, Wind change in boundary layer according to "clear sky" radar reflection, *Meteorologiya i gidrologiya*, No. 8, 1967.

Mel'nichuk, Yu. V. Measuring turbulence in precipitation by means of a Doppler radar station, *Izv. AN SSSR. Ser. FAiO* II(7), 1966.

Gorelik, A. G., V. F. Logunov, Determining rate of updraft of air in showers by Doppler radar, *Izv. AN SSSR, FAiO* V(5), 1969.

Radar Characteristics of Shower and Cumulonimbus Clouds and Cloud Systems:

Zhupakhin, K. S., Technique of operational estimate of turbulent motions in clouds, *Trudy GGO*, No. 224, 1968.

Zhupakhin, K. S., Operational statistical analysis of echo-signals from clouds, showers, and thunderstorms, *Trudy GGO*, No. 224, 1968.

Gashina, S. B., I. M. Imyanitov, I. I. Kamaldina, Ye. M. Sal'man, and Ye. V. Chubarina, Relation of radar characteristics of clouds to their turbulent and electrical states, *Trudy GGO*, No. 173, 1965.

Mel'nichuk, Yu. V., Structure of horizontal pulsations of wind in precipitation according to radar data, *Proceedings of III All-Union Conference on Radar Meterology*, Gidrometeoizdat, Moscow, 1968.

Gorelik, A. G. and V. A. Patsaeva, Study of structure of turbulent current by means of radar in near-earth layer of sub-frontal clouds and precipitation, *Proceedings of All-Union Conference on Radar Meteorology*, Gidrometeoizdat, Moscow, 1968.

Gorelik, A. G. and A. A. Chernikov, Some results

of radar study of structure of wind field at altitudes of 50–70 m, *Trudy TsAO*, No. 57, 1964.

Gorelik, A. G., V. V. Kostarev, and A. A. Chernikov, Coordinate-Doppler Wind observation method, *Trudy TsAO*, No. 57, 1964.

Gorelik, A. G. and V. A. Patsaeva, Wind change in boundary layer according to "clear sky" radar reflection, *Meteorologiya i gidrologiya*, No. 8, 1967.

Mel'nichuk, Yu. V. Measuring turbulence in precipitation by means of a Doppler radar station, *Izv. AN SSSR, Ser. FAiO* II(7), 1966.

Gorelik, A. G. and V. F. Logunov, Determining rate of updraft of air in showers by Doppler radar, *Izv. AN SSSR, FAiO*, V (5), 1969.

Zabolotskaya, T. N. and V. M. Muchnik, Development of shower zones according to radar observation data, *Trudy UkrNIGMI*, No. 82, 1969.

Anchugova, R. A., The question of the dynamics of the development of radio-echo from cumulonimbus clouds, *Trudy GGO*, No. 224, 1968.

Borovikov, A. M., V. V. Kostarev, and A. B. Shupyatskiy, Some results of radar observations of evolution of powerful-cumulous clouds and modification results, *Trudy TsAO*, No. 57, 1964.

Brylev, G. B., Ye. M. Vorob'ev, S. B. Gashina, L. G. Kachurin, G. L. Nizdoyminoga, and Ye. M. Sal'man, Some results of observations on the transformation of radar characteristics of powerful cumulus clouds, *Trudy GGO*, No. 243, 1969.

Borovikov, A. M., V. V. Kostarev, I. P. Mazin, et al., *Radiolokotsionnyye izmereniya osadkov* [Radar measurements of precipitation], Geidrometeoizdat, 1967.

Voronov, G. S., Some data from a hail study in the Alazan Valley, *Trudy TsAO*, No. 65, 1965.

Lapcheva, V. F., Correlation of results of radar studies of convective clouds with mechanism of formation of precipitation, *Trudy VGI*, No. 5, 1966.

Sal'man, Ye. M., G. B. Brylev, V. K. Zotov, B. Sh. Divinskaya, and A. A. Fedorov, Joint use of radar and satellite observations in analyzing meso- and macro-scale systems, *Meterologiya i gidrologiya*, No. 2, 1969.

Gashina, S. B. and Ye. M. Sal'man, Pecularities of radar characteristics of thunderclouds,*Trudy GGO*, No. 173, 1965.

Ignatova, R. V., V. A. Petrushevskiy, and Ye. M. Sal'man, Radar Indications of the nature of cloudiness, *Trudy GGO*, No. 173, 1965.

Divinskaya, B. Sh., Field analysis of cloudiness and precipitation according to radar observations, *Trudy GGO*, No. 173, 1965.

Lapcheva, V. F., Assessing hail risk of convective clouds, *Trudy VGI*, No. 11, 1968.

Abshaev, M. T., Results of experimental verification of radar method of indicating hail zones, *Trudy VGI*, No. 14, 1969.

Gashina, S. B. and Ye. M. Sal'man, Statistical features of radar characteristics of convective clouds under different physical and geographical conditions, *Trudy GGO*, No. 243, 1969.

Kulikova, G. I., G. T. Nikandrova, and V. A. Petrushevskiy, Accuracy of measurements of cloud boundaries by radar, *Trudy GGO*, No. 173, 1965.

Petrushevskiy, V. A. and Ye. M. Sal'man, Radar determination of altitude of cumulonimbus clouds, *Trudy GGO*, No. 1958, 1964.

Gashina, S. V. and Ye. M. Sal'man Radar indications of nature of cloud systems and their evolution, *Trudy GGO*, No. 217, 1967.

Sal'man, Ye. M. and S. B. Gashina, Localizing precipitation and hail risk zones by their radar characteristics, *Trudy GGO*, No. 217, 1967.

Ignatova, R. V. and M. A. Romanov Radar determination of cloud characteristics, *Trudy GGO*, No. 217, 1967.

Muchnik, V. M., Differentiation between thunderstorm and hail zones by the radar method, *Sb. "Issledovaniye po fizike oblakov i aktivnym vozdeystviyam na pogodu"* [Coll. "Study of cloud physics and weather modification"], Gidrometeoizdat, Moscow, 1967.

Abshaev, M. T. and Yu. A. Dadali, Radar measurements of water content of cumulonimbus clouds, *Trudy VGI*, No. 19, 1971.

Sal'man, Ye. M., S. B. Gashina and L. I. Kuznetsova, Dependence of radar criteria of dangerous phenomena upon intensity of convection, *Trudy GGO*, No. 261, 1971.

Sal'man, Ye. M. and B. Sh. Divinskaya, Questions of the meteorological effectiveness of a radar system of observations on cloudiness and dangerous weather phenomena, *Trudy GGO*, No. 261, 1971.

Methods of Identifying Hail Zones and Determining the Degree of Risk:

Abshaev, M. T., O. M. Kuchmezov, and A. M. Pinkhasov, Probability-statistical method of indicating hail clouds, *Trudy VGI*, No. 13, 1969.

Abshaev, M. T., Results of experimental verification of radar method of indicating hail risk, *Trudy VGI*, No. 14, 1969.

Lapcheva, V. F., Assessing hail danger and hail content of convective clouds, *Trudy VGI*, No. 11, 1968.

CREATION AND BREAKUP OF CONVECTIVE CLOUDS

Results of Experiments on Breaking up Cumulus Clouds with Loose Powders:

Gayvoronskiy, I. I., L. P. Zatsepina, and Yu. A. Seregin, Result of modification experiments on convective clouds, *Izv. AN SSSR FAiO* 6(3), 1970,

Gayvoronskiy, I. I., L. P. Zatsepina, B. I. Zimin, and Yu. A. Seregin, Use of reagents in powder form for thundercloud modification, [Proceedings of VIII All-Union Conference on Cloud Physics and Modification], Gidrometeoizdat, Leningrad, 1970.

Stimulating Updrafts by Means of Artifically Created Jets Which Trigger Cloud Development:

Trudy IPG [Proceedings of the Institute of Applied Geophysics—institut prikladnoy geofiziki], No. 12, Moscow, 1970.

ELEMENTARY PHYSICAL AND CHEMICAL PROCESSES IN CLOUDS

Experiments with the Use of a Device for Modeling Cloud Processes:

Volkovitskiy, O. A., A set of experimental devices for geophysical studies, *Meteorologiya i gidrologiya*, No. 6, 1965.

Borovskiy, N. V. and O. A. Volkovitskiy, Large aerosol chamber, *Trudy IPG*, No. 7, 1967.

Borovskiy, N. V. and O. A. Volkovitskiy, Test tank for temperature- and pressure-measuring equipment for studies in atmospheric physics, *Trudy IPG*, No. 7, 1967.

Studies of Elementary Processes in Clouds, Physics of Condensation, Coalescence, Freezing and Electrification of

Cloud Elements:

Volkovitskiy, O. A. and L. N. Pavlova, Measurement of transparency of fogs artifically created in experimental apparatuses, *Trudy IPG*, No. 7, 1967.

Volkovitskiy, O. A. and L. N. Pavlova, Eksperimental'noye issledovaniye vliyaniya otnositel'noy vlazhnosti na prozrachnost' sredy [Experimental study of the effect of relative humidity on the transparency of a medium], *Materials from the VII Conference on matters of evaporation, combustion and gas dynamics of dispersed systems*, Odessa, 1967.

Volkovitskiy, O. A., O vozmozhnosti modelirovaniya kondensatsionnoy stadii formirovaniya oblachnogo spektra v kamere[The possibility of modeling the condensation stage of the formation of a cloud spectrum in a chamber], *Materialy VII konferentsii po voprosam ispareniya, goreniya i gazovoy dinamiki dispersnykh sistem* Odessa, 1967.

Volkovitskiy, O. A., Modeling certain cloud processes in a chamber, *Trudy IEM* [Institut elektromekhaniki—Institute of Electromechanics], No. 1, 1969.

Volkovitskiy, O. A. and A. G. Laktionov, Study of initial stage of cloud formation in a chamber, *Izv. AN SSSR, Ser. FAiO*, 5(3), 1969.

Volkovitskiy, O. A. and A. G. Laktionov, On change of sizes of spectrum of fog droplets in a chamber during condensation growth, *Izv. AN SSSR FAiO* 6(3), 1970.

Volkovitskiy, O. A. and Yu. S. Sedunov, Toward a computation of the concentration of droplets and of maximum supersaturation at the initial stage of cloud formation, *Meteorologiya i gidrologiya*, No. 11, 1969.

Levin, L. M. and Yu. S. Sedunov Some questions on the theory of condensation nuclei, *DAN SSSR 170*, No. 1, 1966.

Levin, L. M. and Yu. S. Sedunov, Some questions about the theory of atmospheric condensation nuclei, *Trudy IPG*, No. 9, 1967.

Buykov, M. V., Kinetics of heterogeneous condensation during adiabatic cooling, *Kolloidnyy zhurnal* [Colloid Journal] 28(2), 1966.

Buykov, M. V., Kinetics of heterogeneous condensation during adiabatic cooling. Kinetic regime of droplet growth, *Kolloidnyy zhurnal*, 28(5), 1966.

Dekhtyar, M. N. and M. V. Buykov, Toward a theory of stratiform cloudiness. Evolution of a cloud in nonturbulent atmosphere, *Trudy UkrNIGMI*, No. 70, 1968.

Sedunov, Yu. S., Kinetics of initial stage of condensation in clouds, *Izv. AN SSSR, Ser. FAiO*, 3(1), 1967.

Laktionov, A. G., On the relation of condensation activity of cloud nuclei to their size, *Izv. AN SSSR*, 3(1), 1967.

Laboratory Investigations of Action of Crystallized Regents, Properties of Crystalline and Drop Fogs, Norm of Flow Rate of Reagents:

Bodunova, L. I., L. P. Zatsepina, and A. D. Solov'ev, Laboratory studies of interaction of particles of undissolved substances in a water aerosol, *Trudy TsAO*, No. 65, 1965.

Plaude, N. O., Some questions about quantitative measurements of characteristics of ice-forming aerosols, *Trudy TsAO*, No. 65, 1965.

Plaude, N. O., and A. D. Solov'ev, Using particles with large specific surface for cloud and fog modification, *Trudy TsAO*, No. 65, 1965.

Piotrovich, V. V., Phlorglucinol, a crystallizer for water droplets in supercooled fog and overcast, *Trudy GGO*, No. 186, 1966.

Byskov, N. V., N. N. Yartseva, and A. V. Bronberg, A study of the ice-forming activity of aerosols of metaldehyde and phloroglucinol, *Trudy GGO*, No. 186, 1966.

Gayvoronskiy, I. I., N. O. Plaude and A. D. Solov'ev, Artificial ice-forming aerosols, *Meteorologiya i gidroligiya*, No. 10, 1967.

Plaude, N. O. Study of the ice-forming properties of aerosols of silver iodide and lead iodide, *Trudy TsAO*, No. 80, 1967.

Plaude, N. O., and A. D. Solov'ev Analiz nekotorykh zakonomernostey l'doobrazovaniya na chastitsakh aerozolya neorganicheskikh veschestv [Analysis of certain regularities of ice-formation on particles of an aerosol of an inorganic substance], *Issledovaniya po fizike oblakov i aktivnomy vozdeystviyu* [Studies in cloud physics and modification], Gidrometeoizdat, Moscow, 1967.

Gromova, T. N., N. V. Gliki, and P. N. Krasikov, The effect of impurities of surface-active substances on the ice-forming effectiveness of solutions of phloroglucinol, silver iodide, and lead iodide, *Trudy GGO*, No. 186, 1966.

Preobrazhenskaya, Ye. V., On the interaction of finely dispersed powders of ion-exchange resins with water aerosol and water vapor, *Trudy GGO*, No. 224, 1968.

Gromova, T. N. and Yu. I. Sumin, The use of cuprous sulfide for modification of supercooled convective clouds, *Trudy GGO*, No. 224, 1968.

Kazankova, Z. P. and Kh. Kh. Medaliyev, Interaction of aerosol particles of ice-forming substances and supercooled water droplets, *Trudy VGI*, No. 13, 1969.

Zhikharev, A. S., Complex apparatus for laboratory study of dispersion of ice-forming reagents by explosion, *Trudy VGI*, No. 14, 1969.

Gromova, T. N., and V. T. Lenshin, On supplying dry ice during modification of convective clouds for the purpose of causing precipitation, *Trudy GGO*, No. 262, 1971.

Vychkova, N. V., T. N. Gromova, and Yu. P. Sumin, Ice-forming properties of cuprous sulfide as a reagent for modification of supercooled clouds, *Trudy GGO*, No. 262, 1971.

Stakevich, D. D. and T. S. Uchevatkina, Norms (rates?) for dispensing ice-forming reagents to cause artificial precipitation from convective clouds, *Trudy GGO*, No. 262, 1971.

Mechanism of Formation of Crystals on Crystallization nuclei:

Kachurin, L. G. and V. G. Morachevskiy, *Kinetika fazovykh perekhodov vody v atmosfere* [Kinetic of phase transitions of water in the atmosphere], Izd. LGU [Leningrad State University Publishing House], 1965.

Gliki, N. V., Simplest types of crystallization of super-cooled water droplets, *Kristallografiy* 2(5), 1966.

Borovikov, A. M., I. I. Gayvoronskiy, Ye. G. Zak, V. V. Kostarev, I. I. Mazin, V. Ye. Minervin, A. Kh. Khrgian, and S. M. Shmeter, *Fizika oblakov* [Cloud physics], Gidrometeoizdat, Leningrad, 1961.

Regularities in growth of individual crystals and droplets:

Gliki, N. V., Simplest types of crystallization of super-cooled water droplets, *Kristallografiya* 3(5), 1966.

Aleksandrov, E. L., L. M. Levin, and Yu. S. Sedunov, Condensation growth of droplets of a solution, *Izv. AN SSSR, Ser. FAiO*, No. 8, 1967.

Aleksandrov, E. L., L. M. Levin, and Yu. S. Sedunov, Condensation growth of droplets on hydroscopic nuclei, *Trudy IEM*, No. 6, 1969.

Mazin, I. P., Toward a theory of formation of size spectrum of particles in clouds and precipitation, *Trudy TsAO*, No. 64, 1965.

Smirnov, V. N., Rate of aggregation and condensation growth of aerosol particles, *Trudy TsAO*, No. 92, 1969.

Stochastic Theory of Condensation:

Levin, L. M., Studies on the physics of coarsely dispersed aerosols, *Izv. AN SSSR*, 1961.

Levin, L. M. and Yu. S. Sedunov, Gravitational aggregation of charged cloud droplets in a turbulent flow, *Pure and Applied Geophysics* **64**(2), 1966

Smirnov, V. N. Rate of aggregation and condensation growth of aerosol particles, *Trudy TsAO*, No. 92, 1969.

Levin, L. M. and Yu. S. Sedunov, On turbulent-gravitational aggregation of cloud droplets, *DAN SSSR* **164**(3), 1965.

Levin, L. M. and Yu. S. Sedunov, Turbulent-electrostatic gravitational aggregation of cloud droplets, *Trudy IPC*, No. 9, 1967.

Belyaev, V. N., On the evolution of the condensation spectrum of cloud droplets, *Izv. AN SSSR Ser. FAiO* **3**, 1967.

Levin, L. M. and Yu. S. Sedunov, The theoretical model of the drop spectrum formation processes in clouds, *Pure and Applied Geophysics* **69**(1), 1968.

Sedunov, Yu. S., Fine structure of clouds and its role in formation of cloud droplet spectrum, *Izv. AN SSSR, Ser. FAiO* **1**(7), 1965.

Levin, L. M. and Yu. S. Sedunov, Stochastic condensation of drops and kinetics of spectrum formation, *J. Rech. Atm.*, No. 2–3, 1966.

Levin, L. M. and Yu. S. Sedunov, Kinetic equation to describe micro-physical processes in clouds, *DAN SSSR*, **170**(2), 1969.

Sedunov, Yu. S., Some questions about the kinetics of condensation processes in clouds, *Izv. An SSSR FAiO*. **5**(1), 1969.

Quantitative Theory of Processes of

Formation of Crystallization Nuclei, Formation of Crystallization on Zone and its Rate of Spread, Technique for Introducing Reagent, Characteristics of Open Zone:

Trudy UkrNIGMI, No. 74, 1968.

Trudy UkrNIGMI No. 77, 1969.

Trudy UkrNIGMI, No. 86, 1969.

Trudy UkrNIGMI, No. 95, 1970.

Polovina, I. P., *Vozdeystviya na sloistoobraznyye oblaka* [Modification of stratiform clouds], Vol. 1, Gidrometeoizdat, Leningrad, 1971.

Trudy UkrNIGMI, No. 99, 1970.

Belyaev, V. I., V. V. Vyal'tsev, and I. S. Pavlova, Experimental weather modification by seeding fog with dry ice, *Izv. AN SSSR FAiO* **2**(6), 1966.

Trudy IPG, No. 1, 1965.

Trudy TsAO, No. 80, 1967.

Trudy TsAO, No. 89, 1969.

Trudy TsAO, No. 90, 1969.

Trudy TsAO, No. 96, 1970,

DISSIPATION OF SUPERCOOLED CLOUDS AND FOGS

Study of Conditions Permitting Fog Dissipation:

Polovina, I. P., *Vozdeystviya na sloistoobraznyye oblaka* [Modification of stratiform clouds], Vol. 1, Gidrometeoizdat, Leningrad, 1971.

Experiments on Clearing Large Areas (on the Order of 10,000 km²) of Overcast Due to a Change in the Radiation Balance:

Belyaev, V. I., V. V. Vyal'tsev, and I. S. Pavlova, Experimental weather modification by seeding fog with dry ice, *Izv. AN SSSR FAiO* **2**(6), 1966.

MODIFICATION OF HAIL PROCESSES

Results of Studies of Processes of Formation of Hail Cloud, Growth of Hail and its Transformation; Development of Technique for Modifying Hail Processes and Results of Experimental Work:

Trudy VIII Vsesoyuznoy konferentsii po fizike oblakov i aktivnym vozdeystviyam [Proceedings of VIII All-Union Conference on Cloud Physics and Modification], Gidrometeoizdat, Leningrad, 1970.

Trudy VII Mezhvedomstvennoy komissii po fizike oblakov i aktivnym vozdeystviyam na pogodu [Proceedings of VII Interdepartmental Commission on Cloud Physics and Weather Modification], Moscow, 1967.

Issledovaniya po fizike oblakov i aktivnym vozdeystviyam na pogodu [Studies in Cloud Physics and Weather Modification], Gidrometeoizdat, Moscow, 1967.

Trudy SANIGMI [Sredneaziatskiy nauchno-issledovatel'skiy gidrometeorologicheskiy institut—[Central Asian Hydrometeorological Institute], No. 3l (46), Leningrad, 1967.

Trudy TsAO, No. 65, 1965.

Voronov, G. S., Taking observations on several atmospheric phenomena at weather stations, *Meteorologiya i gidrologiya,* No. 1, 1967.

Voronov, G. S., I. I. Gayvoronskiy, B. N. Leskov, and Yu. A. Seregin, Hail protection experiment in the Moldavian Soviet Socialist Republic, *Meteorologiya i gidrologiya*, No. 7, 1967.

Sokol, G. P. Some results of an estimate of effectiveness of modification of hail-risk clouds in the Gissar Valley, *Meteorologiya i gidrologiya*, No. 1, 1967.

Trudy VGI, No. 5, 1966.

Trudy VGI, No. 11, 1968.

Trudy VGI, No. 13, 1969.

Trudy VGI, No. 14, 1969.

Trudy VGI, No. 17, 1970.

Trudy VGI, No. 20, 1972.

AUGMENTATION OF PRECIPITATION FROM CLOUDS AND CLOUD SYSTEMS

Results of Modifying Frontal Cloud Systems and Air-Mass Clouds by Means of Dry Ice:

Trudy UkrNIGMI, No. 89, 1970.

Trudy IEM, No. 3, 1968.

Litvinov, I. V., Redistribution of precipitation by modifying clouds with cold reagents, *Meteorologiya i gidrologiya*, No. 9, 1967.

Increasing precipitation from cumulus and powerful-cumulus clouds over a ukrainian test area:

Trudy UkrNIGMI, No. 92, 1970.

Trudy UkrNIGMI, No. 103, 1971.

EXTINGUISHING FOREST FIRES BY CLOUD MODIFICATION

Results of First Experiments Showing Practicability of Work on Extinguishing Forest Fires by Stimulating Artificial Precipitation over Fire Regions:

Trudy GGO, No. 262, 1971.

WATER RESERVES OF CLOUD SUITABLE FOR MODIFICATION

Studies of Water Reserves of Seedable Clouds over Various Regions of the USSR:

Leskov, B. N., L. N. Mogila, and I. P. Polovina, On the possible amount of artificial precipitation from cloudy layers of "dry" fronts, *Trudy UkrNIGMI, izd.* No. 99, 1971.

Leskov, B. N., Preliminary results of correlation of amount of artificial precipitation and water reserves of seeded frontal clouds, *Trudy UkrNIGMI, izd.* No. 106, 1971.

Chuvaev, A. P., A study of resources of artificial augmentation of precipitation from convective clouds in regions with insufficient humidity, *Trudy GGO*, No. 186, 1966.

Litvinov, I. V., and Yu. A. Ruzheynikov, Frequency of clouds near Moscow suitable for artificial modification, *Trudy IEM*, No. 3, 1968.

Mamina, Ye. F. and Ye. K. Fedorov, On the water balance of cloud systems, *Izv. AN SSSR, ser. Geofiziki*, No. 5, 1967.

Minervin, V. Ye., Water content of clouds, *Trudy TsAO*, No. 64, 1965.

Leonov, M. P., The matter of sources of water for artificial clouds in cloud modification, *Trudy UkrNIGMI*, No. 53, 1966.

Leonov, I. P. and V. I. Perelet, *Aktivnyye vozdeystviya na oblaka v kholodnoye polugodiye* [Cloud modification during the cold part of the year, Gidrometeoizdat, 1967.

Leonova, M. P. and I. V. Koketaktinova, Preliminary data on water content and precipitation of cyclones. Sbornik Meteorologiya Klimatologiya i gidrologiya [Collection, Meteorology, Climatology and Hydrology], Izd KGU [Kazan State University Publishing House—Izd. Kazanskiy gosudarstvennyy universitet], 1969.

Leskov, B. N. and I. P. Polovina, Estimating the possible quantity of additional precipitation from clouds during the cloud part of the year, *Sbornik po fizike oblakov i aktivnykh vozdeystviy* [Collection, Cloud Physics and Modification], Moscow, 1969.

Polovina, I. P. *Vozdeystviye na vnutrimassovyye oblaka slozhnykh form* [Modifying air-mass clouds of complex forms], Gidrometeoizdat, Leningrad, 1971.

ESTIMATING THE EFFECTIVENESS OF CLOUD MODIFICATION

Estimating Effectiveness of Cloud Modification Experiments and Monitoring of Results of Modification:

Prikhot'ko, G. F., Estimating the effectiveness of artificial precipitation from convective clouds, *Trudy UkrNIGMI*, No. 61, 1966.

Osipova, G. I. and Yu. S. Fridman, The matter of estimating the effectiveness of cloud modification aimed at increasing rainfall, *Trudy GGO*, No. 156, 1964.

Leonova, M. P., Estimating modification of supercooled clouds, *Meteorologiya i gidrologiya*, No. 6, 1969,

Korniyenko, Ye. Ye., Statistical estimate of effectiveness of modifying convective clouds to increase rainfall, *Trudy UkrNIGMI*, No. 74, 1967.

Kotarin, V. S., Possible effectiveness of modification of lower clouds (fogs), *Trudy GGO*, No. 186, 1966.

Vol'nets, Kh. M., A. A. Levenko, M. L. Markovich, and V. M. Muchnik, Radar observation as a method for studying modification of supercooled stratus clouds, *Meteorologiya i gidrologiya*, No. 10, 1963.

Buykov, N. V., On estimating agricultural effectiveness of artificial redistribution of rainfall, *Trudy UkrNIGMI*, No. 92, 1970.

Gayvoronskiy, I. I. and B. I. Zimin, On monitoring the results of modification of storm processes, *Trudy TsAO*, No. 95, 1971.

11 | Progress of Hail Suppression Work in the USSR

G. K. SULAKVELIDZE

B. I. KIZIRIYA

V. V. TSYKUNOV

Means and techniques for modifying hail processes which have been devised over the past two decades in the USSR, based on a large body of scientifically researched and experimentally designed work, have been used widely for the protection of agricultural crops from hail damage. Hail suppression work has been of a fully applied nature since 1967; it is done to order, under contract with interested agricultural organizations. In 1972 the total extent of hail-protected territory was about 4 million hectares. According to the estimates of a number of agricultural, state insurance, and scientific organizations, the USSR Hydrometeorological Service, and others, hail damage has been cut by a factor of 3 to 5, and the value of the crops preserved from destruction amounts to tens of millions of rubles.

G. K. Sulakvelidze is a professor, Hydrometeorlogical Institute Tbilisi, Union of Soviet Socialist Republics.

B. I. Kiziriya is Director, Officer of Weather Modification, Hydrometerological Service, Moscow, Union of Soviet Socialist Republics.

V. V. Tsykunov is associated with the Institute of Experimental Meterology, Obninsk, Union of Soviet Socialist Republics.

In the USSR systematic studies of hail processes and the quest for a means of modifying these processes were begun in the second half of the 1950s. This work was done by scientific institutes of the USSR Hydrometerological Service and the Institute of Geophysics of the Academy of Sciences of the Georgian Soviet Socialist Republic. Among the hydrometerological scientific institutes during various stages of the studies, and later in the implementation of results obtained by leaders in the field, were the High Mountain Geophysical Institute (Vysokogornyy geofizicheskiy institut) (VGI), the Central Aerological Observatory (Tsentral naya aerologicheskaya observatoriya, TsAO (the Transcaucasian Hydrometeorological Institute), Zak-NIGMI (Zakavkazskiy gidrometeorologicheskiye institut), and the Central Asian Regional Hydrometeorological Institute (SARNIGMI, formerly SANIGMI) (Sredneaziatskiy regionalniy gidrometeorologicheskiye institut).

The groundwork for the study of hail processes was laid by studies of the conditions for formation of convective clouds and rainfall,[1, 17, 36, 38] artificial and natural

Table 11.1   Sizes of Hail Zones

Source	Study Region	Study Period	Hail Zone Volume (km^3)
TsAO	Moldavian SSR	1964–1970	2–40
IGAN	Georgian SSR	1961–1970	10–50
VGI	Armenian SSR, Northern Caucasus	1961–1969	3–15
USA	Eastern Colorado	1964–1967	15–20

crystallization of cloud particles,[15, 19, 32, 35] the thermodynamic conditions for the origin and structure of convective currents,[47, 38] size distribution of cloud and rain droplets,[6, 25, 29, 50] mechanics of aerosols and the aggregation growth of cloud droplets,[13, 24, 30, 37] radar parameters of clouds,[7, 8, 26, 48] and sizes of hail paths and amounts of damage inflicted by hail on agriculture.[14, 40] (Here and in the following passages, only a portion of the many articles extant on a given subject are cited.)

During the period 1956–1958, existing concepts about the mechanism of hail formation were contradicted by data from experimental observations. For a successful solution of the problem it was necessary to resolve three basic problems: (*a*) define more accurately the mechanism by which hail is formed; (*b*) find some step in it, an artificial change that could lead to the prevention of hail incidence, or at least a reduction of the size of hail that falls;

and (*c*) find the technical means of intervening in the natural course of the hail formation process.

CONCEPT OF THE MECHANISM OF HAIL FORMATION

Tables 11.1, 11.2, and 11.3 show some characteristics of hail processes, as obtained by the VGI, TsAO and IGAN (Institute of Geophysics, Academy of Sciences, Georgian Soviet Socialist Republic—Institut geofiziki Akademii nauk Gruzinskoy SSR). Results of observations carried out in the United States are also given, for comparison. It is evident from the tables that hail zones of comparable sizes and approximately equal rates of travel are observed in regions with different physical, geographical, and climatological conditions; lifetimes of hail clouds and growth of hail are also identical. The data in the tables on the rate

Table 11.2   Rate of Travel of Hail Zones

Source	Study Region	Study Period	Rate of Travel of Hail Zones (km/hr)
TsAO	Moldavian SSR	1964–1970	50
IGAN	Georgian SSR	1961–1970	30–50
VGI	Armenian SSR, Northern Caucasus	1961–1970	10–15 during air mass processes, 15–30 during frontal processes
USA	Eastern Colorado	1964–1967	10–15

Table 11.3   Period of Existence of Hail Clouds and Growth of Hailstones

Source	Study Region	Study Period	Lifetime of Hail Cloud (hr)	Time of Growth of Hailstones (min)
TsAO	Moldavian SSR	1964–1967	—	10–12, less commonly 30
IGAN	Georgian SSR	1962–1967	1.0–3.0	15–20
VGI	Armenian SSR Northern Caucasus	1962–1967	1.0–2.5	5–15
USA	Eastern Colorado	1964–1967	1.0–2.0	—

of travel of hail zones and volume of hail zones are average ones. In isolated instances hail zones with characteristics differing significantly from those shown can be observed (for example, the rate of travel can exceed 100km/hr).

The ZakNIGMI, VGI, IGAN, and TsAO have carried out detailed studies of the trajectories of motion of hail zones, of hail paths, and of the spectrum of hail. It would seem that the average value for the diameter of hailstones is equal to 1.0 to 2.0 cm, and the spectrum of hail can best be described by a gamma distribution or by the Khrgian-Mazin formula.[23, 40]

All this indicates, apparently, that the mechanism of formation and growth of hailstones is identical for different regions and is determined principally by the thermodynamic state of the atmosphere.

A working model of a cloud was created at the VGI from 1960 to 1962 according to which the speed of updraft in the cloud increases with altitude, attaining maximum magnitude in the upper part of the cloud, after which it begins to decrease.[41] Such a distribution of updraft velocities creates the conditions for accumulation of large drops in the upper part of the cloud and formation of a zone of increased water content. Hail growth occurs principally on embryonic particles near the top of the cloud because of the accumulated supercooled water in a zone of

increased water content (zone of accumulation). Embryonic hailstones are formed as a result of the freezing of large drops. IGAN scientists developed and completed this model after having shown that in certain cases several accumulation zones can simultaneously be found in a cloud.[9] In this case the explanation of the multilayer structure of hailstones, which caused considerable difficulty in the VGI model, is significantly simplified.

Starting from these concepts, approximative formulas were suggested which made it possible to estimate the possible radius of hailstones R_m and the maximum quantity of water $Q_m{}^3$ which can be stored in a cumulus cloud at a given stratification* of the atmosphere, the stratification determining the magnitude of the maximum rate of updrafts W_m:

$$R_m \leq \frac{W_m{}^2}{\gamma^2} \qquad (11.1)$$

$$Q_m \leq \frac{(W_m{}^2 - V_c{}^2)\bar{p}}{{}^2g}, \qquad (11.2)$$

where γ is a constant equal to 2.2×10^3 cm/sec², which depends on the terminal gravitational speed of large drops or hail; V_c is the critical fall speed of drops at

---

* In meteorological usage, stratification refers to the vertical distribution pressure, temperature, and humidity [Translator's note].

which their breakup occurs (Lenard effect); g is the acceleration of the force of gravity; and $\bar{\rho}$ is the average air density in the accumulation zone.

Several listed references[18, 27, 33, 34] set forth the structure of a mathematical model of a cumulus cloud. The results obtained agree with field data on the first stage of development of a cumulus cloud. The change in the magnitude of the updraft speed in space and time, and also the field of temperature and water content, gives a picture which approximates the data from observations up to the onset of water accumulation in the cloud.

The general inadequacy which these theoretical treatments have in common is that they cannot allow for water accumulation or for phenomena associated with the process of water drop freezing; thus a complete detailed model of a developed "raining" cloud has still not been obtained.

At the same time, however, these investigations confirmed the accuracy of suggestions made earlier about the change of rate of updrafts with altitude and indicated further ways of constructing the mathematical model of a convective cloud.

It is no doubt possible to gather a multitude of logically correct and, it would seem, physically sound schemes of the formation of hail, supported by laboratory experiments and theoretical computations. The whole question is, how well are these schemes verified under field conditions of hail formation and incidence in various climatic and geographical regions. Unfortunately, such verification often gives negative results. This is apparently because we are as yet unable, either in laboratories or in theoretical computations, to construct an exhaustive model of such a complex and multifaceted process as the formation of heavy rain and hail in cumulonimbus clouds. At the present stage of the work, therefore, we must exclude from consideration certain incidental factors which are in themselves of considerable interest, and attempt to ascertain the causal link between basic atmospheric parameters (which govern the formation of precipitation) and the size of hail and amount of precipitation, and then after that try to verify one means or another for the dependence established by means of observations under field conditions.

To verify the accuracy of the VGI concepts about the mechanism of hail formation, a method of hail prediction was developed.[16, 42] The source data chosen were radiosonde data with corrections for advection of air masses according to altitude up to the moment maximum instability developed, that is, considering the predicted stratification of the atmosphere. Concepts about the slice method are used to calculate the magnitude of the maximum updraft velocity, and to determine the altitude of the accumulation zone and the temperature at its upper and lower boundaries, the radius of hail, and the quantity of incident precipitation. Melting of hail during its descent in the warm part of the atmosphere is considered in these computations. The VGI method of hail prediction was tested at the USSR Hydrometeorological Center and during hail suppression projects; verification of the prediction amounted to greater than 90 percent. Moreover, when rates of updrafts in cumulus clouds are calculated according to the means mentioned, they coincide very accurately with observational data. The VGI method of hail prediction described is presently in wide practical use in forecast units and on hail suppression projects.

It should be noted that although the VGI model does not lay claim to great rigor in describing isolated details of the mechanism of hail formation, it evidently provides a correct concept of the general nature of this process and makes it possible, just as experiments have made it possible, to predict the average maximum diameter of hail falling with an accuracy up to 0.5 to 0.8 cm.

After an explanation of the basic principles and conditions of the formation and growth of hail, it became possible to identify the following conditions of stratification of the atmosphere as necessary for hail formation:

1.   The thermodynamic state of the atmosphere should be such as to cause an updraft speed in the cloud varying with altitude and attaining a maximum value roughly just below the top of the cloud, and then decreasing toward its top. In certain instances there may be several such zones with an increased rate of updraft. For hail formation the maximum updraft speed must exceed 15 m/sec.[40]

2.   The velocity distribution in a convective cloud is such as to create conditions for the formation of an accumulation zone where accumulation of a solid or liquid large-drop fraction occurs. The accumulation zone is located above the level of maximum updraft velocity. The water content in the accumulation zone can reach 20 g/m³ or more.

3.   Hail grows basically at the expense of the supercooled large-drop fraction in the accumulation zone, and also at the expense of the solid particles; therefore conditions for hail formation and growth arise only when stratification of the atmosphere leads to the development of an accumulation zone with a temperature between 0 and $-20°C$ at the height of the lower boundary. At other temperatures at the lower boundary of the accumulation zone, as computations and observations have shown, hail growth conditions disappear. The temperature is taken inside the cloud.

4.   Under the conditions mentioned in the preceding three points, there are two situations in which precipitation from convective clouds can fall: (a) when the ascertained rate of descent of a large particle affected by the force of gravity exceeds the maximum updraft rate—in this case it is most common for solid precipitation (hail or graupel) to fall, and its maximum radius can be computed with Equation 11.1; and (b) when the concentration and total weight of the system of drops and solid particles in the accumulation zone reach certain magnitudes, conditons are then created for their descent as updrafts are weakened or suppressed. Using known assumptions, the magnitude of the corresponding water content in the accumulation zone can be calculated by means of Equation 11.2.

5.   A raining cumulus cloud is a generator which condenses water vapor contained in the air, repeatedly stores it in the accumulation zone, and pours it onto the earth. This process goes on for as long as the cloud exists.[12, 40, 42] During this time the water in the cumulus cloud is completely replenished four to five times and sometimes more, depending on the extent of the layer between the condensation level and the level at which the updraft speed attains its maximum value, on the size of the temperature gradient in this layer, and on the temperature at the condensation level. The first two parameters also determine the magnitude of the maximum speed of the updraft.

The energy of a cumlonimbus cloud in which hail forms is so great that direct intervention in this process would scarcely seem either possible or advisable. Therefore, in modifying convective clouds to prevent the fall of hail it was necessary to find another means of converting the cloud into some other state, with modification in mind. This state is achieved by conversion of the potential energy of unstable equilibrium which has been stored in the cloud itself, and in particular, the energy of water-to-ice phase transitions in the supercooled large-drop zone.

Such basic concepts about the mechanism of hail formation comprised the basis on which principles for the modification of hail processes were suggested.

MODES OF INTERVENTION IN HAIL PROCESSES AND SOME SUPPLEMENTS TO WORK ON MODELING A HAIL CLOUD

On the basis of concepts about the mechanism of hail formation, a method of intervening in this process was developed.

According to the concepts developed at the VGI and IGAN, there is sufficient basis for accepting the continuous and uninterrupted artificial crystallization of all supercooled fractions in the hail formation part of the cloud as an efficient means for preventing the growth of hailstone embryos. Computations show that the total conversion to ice of a supercooled cloud requires that the concentration of crystallization nuclei active at a given temperature must reach at least 10^6 to $10^7/m^3$. The artificial creation and continuous maintenance of such a concentration of crystallization nuclei over the entire hail-risk period over the territory to be protected is not practicable, since it would require the dispensing of an enormous quantity of reagent.

In view of this, significant interest was given to the idea of artificially increasing the concentration of ice particles with volume comparable to the volume of natural hailstone embryos, making the artificial particles therefore able to compete with the natural ice particles in capturing supercooled cloud droplets.

Such artificial multiplication of hailstones should lead to an increase in the total mass of the ice fraction and to a slowing of the rate of growth of the

average volume of the individual particles in this fraction.

Actually, if the quantity of precipitation in a cumlonimbus cloud is of limited magnitude according to Equation 11.2, and if the hailstone growth occurs principally because of the large-drop fraction in the accumulation zone, then according to Equations 11.1 and 11.2, increasing the number of hailstone embryos will decrease the final hailstone diameter.

The limiting radius of a hailstone R_α at the 0° isotherm level after such action is taken is determined from the following expression:

$$R_\alpha = R_n \cdot \sqrt[3]{N_n/N_\alpha}, \qquad (11.3)$$

where R_n and N_n are the radius of the hail and the number of hailstone embryos in the natural process; N_α is the number of artificial hailstone embryos. Experimental data have shown that by using this method of modification, it is possible to decrease the diameter of hailstones by a factor of approximately 4 with a reagent dosage of 50 to 100 g/km³ in the hail growth zone.

Such a modification method is employed only when hail growth is occurring because of accumulated water; therefore R_m from Equation 11.1 can be substituted for R_n in Equation 11.3. We note that the results of radar studies show that the principal hailstone growth occurs in the zone of large water content.[3, 9] This is corroborated in Reference 20 and, moreover, the many layers of which hail is composed are explained not only by the presence of alternating zones of increased water, but also by patterns of change in such parameters as the temperature of the cloud medium, the rate of flow, the size of drops, etc.

In an infinitely extended cloud, either with an unlimited prolongation of intense updrafts or of continuously replenished cloud water content, both natural and artificial hailstone embryos can theoretically grow to as large a size as desired, and then artificial modification would lead to intensification of the hail formation process. In actuality, however, under natural conditions the vertical geometrical dimensions and lifetimes of convective clouds, as well as their updraft velocities and the duration of existence during release of convective instability, are rather limited.

Moreover, the accumulation of continuously growing ice pellets, which unlike liquid drops are not subject to the Lenard breakup effect, will still favor their fall as a group even when the rate of descent of each individual particle is substantially less than the rate of the updraft. This fall, it is likely, will facilitate the suppression of updrafts and the formation of downdrafts, leading to breakup of the convective cell involved. Therefore it is in theory always possible to select a concentration of ice pellets that will make it impossible for hailstone embryos to grow to dangerous sizes in real clouds.

Along with this, the increase in number of hailstone embryos inevitably leads to the involvement of new masses of supercooled cloud droplets in the act of accretion, and the increase in the average radius of the total collection of hailstone embryos will be described by a considerably more complicated process than would be anticipated according to Equation 11.3.

I. M. Yenukashvili[49] describes the kinetics of the accretional growth of the hailstone ensemble by a system of integral-differential equations, the approximative solution of which makes it possible to obtain an expression for computing the mean cubic radius of hailstones as a function of time, dependent on the initial values of the concentration and the sizes of droplets and hailstone embryos. The corresponding computations indicate that if the purpose is to substantially decrease the radius over a time characteristic of the duration of hail zones, it is necessary to increase the natural concentration of hailstone embryos (1 to $10/m^3$) by two orders of magnitude; this result coincides with data obtained according to Equation 11.3. A. I. Kartsivadze[22] shows that when the number of hailstone embryos is artifically increased, the number of hailstones of dangerous size falling (which have not melted before striking the earth's surface) will be less than for the natural process. With allowance for the natural concentration of hailstone embryos and the number of artificial crystallization nuclei that are activated as embryos which compete with natural ice pellets, the IGAN believes that to obtain a positive effect, $10^5/m^3$ active crystallization nuclei are required; this means a dosage of 0.1 to 2.0 kg reagent/km^3 at an ice-forming activity of approximately 10^{12} particles from 1 g of reagent. The VGI recommends lower dispersing rates: 0.1 to 0.5 kg/km^3 for a reagent with the same activity characteristics.

The matter of the spread of an aerosol in the open atmosphere as a result of diffusion has been taken up by Chotorlishvili[11]; equations were obtained giving not only the rate of spread of an aerosol from a point (shell) and from a linear (rocket dispersing reagent along its trajectory) source, but also determining the volume of the cloud in which AgI seeding will result in 10^4 hailstone embryos/m^3 Allowing for a coefficient of capture between the aerosol particles of $E = 6.4 \times 10^{-6}$ m^3/sec., a concentration of cloud droplets with which aerosol particles aggregate of $N = 2 \times 10^{-2}/cm^3$, a coefficient of turbulent diffusion of $K = 10^6 m^2/sec^1$, and using a quantity of 100 g of reagent in the shell and 1000 g in the rocket, it was ascertained that 30 sec after explosion a concentration of artificially obtained hailstone embryos equal to $10^{-2}/cm^3$, fills a sphere with a diameter of 700m, which is in good agreement with the data of Reference 5l; for a linear path (rocket) 30 sec after seeding, a similar concentration will be observed in a cylinder with a radius of 300 m and at a length equal to the length of the trajectory. These computations do not allow for change of the seeded volume of the cloud because of translation of the aerosol (convections) which could somewhat increase the volume of dispersal. In general the theoretical computations presented agree with field observation data, starting from which the dosage during modification of a cloud is taken equal to 0.1 to 0.3 kg/km³ at an ice-forming activity of $\sim 10^{12}$ particles with 1 g of reagent.

According to some estimates, a significant increase in the concentration of crystallizing aerosol in the hail zone, composed of supercooled drops, hailstone embryos, and fine ice crystals, can lead to acceleration of the growth of large hail and increase of its concentration. This contradicts the conclusion about the fact that increasing the number of hailstone embryos cannot lead to growth of the concentration of large hailstones. The apparent contradiction is explained by the fact that in these estimates only aggregation processes between hailstone embryos and supercooled droplets are considered; growth of hail particles owing to the solid fraction is not taken into consideration. Thus it is assumed that the overseeding can lead to increase of both concentration and final diameter of the hailstones. It is possible that certain unsuccessful attempts at modifying hail processes involve this phenomenon, although an analysis of the results of modification still does not make it possible to prove this conclusively. In any event, it is very important, in the light of these results, to ascertain not only the lower, but also the upper limit for reagent dosage introduced into clouds during hail modification. Apparently the dosages suggested by the VGI, IGAN, and TsAO satisfy these requirements, as confirmed both by results of field experiments and by the calculations carried out.

Medaliyev[31] has shown that when crystallizing reagent is introduced into a medium with drops whose temperature is higher than the threshold of their crystallization, the reagent particles deposited on drops cause freezing of drops at significantly lower temperatures than the (threshold) temperature of crystallization of drops affected by the given reagent Thus when silver iodide is introduced into a medium with drops at $t = -2$ to $-3°C$, and if the drops are subsequently supercooled, they freeze only at $t = -15$ to $-18°C$, although the crystallization temperature for silver iodide, as is known is $t = -10°C$. This circumstance is taken into consideration in choosing the altitude (level) at which reagent should be introduced into a cloud in order to obtain the maximum modification effect.

Another hail modification method was developed at the ZakNIGMI, which carried out studies on hail formation phenomena for Transcaucasia in general, and for eastern Georgia in particular from 1956 to 1962.[2] This method is based on simultaneous use of hygroscopic and crystallizing reagents. Particles of sodium chloride (table salt) are introduced into the cloud as the hygroscopic reagent; the diameter of the particles amounts to 5 to 10 μ. They are seeded into the warm part of the cloud near its base and at the middle level in the zone where radar reflections increase. The crystallizing reagent used is silver iodide, seeded into the supercooled part of the hail-risk cloud, in the zone of maximum radar reflections. Introducing hygroscopic particles, which facilitate increased condensation and aggregation growth of drops, leads to an increase in the number of large drops; it intensifies the processes of cloud formation and buildup of liquid water in the warm part of the cloud. Drops can grow to such a size that they begin to fall and do not reach the zone of great vertical velocities; that is, a process of premature scavenging of the lower part of the cloud will be observed, resulting in a significant decrease in the entry of liquid water from the lower warm part of the cloud into the upper supercooled zone. A portion of the large drops formed on the artificially introduced hygroscopic particles, depending on the nature of the vertical currents, can be transported into the supercooled part of the cloud, which increases the concentration of potential hailstone embryos. Introducing particles of silver iodide into the hail growth zone can produce crystallization of the supercooled drops, which will make the growth of large hailstones less likely.

In this kind of modification the concentration of giant salt particles (NaCl dosage) is kept within limits of 0.5 to 1.0 kg/km³, depending on the intensity of the processes. This concentration of salt particles corresponds approximately to the concentration found under natural conditions at the base of thunderclouds which are generated over the ocean in the summer and which do not, as a rule, produce precipitation in the form of hail.

The methods of modifying hail processes which have been suggested make it possible to limit the growth of large hailstones by creating competing embryos or decreasing the water content in the hail growth zone. These methods do not make it possible to alter the size and concentration of the hailstones already formed. Therefore seeding crystallizing reagent into a hail zone where there is already a large concentration of large hailstones does not prevent hail damage, and the hail formed in the cloud does reach the surface of the earth.

BASIC TECHNOLOGICAL MEANS FOR MODIFICATION OF HAIL PROCESSES

In order to modify hail processes, it was first necessary to solve the following problems: selecting the most effective active materials (reagents) for creating artificial hailstone embryos and finding ways to disperse them; and developing a method for indicating hail danger and a means of delivering reagents into clouds.

At the VGI the crystallizing reagents used were silver iodide and lead iodide, seeded by explosion. [31, 51.] The IGAN and TsAO developed pyrotechnical means for delivering the reagents mentioned, which were seeded during combustion along the flight path of an anti-hail rocket.

Two means were used to solve the second problem—indication of hail risk—by means of radar. One means (VGI) was to measure the magnitude of a signal reflected from a cloud on one or two wavelengths.[40, 48] By using the Rayleigh and Mie scattering laws and radars with two wavelengths (3.2 and 10 cm), it is possible to estimate the diameter of hailstones from 0.8 to 3.5 cm in size. The diameter of hailstones reaching the surface of the earth is determined by considering melting of hail in its descent below the zero isotherm. Because of the nature of scattering and absorption on the 3.2-cm wavelength, this method gives a systematic error which leads to a computed hailstone diameter which is too high compared with the actual diameter. The error in determining the size of hailstones amounts to 0.5 to 0.8 cm. Further studies have shown that two basic parameters completely describe the hail-bearing properties of a cloud: ratio of reflectivity at the 10.0 and 3.2-cm wavelengths and the temperature at the top of the zone.

To estimate the hail risk of a cloud, the IGAN and TsAO used a number of radar characteristics of the hail zone.[1, 8, 44] Among these characteristics are height of the radar echo above sea level H_m, temperature at this altitude t_m, thickness of the radar reflection zone ΔH, magnitude of maximum radar reflectivity η_m, altitude at which reflectivity reaches maximum value H_{η_m}, temperature at this level

t_{η_m}, ratio of the thickness of the super-cooled part of the radar echo to the warm part h_-/h_+, vertical thickness of the zone of increased reflectivity ΔH_η, at the boundary of which the radar reflectivity is one order of magnitude less than η_m, and temperature at the top of the zone of increased reflectivity $t_{\Delta\eta}$. Each of the parameters has two critical values, unambiguously separating hail and nonhail zones, and in the range between it is possible to ascertain the hail risk probability of the cloud. In order to reduce the region of ambiguity, the average value of the probability $\bar{P}$ was determined according to the separate parameters and the value of the combined radar indication of hail risk K was determined, which made it possible to cut the range of ambiguity almost in two. It was shown that both the separate parameters and the combined indicator K have a seasonal plot. Moreover, the values of the radar parameters are refined by allowing for the special features of individual physical and geographical regions (see Table 11.4).

At the present time radars are being used for the calculation of reflected signals for two wavelengths (10.0 and 3.2 cm). With such a radar the hail zone is determined by the nature of the differential signal entering the amplitude vs. distance display ("A"-scope). The IGAN has developed a calculating device into which the magnitudes of radar parameters of a convective cloud are put. The device computes the probable hail risk of a cloud.

From 1953 to 1958 experiments were done in the northern Caucasus and in the Alazan Valley on delivering reagent from the ground by means of pilot balloons sent to an altitude of 1 to 2 km with a mixture of red phosphorus, silver iodide, and gunpowder, ignited at altitude. Such a method of supplying reagent seems rather ineffective on three counts: a negligible portion of the crystallizing reagent goes into the cloud; the reagent is deactivated by the effect of solar radiation as it travels from earth to cloud; and the particles of reagent deposited on droplets in the warm part of the cloud cause a crystallization effect at

Table 11.4   Criterial Values of Radar Parameters

Cloud	Region	\multicolumn{10}{c}{Values of radar parameters}
		H_m (km)	t_m (°C)	ΔH (km)	h_-/h_+	H_{η_m} (km)	t_{η_m} (°C)	ΔH_η (km)	$t_{\Delta\eta}$ (°C)	$\bar{P}$	K
Non hail	Alazan Valley (Georgia)	6.8	−18	6.2	1.1	2.0	±2	2.8	−1.5	0.15	0
	Moldavia	5.0	−15	4.0	1.0	1.9	±5	2.3	1.5	—	0
Hail	Alazan Valley[a] (Georgia)	11.8	−54	10.6	3.6	7.4	−30	8.2	−34	0.75	1.0
	Moldavia	11.0	−50	9.7	3.8	6.3	−12.5	6.7	27.5	—	1.0

[a]Values of parameters for the Alazan Valley of the Georgian SSR were obtained by the IGAN from data of 9 years of observations (1962–1970), with 400 hail and 3000 heavy shower (non-hail) zones.

significantly lower temperatures (-15 to $-16°C$) than they would if dispersed in the supercooled part of the cloud.

From 1956 to 1959 attempts were made in the same region to modify clouds by seeding reagents from an airplane. The reagent was seeded into the cloud by firing special shells containing silver iodide. Dry ice was also used; granules of CO_2 were crushed in and sprayed from a special device on the airplane. In both these instances the reagent was seeded directly into the cloud and gave a good artificial crystallization effect. However, using this method for hail cloud modification did not succeed, for a number of reasons. For example, great velocities of updrafts and turbulence, significant electrical discharges, and large hailstones in a cloud can disable an airplane. Moreover, even if there are airplanes sturdy enough for these conditions, operationally it is very difficult to seed all hail zones over a large territory, not to speak of the need to use a whole squadron for this purpose, and the difficulties of coordinating it from the ground under complex meteorological circumstances.

For operational delivery of a certain amount of reagent into a given portion of a cloud, and to safeguard human lives and property values, nonbrisance (nondetonating) artillery shells "El'brus-2" and "El'brus-3" and PGI "Oblako" and "Alazani" rockets were developed. After the coordinates of the hail zone are determined by means of radar, they are transmitted to antiaircraft gun or rocket fire positions, and from these positions the appropriate anti-hail means are sent into the cloud. Each shell is furnished with 50 to 300 g of reagent. When the shell explodes, the reagent is dispersed. When a shell is used to introduce reagent into the cloud, scattering does not exceed 0.5 km. Reagent is dispersed along the flight path of the rocket as the result of the burning of pyrotechnic charges. The "Oblako"

rocket lofts about 5 kg of pyrotechnical material, the "Alazani" rocket 1 kg. The triggering devices for the PGI and "Oblako" have four directional bodies; the "Alazani" devices have 12 directed individual lines-of-fire which increase the effectiveness of modification considerably by dispersing up to 12 kg of pyrotechnical material along different paths simultaneously.

The VGI and ZakNIGMI use shells, while the IGAN and TsAO use rockets.

Radar measurement data were used to estimate the results of modification, during the progress of work at the IGAN and VGI.[26, 40] There are two means for reaching such an estimate. The first (VGI, IGAN) is based on considering the change of radar parameters before and after modification. An analysis of these data has shown that within the first 5 min after modification is begun the radar parameters continue to grow and in some cases even do so with greater speed than before seeding. This is probably explained by the effect of release of the latent heat during artificial crystallization and increase of the concentration of relatively large ice particles. In the time interval from 5 to 10 to 15 min after seeding, the parameters reach their maximum values, after which they decrease rapidly, which is a positive effect of the seeding.

ORGANIZATION AND DEVELOPMENT OF HAIL SUPPRESSION OPERATIONS

Hail suppression work was carried out from 1961 to 1963 in the Alazan Valley (Georgia) and in the central piedmont region of the Northern Caucasus, in the Kabardino-Balkar territory. This work was done experimentally, with the aim of suppressing the incidence of hail on a small territory and reducing the losses due to hail damage. The results were encouraging, and demonstrated the feasibility of effective hail suppression by means of anti-hail rockets and shells. In 1964, with the participation of units of the USSR Hydrometeorological Service, the Caucasus Hail Suppression Project (KKPE) was set up, taking in projects of the Management of the Hydrometeorological Service of the Armenian Soviet Socialist Republic, the VGI, and ZakNIGMI, and some later projects of the Management of the Hydrometeorological Service of the Azerbaijan Soviet Socialist Republic. The Anti-Hail Service of the Ministry of Agriculture of the Georgian Soviet Socialist Republic, already organized in 1961, carried out successful hail suppression work. The areas in various regions of the USSR protected against hail were gradually extended, and by 1965 they amounted to about 700,000 ha.

During these same years a considerable volume of research and experimental work was performed—devising, improving and adopting radar methods for indicating hail clouds (one-wavelength and two-wavelength methods) and also modification means (hail suppression devices like the "El'brus-2" and the "Oblako" rocket); testing the ZakNIGMI method; and perfecting the technology for carrying out modification.

The remarkable successes achieved in anti-hail protection are attracting more and more attention from agricultural organizations, and the volume of hail suppression work is continuing to increase. In 1967 a special unit of the USSR Hydrometeorological Service was created, the Department of Modification, to organize and coordinate work on preventing hail damage, as well as other work in the area of weather modification. The general extent of areas protected from hail amounted to more than 3.5 million hectares in 1971. Some 28 hail suppression units carried out about 674 modification operations, as a result of which more

than 1517 hail and hail-risk zones were seeded. Losses from hail were reduced by more than 70 percent.

At the present time, hail suppression work is being carried out in the Northern Caucasus (Russian Soviet Federative Socialist Republic), in the Crimea, and in the Transcaucasian, Moldavia, Uzbek, and Takjik Republics.

Hail suppression work is organized in response to requests from agricultural units by analyzing data on damages inflicted on crops by hail, and then having representatives from the "client" and the "operator" (the hail suppression project) jointly determine the territory to be protected. Top priority for protection from hail damage is given regions where hail inflicts systematic damage on agriculture. Where conditions are otherwise equal, preference is given to regions with perennial types of plantings (such as vineyards and fruit orchards).

An area of 100,000 to 120,000 ha can be protected by one hail suppression unit. Experiment has shown that it is difficult to protect a large area with one unit. With too large an area, the command post (the radar) of the unit is unable to monitor all the clouds and carry out seeding of hail zones. Also, with large areas the distance between the radar and a cloud at the boundary of the territory to be protected can be more than 35 km. In this case the error in determining the exact location of the hail zone, and the degree of hail risk, can become very large.

The territory to be protected, then, is so chosen as to permit treatment of clouds before they reach its boundary. The distance of such advance treatment depends on the travel speed and direction of the clouds; it can amount to 2 to 10 km. Reservation of a zone of preliminary processing is necessary if all hail clouds are to be seeded prior to their entry over territory to be protected.

Several hail suppression units (but not

more than 4 to 5) work together on an anti-hail project, and one of the units acts as the lead unit which directs the others and coordinates their operations. Each of the units consists of a modification group, a radar group, a radiocommunications group, an artillery or rocket group, a group for monitoring results of the modification, and a group to provide materials and technical support. In addition, the lead unit has an aerosynoptic (synoptic meteorology) group, and a group handling coordination with the aviation unit.

The lead unit can consist of from 50 to 60 people, and a subordinate unit has from 25 to 30 people. This is not counting the people stationed at posts where shells are fired from antiaircraft guns, or rockets are fired. When 100,000 to 120,000 ha are to be protected, a back-up of five to seven guns or 10 to 12 rocket devices may be required, depending on the configuration and relief of the territory to be protected. Depending on the complexity of the operations, there can be from three to five people at each seeding post. These posts receive their orders from the command post via radiocommunications or, where it is possible, telephone. The lead unit maintains radiocommunication with subordinate units and with the main aviation units (for example, with the central airport) controlling flights in the region of operations. This is necessary if modification is to be carried out safely.

Every morning there is a staff meeting of supervisory personnel in the hail suppression units. Chiefs of groups report on the operational preparedness (status of the radar posts, communications conditions, quantity of shells or rockets at posts, etc.). If a hail situation is predicted, responsibilities are assigned for modification during the 24-hr period; an additional radiosonde is released for a more exact definition of stratification of the atmosphere, and determination of altitudes at which seeding will be done. The

readiness of all units for action is verified; in the case of a nonhail situation, however, the duty alert is much less strictly interpreted.

In a modification alert, prior to emergence of clouds the radar is switched on every 2 hr and cloudiness within the radius of coverage of the radar is scanned on the 3.2-cm wavelength. Once the zones of reflection from the clouds are detected, the modification group calls the command post, the altitudes of the reflection zones are determined, and the maximum magnitudes of reflectivity, as well as the speeds and directions of motion of the zones are determined. The modification posts (firing points) are notified.

When hail clouds approach within firing distance of the territory to be protected, or when zones in which hail formation is possible develop over the territory to be protected, the chief of modification decides whether or not modification will be carried out. Before modification is performed, the hail risk of the cloud is estimated, the hail zone and its coordinates are determined, initial data are computed to find the zone with the nearest gun or rocket apparatus, and commands are relayed to gun or rocket installations. All data about the operation are recorded at the command post—time of detection and treatment of zones, their radar parameters, and coordinates.

Immediately after modification, the monitoring group leaves by helicopter for a tour of the area, where they use a plotting board to map the regions of hail fall, its size, the number of hailstones per unit of area, etc. If there is agricultural damage inflicted on the protected territory, they note the type of crop, the degree and nature of damage to it, the area of damage, etc. At the morning meeting on the day after modification, by comparing the operational plotting board with results of the tour, conclusions are drawn as to the effectiveness of modification, a critique of the work is carried out, and errors occurring during the operation (if any) are noted, to prevent their recurrence.

SOME SPECIAL ASPECTS OF HAIL SUPPRESSION OPERATIONS

Before carrying out hail suppression work over a large territory it was first of all necessary to protect the people there from the fall or explosion of shells and rockets, and also from injuries which could occur as a result of falling rocket or shell fragments after the reagent was dispersed.

With this in mind, as has been mentioned, special rockets (PGI, "Oblako," "Alazani") and shells ("El'brus-2" and "El'brus-3") were developed. Their technological features provide a high safety level of use. Specifically, the PGI and "Alazani" rockets were designed to destroy themselves by blowing up in mid-air after discharge of the active substances they were carrying. An explosive substance placed in the rockets ensures that they will disintegrate into small fragments which present no danger. The "Oblako" rocket has a parachute system to lower the casing to the ground slowly after the active substance has been discharged. "El'brus" type shells are made of a special alloy and blow up into very small particles when they explode.

When hail suppression work is being performed, precautions for ensuring the safety of aircraft flying in the operational region must be strictly observed. In this connection, instructions have been developed, in cooperation with the aviation units, for all regions setting forth the specific responsibilities of the aviation and anti-hail units. Adherence to the rules and conditions in these instructions guarantees safe, troublefree operations in the aviation and anti-hail units. However, it should be noted that for reasons of safety the aviation units have the right to veto the firing or launching of rockets at any time, even in an obvious hail situation.

Clearly, in the future, when new units are sent to regions with heavier concentrations of flight paths and greater frequency of flights, the possible savings resulting from hail suppression will have to be weighed against the inconvenience of extended flight paths and possible down time of aircraft, before the operations can be adjusted to allow for both these factors.

Operational modification of hail processes, when it is carried out over large areas, involves tremendous difficulties, and its success depends basically on the competence and coordination of the project members, the reliability of the hardware, intelligent organization of the work, and properly timed suspension of civil flights over the territory to be protected. When this work began, and the area brought under protection from hail was not too large (about 100,000 ha), hail incidence was almost totally prevented. But when the area to be protected grew to several hundred thousand hectares or more, the circumstances mentioned sometimes led to breakdowns in the operations, and the intrusion of hail damage into the territory to be protected.

The hypothesis was expressed that although anti-hail work is justified under different physical and geographical conditions and over large areas, there are still some rare kinds of conditions causing hail which apparently cannot be explained on the basis of the proposed hail cloud model. In this regard, Academician Ye. K. Fedorov, head of weather modification operations, wrote at the beginning of 1966: "A great deal of uncertainty undoubtedly remains about the process of hail formation. The unsuccessful experiments, rare as they are, attest of this. A great deal of work is still to be done before this interesting and persuasive hypothesis can be regarded as a solid theory."

During operations each instance of intrusion of hail damage onto a protected territory was subjected to a thorough analysis, in order to explain the reasons for unsuccessful experiments and take measures to prevent their recurrence. The following main causes for the intrusion of hail damage were found:

1. Insufficient accuracy in the determination of parameters and location of hail zone and time for seeding it with reagent.

2. Poor coordination with the aviation unit, and its veto on seedng clouds.

3. Breakdown of radiocommunications between command post and action posts due to hail conditions and intense rainfall.

4. Strong absorption of radar radiation on the 3.2-cm wavelength during heavy showers and screening of hail zones.

5. Drop in sensitivity of the radar at the time of cloud development, or the radar going out of order because of prolonged and intensive use.

6. Defects in (surveyed) interconnection and mutual orientation of guns or rocket installations and radar.

7. Personnel who are not well enough qualified, especially at action posts; inaccurate or mistimed implementation of seeding orders received from command post.

As a result of maximum follow-up on the trouble spots found, it was possible to cut intrusion of hail damage onto protected territory significantly. At the present time it has been possible to reduce losses from hail damage by an average of 70 to 80 percent. However, strictly speaking, it cannot be asserted that all the cases of hail incidence noted could have been prevented if modification had been done in strict accord with instructions. We cannot exclude the possibility that hail would still have fallen even if the modification had been carried out according to instructions in all these cases. Still, what can definitely be said is that in instances when cloud modification was carried out in accordance with the existing instructions, hail incidence was prevented.

ASSESSING THE RESULTS OF MODIFICATION

As usually understood, the results of modification of hail processes include a decrease in the area of hail incidence and a decrease in its intensity, and consequent economic effectiveness of hail suppression work.

However, as the protected areas are extended and the increased volume of work discharges tons of reagent into the atmosphere annually, questions other than those of economic effectiveness must be answered, for example: first, didn't the operations to modify hail processes increase the content of AgI and PbI_2 in the lower layers of the atmosphere and in the ground and water beyond the limits consistent with good health? Second, didn't the introduction of a large amount of reagent into atmosphere lead to a change in the natural course of meteorological processes; specifically, wasn't there an alteration in the amount of precipitation in the modification region and adjacent regions?

Starting in 1960 a study was made of the reagent content in the precipitation which fell, and also in air, water, and soil samples in both the modification region and other outlying regions. The results of observations carried out in the northern Caucasus, where the protected areas are most compactly situated on a territory of about 1 million hectares, and in other modification regions, have shown that the content of lead and silver in the atmosphere, standing and running water, and the soil is not increased, and in absolute magnitude is far less than that found in industrial areas. No increase of the content of these elements in the modification region during the year of operations was noted by health inspection units, either.

The matter of changing the quantity of precipitation in regions where hail suppression work is done is of undoubted interest. According to TsAO data for conditions in the Moldavian region, modification of hail processes leads to some increase in rainfall.[45] To clarify this matter, two areas (polygons) in the northern Caucasus were chosen, located in approximately identical physical and geographical conditions. Hail process modification was carried out over one of them (the test polygon), and the second (the control polygon) was so far from the first that the atmospheric processes over it were left essentially unaffected by reagent seeding done over the first area. Each area had 10 weather stations which had been taking observations for 25 years, 17 years before, and 8 years during modification work. A comparison of the amount of precipitation falling on the test and control areas showed that during a year of modification the amount of precipitation falling on the target area from April to September was increased not less than 15 percent (with 95 percent statistical reliability).

Now it is difficult to say whether or not such an increase in the amount of precipitation incident on the protected territory leads to a change in the pattern of the incidence of precipitation in neighboring regions. At least, the results of work on artificially inducing precipitation carried out in the Ukraine[28] have shown that an increase in precipitation in the operational region of 10 to 15 percent does not lead to a decrease of precipitation in neighboring regions. A more complicated question is the one as to whether or not

doing hail suppression work over a protected territory leads to an alteration in the development of hail processes beyond the limits of the protected territory. Complete answers to this question are not possible at the present time.

In answering the question of which territories should receive modification of hail processes, the basic criteria were reduction of hail damage, and economic savings, attributable to it. Given the applied nature of anti-hail work, randomized experiments to determine the magnitude of the effect are ruled out, since in approximately half of all cases a fall of hail will occur on the protected territory. Therefore, in order to determine the effectiveness of this work, it is necessary to consider the plot of parameters characterizing hail processes over an extended period (10 to 30 years) prior to the beginning of operations, and during the operational period. Control area methods are usually used. The basic parameter is taken to be the area of hail damage, although the number of hail cases and the degree of damage to agricultural crops are also considered. One means of assessing the effectiveness of hail suppression work is given below in general form.

Such an assessment requires a consideration of data from a period of many years (not less than 10) prior to the start of protection, concerning the average area of agricultural crops damaged by hail on the protected ($\bar{S}_0$) and control ($\bar{S}_k$) parts in percents to the entire seeded area. Comparability of these parts is maintained in all characteristics: agricultural technology, climatic, etc. Data from divisions of the State Insurance (Gosstrakh) were used to determine the areas of crops damaged by hail on protected (S_0) and control (S_k) parts in a year of protection.

If it is assumed that the chronological change in the area of hail damage is slight (at least, significantly less than the decrease in the area of hail damage anticipated as a result of protection), then the effectiveness can be determined from the relationship

$$E_1 = \frac{\bar{S}_0}{S_0}. \qquad (11.4)$$

The magnitude E_1 shows how many times less an area is damaged by hail on the protected part in comparison with average values in years prior to protection.

However, the hail damage area in years before protection varies significantly. For example, the coefficient of variation of hail damage area for the Moldavian region δS = 50 to 70 percent,[46] for the Gissar Valley of the Tadjik Soviet Socialist republic δS = 50 to 100 percent,[5] and for the territory of the Northern Caucasus δS = 50 percent.[43] Therefore, if it is found that E_1 < 1 or E_1 > 1, then increase or decrease of hail damage area in a protection year on the protected part can be spoken of only on the average. Natural fluctuations in hail processes lead to the fact that when Equation 14.4 is used to determine the magnitude of the effect there will be a large error, at any rate not less than the absolute magnitude of the coefficient of variation of hail damage area. One way of allowing for natural fluctuations is to compare hail damage areas on protected and control parts. Assuming that the association between damaged areas in both parts can be described by an equation of linear regression form

$$S_0 = S_k \cdot \left(\frac{\bar{S}_0}{\bar{S}_k}\right), \qquad (11.5)$$

we obtain the following equation for the magnitude of the effect

$$E = \left(\frac{\bar{S}_0}{\bar{S}_k}\right) \cdot \left(\frac{S_k}{S_0}\right). \qquad (11.6)$$

In the case in which E = 1, it is most probable that in a protection year the area damaged by hail on the protected part will be the same as if modification were not carried out, and on the average, any modification effect will be absent.

When E < 1, it is most likely that modification over the protected part will lead to an increase in the hail damage area.

When E > 1, it is most likely that possible areas of hail incidence in a protection year will be less on the protected part as a result of modification.

It is not hard to be persuaded that if the relationship between the damage areas on the protected and control parts is weak (coefficient of correlation $\rho \approx 0$), then Equations 11.4 and 11.6 will determine the magnitude of the protective effect with practically identical errors (tens of percents). A notable gain factor in the accuracy of estimating the effect can be obtained with Equation 11.6, if $\rho \approx 0.7$ to 0.8 or greater. At ρ = 1 Equation 11.6 determines the effect with zero error. Under actual conditions the connection noted is weak; on the average the coefficient of correlation is equal to $\rho \simeq 0.3$ to 0.4, and does not exceed 0.6 to 0.7. Consequently, at the present time the use of control areas cannot materially increase the accuracy of determination of the magnitude of the effect. Equations 11.4 and 11.6 determine the magnitude of the effect of modification according to the reduction in the hail damage area. At the same time, considering the great fluctuations of hail processes over space and time, the question arises: is the decrease observed due to the anti-hail protection?

We shall denote as S_m the minimum area (in percents) of agricultural crops damaged by hail on protected territory in a year prior to operations. Using the equation

$$E_2 = \frac{S_m}{S_0}, \qquad (11.7)$$

we can estimate the area of crops damaged by hail in a year of protection, in comparison with the minimum hail-damaged area in a year before protection. If there are enough years of observations in years before protection, and the magnitude is determined with sufficient reliability, then the following conclusions can be drawn.

At $E_2 > 1$, the decrease seen in hail damage area in a protected year is almost certainly the result of hail suppression. If $E_2 \leq 1$, then the observed decrease in hail damage area can with some (not much) probability be explained by the action of natural fluctuating processes. The more E_2 differs from unity, the greater this probability.

A rigorous solution of the question as to the presence or absence of a hail suppression effect can be obtained if the function of overall seasonal distribution of hail damage areas is known. In this case determining the effect amounts to determining the probability of the chance of the hail damage area value observed in a protection year.

Our confidence in the accuracy of an estimate of hail cloud modification effectiveness can be increased by increasing the period of time over which areas of hail damage to crops on protected and control areas are determined, prior to the beginning of hail suppression work, and while the protected and control crop areas are being increased.

A shortcoming of this method is the absence of data on hail incidence in preceding years on nonagricultural areas (forests, pastures, etc). Apparently a significant increase in the area of crops on protected and control territories to some degree compensates for this defect.

Buknikashivili and associates[10] set forth a method for computing the amount and value of agricultural produce preserved from hail damage as a result of modification; the amount of produce saved can be determined from the equation.

$$P_y = \left\{ \left(\frac{\bar{S}_0}{\bar{S}_k} \right) \left(\frac{\bar{b}}{\bar{a}} \right) S_k' b - S_0' a \right\} \cdot y, \quad (11.8)$$

where $\bar{a}$ and $\bar{b}$ are the average weighted percent of damage to crops on protected and control areas, a and b are the percent of damage to crops in a protected year on protected and control parts, y is the average, over several years, yield of crops in tons per hectare. Each crop is computed separately (in this case $\bar{S}_0$, $\bar{S}_k$, $\bar{a}$, and $\bar{b}$ are taken for the specific crop, not for the entire tilled area).

If the cost c of produce taken from one hectare is substituted for y in Equation 11.8, then the equation

$$P_c = \left\{ \left(\frac{\bar{S}_0}{\bar{S}_k} \right) \cdot S_k' \cdot b - S_0' \cdot a \right\} \cdot c \quad (11.9)$$

will give a conditional saving as the result of hail suppression work for a specific crop, expressed in rubles.

Equations 11.8 and 11.9 make it possible to draw a conclusion as to the economic utility of using hail suppression methods, depending on the type of crop and the average distribution of hail processes during the growing season.

It is clear that the saving computed in this way cannot be used to compare different methods of hail process modification used in different regions. Actually, P_y and P_c depend not only on the absolute decrease of hail damage areas and degree of crop damage, which also characterize the effectiveness of the method, but also upon the crop yield and the price of the agricultural produce. Therefore modification methods with an identical effectiveness, used in regions with an identical hail damage intensity, but different crop patterns, give different values of P_y and P_c.

Equations 11.8 and 11.9 determine only the mean-statistical magnitude of the effect; they do not allow for fluctuational processes. The possible value of the total hail damage area estimated according to Equation 14.5 is determined from the root mean square error on the order of 50 to 80 percent. When hail damage areas of separate crops are estimated by Equation

Table 11.5   Mean Decrease of Hail Damage Areas in Hail Suppression Regions of the Northern Caucasus (VGI) and Georgia (ZakNIGMI, IGAN) for the Years 1966–1970

		1966	1967	1968	1969	1970
Total area of protected	VGI	615	890	785	890	960
territory (ha × 1000)	IGAN	220	320	460	460	460
	ZakNIGMI	50	80	110	150	200
Average decrease in hail	VGI	90	50	87	99	62
damage area (%)	IGAN	76	82	67	69	88
	ZakNIGMI	—	96	91	94	87

14.5, errors will be significantly large (on the order of 100 to 150 percent). Errors of as yet unstudied magnitude will also occur in the determination of the degree of crop damage on protected territory (in the absence of modification). Roughly, it can be considered that the mean square error of determining the degree of damage will equal 20 to 40 percent.

Actually, observed values of areas and degree of crop damage on protected and control territories will also be found with errors ("instrumental") approximately equal to 10 to 20 percent.

In conclusion, as an example, Table 11.5 gives the results of an estimate of the effectiveness of hail suppression work from 1966 to 1970, according to data from the VGI, ZakNIGMI, and IGAN for regions of the northern Caucasus and Georgia.

From this table it is evident that during the period shown, the average decrease of the area on which crops were damaged by hail amounts to 80 percent. Approximately the same result is obtained for other regions protected from hail.

CONCLUSION

As a result of the sum total of the research work done in the USSR, there are now three methods of modification. The first two were developed by the IGAN, TsAO,

and VGI, and are based on the use of hail suppression rockets and shells carrying crystallizing reagents; they differ little among themselves, and they work on the principle of artificially altering the conditions for accretional growth of hailstone embryos in the supercooled part of the cloud. Both these methods have been used for more than 10 years over a significant area, in recent years in excess of 3 million hectares. Putting both methods into practice has shown they are approximately equally effective, and decrease losses from hail damage by 70 to 80 percent.

The hail modification method worked out by the ZakNIGMI is based on using shells to introduce hygroscopic and crystallizing substances into the cloud to change conditions for condensational and accretional growth of hail particles in the warm and supercooled parts of the cloud, respectively. For 6 years this method has been employed successfully in the protection of crops from hail damage.

Despite unequivocal successes in hail protection regions, a good many questions remain unanswered in regard to the process of hail formation and techniques for conducting hail suppression work. This is attested to by isolated unsuccessful attempts, and instances of intrusion of hail damage on protected territory. Unfortunately, we cannot as yet say exactly what caused these failures: failures in equipment, failure of the operations chief to modify certain hail processes effectively, or the fact that some mechanisms of hail formation depart to some degree from the concepts which have come into acceptance.

In this connection, it would be of interest to perform further studies of the processes leading to the formation and precipitation of hail. In particular, better studies of convective currents in and around the cloud are needed. It is possible that the results of such studies would enable us to improve the way we seed clouds with reagents. The matter of the extension of the artificial crystallization zone in the cloud has not been pursued to a conclusion; explaining this would make it possible to refine the reagent dosage used during modification. Another subject which is not devoid of interest is the study of the effects of the explosion and products of disintegration of anti-hail shells or rockets on the microstructure and dynamics of development of convective clouds.

REFERENCES

1. Bachiashvili, A. D., R. I. Doreuli, A. I. Kartsichvadze, et al., Some characteristics of radar reflections from rain clouds and hail, *Tr. Vses. soveshch. Akt. Vozdeistv. gradov. protsessy* [Proceedings All-Union Conference on Modification of Hail Processes], Tbilisi, 1964.

2. Bartishvili, B. S., V. P. Lominadze, and Sh. Gudushauri, The matter of combination modificatin of warm and supercooled parts of cloud to prevent hail incidence, *Tr. ZakNIGMI* 21(27), 1967.

3. Bibilashvili, N.Sh., V. F. Lapcheva, and G. K. Sulakvelidze, Special features of aggregation growth of hailstones related to change of updraft velocity with altitude, *Izv. Akad. Nauk SSSR, Ser. Geofiz.* 4, 1960.

4. Bibilashvili, N. Sh. V. F. Lapcheva, and G. K. Sulakvelidze, Water content in rain clouds and some questions of forecasting rainfall, *DAN* 11, 3, 1960.

5. Bokova, P. A. and N. N. Butov et al. Some results of hail suppression work in the Gissar Valley in 1964–1966, *Tr. SANIGMI* 31(46), 1967.

6. Borovikov, A. M., I. I. Gayvoronskiy, Ye. G. Zak, et al., *Fizika oblakov* [Cloud Physics], Leningrad, 1961.

7. Borovikov, A. M., V. V. Kostarev, A. B. Shupyatskiy, Some results of radar observations of the evolution of cumulus congestion and cumulonimbus clouds, and the results of modification, *Proceedings All-Union Conference on Modification of Hail Processes*, Tbilisi, 1964.

8. Borovikov, A. M., R. I. Doreuli, A. I. Kartsichvadze et al., Radar characteristics of hail clouds, *Sb. trudov IGAN "Fizika oblakov" Coll. IGAN Proc. Cloud Physics XXV*, No. I, 1967.

9. Bukhnikashvili, A. V., I. I. Gayvoronskiy, A. I. Kartsivadze et al., Technique for modification of hail processes and results of experiments carried out in the Alazan Valley, *Proceedings All-Union Conference on Modification of Hail Processes*, Tbilisi, 1964.

10. Bukhnikashvili, A. V., A. I. Kartsivadze, B. I. Kiziriya, and G. G. Todua, Organization of experimental protection from hail damage in the Alazen Valley, *Proceedings All-Union Conference on Modification of Hail Processes*, Tbilsi, 1964.

11. Chotorlishvili, L. S., Progapation of inert impurities in clouds away from linear source, *Soobshcheniye AN Gruz. SSR* 62, 1, 1971.

12. Fedorov, Ye. K., *Aktivnoye vozdeystviye na meteorologicheskiye protsessy. Meteorologiya i gidrologiya za 50 let Sovetskoy vlasti* [Modification of Meteorological Processes. Meteorology and Hydrology During 50 Years of Soviet Rule], Moscow, 1967.

13. Fuks, N. A., *Mekhanika aerozoley* [Mechanics of Aerosols], Moscow, 1955.

14. Gigineyshvili, V. M., *Gradobitiya v Vostochnoy Gruzii* [Hail Damage in Eastern Georgia], Leningrad, 1960.

15. Gliki, N. V., A. A. Yeliseyev, and M. N. Marchenko, Conversion of cloud droplets into ice crystals, *DAN* 143, 5, 1962.

16. Glushkova, N. I., and V. F. Lapcheva, The matter of forecasting rain and hail occurring in air-mass cumulus congestus clouds, *Trudy El'brusskoy expeditsii* (Proc. El'Brus Project) 2(5), 1961.

17. Guniya, S. U., *Grozovyye protsessy v usloviyakh Zakavkaz'ya* [Thunderstorm Processes under Transcaucaian Conditions], Leningrad, 1960.

18. Gutman, L.N., *Vvedeniye v nelineynuyu teriyu mezmeteorologichskikh protsessov* [Introduction to Nonlinear Theory of Mesometeorological Processes], Leningrad, 1969.

19. Kachurin, L. G. and V. G. Morachevskiy, *Kinetika fasovykh perekhodov vody v atmosfere* [Kinetics of Phase Transitions of Water in the Atmosphere], Leningrad, 1965.

20. Kachurin, L. G., V. I. Bekryaev, B. M. Vorob 'ev et al., Some questions about phase transitions of water in the atmosphere, applicable to the problem of controlling hail processes, *Trudy VGI,* No. 14, 1968.

21. Kartsivadze, A. I., Experimental modification of hail processes in the Alazan Valley, *Sb. trudov IGAN "Fizika oblakov"* **XXV**(1), Tbilisi, 1967.

22. Kartsivadze, A. I., Estimate of effect of concentration of ice nuclei on formation of hail of dangerous sizes, *Sb. trudov IGAN "Fizika oblakov,"* **XXV**(1), Tbilisi, 1967.

23. Kartsivadze, A. I. and P. I. Makharashvili, Some data on the physical characteristics of hailstones, *Sb trudov IGAN "Fizika oblakov"* **XXV**(1), Tbilisi, 1967.

24. Khorguani, V. G., Some questions about the nature of motion and aggregation of cloud particles, *Proceedings All-Union Scientific Conference on Hail Process Modification,* Tbilisi, 1964.

25. Khrgian, A. Kh. and I. P. Mazin, Size distribution of drops in clouds, *Trudy TsAO,* 7, 1952.

26. Lapcheva, V. F., Results of radar study of zones of reflections from convective clouds, *Proceedings All-Union Conference on Modification of Hail Processes,* Tbilisi, 1964.

27. Lebedev, S. L., Three-dimensional nonstationary model of atmopheric convection cell with clouds, *Izv. An SSR Fao* **2**, 1, 1966.

28. Leskov, B. N. and N. P. Polovina, Modification of a cloud during cold period of year to increase precipitation, *Proc. VIII All-Union Conference on Cloud Physics and Modification,* Leningrad, 1970.

29. Levin, L. M., Function of size distribution of cloud and rain droplets, *DAN* **94**, 6, 1954.

30. Levin, L. M., *Issledovaniya po mekhanike grubodispersnykh aerozoley* [Studies in the Mechanics of Coarsely Dispersed Aerosols], Moscow, 1961.

31. Medaliyev, Kh. Kh., Ice-forming properties of a silver iodide aerosol obtained by explosion, *Uch. zap. KBGU, Nal'chik* **XVI**, 1962.

32. Nikandrov, V. Ya., *Iskusstvennyye vozdeystviya na oblaka i tumany* [Cloud and Fog Modification], Leningrad, 1959.

33. Robitashvili, G. A., Stationary spatial model of cumulus cloud, *Trudy ZakNIGMI* **16**(22), 1964.

34. Robitashvili, G. A., Asymmetrical solution of nonstationary problem of cumulus clouds, *Trudy ZakNIGMI* **25**(31), 1969.

35. Shifrin, K. S. and A. Ya. Pelerman, Kinetics of crystallization of clouds, *Izv. AN SSSR, ser. geofiz.,* 6, 1960.

36. Shishkin, N. S., Study of process of formation of summer precipitation and storm electricity, *UFN* **45**(3), 1951.

37. Shishkin, N. S., *Oblaka, osadki i grozovye elektrichestvo* [Clouds, Precipitation, and Storm Electricity], Moscow, 1954.

38. Shmeter, S. M., Stages in the development of cumulonimbus clouds and features of distribution of meteorological parameters in their zone, *Tr. TsAO* No. 53, 1962.

39. Sokol, G. P., Some results of assessing effectiveness of modification of hail clouds in the Gissar Valley, *Meteorol. i Gidrol.,* 1, 1967.

40. Sulakvelidze, G. K., *Livnevyye osadki i grad* [Rainfall and Hail], Leningrad, 1967.

41. Sulakvelidze, G. K., N. M. Bibilashvili, and V. F. Lapcheva, Technique and physical bases for modification of hail clouds, *Meteorol. i Gidrol.,* 12, 1965.

42. Sulakvelidze, G. K., N. I. Glushkova, and L. M. Fedchenko, *Prognoz grada, groz i livnevykh osadkov* [Forecasting Hail, Storms, and Rainfall], Leningrad, 1970.

43. Sulakvelidze, G. K. and V. F. Lapcheva, Review of work on cloud process modification, *Tr. VGI,* No. 13, 1969.

44. Voronov, G. S. and I. I. Gayvoronskiy, Radar studies of hail processes in Moldavia, *Meteorol. i Gidrol.,* 4, 1969.

45. Voronov, G. S. and I. I. Gayvoronskiy, Results of modification of hail processes, *Trudy TsAO,* No. 100, 1970.

46. Voronov, G. S. I. I. Gayvoronskiy, and Yu. A. Seregin, Studies and modification of hail processes, *Trudy VIII Vsesoyuzhnoy konferentisii po fizike oblakov i aktivnym vozdeystviyam* [Proceedings VIII All-Union Conference on Cloud Physics and Modification], Gidrometeoizdat, Leningrad, 1970.

47. Vul'fson, N. I., Issledovaniya konvektivnykh dvizheniy v svobodnoy atmosfere [Studies of convective motions in the free atmosphere], *AN SSSR,* Moscow, 1961.

48. Yefimov, V. Ye., V. F. Lapcheva, and G. K. Sulakvelidze, Determining hail zones by a radar method, *Meteorol. i gidrol.,* 10, 1963.

49. Yenukashvili, I. M., Effect of change of updraft velocity with altitude on kinetics of aggregation in spatially inhomogeneous clouds, *Coll. Proc. IGAN "Cloud Physics"* **XXV**(1), 1967.

50. Zaytsev, V. A., Sizes and distribution of droplets in cumulus clouds, *Trudy GGO* **13**(75), 1968.

51. Zhikharev, A. S., G. B. Myakon'kiy, G. K. Sulakvelidze, and V. G. Khorguani, Study of ice-forming activity and degree of dispersion (particular size) of reagents obtained by explosion under field conditions, *Proceedings VIII All-Union Conference on Cloud Physics and Modification,* Leningrad, 1970.

12 | Cloud Seeding in Australia

E. J. SMITH

Much of Australia, the driest continent, consists of desert where the rainfall is sparse and unreliable. The idea of stimulating extra rain has always appealed to Australians, and ancient "rainmaking" ceremonies play a part in the culture of the aborigines. In this environment it is natural that an enthusiastic reception met the news in 1946 that Vincent Schaefer in Schenectady, New York, had dropped dry ice into a supercooled cloud causing it to change to ice crystals,[27] thus raising a hope that to stimulate the elusive "extra rain" might at last be possible.

On February 5, 1947, the area inland from Sydney, New South Wales, was covered with deep cumulus clouds, which all appeared to be similar in type and size.[22] Dry ice was dropped into one of them (Figure 12.1). Within a few minutes rain started to fall from the "seeded" cloud and the top began to grow explosively; its appearance 13 and 40 min after seeding is shown in Figures 12.2 and 12.3. The shower grew in size and intensity, lasted for some hours and brought ½ in. of rain over an area of something like 50 mi². Surrounding clouds gave no rain. This is believed to be the first documented case anywhere of rain, stimulated by man, falling to the ground in appreciable quantity and also the first case in which dynamic cloud growth followed seeding.

This experiment marked the beginning of a program of research, which has continued to the present day, by the Commonwealth Scientific and Industrial Research Organization. This is an agency of the Government of the Commonwealth of Australia, established to undertake scientific research to benefit primary and secondary industry. The program includes theoretical, laboratory, and airborne investigations in fundamental cloud physics and also experiments in weather modification. Because lack of rain is by far Australia's biggest "weather" problem, most of the work on "modification" has been concerned with seeding clouds for the purpose of increasing the rainfall. This chapter gives a summary of the Australian weather modification experiments and their results, the physical principles on which they are based being described in other chapters.

EXPERIMENTS WITH DRY ICE

"Dry ice" is frozen carbon dioxide: its temperature is $-80°C$ or colder. If a piece the size of a pea is dropped from an aircraft into a supercooled cloud it falls about 1 or 2 mi before evaporating, and a large number of ice crystals forms in the

---

E. J. Smith is with the Commonwealth Scientific and Industrial Research Organization, Division of Cloud Physics, Epping, NSW, Australia.

wake. Each crystal is potentially capable, if conditions are ideal, of growing into a large snowflake at the expense of the cloud drops and melting to form a raindrop when it falls to levels where the temperature is warmer. This is an attractively simple principle, and during the period 1947 to 1950 it was tested in a series of experiments performed near Sydney.[34]

In these early experiments the principle was, briefly, to drop about 100 lb of dry ice into a suitable cloud and watch what happened to it immediately afterwards, using where possible adjacent untreated clouds for comparison and using any observational instruments such as radar which happened to be available. These experiments lacked refinements which have been subsequently introduced, such as randomization and computer modeling which are described in other chapters.

Nevertheless, their results may still be of interest because although they may not form the basis for rigorous conclusions there is no reason to doubt their validity.

One series of experiments was performed on cumulus clouds; most of them were close to other clouds of similar type and characteristics and in no case was any rain falling from nearby clouds.

The chance of success depended on the cloud-top temperature (Figure 12.4). When this was warmer than $-8°C$ half the treated clouds rained and half did not; when it was colder almost all of them rained.

The time that elapsed between the seeding and the start of rain was about 10 min for shallow clouds, but for deeper ones it took longer before the rain appeared below the cloud base.

These simple observations were

Figure 12.1   February 5, 1947. Inland cumulus clouds before dry ice was dropped.

Figure 12.2   Cloud growth 13 min after seeding.

consistent with the dry ice causing the rain, on most occasions at least, and the general conclusion drawn from these experiments was that dry ice appeared to be an effective agent for stimulating rain from supercooled cumulus clouds; further, during these experiments it was observed that clouds suitable for the use of this technique were of reasonably frequent occurrence in southeast Australia.

From a more practical point of view, however, this method looked less attractive at that time. If attempts had been made to stimulate rain over a large area it would have been necessary to transport large quantities of dry ice to the altitudes of the cloud tops, which in this area in summer extend to about 25,000 ft. This

would have required expensive military-type aircraft. Moreover, suitable clouds occurred at unpredictable intervals, necessitating the storage of large quantities of dry ice at a time when storage facilities were not readily available. Thus it appeared that to stimulate a useful increase of rain by the use of dry ice might be possible but would be economically doubtful. It was therefore desirable to investigate possible alternatives.

THE USE OF SILVER IODIDE

Dry ice affects supercooled clouds by causing ice crystals to form in them. It should be possible to cause the same effect in another way: Vonnegut had sug-

Figure 12.3 The fully developed anvil 40 min after seeding.

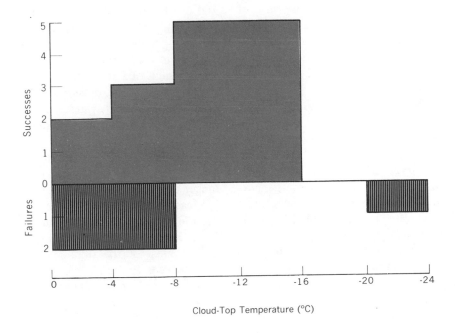

Figure 12.4 Dry ice seeding, cumulus clouds. Distribution of successes and failures with cloud-top temperature.

Figure 12.5   Silver iodide smoke generator for use on aircraft.

gested that silver iodide and other chemicals can be prepared in the form of nuclei on which ice crystals can grow.[42] If these nuclei were introduced into a suitable cloud and caused the formation of ice crystals, they should grow, and eventually become raindrops, just as they do when ice crystals are caused to form by dry ice.

Around 1948–1952 there were many attempts at "rainmaking" by this method, the silver iodide being released in most cases in the form of smoke from generators on the ground in the hope that it would rise into the supercooled levels of suitable clouds. Two preliminary experiments of this type were performed in Australia, but no results could be detected in the rain near the smoke generators. This raised questions as to how long it took the smoke to rise to appropriate levels, and during this interval how far and in what direction it traveled with the wind, and whether it retained its effectiveness. Doubts arose on the first score because in much of Australia the freezing level is high and the terrain is relatively flat, so the smoke must rise through a considerable height range, and on the second because the silver iodide (which resembles the material used in photographic emulsions) might be altered by exposure to daylight, and perhaps no longer act as an ice-crystal nucleus.

Measurements were therefore made of the life and trajectory of smoke released from the ground.[31, 32] Under the test conditions most of the smoke remained at low altitudes for some hours and lost its nucleating power so rapidly that by the time it reached cloud height it was mostly ineffective. Further, its trajectory was unpredictable, making it difficult to know where to look for effects on the rainfall, if there were any.

It appeared that, for general use in Australian experiments, releasing the smoke from the ground had its problems, and it would be better to release the smoke directly into the clouds from an aircraft. The nucleation could then proceed at a known position without there being time for appreciable decay to take place. With cumulus clouds it might be sufficient to fly at cloud base, relying on updrafts to distribute the smoke through the cloud, but with stratiform clouds, in which strong updrafts cannot be expected, it would be necessary to fly through the cloud at supercooled levels while releasing the smoke. Appropriate smoke generators for use on aircraft were therefore designed and tested[33] (Figure 12.5) and were used in the experiments described below.

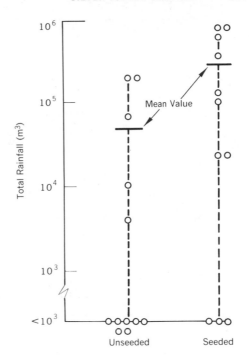

Figure 12.6 Total rainfall from isolated cumulus clouds, unseeded and seeded with silver iodide.

SILVER IODIDE EXPERIMENTS ON SINGLE CLOUDS

Preliminary trials[48] were first performed in which silver iodide smoke was released from an aircraft into single supercooled clouds with the intention of stimulating rain. The results were encouraging, and fully randomized series of trials was undertaken.[3] In each of the experiments of this series a cumulus cloud was selected which complied with a fixed specification: it had to be supercooled with top colder than −10°C, reasonably isolated, deep, of long duration and not leaning sideways too much, and not within 20 mi of any other cloud which was either raining already or glaciated (i.e., the top consisting of ice crystals). The cloud was then either seeded or not seeded according to a random sequence (similar to spinning a coin): in either case the seeding aircraft

flew under the cloud, and if it was to be seeded the silver iodide smoke was released into updrafts. The subsequent history of the cloud was then observed, and any rain that fell from it was measured by means of an impactor mounted on an aircraft, the crew of which did not know if the cloud had been seeded or not.

The total rainfall from each cloud is displayed in Figure 12.6, seeded and unseeded cases being separated. More rain fell from the seeded clouds than from the unseeded ones; there was very little likelihood that such a large difference could have arisen by chance, and the conclusion was drawn that the seeding had increased the rainfall.

In these experiments the mean rainfall from a seeded cloud exceeded that from an unseeded cloud by 2×10^5 m³, or well over 100 acre-ft. Put in another way, the

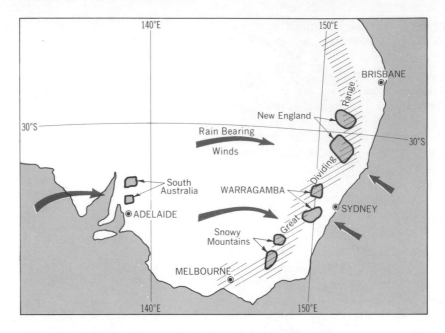

Figure 12.7   Location of cloud seeding experiments in southeastern Australia.

seeded clouds yielded several times more rain than the unseeded ones.

Of course it could not be assumed that a similar increase could be stimulated by the same method in the rainfall over an area. Only some of the rain is likely to fall from clouds of this type, and other clouds may not react so well to seeding; further, if rain is stimulated from one cloud it may affect the behavior of other clouds. To find out what effects can be caused on the rain over an area a different type of experiment is needed.

AREA EXPERIMENTS

In 1955 it seemed that it should be quite straightforward to seed clouds over areas of Australia and to measure the rainfall increases produced. After 16 years' work in this field the position can be summed up as follows.

1.   To measure the effects of seeding over an area is much more difficult than was expected.

2.   In selected climates, seeding clouds with silver iodide has substantial effects on the rain over an area.

3.   The effects are not simple: seeding may either increase or decrease the rain, and not necessarily only in the desired area.

4.   The job of finding out what cloud seeding can do and how best to do it is by no means finished, but the indications are that selected applications can be very rewarding.

This section describes the detailed work leading to these conclusions; the casual reader may prefer to omit it.

EARLY EXPERIMENTS

During the period 1955 to 1963 four experiments were carried out in Australia[28] in the locations shown in Figure 12.7. The purpose in each case was to find out if rain over the specified area could be increased by seeding clouds with silver iodide released from an aircraft.

Time during each experiment was divided into periods of about 14 days' duration, an exception being the Warragamba experiment where the period length was 1 day. Each experiment used two areas of 1000 to 3000 mi², the rainfall in each being measured by a network of 30 to 150 gauges.

In the Snowy Mountains experiment the two areas were used as target and control, respectively, and during any one period a random process determined whether clouds over the target area should or should not be seeded. In the other experiments, clouds over one or the other area were seeded, a random process determining which of the two would be used as the target during any period.

In other respects the experiments were similar. The purpose was to establish whether the rainfall over a specified area could be increased. The operational objective was to seed as many as possible of the deep, supercooled clouds passing over the area, cumulus clouds being seeded at the base and stratiform clouds at the −5 to −10°C level.

The duration of these experiments was 3 to 6 years; they operated continuously except that there was provision for suspending them when there was too much rain (the decision being made by an impartial "referee" who did not know the seeding sequence), and some of them were shut down in specified seasons, for example, in South Australia in summer, when there are very few clouds, and in New England during the wheat harvest, when rain is not desired.

The prevailing rain-bearing winds are shown in Figure 12.7. The Snowy Mountains and New England experiments were both situated on the western slopes of the

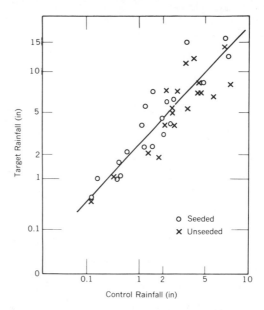

Figure 12.8 Snowy Mountains experiment. First 2 years' results. Period rainfalls in target and control areas.

Great Dividing Range and were affected mainly by continental air masses. The South Australian experiment mostly took place in maritime air while that in the Warragamba catchment near Sydney experienced weather of both types.

Before the experiments started it seemed likely that the results would vary from one to the next, for example, due to the differing types of cloud which one might expect to find in the different areas. It was thought likely that experiments in the areas with more continental air masses (New England, Snowy) might give the best results as more of the rain should form by the ice crystal process than in more maritime climates.

The early results supported these expectations. The first year's operation in Warragamba and in South Australia, with maritime climates, gave no indication that seeding had had any effect on the rainfall. However, initial results in the Snowy Mountains and in New Eng-

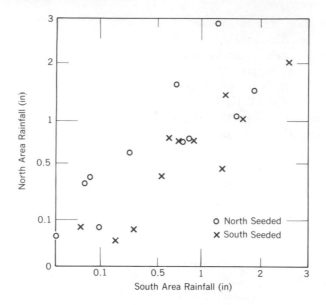

Figure 12.9   New England experiment. First year's results. Period rainfalls in north and south areas.

land were most promising. Figure 12.8 shows a plot of the target and control rainfalls in the first 2 years' operation of the Snowy experiment. Each point represents the rainfall for a period (about 12 days) in the two areas. The circles representing seeded periods are clearly displaced above the crosses (unseeded periods), as they should be if seeding increased rain in the target area: the extent of the displacement correspond to an increase of 26 percent and the results are statistically significant; that is, there is very little likelihood that such an increase could have arisen by chance. Similar results shown (Figure 12.9) for the first year's operation of the New England experiment are even more convincing: they correspond to a rainfall increase of over 30 percent at an even better significance level. At this stage the results appeared to be extremely heartening: the method of seeding clouds stimulated rain from individual clouds and appeared to cause a substantial rainfall increase in selected climates but not in others. However, when the experiments

continued the results varied from year to year and showed a consistent tendency to deteriorate, as shown in Figure 12.10. The results of the initially successful experiments in New England and the Snowy Mountains appeared to die out after a few years, while even the results of the other experiments that started poorly seemed to get worse.

The reason for this apparent deterioration of the results of seeding presents a puzzle that is not yet solved. We wonder if the effects may last for a long time after the seeding stops, perhaps into periods when no seeding was supposed to have occurred. Or perhaps the results may gradually, over a long time, spread out from the target area and contaminate the control areas. In either case the ability of the experiment to detect the effects of seeding would decrease even if the effects were still occuring. Whatever the reason, it was not foreseen in the design of these experiments, and it is clear that in future experiments provision should be made to investigate this effect if it occurs and to make sure that it does not prevent the true results of seeding being detected.

Other limitations of these experiments became apparent only when their results were being analyzed. They were designed to detect changes in the mean rainfall. This would be appropriate if the changes were always of the same type (e.g., always increases). However, during these experiments it appeared that the seeding sometimes increased the rainfall and sometimes reduced it. Thus rain in the target areas during the experiments was more variable than it had been before they started, and the variability of rainfall depended on the seeding even in areas where the mean rainfall was unaffected.[28] Obviously it was necessary to investigate the conditions in which seeding might stimulate increases and decreases. Practical seeding attempts might then be concentrated on the best conditions, and investigations could be made into how to improve the seeding technique for other conditions. As an example, it seemed that the effects of seeding varied with cloud top temperature. In the New England experiment the results suggested a substantial increase in rainfall on days when cumulus clouds with tops colder than $-10°C$ were seeded, but there appeared to be no change in rainfall on days when the seeded clouds were warmer. In Warragamba, it appeared that there were increases from cold clouds and decreases

when warmer clouds were seeded.[28] However, this sort of investigation was not foreseen in the experimental design, and cloud observations were made in the target area only: firm conclusions cannot be drawn from a comparison of rain in one area from clouds whose characteristics were observed, with rain in a control area falling from clouds of unknown characteristics.

Clearly it was necessary to plan future experiments in such a way that in addition to detecting changes in the mean rainfall over a long time they would allow proper investigation of effects of seeding which vary with conditions.

These early experiments were designed, as shown in Figure 12.7, each with a pair of topographically similar areas aligned north-south so as to be across the prevailing wind. It was hoped that the rainfall, which might vary from time to time, and at given times might vary in an east-west direction, would usually be similar as between each pair of areas. The rain in an unseeded control area would then be a good guide to how much rain there should have been in the target area if it had not been seeded. This hope was only partially fulfilled: rainfall gradients occurred in a north-south direction as well as east-west, and the variability of the north-south gradient reduced the value of the rain in the control area as a "yardstick" for that in the target. Future ex-

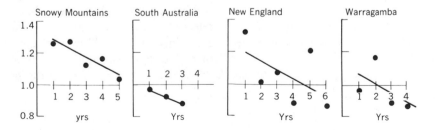

Figure 12.10   Variation of observed result with time in each experiment. Ordinate is rainfall increase factor associated with seeding.

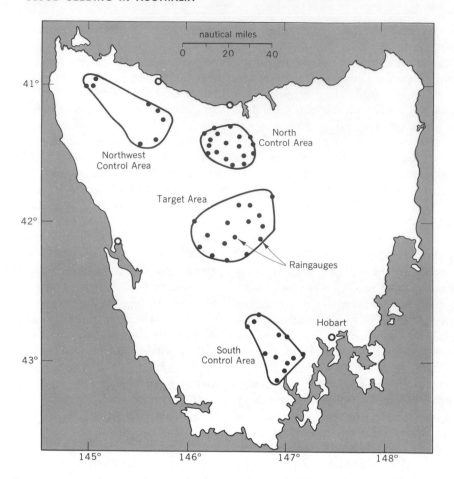

Figure 12.11   Experimental areas in Tasmania.

periments should make provision for allowing for rainfall gradients in both directions.

Summing up, we learned that future experiments on the results of cloud seeding on the rainfall over areas should be designed in such a way that they allowed for the investigation of certain effects, and were not prevented from achieving their objectives if these effects occurred.

The effects were as follows: (*a*) deterioration of apparent results with time, (*b*) variability of results of seeding with conditions, and (*c*) variability of rainfall gradients.

THE EXPERIMENT IN TASMANIA

A cloud seeding experiment was started in Tasmania in 1964.[30] It was of a generally similar type to those described in the previous section, in that it was intended to investigate the effects of seeding clouds with silver iodide smoke released from an aircraft on rainfall over a designated area. However, the experimental design was improved in view of the suggestions arising from the previous section.

The location of the experiment is illustrated in Figure 12.11. The target was a hydroelectric catchment area on a plateau, and there were three designated control areas which were never seeded. This made it possible to allow for the effects of rainfall gradients in all directions.

Time during the experiment was divided into pairs of periods of about 12 days' duration; clouds were seeded during half the periods, selected on a random basis.

An aircraft was used both for seeding and for making observations of clouds and of various aspects of the weather in the target area; these observations were made in both seeded and unseeded periods. This made it possible to compare rain on days when clouds with certain observed characteristics were seeded, with rain on days when clouds observed to be similar were not seeded, without any bias being involved.

The seeding schedule was carried out in alternate years only: during the intervening years no seeding was performed but the rainfall measurements were continued. In these circumstances, if the observed results tended to deteriorate after a long time, as in the earlier experiments, it was hoped that the off years might allow the areas to recover from this effect—whatever it is —and perhaps give better overall results than would be achieved with continuous operation. Further, by observing the way in which the effects of seeding built up and died away after the seeding started and stopped, it was hoped to gain some insight into the processes which had caused the deterioration.

The first stage of this experiment took place in the four operational years 1964, 1966, 1968, and 1970. The preliminary result was that the rainfall in the target areas was increased by seeding during autumn and winter, the effect being estimated as 15 to 20 percent at a satisfactory significance level: the value of this increase is much more than enough to justify the cost of the experiment. No changes were detected in spring and summer.

Detailed analysis of this type of experiment takes a long time and it is not yet complete, but when it is, it should give information, for example, on the meteorological conditions and cloud types when the most and least favorable results of seeding occurred: this should be of great help in planning future seeding experiments and operations. Until the analysis is complete we cannot say if the results throw any light on questions such as how long the effects of seeding last; but at least this experiment does not seem to have suffered from deterioration of results such as plagued previous experiments, possibly because the areas were allowed to "recover" during the off years.

The results of the first stage of this experiment were regarded as so promising that the Hydro-Electric Commission of Tasmania started a second 4-year stage in 1971. This is essentially similar to the first stage, except that clouds are seeded in two periods out of three, whereas in the first stage the proportion was one in two. In this way it is hoped to stimulate more rain while still providing enough information to confirm and supplement the results of the first stage. The first year's rainfall results (1971) are consistent with those of the first stage in suggesting a substantial rainfall increase in autumn and winter.

OPERATIONAL CLOUD SEEDING

While the Commonwealth Scientific and Industrial Research Organization (CSIRO) has been conducting cloud seeding research, Australia has suffered from its customary droughts, periods of forest-fire danger, and similar emergencies. In addition, there are large areas

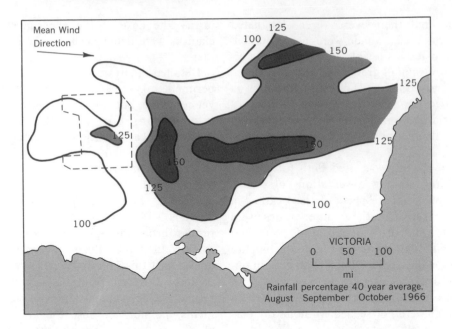

Mean Wind Direction

125
100
150
125
125
150
150
125
125
100
100

VICTORIA
0   50   100
mi
Rainfall percentage 40 year average.
August September October 1966

Figure 12.12   Rainfall pattern during seeding operations in Victoria. Dashed line indicates target area.

where increased rain is desirable as a means of improving the normal climate. These circumstances periodically caused the question to be raised of whether cloud seeding should be put to practical use.

At the present "state of the art" the effects of seeding in given circumstances are unpredicatable, if only because it is not possible to predict what clouds will be encountered. Nevertheless it seemed, in some cases, that there was reason to hope that seeding might stimulate extra rain and that, if it did, it would be worth more than the cost of trying. At first this was done by diversion of research staff to operational seeding, but the limitations of this approach, while obvious in principle, proved even more disruptive in practice. Further, it appears that the legal position in Australia is that responsibility for practical attempts to alter the weather rest with the State Governments rather than the Commonwealth.

The CSIRO. therefore set up "Courses of Instruction" in cloud seeding techniques. These were attended mostly by two types of customer: first, by senior officers of State Government Departments, who wished to be informed as to what cloud seeding had to offer and what was involved; those who were suitably impressed sent, to the next course, junior officers to learn the job. As a result of these courses cloud seeding programs have been implemented by all the Australian States as the need has arisen, the aircraft being hired and operated by the States under the direction of graduates of the CSIRO courses, while CSIRO acts only in an advisory capacity. A special case is in Tasmania where a State Government instrumentality is currently conducting the second stage of an experiment, the first stage of which was conducted by CSIRO.

ASSESSMENT OF OPERATIONS

Cloud-seeding experiments and operations are distinguished from each other by the intended end-product, which in the case of experiments is information while in the case of operations it is rain. Nevertheless, investigators conducting experiments usually hope that their efforts will produce useful rain as well as information, while instrumentalities conducting operations are often called upon, for example, by State Treasuries, to give an account of how much rain their activities (and expenditure) have produced. The latter situation raises problems. It is difficult to detect artifically induced changes in rainfall: in a carefully controlled experiment such as that described in the Tasmania Experiment, with several control areas, randomized seeding on only half the opportunities, and measurement of all sorts of physical parameters, it takes several years to be sure that a change of, say, 15 percent is really caused by seeding and is not just a chance product of the natural variability of rainfall. In an operation in which clouds are simply seeded without these aids to assessment, to detect with confidence a similar change in rainfall caused by seeding would at best take much longer and is often impossible: detection in a short time would be possible only if the change were very large.

Nevertheless, the cost of operating aircraft is high, the value of extra rain is even higher, and treasuries are insistent, so attempts must be made to assess the results of operations; however, all concerned should recognize that simple assessments can at best give only a rough indication of what is going on. They cannot record the details over short periods and they cannot provide conclusive proof of anything.

Assessments have been undertaken in this spirit of some of the Australian operations. For example, clouds were seeded over a specified area of Victoria in 1966.[1] Figure 12.12 shows contours of rainfall in the relevant months expressed as a percentage of the long-term averages. This shows that there was more rain than usual in part of the target area, but even more so in an unexpectedly large area extending down the prevailing westerly wind for some hundreds of miles, while rainfall was less than normal elsewhere. Can seeding clouds have increased rain over a larger area than that intended? This suggestion has also emerged from several other operations both in Australia and overseas.[1] Seeding continued in Victoria and very similar rainfall patterns were seen in 1967 and 1968, supporting the idea of increases extending a long way downwind. However, in 1969 and 1970 the seeding continued as before but the rainfall patterns downwind showed a depressing decrease where previously there had been an increase.

The question of what effect seeding really had on the rainfall of Victoria is still unsolved: we do not know if these patterns indicated variations in the results of seeding or simple variations in the natural rainfall. This again illustrates the fact that in operations without controls, rainfall changes can be measured with confidence only if they are very large or go on for a very long time. It is clear that we need some good new ideas as to how to assess the results of cloud seeding operations in a reasonable time.

CLOUD PHYSICS RESEARCH

The weather modification work in Australia has been supported, from its beginning in 1947 to the present, by a substantial program of research into the physics of clouds and rain.

Some of this work has concerned physical aspects specific to the climate of Australia, but most of it has been directed toward fundamental aspects not previously investigated.

Throughout this program, considerable stress has been laid on measurements from aircraft, rather than on the ground, when they are more relevant to the processes involved in real clouds.

Only a brief summary of some of the highlights is possible here.

Cloud Condensation Nuclei

Measurements of the concentrations of cloud nuclei have been made in many situations in the air and at the ground over the range of supersaturations important in cloud formation.[37] Direct comparison of observed cloud droplet concentrations with predictions based on observed cloud nucleus concentrations has shown satisfactory agreement.[38] Concentrations of nuclei have been found to be high in times of drought and to be strongly influenced by such activities as the burning of sugar cane. Annual and diurnal variations have been observed at a ground site.[41] Sizing techniques have shown that the particles involved must be very small, with radii not much larger than 10^{-6} cm,[40] while their instability to heating suggests that they are ammonium sulfate rather than sodium chloride.[39]

Cloud Droplet Spectra and Related Aircraft Measurments

Extensive measurements of temperature, humidity, and vertical velocity have been made in conjunction with cloud droplet spectra obtained by exposing sooted slides for a short accurately determined time[15] and of cloud liquid water content using a paper tape device.[47] This has allowed checks to be made on theories of the formation and evolution of convective clouds and on the degree of reliability of the instruments. As an illustration of an application of the measurements, observations on predominantly warm isolated maritime cumulus have shown a tendency to bimodal size distributions together with a relative uniformity across the cloud in the shape of the spectrum at a constant level.[46] This provides some information on the mixing of a cloud with its environment, a problem of great importance in the rapidly expanding subject of numerical modeling of cloud growth.

A DC-3 aircraft, instrumented for cloud physics research, is shown in Figure 12.13.

Figure 12.13   DC-3 aircraft instrumented for cloud physics research.

Figure 12.14 Laboratory investigation of coalescence: water droplets collide and coalesce in a vertical wind tunnel.

Coalescence

The theory of the coalescence process of rain formation was developed[13] and led to the studies of liquid water content, droplet spectra, and updraft velocities mentioned in the last section. The spectra of raindrops were also measured at the ground,[9] while extensive observations were also made of collection efficiencies of drops of different sizes suspended in a vertical wind tunnel[36] Figure 12.14). It is interesting to note that the effect of random collisions on broadening the droplet spectra[35] was discussed long before it became a generally accepted concept.

Natural Ice Nuclei

Several techniques have been developed for counting the nuclei active in ice formation in the atmosphere. These include mixing and expansion chambers using supercooled sugar solutions for detecting the ice crystals[4, 44] and membrane filters (Figure 12.15).[8] Although there are some inherent difficulties with all these techniques it has been established that in Australia, although concentrations of nuclei may fluctuate by two orders of magnitude over a few hours, the background level at temperatures warmer than $-20°C$ is relatively constant. In other

Figure 12.15   Ice nuclei on membrane filters are easily seen and counted after ice crystals have grown on them.

countries nuclei from dust storms and other surface sources are said to predominate, but Australian experience in which jet streams and pressure patterns seem to have more relevance (Figure 12.16)[7] suggests that the stratosphere is a more important source. Meteoritic material was shown[6] to be able to provide sufficient ice nuclei to explain observed concentrations in the troposphere. A

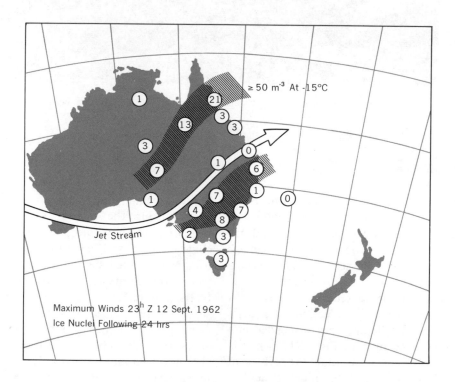

Figure 12.16   The ice nucleus concentration is low under a jet stream but higher on either side.

positive correlation between rainfall and ice nucleus concentrations has been established for several places.[7]

Artificial Ice Nuclei

Silver iodide has been the most intensively studied of the artificial ice nuclei although other interesting substances such as steroids[21] and metaldehyde[20] have received attention. Since silver iodide has been used almost exclusively for cloud seeding, considerable effort has been devoted to developing and testing satisfactory nucleus generators.[33] The decay of the ice-nucleating properties of silver iodide smoke has been studied.[32] It depends on the method of generation and can be very rapid in daylight, but does not occur at all at night.

Ice Nucleation Theory

The theory of heterogeneous nucleation has been developed in a form useful for comparison with experimental results, with a minimum of adjustable parameters.[18, 19] The main factors involved have been revealed by extensive experiments on silver iodide's special nucleating ability, and its dependence on the chemistry of production and the methods of observations.[16] Some new evidence on the nature of nucleation from the gas phase has come from study of "preactivation" of the nucleating properties of various substances.[17]

Ice Crystals in Clouds

Ice crystals have been detected in natural clouds by a replication impaction technique and by impressions on thin foils. In general the concentrations found in strati-

form clouds are consistent with the concentrations of ice nuclei, but in maritime cumulus clouds ice crystals may be several orders of magnitude more numerous than ice nuclei.[24, 23]

Secondary production of ice crystals on the freezing of water drops has been found to occur[14] but is insufficiently frequent to explain the observations, so that reasons for the discrepancy between crystal and nucleus concentrations are still being sought.

Silver in Rainwater

Attempts at the physical verification of cloud seeding experiments have included the development of a means of detecting minute traces of silver in rainwater.[43] Silver was detected much more frequently in areas where the clouds were seeded than in control areas where the clouds were not seeded.

Radar Meteorology

Radar equipment both on aircraft and the ground was used to study rainfall and contributions were made to the early stages of the science of radar meteorology.[10]

Factors Influencing Rainfall

Rainfall, and the chance of increasing it, may be affected by many surprising influences. The possibility that meteor showers provide ice nuclei leading to singularities in rainfall on certain calendar dates has been widely studied.[12] While statistical treatments have suggested the reality of some singularities,[26] in general, proof of their reality is an intractable problem. No adequate experimental evi-

Figure 12.17 Cumulus cloud seeding with pyrotechnics: 1 min before seeding; 7 min after seeding; 38 min after seeding; 79 min after seeding; 42 min after seeding (below cloud).

dence of a direct link between meteor showers and the supposed singularities has yet been found. The influence of the moon on rainfall[2] is easier to demonstrate but varies considerably from one place to another. Again the physical mechanism is not clearly understood. The possibility that industrial[25] and agricultural[45] activities may influence rainfall has also been investigated.

Numerical Modeling of Cloud Behavior

An accurate numerical prediction of cloud behavior based on correct physical principles is valuable in deciding whether seeding has made a difference to clouds, and the aircraft measurements described earlier in this section are being applied towards that end.[46] In addition, the value of modeling is illustrated by some experiments in fog dispersal.[5] A pilot experiment produced encouraging but not conclusive results. A modeling program[49] later showed that the differences between treated and untreated fogs could only be marginal, thus saving the effort involved in a larger test.

FUTURE EXPERIMENTS

It was concluded at an early data that, for general use in Australian conditions, silver iodide smoke should be released from aircraft, and for many years research has been based on this method. However, there are exceptions to most rules, and in specially favorable areas we should test the effects of releasing the smoke from the ground. But other methods are possible and should be investigated. Ice nuclei can be generated by pyrotechnics which can be dropped into cloud tops. Preliminary experiments[29] suggest that this method has promise in

the local conditions: photographs of a cloud seeded by this method (Figure 12.17) show an intriguing resemblance to those (Figure 12.1 to 12.3) of a cloud seeded over two decades earlier with dry ice. Perhaps dry ice should be tried again: the original objections are not as vaild as they were 20 years ago, as relatively cheap aircraft can now fly high enough, and dry ice is now much easier to store. Dry ice does have advantages over pyrotechnics, not the least being that there is no fear that it may set fire to a drought-stricken countryside. When testing any of these methods full advantage should be taken of the experimental facilities which have recently become available, such as computer modeling.

Finally, much of the rain in Australia falls from clouds which do not reach up to the freezing level: no ice crystals can be involved in the process of rain formation. Preliminary experiments[11] suggest that these clouds too can be caused to yield extra rain, and other suggestions as to how to achieve this end may well emerge from a continuing program of investigation in fundamental cloud figures.

It is clear that in the dry climate of Australia, research workers face a continuing challenge to explore the possibilities of stimulating more rain.

REFERENCES

1. Adderley, E. E. and E. K. Bigg, The downwind pattern of rainfall following cloud seeding in western Victoria, Australia, *Transactions of the Seminar on Extended Area Effects of Cloud Seeding,* Santa Barbara, Vol. 1, 1971.

2. Adderley, E. E. and E. G. Bowen, A lunar component in precipitation data, *Science* 7, 749, 1962.

3. Bethwaite, F. D., E. J. Smith, J. A. Warburton, and K. J Heffernan, Effects of

seeding isolated cumulus clouds with silver iodide, *J. Appl. Meteorol.* **5,** 513, 1966.

4. Bigg, E. K. A new technique for counting ice-forming nuclei in aerosols, *Tellus* **9,** 394, 1957.

5. Bigg, E. K., J. L. Brownscombe, and W. J. Thompson, Fog modification with long-chain alcohols, *J. Appl. Meteorol.* **8,** 75, 1969.

6. Bigg, E. K., and J. Giutronich, Ice nucleating properties of meteoritic material, *J. Atmos. Sci.* **24,** 46, 1967.

7. Bigg, E. K., and G. T. Miles, The results of large-scale measurements of natural ice nuclei, *J. Atmos. Sci.* **21,** 396, 1964.

8. Bigg, E. K., S. C. Mossop, R. T. Meade, and N. S. C. Thorndike, The measurement of ice nucleus concentrations by means of millipore filters, *J. Appl. Meterol.* **2,** 266, 1963.

9. Bowen, E. G., and K. A. Davidson, A raindrop spectrograph, *Quart. J. Royal Meteorol. Soc.* **77,** 445, 1951.

10. Bowen, E. G., Radar observations of rain and their relation to mechanisms of rain formation, *J. Atmos. Terr. Phys.* **1,** 125, 1951.

11. Bowen, E. G., A new method of stimulating convective clouds to produce rain and hail, *Quart. J. Royal Meteorol. Soc.* **78,** 37, 1952.

12. Bowen, E. G., The influence of meteoritic dust on rainfall, *Australian J. Phys.* **6,** 490, 1953.

13. Bowen, E. G. The formation of rain by coalescence. *Australian J. Sci. Res.* A3, 193, 1950.

14. Brownscombe, J. L. and N. S. C. Thorndike, Freezing and shattering of water drops in free fall, *Nature* **220,** 687, 1968.

15. Clague, L. F., An improved device for obtaining cloud droplet samples, *J. Appl. Meteorol.* **4,** 549, 1965.

16. Edwards, G. R. and L. F. Evans, Ice nucleation by silver iodide. I. Freezing vs. sublimation, *J. Meteorol.* **17,** 627, 1960. II. Collision efficiency in natural clouds, *J. Meteorol* **18,** 760, 1960. III. The nature of the nucleating site, *J. Atmos. Sci.* **25,** 249, 1960.

17. Edwards, G. R. and L. F. Evans, The mechanism of activation of ice nuclei, *J. Atmos. Sci.* **28,** 1443, 1971.

18. Fletcher, N. H., *The Physics of Rain Clouds,* Cambridge University Press, 1962.

19. Fletcher, N. H., Active sites and ice crystal nucleation, *J. Atmos. Sci.* **26,** 1266, 1969.

20. Fukuta, N., Ice nucleation by metaldehyde, *Nature* **199,** 475, 1963.

21. Head, R. B., Steroids as ice nucleators, *Nature* **191,** 1058, 1961.

22. Kraus, E. B. and P. Squires, Experiments on the stimulation of clouds to produce rain, *Nature* **159,** 489, 1947.

23. Mossop, S. C., R. E. Cottis, and B. Bartlett, Ice crystal concentration in cumulus and stratocumulus cloud, *Quart. J. Royal Meteorol. Soc.* **98,** 105, 1972.

24. Mossop, S. C. and A. Ono, Measurements of ice crystal concentrations in clouds, *J. Atmos. Sci.* **26,** 130, 1969.

25. Ogden, T. L., The effect on rainfall of a large steelworks, *J. Appl. Meteorol.* **8,** 585, 1969.

26. O'Mahony, G., Singularities in daily rainfall, *Australian J. Phys.* **15,** 301, 1962.

27. Schaefer, V. J., The production of ice crystals in a cloud of supercooled water droplets, *Science* **104,** 457, 1946.

28. Smith, E. J., Cloud seeding experiments in Australia, *Proceedings of the 5th Berkeley Symposium on Mathematical Statistics and Probability,* Vol. 5, 1967, p. 161.

29. Smith, E. J., Cloud-seeding with pyrotechnics in Australia, *Preprints, Second National Conference on Weather Modification, Santa Barbara,* 1970, p. 186.

30. Smith, E. J., E. E. Adderley, L. Veitch, and E. Turton, A cloud-seeding experiment in Tasmania. *Proceedings of the International Conference on Weather Modification, Canberra,* 1971, p. 91.

31. Smith, E. J. and K. J. Heffernan, Airborne measurements of the concentration of natural and artificial freezing nuclei, *Quart. J. Royal Meteorol. Soc.* **80,** 182, 1954.

32. Smith, E. J. and K. J. Heffernan, The decay of the ice-nucleating properties of silver iodide released from a mountain top, *Quart. J. Royal Meteorol. Soc.* **82,** 301, 1956.

33. Smith, E. J., J. A. Warburton, K. J. Heffernan, and W. Thompson, Performance measurements of silver iodide smoke generators on aircraft, *J. Appl. Meteorol.* **5,** 292, 1966.

34. Squires, P. and E. J. Smith, The artificial stimulation of precipitation by means of dry ice, *Australian J. Sci. Res.* **A2,** 232, 1949.

35. Telford, J. W., A new aspect of coalescence theory, *J. Meteorol.* **12,** 436, 1955.

36. Telford, J. W., N. S. Thorndike, and E. G. Bowen, The coalescence between small water drops, *Quart. J. Royal Meteorol. Soc.* **81**, 241, 1955.

37. Twomey, S., The nuclei of natural cloud formation, Parts 1 and 2, *Geofis. Pura Appl.* **43**, 227, 1959.

38. Twomey, S. and J. Warner, Comparison of measurements of cloud droplets and cloud nuclei, *J. Atmos. Sci.* **24**, 702, 1967.

39. Twomey, S., The composition of cloud nuclei, *J. Atmos. Sci* **28**, 377, 1971.

40. Twomey, S., Measurement of the size of natural cloud nuclei by means of nuclepore filters, *J. Atmos. Sci.* **29**, 318, 1972.

41. Twomey, S. and K. A. Davidson, Automated observations of cloud nuclei, September 1969–August 1970, *J. Atmos. Sci.* **28**, 1295, 1971.

42. Vonnegut, B., The nucleation of ice formation by silver iodide, *J. Appl. Phys.* **18**, 593, 1947.

43. Warburton, J. A. and C. T. Maher, The detection of silver in rain water. Analysis of precipitation collected from cloud-seeding experiments, *J. Appl. Meteorol.* **4**, 560, 1965.

44. Warner, J., An instrument for the measurements of freezing nucleus concentration, *Bull. Obs. Puy de Dome* **2**, 33, 1957.

45. Warner, J., A reduction in rainfall associated with smoke from sugar cane fires—an inadvertent weather modification? *J. Appl. Meteorol.* **7**, 247, 1968.

46. Warner, J., The microstructure of cumulus cloud. Parts 1 and 2, *J. Atmos. Sci.* **26**, 1049 and 1272, 1969.

47. Warner, J. and T. D. Newnham, A new method of measurement of cloud water content, *Quart. J. Royal Meteorol. Soc.* **78**, 46, 1952.

48. Warner, J. and S. Twomey, The use of silver iodide for seeding individual clouds, *Tellus* **8**, 453, 1956.

49. Warner, J. and W. G. Warne, The effect of surface films in retarding the growth by condensation of cloud nuclei and their use in fog suppression, *J. Appl. Meteorol.* **9**, 639, 1970.

13 | Rain Stimulation and Cloud Physics in Israel

A. GAGIN

J. NEUMANN

RAIN STIMULATION EXPERIMENTS IN ISRAEL AND ISRAEL'S WATER POTENTIAL

History

Cloud seeding activities in Israel began in 1948. At first individual clouds were seeded with dry ice from a plane of the Israel Air Force and with the partial financial assistance of the Israel Ministry of Agriculture. However, financial conditions soon forced us to switch over to AgI ground generators.* All the activities were planned and supervised by a self-

---

* Little did we realize at the time that the AgI smoke from conventional ground generators placed over flat terrain is not very likely to reach the clouds in adequate concentration. We came to learn at a later date that in order to ensure a sufficient concentration of AgI particles in the clouds, the output of ground generators must be greatly enhanced. But since we deal here with ground generators, a greatly increased output might create environmental problems. This need for an adequate concentration in clouds was the principal factor for deciding in favor of seeding by aircraft in the 1961–1967 project in Israel. Additionally, a number of ground generators were used (see below).

A. Gagin and J. Neumann are with the Department of Atmospheric Sciences, The Hebrew University of Jerusalem, Israel.

constituted unofficial committee, which called itself the "Rain Committee" and was composed of scientists and government officials.

Perhaps the most significant outcome of the first 3 years of activities was that we learned to appreciate the great difficulties connected with a critical evaluation of the results of seeding. It was this realization that led us early in 1951 to advance the idea of the need for a randomization of the experiment. The idea was approved by the Rain Committee and a randomized experiment—although on a very small scale—was put into operation in 1952. In all probability, it was the first randomized cloud seeding project in the world. A brief account of it was given in an invited paper at a conference in Tucson, Arizona in 1956.[44, 48] Admittedly, the statistical efficiency of the design of the project was low. Its greatest merit was in the introduction of randomization in rain stimulation trials.

Except for the years 1953 through 1956, when the Lasker Foundation of New York sponsored and financed a partially randomized seeding project in parts of the country, the above small randomized scheme continued throughout the 1950s. The Israel Government was not ready at the time to support a major experiment. In those years when an Israel Government official asked for the opinion of his colleagues in the U.S. Government service, the reply was usually skeptical. . . . Under the circumstances, the small randomized project could not have continued without the efforts of E. D. Bergmann, Chairman of the "Rain Committee."

The growing realization over the 1950s that Israel must find new water resources prepared the ground for a change in the attitude of the official water organizations of the country. In 1960 an "International Conference on Science in the Advancement of New States" was convened in Rehovot, Israel, by Abba Eban, then Minister of Education and Culture and President, Weizmann Institute of Science, Rehovot.[27] Among the invited speakers were E. G. Bowen, Radiophysics Division, CSIRO, Australia, and L. J. Battan, Institute of Atmospheric Physics, University of Arizona, Tucson, Arizona. Both spoke of cloud seeding experiments. Bowen was so positive about the promises of seeding of "cold" clouds with AgI (the text of the staid Proceedings does not quite reflect the enthusiasm of the verbal presentation) that his words convinced heads of Israel's water organizations (A. Wiener, TAHAL—Water Planning for Israel, in particular) of the worthwhileness of a serious major project. The ancient proverb, "No one is a prophet in his own country" (or "in his city," as the Hebrew form of the old proverb puts it), proved true once again. What the recommendations of local scientists could not achieve over a period of 12 years was accomplished by the foreign scientists.

This major project was to be financed by the Ministry of Agriculture, responsible for the development of water resources, and administered by the Mekorot Water Company through its subsidiary, the Electrical and Mechanical Services (SHAHAM), Ltd. The 12-year-old "Rain Committee" was made an official organ of the Israel National Council for Research and Development and entrusted with the scientific planning and supervision of the experiment. In actual fact, the statistical design of the randomized experiment was the work of K. R. Gabriel of the Statistical Laboratory* at the Hebrew University of Jerusalem. A decision was also made concerning the initiation of a program of cloud physics research connected with the seeding project. The conduct of that research was put into the hands of the Hebrew University Department of Meteorology (renamed Department of Atmospheric Sciences in 1972) Cloud Physics Laboratory, headed by A. Gagin.

Since by 1960 doubts as to the usefulness of AgI smoke generators placed over flat terrain were rather strong, a decision was made (aided by E. G. Bowen) to use aircraft equipped with AgI burners, the seeding flights to take place at cloud base level. The aircraft were to be equipped with instruments in order to study the physics of clouds.

The statistical design of the randomized experiment was of the type designated "crossover" and involved *two* experiment areas referred to by the names "North" and "Center" (see Figure 13.1).

The actual seeding operations began in February 1961 and continued for a total of 6½ rainfall seasons (the rainfall season of Israel is much like that of California, from about November to April, or "Mediterranean climate"), that is, until spring of 1967.

During this first major experiment only a fraction of the catchment area of Lake Tiberias (the "Sea of Galilee" of the New

* Now called Department of Statistics.

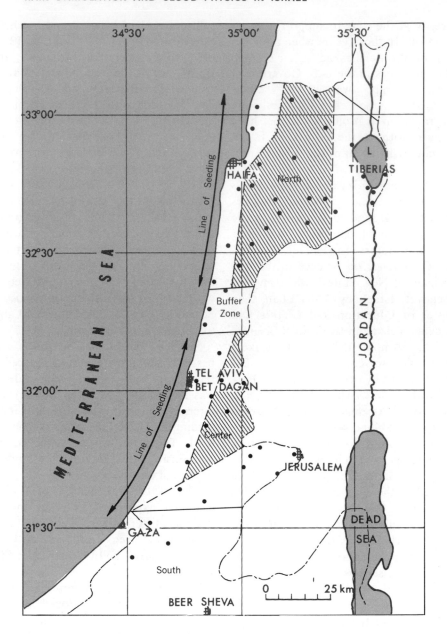

Figure 13.1   Map of Israel, showing the two experimental areas and their interior parts (shaded), for the 1961–1967 experiment. Dots indicate location of rainfall gauges used in the statistical analysis.

Testament), the principal water reservoir of Israel's Main Water Carrier, could be seeded, the rest of the catchment being in Lebanon and Syria. Because of the importance of that catchment and lake for Israel's water resources, a second major

randomized experiment was started in 1969. The primary aim of the new project is to explore the possibility of increasing the rain over the catchment, a major fraction of which has become accessible to seeding since June, 1967. As in the first experiment, there are two experiment areas, North and South. These areas are larger than those of the first experiment. Again we can use the randomized crossover scheme. However, the design of the second experiment is much more versatile. More subareas can be compared now and, in particular, results for the North can be examined separately using control areas whose rainfall is highly correlated with that of the North. Some of the design features are such that we expect to obtain for the North significant, and possibly highly significant, results, positive or negative, within a shorter period than the length ($6\frac{1}{2}$ seasons) of the first experiment.

Finally, parallel with the above areal seeding, a single-cloud seeding study is in progress accompanied by an intensive study of the physical properties of cumulus clouds of Israel. The latter effort has already produced a number of interesting and important results, which have advanced our understanding of the precipitation-forming processes in these clouds.

The Water Potential of Israel*

Several estimates were made over the years by different individuals of Israel's water potential. Most of these agree that it amounts to about $1.5 \times 10^9/m^3/year$. Moreover, since most of that potential is generated in the north while the south is arid, or nearly so, a need exists for transporting the water from north to south. The idea of the Water Carrier first appears in the visionary book of Theodor Herzl,[30] the founder of the World Zionist Organization. A more concrete form of

the idea was set forth in a book by Walter C. Lowdermilk,[39] the American soil and water conservationist. The engineering details were worked out by James B. Hays[29] who was formerly a project manager of the Tennessee Valley Authority. Interestingly enough, he called his major report "T.V.A. on the Jordan."

As was stated earlier, the principal source of Israel's Main Water Carrier is Lake Tiberias, which is situated in the Jordan Valley (Figure 13.1), approximately 210 m *below* mean sea level. Water is pumped up from there to a height of 150 m *above* mean sea level and sent southward. The Carrier was started up in 1964. About $3.5 \times 10^8 m^3$ of water is pumped annually from the Lake and transported south.

It was quite clear already in the 1950s that the aforementioned water potential is inadequate for the development and settlement of the country and that, therefore, ways and means are to be found for augmenting the potential. Unorthodox methods such as evaporation reduction (Lake Tiberias alone evaporates about 270 million $m^3/year$—equivalent to over $\frac{1}{6}$ of the potential*),[47] evapotranspiration reduction, recycling of sewage water, desalination of brackish and seawater, and artificial rain have all been considered. (We ignore such ideas as "painting the country black" in order to increase convection and rainfall; see Katz and Gagin.[35]) Evaporation reduction efforts do not appear promising; evapotranspiration reduction experiments are still in the early stages. Recovery of sewage water has begun. Open oxidation ponds do occasionally give rise to unpleasant smells; closed recovery systems require large investments, but the water

---

* The writers are indebted to E. Kally, TAHAL— Water Planning for Israel, for information provided by him. Some of the information is published in his book (in Hebrew), *The Struggle for the Water*.

* Of course, the potential is the annual water mass obtained *after* deducting from the annual rainfall the various losses such as the evaporation from Lake Tiberias, etc.

needs of the country will force the water organizations to make the investments.

A good deal is being done in Israel in the area of desalination,[40] but the quantities involved at this time are relatively small and are, in the main, supplying local needs of some settlements in the Negev (desalination of brackish waters) and of the township of Eilat (seawater). The idea of a major dual atomic plant, producing both electricity and desalted water, has also been entertained, but the investments required are so large that the matter is still in the study stage.

Artificial increase of rain, if feasible, would be a rather inexpensive way toward augmenting water resources, Assuming that cloud seeding is carried out over Israel north of Beer Sheva (south of Beer Sheva seeding opportunities are less frequent) and assuming a 10 percent increase in rainfall through seeding, on the average (we note that the results of the 1961–1967 experiment indicate a 15 percent increase on days seeded), the result being statistically significant and even highly significant, an addition of over 500 million m^3/year is expected, on the average. Allowing for evaporation and other losses, we can expect that 300 or, perhaps, 400 million m^3/year would accrue to Israel's water potential (about as much as is transported yearly at present by the Main Water Carrier). That is, assuming a conservative* 10 percent increase, the water potential of Israel could be augmented by $\frac{1}{5}$ and more. The cost of that additional water—if real—is extremely low: about $\frac{1}{30}$ or even $\frac{1}{60}$ of the next cheapest method for developing new water resources.

---

* Conservative if we base our figures on the statistical results of the 1961–1967 experiment. We must realize, however, that at this stage the quantitative results are based on statistical evaluation. Some of the difficulties encountered in purely statistical evaluations are discussed in a subsequent section of this chapter.

The decisions facing Israel in a very few years are grave. Already the pumping facility of the Main Water Carrier at Lake Tiberias is operating close to its maximum capacity. If the results of the second randomized cloud seeding experiment prove successful, the pumping facility will have to be increased, which means a rather large capital outlay. Here is the dilemma:

Suppose the statistical results of the second experiment prove positive and the investment in the enlarged pumping facility is made. Should these statistical results prove misleading (whether we discover this or not), not only is an investment wasted but, possibly, a water shortage may be ahead. Alternative water resources must be developed to avoid such a shortage but the country cannot very well afford investments in different alternatives.

If, on the other hand, we do not believe in the statistical results or if the statistical results are negative and we accept them as such, although in actual fact there is an increase through seeding, then we shall let the artificially added waters go to waste since we shall not have the adequate pumping facilities: these waters will flow down to the Dead Sea and become highly saline (over 20 percent salinity).

STATISTICAL ASPECTS OF THE EXPERIMENT: DESIGN, RESULTS, AND PROBLEMS

Statistical Design of the Experiment

As was pointed out earlier, the 1961–1967 Israeli randomized cloud seeding project was designed as a "crossover" experiment. To be most effective, the design requires two experiment areas whose rainfall is well correlated in a positive sense. The scheme was first introduced in Australia[1]; in its Israeli application, Gabriel[16] introduced a number of improvements. A time unit is to be chosen for the experiment: in Israel it was the 24-hr day* and consequently we will refer to the time unit, briefly, as a "day." In the design only one of the two experiment areas is seeded on any one day, the area being determined in a random manner. On the same day the second area serves as a "control" area. During the next day of seeding again only one of the two areas is seeded, the "choice" of the area of seeding being random at all times. For the test statistic of the experiment, to measure the effect of seeding, the root-double-ratio (RDR)

$$R = \left(\frac{N_s}{C_u} \cdot \frac{C_s}{N_u} \right)^{1/2} \qquad (13.1)$$

was selected. In Equation 13.1, the subscript s refers to days seeded, u to days unseeded.† N and C, respectively, stand for the average rainfall of the two areas (the letters N and C are the initial of the two areas of the 1961–1967 experiment, North and Center). The great merit of the crossover design and, especially, that of the test statistic is that it eliminates, to some extent, the troublesome and misleading effects of the *natural* fluctuations in rainfall. These fluctuations are or can be of an order of magnitude greater than, or at least similar to, those of artificial increases which we believe are possible, and can therefore mask the latter unless the statistical design of the experiment is "efficient." It will be seen in Equation 13.1 that if the rainfalls of the two experiment areas are closely correlated in a positive sense, then a "natural" fluctuation in the

rainfall of area N (or C) will be neutralized, to a degree, by the more or less parallel fluctuation on the same day (N_s and C_u represent *the same set of days* and the same applies to C_s and N_u) in C (or N). Moreover, the crossover design avoids the direct use of historical rain data. This important feature prevents basing our estimates on past relationships between the two areas when, in actual fact, the relationship during the experiment period might be different.

The following comparison illustrates the "efficiency" of the crossover design. In some of the randomized experiments, for example, in Project Whitetop,[8] the so-called single-area design was used. In this design a single area serves on some days (or time units) selected at random as the experiment area, on others as the control area. For test statistic, one can use the ratio

$$\frac{S - U}{U} = \frac{S}{U} - 1, \qquad (13.2)$$

where S = average precipitation for days seeded, and U = the same for days unseeded. The inefficiency of the single-area design is shown by a recent study of Neumann and Shimbursky.[50] In the 1961–1967 Israeli crossover experiment involving the two experiment areas designated North and Center, the RDR amounted to 1.15, that is, a 15 percent increase is indicated for days of seeding* and the statistical significance of that result was estimated (on the basis of a computer experiment shown by Gabriel in his Table 4[18]) as 0.9 percent. If we now look at the North area as a single-area experiment, the ratio of Equation 13.2 is 0.15; that is, again a 15 percent increase is indicated, but the level of significance

---

* To be precise, the 24-hr day was adopted from November, 1961, that is, for about 6 out of 6 ½ seasons of the experiment the time unit was the 24-hr day. See Table 1 in Gabriel.[18]

† "Days seeded" stands for days *allocated* randomly to seeding irrespective of whether the day was actually seeded or not.

---

* As was pointed out earlier, results for "days of seeding" include days allocated to seeding irrespective of whether the day was actually seeded or not.

now is about 16 percent.[50] Here and in subsequent sections, the tests adopted are one-sided. What is tested is the hypothesis of a null effect versus the possibility of enhancing the rains through seeding.

Despite the great superiority of the crossover scheme, it has its practical shortcomings. We have seen above that it is desirable for the rainfalls of the two experiment areas to be highly correlated. This will often mean that the two areas are to be situated near one another. Hence on some days, the winds may carry the seeding material intended for experiment area A to B where B is not to be seeded on those days and vice versa. We note in passing that the adoption of a buffer zone (see Figure 13.1) between the two experiment areas reduces, to a degree, the possibility of "cross-contamination."

Another problem, qualitatively common to both the RDR and the ratio shown in Equation 13.2, is bias,[15] a fact which leads to an overestimate of the effect studied by the ratios in question. Gabriel, however, has pointed out that the bias becomes smaller as the number of experiment days is increased. In the case of the single-area ratio, Equation 13.2, the reduction in bias as the number of experiment days is enlarged is shown in Table 1 of a paper by Neumann and Shimbursky.[50]

Another problem common to many randomized designs is that a fraction of the total number of experiment days is not seedable. For instance, on some days either there are no clouds at all or the clouds are small and perhaps have no supercooled top at all. Clearly, there is no point in seeding on such days. Yet statistical significance testing of the results of seeding is only meaningful if the data of *all* days randomly allocated to seeding, whether actually seeded or not, are considered together. Since seeding is actually carried out when suitable clouds are present in the area designated for seeding, it is biased in favor of days when natural conditions may promote the formation of rain. Restricting the analyses to days actually seeded would have rendered us similar to a character in a Thomas Mann novel who would go daily to the railway station, give a signal every time a train started off, and believe that the train was put into motion at his signal. On the other hand, if the days totally unsuitable to seeding are retained, and especially if they are numerous, then they are likely to blur and even "blot out" the effect of seeding in the statistics.

Gabriel's solution to the problem was in using the buffer zone between the two experiment areas as an *objective indicator* of "natural" conditions in the experiment areas independent of the effects of seeding. Since the rainfall of the buffer is rather closely correlated with that of the experiment areas, it can serve as a good indication of "natural" conditions. On this basis, the following definition was adopted for the experiment (or rain) days: any day in the buffer when there occurs a measurable amount of rain is an experiment day, or rain day, in the experiment areas from the point of view of the design of the experiment. In this manner, about 38 percent of the days of the seeding season were eliminated. Only on about 2 percent of the total number of days during the seeding seasons of the 1961–1967 project did a measurable rain fall in the experiment areas when there was no measurable rain in the buffer. Thus the total loss of 2 percent of the seedable days was very small compared with the gain accrued by eliminating unsuitable days from seeding in an objective manner and the associated gain in the ratio of signal to noise.

Statistical Results of the 1961–1967 Experiment

The statistical results of the 1961–1967 experiment are discussed in various papers by Gabriel. The most comprehensive report of the statistical results are in his 1970 report,[18] although some aspects, for example, that of the design, are discussed in greater detail in earlier reports and papers, the list of which is too long to quote here. Nevertheless, we think it worthwhile to refer to some papers by Gabriel[16, 17] and to another paper by Neumann, Gabriel, and Gagin.[49]

At the outset of the experiment, the Rain Committee approved that the results of seeding would be assessed by the RDR, Equation 13.1. For an estimate of the statistical significance of the effect of seeding, Gabriel proposed the Wilcoxon-Mann-Whitney (WMW) non-parametric test. The crossover design was applied not only to the entirety of the North and Center experiment areas but also their inland sections somewhat downwind from the line of seeding.* These inland areas are referred to as "interiors" or "interior areas." We anticipated that the major effect of seeding, if any, should occur at some distance downwind, and hence the interior areas should show the largest effect. In fact, our estimate is that the major effect of seeding should occur roughly half an hour after the introduction of the seeding material at cloud base level. This estimate is based on theoretical calculations of

growth of precipitation elements summarized later. In terms of wind, half an hour is equivalent to 15 to 25 km horizontal distance; that is, we anticipate the maximum effect to take place at that distance downwind from the line of seeding. The location of the interior areas is indicated in Figure 13.1.

For the entire experiment areas of the 1961–1967 experiment, the RDR is about 1.15 (= 15 percent increase). The standard error of the RDR is such that the indicated increase is 15 ± 6 percent and the WMW test yields a significance of 5.4 percent. For the interior areas, the parallel results are close to 22 ± 7 percent, the WMW test being significant at the 1.3 percent level; that is, for the interior areas the positive effect is highly significant.

It is noted that the first "official" evaluation procedure had the shortcoming that the significance test did not relate directly to the RDR. This defect was corrected by Gabriel at a later stage of the experiment. He undertook *computer experiments* with the rain data of the two areas. In these, the data were permuted in a random manner and for each such computer experiment an "RDR" was obtained. On carrying out a large number of permutations, one can get a distribution of the "RDR's" in random experiments. By relating the RDR actually obtained in the field experiment to the computer experiments, one can judge the statistical significance of the actual RDR. Such a procedure is much more "powerful" than the aforementioned WMW test since much greater use is made of the data. Additionally, the test of significance relates directly to the RDR.

Gabriel and Feder[19] have studied the asymptotic distribution of the RDR for computer experiments involving an increasing number of rain days (or rain

---

* In Israel the winds on rain days practically always have a westerly component; this is especially true for the winds aloft, say, at 950-mb and at heights above it. It will be recalled that aircraft seeding took place at about cloud base level, which for cumulus clouds is usually at 700 to 800 m, or at nearly 900 mb.

"time units").* They were able to show that the distribution is asymptotically normal as the number of rain days is enlarged, and this provides a way (mentioned in the foregoing paragraph) of determining the statistical significance of a field experiment. The theoretical derivation of the asymptotic distribution is based on some assumptions which, we believe, are not restrictive.

With the aid of the above computer-randomization experiments Gabriel finds that the RDR for the entire areas is 1.153 ± 0.064 and for the interiors 1.218 ± 0.067.[18] The pertinent one-tailed significance levels are, respectively, 0.9 and 0.2 percent. Thus the randomization experiments indicate highly significant results both for the entire and the interior areas. Nevertheless, some problems remain.

Other Evaluations of the Overall Results

In addition to the evaluations described, Gabriel has studied the results of a procedure involving "historical" rain data.[18] Moreover, Dalinsky and Wurtele have studied the data and published results of their investigations.[11, 60] Generally, the test statistics considered vary from author to author. The "precise" results depend to some extent on the test statistic, but all three investigators agree that seeding apparently increased rainfall, and the results are found to be statistically significant or highly significant.

Anatomy of the Results

In discussions with hydrologists the following interesting and important

* Some criticism of their paper was put forward by a referee of a paper by Neumann and Shimbursky.[50] This criticism was more of a formal than of a practical nature; Gabriel's results remain unchanged. The matter is discussed in the paper by Neumann and Shimbursky.

problem emerged: assume we accept the statistical indications of an increase of 15 percent through seeding as real, and suppose that this increase is made up of *large* enhancements on a relatively few days, while on a good number of other days we actually cause a decrease. Then since on days of large rainfall, natural and artificial, some, and possibly a great deal, of the rain will be lost irretrievably through flood runoff, while on some other days the rainfall is supposedly reduced at times when runoff losses would be nil or small, seeding may, in fact, do more damage to water supply than benefit, despite the overall "addition" to rainfall.

The intensive physical studies conducted by us in conjunction with the seeding project have contributed greatly to our understanding of the rain-formation processes. We believe that in "continental"-type of cumulus clouds, typical of Israel's winter season, an "overseeding" is not reasonably possible. The problem was also studied statistically by both Gabriel and by Wurtele.[18, 60] Their methods differ, but the two authors agree that important negative effects are not supported by the data.

Another problem of concern was as follows: is the indicated overall increase made up of large increases on a few days with no increases of any importance on other days? The views on this matter varied until the first experiment was concluded in 1967. Gabriel in 1970 studied the problem in great detail. He found positive effects of varying intensity in the great majority of categories and subcategories of days of the different classifications. The indications are that the daily effects may have been proportional to the amounts of "natural" rainfall, up to a limit; see his Tables 5 and 6.[18]

One other aspect of interest is the possible occurrence of long-distance effects of seeding. Large apparent positive downwind effects have been reported for the case of the Swiss Hail Prevention Experiment[51] and for the experiment in the Climax area of Colorado[26]; large apparent negative effects downwind, upwind, and lateral to the mean wind have been reported for Project Whitetop.[37]

In the case of the Israeli experiment, there are insufficient rain data available to us for the Kingdom of Jordan to examine the matter of long-distance effects for all the seasons of the experiment. However, Table 3 in Gabriel's 1970 study suggests a decrease in the apparent positive effect of seeding east of the "interior areas" (Figure 13.1). It is precisely this apparent decrease that gives prominence to the "interior areas." See Table 13.1 below.

Our experience with rainfall data suggests that the above-mentioned apparent increases or decreases at large distances from the line of seeding could well be manifestations of the great natural fluctuations in rainfall, possibly modified somewhat by the effects of seeding. Both in the case of large distances and in the cases usually studied (i.e., not too large distances), it is very difficult to arrive at unequivocal answers by purely statistical methods. It appears to us that the statistical techniques applied so far are inadequate to cope with the great variability of rainfall. Indeed, some of the negative results for Project Whitetop reported earlier by Neyman and ascribed by him to seeding are now tentatively interpreted by him and by his collaborators[38] as representing rainfall fluctuations.

Some Problems of the 1961–1967 Experiment

It was observed by Wurtele in 1971 that in the 1961–1967 experiment rainfall in the Buffer, which was not to be seeded, was greater on C-allocated days (= days allocated to seeding in the Center) than on N-allocated days (Table 13.1).[60]

We have already mentioned in a footnote that on rain days in Israel the winds tend to be from the west. A more detailed study for the country north of the Negev indicates that on the above set of days the winds in the first few kilometers, and especially in the layer above the first few hundreds meters, are rather often from the southwesterly quarter. Under these circumstances it was tempting to assume that on days allocated to seeding in the Center, which lies just south of the Buffer, AgI particles were carried by the winds to the Buffer. Based on that premise, the larger rainfall in the Buffer on C-allocated days would be caused by an unintentional seeding of the Buffer.

The foregoing view is weak because rainfall in the Buffer was also greater on the C-allocated days, which were not actually seeded. The difference between these two sets of unseeded days can only be caused by the wild fluctuations in rainfall. But the difference does raise the question: could it be that the heavier rainfall of the Buffer on C-allocated and actually seeded days, as compared with N-allocated and actually seeded days, was also a matter of random fluctuations in rainfall rather than a matter of wind transport of seeding particles from the Center?

E. Shimbursky has completed, at the time of writing this chapter, an investigation involving rain and wind data for the 1961–1967 experiment. The winds are for the 850-mb level, about 1500 m msl, somewhat above cloud base level, as measured at 1400 hours LST daily at Bet Dagan by the Israel Meteorological Service. Bet Dagan (see Figure 13.1) is situated slightly to the north of the central latitude of the Center area. It is at a distance of about 35 km from the Buffer. Its wind-aloft directions are presumably representative for flow condi-

Table 13.1   Mean Daily Rainfall (mm) in the 1961–1967 Israeli Randomized Cloud Seeding Experiment on Days for Which Wind-Aloft Data are Available[a]

	Allocation of Seeding to					
	North, Seeded and	Center, Unseeded	North		Center	
Area or Subarea			Seeded,	Unseeded,	Seeded	Unseeded,
	197 Days	167 Days	111 Days*	86 Days†	91 days*	76 Days†
North						
Total	8.69	7.98	13.49	2.48	11.75	3.47
Coast	8.56	7.82	13.02	2.81	11.10	3.89
Interior	9.96	8.87	15.49	2.82	13.07	3.83
East	6.21	6.15	9.85	1.50	9.01	2.72
Buffer	6.96	9.12	10.84	1.95	14.23	3.00
Center						
Total	6.71	8.23	10.38	1.98	12.89	2.64
Coast	6.64	8.14	10.30	1.80	12.78	2.58
Interior	7.07	9.47	11.14	1.82	14.58	3.34
Southeast	6.59	7.03	9.95	2.25	10.96	2.31

[a] Total number of experiment (or "rain") days: 364. Number of rain stations: North, ca. 26; Buffer, 3; Center, ca. 18.

* Actually seeded

† Allocated to seeding but not actually seeded.

tions along the Center-Buffer boundary zone. A potentially serious shortcoming of the investigation is that it is based on winds at one fixed hour of the day and does not allow for changes in direction throughout the day.*

The investigation classifies the days into three groups: (1) winds from the southwesterly quarter (181° to 270°), (2) winds from the northwesterly quarter (271° to 360°), and finally (3) winds from the easterly semicircle. It is the days in group 1 when the Buffer and, at times, when the winds are close to being sou-

* On the other hand, a study involving more than once-a-day wind measurements will encounter the difficulty that at most stations the rain is only measured once a day.

† One could have divided the southwesterly quadrant into two or more sectors, but then relatively few days will be associated with each sector. In view of the great variability of rain, we are hesitant to work with low numbers of rain days.

therly,† that the North might have been contaminated by AgI smoke from the Center. This group comprises about ⅔ of all rain days (see Table 13.2) and is thus the most important group. Seeding the North on days of group 2 should have affected the Buffer and the Center less often both because of the fewer number of days involved and because of the fact that the Mediterranean coast runs from SSW to NNE.

Wind data were available for 364 "rain days" out of a total of 381 "rain days" of the experiment. Table 13.2 lists the results of the computations, for the 364 days in question and for the days when the winds were from either the southwesterly or the northwesterly quarter. The significance levels, both for single-area ratios (Equation 13.2) and for RDR's have been estimated in each case from 1000 Monte Carlo computer experiments, that is, random permutations, using the 1961–1967 data of the area(s) of concern. We have marked by the symbol ≠ the cases where the significance levels are poorer than 5 percent, to which we should give little weight in our considerations. In preparing the table, the assumption has been made that the Buffer is seeded on days when the Center is seeded.

Before listing the salient features of Table 13.2 we wish to point out the following. Let us assume that seeding increases the rain. If we deal with the single-area ratio (Equation 13.2) this ratio will be large if the area is not inadvertently contaminated on days not to be seeded. Inadvertent contamination of the area on days not to be seeded will have the effect of decreasing the ratio. Parallel considerations apply to the RDR of Equation 13.1. Random fluctuations can, of course, render the ratios large or small. We "guard" ourselves against random fluctuations by using

statistical significance tests. Thus large ratios can mean large increases in rain and proper conduct of experiment (no contamination) or be the result of rainfall variability or a suitable combination of the aforementioned factors; small ratios can be caused either by small increases, contamination, random fluctuations, or a combination of them.

Some of the salient features of Table 13.2 are as follows:

1. On SW days (= days with winds from the southwesterly quarter), the SAR (= single-area ratio) for the Buffer is high (1.32, and that at a reasonable level of significance, 4.7 percent). In contrast, on NW days, the parallel ratio is but 1.05 and insignificant.

2. On SW days, the SAR for the North is low (1.04), whereas on NW days (when the North, presumably, could not have been contaminated on C-days) it is very high and not significant (6 percent).

Table 13.2   Root-Double Ratios (RDR) and Single-Area Ratios (SAR) and Statistical One-Tailed Significance Levels (in Parentheses) for the 1961–1967 Israeli Randomized Cloud Seeding Experiment for Days with Wind-Aloft Data[a]

	All Wind Directions 364 days	SW days, 243 days	NW days, 81 days
RDR			
North-Center	1.15 (2%)	1.10 (11%) ≠	1.35 (1%)
North-Buffer	1.19 (0.1%)	1.17 (1%)	1.26 (1%)
SAR			
North	1.08 (v.l.s.) ≠	1.04	1.52 (6%) ≠
Buffer	1.31 (7%)	1.32 (4.7%)	1.05
Center	1.22 (11%)	1.16 (v.l.s.) ≠	1.20 (v.l.s.) ≠

Note: It is assumed for the purposes of the table that the Buffer was seeded (inadvertently) on days allocated to seeding in the Center.

[a] SW days = Days with winds at the 850-mb level from the southwesterly quarter (181°–270°). NW days = The same, but winds from the northwesterly quarter (271°–360°). Data for winds from the east not quoted; symbol ≠ emphasizes that significance level is low. v.l.s. = Very low significance.

3. On NW days, the RDR for North versus Center is high (1.35) and highly significant (1 percent), whereas on SW days the RDR is much lower (1.10) and insignificant.

4. On NW days, again, the RDR for North versus Buffer is rather high (1.26) and highly significant (1 percent).

All the above-listed features would be consistent with the assumption that the Buffer was contaminated on SW days when the Center was seeded. But, as was pointed out earlier, Table 13.2 is based on once-a-day winds and these winds are not necessarily valid for all hours of seeding. Moreover, the finding "consistent with the assumption, etc." is no proof.

It will be seen in the next section that preliminary results of the second experiment show a number of features paralleling those of the 1961–1967 project, a fact which lends support to the interpretation that the findings with respect to the experiment areas for the 1961–1967 experiment are not mere statistical flukes. Moreover, the meteorological-physical research carried out by us over the past few years in clouds and in laboratory, as well as studies of a more theoretical nature, suggest that AgI seeding of the "continental"-type of cumulus clouds of Israel's winter season (a reference to this cloud type has already been made earlier) should lead to a stimulation of precipitation from these clouds.

In the early stages of the 1961–1967 experiment we could not be sure that we would be able to advance our understanding of the physical processes in cumlus clouds to any great extent, and therefore we necessarily put our faith into statistical methods. The experience of the past few years with purely statistical evaluations and the progress in physical research have reversed our early attitude.

Preliminary Results of the Second Experiment

At the time of writing this chapter (April, 1972), we are close to the end of the third season of seeding in the framework of the second experiment (1969–1970). Results of physical measurements and of statistical evaluations are available for the first two seasons. These results, and especially the statistical ones, are necessarily preliminary; more seasons of experiment are needed.

The statistical studies involve not only rain data, including those for the experiment, control, and buffer areas, but also cloud top data as obtained with the aid of the 3-cm weather radar of the Israel Meteorological Service which participates in the experiments.

The RDR North versus South is about 1.10; the single-area ratio, Equation 13.2, for the North alone is over 1.2 while for the South it is less than 1. However, none of those results is significant as yet. On the other hand, the use of the control area associated with the North experiment area (the correlation coefficient between the daily rain data of the two is .87) already indicates an increase of rain in the North on days of seeding at a significant level of 4 percent, and this after two seasons (185 experiment, or "rain" days). If the rain data are coupled with cloud top temperature data through a multiple, two-variable regression, the significance is improved to 3 percent. The rise in significance occurs in spite of the fact that the number of rain days for which we have cloud top data as well as rain data is 105 against the 185 days for the rain data alone. Unfortunately, we do not have control areas for the South for which parallel computations could be carried out.

Some of the salient features of the 1961–1967 experiment reappear in the

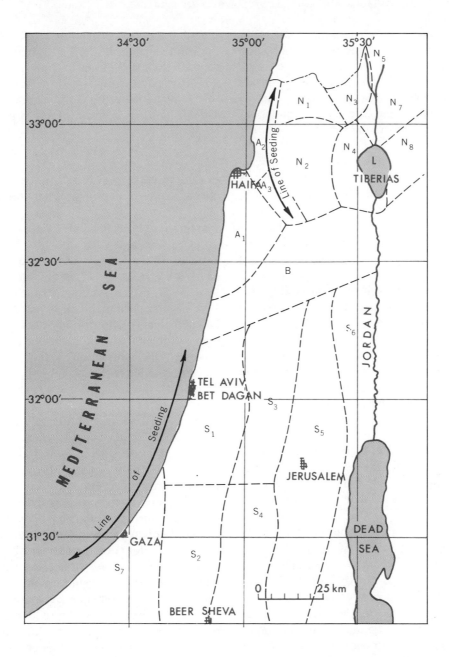

Figure 13.2 Map of Israel, showing the two experimental areas and their subareas (N_i, S_i), the control areas (A_i), and the buffer zone (B), for the second experiment (beginning in 1969–1970).

1969–1970 experiment. For instance, the "interior" areas of the North again show a larger apparent increase than the rest of the experiment area. These subareas are now located somewhat more inland (and over a part of the catchment area of Lake Tiberias) than in the case of the 1961–1967 project. We note that the seeding line for the North of the current experiment (see Figure 13.2) is slightly landward from the Mediterranean coast, whereas in the earlier project the seeding line was offshore (Figure 13.1). It is reasonable to speculate that the shift to a more inland location of the subareas of maximum enhancement is connected with the shift in the location of the seeding line as it takes some finite time for the AgI to produce its effects.

Finally, we wish to point out that although the single-area ratio for the *combined data* of the $6\frac{1}{2}$ seasons for the Center area of the 1961–1967 experiment was 1.15 (practically equal to the RDR of North versus Center), in actual fact, in 4 seasons out of the $6\frac{1}{2}$ seeding seasons that ratio was less than 1 (the North had in only 1 out of the $6\frac{1}{2}$ seasons a single-area ratio less than 1). We wonder if these facts are related to the finding of the last few years that cloud tops in the area now designated South (which incorporates the Center of the 1961–1967 project), are, on the average, about 4°C warmer. In the South, the median of cloud top temperatures is just at about −16°C, whereas in the North it is at about −19.5°C. To the best of our understanding, the most suitable conditions for AgI seeding are offered by supercooled clouds whose top temperature is at −18 or −19°C and these conditions are more often met in the North than in the South.

The relatively large cloud top temperature differences between the North and the South, over distances between 50 and 200 km (or 250 km at the most), draw fresh attention to the well-known fact that Israel is located in a region of large climatic "gradients." Nevertheless, it came as a surprise to find that over a relatively short distance one should have such great cloud top differences.

CLOUD PHYSICS ASPECTS OF THE ISRAELI CLOUD SEEDING EXPERIMENT

Statement of Problems

In view of the lack of any basic cloud physics studies in Israel before 1961, the 1961–1967 cloud seeding experiments were begun under the assumption that in Israel the most effective mechanism for precipitation formation is the ice crystal process. The assumption rested on the following simple reasoning.

In Israel the rainfall season is the cold season of the year; the freezing level drops, at least on days of precipitation, to about 2 km. With cloud bases at about 700 to 800 m and cloud tops at 4 to 6 km, these major cumulus clouds have the greater part of their vertical extent in air layers whose temperature is below the freezing point. Although the earlier assumption concerning the importance of ice crystals for the formation of precipitation was reasonable, it nevertheless required substantiation. This situation, where a major experiment is launched without adequate fundamental research, is no exception to the rather general rule that financial support for such research becomes available (and even then on a limited scale) only when the need for applied research requires it. Consequently, at the outset, the cloud physics studies that accompanied the seeding project were directed mainly to test the hypotheses underlying seeding with AgI crystals. These studies had to

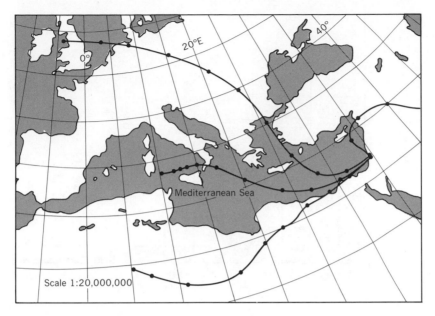

Figure 13.3   Typical trajectories of air masses arriving at the Israeli Mediterranean coast on rain days. The marked segments, on the lines, are 12-hr intervals. The trajectories refer to 850-mb level (about 1500 m).

clarify the following problems:

1. What is the predominant mechanism for the formation of precipitation in cumulus clouds of Israel? Are the formation of ice crystals and their subsequent growth essential factors?

2. Are natural ice nuclei solely responsible for the formation of ice crystals in clouds? Are there conditions when a deficiency of such nuclei is responsible for a delay, and even a failure, in the initiation of precipitation?

3. Will the introduction of artificial ice nuclei into clouds and the associated increased concentrations of ice crystals at temperatures higher than in natural processes lead to an increase in precipitation?

4. Can we reach a better understanding of the quantitative aspects of seeding? (The importance of that problem

grew as the work progressed and the statistical findings suggested a variable effect on different days.)

5. Could an overall positive effect on days of seeding over a long period be the result of the "superposition" of positive, negative, and nil effects?* Can we identify the conditions when the effects of seeding are most favorable?

As we felt that any answers to these problems and any computational evaluation of seeding effects should be based on the development and use of a realistic cloud model, a substantial part of our research effort was to be directed at the study in detail of the local rain clouds. The importance of these studies gained appreciation in the course of the experiment as we realized that no statistical evaluation will be of real value unless these results are substantiated by detailed physical considerations.

The summary of our research presented below results from the studies performed in this spirit.

* The statistical investigations relating to the problems raised here are mentioned in an earlier section of this chapter.

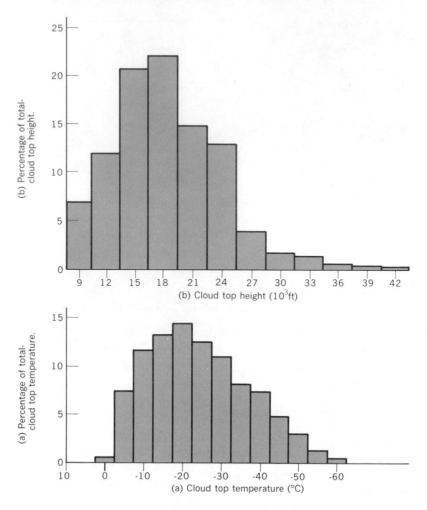

Figure 13.4 Overall average frequency distribution of cloud top heights and temperatures on rain days in Israel.

General Meteorological Outline

As was pointed out earlier, in Israel the rainfall season is limited, for all practical purposes, to the period November through April, that is, the winter season. Therefore, this period forms the cloud seeding season.

The great majority of Israel's rain-bearing clouds are cumulus clouds associated in bands that move in a general west-to-east direction. These bands are embedded in a cyclonic flow having its center typically in the northeastern Mediterranean, and are located in both the pre- and postfrontal air masses. An illustration of the four main trajectory types for air masses crossing the Israeli coast on rainy days is given in Figure 13.3, where the trajectories refer to the 850-mb level (about 600 to 700 m above usual cloud base; see next paragraph). We note in the diagram the relatively brief over-water track of these air masses.

Cloud bases on rainy days are, in general, at about 700 to 800 m (about 2500 ft) msl with an average temperature of 9°C. Figure 13.4 gives the representative cloud top height and temperature distributions on these days. As is obvious, the major part of the vertical extent of these cumuli is above the altitude of the 0°C isotherm which is often under 2 km (about 6000 ft). Both the median and the mode of the cloud top temperatures are close to −20°C.

Parameters Measured and Research Equipment

Selection of the atmospheric and cloud parameters to be studied was made on the basis of their potential importance for

local rain formation processes:

1. Cloud condensation nuclei (CCN) spectra were obtained by utilizing a thermal diffusion chamber.[24] All measurements were taken during about two winter seasons, at just below cloud base before measurement of cloud droplet characteristics in the clouds. Additional, complementary measurements were performed extensively at a mountain station near Jerusalem.[56]

Most of the research equipment connected with items 1 through 7 and details of our airborne facilities are illustrated in Figures 13.5 through 13.12.

2. Cloud droplet size distributions, at

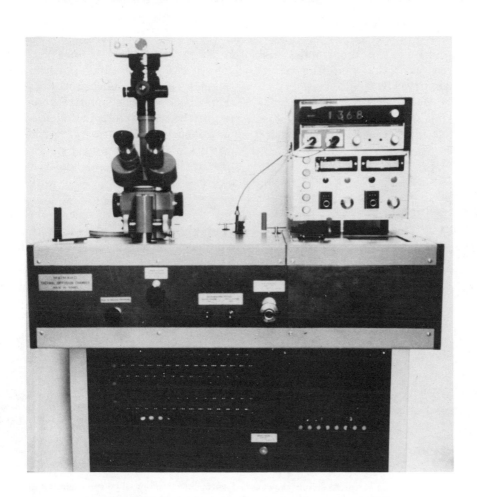

Figure 13.5 Thermal diffusion ice nuclei counter.

Figure 13.6   Research aircraft—general view.

various altitudes in the clouds, were measured with the sampling device described in 1965 by Clague.[10]

3.   Ice nuclei were sampled at below cloud base altitude, using 0.45-μ membrane filters, a method of ice nulcei

counting. Each set of measurements consisted of at least four separate filter samples which were later processed in the thermal ice nuclei counter (Figure 13.5) to give the temperature dependence curve of ice nuclei concentrations.[23] Again, a network of up to six ground stations was operated to sample continuously ice nuclei at -15 and $-20°C$. This work went on for four winter seasons.

4.   Ice crystal data were obtained through the use of an airborne, continuous, Formvar replicator described

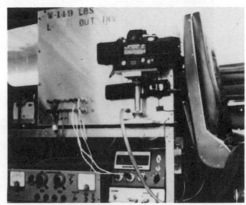

Figure 13.7   Interior of research aircraft showing the ice crystal replicator (right) and recorder (left).

Figure 13.8   Interior of research aircraft showing cloud condensation nuclei counter.

Figure 13.9 Large hydrometeor aluminum foil impactor.

Figure 13.10 Rack carrying ejectable pyrotechnic devices.

by Patrich and Gagin[53] (see Figure 13.7). The samples were taken during multiple traverses at altitudes of around 300 to 600 m (1000 to 2000 ft) below cloud tops. A total of about 70 clouds were sampled so far, in the range of cloud top temperature of -8 to $-25°C$.

5. Large cloud particles $\geq 250\ \mu$ in diameter were sampled for their concentrations and sizes through the use of a continuous aluminum foil impactor (CSIRO design) (see Figure 13.9).

6. Temperature was measured directly from the aircraft using a fine platinum-resistance, vortex-type thermometer. Simultaneous measurements of vertical air temperature profiles performed with this thermometer and a

Figure 13.11 Research aircraft showing ice crystal replicator and liquid water content sensors.

Figure 13.12   Seeding aircraft.

radiosonde have shown agreement to within 0.5°C.

7.   A 3-cm weather radar of the Israel Meteorological Service equipped, inter alia, with an RHI facility, was operated continuously during the three winters 1969–1972 to measure all echo tops present on a total number of 107 days. Prior to this period, radar studies were confined to the detection and measurement of the characteristics of first-precipitation echoes of local clouds.

The Colloidal Stability of Local Cumulus Clouds—Studies of the Factors Governing the Nature of Rain-Forming Mechanisms

Any effective artificial intervention in the microphysical processes in clouds, aimed at the stimulation of precipitation, should be based on a detailed knowledge of the microphysical properties of the local clouds. Although our present understanding of the basic processes responsible for the initiation of rain elements is far from being perfect, it is widely accepted that they can be related to two basically different, alternative mechanisms. These elements, whose existence in the clouds in *sufficient numbers* constitutes a necessary condition for the efficient formation of precipitation, are thought to originate either (1) through the initial formation of ice crystals which subsequently grow by vapor diffusion and riming, or (2) through the rapid condensation of cloud drops to be followed by a fast enough process of collision-coalescence growth of the drops. The efficiency of the latter process will depend on whether or not the condensation stage produced a sufficiently wide spectrum of cloud droplets.

These two processes are often in competition with each other. The nature of the mechanism that will dominate finally will be determined primarily by the microphysical structure of the cloud. The failure to recognize the dominant mechanism may result in the use of the wrong seeding technique and in undesirable negative effects or in no effects at all on precipitation. Thus the primary aim of our research was to evaluate the prospects of cloud seeding through the studies.

The Evolution of the Cloud Droplet Spectrum. Extensive measurements of CCN spectra for two winter seasons[56] have revealed the significant continental aerosol content of the air masses arriving at the eastern Mediterranean coasts. A typical daily variation of the CCN content of these air masses is displayed in Figure 13.13. In Figure 13.3 typical trajectories of the large-scale flow regime on rainy days in Israel were presented. As can be seen from the latter diagram the air masses, originating either in the European continent or the North African desert, stay for too short periods over the sea for the aerosol-removing processes to change significantly their continental nature. Thus the CCN spectrum obtained on April 14, 1971, which is typical for air masses staying over North Africa prior to their arrival in Israel, is as follows:

$$n = 1290S^{0.42} \qquad (13.3)$$

as compared to our average CCN

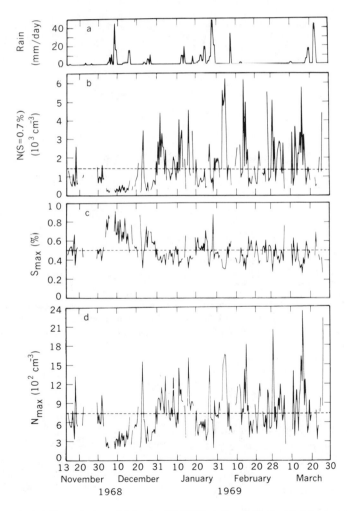

Figure 13.13 Typical daily variation of (*a*) rain, (*b*) CCN at S = 0.7 percent, (*c*) supersaturation, and (*d*) computed cloud drop concentration.

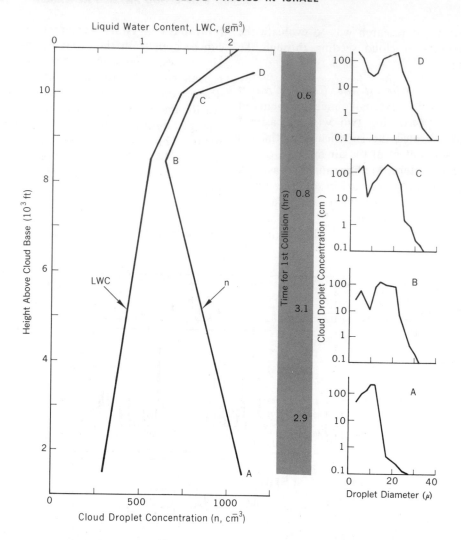

Figure 13.14 The variation with height above cloud base of the cloud drop spectrum (right), the liquid water content (LWC), and cloud drop concentration (n) in a cumulus cloud on April 14, 1971. (For other details on this cloud, see Table 13.3, No. 1.) The time for the first collision in the various spectra is given in the middle column (after Gagin[22]).

spectrum $n = 1788\ S^{0.68}$, where $n =$ number of the CCN/cm³ of air and $S =$ supersaturation in percent with respect to water.

Using the expression developed by Twomey[57] for the maximum concentration (N_{max}) of CCN activated to form cloud droplets, namely,

$$N_{max} = C(S_{max})^k$$

$$= C^{2/(k+2)} \left(\frac{6.87 \times 10^{-2} V^{3/2}}{kB(3/2,\ k/2)} \right)^{2/(k+2)}$$

$$(13.4)$$

(B is the beta function), inserting in Equation 13.4 the values $C = 1290/cm^3$ and $k = 0.42$ from Equation 13.3, and assuming the updraft velocity at cloud base to be $V = 2.5$ m/sec (in conformity with Sulakvelidze's[55] measurements of vertical velocity variation with height in similar clouds), the predicted concentration of cloud drops at cloud base works out to be about $890/cm^3$. Figure 13.14 displays the details of the variation of cloud drop concentration, as well as the liquid water content and droplet size spectrum with height above the base of a cumulus cloud sampled on the same day (April 14, 1971). If instead of using a value of 2.5 m/sec for the base updraft velocity, a value of 4 m/sec is inserted in Equation 13.4 for V, then the predicted value of N_{max} obtained is $1010/cm^3$, which is somewhat closer to the observed value of $1100/cm^3$. In the

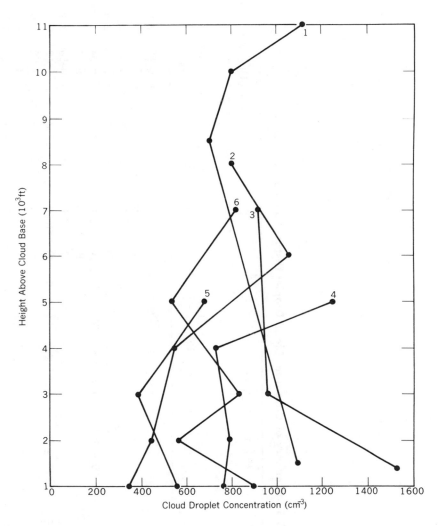

Figure 13.15 The variation with height of cloud drop concentration in the cumulus clouds described in Table 13.3. The clouds are numbered in agreement with the details of Table 13.3 (after Gagin[22]).

Table 13.3 Summary of Data from Cloud Droplet Distribution Measurements

Type and Date	Cloud Base T (°C)	Cloud Base H (kft)	Cloud Top T (°C)	Cloud Top H (kft)	CCN (cm⁻³)	Traverse No.	Traverse H (kft)	Traverse T (°C)	N_{max} (cm⁻³)	$\bar{N}$ (cm⁻³)	LWC (g⁻³)	$\bar{a}$ (μm)	Q_1 (μm)	Med (μm)	Q_3 (μm)	Sample Total No.	Sample Portic Bimodal
1. Cu April 14, 1971	9.5	3.5	−13.5	14	$n = 1290\,S^{0.42}$	A	5.0	6.2	1085	730	0.60	9.7	5.4	9.0	10.8	3	1
						B	12.0	−8.0	631	575	1.10	10.7	5.5	11.4	13.2	5	4
						C	13.5	−11.0	800	670	1.47	12.0	5.0	12.7	15.7	3	3
						D	14.0	−12.0	1114	970	2.18	12.5	5.5	12.0	16.9	2	2
2. Cu April 23, 1971	5.5	4.0	−10.0	13	$n = 740\,S^{0.46}$	A	5.0	4.8	352	—	0.04	6.3	3.8	5.4	6.8	1	0
						B	6.0	2.5	454	—	0.17	7.7	4.3	5.8	9.5	1	0
						C	8.0	−1.9	544	—	0.50	8.4	4.0	6.7	10.4	1	0
						D	10.0	−5.2	1048	1021	1.12	10.7	4.0	11.2	14.2	2	2
						E	12.0	−8.2	803	730	1.02	10.8	4.9	11.4	15.0	3	3
3. Cu March 27, 1971	2.6	5.0	−12.0	13	$n = 1280\,S^{0.41}$	A	5.5	2.2	1567	1371	0.11	4.8	2.4	4.0	4.9	4	0
						B	6.5	−0.1	1550	1520	0.21	6.1	4.1	4.9	5.8	3	0
						C	8.0	−6.2	872	—	0.32	8.6	6.2	7.3	8.8	1	0
						D	12.0	−12.0	913	903	0.94	10.7	4.5	11.4	i3.2	2	2

		$\bar{d}$														
4. Cu March 27, 1971	4.8	4.0	−9.1	11	$n = 1280\,S^{0.41}$	A	5.6	1.5	795	—	0.19	7.0	4.8	6.5	7.2	1	0
						B	6.0	−0.6	799	695	0.19	7.4	5.6	6.4	7.2	3	0
						C	8.0	−4.0	744	—	0.34	8.1	5.5	7.1	10.4	1	0
						D	9.0	−6.2	1245	1191	0.77	9.6	5.2	9.8	11.0	3	3
5. Cu April 14, 1971	7.8	4.0	−10.0	13	$n = 1290\,S^{0.42}$	A	6.0	4.0	551	540	0.36	8.8	4.1	6.9	11.6	3	2
						B	7.0	2.3	389	—	0.29	9.5	5.0	7.2	11.9	1	1
						C	9.0	−0.6	685	560	0.42	9.9	4.6	8.2	12.8	4	4
6. Cu March 19, 1971	7.5	3.0	−3.4	11	$n = 1070\,S^{1.67}$	A	5.0	4.0	890	515	0.33	8.0	4.4	7.7	8.6	3	1
						B	6.0	1.5	550	410	0.24	7.8	6.1	7.8	10.4	3	1
						C	7.0	−0.3	853	839	0.59	9.3	4.2	8.3	12.2	3	3
						D	9.0	−3.0	510	495	0.39	9.8	5.3	9.4	12.2	2	1
						E	10.0	−3.4	810	520	0.71	9.9	5.2	9.6	13.4	2	2
7. AC March 25, 1971	11.6	7.0	−1.5	12	1500 ft: $n = 1600\,S^{1.60}$ 4000 ft: $n = 921\,S^{1.61}$	A	7.0	4.0	973	781	0.12	7.0	4.2	4.9	6.8	5	0
						B	10.0	0.0	537	494	0.28	9.0	4.3	7.0	12.0	5	2

[a] The average value ($\bar{d}$), first quartile (Q_1), median (Med), and third quartile (Q_3) of the droplet distribution are given in terms of droplet diameters.

forthcoming calculations a vertical velocity profile was assumed for this cloud having a value 2.5 m/sec at cloud base and a maximum value of 7 m/sec at about 6000 ft above cloud base. This averages around 5 m/sec for the whole cloud, again in conformity with Sulakvelidze's data.[55] As this cloud is of a depth and dimensions near to the modal values for our precipitating clouds (Figure 13.4), and since the air mass-aerosol properties and trajectory are quite representative for a typical rainy day, it seems desirable to attempt shedding some light on the nature of the mechanisms that may be responsible for rain formation in such clouds.

As to the concentration of cloud droplets, it is easily seen that not only in this cloud but in all the others (Table 13.3) the variation with height of droplet concentration does not show any trend of a systematic nature, and increases as well as decreases may occur, such that the overall changes with height are not significant (Figure 13.15). This is in agreement with Warner's findings.[59] However, both liquid water content and the droplet size spectrum vary significantly with height. The extremely small average droplet diameters as well as the narrow spectra can simply be the result of the high concentrations of droplets found in our clouds, very much in agreement with Warner's studies.[59]

We now turn to the obvious evolution and broadening of the cloud droplet spectrum. In view of the latest studies on the collection efficiencies of small drops (radii $\leq 15\ \mu$) of Davis and Sartor[12] and Hocking and Jonas,[31] it seems worthwhile to investigate to what extent collision-coalescence processes may be responsible for the observed broadening of the spectrum and, later, for the production of precipitation elements. In order to do this, let us consider Equation 5 in Bartlett's 1970 paper for the time (Δt) it

takes before the first collision between the largest drop ("collector drop") of radius a_1 in the spectrum and the drops having $\frac{1}{2}$ the radius of the collector drop.[2] As Bartlett shows, for this radii ratio the collision rate is at maximum. Thus

$$\Delta t = \frac{1.45 \times 10^{-7}}{E \cdot N \cdot a_1{}^4}, \qquad (13.5)$$

where E is the collection efficiency of the drop a_1 for drops having radii $a_2 = a_1/2$ and concentration N. The results of this computation are given in Figure 13.14 for each of the four levels A, B, C, and D. Bearing in mind that for an average updraft velocity of $V = 5$ m/sec the time available for the cloud drop spectrum at point A (Figure 13.14) to produce the observed spectrum at the highest point D is 670 sec, the time of 2.8 hr for the first collision, computed for spectrum A, excludes the possibility that collision-coalescence processes can have any significant effect in such clouds. This statement holds also for the other levels in this cloud.

It will be in order to add before passing on that Bartlett's Equation 5 is founded on the assumption of a uniform distribution of droplets in the cloud. This, of course, is an unrealistic premise. Noah Chodes of the Hebrew University, Department of Atmospheric Sciences' Cloud Physics Laboratory has most recently undertaken a recalculation based on the assumption of a random reduction in first-collision times but little else of consequence. In the case of spectrum A, Table 13.3, the droplets are small and are not "collected" due to the low collision efficiencies of small droplets. Taking the other extreme of the spectrum, namely, spectrum D, there occurs some broadening of the spectrum through collisions, but the concentration of the "newly" gained large droplets is very low. Also, it transpires that for a significant broadening of the spectrum and the formation of precipitation elements, the process should continue (for spectrum D) for at least 900 sec more. If we make the assumption of an average updraft of 5 m/sec (and we believe that this is a conservative assumption), the additional time required implies that the cloud must extend 4.5 km farther, or the total height must be about 9 km. Such a cloud thickness is seldom observed in our part of the world. In any case, the calculations show that pure condensational growth alone can produce larger droplets than the collision process (assuming that the theories of collisional growth available at present are sufficiently near the truth), but even these droplets, which are supposed to have grown through condensation, are far too small to count as "precipitation elements."

It is appropriate at this stage to investigate the effect of condensation on the evolution of the spectrum. Squires[54] gave the expresssion (Equation 13.6) for the supersaturation (S) in a cloud, assuming that at that stage the drops are growing by vapor diffusion and are well beyond the activation stage at cloud base:

$$S = \frac{6.7}{\Sigma r}(0.0239V + 1.72 \times 10^{-6}n),$$

$$(13.6)$$

where S is given in percent, n is the number of drops per gram of air, Σr is the sum of their radii, and V is the updraft velocity in cm/sec. Assuming again an average updraft velocity of 5 m/sec and the spectrum at point A, we obtain for the supersaturation the value $S = 0.12$ percent, a perfectly reasonable value in view of the result given by Warner[58] that the median supersaturation is 0.1 percent in cumuli of the size discussed above. Now, as Warner[59] suggested, the growth by condensation with time (t) of a given concentration of droplets of diameter d_1 to diameter d_2, is given by the following

approximate equations (r and d in μ):

$$S = r\frac{dr}{dt} \quad \text{or} \quad d_2{}^2 - d_1{}^2 = 8St. \quad (13.7)$$

Applying this calculation to the cloud demonstrated in Figure 13.14, the drops at a concentration of 0.1/cm³ and diameter $d_1 = 28\ \mu$ (spectrum A) will grow in the cloud having a supersaturation of 0.12 percent in 670 sec (the time that elapses between points A and D in the cloud) to the size d_2 given by

$$d_2{}^2 = (28)^2 + 8 \times 0.12 \times 670 - 1427,$$

$$d_2 = 37.8\ \mu$$

and indeed, this is exactly the size attained by these drops, as can be seen in Figure 13.14, spectrum D. Thus the broadening of the spectrum can be attributed to condensation alone. This and the exclusion of the effectiveness of collision-coalescence processes suggest that an alternative mechanism should be responsible for rain formation in these clouds. The validity of this conclusion can definitely be extended to all clouds sampled and summarized in Table 13.3. Although the total sample of clouds in which cloud droplet measurements were made is small, the air mass trajectory analysis done for each of the five days in question suggests that these measurements can be regarded as representative of most of our rainy days. Moreover, the CCN spectra for these days (Figure 13.13) are well within the typical range of values that can be taken as representative for our air masses.[56]

We tend to conclude that all-water processes, that is, condensation and subsequent collision-coalescence, may play a negligible role in the process of rain formation in the local clouds. From this important conclusion we infer that the dominant rain-forming mechanism is the second, alternative, mechanism, that is, the "cold" ice crystal process.

The Role of Ice Phase in Local Rain-Bearing Cumulus Clouds.   In view of the lower vapor pressure of ice compared with supercooled water at the same temperature, it was suggested as early as 1935 that ice crystals will grow to the size of precipitation elements much faster than water drops. In fact, it was thought at the time that ice crystals are responsible, in general, for the formation of precipitation. Although the latter statement is no longer held true, we have shown in the previous section that the known alternative rain-formation mechanism, namely, the collision-coalescence process,* seems to be ineffective in our cumulus clouds, and very probably this is true for all continental-type cumulus clouds. We tend to conclude, therefore, that the ice crystal mechanism is an essential step in the generation of rain in local clouds. Since our seeding techniques are aimed at creating ice crystals, it is appropriate to study the effects produced by the increased concentration (enhanced by seeding) of ice crystals.

For an efficient initiation of precipitation in supercooled clouds, ice crystals should in these clouds (a) exist in sufficient concentrations, and (b) enjoy conditions that will make them grow to the sizes of precipitation elements within the lifetimes of the clouds.

In regard to (a), ice crystal formation in natural supercooled clouds has long been attributed to heterogeneous nucleation by ice-forming nuclei. The common assumption underlying cloud seeding techniques relies, for its effectiveness, upon a deficiency of ice particles in the clouds resulting from a corresponding lack of active atmospheric ice nuclei. Until recently, surprisingly little was

* That is to say, the collision-coalescence processes as the published literature describes it. It appears, however, that the present state of the theory is far from satisfactory.

known about the variation with temperature of ice crystal concentrations in clouds. The recent studies of Mossop[45] are an excellent documentation of an anomaly on which data has been accumulating for the last 20 years. These studies show that quite frequently ice crystal concentrations in maritime clouds may exceed those of ice nuclei by factors of the order of 10^3 to 10^4. Cases have been reported[25, 32, 46] also in which the existence of good correspondence between ice crystal and ice nuclei concentrations has been detected. Mossop[45] draws attention to an earlier postulate by Findeisen[14] that there may be differences in the propagation of the ice phase, between clouds having maritime-like microstructure and clouds having so-called continental-like properties. The high degree of continentality of the clouds reported here has been clearly shown in the preceding section. Figure 13.16, based on 70 cloud penetrations during two winter seasons, displays the overall relationship between ice crystal and ice nuclei concentrations in these continental cumulus clouds. Here, as before, a relatively good agreement exists between the two.

Furthermore, both Mossop[45] and Koenig[36] have reported very rapid glaciation rates in maritime-type cumulus clouds at relatively warm temperatures. Mossop[45] suggested a glaciation rate of about one order of magnitude increase in concentration every 8 min. Koenig[36] reported rates of about three orders of magnitude increase in 5 min. Consecutive penetrations in our clouds, up to four in a row, have not revealed any systematic variation with time of ice crystal concentrations. There are, however, discrepancies, especially in the temperature range of -14 to $-20°C$, in which, on the average, our ice crystal concentrations exceed those of the average ice nuclei concentrations by about a factor of 5. Such differences may be the result of the "filtering" effect of the cloud on ice nuclei. Assuming an average updraft velocity of 5 m/sec and a cloud lifetime of 3600 sec, a total volume of about 4.5 times the cloud volume (the cloud being 4 km deep) will be "filtered." It is suggested, therefore, that this effect may account for at least a part of the above mentioned discrepancies.

On a smaller number of flight traverses (14 cloud penetrations), the aluminum-foil impactor for drops was operated simultaneously with the ice crystal replicator. The data on large water-drop (≥ 250 μ diameter) concentrations suggests that such large drops are totally absent in clouds whose tops are warmer than $-12.5°C$. In colder clouds with top temperatures in the range of $-12.5 > T > -23°C$, the concentration of such drops is less than 0.1/liter. In contrast, both Mossop and Koenig[36, 45, 46] em-

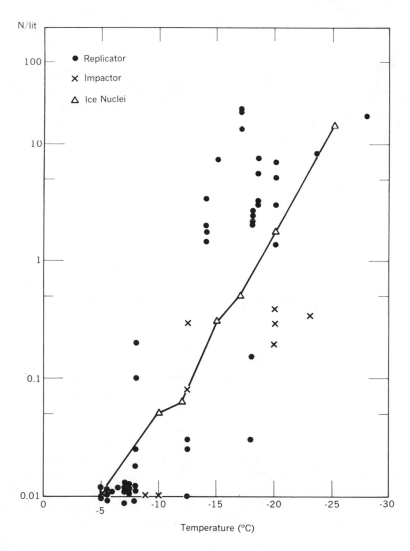

Figure 13.16   The relationship betweeen ice crystal concentration and the average ice nuclei concentration as a function of temperature. Ice crystal concentrations are related to pertinent actual cloud top temperatures (after Gagin[22]).

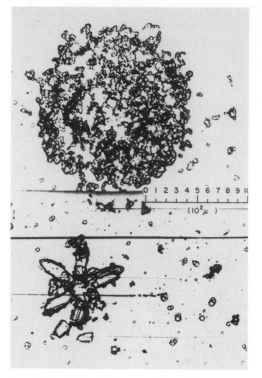

Figure 13.17 Typical replicas of a low-density graupel resulting from the rapid riming of dendritic or stellar ice crystals shown below (after Gagin[22]).

until the graupel phase is achieved. This chain of events can in fact be inferred from our observations. Ice crystals that have not gone through the dendrite, crystal-type regime seem to experience very little riming. However, once they enjoy the temperature domain of rapid dendritic growth and attain a size of several hundred microns, riming produces rapidly the very low density graupel shown in Figure 13.17. The low density of those graupels, indicative of the major role that riming by small droplets plays, is obvious in this picture, as these graupels shatter upon impaction on the replicator film. It seems that because of the higher collection efficiency of the branches of dendritic or stellar crystals, they bring about a rapid transition to the graupel stage. This rapid transition is typical of our clouds.

Both the high colloidal stability of local clouds and the relatively good correspondence between the concentrations of ice nuclei and of ice crystals, which so clearly emphasize the lack of sufficient ice crystal concentrations in clouds with top temperature warmer than $-13°C$, present almost ideal conditions for seeding the clouds with heterogeneous ice nuclei.

Furthermore, the absence of rapid glaciation mechanisms at higher temperatures, which are so typical of maritime clouds, suggests that ice nucleus measurements are important for the evaluation of the possible effects of cloud seeding. Figure 13.18 displays the typical daily variation of ice nuclei active at -15 and $-20°C$. One important conclusion which can immediately be drawn, on the basis of these measurements and the output rates of our seeding technique, is that the possibility of overseeding of the local cumuli can be rejected. It is recalled that overseeding is believed to be responsible for negative effects on rainfall in some other cloud seeding experiments.

phasize the existence of large drops, in concentrations of a few hundred per liter, prior to the inception of the rapid glaciation processes that they observed. Thus it is appropriate to suggest that the clouds described above are of a different category than that of the clouds investigated by Mossop and Koenig.

This absence of large cloud drops in clouds with summit temperatures warmer than $-12.5°C$ and the relative scarcity of these elements in colder clouds are not at all surprising in view of the inferred inefficiency of the collision-coalescence processes in these clouds. We tend, therefore, to speculate that precipitation elements in our clouds form primarily through the process of nucleation of ice crystals that initially grow by water vapor diffusion and subsequently by riming

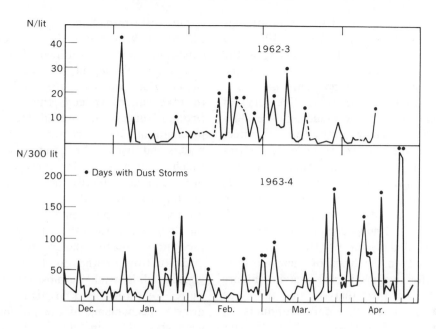

Figure 13.18 Typical daily average counts (N) of ice nuclei at Jerusalem. Counts are at $-20°C$ for 1962–1963 and at $-15°C$ for 1963–1964 (after Gagin[20]; Neumann et al.[49]).

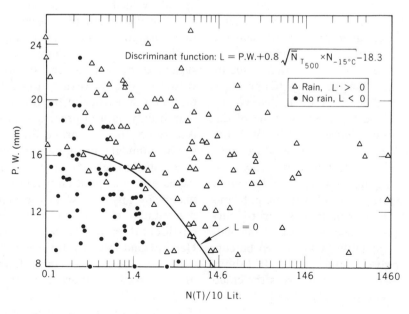

Figure 13.19 Discriminant function separation between the occurrence and nonoccurrence of rain, as a function of daily values of effective ice nuclei concentration and precipitable water (P. W.) (after Gagin[30]; Neumann et al.[49]).

It is of interest to describe an attempt made by Gagin as early as 1965[20] to evaluate the role of ice nuclei in the formation of precipitation. In view of the observed good correspondence of ice nuclei concentration with ice crystal concentration in the clouds, this evaluation is especially instructive. A direct comparison between ice nuclei concentrations and precipitation amounts has suggested no relationship whatsoever. On the other hand, a rather sharp discrimination has been found to exist between the occurrence and nonoccurrence of precipitation on a daily basis.* Days were characterized on the basis of their moisture and effective ice nuclei content (Figure 13.19). In this analysis moisture was represented by the daily amounts of total precipitable water. The effective number of ice nuclei (N_{eff}) was computed from

$$N_{eff} = \frac{N_{-15}}{\bar{N}_{-15}} \bar{N}_{T500}, \qquad (13.8)$$

where N_{-15} is the measured daily average concentration of ice nuclei active at $-15°C$/liter of air, N_{-15} the general average (i.e., average for all seasons investigated) value of this concentration, and $\bar{N}_{T500}$ the computed average value of concentration for the daily temperature of the 500-mb level, taken as the average height of cloud tops. At the time of the preparation of Figure 13.19 in 1969, the cloud top height and temperature data were not available. However, as can be seen from Figure 13.4, both modal values of the average cloud top height and temperature distributions seem to correspond quite closely to those of the 500-mb level. The latter is known to be correspondingly around 18,200 ft (about 5460 m) and $-17.5°C$, whereas the cloud tops are about $-17°C$ at 18,000 ft (about 5400 m).

---

* For a discussion of the discriminant function analysis, see pp. 118–122 in Panofsky and Brier.[52]

The discrimination, as presented in Figure 13.19, suggests that a necessary condition must be fulfilled; that is, a certain critical concentration of ice nuclei should exist if initiation of precipitation is to take place. Indeed, the statistical findings suggest an overall increase in precipitation under seeding. In order to understand better the conditions under which seeding would be most effective, one should also identify the other complementary conditions which would carry the process of precipitation formation from the stage of initiation to that in which the precipitation reaches the ground.

Evaluation of the conditions under which seeding would be most effective requires quantitative knowledge of the growth mechanisms of the precipitation elements. Unfortunately, at the present state of knowledge, one can only hope to consider this problem qualitatively.

In cumulus clouds, which form the majority of precipitating clouds in this region, the relationship between the sizes of clouds and the magnitudes of their updraft velocities imposes some restrictions on the length of time available for rain formation—the deeper the clouds, the longer the time available. It has been suggested by various investigators that seeding would result in initiating precipitation earlier in the life of a cloud than if only natural processes took place. According to this proposition, seeding would stimulate precipitation from relatively shallow clouds, and the artificial increase in precipitation would be caused mainly by an increase in the number of precipitating clouds. However, recent experimental evidence put forward by Hallett[28] and later by Jayaweera,[33] and a number of limitations on the effectiveness of AgI seeding show that the above proposition is not generally valid.

Let us examine the advantages gained by the introduction of AgI particles in cumulus clouds. On adopting the view that a concentration of one ice crystal/10 liters of air[43] is the critical concentration necessary for an efficient precipitation-forming mechanism, and on the basis of our extensive measurements of raindrop concentration in natural shower conditions indicating an overall average of 1.4 drops/10 liters of air, then, as an average for this region, the above critical concentrations would be achieved at about −14°C (see Figure 13.16). Since AgI has a threshold temperature of −5°C, it would seem logical to expect that AgI seeding at times of deficiency in natural ice crystals would stimulate the formation of precipitation which otherwise would not have been initiated

naturally. It still remains to explore whether ice crystal formation in sufficient concentration at −5°C, due to seeding, is indeed necessary for a more efficient precipitation-forming process. In order to do this, let us now consider the growth mechanism of precipitation elements.

According to the theory of precipitation formation by the ice crystal mechanism, the process of a precipitation-element formation is initiated by nucleation and the formation of an ice crystal. The crystal will grow by diffusion-sublimation until it becomes large enough for riming to take place. The continued process of riming will lead eventually, and at a

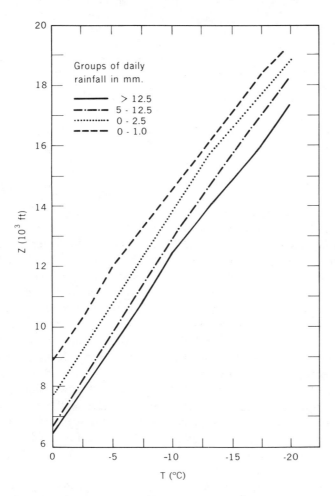

Figure 13.20 Temperature-height relationship for groups of days having different daily rainfall. Z = height (after Neumann et al.[49]; Gagin[21]).

488 RAIN STIMULATION AND CLOUD PHYSICS IN ISRAEL

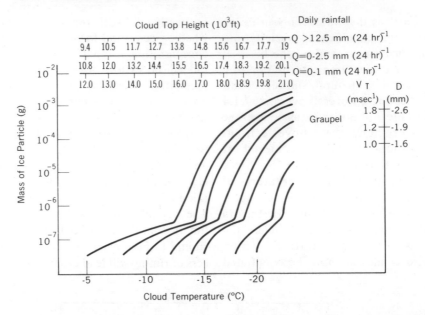

Figure 13.21 Mass of ice particles as a function of height and temperature in a cumulus cloud. Growth computation takes into account sublimation as well as riming. Graupel riming is presented for a cumulus cloud having a maximum liquid water content of 1.2 g/m³. Vertical velocity profile has a corresponding maximum of 7 m/sec¹.

fast pace, to the formation of a graupel. From then on, growth will take place by riming and coalescence, according to the position of the element relative to the melting level.

Hallett[28] and Jayaweera[33] presented estimates of the rate of growth of sublimating ice crystals. Hallett's experimental evidence suggests that the relative growth rates depend only on temperature, whereas ·Jayaweera's computations suggest a dependence on both temperature (T) and time (t). It was decided, therefore, to use in the present analysis Jayaweera's more general proposition. As can be easily seen from this work the integration of the mass of an ice crystal as a function of temperature and time in the first 300 to 400 sec of its life is made easy, as it can be treated to behave as a sphere. The relative position of the growing ice crystal in the cloud can be determined, once the temperature and updraft velocity are known functions of

height. Thus the mass (m) as a function of height in the sublimation zone can be computed from

$$m_z = m_0 + \frac{z - z_0}{\Delta U}\left(\frac{dm}{dt}\right)_{T,t},$$

where z is the height of the upper boundary of the layer (usually 1°C wide), z_0 the height of the lower boundary, and T and ΔU the mean temperature and relative updraft velocity for this layer respectively. Mean curves for temperature versus height have been determined for different groups of days (see Figure 13.20). Riming was computed on the basis of the collection of cloud water by ice crystals with an average collection efficiency $E = 0.5$ for disk-type crystals having coalescence cross-section of $\pi d^2/4$, and $E = 1$ for dendrites. However, the coalescence cross-section for dendrites was taken as $\pi d^2/8$, where d is the diameter of the crystal.[34] The vertical liquid water content profile was taken, on the basis of our direct measurements, to have on the average a maximum of 1.2 g/m³ at about 1 km below cloud top and 0.2 g/m³ at 0.3 km above cloud base. In accordance with Douglas's criterion for such clouds,[13] the threshold mass for riming was taken as 3.2×10^{-7} g. This value seems reasonable in view of the frequent occurrence in our clouds of plane dendrites or stellar crystals in the range 200 to 400 μ with little or no riming at all (see Figure 13.17). In view of the absence of suitable data with regard to updraft velocity, the assumption was made that vertical velocities were similar to those reported by Sulakvelidze, Bibilashvili, and Lapcheva in similar 4 to 5 km deep clouds,[55] that is, 2.5 m/sec at cloud base and a maximum of 7 m/sec at 3.5 km above cloud base.

In Figure 13.21 the abscissas are for either height or temperature corresponding to the three groups of days shown in Figure 13.20. Mass or the corresponding terminal velocity for graupel is presented as the ordinate. Ludlam's criterion for the minimum height of the clouds for shower formation was adopted.[41]

The results pertaining to the growth of solid precipitation elements in such cumulus clouds are summed as follows:

1. The contribution of ice crystal growth by sublimation in the temperature interval -5 to $-12°C$ to the total mass of the precipitation elements is negligible.

2. Because of the very rapid growth at around $-14°C$, and the dominant influence of riming, it seems immaterial whether nucleation takes place at -5, -10, or even $-12°C$. The resultant mass at $-20°C$ will be practically the same.

3. The occurrence of riming on quasispherical graupel, which is critically necessary for efficient and fast precipitation formation, requires dendritic growth at the $-14°C$ zone.

4. For an efficient process of shower formation in cumulus clouds, the tops of the cumulus clouds must attain the following minimum heights: 17,500, 19,000, and 19,6000 ft for a liquid water content of 1.2 g/m³ in the riming zone for the respective three rainfall intensity groups plotted in Figure 13.20 and the top of Figure 13.21.

These calculations are based on Ludlam's criterion[42] that solid precipitation elements must have a fall velocity of about 1 m/sec in the vicinity of cloud tops for efficient shower formation. Actually, a fall velocity of 1.1 m/sec has been adopted in the present study.

Israel's Cumulus Clouds and Clouds of Project Whitetop and Arizona's Santa Catalina Mountains

Our studies of the physics of Israel's winter cumulus clouds, reported earlier, indicated that the all-water processes of condensation and collision-coalescence of droplets are unable, judged by the present status of the theories of droplet growth, to account for the formation of rain in these clouds. In contrast, the investigations described show that the presence or stimulation of generation of the ice phase can explain it, based again on the present state of the appropriate theories. Furthermore, the various statistical analyses set forth show that there was a statistically significant increase in precipitation under seeding, though some problems still remain.

Our findings raise the following intriguing questions: if the results of the Israeli experiment are positive, why were the results of Project Whitetop[8] and those of Arizona Program II[5] either negative, or at least not supporting, the hypothesis of enhancement under seeding? In all these experiments the clouds were mainly, or exclusively as in Arizona, cumulus clouds and the seeding agent AgI smoke dispensed from burners fitted to aircraft.

Despite the aforementioned similarities, we note at the outset that the clouds involved in the Israeli experiment are *winter* cumulus clouds associated, in most cases, with cold fronts and pre- and postfrontal air masses. Both the Whitetop and Arizona clouds of concern here were *summer* cumulus clouds. The Arizona clouds differ even from those of Whitetop insofar as the former were "convective" cumulus clouds.

A comparison between Israel's winter rainbearing cumulus clouds and the summer cumulus clouds developing on Arizona's Santa Catalina Mountains was made some years ago by Gagin.[21] In that paper the author studies relationships between liquid water contents (LWC) in clouds, precipitable water in the air, environmental lapse rate, and temperature on the one hand, and daily precipitation amounts on the other, in Israel. The relationship between the daily rainfall amounts and the air temperature as a function of height is plotted in Figure 13.20, which has been taken from the 1965 Gagin paper. The study shows that there are but rather minor differences between the values of LWC, precipitable water, and lapse rate for the different daily precipitation-amount groups. Air temperature, however, turns out as a decisive factor through its control of the activation process of natural ice nuclei. Thus both this earlier study and the more recent and detailed investigations described in the present chapter, based on an intensive study of the microphysics of Israel's winter clouds, indicate that the ice phase is of vital importance to precipitation formation in these clouds.

There are no comparable cloud microphysical data available for Arizona's Santa Catalina Mountains, but Battan[4] summarizes his conclusions concerning the rain-generating mechanism in his clouds as follows:

... One must also admit the possibility that the ice nuclei, in this and the other experiments as well, did, in fact, enter the cloud, but that they did not play an important role in governing how much precipitation reached the ground. The view that seeding might not have an important effect on rainfall from certain clouds is supported by analyses which concluded that the dominant precipitation mechanism in the convective clouds of Arizona is the coalescence mechanism.[5]

Turning now to Project Whitetop, the associated physical researches indicate that the majority of cumulus clouds ("cumulus congestus") examined contained precipitation particles larger than 250μ in diameter in the upper levels sometime during the life cycle of the clouds,[9] and exhibited rather large glaciation rates involving ice crystal concentrations orders of magnitude in excess of ice nuclei concentrations.[7, 36] None of these phenomena is found in Israel's winter clouds which, in fact, are of a continental type, whereas the Whitetop clouds were of a maritime type. Judged by our present understanding of the rain-forming processes, the introduction of large numbers of artificial nuclei into clouds that develop naturally high glaciation rates is likely to inhibit the precipitation-forming process.

We thus conclude that the physical characteristics and processes of Israel's winter cumulus clouds are very different from those pertaining to the clouds of Project Whitetop and Arizona Program II, and these differences are likely to be responsible for the different apparent responses of these clouds to the same treatment.

CONCLUSIONS

1. The high colloidal stability of the local continental cumulus clouds, indicated by the narrowness of the cloud drop-

let radii spectra and the apparent inefficiency of the collision-coalescence mechanism at the droplet sizes involved, suggests that ice crystals are essential for the formation of precipitation in these clouds. This and the absence of ice crystal multiplying mechanisms provide a physical basis for cloud seeding with artificial ice nuclei in Israel.

2. The clear correspondence between ice crystal and ice nuclei concentrations suggests that often conditions exist where supercooled clouds are deficient of such concentrations of ice crystals necessary for promoting an efficient rain-forming process. The discriminant function analysis (Figure 13.19) indicates that ice nuclei have a marked influence on the initiation of precipitation.

3. Because of the rapid growth of ice crystals at a temperature of about $-14°C$, it does not seem necessary to use cloud seeding agents which produce nucleation at a relatively high temperature. On the other hand, at temperatures below $-20°C$ the atmosphere in this country appears to have (see Figure 13.16) sufficient natural nuclei that become active at the low temperatures in question, so that the addition of artificial nuclei to these cold tops would serve no useful purpose. Judging by (a) our measurements of the concentrations of raindrops that average 0.14/liter (never exceeding the value of 0.6/liter), and (b) the definite scarcity of ice crystals in clouds warmer than $-12°C$, it appears that frequently conditions exist when natural processes will not provide the concentrations of ice crystals needed for ideal growth conditions.

4. Once cumulus clouds build up in an atmosphere deficient of suitable nuclei, they must attain for shower formation a vertical extent with tops colder than $-15°C$. In fact, to be favorably affected by AgI seeding, they must have at their tops temperatures around $-16°C$ (for clouds with a liquid water content of 1 to 2 g/m^3, which ap-

pear to be typical for Israel). The dependence of daily precipitation amounts on temperature (see Figure 13.20) is probably caused by the fact that the colder the atmosphere, the lower is the altitude of the $-14°C$ level. Since cloud base temperatures were found to be practically the same on all rainy days, the likelihood of precipitation formation in clouds having their tops at one and the same altitude is greater for clouds forming on cold days.

5. The authors believe that the above conclusions and the observed nature of local cumulus cloud top height and temperature distributions (having modal values of $-20°$ and 5450 m or 18,000 ft), provide a satisfactory basis for cloud seeding as suggested also by the significance of the positive effects of seeding obtained by statistical analyses, except that these analyses still leave some problems unanswered.

These conclusions imply that seeding can produce on any one day large differential effects that depend mainly on the specific nature of cloud top height and temperature distributions.

The existence of clouds that are favorable to seeding, along with other clouds that contribute a substantial amount to the total daily rainfall and that are not affected at all or even may respond negatively, imposes a serious complication to the statistical analysis additional to the many other known complications.

It also appears that, when evaluating the possible effects of seeding on a geographical basis, a better understanding of the results and prospects of cloud seeding may be achieved only if an adequate effort is invested in studies of a physical nature which provide the basis for statistical analyses. The key to the improvement and optimization of our seeding techniques is to be found only in a clear physical recognition of cloud seeding opportunities. Significant progress can be expected only when physically sound hypotheses are presented for testing.

ACKNOWLEDGMENTS

The number of organizations and individuals whose support, moral, scientific, financial, and/or administrative, have been vital, is considerable.

D. Zahavi and J. Neumann initiated the experiments in 1948. Appreciation is due to members of the "Rain Committee" and in particular to E. D. Bergmann, Chairman (1948–1968) of the Committee, for their loyal support during the very difficult 1950s when few other people had faith in the worthwhileness of the experiments. Without Professor Bergmann's efforts the small randomized experiments of the 1950s could not have survived.

A. Wiener, Director-General, TAHAL-Water Planning for Israel, brought about in 1960 the change in the attitude of the official organizations: In 1961 the Israel Ministry of Agriculture undertook to finance a major randomized project. Since then the Mekorot Water Company has been the host organization of the project and its subsidiary SHAHAM-Electrical and Mechanical Services, Ltd., has been responsible for the administration of the experiments. The individuals who have been most closely connected with the project are S. Kantor (Mekorot), M. Cohen (first with SHAHAM, subsequently with Mekorot), Ch. Molad, and M. Nave (SHAHAM).

K. R. Gabriel, Department of Statistics, the Hebrew University of Jerusalem, designed the 1961–1967 experiment and carried out important statistical analyses and other studies. Since 1970 his work has been carried on by J. Neumann and, especially, by E. Shimbursky. Zivya Wurtele (Los Angeles) and P. Dalinsky

(TAHAL) contributed to the statistical studies, while E. Kally of TAHAL studied the water-potential aspects of rain stimulation in Israel.

On the meteorological-physical side, the project was carried forward by the Cloud Physics Laboratory, headed by A. Gagin, Dept. of Atmospheric Sciences, The Hebrew University of Jerusalem. He was assisted by M. Aroyo, E. Hinkis, J. Patrich, B. Terliuc, and J. Twena. Further, credit is due to K. Rosner for his work as Project Meteorologist.

A profound debt of thanks is due to the Israel Meteorological Service for participating in the project. Special mention should be made of the interest and assistance of M. Gilead, formerly Director of the Service, and of S. Jaffe.

Finally, we wish to thank the Cloud Physics Group of CSIRO, Australia, for encouragement, scientific exchanges, and instrumentation. Discussions, in letters and in person, with E. K. Bigg, E. G. Bowen, S. C. Mossop, E. J. Smith, and J. Warner have been most helpful.

REFERENCES

1. Adderley, E. E. and S. Twomey, An experiment on artificial stimulation of rainfall in the Snowy Mountains of Australia, *Tellus* **10**, 275–280, 1958.
2. Bartlett, J. T., The effect of revised collision efficiency for small drops, *Quart. J. Royal Meteorol. Soc.* **96**, 730–738, 1970.
3. Battan, L. J., Precipitation modification in arid regions, *Science and the New Nations*, Basic Books, New York, 1961, pp. 164–167.
4. Battan, L. J., Relationship between cloud base and initial radar echo, *J. Appl. Meteorol.* **2**, 333–335, 1963.
5. Battan, L. J., Silver-iodide seeding and rainfall from convective clouds, *J. Appl. Meteorol.*, **5**, 669–683, 1966.
6. Bowen, E. G., Cloud seeding, *Science and the New Nations*, Basic Books, New York, 1961, pp. 161–163.
7. Braham, R. R., Meteorological bases for

precipitation development, *Bull. Am. Meteorol. Soc.* **49,** 343–353, 1968.

8. Braham, R. R. and J. A. Flueck, Some results of the Whitetop experiment, *Preprints of Papers, Second National Conference on Weather Modification, Santa Barbara, Cal.,* Am. Met. Soc., 1970, pp. 176–179.

9. Brown, E. N. and R. R. Braham, Precipitation particles measurements in cumulus congestus, *J. Atmos. Sci.* **20,** 23–28, 1963.

10. Clague, L. F., An improved device for obtaining cloud droplet samples, *J. Appl. Meteorol.* **4,** 549–551, 1963.

11. Dalinsky, P., Evaluation of rain stimulation activities in Israel 1961/62-1967/68, Tel-Aviv, Tahal-Water Planning for Israel, 1969, 12 pp.

12. Davis, M. H. and J. D. Sartor, Theoretical collision efficiencies for small drops in Stokes flow, *Nature* **215,** 1371–1372, 1967.

13. Douglas, R. H., Growth by accretion in ice phase, *Geophys. Monogr.* (AGU) **5** *Physics of Precipitation,* 1960, pp. 264–270.

14. Findeisen, W., *Ergebnisse von Wolken und Niederschlagsbeobachtungen bei Wetterekungsfluegen ueber See. Forsch. Erf. d. RWD, Reihe B,* No. 8, 1942. (Also, *Ber. Deut. Wetterdienstes U.S. Zone,* 1, 1947, pp. 21–29.)

15. Flueck, J. A., The expectation of a ratio of non-negative random variables and some present meteorological problems (abstract), *Technometrics* **10,** 421–422, 1969.

16. Gabriel, K. R., Statistical design of an artifical rainfall stimulation experiment, *Le Plan d'Expériences,* Paris, C.N.R.S., 1963, pp. 147–163.

17. Gabriel, K. R., The Israeli artificial rainfall stimulation experiment. Statistical evaluation for the period 1961–65, *Proceedings, Fifth Berkeley Symposium on Mathematical Statistics and Probability,* Vol. 5, *Weather Modification Experiments,* University of California Press, Berkeley, 1966, pp. 91–114.

18. Gabriel, K. R., The Israeli rainmaking experiment 1961–67, final statistical tables and evaluation (Tables prepared by M. Baras), Tech. Rep., Jerusalem, Hebrew University, 1970, 47 pp.

19. Gabriel, K. R. and P. Feder, On the distribution of statistics suitable for evaluating rainfall stimulation experiments, *Technometrics* **11,** 149–160, 1969.

20. Gagin, A., Ice nuclei, their physical characteristics and possible effect on precipitation

initiation, *Proceedings, Tokyo-Sapporo International Conference on Cloud Physics,* 1965, pp. 155–162.

21. Gagin, A., Studies of the nature of precipitation mechanisms for the physical evaluation of cloud seeding experiments, *Proceedings, Toronto International Conference on Cloud Physics,* 1968, pp. 730–734.

22. Gagin, A., Studies of the factors governing the colloidal stability of continental cumulus clouds, *Proceedings, Canberra International Weather Modification Symposium,* 1971.

23. Gagin, A. and M. Aroyo, A thermal diffusion chamber for the measurement of ice nuclei concentrations, *J. Rech. Atmos.* **4,** 115–122, 1969.

24. Gagin, A. and B. Terliuc, A modified Wieland-Twomey thermal diffusion cloud nuclei counter, *J. Rech. Atmos.* **3,** 73–77, 1968.

25. Grant, L. O., The role of the ice nuclei in the formation of precipitation, *Proceedings, Toronto International Conference on Cloud Physics,* 1968, pp. 305–310.

26. Grant, L. O., Some preliminary analyses to explore the possibility of extended area effects from the Climax seeding experiment, *Transactions, Seminar on Extended Area Effects of Cloud Seeding, Santa Barbara, Feb. 1971,* pp. 103–120.

27. Gruber, R, Ed., *Science and the New Nations,* Basic Books, New York, 1961, 314 pp.

28. Hallett, J., Field and laboratory observations of ice crystal growth from the vapor, *J. Atmos. Sci.* **22,** 64–69, 1965.

29. Hays, J. B., *T.V.A. on the Jordan: Proposal for Irrigation and Hydro-Electric Development in Palestine,* Public Affairs Press, Washington, D.C., 1948, 114 pp.

30. Herzl, T., *Alteneuland,* Seemann, Leipzig, 1902, 343 pp.

31. Hocking, L. M. and P. R. Jonas, The collision efficiency of small drops, *Quart. J. Royal Meteorol. Soc.* **96,** 722–729, 1970.

32. Isono, R., Variations of the ice nuclei concentration in the atmosphere and effects on precipitation, *Proceedings, Tokyo-Sapporo International Conference on Cloud Physics,* 1965, pp. 150–154.

33. Jayaweera, K. O. L. F., Calculations of ice crystal growth, *J. Atmos. Sci.* **28,** 728–736, 1971.

34. Jiusto, J. E., Crystal development and glaciation of a supercooled cloud, *J. Rech. Atmos.* **5,** 69–85, 1971.

35. Katz, E. J. and A. Gagin, Evaluation of a convection theory for the inducement of rain in Israel, Jerusalem, Dept. of Meteorology, Hebrew University, 1964, 35 pp. (Mimeographed report).

36. Koenig, L. R., The glaciating behavior of small cumulonimbus clouds, *J. Atmos. Sci.* **20**, 29–47, 1963.

37. Lovasich, J. L., J. Neyman, E. L. Scott, and M. A. Wells, Further studies of the Whitetop cloud-seeding experiment, *Proc. Natl. Acad. Sci.* **68**, 147-151, 1971.

38. Lovasich, J. L., J. Neyman, E. L. Scott, and M. A. Wells, Hypothetical explanations of the negative apparent effects of cloud seeding in the Whitetop experiment, *Proc, Natl. Acad. Sci.* **68**, 2643–2646, 1971.

39. Lowdermilk, W. C., *Palestine, Land of Promise*, Harper, New York, 1944, 236 pp.

40. Levite, G. E. Ed., Developments in desalination technology in Israel, *Proceedings, Eighth National Symposium on Desalination, Ein Boqeq, 14-15 March, 1971*, National Council Res and Dev., Jerusalem, 1971, 163 pp.

41. Ludlam, F. H., The production of showers by the growth of ice particles, *Quart. J. Royal Meteorol. Soc.* **78**, 543–545, 1952.

42. Ludlam, F. H. and P. M. Saunders, Shower formation in large cumulus, *Tellus* **8**, 424–442, 1956.

43. Mason, B. J., *The Physics of Clouds*, Oxford University Press, 1957, 481 pp.

44. McDonald, J. E., Scientific basis of weather modification, *Science* **124**, 86–87, 1956.

45. Mossop, S. L., Concentration of ice crystals in clouds, *Bull. Am. Meteorol. Soc.* **51**, 474–478, 1970.

46. Mossop, S. C. and A. Ono, Measurements of ice crystal concentration in clouds, *J. Atmos. Sci.* **16**, 130–137, 1969.

47. Neumann, J., Energy balance and evaporation from sweet-water lakes of the Jordan Rift, *Bull. Res. Council Israel* **2**, 337–357, 1953.

48. Neumann, J., Artificial stimulation of precipitation: A report of activities in Israel, *Conference on the Scientific Basis of Weather Modification Studies* (Collection of extended abstracts), University of Arizona, Tucson, 1956, 3 pp. (Each abstract is paginated separately.)

49. Neumann, J., K. R. Gabriel, and A. Gagin, Cloud seeding and cloud physics in Israel: Results and problems, *Proceedings, International Conference "Water for Peace,"* Washington, D.C., May 1967, Vol. 2, pp. 375–388.

50. Neumann, J., and E. Shimbursky, On the distribution of a ratio of interest in single-area cloud seeding experiments, *J. Appl. Meteorol.* **11**, 370–375, 1972.

51. Neyman, J., H. L. Scott, and M. Wells, Influence of atmospheric stability layers on the effect of ground-based cloud seeding, I. Empirical results, *Proc. Natl. Acad. Sci.* **60**, 416–423, 1968.

52. Panofsky, H. A. and G. W. Brier, *Some Applications of Statistics to Meteorology*, Pennsylvania State University, College of Mineral Industries, University Park, Pa., 1958, 224 pp.

53. Patrich, J. and A. Gagin, Ice crystals in cumulus clouds - preliminary results, Int. Conf. Meteor., A.M.S.-I.M.S., Tel-Aviv, Nov. 30–Dec. 4, 1970, 15 pp. *Collected Papers*, Rept. No. 4, Cloud and Rain Physics Laboratory, Dept. of Meteorology, The Hebrew University of Jerusalem, 1971.

54. Squires, P., The growth of cloud drops by condensation, *Australian, J. Sci. Res.* **45**, 59–86, 1952.

55. Sulakvelidze, G. K., N. Sh. Bibilashvili, and V. F. Lapcheva, *Formation of Precipitation and Modification of Hail Processes*, Israel Program for Scientific Translations, Jerusalem, 1967, pp. 6–56.

56. Terliuc, B. and A. Gagin, Cloud condensation nuclei and their influence on precipitation, *J. Appl. Meteorol.* **10**, 474–481, 1971.

57. Twomey, S., The nuclei of natural cloud formation. Part II. The supersaturation in natural clouds and the variation of cloud droplet concentration, *Geofis. Pura Appl.* **43**, 243–249, 1959.

58. Warner, J., The supersaturation in natural clouds, *J. Rech. Atmos.* **3**, 233–237.

59. Warner, J., The microstructure of cumulus clouds. Part I. General features of the droplet spectrum, *J. Atmos. Sci.* **26**, 1049–1056, 1969.

60. Wurtele, Z. S., Analysis of the Israeli cloud seeding experiment by means of concomitant meteorological variables, *J. Appl. Meteorol.* **10**, 1185–1192, 1971.

E
SEVERE STORMS

Is it possible now to tame hurricanes? Can we reduce lightning and hail from thunderstorms?

Maybe the most important problem to be tackled in weather modification is taming severe storms. Rain augmentation may not be too important in the long run, because we may have to solve our water shortages in more dramatic ways, but the ability to decrease damage due to hail, lightning, and hurricanes by weather modification is very important. I deliberately did not mention tornadoes because, although they are very damaging, no one now has any good ideas about how to modify them beneficially. Chapter 16 discusses our state of knowledge of tornadoes, but really can not treat seriously the matter of tornado modification.

Probably the most exciting idea in all of weather modification work is the possibility of taming hurricanes. Suppose we could cut the winds in these storms by 30 percent and thereby cut wind damage in half. That is an important prospect for the United States, Japan, India, the Philippines, Mexico, China, and several other countries subject to large losses from these violent storms.

There are two ongoing programs in the United States on lightning suppression, both discussed in Chapter 17, that look promising but need more work before either technique is ready for operational use. Whether AgI seeding reduces lightning strokes has not been verified to the satisfaction of most scientists yet, and tests on chaff seeding have really only started.

Although no major success has been achieved yet in severe storm modification, there are lots of good prospects and almost certainly some of them will pay off in a few years.

14 | Hurricane Modification

R. CECIL GENTRY

Hurricanes are the most destructive of natural phenomena. An earthquake may take an enormous toll in life and property, but it does so with a dislocation that lasts only a minute or two. Tornadoes are the most violent storms on earth, but they are small and short-lived, and do their destructive work along a narrow track. The extratropical cyclones associated with our winter storms are the largest events of that type in the atmosphere, but they are usually milder than their smaller, tropical cousins.

In their combination of violence and duration, hurricanes are unsurpassed. They attack a coastal sector involving thousands of square miles, control the atmosphere over synoptically large areas, and whip the people and communities in their path with a deadly combination of high winds, heavy rains, and storm tides. Even when the storm itself has passed, the stricken area is left isolated and faced with difficult problems of reconstruction; the human loss paid to these storms is, of course, incalculable.

For scientists concerned with weather modification, hurricanes are the largest and wildest game in the atmospheric preserve. Moreover, there are urgent

reasons for "hunting" and taming them. Although improved techniques of hurricane detection and timely warning have greatly reduced the number of deaths caused by hurricanes in the United States, they have not been able to halt the rising loss of property (Figure 14.1).*

Damages from Hurricanes

Average annual cost (before Hurricane Agnes, 1972)† of hurricane damage is now about $450 million. During the 1960s, two hurricanes, Betsy (1965) and Camille (1969), caused more than $1.4 billion damage each, and losses for the period 1965–1969 in current dollars exceed $3 billion. In 1970, Hurricane Celia caused $454 million damage, the sixth highest hurricane-damage figure in United States history. Even in 1971, when only moderate hurricanes affected the United States, damage exceeded $200

R. Cecil Gentry is Director of the National Hurricane Research Laboratory, Environmental Research Laboratories, National Oceanic and Atmospheric Administration, Miami, Florida.   .

* The data in Figure 14.1 illustrate that damages from hurricanes in the United States have been increasing since 1915. In the top half of the illustration damages are summarized by 5-year periods. The statistics have been adjusted to a 1957–1959 base using the Department of Commerce Composite Cost Index for Construction.[2]

† Damage by Hurricane Agnes was due mainly to river floods caused by a complex low-pressure area developed over inland regions long after hurricane force winds had subsided. Damage in the United States exceeded $3 billion.

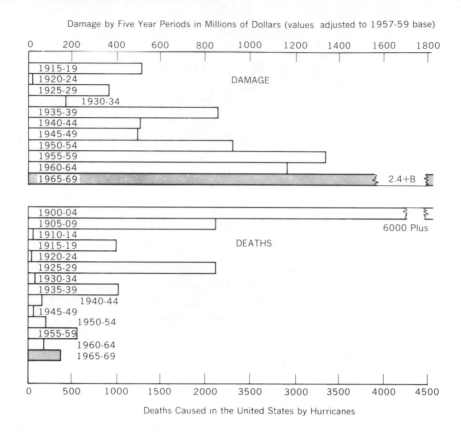

Figure 14.1 Trends in losses from hurricanes in the United States summarized by 5-year periods. Damage statistics have been adjusted to the 1957–59 Department of Commerce composite cost index for construction.

million. The additional loss in human suffering is difficult to evaluate. Clearly, any weather modification techniques which could be used to mitigate the destructive aspects of these storms would be of great general benefit.

What are the damaging elements? Wind, storm surge, and rain-caused flooding. In most United States hurricanes, the largest part of the damage is caused by either the wind or the storm surge, which is caused by the wind.

Damages from Winds

The force of the wind varies with the square of the wind speed. This means a

wind of 50 mps exerts four times as much force as a wind of 25 mps. Similarly, a reduction of the maximum winds of about 10 percent would mean about a 20 percent reduction in wind force. This suggests that such a reduction in wind speed could be equated with at least a 20 percent reduction in damages caused by the winds.

Damages from Storm Surge

Storm surge refers to the rise of water along a coast as a hurricane moves inland. This includes the rise in sea level plus wave action superimposed on the mean water level. For mature hurricanes along the United States coast this rise can vary from 1.5 to 8 m. For enclosed bays the rise can be even higher.

Over the deep ocean, hurricane-driven wind waves may be 15 m or higher, but the mean water level will usually be less than 1 m higher than that outside the storm. In shallow water along the coast the waves are greatly damped, but the rise in sea level can be quite dramatic. Also, the differential rise in sea level along a coastline can set up strong currents which erode beaches and sweep away even strong structures. This combination of rising sea level, battering waves, and strong currents constitutes an almost irrestible force, and one that does great damage along American coastlines.

The height of a storm surge is a function of wind speed, but it is also greatly affected by the slope of the continental shelf, that is, the bottom topography features. For example, where the depth drops off sharply to 100 m not far off the beach, the rise in mean sea level as a hurricane approaches will be small. Other factors also complicate the storm-surge computation. A modification experiment that reduced the wind speed, without moving the belt of maximum winds along the coast, would reduce the storm surge. On the other hand, if an experiment reduced the maximum winds but developed a new maximum at another location, the surge could either increase or decrease, depending largely on bathymetry or bottom topography. This storm surge problem is being studied with the aid of models for predicting the height of the storm surge, and the solution is not yet clear-cut.[14]

Damage from Hurricane Rains

Hurricane rains and the widespread, often sudden, flooding they produce are a major damaging element. In a typical tropical cyclone, 15 to 30 cm of rain will fall over much of the area traversed by the storm. When the hurricane moves inland, and particularly when it moves into a mountainous region, this rainfall can cause floods which take a large toll in life and property.

The most expensive hurricane in history before 1965 was Diane (1955), which caused little damage as it came ashore in North Carolina; however, long after the storm winds had ceased to be dangerous, the hurricane brought rains to Pennsylvania, New York, and New England, killing 200 people and doing an estimated $750 million in property damage.

Hurricane Camille (1969), after Agnes (1972) the most destructive storm ever to strike the United States, caused $1280 million damage along the northern Gulf of Mexico with its winds and storm surge. Further inland, wind became negligible; however, as the storm moved eastward across Kentucky, Virginia, and West Virginia, its torrential local rains caused a series of disastrous flash floods along the eastern slope of the Alleghenies, and property loss in these areas was more than $140 million. Agnes (1972) caused virtually no damage when it went ashore, but later caused floods that have made it the most destructive hurricane in United States history. Unfortunately, the modification experiments presently designed are not expected to affect the hurricane rains significantly.

Variation of Damage Related to Variation of Wind Speed

Most of the damage caused by hurricanes in the United States is caused by either the hurricane winds or the storm surge. We cannot be certain that modifying the wind field will reduce the storm surge, but the surge does vary with the wind, so there is a good chance that reducing the wind will reduce the storm surge in many cases. In any event, a large amount of damage is caused by the winds directly. In hurricane Celia (1970), for example, damages exceeded $450 million and nearly all were caused by winds even though the storm surge on the coast near Corpus Christi, Texas, was greater than 2.8 m.

In previous sections, it was suggested that the damage might vary with the force of the wind which varies as the square of the wind velocity. The relationship between hurricane intensity and property damage, however, is a complex one. First, the amount of damage is a function of the value of property affected by the storm: a very intense storm sweeping a desolate area will cause little damage. Second, the value of property constructed in many locations has changed greatly through the years. And finally, damage statistics from past storms have not been compiled in a manner that facilitates a solution of this problem. Accordingly, there is still no reliable, statistical method of forecasting what a given storm will do to a given area. Some investigators, however, have used the damage statistics from previous storms to solve the following relationship:

$$D = cv^n,$$

where D is the damage, c and n are constants, and v is the maximum wind speed. The interesting quantity here is n. If the damage varies with the wind force, n will equal 2. Values found from analyzing the damage statistics give n as being between 2 and 6[13]; however, to obtain these solutions, investigators have had to adjust the damage statistics for such things as changing populations and inflation in building costs.

If better data should confirm that n is approximately 3, this would mean a reduction of 10 percent in the maximum winds could result in a reduction of about 30 percent in the damage. For storms in which the wind damage exceeded $450 million (e.g., Hurricane Celia, 1970, and Hurricane Camille, 1969), such a reduction could be quite significant.

In this chapter the following is discussed: structure and energetics of hurricanes insofar as they are pertinent to the modification experiments, history of hurricane modification work, descriptions of seeding experiments and an explanation of why they should cause beneficial modification, future prospects, and special problem areas.

STRUCTURE AND ENERGETICS OF HURRICANES RELATED TO MODIFICATION EXPERIMENTS

Each summer and autumn, many areas of disturbed weather move across the tropical oceans. Sometimes, when conditions are right, one of these will draw additional energy from the warm, moist air masses of those latitudes, organize itself into a heat engine, and run for days, sometimes weeks, fed by the atmospheric energy source. When the disturbance becomes a tropical cyclone with winds of at least 32 m/sec blowing counterclockwise (in the northern hemisphere) around a low-pressure center, it is called a hurricane, at least in the United States; it is also called a tropical storm if its maximum winds are less than 32 m/sec.

A more comprehensive discussion of the physical processes is given in Chapter 15, but a short, qualitative description may help explain the proposed hypotheses for hurricane modification. In areas of disturbed weather, the wind field is usally perturbed and stronger winds tend either to detour around a center of relative calm or to form a closed circulation. In such a circular wind field, a number of forces operate on the air particles, as shown in Figure 14.2, where P represents the pressure gradient force which operates from high to low pressure. Co is the Coriolis force, Ce is the centrifugal force, F is the friction, and V is velocity. The Coriolis force in meteorology arises solely from the earth's rotation, and is an apparent force on moving particles that acts normal to the velocity to "deflect" motion to the right in the Northern Hemisphere, and to the left in the Southern Hemisphere. The Coriolis force is proportional to the speed of the particle, that is, $Co = fV$, where V is the wind speed, $f = 2\Omega \sin \phi$, Ω is the angular velocity of the earth, and ϕ is latitude.

The centrifugal force, an apparent force in any rotating system, deflects masses radially outward from the axis of rotation. It is expressed as

$$Ce = \frac{V^2}{R}$$

where R is the radius of curvature of the path.

Friction is a decelerating force. In balanced horizontal flow with no friction,

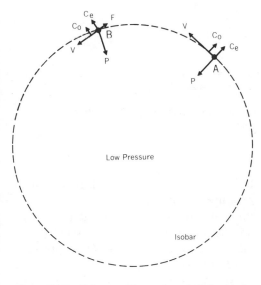

Figure 14.2 Balance of forces for winds in circular flow. P is the presure gradient, Co is the coriolis force, Ce is the centrifugal force, F is friction, and V is velocity.

the air flows parallel to the isobars or curves of equal pressure on a level surface. This is illustrated at point A in Figure 14.2, where the direction of motion is along the isobar. Where there is friction, however, a balance of forces is possible only if the air flows across the isobars toward lower pressure, as shown at point B in Figure 14.2. This results in air at low levels in the disturbance to spiral toward the center of low pressure. As the air converges it must rise and pressure on it decreases, causing the air to expand and cool. Because the amount of water a volume of air can hold is a function of the temperature, vapor condenses in the now-saturated, rising cooling column. This change of phase of water releases latent heat of condensation equivalent to that required to convert the liquid water into steam, approximately 600 cal/g of water condensed. This large quantity of heat is now available to warm the air or to do the work of expanding the air volume.

If the circulation becomes so organized that a layer of divergent flow at high levels is associated with the layer of converging currents at low levels, the temperatures rise at the upper levels and the pressure falls at sea level as much of the rising air is evacuated from the area. Warming results from the latent heat released in the ascending currents and the heating caused by compression in the descending currents. This warming develops a pool of relatively warm air in the middle and upper levels of the troposphere (e.g., between 3 and 14 km). Figure 14.3 shows the temperature distribution including a warm pool of air for a cross-section through Hurricane Hilda (1964). Since warm air is less dense than cool air, this accumulation of warm air through a great depth of the atmosphere results in the low sea level pressure at the center. (This assumes that a pressure surface is level somewhere above the top of

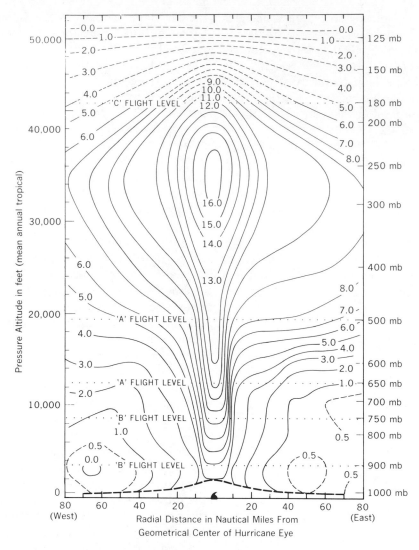

Figure 14.3 Vertical cross-section of temperature anomalies (from mean tropical conditions) for Hurricane Hilda, October 1, 1964 (after Hawkins and Rubsam[12]).

the storm.) The lowering of the pressure at the center increases the force P in Figure 14.2, and if other conditions are right, results in a greater inflow of air, more condensation, and more warm air.

Eventually a state of relative equilibrium is reached in which the mature hurricane has the temperature, pressure, and wind distributions shown schematically in Figure 14.4a. Here the temperature graph represents the radial variation typical for levels between about 3 and 14 km. The pressure distribution is for sea level and illustrates the low pressure at the center and the larger gradient of pressure in the vicinity of the eye wall. (The pressure gradient is the pressure difference between two points divided by the distance between them. The pressure gradient force, P, in Figure 14.2 is proportional to the pressure gradient.) Since the pressure gradient is the forcing function for the winds, the maximum wind speeds are associated with the maximum in pressure gradient which in a mature hurricane usually occurs in the wall clouds not far from the center of the storm.

At least two fundamentals established in recent years by studies of these and other features of tropical cyclone structure and maintenance suggest possible avenues for beneficial modification. (1) The transfer of sensible and latent heat from the sea surface to the air inside the storm is necessary if a hurricane is to reach or retain even moderate intensity. (2) The energy for the entire synoptic-scale hurricane is released by moist convection in highly organized convective-scale circulations located primarily in and around the eye of the storm and in the major rain bands. The first principle explains the observation that hur-

ricanes form only over warm tropical waters and begin dissipating soon after moving over either cool water or land, neither of which provides a flux of energy to the atmosphere sufficient to keep the storm at full intensity. The second accounts for the low percentage of tropical disturbances that reach hurricane intensity. If a warm sea with its large reservoir of energy was the only requirement for hurricane development, we would have more than 10 times as many hurricanes as normally form.

During the 1969–1971 seasons, 295 tropical waves were tracked in the Atlantic and adjacent areas where sea surface temperatures were warm enough for hurricane genesis, but only 32 of these areas developed sufficiently to become named tropical storms.[6-8] Of these, only 18 reached full hurricane intensity (maximum winds greater than 32 mps).

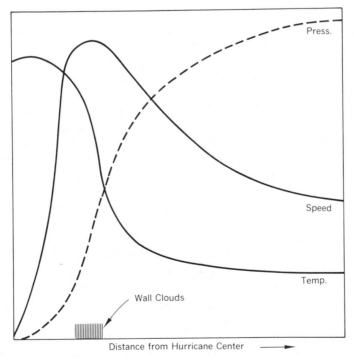

Figure 14.4a Schematics of radial profiles of temperature, pressure, and wind speed for a mature hurricane. The temperature profile is applicable for levels between 3 and 14 km. The pressure and wind speed profiles apply to levels near the surface.

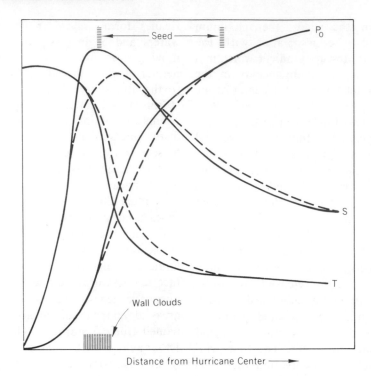

Figure 14.4b   Same as Figure 14.4*a*, except possible effect of a modification experiment is shown.

Apparently there is a limit to the number of ways in which the convective and synoptic scales can interact to achieve optimum utilization of the energy flowing upward from the ocean; then it is not surprising that few tropical disturbances intensify and become major hurricanes.

Both the findings suggest possible field experiments which may modify a hurricane beneficially. On the basis of the first, we may attempt to reduce the flux of energy from the sea surface to the atmosphere, for example, by techniques for inhibiting evaporation. On the basis of the second, we may try to modify the rate of release of latent heat in the small portion of the total storm occupied by the organized active convective-scale motions in a manner that redistributes heating to produce a weakening of the storm. The success in the last few years of experiments to modify cumulus clouds (Chapter 6) suggests that the seeding of hurricanes

is a promising approach. If we can modify the clouds in the hurricane, we may modify the storm—that is, by modifying interactions between the convective (cloud) scale and the synoptic (hurricane) scales, we can affect the storm's intensity.

Consideration of Figure 14.4*a* suggests ways of doing this. Since the damage caused by hurricanes is largely a function of the wind speed, one can achieve beneficial modification of hurricanes if one can effect a reduction of the maximum winds in the storm. Since the forcing function for the winds is the pressure gradient, one might attempt to alter the wind by modifying the pressure gradient. The pressure gradient at sea level in turn is related hydrostatically to the temperature field integrated through the atmosphere. One then might try to modify the winds in a hurricane by seeking adjustments in the temperature field by reducing the contrast between temperatures in the core of the storm and those outside the storm, either by cooling air in the core or warming air outside the core. One such hypothetical redistribution is illustrated in Figure 14.4b.

Various suggestions have been made as to means of changing the temperature field. One is to modify the development of the warm core. Since development of the warm core is greatly influenced by the amount of latent heat available for release in the air columns rising near the center of the storm, one may affect the temperature by reducing the water vapor in the rising columns. Studies of hurricanes indicate that a significant part of the vapor condensed in these rising columns is evaporated from the ocean inside the high wind area of the hurricane.[16, 19] This has led to suggestions that one spread a film over the ocean to reduce evaporation. Theoretical models of hurricanes[18, 21, 28] all indicate that inhibiting evaporation of water from the ocean to the atmosphere inside the hurricane would indeed reduce the intensity of the storm. Unfortunately, we know of no film capable of suppressing the evaporation that is also strong enough to persist in the high waves and strong winds of the hurricane.

Others have suggested cooling the water of the hurricane or dropping cold materials in the air near the center of the storm. All suggestions for reducing the temperature of the water so far evaluated (e.g., mixing the water to bring cold water from lower levels to the sea surface) require such great expenditure of energy that they are impractical. Likewise, inserting cool substances in the air near the center of the storm require either more aircraft or ships to transport the material than can practically be accommodated in the dangerous environment of the hurricane.

Modifying the radiation mechanism at the top of the hurricane has also been advanced, and usually involves the distribution of material of various radiative properties on top of the clouds at selected locations about and at various distances from the center of the hurricane. It has been argued that such materials might induce a change in the temperatures in the upper portions of the hurricane. These suggestions have not yet been evaluated fully either from the viewpoint of the practicality of distributing suitable material, or from the theoretical viewpoint of what effect such changes would have on hurricane intensity.

Finally we come to suggestions for modifying hurricanes which appear logistically practical and offer hope for success. These relate to modifying the mechanism by which convective processes in the eye wall and the rain bands of the hurricane distribute heat through the storm. Water vapor is condensed and the latent heat released in the convective clouds; if one could change the pattern of these convective clouds, one might affect the distribution of heat in a hurricane. This is the approach adopted in the Stormfury experiments.

HISTORY OF HURRICANE MODIFICATION EXPERIMENTS

The first known attempt to modify a hurricane was made on October 13, 1947, when crews of Project Cirrus* dropped dry ice (Dry Ice is a trade name for solid CO_2) in the thin stratified clouds outside the wall clouds of a hurricane east of Jacksonville, Florida. This was a "Let's try it and see what happens" type of experiment. The idea was that the dry ice would cause freezing of the supercooled water in the hurricane clouds and through

* Project Cirrus was a pioneer experiment in weather modification by cloud seeding conducted under federal sponsorship by the General Electric Company (Chapter 1).

Table 14.1   Results of Experiments in Seeding Hurricane Clouds Near the Eye Wall

Name	Date	No. of Seedings	Silver Iodide Used[a] (No./kg)	Approx Max Wind Speed Change (%)
Hurricane Esther	Sept. 16, 1961	1	8/35.13	−10
Hurricane Esther	Sept. 17, 1961	1	8/35.13	0[b]
Hurricane Beulah	Aug. 23, 1963	1	55/219.96	0[b]
Hurricane Beulah	Aug. 24, 1963	1	67/235.03	−14
Hurricane Debbie	Aug. 18, 1969	5	976/185.44	−30
Hurricane Debbie	Aug. 20, 1969	5	978/185.82	−15

[a] Values in column are for total number of units and total kilograms of silver iodide used each day (based on records kept by Sheldon D. Elliott, Jr.). Test results indicate the smaller seeding pyrotechnic units make more efficient use of the silver iodide.

[b] Pyrotechnics dropped outside seedable clouds.

Note: In addition, a hurricane was seeded Oct. 13, 1947 and Hurricane Ginger was seeded Sept. 26 and 28, 1971. The clouds seeded in these storms were far different and the seedings were done in a different fashion than for the storms listed above. See the text for further comments about these results.

the release of latent heat might cause some change in the hurricane. Langmuir in 1948 wrote: "Visual observation of the seeded area showed a pronounced modification of the cloud deck seeded. No organized trough was observed; rather the overcast previously observed appeared as an area of widely scattered snow clouds. The disturbed area covered perhaps 300 square miles. No convective activity was seen to follow the seeding process at any time during the mission."[15]

Unfortunately the experimenter did not have the equipment to make and record measurements in the clouds before and after the seeding, or to make extensive observations of the changes in hurricane structure with time. Because of failure of navigation aids, the crews did not penetrate the wall clouds either to the eye or to the intense convective area in the principal rain band of the storm. Either shortly before or after the seeding, the hurricane track changed from north-northeast to westerly and the storm later moved inland over Savannah, Georgia.

Judging from results of seeding experiments in recent years, and from our increased knowledge of the behavior of hurricanes, it seems very unlikely that the 1947 seeding could have had much affect on the hurricane except for the seeded clouds. Mook and his associates,[17] who collected all obtainable data for an analysis of the hurricane and its track, concluded that the change in course began before the seeding took place.

R. H. Simpson proposed in 1961[26] that hurricanes might be modified by introducing freezing nuclei into the massive wall clouds around the eye of a hurricane. At about the same time, Pierre St. Amand and his associates at the Naval Weapons Center, China Lake, California developed pyrotechnic generators, which made it practical to introduce very large quantities of silver iodide crystals into clouds within a few minutes. Since silver iodide crystals are effective freezing nuclei, this provided a means of causing the supercooled water in the clouds to freeze rapidly. A group from the Weather Bureau and the Navy seeded Hurricane Esther on September 16, 1961, and again on the next day. Results (see Table 14.1) were encouraging if nonconclusive.[26]

But the prospects of beneficial hurricane modification were sufficiently encouraging that Project Stormfury was formally organized by the Weather Bureau and the Navy. The agreement has been renewed and revised several times and continues today as a cooperative effort of the National Oceanic and Atmospheric Administration in the Department of Commerce and the Navy and the Air Force in the Department of Defense. Stormfury is also strongly cooperative with other government agencies and several universities.

In 1962, no hurricane considered suitable for seeding was available. Extensive data were collected in a number of hurricanes, and the Stormfury forces actually flew into Hurricane Ella before deciding that the wall clouds were not developed well enough for the experiment.

In 1963, Hurricane Beulah was seeded on August 23, and again on August 24. The results (Table 14.1) were again described as encouraging but not conclusive.[27]

No operations were planned for 1964. By 1965, Dr. Joanne Simpson had become Director of Project Stormfury and the design of the experiment was changed to provide for massive seedings at 2-hr intervals for a total of 5 seedings/day. Unfortunately, no suitable storm moved into the area set aside for the seedings during 1965 or 1966.

The author became director of Project Stormfury before the 1967 season. Even though the areas approved for seeding were extended, no eligible storm occurred in either 1967 or 1968.

Theoretical and experimental hurricane research continued through these years, however, with results that produced

changes in the original design of the multiseeding experiment. Furthermore, improved pyrotechnic silver iodide generators were developed before the 1969 season. The frustration of 4 years without opportunities may not have been in vain: apparently minor changes to improve the design of the seeding experiment may have been the difference between success or failure.

Hurricane Debbie was seeded five times within 8 hr on August 18, and again on August 20, 1969. The maximum winds measured and recorded by highly instrumented aircraft flying at 3600 m decreased from 50 mps before the first seeding on August 18 to 35 mps 5 hr after the fifth seeding, a 30 percent reduction. The maximum winds, which had meanwhile returned to the original levels, decreased on August 20 from 51 mps before the first seeding to 43 mps within 6 hr after the fifth seeding, or by about 15 percent.[9, 10] These results were so encouraging that a greatly expanded research program was planned.

The research has continued to go well, and nature has continued to be uncooperative. No storm was available for experimentation in 1970, nor was one suitable for the big experiment on the clouds surrounding the wall clouds in 1971. Hurricane Ginger (1971) did move into an eligible area, and clouds in rain bands far removed from the center of the storm were seeded twice on September 26 and four times on September 28. In this storm, the clouds had low seedability (see Chapter 6), contained little supercooled water, and were in a hurricane wind field that was already relatively weak and diffuse. The storm did not change in intensity following the seeding on either day, and in each case the changes were largely caused by natural forces. The seeding did not appear to affect the hurricane very much other than to cause changes in the seeded clouds.[11]

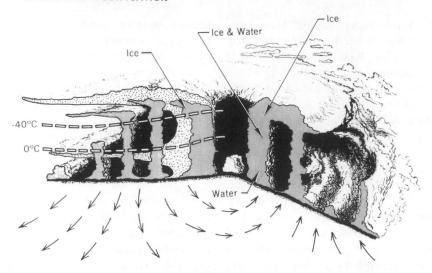

Figure 14.5 Schematic cross-section for a hurricane illustrating the eye, the wall clouds, rain bands, inflow winds, clouds in the outflow layer, and supercooled water in the clouds.

THE STORMFURY EXPERIMENT

Hurricanes differ greatly in structure, in energy processes, and in their life history. The type of storm discussed thus far (Figure 14.4) is one that would have developed over the warm tropical oceans at latitudes less than 25°, whose eye and eye wall were well-developed, and whose maximum winds were located in the eye wall between 10 and 50 km from the center of the storm. Figure 14.5 is a schematic cross-section of such a hurricane showing low-level convergent circulation, the eye, wall clouds, rain bands, the clouds in the outflow layer, and the presence of supercooled water in the clouds.

This supercooled water in the convective clouds of the hurricane provides an opportunity for modifying the convective processes and, in turn, the interaction between the convective and hurricane scales of circulations. In clouds, water drops do not automatically freeze when the temperature is below O°C (see Chapter 6). Freezing nuclei are needed to induce freezing unless the temperature is as low as −40°C. Ice crystals make ex-

cellent freezing nuclei, and there are many other natural substances that serve this purpose. In hurricanes, however, there often seems to be a deficiency of freezing nuclei that are active at temperatures higher than about −25°C. If one introduces freezing nuclei into clouds, the supercooled drops will freeze. The phase changes from water to ice causing a release of the latent heat of fusion (approximately 80 cal/g) which is equal to the heat that would be required to melt the ice. This provides a mechanism by which one can add heat to the air. Study of hurricane structure and, in particular, results from experiments with hurricane models (Chapter 15), suggest that the addition of heat in certain sectors of hurricanes may cause a reduction in the maximum winds of the storm.

Besides the release of the latent heat of fusion initiated by freezing of supercooled water drops, there is another means of selectively adding heat or redistributing heat in storms. In Chapter 6 the seedability of clouds is discussed. It has been shown that, given certain vertical temperature and humidity distributions, clouds will grow more if seeded than they would have grown naturally. Calculations made by Sheets in 1969[24] suggest that hurricane clouds have some seedability and that the increased buoyancy induced by release of latent heat of fusion in the tops of seeded clouds would cause greater growth of these clouds. As the clouds grow, more water condenses, releasing much greater quantities of latent heat. Calculations with theoretical models suggest that heat released by this condensation mechanism is many times* that of

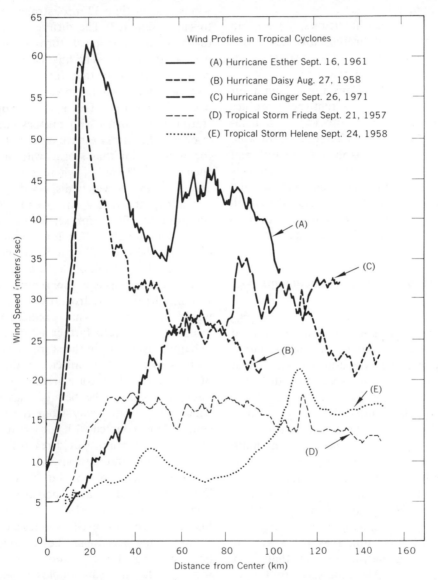

Figure 14.6 Wind speed radial profiles for a number of tropical cyclones. In profiles *A* and *B* the kinetic energy is concentrated near the center of the storm, while in the other three it is dispersed over a much larger area.

* The latent heat of condensation, approximately 600 cal/g, is about 7½ times the latent heat of fusion.

the heat released when the existing super-cooled water is frozen,[23] and that the proper area in which to enhance heating in the storm is the area outside the radius of maximum vertical motion (the eye wall). In a mature hurricane this usually coincides with the belt of maximum winds (see Chapter 15).

There are other hurricanes, however, in which the wind profiles are vastly different. There are also tropical cyclones which never reach hurricane intensity and remain tropical storms throughout their life. Some widely different speed profiles are illustrated in Figure 14.6. Modification of storms with different types of wind structure will probably require different techniques. The modification experiments that have been conducted thus far and that have seemed to be successful have all been on storms whose wind speed profiles resemble A or B in Figure 14.6.

Explanation of Seeding Experiments

The goal of all seeding experiments is to reduce the maximum winds in the hurricane. Since energy is added when a cloud is seeded, one can expect to reduce the maximum winds in a seeded hurricane only by distributing the energy over a larger area. Thus the goal of the seeding experiments is to reduce the maximum winds by dispersing the energy that is normally concentrated in a relatively narrow band curving around the center of the storm (profiles A and B in Figure 14.6). This concept is illustrated in Figure 14.7. The diagram compares total kinetic energy in a hurricane with that in a tropical storm. The wind speed profile for the hurricane is a theoretical one, but the small circles show how well it fits the data for at least one hurricane. The horizontal and vertical scales for the diagram are V^2 and R^2, so that unit area on the diagram is proportional to kinetic energy. The total area below the two graphs is essentially equal. Thus the total kinetic energy in the tropical storm whose maximum winds were less than 23 mps is essentially the same as the total kinetic energy in a hurricane with maximum winds greater than 32 mps, strong enough to cause damage. The difference is that the winds in the tropical storm are spread over a larger area. The hurricane modification experiments thus far designed all have as their goal to spread the energy over a larger area. This is the reason for stating that storms with profiles such as A or B in Figure 14.6 have more potential for modification than storms with profiles such as C, D, or E, where the kinetic energy is already widely dispersed.

The hypothesis first advanced by R. H. Simpson in 1961[26] for seeding a storm with silver iodide called for seeding the clouds in the eye wall of a hurricane. He proposed that release of the latent heat of fusion would change the temperature, which, through hydrostatic relations, would change the pressure gradient at sea level. His hypothesis further assumed that dynamic instability in the storm would be such as to cause the particles in the area of the weakened pressure gradient to flow outward, causing the belt of maximum winds to migrate away from the storm center. This, coupled with conservation of angular momentum, would result in a reduction of maximum winds.

Measurements of the liquid water in hurricane clouds have increased considerably since the 1961 experiments but are probably still inadequate to answer many questions. They do suggest, however, that there is too little liquid water in the hurricane clouds above the freezing level to give the effect hypothesized when the clouds are seeded. Theoretical studies and seeding experiments with cumulus clouds (Chapter 6) indicate, however, that seeding can do more than release the latent heat of fusion. By stimulating condensation seeding may release many times more heat than originally hypothesized. Furthermore, calculations with theoretical models of hurricanes (Chapter 15) indicate that a new eye wall can be developed by encouraging growth of clouds in the inflowing air several kilometers outward from the old eye wall. This would have the effect of producing a new eye wall radially outward from the old, through which the inflowing air would ascend to the outflow layer at the top of the storm. Again, from considerations of partial conservation of angular momentum, the farther outward from the center of the storm that the air ascends, the less will be the wind speeds. Thus

developing a new eye wall by seeding clouds radially outward from the old eye wall could provide a mechanism for reducing the maximum winds in a hurricane.

The portions of the hurricanes seeded in the various experiments in Hurricanes Esther, Beulah, and Debbie are similar even though there has been a gradual evolution in the hypothesis for the seeding experiment. In the 1961 and 1963 experiments, the apparently successful seedings extended radially outward from approximately the inner edge of the eye wall. The seedings in the 1969 experiments were radially outward from the radius of maximum winds, which was about 3 mi beyond the inner edge of the

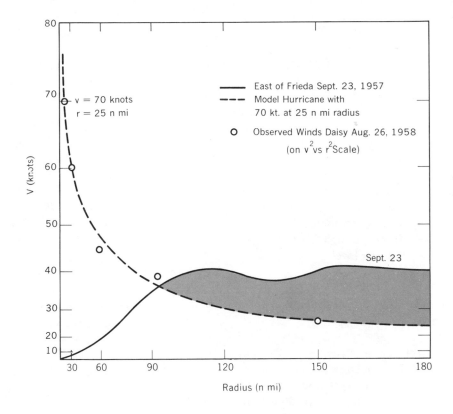

Figure 14.7 Profiles of wind speed against radius of Tropical Storm Frieda (September 23, 1957), and for hypothetical hurricane with a profile that is quite similar to that of Hurricane Daisy (August 26, 1958)—see small circles. Vertical and horizontal scales are such as to make area under curves vary proportionally to kinetic energy (after Riehl and Gentry[20]).

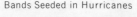

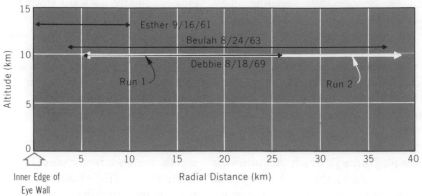

Figure 14.8 Radial extent of the clouds seeded in various hurricane experiments. In Hurricane Debbie (1969) there were five seeding runs of varying lengths on each of the days. Runs 1 and 2 of August 18 illustrate the range.

eye wall. There was, however, considerable overlap of the areas seeded (see Figure 14.8). For this reason, results from the 1961 and 1963 experiments can supplement those of 1969. The fact that in each of the earlier experiments only one seeding was made while the latter had five seedings probably introduces more variability in the results than does the difference in the areas seeded. In Hurricane Beulah, especially, the area corresponded closely to those seeded in Hurricane Debbie. All the seedings used pyrotechnics which produced silver iodide nuclei approximately from the flight altitude down through the freezing level.

Pyrotechnics Used for Seeding of Hurricanes

Silver iodide seeding by Project Stormfury has been by pyrotechnics dropped from airplanes flying across the storm at levels mostly above 30,000 ft. The pyrotechnics were developed by the group at the Navy Weapons Center, China Lake, California, under the leadership of Dr. Pierre St. Amand. A number of mixtures have been used, but the one used in the seeding of hurricane Debbie and the one that is still used primarily for seeding the clouds radially outward from the radius of maximum winds is the LW 83 formulation whose mixture is given in Table 14.2.[4]

The mixture is packaged in the M112 photoflash cartridge that is standard for United States military organizations. The compound weighs 290 mg and the silver iodide output is 190 g/unit. Each gram of mixture produces between 10^{12} and 10^{14} freezing nuclei that are effective at

Table 14.2 LW 83 Formulation (Stormfury I)

Component	Percent by Weight
$AgIO_3$	78
Al	12
Mg	4
Binder	6
	——
	100
AgI Output (calculated)	65% (190 g/unit)

temperatures commonly found in hurricanes at and below the drop altitude.[3, 4] The pyrotechnic is designed to be released at 10 to 11 km, and to burn for 120 sec or about 6 km.

In the experiments designed to develop a new eye wall at greater radius, the seeding aircraft crosses the storm from the left rear quadrant to the eye and through the right front quadrant (Figure 14.9). After the aircraft crosses the radius of maximum winds, seeding drops are started and last for 30 to 40 km. During the course of this run approximately 200 of the pyrotechnics are dropped, producing a curtain of silver iodide crystals 25 to 40 km long and 6 km deep. The crystals (freezing nuclei) are transported in a circular path by the winds blowing around the storm center. There are considerable turbulence and strong horizontal shears in this layer, so that in about 2 hr the silver iodide should be dis-

tributed in an annulus around the hurricane center with inner edge near the center of the eyewall and extending radially outward for about 30 km. At the present time, it is not known what percentage of the silver iodide crystals blow out the top of the hurricane or fall out the bottom in raindrops. The remainder, however should be fairly distributed around the storm within 3 hr.

Monitoring the Storm

The most difficult job in a hurricane modification experiment is determining whether the seeding causes changes in the storm. By comparison, it is relatively simple to drop the pyrotechnics, and it is not much more difficult to determine how the intensity of the storm changes with time. The really difficult problem is determining whether changes are caused by

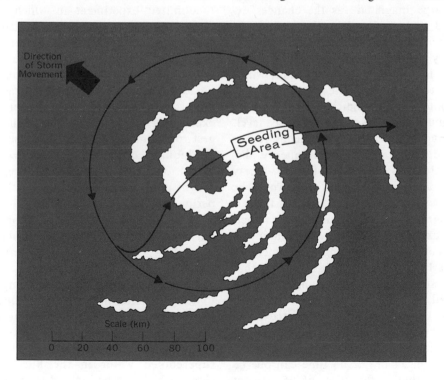

Figure 14.9 Track of seeding aircraft showing seeding area relative to eye and wall clouds of a hurricane.

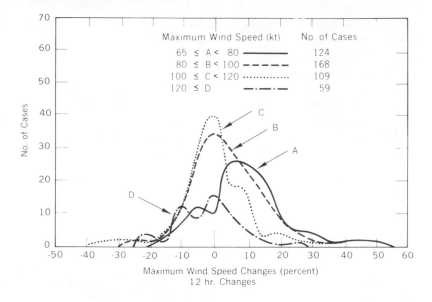

Figure 14.10   Frequency diagram of changes in maximum winds of hurricanes. (After Sheets[25]).

the seeding. The reason for this is that the natural variability of hurricanes is nearly the same magnitude as the changes expected to be induced by the seeding experiment. Figure 14.10 represents the frequency of wind speed changes in hurricanes for 12 hr.[25] The maximum winds used to calculate the changes were computed by an empirical formula from minimum pressures measured at sea level. Although this can give erroneous results in individual cases, the frequency distribution is believed to be fairly representative of changes that would be obtained from measured winds if the latter were available in sufficient number. The expected reduction in maximum winds from the seeding experiment is about 10 to 15 percent. It is obvious that many storms have large changes that are caused strictly by natural variations. There have been well-documented cases where the maximum winds changed up to 30 percent in 12 hr and up to 50 percent in 24 hr. At no time can one be positive that a reduction of 15 percent was caused by seeding independently of natural changes.

One alternative for handling the difficult evaluation problem is to design a randomized experiment in which some storms selected by chance are seeded and some are kept for controls. Afterwards the seeded and control populations could be compared to see which had the greatest reduction in maximum winds. Unfortunately, this requires a rather large experimental population. One of the biggest problems Stormfury has faced is inadequate opportunities for experimentation. The Project, therefore, has sought other means of evaluating the experiment.

Through the development of theoretical hurricane models which are capable of simulating the modification experiment (Chapter 15), it has been possible to calculate the expected effect of seeding and to design a better means of evaluating the experimental results. The present Stormfury experiment is based on the following set of hypotheses. (1) There is supercooled water in the hurricane's clouds which can be induced to freeze by the introduction of freezing nuclei. (2) The freezing will cause release of heat, an increase in the temperature, and in some cases increased growth of the clouds. (3) A change of temperature will result at various locations in the storm. (The largest temperature changes are not necessarily in the seeded clouds according to results from the model calculations.) (4) The pressure distribution will be altered. (5) A new eye wall will develop radially outward from the old one. (6) The belt of maximum winds will move radially outward from the old original maximum and the maximum winds will decrease. All these hypotheses suggest measurements which can be made to help evaluate the experiment, such as (1) the number of ice nuclei and ice particles in the cloud before and after seeding, (2) radar monitoring of the changes in structure and size of the seeded clouds, (3) evolution of size and shape of the eye, (4) temperature and pressure gradients at various levels, and (5) wind distribution

across the storm at frequent intervals. In particular, these measurements can be made often enough to determine if the variations are in phase with the seedings and if they come in the right sequence, at the right time, and in the right place to be reasonably accounted for by the modification experiments. If all the data just mentioned could be accurately and completely measured, one should be able to determine with only a few experiments whether the hurricane was modified by the experiment or by natural forces. Unfortunately, limitations on the number of aircraft and, especially, on aircraft instrumentation, have prevented Stormfury from obtaining all these data in the past.

The principal data used in evaluating the modification experiments thus far have been the measured winds and the observations of the size of the eye.[1, 9, 10] In any case, many well-instrumented air-

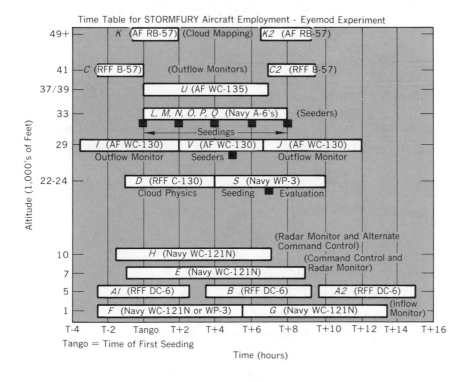

Figure 14.11   Flight schedules for monitoring and seeding aircraft for planned experiments on hurricanes.

craft are needed to collect sufficient data to properly evaluate the results of a modification experiment. Furthermore, these data are needed from a few hours before the first seeding to establish the base state of the hurricane until about 6 hr after the last seeding. Figure 14.11 shows the schedules of the monitoring and seeding aircraft planned for the experiment. Although fewer aircraft would be needed if they all had the proper instrumentation, this figure illustrates the efforts being made to collect the needed data.

FUTURE PROSPECTS, SPECIAL PROBLEM AREAS, AND PLANS

Evaluation of Experiments

A statement of the true probability of achieving beneficial modification is greatly complicated by the few experimental cases we have had and by the large natural variability of hurricanes. Figure 14.10 shows the natural variability of winds, and illustrates why the results from Hurricane Debbie were so encouraging. The winds in that storm were reduced following the seedings on the two days by 30 and 15 percent, respectively. Each of these values is greater than the standard deviation of the data used in the graph for the "100-knot storms" in Figure 14.10. Additional details of Hurricane Debbie results are given in Figures 14.12 and 14.13.

If one considers the sequence of events in the Hurricane Debbie (1969) experiment, the results are quite impressive: seeded five times on August 18, 1971 and the wind speeds decreased 30 percent; no seeding August 19 and the storm in-

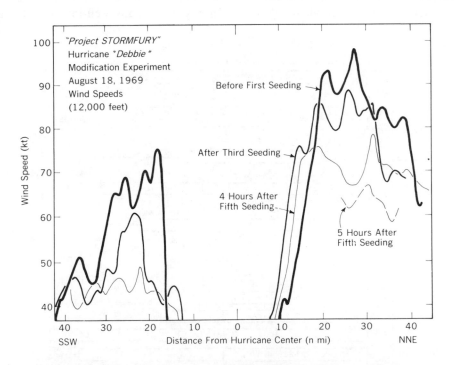

Figure 14.12 Changes with time of wind speeds at 3600 m in Hurricane Debbie on August 28, 1969. (Lines have been smoothed. The winds were measured by aircraft flying across the storm from south-southwest to north-northwest or the reciprocal tract. Profiles are given which show the wind speeds before the first seeding, after the third seeding, and after the fifth seeding (Copyright 1969 by the American Association for the Advancement of Science) (reprinted by permission, from Gentry[9]).

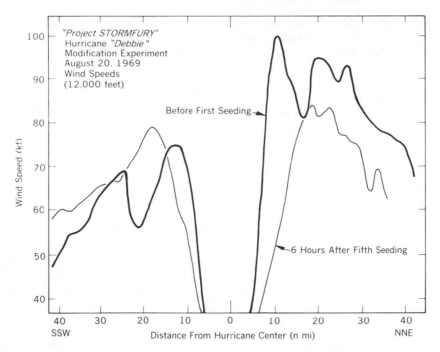

Figure 14.13 Same as Figure 14.12 except that the wind speed profiles are for August 20, 1969 (Lines have been smoothed). (Copyright 1969 by the American Association for the Advancement of Science) (reprinted by permission, from Gentry[9])

tensified; seeding of the inner* wind maximum on the first four seedings on August 20 and it decreased, and seeding of the outer wind maximum on the fifth seeding and the winds decreased 15 percent. From studies of previous storms it seems unlikely that all these things could happen by chance as often as one time in 50.

One still must consider that natural forces may have caused the changes that occurred in Hurricane Debbie; these have been considered[10] and may have contributed. This is especially true on August

* There were two wind maxima on August 20, 1969 (see "before first seeding" wind speed profile in Figure 14.13). Theory suggests that the most effective belt in which to seed is outward from the radius of maximum winds. To seed radially outward from the inner maximum means to span the outward maximum with the seeding. The theoretical model suggests this can cause intensification (see Chapter 15). During the first four seedings, the outer maximum did increase slightly. The fifth seeding was done radially outward from the outer maximum and the winds decreased.

18, 1969. Unfortunately, the data are inadequate to determine satisfactorily whether the changes were induced by the seeding, by natural forces, or by a combination of the two.

The Hurricane Debbie results are somewhat supported by results from attempts to simulate the modification experiment with the theoretical hurricane model.[23] Although the models are not capable of completely simulating the modification experiment, the model results suggest that the wind speeds should have decreased 4 to 8 percent. Some of the other theoretical models in which the modification experiments were simulated in a different fashion have given contrary results.[5, 28]

Howard, Matheson, and North[13] have applied Bayesian statistics to all the data on past experiments to calculate the probability of changes in the next hurricane experiments. In their work, they have included the effect of natural varia-

Table 14.3 Probability Assignments to 12-Hr Wind Change. Discrete Approximation With Five Outcomes

Interval of Changes in Maximum Sustained Wind	Probability that Wind Change Will be Within the Interval	
	If Seeded	If Not Seeded
Increase of 25% or more	0.038	0.054
Increase of 10 to 15%	0.143	0.206
Little change +10 to −10%	0.392	0.480
Reduction of 10 to 24%	0.255	0.206
Reduction of 25% or more	0.172	0.054

bility and have recognized that a storm may intensify following a seeding even if one assumes that the effect of the seeding is to cause a decrease in wind speed. This could happen if the seeding effect were −15 percent and the natural variability for the particular case were +30 percent. Their results, summarized in Table 14.3, strongly suggest that seeding should result in reduced intensity.

The difficulty of getting frequent experiments in hurricanes emphasizes the need for developing improved techniques for determining the probability from a limited number of experiments. With the present state of our resources, this probably means collecting more comprehensive data in hurricanes and the development of improved theoretical models to indicate which data are most sensitive to changes induced by the seeding experiments.

Limitation on Experiments

One of the reasons for being unable to conduct more experiments in hurricanes

is the need to restrict the area for conducting the experiments. Until the present, the experiments have largely been conducted by aircraft operating either from Puerto Rico or Florida. The need to monitor the storms requires that the aircraft remain on station for several hours. This means that the experimental storms must be within about 700 mi of an operating base. There are, however, restrictions on experimenting close to land. These restrictions have two motivations: (1) the data for evaluating the experiment should be collected before proximity of land modifies the storm; and (2) so long as the program is experimental the Government prefers not to seed any storm which might quickly move inland and cause either great damage or loss of life. This is based not on fear that the seeding would cause intensification but on concern for the thus far unresolved legal and sociological problems such an event would raise.

Howard, Matheson, and North[13] have performed a decision analysis study of the hurricane modification program and have concluded that all suitable hurricanes should be eligible for seeding if there were no issue about government liability. At the present time, the legal aspects have not been sufficiently resolved (see Chapter 21).

Effects of Ecology

In conducting any experiment in artificial weather modification, one must consider the possibilities of harmful side effects on the environment. Most of these concerns relate to possible effects on the rainfall, general circulation of the atmosphere, plant growth, and of the seeding agent on plant and animal life.

Many areas of the world depend on rainfall from tropical cyclones to alleviate drought, nurture crops, fill reservoirs, restore ground water, and satisfy other water requirements. The question is often raised whether modifying a hurricane will significantly affect the rainfall patterns. The data collected in previous modification experiments are inadequate to answer this question, but they have not indicated that the experiments had any effect on the rainfall. The models[22] suggest that the seeding will cause a small increase in the total rainfall in the storm, but that the rainfall will be more widely distributed.

Those who are concerned about the rainfall also ask if the experiments will change the course of the storm and cause it to take the needed rain to an entirely different area. Most successful techniques used in predicting hurricane tracks suggest that the modification could have very little, if any, effect on the direction of motion of the storm. The experiments thus far conducted do not show any identifiable effect. This problem can best be studied with a full three-dimensional model of a hurricane that includes interaction with the environment. Work is underway on such a model, but it is not yet perfected (Chapter 15).

The proposed hurricane seeding experiments are not expected to have any noticeable effect on the general circulation of the atmosphere. The experiments are not expected to destroy hurricanes, change their tracks, or have much effect on the rainfall, and the experiment would have to cause significant changes in one of these areas to affect the general circulation.

A reduction in the maximum winds should result in less damage to trees and shrubs, so that the effects of modifying the storm should be beneficial in this regard. A possible exception might be in areas that depend on periodic visits of tropical cyclones for pruning of excess growth.

The agents used for seeding the hurri-canes have consisted primarily of silver iodide and certain by-products that are not known to have any undesirable affect. Evidence presented so far has not shown that silver iodide in the concentration used in the experiments has any harmful effect. Research continues on this problem.

Future Plans

Plans of Project Stormfury based on evidence of the type summarized in this chapter suggest that the following should be emphasized in future activities of the project:

1. Increased efforts to improve theoretical models.
2. Collection of data to further identify natural variability in hurricanes.
3. Expanded research—both theoretical and experimental—on physics of hurricane clouds and interactions between the cloud and hurricane scales of motion.
4. More field experiments on tropical cyclones at every opportunity.
5. Tests of other methods and material for seeding.
6. Further evaluation of other hypotheses for modifying hurricanes.
7. Development of the best procedures to maximize results of field experiments.

ACKNOWLEDGMENTS

The author gratefully acknowledges the contributions of the reconnaissance crews of the Navy, Air Force, and the Research Flight Facility of NOAA in the field experiments of Project Stormfury, and the research contributions of his colleagues in the National Hurricane Research Laboratory, the Navy, and various universities. Without their work, this chapter would not have been possible.

The development of the current hypothesis for hurricane modification has been an evolutionary process and was facilitated by the contributions of many people. Many of these are noted in the references. Special thanks are due to Doctors Stanley L. Rosenthal and Harry F. Hawkins of the National Hurricane Research Laboratory and to Professors Noel E. LaSeur, Charles L. Hosler, James McDonald (deceased), and Jerome Spar, all members of the Stormfury Advisory Panel.

REFERENCES

1. Black, Peter G., Harry V. Senn, and Charles L. Courtright, Airborne radar observations of eye configuration changes, bright band distribution, and precipitation tilt during the 1969 multiple seeding experiments in Hurricane Debbie, *Monthly Weather Rev.* **100**(3), 208–217, 1972.

2. *Construction Review*, Bureau of Domestic Commerce, Department of Commerce, 1970, Table E1.

3. Eliott, S. D., Jr., STORMFURY seeding pyrotechnics, 1969, *Project STORMFURY Annual Report, 1969*, U.S. Department of Commerce and U.S. Department of the Navy, D1–5, 1970.

4. Eliott, S. D., Jr., Roger Steel, and William D. Mallinger, STORMFURY pyrotechnics, *Project STORMFURY Annual Report, 1968*, U.S. Department of Commerce and U.S. Department of the Navy, B1–11, 1969.

5. Estoque, Mariano A., *Hurricane modification by cloud seeding*, Final Report NOAA Grant E-22-11-71(G), University of Miami Division of Atmospheric Science, 1972, 13 pp.

6. Frank, Neil L., Atlantic tropical systems of 1969, *Monthly Weather Rev.* **98**(4), 307–314, 1970.

7. Frank, Neil L., Atlantic tropical systems of 1970, *Monthly Weather Rev.* **99**(4), 281–285, 1971.

8. Frank, Neil L., Atlantic tropical systems of 1971, *Monthly Weather Rev.* **100**(4), 268–275, 1972.

9. Gentry, R. Cecil, Hurricane Debbie modification experiments August, 1969, *Science* **168**, 473–475, 1970.

10. Hawkins, H. F., Comparison of results of the Hurricanes Debbie (1969) modification experiments with those from Rosenthal's numerical model simulation experiments, *Monthly Weather Rev.* **99**: 427–434, 1971.

11. Hawkins, H. F., Kenneth Bergman, and R. Cecil Gentry, Report on seeding of Hurricane Ginger, *Project STORMFURY Annual Report, 1971*, U.S. Department of Commerce and U.S. Department of the Navy, 1972.

12. Hawkins, H. F. and D. T. Rubsam, Hurricane Hilda, 1964. II. Structure and budgets of the Hurricane on October 1, 1964, *Monthly Weather Rev.* **96**(9), 617–636, 1968.

13. Howard, Ronald A., James E. Matheson, and D. Warner North, The decision to seed hurricanes, *Science*, **176**(4040), 1191–1202, 1972.

14. Jelesnianski, C. P. and A. D. Taylor, A preliminary view of storm surges, before and after storm modification, NOAA Tech Memo ERL WMPO-3, U.S. Dept. of Commerce, NHRL, Miami, Florida, 1973, 33 p.

15. Langmuir, Irving, The growth of particles in smoke and clouds and the production of snow from supercooled clouds, *Proc. Am. Phil. Soc.* **92**(3), 167–185, 1948 (Daniel F. Rex quoted about hurricane, p. 184).

16. Miller, B. I., A study of the filling of Hurricane Donna (1960) over land, *Monthly Weather Rev.* **92**, 389–406, 1964.

17. Mook, Conrad P., Eugene W. Hoover, and Robert A. Hoover, An analysis of the movement of a hurricane off the east coast of the United States, October 12–14, 1947, *Monthly Weather Rev.* **85**, 243–250, 1957.

18. Ooyama, K., Numerical simulation of the life cycle of tropical cyclones, *J. Atmos. Sci.*, **26**, 3–40, 1969.

19. Palmén, E. and Herbert Riehl, Budget of angular momentum and energy in tropical cyclones, *J. Meteorol.*, **14**, 150–159, 1957.

20. Riehl, Herbert and R. C. Gentry, *Analysis of Tropical Storm Frieda, 1957, A Preliminary Report*, National Hurricane Research Project Report No. 17, U.S. Department of Commerce, Weather Bureau, 1958, 16 pp.

21. Rosenthal, S. L., A circularly symmetric primitive-equation model of tropical cyclone development containing an explicit water vapor cycle, *Monthly Weather Rev.* **98**, 643–663, 1970.

22. Rosenthal, S. L., A circularly symmetric primitive-equation model of tropical cyclones and its response to artificial enhancement of the convective heating functions, *Monthly Weather Rev.* **99,** 414–426, 1971.

23. Rosenthal, S. L. and M. S. Moss, Numerical experiments of revelance to Project STORMFURY, *NOAA Tech. Memo. ERLTMNHRL 95,* U.S. Department of Commerce, National Hurricane Research Laboratory, Miami, Florida, 1971, 52 pp.

24. Sheets, R. C., Computations of the seedability of clouds in a hurricane environment, *Project STORMFURY Annual Report, 1968,* U.S. Department of Commerce and U.S. Department of the Navy, E1-5, 1969.

25. Sheets, R. C., Application of Bayesian Statistics for STORMFURY Results, *Project STORMFURY Annual Report, 1969,* U.S. Department of Commerce and U.S. Department of the Navy, H1-16, 1970.

26. Simpson, R. H., M. R. Ahrens, and R. D. Decker, A cloud seeding experiment in Hurricane Esther, 1961, *National Hurricane Research Project Report No. 60,* U.S. Department of Commerce, Weather Bureau, Washington, D.C., 1963, 30 pp.

27. Simpson, R. H. and J. S. Malkus, Experiments in hurricane modification, *Sci. Am.,* **211**(6), 27–37, 1964.

28. Sundquist, Hilding, Model tropical cyclone behavior in experiments related to modification attempts, *Tellus* **XXIV**(1), 6–12, 1972.

15 | Computer Simulation of Hurricane Development and Structure

STANLEY L. ROSENTHAL

Tropical cyclones form only over the oceans. They are accompanied by violently rotating winds and torrential rain. Observational studies show that the energy required to drive these systems is derived from the latent heat of condensation released in tall convective clouds around the center.

These storms form in large-scale disturbances of the atmospheric flow patterns over the tropical oceans. Often the lower portions of these disturbances are cold core in the sense that their rain areas (as observed on a scale much larger than that of the individual cumulus cloud) are colder than surrounding areas.[18] Since the rain areas are usually associated with low pressure, the horizontal pressure gradients increase upward because the falloff of pressure with height is more rapid in cold air than in warm air.

Some of these cold-core systems develop warm cores with the transformation first evident at upper tropospheric levels,

then spreading downward. Although the detailed processes responsible for this transformation are not known, the change is clearly attributable to the release of latent heat in the rain area.[26] The conversion proceeds slowly and may require several days for its completion. Once the warm core has been established, the chances of intensification (in terms of increased winds) become more likely.[18]

Most of these warm-core systems never attain hurricane intensity. The few cyclones that do reach hurricane intensity begin to deepen rapidly at some point in time and can evolve from a system with strongest winds of perhaps 30 knots to a hurricane with winds of 70 to over 100 knots within 12 to 24 hrs. This impulsive deepening can occur soon after the formation of the warm core or it can take place days later.

As intensification to the hurricane stage occurs, the strong winds form a tightband around the vortex center and pressures at the storm center drop rapidly. The characteristic ring of cumulus convection around the vortex center (the eye wall) seems to form just before the onset of rapid intensification[10]

Stanley L. Rosenthal, is Chief, Theoretical Studies, National Hurricane Research Laboratory, National Oceanic and Atmospheric Administration, Coral Gables, Florida.

and is well organized by the time the storm reaches a quasisteady mature stage. The eye and eye wall typically occupy less than 5 percent of the total storm volume (with an average radius of about 15 nmi) but evidence indicates that the processes that take place there are largely responsible for the structure and maintenance of the entire system.[10] The cumulonimbus clouds in the eye wall may extend upward to 50,000 ft and there merge with extensive cirrus shields. In the mature stage, several bands of cumulonimbi spiral into the eye wall (the well-known rain or spiral bands).

Temperatures show only small irregular gradients outside the eye wall. Warmest temperatures are formed inside the eye with largest gradients occurring through the eye wall. In terms of temperature departures from average tropical conditions (anomalies), the largest positive values occur near an elevation of 40,000 ft over the eye. These anomalies decrease downward and outward.[7, 10]

In the low and middle troposphere, pressures decrease rapidly as one approaches the eye and strongest pressure gradients are found in the eye wall region.[10] Only slight variations of pressure with azimuth are apparent. At upper tropospheric levels, the horizontal pressure gradients decrease rapidly with elevation in response to the strong positive temperature anomalies.[7, 10]

Winds spiral inward at low levels and outward at upper tropospheric levels. In the typical case, little inflow or outflow is found at midtropospheric levels.[7, 10, 19] Strongest tangential winds are found near the region of maximum pressure gradient and these decrease little with height in the lower troposphere where pressure gradients also vary little with height. In the higher troposphere, where pressure gradients decrease rapidly with height, winds respond in the same manner and

there is usually little evidence of the system at tropopause heights.[10]

The rainfall rates associated with mature hurricanes indicate that vast amounts of latent heat (10^{13} cal/sec according to Palmén and Riehl)[14] are released and transported upward through the action of the cumulus clouds. Only 2 to 4 percent of this released latent heat is converted to kinetic energy; most of the released latent heat is converted to potential energy and exported from the hurricane by the upper tropospheric outflow branch.[14, 19]

The inflow of mass to the hurricane in the low troposphere is very nearly equal to the upper tropospheric outflow.[14] The inflowing air moves toward lower pressure and hence has work done on it by the pressure-gradient force. This constitutes the mechanism whereby potential and internal energy are converted to kinetic energy; however, in the upper troposphere, the inward directed pressure gradient must be smaller than that of the inflow layer. If this were not the case, the outflow would convert kinetic energy to potential energy at the same rate that the inflow generates kinetic energy, and there would not be a net generation of kinetic energy to maintain the storm against frictional dissipation. As already noted, the warm core of the hurricane produces the required reduction of pressure gradient with height. The prime role of the condensation heating is then to develop and maintain the storm's warm core.

THE THEORETICAL PROBLEM AND A STRATEGY FOR ITS SOLUTION

The Hurricane Problem in Relation to Basic Physical Laws

The theoretical meteorologist is faced with the problem of providing an explanation of the events described in the introduction and ultimately a method for predicting their occurrence based on the pertinent physical laws. These are the same physical laws that, in principle, govern all atmospheric motions: (1) Newton's second law of motion expressed for fluids in the form of the Navier-Stokes equations, (2) the first law of thermodynamics, (3) the law of mass conservation, (4) the Charles-Boyle equation of state for an ideal gas, and (5) a conservation law for water substance.

The mathematical expressions for these laws are usually referred to as the "hydrodynamical equations." They comprise a system of seven equations that are mathematically complete for seven dependent variables (pressure, temperature, density, three velocity components, and humidity), provided that diabatic and frictional effects are either known or can be deduced from the basic seven variables. The independent variables of the system are the space coordinates and time. With boundary conditions, the hydrodynamical equations comprise an initial-value problem that determines the future state of the atmosphere given an initial state.

A Method for Solution of the Hydrodynamical Equations

Unfortunately, six of the hydrodynamical equations are highly nonlinear, partial differential equations and solutions in closed form can be obtained only for certain highly idealized flows. Only with the development of high-speed digital computers has it become feasible to obtain solutions pertinent to a wide variety of meteorological problems through application of the techniques of numerical integration.

L. F. Richardson in 1922 seems to have been the first to observe that the form of

the hydrodynamical equations renders them particularly amenable to numerical solution as an initial-value problem.[16] This arises from the fact that local time derivatives (time rate of change at a given point in space) of each dependent variable can be expressed in terms of space derivatives of the same set of variables. If the initial values of the dependent variables are specified on a network of grid points, the space derivatives may be approximated by finite difference techniques. Once this has been achieved, the system of partial differential equations is converted to a system of ordinary differential equations (one for each dependent variable and for each grid point) in which only time derivatives appear. Techniques of numerical integration may then be employed to obtain values of the dependent variables at future times on the array of grid points.

Certain restrictions must be considered if the numerical solution is to approximate the solution to the differential equations. (The interested reader may refer to Richtmyer and Morton[17] for detailed discussions of these matters.) Given the spatial separation between grid points, the maximum allowable time step for the numerical integration is directly proportional to the distance between grid points and inversely proportional to the phase speed of the most rapidly propagating wave whose solution is contained in the hydrodynamical equations. If the time step exceeds this critical value, random errors (such as those produced by rounding off), grow exponentially with time and quickly make the calculation meaningless. This phenomenon, known as "linear computational instability," can occur in the absence of nonlinear terms and was first discovered through numerical studies of the linear wave equation. A second type of computational instability can arise from truncation errors in certain nonlinear terms and requires special care in the finite difference formulations of these terms.

The Design of Numerical Models

The hydrodynamical equations, in their complete form, contain solutions corresponding to acoustic and gravity waves as well as solutions pertinent to meteorological events. Although the effect of the acoustic waves on the meteorological solutions is insignificant, their very presence in the equations would require the numerical integration to proceed with extremely small time steps to preserve linear computational stability (acoustic waves propagate at speeds an order of magnitude greater than do meteorological systems). It therefore becomes an economic necessity to modify the hydrodynamical equations in such a way that acoustic waves are eliminated. This may be achieved by replacing the vertical component of the equation of motion with a statement of hydrostatic equilibrium.[25] Based on order of magnitude considerations, this approximation is entirely valid if the horizontal dimensions of the phenomena of interest are much larger than their vertical dimensions.[15] In the meteorological literature, the hydrodynamical equations with this approximation are known as the "primitive equations."

Economics usually dictates further simplifications. For different problems, these may include elimination of some or all gravity waves, neglect of some or all diabatic effects, neglect of some or all frictional effects, assumptions of certain types of symmetry, and so on. These simplifications must proceed in an orderly fashion with a priori information concerning the nature of the solution desired and the order of magnitude of neglected effects in relation to the effects that have been retained. If, for instance, we wish to obtain

solutions relevant to hurricanes, we cannot neglect condensation heating since it drives the storm, nor can we neglect centrifugal terms since they are among the largest terms in the equations of motion. On the other hand, the assumption of circular symmetry is reasonable as a first approximation and the elimination of certain gravity waves is probably also justifiable.[13]

The term "numerical model" denotes a suitably simplified version of the hydrodynamical equation that has been cast in finite difference form for the purpose of numerical integration. A "numerical experiment" consists of the determination and examination of a solution of the model equations for some chosen set of initial conditions.

Spacing between grid points is determined by the dimensions of the physical phenomenon to be investigated. Once this spacing has been selected, those motions with scales smaller than the grid-point separation (usually referred to as "subgrid" motions) are eliminated, and those motions with dimensions slightly larger than that of the grid spacing are severely distorted through truncation error. At this point, the investigator must determine whether or not the subgrid motions have important interactions with the motions under investigation. If this is indeed the case, a method must be devised through which the effects of the subgrid motions may be represented with knowledge of only the larger-scale motions. Such a method is known as "parameterization" and is the finite difference analogue to a turbulence theory.

Horizontal grid spacing on the order of 10 km appears adequate to resolve the structure of the hurricane vortex. However, cumulus clouds have horizontal scales ranging from a few hundred meters to a few kilometers. Thus, if 10 km resolution is adopted, the clouds that provide the energy to drive the storm become subgrid motions. If, on the other hand, we decide to make the resolution sufficiently fine so that at least the largest cumulus clouds can be resolved, a grid spacing on the order of 100 m would be required. It would then not be economically feasible to span several hundred kilometers of radial distance from the storm center.

The cumulus clouds must, therefore, be parameterized and this has proven to be the most difficult problem in the design of hurricane models. Although no widely accepted technique exists, a number of techniques provide reasonable results and are therefore useful.

Guidance from Linear Analysis

By the time a numerical model is ready for machine coding, the hydrodynamical equations have been substantially modified through simplifications and parameterizations. The danger always exists that the representation of physical processes, important with respect to the phenomena under investigation, have inadvertently been destroyed or severely distorted. After the machine program has been prepared and calculations have been performed, the numerical solutions will ultimately allow the investigator to discover whether or not he has erred along these lines, since the solutions will not realistically simulate the natural event. With determined effort and ingenuity, the sources of difficulty can ultimately be discovered and corrected. This is a tedious and inefficient procedure that often becomes quite expensive in terms of man hours and machine time. Although this stage of the investigation cannot be avoided entirely, its magnitude can be substantially reduced in many cases through application of a standard technique of theoretical analysis.

This technique, particularly useful when the modeling effort is aimed at an explanation of the intensification of a meteorological system, is referred to as "linear analysis" or "the perturbation method." The analysis consists of two steps. First, the nonlinear model equations, through systematic procedures, are simplified to take the form of *linear* partial differential equations. The linearization is dependent upon the assumption of a "base state" that is invariant in time and represents a simple and known solution of the nonlinear equations. The meteorological system under study is then assumed to be a small-amplitude, space- and time-dependent perturbation on the base state. Terms in the nonlinear model equation that contain products of the perturbation quantities are considered higher-order and negligible.

The second part of the analysis is solution of the linear equations. The solutions are then scrutinized to determine the conditions under which the perturbation will grow (be unstable), decay (be stable), or remain unchanged (be neutral). In most cases, stability is dependent on scale. The scale with the largest rate of growth is assumed to be that which would evolve in the real atmosphere if random perturbations were superimposed on an atmospheric situation similar to the base state. Although it cannot be proved with rigor that the scale with the largest growth rate (the preferred mode), as predicted by linear theory, will be dominant in a solution of the nonlinear equations, experience has shown that this is almost always the case.

If the preferred mode does not resemble the atmospheric phenomenon in spatial scale, structure, and growth rate, it is likely that the nonlinear solutions will also be unrealistic. Since the linear equations are usually solved in closed form, the sources of the unrealistic features, if any, may be searched out far more systematically than is the case for the nonlinear numerical model.

A THEORY FOR HURRICANE DEVELOPMENT

Some Elementary Cloud Dynamics

To understand the problems that arise in the design of hurricane models, it is necessary to review some elementary cloud dynamics. Consider air parcels that rise adiabatically. If these parcels become less dense than their environment, buoyant forces will accelerate them upward and a situation exists that meteorologists refer to as "static instability." Since the pressure of the parcel will be very nearly the same as the environmental pressure, the parcel must become warmer than the environment to become buoyant.

By simple manipulation of the first law of thermodynamics (making use of the fact that the parcel's environment is very nearly in a state of hydrostatic equilibrium), it may be shown that a *necessary* condition for static instability is

$$\frac{\partial T}{\partial Z} < - \frac{g}{c_p}, \qquad (15.1)$$

where g is the acceleration of gravity, c_p is the specific heat capacity at constant pressure for air, T is air temperature, and Z is height above some reference level (usually mean sea level). The physical significance of condition 15.1 is rather simple to grasp. Air that rises adiabatically cools at the rate g/c_p (dimensions of degrees per unit length) as internal energy is converted to potential energy. The quantity g/c_p is known as the "dry adiabatic lapse rate" and, numerically, is about $1°C/100$ m. A parcel rising with this lapse rate can become buoyant only if the ambient atmosphere cools with

height more rapidly than g/c_p. It is extremely important to note that Condition 15.1 is satisfied in the atmosphere only in very limited regions and under very special circumstances.

If the parcel contains water vapor, and if it rises and cools adiabatically until the saturation vapor pressure at the parcel's temperature is reduced to the actual partial pressure of the water vapor (this point is known as the "lifting condensation level"), the criterion for static instability becomes significantly different from condition 15.1. The rate of temperature decrease with further ascent will be far smaller than the dry adiabatic rate (in the lower troposphere, about half) because the latent heat released as water vapor condenses will counteract the adiabatic cooling. This is called a "moist adiabatic" or "pseudoadiabatic" process. Clearly a parcel rising moist adiabatically is much more likely to become buoyant than is a dry adiabatic parcel. It is then clear that a *necessary* condition for static instability of moist adiabatic ascent is

$$\frac{\partial T}{\partial Z} < - \gamma_m, \qquad (15.2)$$

where $\gamma_m < g/c_p$.

Condition 15.2 is not at all uncommon in the atmosphere. Indeed, it is satisfied virtually everywhere over the tropical oceans in the lower half (by mass) of the troposphere. Such regions are known as "conditionally unstable." If air in a conditionally unstable atmosphere is forced to rise beyond its lifting condensation level (as might be the case if the horizontal motions of the atmosphere produced mass convergence close to the sea surface), the rising air may become buoyant. When this takes place, further ascent will not be in the form of a uniform large-scale (say, the scale of the horizontal convergence that forces the ascent) current. The ascending air must break up into narrow convective columns (cumulus clouds) with large intervening areas. Observations indicate that even in the inner core of hurricanes the aerial coverage of active updrafts is less than 10 percent.[11] Stated in different words, the preferred mode for condtionally unstable atmospheric perturbations is that which has the dimensions of a cumulus cloud.

The First Hurricane Models

The difficulties that arose in the early hurricane models were related to the differences between dry adiabatic and moist adiabatic motions and to the failure of the modelers to recognize that the moist adiabatic process takes place only in clouds and that active cumulus clouds occupy only a small percent of the area even in the inner core of a hurricane.

A numerical model predicts and makes use of a vertical velocity that is an average over a grid interval (about 10 km for hurricane models). In the real atmosphere, an average over such a large interval would have contributions from large cloud updrafts over a few percent of the area and perhaps gentle downdrafts over most of the area. It is, therefore, clear that the average vertical motion is a fictitious quantity since no air moves with this velocity. However, the first hurricane models assumed conditionally unstable pseudoadiabatic ascent with the mean motion. Since the preferred modes with static instability are narrow convective columns and since the smallest scale that the numerical model can "see" is that with wavelength of twice the grid interval, the results of the numerical experiments showed unchecked development of two-grid interval waves and these perturbations completely obscured the hurricane-scale motions. In essence, the modelers were attempting to obtain hurricane

formation from the release of an instability that is ressponsible only for cloud development.

Conditional Instability of the Second Kind (CISK)

Amplification of the warm-core prehurricane vortex to the hurricane stage proceeds rapidly once started. This suggests the release of some atmospheric instability. Earlier investigators thoroughly studied the applicability of known instabilities to the hurricane problem. These efforts (see summaries by Spar,[24] Alaka,[1] Kuo,[8] and Yanai,[27]) were largely unrewarding. A new instability was first suggested by Charney[5] and later elaborated upon by Charney and Eliassen[6] and Ooyama.[12, 13]

A basic feature of the hurricane, as well as the prehurricane warm-core disturbance, is enormous rainfall over that portion of the storm that shows horizontal mass convergence close to sea level. Indeed, rainfall rates in mature storms are very closely equal to the water-varpor convergence in the lowest kilometer of the atmosphere. This suggested to Charney and Eliassen[6] and to Ooyama[12] that the large-scale vortex and the cloud system act in cooperation. The clouds supply the heat energy needed to drive the vortex. The vortex, by providing low-level water-vapor convergence, organizes and maintains the cloud system. Linear analyses performed by Ooyama[12] and Charney and Eliassen[6] showed that this cooperative mechanism could lead to instability of the large-scale vortex. The instability was called "conditional instability of the second kind" by Charney and Eliassen and is known in the meteorological literature by the acronym "CISK."

CISK has been the dominant mechanism of all hurricane models that have provided realistic results. It is, therefore, important to discuss its various aspects in sufficient detail to gain a comprehensive qualitative feeling for the physical processes involved. The water-vapor budget for an atmospheric column (extending from the sea surface to the top of the atmosphere) may be written in the following schematic form:

$$\frac{\partial W}{\partial t} = \text{Inflow} + \text{Evaporation}$$

$$- \text{Precipitation,} \quad (15.3)$$

where W is the mass of vapor contained by the column, Inflow is the vertically integrated water-vapor convergence, Evaporation is the evaporation from the ocean surface on which the atmospheric column rests, and Precipitation is the mass per unit time of solid and liquid water falling out of the column and reaching the sea. We have here neglected condensation products stored in clouds. This is valid on a long enough time scale since if the liquid does not rain out, it will evaporate.

From Equation 15.3, it is clear that the atmospheric humidity would decrease in heavy rainfall if the evaporation or the inflow or both were not as large as the precipitation. This is not what is commonly observed in the atmosphere where $\partial W/\partial t$ is either positive or near zero in rainy areas. Consequently, the precipitation loss must be nearly balanced by the evaporation and inflow. Budget studies for mature and immature hurricanes[19] indicate that the inflow of water vapor is an order of magnitude greater than evaporation. It is clear, therefore, that horizontal convergence must supply the bulk of the water vapor precipitated by the cumulus convection. Furthermore, the water-vapor mixing ratio decreases almost exponentially with altitude so that

large amounts of water-vapor convergence can be achieved only through inflow at low levels.

Consideration of the cloud dynamics also indicates that the inflow must be at low levels if cumulus clouds are to be formed. We have already noted that the lower tropospheric portion of the marine tropical atmosphere is conditionally unstable. This is only a *necessary* condition for buoyancy of rising air; it is not *sufficient*. Ooyama[13] showed that in the mean tropical atmosphere, parcels must begin their ascent from the lowest kilometer of the atmosphere if they are to become buoyant. Although the vertical distributions of temperature and humidity in a tropical cyclone will not be that of the mean tropical atmosphere, it is generally the case, even in conditions of disturbed weather, that parcel ascent can become buoyant only if it begins from close to sea level.

There is still another factor that limits the origin of buoyant cloud parcels to the lowest kilometer of the tropical atmosphere. Temperatures within cumulus clouds are significantly less than those to be expected from undilute pseudoadiabatic ascent. Cumulus clouds continuously mix with the environmental air as the cloud parcels ascend; the cloud parcels entrain drier and cooler air in this mixing process. Consequently, the buoyancy of cumulus clouds is then much smaller than that of undilute parcels and significant buoyancy can be achieved only if cloud parcels originate at very low levels.

We now turn our attention to the mechanisms that can provide the low-level convergence required to supply the water vapor. Consider a cyclone that at some instant is in a steady state. In the absence of friction, the radially outward directed Coriolis and centrifugal "forces" must be balanced by the inward directed pressure-gradient force. This may be written

$$\frac{v^2}{r} + fv = \frac{1}{\rho}\frac{\partial p}{\partial r}, \qquad (15.4)$$

where v is the tangential component of wind (positive when counterclockwise), r is radius, ρ is density, p is pressure, $f = 2\Omega \sin\Psi$, Ω is the earth's angular velocity, and Ψ is the latitude. This equation is known as the "gradient-wind" equation and the value of v that can be calculated from it is known as the "gradient wind." For the hypothetical situation under consideration, the actual winds and the gradient winds are the same.

Supppose that an air parcel is displaced inward. In the absence of friction, the parcel would conserve its absolute angular momentum with respect to the vortex center. Since the absolute angular momentum may be written

$$M_a = rv + \frac{fr^2}{2}, \qquad (15.5)$$

the tangential velocity of the parcel would increase. Since M_a increases outward for virtually all atmospheric vorticities, the parcel displaced inward would arrive at smaller radii with values of v greater than the local gradient wind. The pressure-gradient force would then be too small to balance the centrifugal and Coriolis forces associated with the displaced parcel. The parcel would, therefore, be accelerated outward and would be unable to penetrate toward the vortex center.

If frictional torques were to reduce the parcel's absolute angular momentum rapidly enough so that the parcel arrived at smaller radii with tangential velocities smaller than the local gradient winds, the parcel would be accelerated inward and penetration would be possible. Since frictional forces are the largest near the air-sea interface, this mechanism seems well suited to provide the low-level inflow required to support the cumulus convection.

A second way in which the balance of forces given by the gradient-wind equation can be upset is to increase the inward pressure-gradient force. In the absence of friction, this can occur only through the action of differential (with radius) heating. However, since the condensation heating under consideration here is dependent on the existence of the low-level inflow, it cannot be the source of this inflow and, consequently, surface friction becomes the explicit mechanism for production of the low-level convergence.

A qualitative picture of CISK emerges from the previous paragraphs. Suppose a weak cyclone (in more or less gradient balance) is present over the tropical oceans. A frictionally induced inflow will then develop at low levels. In those regions where the inflow is convergent, mass will be transported upward in cumulus clouds. The latent heat released by the clouds will allow the inner portions of the vortex to warm and thus the pressure will be lowered. This increases the inward-directed pressure-gradient force and thereby increases the inflow. This in turn increases the cumulus convection and the system may intensify further. Since angular momentum is also transported inward, the tangential winds increase. In the upper troposphere, the mass that has been transported upward is evacuated by an outward branch of the transverse circulation.

Intensification of the cyclone does not proceed without limit. The temperature increases in the central portion of the vortex are large and concentrated at upper levels.[10] This concentration acts to reduce the convective instability as measured by the reduction of temperature with height. When the vertical profile of temperature becomes similar to the pseudoadiabatic profile, cumulus activity is unlikely to provide further warming and thus further reductions of pressure do not occur. The system is then likely to reach a steady mature state or even to begin a period of decay.

CISK, as outlined above, constitutes an attractive hypothesis for hurricane development. The arguments have, however, been entirely qualitative. A quantitative examination is dependent on the availability of a parameterization for the cumulus activity. It is important to note that the qualitative concept of CISK is independent of the parameterization.

QUANTITATIVE ASPECTS OF CISK: THE PROBLEM OF CUMULUS PARAMETERIZATION

Although the release of latent heat in cumulus clouds is an obvious source of energy for the larger-scale motions of the atmosphere, we have little understanding of the mechanisms by which the energy is transmitted from cloud to ambient atmosphere. Theoretical considerations by several investigators indicate that dry adiabatic subsidence between the cloud towers may be the primary mechanism for raising the macroscale temperature. Even large cumulonimbus clouds, because of entrainment, are only a few degrees warmer than their environment. Furthermore, the active clouds cover only a few percent of the rainy area. Gentle downdrafts may cover most of the region. If we use the subscript "E" for environmental values, the subscript "c" for cloud values and the symbol "λ" for the fraction of the area covered by cloud,

$$\bar{T} = \lambda T_c + (1 - \lambda) T_E. \quad (15.6)$$

Since $\lambda \ll 1$ and $T_c \approx T_E$,

$$\bar{T} \approx (1 - \lambda) T_E \approx T_E \quad (15.7)$$

and

$$\frac{\partial \bar{T}}{\partial t} \approx \frac{\partial T_E}{\partial t}. \quad (15.8)$$

With this view, Ooyama[13] argues that the bulk of the condensation heating, as a cloud parcel rises, is coverted to potential rather than internal energy and is largely exported by the upper-level outflow (this is consistent with the empirical budget study by Palmén and Riehl[14]). According to this hypothesis, it is only when parcels subside that potential energy is converted to internal energy in significant amounts.

These processes, if correct, are clearly complex with subtle interactions. If it were necessary to represent all these details of the cloud ensemble to compute the temporal variations of the macroscale temperature, the progress in modeling and understanding hurricanes that has taken place over the last 10 years could not have been achieved. Experience indiates that much can be achieved even with a rather crude parameterization based largely on intuition and empiricism.

When the hydrodynamical equations are averaged over a grid interval eddy correlation terms arise that are a result of subgrid motions. These terms are, in part, produced by cumulus convection. The cumulus parameterization problem involves replacing these portions of the eddy terms with virtual sources and sinks of heat, moisture, and momentum that can be calculated from the macroscale data. With regard to heat, the problem is frequently considered in two parts: (1) the vertically integrated heating in an atmospheric column from the sea surface to the top of the atmosphere, and (2) the distribution of this heating with height.

The first quantitative studies of CISK circumvented the second aspect through the design of models that represented the temperature at only one midtropospheric level. It was also assumed that cumulus convection occurs only in the presence of convergence in the frictional boundary layer. It was further assumed that the vertically integrated condensation rate was equal to the frictional convergence of water vapor. This assumption presumes that all water vapor entering the system ascends in clouds, condenses, and is precipitated. None of the vapor is either stored in the atmosphere or transported upward and outward. Although this is a reasonable assumption for the mature hurricane, a substantial fraction of the inflowing water vapor must be used to increase upper tropospheric humidities in the incipient storm. It is expedient to ignore this complicating aspect since the difficult task of formulating and integrating an equation for the prediction of atmospheric humidity may be avoided.

The parameterization in these first CISK models also assumed that all convective clouds reach the upper troposphere. Ooyama argues that the smaller convective elements probably contribute little to the hurricane's energetics.[13] This is probably correct for the mature storm. In the incipient stage, however, the smaller clouds provide upward transports of water vapor that may well be of importance for development of the deep moist layer[18] that seems to be necessary for the intensification of tropical cyclones.

A SURVEY OF RESULTS FROM CIRCULARLY SYMMETRIC HURRICANE MODELS

Basic Characteristics of the National Hurricane Research Laboratory Model

In the presentation of sample results, we draw primarily from the work of the National Hurricane Research Laboratory (NHRL).

The model provides for seven labels of vertical resolution as shown by Table 15.1. The grid points are thus spaced irregularly with respect to both the height coordinate and the mean mass stratification of the tropical atmosphere (last column of Table 15.1). The irregular

spacing provides better resolution (by mass) in the lower tropospheric inflow layer and in the upper tropospheric outflow layer. Coarser resolution is present at middle levels where vertical gradients in hurricane tend to be small.

Parameterization of Cumulus Convection

The cumulus parameterization is based on a theory by H. L. Kuo.[9] The heating at a given level is proportional to the difference between the temperature of a parcel undergoing pseudoadiabatic ascent and the actual macroscale temperature at that level. Kuo's (1965) technique has not been widely accepted by meteorologists. In his derivation, Kuo assumes the clouds to have the temperature and humidity of the pseudoadiabat followed by a parcel that ascends from sea level without entrainment. The clouds, according to Kuo's derivation, impart their properties to the macroscale through instantaneous horizontal mixing with the surrounding cloud free environment.

In our earlier discussions, it has been noted: (1) because of entrainment, clouds

Table 15.1 Heights and Mean Pressures of the Information Levels[a]

Level	Height (m)	Mean Pressure (mb)
1	0	1,015
2	1,054	900
3	3,187	700
4	5,898	500
5	9,697	300
6	12,423	200
7	16,621	100

[a] The mean pressures are approximate and are based on a mean hurricane season sounding.

differ in temperature from their environment by much less than does the pseudoadiabat; (2) the aerial coverage of active updrafts is only a few percent; and (3) the actual mechanism for macroscale warming may well be dry adiabatic subsidence between the cumulus towers. In view of this, it is easy to see why Kuo's technique has been subject to criticism.

Despite these criticisms, the technique gives realistic results for the hurricane problem and evidence exists that it may also be useful for other problems of the tropics. Although the physics called upon to justify the derivation are questionable, the final equations are qualitatively acceptable in that the effect of the cumulus is to drive the vertical distributions of moisture and temperature toward those appropriate to moist adiabatic parcel ascent from mean sea level. In nature, this limiting case is approached only in the inner rain areas of hurricanes[18]; however, there is ample observational evidence that cumulus convection associated with large-scale disturbances tends to produce this type of thermal structure when it is of sufficient duration. Based on observations, therefore, and for the sake of proceeding with a study of the macroscale dynamics, Kuo's technique serves as a useful interim device. Mathematical details of the NHRL model can be found in a 1970 paper.[21]

A Typical (Control) Experiment

For this experiment (S-35) the radial limit of the computational domain is 440 km. The sea temperature is 2°K greater than the initial sea-level air temperature (the latter is initially horizontally uniform for all experiments). Since the sea-level air temperature varies with time according to the thermodynamic equation, the air-sea temperature difference varies both with radius and time as the model

Table 15.2   Initial Values of Relative Humidity at the Information Levels

Level	Height (m)	Relative Humidity (%)
1	0	90
2	1,054	90
3	3,187	54
4	5,898	44
5	9,697	30
6	12,423	30
7	16,621	30

hurricane evolves. Radial resolution for the basic experiment is 10 km and the time step is 2 min.

The initial conditions represent a weak, warm-core vortex in gradient balance. The maximum wind is 7 m/sec located at a radius of approximately 250 km. The mixing ratio at the initial instant is horizontally invariant and nearly equal to that of the mean tropical atmosphere (see Table 15.2 for approximate initial values of relative humidity).

Figure 15.1 summarizes the control's life cycle in terms of central pressure and maximum wind at sea level. The initial conditions are clearly arbitrary, and therefore the early portions of the solution are not particularly significant. The rather long "organizational" period (168 hr) required for the vortex to begin intensification is easily altered by changing such arbitrary parameters as the scale and intensity, or both, of the initial vortex, the size of the computational domain, and the lateral boundary conditions. The values of some rather poorly defined physical parameters also strongly effect the organizational period.

The rapid intensification of the storm to the mature stage is fairly typical of real hurricanes. Indeed, cases exist where deepening of this magnitude has occurred

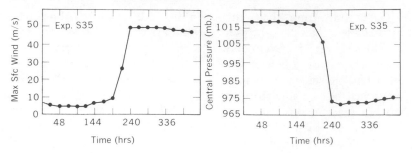

Figure 15.1 Results from control experiment (S-35). Left, maximum wind at sea level as a function of time. Right, central pressure as a function of time.

in as little as 12 to 24 hr. The long, quasi-steady, mature stage is similar to that of hurricanes that remain over the ocean without encountering cold surface waters or unfavorable surrounding flow patterns.

The organizational period for a developing model storm does not terminate until a concentrated vertical-velocity maximum (eye wall) develops at a radius smaller than that of the low tropospheric wind maximum. In the introduction, we saw that this is also the case for real storms. Once this feature appears, the model storm deepens rapidly and, at the same time, the eye wall and wind maximum migrate inward. The wind maximum, however, moves more rapidly than the eye wall. Invariably development ceases when the eye wall and wind maximum become nearly coincident.

The thermal structure evolves in such a way that lapse rates in the rain area approach that of a pseudoadiabat during the period of rapid development.[20] In the mature stage, significant conditional instability is found only in the surface friction layer. Figures 15.2 through 15.6 illustrate structural features representative of the mature stage. The strongest winds (Figure 15.2) are concentrated in a narrow zone near the storm center. The strongest vertical motions (Figure 15.3) and condensation heating, as represented by the rainfall rate (Figure

15.4), are also concentrated in an extremely narrow zone that extends from slightly outside to slightly inside the wind maximum. The radial motions (Figure 15.5) show virtually all the storm's inflow to occur in the lower kilometer of the atmosphere, whereas significant outflow is found only in the high troposphere. The transverse circulation is then characterized by inflowing air very close to the surface, ascent in the vicinity of the wind maximum, and outflow at high levels.

Close to the sea surface, the tangential wind is controlled by the opposing effects of angular momentum conservation as air spirals inward and loss of angular momentum to the ocean through surface drag. This then explains the close association of the eye wall and the wind maximum, since it is in the eye wall region that large centrifugal and Coriolis effects prevent further inward penetration of the air and hence force ascent to take place. Inflow is concentrated close to the sea surface because the rotational stability of the mature vortex does not allow inward penetration unless absolute angular momentum (following a parcel) is rapidly dissipated. Surface drag provides the only sink of angular momentum sufficient to allow significant penetration. The outflowing branch of the circulation in the high troposphere is largely controlled by conservation of absolute angular mo-

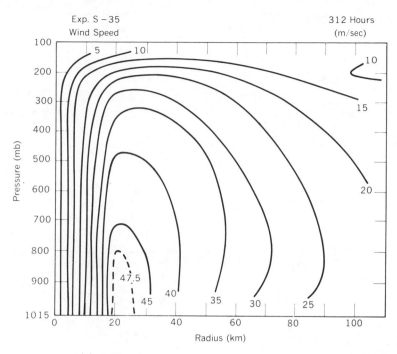

Figure 15.2 Cross-section of wind speed at 312 hr of experiment S-35. Isotachs are labeled in m/sec.

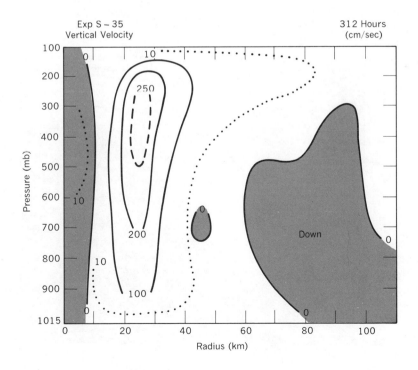

Figure 15.3 Cross-section of vertical velocity at 312 hr of experiment S-35. Isotachs are labeled in cm/sec. Hatched areas indicate negative values (subsidence).

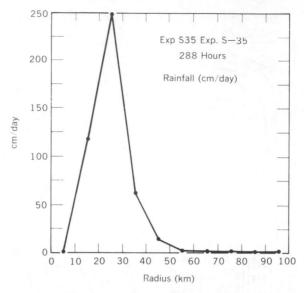

Figure 15.4 Radial profile of rainfall rates at 288 hr of experiment S-35.

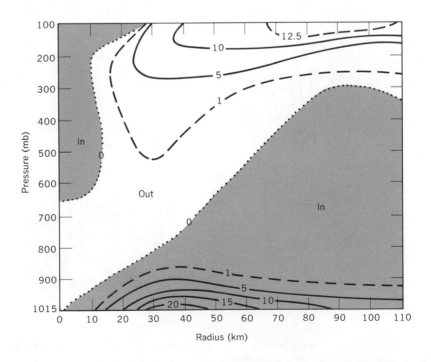

Figure 15.5 Cross-section of radial velocity at 312 hr of experiment S-35. Hatched areas depict inflow.

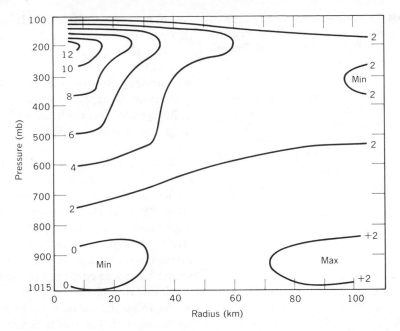

Figure 15.6 Cross-section of temperature anomalies at 312 hr of experiment S-35. Isotherms are labeled in degrees K.

mentum since frictional forces are small at these levels.

The fact that the wind maximum and the isotachs inward of the maximum (Figure 15.2) are essentially vertical is a consequence of the control of the midtropospheric wind field by vertical transports of absolute angular momentum. The model very nicely provides a subsiding eye (Figure 15.3) at the storm center.

Tables 15.3 and 15.4 make some comparisons between the energy budgets of the mature stage of the model and that of Hurricane Daisy (1958). The data for Hurricane Daisy are taken from Riehl and Malkus.[19] In view of the observational uncertainties in the empirical estimates and the natural storm to storm variability (compare estimates by Riehl and Malkus[19] with those of Hawkins and Rubsam[7]) the agreement between the model values and those for Hurricane Daisy is certainly satisfactory. Detailed energy budgets as functions of time and space for model hurricanes may be found in previous publications.[20, 21]

If the rainfall rate given by Table 15.3 is converted to a mass of condensed water per day, multiplied by L to obtain energy released per day, and divided into the rate

Table 15.3 Sensible Heat Flux and Evaporation Rates for Hurricane Daisy (1958) and Experiment S-35 at 312 Hr.

	Average Values from Model for Radial Interval 0–100 km at 312 hr	Average Values for Hurricane Daisy for Radial Interval 0–80 nmi on Aug. 27, 1958
Sensible heat flux (cal/cm²−sec)	1.5×10^3	2.9×10^{-3}
Evaporation (cm/day)	1.8	2.3
Rainfall (cm/day)	22	

Table 15.4 Kinetic Energy Generation and Dissipation by Surface Drag Friction for Hurricane Daisy (1958) and Experiment S-35 at 312 Hr.

	Integrated Values from Model for Radial Interval 0–100 km at 312 hr	Integrated Values for Hurricane Daisy for Radial Interval 0–60 nmi on Aug. 27, 1958
Kinetic energy production (10^{14} KJ/day)	5.8	4.6
Kinetic energy dissipation by surface friction (10^{14} KJ/day)	2.9	2.1

of kinetic energy production (Table 15.4), the efficiency of the mature storm is found to be about 3 percent, which is in excellent agreement with 2.7 percent estimated by Palmén and Riehl.[14] While the average rainfall rate shown by Table 15.3 is not overly impressive, we note (Figure 15.4) that the eye wall rainfall is an order of magnitude greater than the average.

Detailed examination of the kinetic energy budget (as, for example, in Rosenthal[21]), shows the model storm to be self-contained over the limit of the computational domain; at radii of 400 km and greater, the system transports small amounts of kinetic energy outward.

Sensitivity Experiments

A number of experiments, otherwise identical, were conducted with 10-km and 20-km horizontal resolution. It was found that reasonably good results could be obtained with 20-km resolution but that certain aspects of the storm structure were substantially improved by use of 10-km resolution. An experiment with 5-km resolution yielded results that differed little from the control experiment discussed above.

A 440-km computational domain has been used extensively in these experiments based on a need for computational economy. With the selection of this rather limited domain, it was realized that lateral boundary conditions would be extremely important. It was clear that the model hurricane could not be treated as a mechanically closed system since this would force the outflow air to sink relatively close to the storm center and the attendant adiabatic warming would inhibit the development of a warm-core system. Empirically, it is well known that upper tropospheric flow patterns that inhibit the escape of outflow air from the near vicinity of tropical disturbances generally are unfavorably for storm development.[18] Lateral boundary conditions of zeros for the vertical component of relative vorticity and horizontal divergence, introduced at the very beginning of our experimental program,[20] have proved satisfactory. Comparisons have been made between experiments with the original boundary conditions and experiments in which the lateral boundary conditions were approximately those of a smooth, insulated wall.

A first experiment with the closed boundary conditions, otherwise identical to the control, was carried to 504 hr and showed continuous decay from the initial state.

The computational domain was then increased to 1000 km. With open boundaries, the effect of increased domain size was merely to allow the model storm to reach its mature state about 24 hr earlier than was the case for the control. With closed boundaries, a peak intensity of 21 m/sec occurred at 384 hr followed by rapid decay. A further increase in domain size from 1000 to 2000 km produced similar, but not identical, mature storms with both sets of boundary conditions. Experiments with closed boundaries showed a marked linear increase of peak winds with domain size of about 16 m/sec per 1000 km.

Since open lateral boundary conditions are more realistic[14] and since these boundary conditions allow a computational economy in that growing disturbances may be obtained with relatively small computational domains, our discussions for the remainder of the chapter are limited to this case.

Strengthening the initial tangential wind without changing its geometry (scale) decreases the length of the organizational period and allows the storm to reach its mature stage more rapidly. Despite this, the intensity and structure of the mature stage seems to be independent of the intensity of the initial vortex.

As pointed out by Charney and Eliassen,[6] the frictional convergence of water vapor is proportional to the surface stress. As the initial wind grows, the stress also becomes greater, and con-

sequently the energy supply for convection is increased.

The overwhelming significance of this mechanism was demonstrated when in the mature stage (288 hr) of the control experiment, the transverse circulation was set to zero while the tangential wind was unaltered. Only a few hours were required to restore the transverse circulation.

Despite the importance of the initial tangential wind in determining the length of the organizational period, Anthes[3] was able to show that mature hurricanes can be generated from initial conditions that contain no organized vortex but only random velocity components.

The Effect of the Water Vapor Content of the Atmosphere

In the real atmosphere, the development of large moisture content to great heights is necessary before rapid intensification of the prehurricane vortex occurs.[18] Since the entrainment of dry air by cumulus clouds acts as a suppressor of both their vertical development and their net condensation, the role of the deep moist layer seems to be to allow the clouds to entrain relatively moist air and thus to grow and provide diabatic heating more

Table 15.5 Experiments in Which the Drag Coefficient (C_D) is Varied During the Immature Stage of the Model Cyclone[a]

Experiment	C_D	Time of Maximum Intensity (hr)	Strongest Sea-Level Wind at Maximum Intensity (m/sec)
Q20	0.2 × control	432	63
Q17	0.75 × control	264	55
Control	Control	240	50
Q3	2 × control	216	40
Q5	5 × control	192	28

[a] The initial data are at hour 168 of the control experiment.

effectively than would otherwise be possible. The cumulus parameterization employed in the model does not contain an explicit representation of entrainment. However, the sense of this process is simulated since, for a given water vapor supply, rainfall varies with atmospheric humidity.

Experiments with the initial humidity given by Table 15.2 have been compared to an experiment where the initial relative humidity was everywhere 90 percent. The latter experiment reached the mature stage about 48 hr earlier than did the former. More dramatically, at 120 hr of the control experiment (see Figure 15.1), which is just before the model storm begins to intensify, an experiment was performed where the humidity was restored to its intial value. Intensification to a mature state never took place despite the fact that the calculation was carried to 408 hr. Other experiments in which other variables (one at a time) were restored to initial values showed only minor differences from the control. It becomes clear, therefore, that the role of the model organizational period, and probably also that of the real prehurricane depression, is the development of a deep, moist core in the storm interior.

The Dual Role of the Drag Coefficient

The primary source of energy for the hurricane is derived from the frictionally induced inflow of water vapor. As already noted, this inflow develops more rapidly when surface drag is greater. Yet increased surface drag also leads to increased dissipation of kinetic energy. The linear studies, therefore, indicate that in the earlier stages of development, the frictional inflow of water vapor is the dominant effect and growth rates increase with drag coefficient. These results cannot, however, be extrapolated to the mature stage of the storm.

To study this, the surface drag coefficient (C_D) was varied during the earlier phase of the model storm. The initial data for these calculations were at hour 168 of the control. The calculations were continued from this point with modifications in C_D (see Table 15.5).

Clearly, the early portions of the experiments follow linear theory. On the other hand, peak intensity is inversely related to C_D. The table also verifies that decreased drag coefficients (experiments Q17 and Q20) lead to smaller growth rates during the early phases of the computations but ultimately to larger peak intensities.

Air-Sea Exchange of Sensible and Latent Heat

Oceanic evaporation and heat added to the atmosphere by contact with the warm ocean surface have long been considered important ingredients in the development and maintenance of tropical storms. Observations indicate that tropical storms

Table 15.6 Experiments that Examine the Relative Importance of Air-Sea Exchanges of Sensible Heat, Latent Heat, and Momentum During the Immature Stage of the Model Cyclone[a]

Experiment	Surface Drag	Sensible Heat Transfer	Evaporation
S35	Control	Control	Control
Q8	Control	None	Control
Q9	Control	Control	None
Q10	None	Control	Control
Q11	Control	Twice control	Twice control
Q18	Twice control	Twice control	Twice control

[a] The initial data are from hour 168 of experiment S-35.

form only over warm ocean waters (T_{sea} > 26°C).

Despite this, the evaporation rates are very small compared to the lateral inflows of water vapor to the hurricane. Also, the exchange of sensible heat between the hurricane and the ocean is only a few percent of the latent heat release.

Ooyama[13] found drastic reductions in the strength of his model hurricane when evaporation and sensible heat transfer were suppressed. He pointed out that at sufficiently large radii, midtropospheric air subsides into the boundary layer. (An example of this feature from our model is shown by Figure 15.3.) This subsidence tends to dry out the boundary layer air since atmospheric humidity decreases

with height. Ooyama argued that unless evaporation from the ocean can raise the humidity to sufficiently large values before the inflowing air reaches the inner region, ascending air parcels will have insufficient buoyancy to maintain the convective activity, and hence the storm will begin to weaken. Ooyama's line of reasoning can be extended to show that evaporation should be far more important than sensible heat flux. These ideas are confirmed by the series of experiments discussed below.

Experiments appropriate to the immature stage of the model storm are summarized in Table 15.6. Results are given by Figure 15.7. During the first 48 hr, the only experiments in this group that depart significantly from the control are those

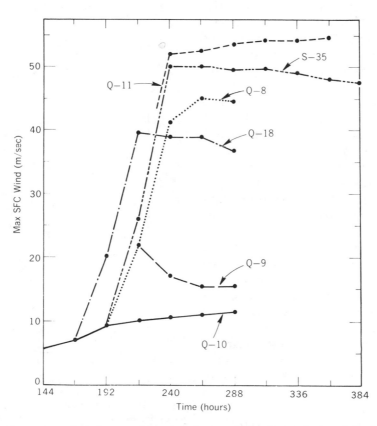

Figure 15.7 Comparison of maximum surface winds for experiments that study the relative importance of air-sea exchanges of sensible heat, latent heat, and momentum during the immature stage of the model cyclone. See Table 15.6 for details.

Table 15.7 Experiments that Examine the Relative Importance of Air-Sea Exchanges of Sensible Heat, Latent Heat, and Momentum During the Mature Stage of the Model Cyclone[a]

Experiment	Surface Drag	Sensible Heat Transfer	Evaporation
S35	Control	Control	Control
Q13	Control	None	Control
Q14	Control	Control	None
Q15	None	Control	Control
Q16	Control	Twice control	Twice control

[a] The initial data are from hour 288 of experiment S-35.

for which surface drag was altered (Q10 and Q18). This again verifies that frictional convergence is the most crucial factor during the early stages of development. Particularly it should be noted that experiment Q18 is nearly identical to Q3 (Table 15.5) in which only surface drag was doubled. In experiment Q11, where evaporation and sensible heat transfer are doubled while surface drag is left as in the control, significant departures from the control are noted only after the mature stage is reached.

Experiment Q8 (no sensible heat transfer) and Q9 (no evaporation) support the extension of Ooyama's argument offered earlier. When the sensible heat flux is suppressed, the model storm develops a peak intensity of 45 m/sec in comparison to 50 m/s for the control. *When evaporation is suppressed, the peak intensity is only 22 m/sec.*

Experiments for the mature stage are listed in Table 15.7 and results are summarized by Figure 15.8. *Suppression of the evaporation (Q14) again leads to a more dramatic response than does cutting off the sensible heat supply (Q13).* In contrast to the immature stage, a greater departure from the control occurs in Q14

(no evaporation) than in the Q15 (no drag). This is caused by the presence of a well-marked friction layer convergence in the initial data. The latter is able to maintain itself for some time after the surface drag is suppressed.

Experiment Q16, where evaporation and sensible heat transfer are doubled, shows the type of response to be expected of a mature hurricane moving over warmer waters.

Despite the qualitative similarity of the wind maxima curves for Q14 and Q15 (Figure 15.8), the structural changes of the model storms are very different. In experiment Q14, where evaporation is suppressed, winds decrease everywhere and the area covered by strong winds decreases in size. In contrast, when the

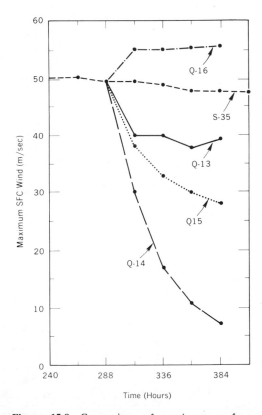

Figure 15.8 Comparison of maximum surface winds for experiments that study the relative importance of air-sea exchanges of sensible heat, latent heat, and momentum during the mature stage of the model cyclone. See Table 15.7 for details.

surface drag is suppressed (Q15), the area covered by strong winds expands as the peak wind decreases. This is because air spiraling inward does not lose momentum to the sea and hence reaches a particular radius with tangential winds greater than those found in the control. The strongest winds, however, decrease because air parcels are unable to penetrate very far into the vortex as a result of excess centrifugal forces that arise from angular momentum conservation.

Nondeveloping Experiments

In published papers, hurricane modelers tend to emphasize experiments that produces hurricanes. However, preceding sections of this chapter have shown that factors such as reduced evaporation and reduced atmospheric humidity inhibit storm development. Anthes[3] showed hurricane development to be extremely sensitive to small variations in sea temperature, vertical temperature gradient, and relative humidity. Reductions of only 2°C in sea temperature were sufficient to prevent hurricane development. Increases of upper tropospheric air temperature of only 2.5°C (thus increasing the static stability) delayed the formation of hurricane force winds while increases of 5°C prevented it altogether.

Anthes[3] found rapid storm formation occurred only with a combination of high humidity, high sea temperature, and relatively small values of static instability. Small unfavorable changes in any of these parameters were sufficient to delay or entirely inhibit hurricane development.

Computations Relevant to Hurricane Modification Through Cloud Seeding by Silver Iodide

A hypothesis by Simpson and Malkus[23] recognized that the hurricane eye wall cloud is located quite close to the region of maximum low-level pressure gradient and further contended that the wall cloud contains significant quantities of supercooled water. According to their hypothesis (Hypothesis I), if this supercooled water were frozen through nucleation by silver iodide crystals, the released heat of fusion would produce temperature increases, and therefore, hydrostatically, pressure decreases near the region of the strongest pressure gradient. If the central pressure did not concomitantly decrease, a reduction in maximum pressure gradient and, in turn, a reduction in wind speed should be the net result. Temperature increases of 2°C were estimated on the assumption of freezing 1 g/m³ of supercooled water in the layer 500 to 150 mb. With the further assumption that the 150-mb surface remains unaltered, hydrostatic arguments were invoked to estimate reductions of 10 to 15 percent in the maximum pressure gradient force.

Difficult questions can be raised concerning both the actual supercooled liquid water content of eye wall clouds and the efficiency of the seeding operation in freezing this supercooled water. The hypothesis is also questionable from other points of view. The estimates of temperature increase to be realized from the released heat of fusion assume a constant pressure process. It is, however, not unlikely that the air would follow an ice pseudoadiabat with substantial amounts of the heat being converted to potential rather than internal energy.

An even more critical consideration is the fact that the eyewall drives the storm's transverse circulation and seeding this region alone would very likely accelerate this circulation, thus providing a more rapid inflow of both angular momentum and water vapor to the eye wall region.

Hypothesis II differs from Hypothesis I in that the latter calls for seeding the eye wall alone, whereas the former suggests seeding either from the eye wall outward or entirely outward from the eye wall. In Hypothesis II, basically, one stimulates convection and ascent at radii greater than that of the eyewall. The region of stimulated convection is intended to compete with the eye wall for the inflowing air. If significant portions of the inflow can be diverted upward at the seeded radii, the angular momentum and water vapor supplies to the original eye wall and wind maximum will be reduced. As a consequence, one would expect the original wind maximum to be reduced and the eye wall convection to be diminished.

Radar data indicate that more than 50 percent of the cumuli within 30 nmi of the hurricane center have tops within the 20,000 to 30,000 ft range. Since, in most hurricanes, very little macroscale inflow or outflow occurs at middle tropospheric levels, the mass transported upward in clouds whose tops are at these levels must subside relatively close to the cloud and thus reduce the net upward mass transport. If through silver iodide seeding these clouds can be induced to grow vertically into the outflow layer, not only would additional latent heat be realized but also, the mass transported upward would be caught in the macroscale outflow and evacuated from the storm core. Consequently, Hypothesis II provides an efficient mechanism for evacuation of the air diverted from the inflow layer. Calculations with cloud models indicate that individual clouds, just beyond the eye wall, can indeed be induced to grow to outflow levels by silver iodide seeding.

Since the hurricane model contains no cloud physics, simulation of silver iodide seeding through the application of basic physical laws, as has been done in cloud models, is not yet possible. The techniques employed are therefore highly pragmatic and presuppose the occurence

of certain cloud physical processes. They have evolved from extensive discussions with those involved in the field program. Space is not sufficient for the presentation of details and the interested reader is referred to a paper written in 1971 by Rosenthal and Moss.[22] The highly idealized nature of the model, and the extremely crude techniques employed to represent silver iodide seeding, preclude direct comparisons of model calculations with particular field experiments. Calculations can be compared validly only against other calculations. Given two proposed tactics for hurricane modification, the model can provide useful information concerning the relative merits of the two tactics but it cannot predict detailed results of an actual field test.

With these limitations in mind, we can summarize the results of a large number of calculations. The computations provide no support for Hypothesis I; model storms consistently intensify with this type of modification. The simulations further indicate that of the potential tactics under Hypothesis II, a seeding operation just radially outward from the eye wall center is optimum. This is primarily reflected in the rapidity of response to the artificial heating rather than in the ultimate reduction of the maximum winds. The latter is more or less the same for all calculations that succeed in building a new eye wall at a larger radius.

The calculations indicate that seeding must be continued at least until a new eye wall is established. On the other hand, prolonged seedings at the same radii, beyond the time that the new eyewall is established, does not appear to be desirable and may well reduce the effect of the earlier seeding.

In all successful calculations, the ultimate effect was to displace the eye wall 10 km (one grid point) radially outward from the original eye wall and to reduce the wind maximum at sea level by 3 to 4 m/sec (6 to 8 percent). Variations in the radius of seeding and the rate of artificial heating altered only the time required for these changes to take place.

Response times decreased markedly with storm age. Intuitively, this appears to be related to an increase of the static stability of the control storm's eye wall as the model storm ages. In addition, aging of the control storm is accompanied by a natural tendency for eye wall rainfall to diminish and for rainfall to spread radially outward.

The increased transverse circulation that results from application of the artificial heating tends to increase winds at relatively large radii as inward transports of absolute angular momentum are increased. This is not a particularly desirable feature of a modification experiment since it may well produce adverse storm-surge effects. On the other hand, these increased winds are vital if the eye wall is to be reformed at a larger radius and the maximum wind is to be reduced. It is the increased centrifugal and Coriolis forces arising from these larger winds that prevent the inflow from reaching the original eye wall and thus forces ascent at a larger radius.[22]

Although virtually all the tactical variations of Hypothesis II ultimately resulted in similar modifications of the model hurricane, those with larger response times are clearly less likely to be effective in a real hurricane. It thus appears that field programs conducted with older storms seeded just beyond the eye wall are most amenable to beneficial modification. It also appears that the seeding operations should be massive since larger heating rates clearly lead to more rapid responses of the model storm.

Since many of the numerical experiments show temporary increases of the maximum wind during the early phases of the operation, it would appear that the policy of not seeding storms whose early landfall is predicted should be continued.

A THREE-DIMENSIONAL MODEL OF THE HURRICANE

General Features

The previous sections of this chapter provide ample evidence to support the contention that circularly symmetric models have simulated the life cycle of tropical cyclones with a large degree of realism and have yielded valuable insight into hurricane dynamics and energetics. With this background, and with ever increasing computer capability, it is reasonable to begin the study of the asymmetric features of the hurricane. Among the more notable of these are the upper tropospheric outflow layer, the rain bands, hurricane motion, and the interactions between the hurricane and nearby synoptic systems.

To incorporate all these features in a single numerical model is an extremely ambitious goal. The model developed at NHRL represents only an isolated stationary vortex and appears to be the logical first step beyond symmetric models.[2, 4] Computational economy has demanded that the model be limited to only three levels in the vertical, a coarse horizontal resolution of 30 km, and a relatively small computational domain of radius 435 km. The results, nevertheless, are encouraging and justify further effort.

Once the assumption of circular symmetry has been discarded, there is no further advantage of working with cylindrical coordinates. The horizontal mesh is a square grid with a uniform spacing of 30 km. The lateral boundary points approximate a circle, and all grid points are contained between radii of 450 and 435 km.

The parameterization of cumulus convection closely follows the scheme used in the symmetric model although some modifications are necessary because of the reduced vertical resolution. At the air-sea interface, the sensible and latent heat fluxes as well as the drag frictional effects are modeled in a fashion similar to that of the symmetric model.

Initial Conditions

Initial conditions consist of an axisymmetric vortex in gradient balance. Minimum pressure is 1011 mb, and environmental pressure on the lateral boundary is 1015 mb, which gives a maximum wind of 18 m/sec at a radius of 240 km. With these initial conditions and utilizing a time step of 45 sec, the experiment to be discussed was executed for 192 hr.

With symmetric initial and boundary conditions, the solutions to the differential equations remain symmetric for all time. However, asymmetrics in the truncation and round-off errors, as well as in the lateral boundaries produce weak asymmetries (on the order of 10^{-10} percent of the symmetric component) in the finite difference equations after the first time step. These perturbations may then grow with time and become a significant part of the total circulation. In a later section the mechanism for this growth is discussed and it is also shown that the initial form of the perturbation is unimportant in determining the final form of the asymmetries.

Some Results

An overall view of the three-dimensional time-dependent structure of a typical experiment is illustrated by Figure 15.9

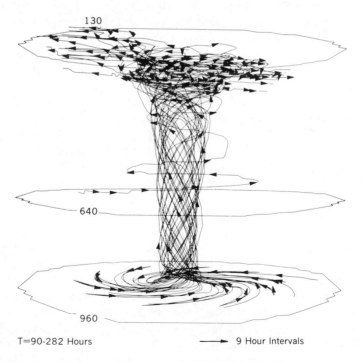

130

640

960

T=90-282 Hours   ⟶ 9 Hour Intervals

Figure 15.9 Particle trajectories calculated over a 9-day period in an experiment. The three levels are labeled in millibars (approximate). All particles start in boundary layer except one, which is started in the middle troposphere.

Streamlines and Isotachs   Level 1½   156 hours

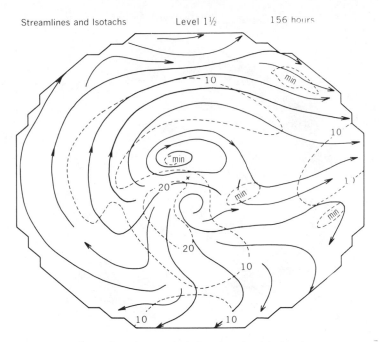

Figure 15.10   Streamline and isotach (m/sec) analysis for upper level during the mature, asymmetric stage.

where the tracks of particles released in the hurricane circulation over an 8-day period are shown. Figure 15.9 reveals a nearly symmetric boundary layer in which air accelerates as it flows inward to the center. Reaching the center, the particles are carried rapidly upward, reaching the outflow layer in about 2 hr. (Note that the macroscale vertical velocities are used to compute the vertical displacements; in reality, a particle would be carried upward in a cumulonimbus updraft in considerably less time.) After the particles reach the outflow layer, they decelerate and move outward in a highly asymmetric, unsteady flow.

Figure 15.10 shows a typical streamline and isotach pattern in the upper level during the mature stage of the experiment. Noteworthy is the anticyclonic eddy located to the "north" of the storm center. The outflow occurs mainly in two jets, in agreement with storms in nature.

Figure 15.11 shows the vertically integrated convective heat release, expressed as centimeters of rain per day.

The semicircle of rainfall rates over 200 cm/day corresponds well to the nonuniform eye wall convective region in real storms[7]; however, Figure 15.11 shows that a rainfree "eye" is not present at this time, but a region of relatively light rainfall occurs at the center of the storm. The absence of an eye is probably caused by the coarse horizontal resolution. The spiral bands, with rainfall rates averaging about 2 cm/day, are approximately 90 km wide at large distances from the center and somewhat wider closer to the center. These bands rotate cyclonically about the storm center and propagate outward at a speed of about 24 knots. Although the outer rain bands in nature apparently propagate outward, the model rate of 24 knots is probably somewhat too high.

The spiral bands in the model are undoubtedly gravity waves modified by latent heat release. However, the mechanism that excites these waves is as yet unknown. There does seem to be an interesting, although obscure, relationship

Convective Rainfall Rates 156 Hours

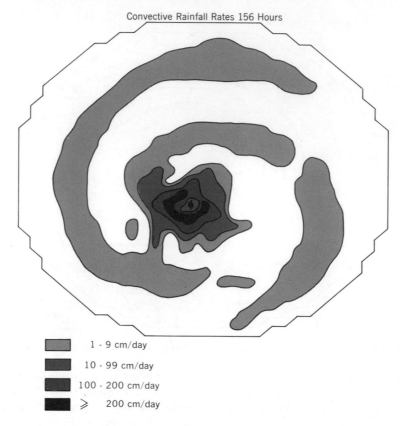

	1 - 9 cm/day
	10 - 99 cm/day
	100 - 200 cm/day
	$\geqslant$ 200 cm/day

Figure 15.11   Rainfall rates (cm/day) during mature, asymmetric stage.

between the bands (most pronounced at the top of the boundary layer) and the asymmetries in the outflow layer. In all experiments, the bands are conspicuously absent until the symmetric flow in the upper levels breaks down. Also, the number of bands (two) seems to be associated with the two jets of upper tropospheric outflow.

Development of the asymmetric features of the model storm is certainly among the most interesting aspects of the calculation. During the early rather symmetric stage, when circular variances of all dependent variables are extremely small, the little variance that does exist is concentrated in azimuthal wave number 4 (four complete harmonics in 360° of azimuth), which is attributable to the irregular shape of the lateral boundary that

has four rather sharp corners. Once the asymmetric stage has been reached, however, azimuthal wave numbers 1 and 2 account for most of the circular variance, and this is in good agreement with observations.

The appearance of these asymmetries raises questions concerning the influence of the initial perturbations, the predominance of wave numbers one and two, and the mechanisms that allow these asymmetric perturbations to grow.

As noted earlier, the initial asymmetry is introduced through truncation and round-off errors and irregular boundaries. Since it is more appealing to deliberately introduce perturbations of known amplitude and variance, the calculation was repeated with the deliberate introduction of asymmetries on the order of 10^{-1}

percent of the symmetric flow (10 orders of magnitude greater than the perturbations due to truncation errors). This was achieved by adding random numbers to the initial winds. Although the asymmetric stage was reached more rapidly than in the case discussed above, the structures of the two model storms were essentially the same. The source of the initial asymmetries appears unimportant in determining the ultimate structure of the asymmetric circulation.

The kinetic energy budget for the asymmetric component of the motion (the eddy component obtained by subtracting the circular mean from the total velocity) showed that the asymmetric (or eddy) kinetic energy was derived almost totally from the kinetic energy of the circularly averaged flow. These eddies continually extract kinetic energy from the mean circulation. The mean kinetic energy is maintained through the low-level cross-isobar flow associated with the mean transverse circulation and the upward transport of kinetic energy by the rising motion near the center of the storm.

THE FUTURE

The ultimate test of nonlinear models is an ability to make skillful predictions from initial conditions that represent an observed state of the atmosphere. Hurricane models have not reached this level of sophistication although valiant attempts have been made in this direction. The models discussed above are to be considered theoretical tools since they employ hypothetical initial data. These models represent some sort of average or typical hurricane. They are idealized to the extent that the storm is stationary and isolated from the surrounding atmosphere.

Models capable of dealing with real hurricanes are under development at a number of institutions. There is ample reason why models of this type have not been developed heretofore. As we have seen, adequate resolution of the inner core of the hurricane requires grid points spaced at about 10 km. On the other hand, if the hurricane is to move and interact with nearby atmospheric disturbances, the computational domain would have to span several thousand kilometers since atmospheric events on this scale determine the storm track.[18] Clearly to do this with grid points spaced at 10 km intervals cannot be justified from an economic point of view.

An expanding mesh, with very fine resolution near the hurricane center, and with the resolution becoming coarser with distance from the center, must be utilized; however, after some time the vortex center will move away from the high-resolution portion of the grid if the fine mesh is fixed in space. The fine mesh must, therefore, be allowed to move with the hurricane center. To achieve this, we must face major unresolved mathematical procedures for linking the coarse mesh to the moving fine mesh.

A second major deterrent to real-data hurricane models has been the lack of adequate meteorological observations over the tropical oceans. At the current time, observations of the atmosphere over the tropical oceans are completely inadequate for specification of the structure of even large-scale (horizontal dimensions on the order of 10^3 km) disturbances. To compound these difficulties, data at several altitudes with about 10-km horizontal resolution in the storm core are also required. Some data of this type are currently gathered by NOAA's research aircraft. However, the number of aircraft would have to be substantially increased to provide adequate data for numerical models if current techniques of observation are retained.

In view of the difficulties summarized in the last few paragraphs, it is easy to understand why hurricane modelers have, for the main part, restricted their efforts to the theoretical tools that have been the subject of this chapter. It now appears that the difficult problems of real-data models must be vigorously attacked before new and significant advances are achieved.

Note added in proof: This survey was written in late 1971 and early 1972. Since that time, rapid and significant progress has been achieved relative to a number of problems cited in the text. Cumulus parameterizations far more sophisticated than those available in 1971 have emerged, primarily through the efforts of A. Arakawa and M. Yanai at the University of California, Los Angeles and K. Ooyama at New York University. Professor Ooyama's theory was tested in a hurricane model by the author, and highly encouraging results were obtained.

A few interesting forecasts of real hurricanes with three dimensional models have been carried out. The active researchers in this work have included B. Miller (National Hurricane Center, NOAA, Miami), M. Mathur, and B. Ceselski (Florida State University). Experimentation with three dimensional theoretical models by R. Madala (Florida State University), Y. Kurihara (Geophysical Fluid Dynamics Laboratory, NOAA, Princeton, N.J.), and R. W. Jones (NHRL, NOAA, Miami) has provided strong background material for proceeding with variable resolution nested grids.

ACKNOWLEDGMENTS

The material of this chapter represents a synthesis of many investigator's efforts at a number of institutions in several nations of the world. The efforts at NHRL could not have proceeded without this background. The reference lists in the papers cited in the bibliography should be consulted for a fuller appreciation of those who have made important contributions. Certainly, the NHRL program was not a one- or two-man achievement. The persistent and determined efforts of Richard A. Anthes, James W. Trout, Robert W. Jones, Burt Morse, Michael Moss, and Walter J. Koss are acknowledged and appreciated by the author.

REFERENCES

1. Alaka, M. A. Instability aspects of hurricane genesis, *National Hurricane Research Project Report No. 64*, U.S. Weather Bureau, Washington, D.C., 1963, 23 pp.
2. Anthes, R. A., The development of asymmetries in a three-dimensioal numerical model of the tropical cyclone, *NOAA Tech. Memo-ERLTM-NHRL 94*, U.S. Department of Commerce, National Hurricane Research Laboratory, Miami, Fla., 1971, 55 pp.
3. Anthes, R. A., Non-developing experiments with a three-level axisymmetric hurricane model, *NOAA Tech. Memo. ERLTM-NHRL 97*, U.S. Department of Commerce, National Hurricane Research Laboratory, Miami, Fla., 1971, 18 pp.
4. Anthes, R. A., S. L. Rosenthal and J. W. Trout, Preliminary results from an asymmetric model of the tropical cyclone, *Monthly Weather Rev.*, **99**(10), 744–758, 1971.
5. Charney, J. G., On the formation of tropical depressions, *Proceedings of the American Meteorological Society Technical Conference on Hurricanes*, Miami Beach, Fla., November 19–22, 1958 (E 1-1)-(E 1-2).
6. Charney, J. G. and A. Eliassen, On the growth of the hurricane depression, *J. Atmos. Sci.* **21**(1), 68–74, 1964.
7. Hawkins, H. F., and D. T. Rubsam, Hurricane Hilda, 1964: II. Structure and budgets of the hurricane on October 1, 1964, *Monthly Weather Rev.* **96**(9), 617–636, 1968.
8. Kuo, H. L., Mechanism leading to hurricane formation, *Proceedings of the Second Technical Conference on Hurricanes, June 27–30, 1961*, Miami Beach, Fla. National Hurricane Research Project Report No. 50,

Part II, U.S. Weather Bureau, Washington, D.C. 1961, pp. 277–283.

9. Kuo, H. L., On formation and intensification of tropical cyclones through latent heat release by cumulus convection, *J. Atmos. Sci.*, **22**(1), 40–63, 1965.

10. LaSeur, N. E., The structure of hurricanes: a survey, *Geofis. Int.* **3**(3 and 4), 111–116, 1963.

11. Malkus, J. S., Recent developments in studies of penetrative convection and an application to hurricane cumulonimbi towers, *Cumulus Dynamics*, Pergamon Press, New York, 1960, pp. 65–84.

12. Ooyama, K., A dynamical model for the study of tropical cyclone development, *Geofis. Int.* **4**(2 and 4) 187–198, 1963.

13. Ooyama, K., Numerical simulation of the life-cycle of tropical cyclones, NSF Grant No. GA-623, Dept. of Meteorology and Oceanography, New York University, New York, 1967, 133 pp. Also: Numerical simulation of the life cycle of tropical cyclones, *J. Atmos. Sci.*, **26**(1), 3–40, 1969.

14. Palmén, E. H., and H. Riehl, Budget of angular momentum and enenergy in tropical cyclones, *J. Meteorol.* **14**(2), 150–159, 1957.

15. Phillips, N. A., Numerical weather prediction, *Advances in Computers,* Vol. 1, Academic Press, New York, 1960, pp. 43–86.

16. Richardson, L. F., *Weather Prediction by Numerical Process,* Cambridge University Press, London, 1922, 236 pp.

17. Richtmyer, R. D., and K. W. Morton, *Difference Methods for Initial Value Problems*, 2nd ed., Interscience, New York, 1967, 405 pp.

18. Riehl, H., *Tropical Meteorology*, McGraw-Hill, New York, 1954, 322 pp.

19. Riehl, H., and J. S. Malkus, Some aspects of Hurricane Daisy, 1958, *Tellus* **13**(2) 181–213, 1961.

20. Rosenthal, S. L., Numerical experiments with a multilevel primitive equation model designed to simulate the development of tropical cyclones: Experiment I, *ESSA Tech. Memo. ERLTM-NHRL 82,* U.S. Department of Commerce, National Hurricane Research Laboratory, Miami, Fla., 1969, 36 pp.

21. Rosenthal, S. L., A circularly symmetric primitive equation model of tropical cyclone development containing an explicit water vapor cycle, *Monthly Weather Rev.* **98**(9), 643–663, 1970.

22. Rosenthal, S. L., and M. S. Moss, Numerical experiments of relevance to Project STORMFURY, *NOAA Tech. Memo. ERLTM-NHRL 95*, U.S. Department of Commerce, National Hurricane Research Laboratory, Miami, Fla., 1971, 52 pp.

23. Simpson, R. H., and J. S. Malkus, Experiments in hurricane modification, *Sci. Am.* **211**(6), 27–37, 1964.

24. Spar, J., A survey of hurricane development, *Geofis. Int.* **4**(2 and 4), 169–178, 1963.

25. Thompson, P. D., *Numerical Weather Analysis and Prediction,* Macmillan, New York, 1961, 170 pp.

26. Yanai, M., A detailed analysis of typhoon formation, *J. Meteorol. Soc. Japan,* **39**(4), 187–214, 1961.

27. Yanai, M., Dynamical aspects of typhoon formation, *J. Meteorol. Soc. Japan* **39**(5), 282–309, 1961.

16 | Tornadoes

ROBERT DAVIES-JONES
EDWIN KESSLER

Among the smallest in horizontal extent of the atmosphere's whirling winds are tornadoes, but they are the most destructive locally. During the last 20 years in the United States, about 113 persons have been killed by tornadoes annually, and the annual property damage has averaged about $75 million. These figures compared with estimated losses caused by lightning, hail, and hurricanes are shown in Table 16.1.

Extreme variability characterizes tornadoes. Most tornado losses are associated with a few storms[21] that utterly destroy the structures in large parts of urban areas, or entire small communities. Especially when structures are poorly vented, higher pressure within may move roofs and walls outward, and the wind may carry them off as their connections weaken (see Figure 16.1). Only buildings of reinforced concrete and rigidly connected structural steel characteristically escape serious structural damage from violent tornadoes. Windows, roofs, and sidings are always vulnerable.[9, 33, 84, 115] Allen Pearson, Director of the NOAA National Severe Storms Forecast Center, has noted that 85 percent of tornado fatalities from 1960 to May, 1970 were produced

by 1 to 1½ percent of reported tornadoes. A severe tornado event leaves a community momentarily stunned and disorganized and draws a response of the magnitude demanded in war.[86]

Even in the relatively small regions of the world favorable for severe thunderstorm development, the vast majority of storms do not spawn tornadoes. Because of the complex processes involved in formation and maintenance, and their high energy density, tornadoes represent an awesome challenge to the would-be weather modifier. Our present state of knowledge allows us only a brief and speculative concluding section on modification; therefore the bulk of this chapter is devoted to reviewing the tornado field.

TORNADO CHARACTERISTICS

Tornadoes are customarily defined as violently rotating columns of air in contact with the ground; their lower portions are nearly always visible as writhing funnels pendant from cumulonimbus clouds or occasionally from shallow cloud shelfs or flanking lines of cumuli which are extensions of thunderstorm cloud systems.[6, 101] The flow within the parent cloud is totally obscured. The funnel may assume various forms, from a thin rope to a thick amorphous mass of black cloud in contact with the ground. A tornado

Robert Davies-Jones is attached to the National Severe Storms Laboratory, National Oceanic and Atmospheric Administration, Norman, Oklahoma, where Edwin Kessler is the Director.

Table 16.1 United States Losses Attributed to Some Weather Phenomena

Type of Storm	Average Annual Deaths	Average Annual Property Damage ($)
Tornado[a]	113	75 million
Lightning	150[b]	100 million[c]
Hail	—	284 million[d]
Hurricane[a]	75	500 million

Source: Kessler.[66]

[a] Based on data from the Environmental Data Service, ESSA; applicable to period 1953–1972.

[b] Estimate based on data from National Center for Health Statistics applicable to period 1959–1965.[138]

[c] Includes property damage by lightning-caused building fires, $30,600,000 in 1967, according to *Accident Facts*, National Safety Council, Chicago, Ill., 1968, 96 pp. Other property loss includes forest fires, aircraft damage, disruption of electromagnetic transmissions and casualties to livestock.[78]

[d] Estimate for period 1958–1967 provided by Stanley Chagnon, Illinois State Water Survey, Urbana, Ill. About 10% of Illinois losses represents property; the remainder is crop damage.[17]

which does not reach the ground is usually called a funnel cloud. Tornadoes are characterized by narrow cores of concentrated vorticity within a background of much weaker vorticity,* and resemble other long thin vortices in this aspect (e.g., the draining bathtub vortex). Similar atmospheric vortices originating in convection are waterspouts (commonly defined as tornadoes over water, although generally less intense) and dust devils, which respond to intense heating of dry surfaces.

* Vorticity is defined and its significance discussed in the section called Mathematical Modeling.

Although a few tornadoes destroy large areas, a typical tornado in the United States is on the ground for only one to two minutes and lightly damages an area 50 yd wide along a path toward the northeast about 1 mi long. In extreme cases, the path is 1 mi wide or 300 mi long, and the record duration is 7 hr, 20 min. Although much damage is probably caused by winds of about 125 mph, the maximum winds of intense tornadoes (never accurately measured) are probably between 175 and 250 mph. Damage has been attributed also to the sudden pressure drop accompanying tornadoes.

Often the tornado funnel is observed to skip along the ground and to move rather erratically; however, tornado movement tends to be dominated by the motion of the parent thunderstorm. Fifty-eight percent of tornadoes move from the southwest, the most frequent direction from which thunderstorms are steered by the (mid-level) wind, and only 4 percent of tornadoes have a westward component of motion.[136] The translation speed averages 35 mph but varies from practically zero with extremely sinuous paths to 70 mph.

Practically all tornadoes rotate cyclonically. Documented cases of anticyclonic ones are rare, but there is no reason to suppose that they do not occur since V. Rossow[103] found nine anticyclonic waterspouts out of 39 whose sense of rotation was determined, and J. H. Golden[51] similarly found six out of 54.

Tornadoes are observed frequently in the forward right quadrant of hurricanes as they cross the American coast. These tornadoes are usually short-lived and less severe. In 1967 Hurricane Beulah spawned 115 tornadoes in Texas. Other more severe tornado outbreaks, such as the 47 Palm Sunday tornadoes of April 11, 1965, in the Midwest, are associated with extratropical cyclones.

Figure 16.1   Effects of a tornado at Jonesboro, Arkansas, May 15, 1968. American Red Cross photo taken on May 17 by Jack Shere.

TORNADO CLIMATOLOGY

Geographical Distribution

Tornadoes occur much more frequently in the interior of the United States than in other parts of the world (Figure 16.2). Tornadoes are most common in the zone of transition between polar and tropical air masses. They are very rare poleward of 60° where the air seldom is warm and moist enough to support them. They are also uncommon in the tropics, although this area of the world has the highest thunderstorm frequency; however, the atmosphere generally does not attain the high potential instability in the tropics that it does in midlatitudes, and consequently, thunderstorms are less severe. Further, the tropical air rarely harbors

the large values of vorticity favorable for tornado formation.

Seasonal Distribution in the United States

During the cold months tornadoes occur most frequently along the Gulf Coast and southeastern states, but during the period between late winter and midsummer the location of maximum activity progresses northward through the central into the northern plain states,[118] and returns south from late summer to fall. Thus states such as Oklahoma have two maxima, but the fall maximum is much smaller than the spring one. The migration of the tornado belt can be explained by noting that tornadoes almost always occur in the warm sector of a low and in air with dew point higher than 53°F.[135] The 50°F monthly mean dew point line lies along the Gulf Coast in December but progresses northward until July, when it covers the whole of the United States east of the high plains. (The thermal contrast between the Gulf of Mexico, which is relatively cold during spring and summer, and the continental land mass plays a vital role in the northward transport of moist air during this period through a monsoon effect.) Simultaneously, the storm track moves northward. Thus tornadoes are rare in the northern states during winter, spring, and fall owing to low dew points, and in the southeastern states during summer owing to the storm tracks being too far north.

About 700 tornadoes per year are reported currently in the United States, with the highest percent (about 20 percent) in May and lowest (3 percent) in January.

Diurnal Distribution in the United States

Tornadoes occur during all hours of the day but a recent study[110] shows that the preferred hours are between 1500 and 1900 local standard time (LST), when 42 percent of all United States tornadoes take place. This afternoon maximum, due to diurnal heating, applies everywhere except near the Gulf Coast where early morning convective activity on the sea breeze front during the period April to July causes the hour of the maximum to be closer to noon. Furthermore, the Gulf Coast has a much flatter maximum with more total occurrences between 0000 and 1200 LST than elsewhere.

TORNADO PREDICTION

As used here, "prediction" or "forecast" refers to phenomena foreshadowed in advance of development, rather than to

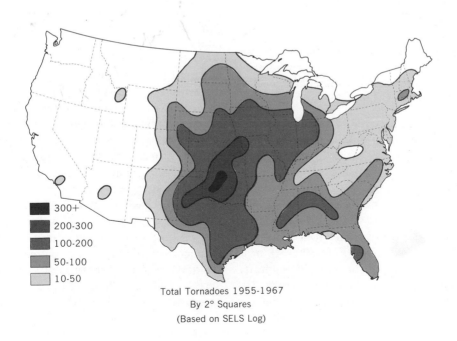

300+
200-300
100-200
50-100
10-50

Total Tornadoes 1955-1967
By 2° Squares
(Based on SELS Log)

Figure 16.2 Distribution of tornadoes in the United States (after Pautz[96]).

description of the expected behavior of entities already developed.

The correlation between thunderstorm severity and tornado occurrence forms the basis of tornado forecasts. Like other weather forecasts, these must start from a description of the present state of the atmosphere. They are less specific than we would like, because our observations are too sparse to describe atmospheric variability on the scale producing the tornado or thunderstorm phenomena. A severe thunderstorm may extend 10 to 25 mi, and exist for about 6 hr, while the distance between primary surface weather stations is about 100 mi and between upper air stations more than 200 mi. Observations are made hourly at the surface stations (more often under special conditions) but usually at only 12-hr intervals at the upper air stations. Even if our knowledge were otherwise adequate to the task, the current weather observing system limits us to indicating the probability of thunderstorms and accompanying tornadoes in regions much larger than the storms.

Forecasters look for the following conditions in forecasting severe thunderstorms and possible associated phenomena (tornadoes, hail, damaging surface winds, severe turbulence)[85, 95, 135]:

1. Deep conditional and potential instability, that is, large lapse rates of temperature and moisture through a great depth.
2. Abundant moisture in the surface layer to a depth of at least 3000 ft. The presence of a dry air mass at intermediate levels (with base at 3000 to 8000 ft), providing the potential for strong downdrafts through evaporative cooling, is also favorable although not absolutely essential.
3. The presence of a stable layer or inversion to prevent deep convection from occurring until the potential for explosive overturning is established.

4. A mechanism to remove the stable layer such as surface heating, dynamic lifting, horizontal advection, or some combination of these factors. Slow lifting ($<$ 15 cm/sec) occurs normally in regions of cyclonic flow, and more rapid lifting ($\leq$ 2 m/sec) is associated with thin zones of low-level convergence such as fronts, troughs, or dew point fronts. Warm air advection below and cooling above the inversion also tend to destabilize the atmosphere.
5. Winds that increase greatly with height with large values in narrow horizontal bands at altitudes above 20,000 ft. Such winds are associated with transition zones between cold and warm air masses, in which severe thunderstorms most frequently form.

The high tornado frequency in West Texas, Oklahoma, and Kansas can be explained by the barrier influence of the Rocky Mountains to the west, and the Gulf of Mexico, a source region for moist tropical air, to the south. The relatively shallow moist Gulf air generally intersects the eastern slopes at a height of 3000 to 5000 ft above sea level. The mountain barrier hinders displacement of the Gulf air by eastward-moving Pacific air. The part of the air mass which comes over the mountains is warmed and dried by compression and surface heating along the lee slopes, generally causing it to leave the surface and overrun the tropical air with a inversion forming at the interface of the two air masses. A common situation (see Figure 16.3) is for a cold trough aloft to approach from the west, induce cyclogenesis, and increase the southerly current of maritime tropical air over the Great Plains. The mountains constrain the air from spreading westward, and the moist surface air warms and deepens. At the same time, the air aloft is cooled by lifting and cold air advection produced by the upper

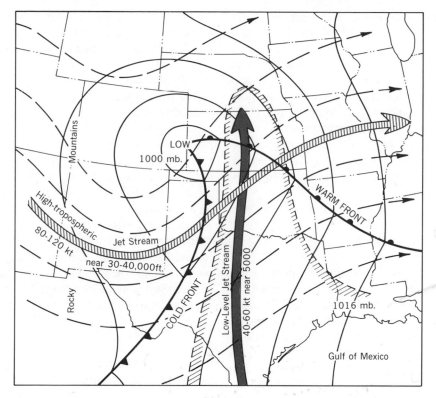

Figure 16.3 Schematic features of a severe weather outbreak. Solid lines are sea-level isobars; dashed lines are streamlines of upper tropospheric flow. Shading outlines general area of moist tongue in lower levels; this is in general associated with the region of potential instability (Reprinted by permission from Newton[91]).

trough. These processes tend to break the inversion and initiate severe convective activity.

Present tornado forecasts usually refer to developments expected to begin from 1 to 7 hr after the forecast is issued in regions of about 25,000 mi². About one-third of tornado forecasts are issued as the immediate consequence of a radar observation of an existing thunderstorm in a suspicious area. About 40 percent of affirmative predictions are correct, that is, are followed by tornadoes somewhere in the forecast box during the forecast period. The incorrect affirmative forecasts divide about evenly between cases without tornadoes and cases with tornadoes outside but near the predicted regions. Since

the climatological expectancy of tornadoes during 6 hr in a randomly selected 25,000 mi² area in eastern and central United States is only about one in 400,* it is plain that affirmative forecasts evidence considerable skill in identifying the meteorological parameters associated with development of severe storms and tornadoes.

Forecasts of severe storms and tornadoes 1 to 7 hr in advance are called "watches." Considering the wide area the forecast covers in relation to an area likely to be affected, the public is only encouraged by a "watch" to remain alert to

* The expectancy of a tornado during 6 hr in a 25,000 mi² area is about $\frac{1}{20}$ near the seasonal and geographical maxima of tornadoes.

further advisories. The forecasts are disseminated by teletype from the National Severe Storms Forecast Center in Kansas City, Missouri to local offices around the country. Occasionally, a local National Weather Service office may issue a modified local forecast that takes special account of peculiar local conditions. Since almost all the communication media subscribe to the teletype service, storm indications are announced quickly to the radio and television public.

TORNADO WARNING

As used here, a "warning" refers to advice issued on a severe phenomenon in progress. When the National Weather Service, through its own action or receipt of a report becomes aware of a tornado in existence, a warning to communities in the extrapolated path of the storm is immediately issued by teletype, and directly by radio and television. In threatened communities, the public may also be warned by actions of local authorities including the sounding of sirens. The warning of at least a few minutes thus provided is credited with halving the loss of life. The greatest loss of life is found commonly in the first community struck by a tornado; downstream locations benefit from longer warning time.

As suggested, observer reports these days are valuably augmented by radar observations. The primary radar network of the National Weather Service has stations spaced 200 to 250 mi apart. When severe storms threaten, the radar screens are monitored continuously. More intense echoes are associated with heavier precipitation and greater likelihood of hail, strong straight-line winds, and tornadoes. The echo from a tornadic storm within about 60 mi of the radar often displays a hook-shaped appendage, as in Figure 16.4. Thus the forecaster's

observation of intense radar echoes and hook-shaped echoes provides a continuous check on visual sightings and damage reports and facilitates issuance of timely warnings to communities lying in the projected path of storms.

During the last 8 years warnings of several minutes to 1 hr or more have been provided in connection with practically all major tornadoes. Usually, however, there are some endangered persons who cannot be reached by the warning system, or whose response to the warnings does not increase their safety.[35]

Effectiveness of the warnings can be judged from the fact that the average annual number of tornado deaths in the United States decreased from 215 during 1916–1958 to 113 for 1953–1972, despite increasing population.

RELEVANT ASPECTS OF THUNDERSTORM CIRCULATION

To understand the tornado, it is necessary to know the larger-scale thunderstorm circulation in which it is embedded. Observations have shown that the preferred region for tornado development is on the right* rear side of the storm, which is also a favored location for rapid convective development. Browning,[12] Newton,[91] and Fankhauser[31] summarized present knowledge of the airflow in severe thunderstorms (Figures 16.5 and 16.6). Warm, moist air flows into the storm on its right flank (usually the southeast side in the Great Plains), rises in a sloping updraft, and leaves the storm largely in the anvil. A mid-level current enters the storm from the rear, is cooled by evaporating precipitation falling out of the updraft, sinks, and undercuts the updraft. A "pseudo-cold front" forms at the edge of this rain-cooled air mass. As

* Directions apply to Northern Hemisphere. Southern Hemisphere storms are mirror images.

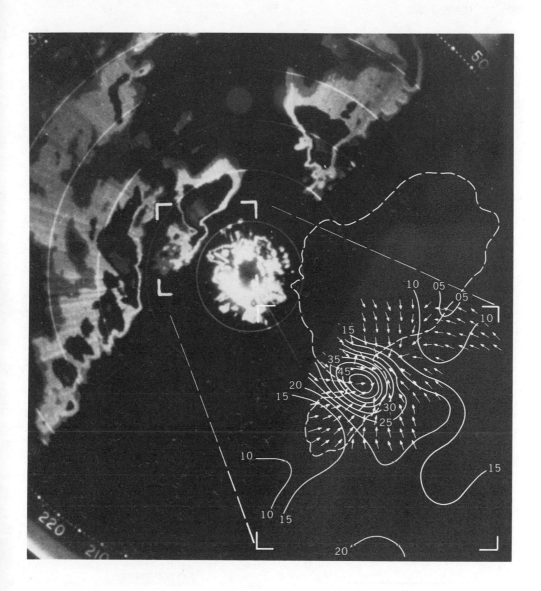

Figure 16.4 10-cm radar display of tornadic storms recorded at Norman, Oklahoma, on April 30, 1970. Range marks denote intervals of 20 nmi. North is toward the top; the radar is located at the center of the range circles. Wind speed in knots and wind direction as recorded by surface network stations at 5-mi intervals are shown in the inset, lower right. A tornado was on the ground near the forward edge of the hook-shaped appendage, close to the center of convergence of the wind field (reprinted by permission, from Kessler[67]).

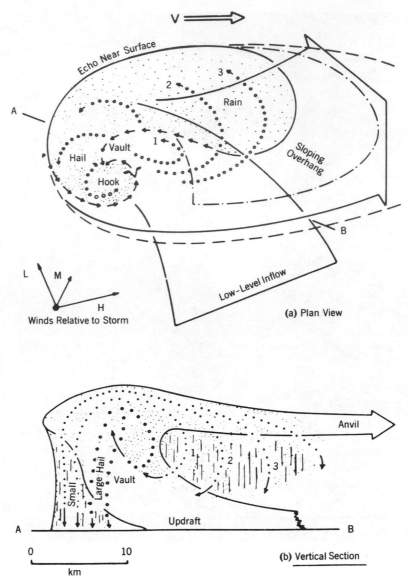

Figure 16.5 Top, plan view of a steady-state severe storm moving with velocity V; bottom, vertical section along line AB. In the plan view, the stippling represents precipitation of varying intensity which reaches the ground, with hail in the denser area. Dash-dotted lines indicate the extent of echo in middle and high levels of the storm. Small circles show precipitation trajectories. Tornadoes, if any, tend to occur near flying V. Broad arrows indicate general inflow in updraft and outflow in anvil. Arrows in plan view show motions of small proturbances seen on edges of low-level radar echoes (after Browning[12]).

it advances, the front lifts the moist potentially unstable air ahead of it, and thereby continuously regenerates the updraft. Probably the passage of such a front is familiar. It is characterized by drastic cooling, a drop in humidity, an abrupt pressure rise and strong wind gusts. Observations by Ward,[124] Ward and Arnett,[131] and an experiment by Ward,[125] have shown a tendency for tornadoes to occur underneath a convective cell along the cyclonic shear

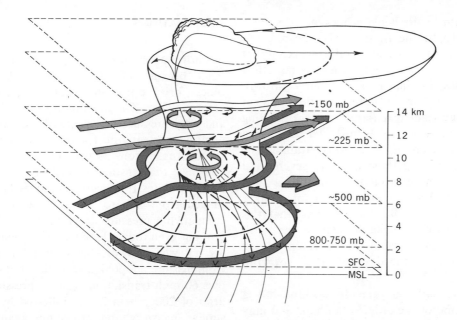

Figure 16.6 Three-dimensional interpretation of the interacting external and internal airflow associated with an individual persistent Great Plains cumulonimbus. The thin, solid inflowing and ascending streamlines represent the history of moist air originating in the subcloud layer (surface to ~750 mb). The heavy dashed streamlines trace the entry and descent of potentially cold and dry middle-level (700- to 400-mb) air feeding downrushing and diverging downdraft. The surface boundary between the inflow and downdraft is shown as a barbed band. The internal circular bands signify net updraft rotation. The shape and orientation of the dividing external bands represent typical vertical shear and character of ambient relative horizontal airflow at middle (~500 mb) and upper (~225 mb) levels. The approximate pressure-height relationship is shown on the left forward corner of the perspective box. The broad flat arrow on the right represents direction of travel (after Fankhauser[31]).

line between inflowing warm moist air from the south and the parent storm's rain-cooled outflow from the northwest. Likewise neighboring heavy rain showers are important in the production of waterspouts, since they, too, tend to form in strong updrafts just ahead of the leading edge of rain-cooled air.[51]

There appear to be two types of tornadoes. The first type form under the flanking line of new cells which continuously develop on the right rear flank of severe thunderstorms and probably mark the pseudo-cold front. In this location F. C. Bates has seen a line of dust whirls with very tenuous connections to the base of the cumuli.[6] Several vortices abort, but one or two often develop into tornadoes. At the time the vortex forms, the cloud tops overhead may be only 12,000 ft, but rapid development presumably ensues. The tornadoes can be a considerable distance (20 km) from the rain shaft of the parent storm. The second type form underneath rotating cumulonimbi (tornado cyclones) and are generally more severe tornadoes. (Perhaps the first type sometimes evolves into the second.) A pendant or hook which forms on the right rear of an echo[38] (see Figure 16.4) identifies tornado cyclones on radar. The hook is composed of precipitation drawn into a cyclonic spiral by the winds. The associated notch in the echo is caused by precipitation free, warm, moist air flowing into the storm.[12]

Storms with hook echoes move significantly to the right of the mean winds in the troposphere, perhaps because of the Magnus effect[38] (this same effect causes a spinning ball to curve through the air) or to continual development on the right rear flank (where a flanking line of new cells forms) and dissipation at the front left.[92] Storms attaining the right moving stage are very severe, and assume a supercell structure rather than the multicell configuration of weaker storms.[12] N. B. Ward showed that this is caused by enhancement of the vertical mass transport as a result of the dynamic stability of the rotating updraft.[127]

Typically, a tornado cyclone has a lifetime of the order of an hour and may spawn several tornadoes, which tend to dissipate after about half a revolution of the tornado cyclone.[37] More than one tornado may be on the ground simultaneously. Case studies of individual tornado cyclones have been made by Fujita,[36] Ward,[126] and Koscielski,[70] among others. Often the tornado cyclones are recognized visually by a cylindrical rotating collar cloud, up to 3 mi in diameter, hanging down below a bell-shaped cumulonimbus (itself rotating) with spiral streaks on its base. In several cases, an eye has been observed in the center, indicating descending motion along the axis of the mother cyclone (this is clearly shown in Kelly Foss's film of the Borger, Texas tornado cyclone of June 13, 1971). Sometimes a low tail cloud extends outward from the collar cloud and rotates around it as it is drawn inward. A new tail cloud then begins to form. Very intense vertical motion is observed near the collar cloud.

To summarize, tornadoes are likely to occur in intense convective regions where the mesoscale circulation shows signs of significant rotation (radar hooks, right moving storms) or strong convergence (merging radar echoes).

OBSERVATIONS

Scientific observation of tornadoes is difficult because of their random occurrence, brief duration, small size, and great violence. The maximum wind speed and total pressure deficit are of great importance to engineers designing vital structures, such as nuclear reactor housings, to withstand tornado passages. Past estimates for severe tornadoes have ranged from 200 to 500 mph with accompanying pressure drops of 0.1 to 0.5 atm but even today these vital parameters are not well known. The Atomic Energy Commission adopted 300 mph rotation plus 60 mph translation, a linear pressure drop of 200 mb in 2 sec, followed by a similar recovery and protection against missiles for its reactor safety criterion,[82] but some current thinking favors lowering these values considerably, especially in low-risk areas.[102]

Tornadoes are characterized by a practically inviscid outer region, where converging fluid rotates increasingly faster by conservation of angular momentum, and a narrow inner core region, where tangential winds fall off to zero at the center (as with the rotation of a solid). A first approximation to the dependence of tangential velocity. v on radial distance, r, is given by the Rankine combined vortex (which does not allow for axial flow or axial gradients)

$$v = v_{max} \frac{r}{r_c} \qquad 0 \le r \le r_c$$

$$v = v_{max} \frac{r_c}{r} \qquad r \le r_c,$$

(16.1)

where v_{max} is the maximum v, and r_c is the core radius. The cyclostrophic relationship, which states that the inward radial pressure gradient force and outward centrifugal force balance one another, is accepted generally as a good approximation outside the ground boundary layer. Thus

$$\frac{\rho v^2}{r} = + \frac{\partial p}{\partial r}, \qquad (16.2)$$

where ρ and p are air density and pressure. Assuming that ρ is horizontally constant (which is not strictly true) and the velocity distribution is as given in Equation 16.1, we may integrate Equation 16.2 to obtain the total pressure drop

$$-\Delta p \sim \rho v_{max}^2. \qquad (16.3)$$

A knowledge of either maximum tangential velocity above the boundary layer or total pressure deficit* allows the other one to be crudely estimated. Tables 16.2 and 16.3 summarize estimates of tornadic wind speeds and pressure drops from the following methods.

Damage Surveys

Engineering analyses of damaged structures are one source of tornadic wind estimates; however, observed damage patterns, open to different interpretations, depend not only on tornado intensity but on many factors such as construction quality, shape, and initial state of each structure, the sequence of failure, localized torques and stresses caused by rapidly changing pressure patterns, the rate at which air leaks out to compensate for decreasing atmospheric pressure, and the density and composition of flying debris. For example, a well constructed house with all windows and doors tightly closed except, say, one open window on the upwind side (maybe broken by a piece of debris) is likely to explode in tornadic winds because of ram pressure effect, while a similar house with lee windows or vents open may suffer only light damage.

---

* Since pressure is almost constant across the boundary layer, Δp can be measured at the surface.

The Lubbock tornado of May 11, 1970 received much attention because of the immense destruction it caused over a wide path. T. Fujita[40] assumed from the disorganized damage pattern that the devastating destruction was caused by smaller "suction spots" rotating around the center of the tornado system. He has since identified the "suction spots" as multiple vortices after observing a system of multiple dust devils.[41] N. B. Ward[129] also made this identification based on a laboratory experiment in which multiple vortices were produced in a single convergence zone. Fujita found the storm's translation speed was 21 mph, and the velocity of "suction spots" around the center was $290/n$ where n was the number of spots. The maximum speed is found by adding these two velocities together with a third velocity, the maximum tangential velocity around each spot's center. He estimated winds over 260 mph in one location.

K. C. Mehta, J. R. McDonald, J. E. Minor, and A. J. Sanger,[83] who also made a study of the damage, disagree with Fujita. They conclude that the observed damage pattern did not require the postulation of "suction spots" and that no evidence of wind velocities over 200 mph at ground level was found. They state that most of the structural damage was caused by winds in the range of 75 to 125 mph. A 13-ton cylindrical fertilizer tank, originally resting on saddle supports, was found half a mile away. The wind speed necessary to slide the tank off its supports was computed to be 163 mph.*

Direct Passages Over Instruments

Anemometers are often blown away in tornadic conditions; in cases where they

---

* According to K. C. Mehta in private correspondence in 1972.

Table 16.2 Maximum Wind Speed Estimates in Tornadoes

Tornado	Method	Wind Speed Estimate (mph)	Rad (m)	Circulation ×10³ (m²/sec)	Remarks	Reference
Cleveland, 1953	Anemometer			47	Closest distance 700 ft	Lewis and Perkins[76]
Worcester, Mass., 1953	Damage	170			Transmission tower	Booker[8]
Sydney, Neb., 1955	Ground marks	484[a,b]		10	Overestimate	van Tassel[119]
Scottsbluff, Neb., 1955	Movie				At 200 ft from center	Hoecker[61]
Bell Co., Tex., 1956	Funnel shape	230[a]	61	39		Glaser[48]
Texas City, Tex., 1957	Damage	285			Water tank	Bigler[7]
Dallas, 1957	Damage	302			Signboard—doubtful	Segner[106]
Dallas, 1957	Damage	217			Railroad freight car	Segner[106]
Dallas, 1957	Movie	170[a]	40	19		Hoecker[62]
Dallas, 1957	Funnel	209[a]				Dergarabedian et al.[25]
Fargo, N.D., 1957	Funnel shape	230[a]				Fujita[36]
Fargo, N.D., 1957	Movie	112[a]		35	At 110 m from axis	Fujita[36]
Newton, Kansas, 1962	Estimate from pressure trace	126		600	Tornado cyclone, 1-mi radius	Ward[126]
Palm Sunday, 1965	Anemometer	151			Peak gust	Fujita et al.[43]
Palm Sunday, 1965	Ground marks	180[b]		23	Tangential velocity 118 mph at 70 m	Fujita et al.[43]
Westcliff, Colo., 1965	Funnel	173[a]	18	9		Dergarabedian[23,24]
Bainsville, Ont., 1965	Funnel	93[a]				Dergarabedian et al.[23,24]
Belvidere, Ill., 1967	Ground marks	210[b]				Fujita et al.[42]
Madera, Calif., 1967	Funnel	102[a]	15	4		Dergarabedian et al.[23,24]
Waterspout, 1967	Movie	144[a]	12	5		Golden[50]
Waterspout, 1967	Funnel	180[a]	12	6		Dergarabedian et al.[23,24]
Charles City, Iowa, 1968	Ground marks	458[a,b]			Overestimate	Waite and Lamoureux[123]
Lubbock, Texas, 1970	Damage	163			Displaced cylindrical tank	Mehta (1972, unpublished)

[a] Does not include translation speed.
[b] Does not include tangential velocity around axis of suction spot.

Table 16.3 Measured Pressure Deficits in Tornadoes[a,b]

Tornado	Pressure Drop (mb)	Distance from Center[c] (m)	Remarks	Reference
Little Rock, 1894	13			Harkness[57]
St. Louis, 1896	82	Very near center	Citizen's barometer reading	Frankenfield[34]
St. Louis, 1896	10	800	Weather Bureau	Frankenfield[34]
Minneapolis, 1904	192		Unofficial citizen's barometer reading (dubious)	Outram[94]
Minneapolis, 1904	19		Weather Bureau	Outram[94]
Sydney, Neb., 1951	16	Fringe of funnel		Harrison[58]
Minneapolis, 1951	14	Fringe of funnel		Harrison[58]
Dyersburg, Tenn., 1952	22	38		Carr[16]
Cleveland, 1953	8	213		Lewis and Perkins[76]
Fargo, N.D., 1957	12		Near center of tornado cyclone	Fujita[36]
Waterspout, 1958	21	Very near center	Waterspout passed over ship	Chollet[18]
Austin, Tex., 1959	5			Wagner[122]
Newton, Kansas, 1962	34	214	Near center of tornado cyclone	Ward[126]
Topeka, Kansas, 1966	21		290 ft from left or NW edge of extreme wind streak	Galway[46]
Oklahoma City, 1970	10	400–800	NSSL mesonetwork station	
Lubbock, Tex., 1970	12	Near center	Not in suction spot	Fujita[40]
Springfield, Mo., 1971	12	Near center		Pearson and Wilson[98]

[a] The effects of instrumental damping on these measurements are unknown.

[b] Tepper and Eggert[115] report additional measured pressure deficits of 6 to 12 mb within ½ mi of tornadoes.

[c] Center here refers to center of extreme damage, which may not coincide with center of vortex.

survive there is always the question of whether they received the storm's full force. The highest value ever recorded was 151 mph at Tecumseh, Michigan, along the southern edge of a 3-mi wide damage path during one of the Palm Sunday tornadoes (April 11, 1965).[43]

The Weather Bureau Office, lying near or in the path of the Lubbock tornado, recorded a wind speed of only 90 mph and a pressure drop of 12 mb. However, the extremes were undoubtedly much higher.

The Newton, Kansas tornado cyclone of May 24, 1962 passed directly over a National Severe Storms Project microbarograph which recorded a pressure drop of 34 mb.[126]

Aneroid barometer falls of 82 mb, during the 1896 St. Louis tornado and of 192 mb during the 1904 Minneapolis tornado were read by citizens as the storms passed over. Equation 16.3 relates these pressure drops to maximum wind velocities of 193 and 296 mph, respectively, assuming that the barometers were in the core of the vortex at the time. It is difficult to assess the validity of these readings, considering they were taken under great stress. The damage associated with the Minneapolis tornado was not apparently severe enough to substantiate the reported pressure deficit.[94]

During the past 9 years, the National Severe Storms Laboratory (NSSL) has maintained a network of 30 to 60 conventionally equipped surface stations in areas of 1500 to 7000 mi² where tornadoes are expected at the rate of about 1/1000 mi² per spring season.[65] Only two stations, however, have been strongly affected by a tornado vortex since 1965. One of the April 30, 1970, Oklahoma City tornadoes passed within ½ mi of a station. A wind gust of 97 knots and pressure drop of 10 mb were recorded.

Photographic Methods

The tornado funnel varies considerably in size and shape. In some photographs it appears smooth (Figure 16.7a) suggesting weak turbulence; in others it appears strongly turbulent (Figure 16.7b). Since waterspouts usually are cylindrical and smooth walled (Figure 16.7c), we are led to search for significant variability in surface roughness or in atmospheric conditions to account for the range of turbulence and shapes.

A. H. Glaser[48] and Dergarabedian and Fendell[23, 24, 25] have shown how the maximum tangential wind speed at a given instant can be estimated from a still photograph of the funnel if certain gross assumptions are made. These include a subcloud layer so well mixed that the mixing ratio is constant and the lapse rate adiabatic, and a flow that is steady, axisymmetric, hydrostatic, cyclostrophic, and adiabatic. Mathematically, we have

$$\frac{\partial p}{\partial z} = -\rho g \qquad (16.4)$$

$$\frac{\partial p}{\partial r} = \frac{\rho v^2}{r}, \qquad (16.5)$$

where the symols have their conventional meanings. Under the above conditions the funnel cloud boundary is an isobaric (constant pressure) surface and is given by the equation

$$\left(\frac{\partial z}{\partial r}\right)_p = \frac{-(\partial p/\partial r)_z}{(\partial p/\partial z)_r} = \frac{1}{g}\frac{v^2}{r}, \qquad (16.6)$$

and hence by

$$z = h - \frac{1}{g}\int_r^\infty \frac{v^2(r_1)}{r_1}\,dr_1, \qquad (16.7)$$

where h is the height of the ambient cloud deck above the surface if variations of v with z are neglected. By further assuming

Figure 16.7a Smooth-walled tornado cloud at Enid, Oklahoma, on June 5, 1966 (ESSA photo by Leo Ainsworth).

Figure 16.7b Turbulent double tornado on April 11, 1965 at Elkhart, Indiana (photo by Paul Huffman).

Figure 16.7c Waterspout over the Florida Keys on September 10, 1969 (ESSA photo by Joseph Golden).

the tangential wind distribution of Equation 16.1, we obtain

$$z = h - \frac{v_{\max}^2}{2g} \frac{r_c^2}{r^2} \qquad \text{for } r > r_c$$

$$(16.8)$$

$$= h - \frac{v_{\max}^2}{2g}\left(2 - \frac{r^2}{r_c^2}\right) \quad \text{for } r < r_c.$$

Thus for cases where the tip of the condensation funnel is not obscured by dust, and is above the ground at height z_o,

$$v_{\max} = \sqrt{g(h - z_0)}. \qquad (16.9)$$

When the condensation funnel is on the ground $v_{\max}$ and r_c can be obtained by matching the theoretical and observed funnels. Glaser also showed that the same results apply for funnels inclined to the vertical if vertical heights and horizontal radii are used rather than slant values.

The main criticism of this method lies in its approximations. Tornadoes do not form under hydrostatic conditions; however, it is claimed that neglecting radial and vertical components of motion leads only to a moderate overestimation of tangential velocity. Approximating the swirling flow by a combined Rankine vortex may be quite poor in view of B. Morton's conclusion from a mathematical scale analysis that v must be a function of height.[88] Undoubtedly the flow is asymmetric and nonsteady. Condensation pressure variations may result from non-adiabatic processes, such as surface heating and evaporative cooling or from mixing ratio variations below cloud base. Despite these difficulties, this method agrees reasonably well with other estimates of the maximum wind speed in the Dallas tornado (see Table 16.2).

The velocity field of a tornado can be deduced from careful analysis of suitable movies. For example, W. H. Hoecker[62] derived tangential and vertical velocities for the 1957 Dallas tornado by timing the motion of cloud wisps, dust, and debris.

He allowed for the vertical fall speed of solid debris, but not for the centrifuge effect. Hoecker found a maximum tangential velocity of 170 mph at a height of 70 m and radius 40 m.

Proximity Soundings

D. K. Lilly[79] and Edwin Kessler[66] have discussed estimation of surface pressure drops and tornadic wind speeds from atmospheric soundings representative of the tornado air mass. The following derivation shows in what sense hydrostatic balance applies.

Along the axis of a vertical, steady, axisymmetric updraft the vertical equation of motion is (neglecting eddy mixing)

$$\frac{\partial}{\partial z}\left(\frac{w^2}{2}\right) = -\frac{1}{\rho}\frac{\partial p}{\partial z} - g. \quad (16.10)$$

Assume an environment in hydrostatic equilibrium, that is,

$$0 = \frac{1}{\rho_e}\frac{\partial p_e}{\partial z} - g, \qquad (16.11)$$

and let $\hat{T}$, $\hat{p}$, and $\hat{\rho}$ be relatively small axial departures of T, p, and ρ from the environmental values. Then Equation 16.10 may be written

$$\frac{\partial}{\partial z}\left(\frac{w^2}{2}\right) = -\frac{1}{\rho_e}\frac{\partial \hat{p}}{\partial z} - \frac{\hat{\rho}}{\rho_e} g. \quad (16.12)$$

By using Equation 16.11 and the equation of state for moist air,

$$p = \rho R T_v, \qquad (16.13)$$

(where T_v is virtual temperature) and integrating from the top of the ground boundary layer ($z = 0$) to some height H, we obtain

$$\left[-\frac{\hat{p}}{p_e}\right]_H^0 = \frac{gH\bar{\hat{T}}_v}{R\bar{T}_{ve}^2}$$

$$- \frac{1}{2R\bar{T}_{ve}}[w^2(H) - w^2(0)], \quad (16.14)$$

where the overbars refer to pressure weighted mean values between $z = 0$ and H. The last term on the right represents dynamic pressure. Only with this term negligible can we obtain surface pressure by integrating the hydrostatic equation between $z = 0$ and H. We can let H be the top of the updraft so that $w(H) = 0$. It is then assumed that the vertical velocity at the top of the boundary layer is not large enough for the $w(0)$ term to contribute significantly to the left side of Equation 16.4.

First consider a nonrotating updraft forming in a conditionally unstable atmosphere. In the lowest levels the air may be negatively buoyant and forced to ascend by low-level convergence, or other means. However, at a certain height this air becomes sufficiently cooled by expansion for condensation to occur, and the rising air rapidly becomes buoyant through latent heat release. (Note, however, that the heat of condensation is not sufficient to prevent net cooling during ascent.) Assuming no mixing with environmental air (believed to be true at least in the lower halves of the cores of strong updrafts), the updraft extends into the lower stratosphere where the environment is isothermal with height and soon becomes negatively buoyant. Because of its inertia, it overshoots its equilibrium level by such a distance that the updraft column and a similar ambient one weigh the same, before falling back down. The result is zero hydrostatic pressure deficit at the ground. Observed pressure deficits of 1 to 2 mb or less underneath nonrotating updrafts support this conclusion.

What then is the fundamental difference between rotating and nonrotating updrafts? Lilly suggests that rotation is very important in determining the shape of the outflow streamlines at the top of the updraft. In a nonrotating convective cell the overshooting of the updraft is necessary to produce sufficient pressure excess in the upper layers to turn the updraft aside and produce an outflow anvil. According to Lilly, air in a rotating updraft proceeds outward and approaches the equilibrium level asymptotically without overshooting. The excess pressure requirement is eliminated since the centrifugal acceleration is sufficient to produce outflow as soon as the radial pressure gradient is relaxed.

This being the case Equation 16.14 estimates the surface pressure drop if H is the height of the top of the updraft. In the nonrotating case $\bar{T}_v$ and hence $\hat{p}$ at $z = 0$ is zero due to the overshoot. Proximity soundings indicate that for the Enid tornado (and others in the same squall line) $h \sim 13$ km, $\bar{T}_{v_e} \sim 260°C$, and $\bar{T}_v \sim 7°C$ for pseudo-adiabatic ascent along the axis. From Equation 16.14 the surface pressure drop is 45 mb (note there should be no pressure differential at the equilibrium level as the fluid above is undisturbed). According to Equation 16.3, the equivalent maximum wind speed is 145 mph.

Although the primary vertical motion must be upward there may be a secondary flow in the core which is down close to and along the axis and up near the core walls. If this is so, the above estimate of central pressure deficit is an underestimate since the descending air becomes hot because of dry adiabatic compression. The maximum pressure drop that can be supported thermodynamically is obtained by assuming a forced, energy-consuming, axial descent from the equilibrium level all the way to the ground. For the Enid tornado we then have $\bar{T}_v \sim 23°C$, $\Delta p \sim 151$ mb, and $v_{max} \sim 262$ mph. The central temperature at the ground would be 140°F. These extreme values probably never occur due to mixing in the core and the possibility that the descending air originates from the high troposphere rather than the low stratosphere.

Finally, we note another effect of rotation on an updraft. Suppose that an updraft is tilted significantly from the vertical. Even in the absence of an overshoot, we still would not expect much of a surface hydrostatic pressure deficit since the updraft air near the top does not overlay the base of the column. However, a vortex column acts as if it has "dynamic walls" because of its great stability against radial displacements, and so the hydrostatic pressure deficit (which depends on vertical height, not slant height) is transmitted along the column (in the same fashion as in a tilted mercury barometer).

Ground Marks

Fujita, Bradbury, and Black[42] identified six characteristic marks or debris patterns left by tornadoes crossing open fields. The cycloidal suction marks (see Figure 16.8) are the most interesting and useful. Ground inspections reveal that they consist of short pieces of corn stubble laid in cycloidal rows about 6 in. high and 5 ft wide.[42, 123] They are indicative of local "suction spots" of convergence rotating around the axis of the vortex system, because a combination of straight line and circular motion yields a cycloidal path. The suction spots are thought to be multiple vortices in a single convergence area.

The locus of a spot depends on the translation velocity, the radial distance between spot and center of the system, and the tangential velocity of spot around center. The first two are generally known from radar data and the observed size of the cycloidal loops. The tangential velocity of the spot, which depends only on

Figure 16.8   Tornado path with suction spots (After Fujita[39]).

the shape of the cycloid and the translation velocity, can thus be computed. However, it is difficult to follow single spot paths because of erasure of part of the tracks by other vortices; in many cases the actual number of spots is not clearly resolvable.

For one of the Palm Sunday tornadoes, spot tangential velocities of 120 mph were estimated; also similarly obtained was a 160 mph estimate for a 1967 Illinois tornado.[42] Van Tassel[119] had earlier assumed that the cycloidal marks were scratch marks made by a solid object being carried around the funnel and obtained a tangential velocity of 484 mph.

According to Fujita this figure should be divided by n, the number of suction spots. Similarly Waite and Lamoureux[123] computed $459/n$ mph for the 1968 Charles City, Iowa, tornado. Fujita[40] inferred $290/n$ mph from his damage survey of the Lubbock tornado.

Estimated distances of the suction vortices from the system center range from 50 m measured from the field markings to 300 m in the Lubbock damage pattern. These agree well with filmed observations of a Fargo tornado, which showed seven cloud pendants appearing at 100 m from the center of a sheared-off funnel whose base was just off the ground.[36] According to Fujita the core diameters of suction vortices are less than 20 m, and three to five suction vortices are often present simultaneously.

Electrical Activity

Generally sferics and visible electrical phenomena accompany tornadic storms, but the dynamical role of electromagnetic phenomena in tornadoes and the uniqueness of tornadoes' electromagnetic signature remain uncertain, even controversial. J. P. Finley's report[32] on 600 tornadoes in 1882 indicated variability of tornado electricity. He listed thunder and lightning observations in 425 associated rainstorms. In 17 cases, luminosity of an apparently electrical origin was noted in the tornado funnel itself, while in 49 cases the absence of electrical indication in the cloud was specifically reported. Interest in electrical theories rose when H. L. Jones reported unusual 160 kHz radiation from a tornadic storm.[64] Vonnegut[120] presented an electrical theory of tornadoes and Brook[10] reported a magnetic variation observed during touchdown of a tornado near Tulsa. Weller and Waite[133] proposed that tornadoes are associated with intense electromagnetic radiation at television frequencies. On the other hand, Gunn,[53] measuring the electrical activity of the tornadic storm that devastated Udall, Kansas on May 25, 1955, found it, "more or less typical of exceptionally active storms." Wilkins' experiments[134] cast doubt on a fundamental dynamical role for electrical phenomena in tornadoes, and Rossow,[103] measuring magnetic fields over numerous waterspouts, found little disturbance. Kinzer and Morgan[68] locating sferics sources in an Oklahoma tornadic storm on June 10, 1967 reported no obvious close connection between lightning areas and tornado locations.

Recently Taylor[114] tentatively concluded from results during one Oklahoma storm season that the best electromagnetic indicator of tornadic activity within 30 to 50 km is the number of bursts per unit time at frequencies in the range 1 to 100 mHz. Apparently a definite correlation exists between high burst rates and nearby tornadic storms, although a few tornadic storms do not have enhanced activity and a few non-tornadic storms have above-average burst rates. To attempt to verify this preliminary conclusion "black boxes" containing 3-mHz receivers and event recorders were placed in the field during the springs of 1972 and

1973 close to but outside the ground clutter of radars, which were used to verify tornadic events. If proved good tornado detectors, the boxes could be used as community warning devices.

Ordinary atmospheric energy sources can produce winds as high as 250 mph. If tornadic winds prove much in excess of this value, we would seek a fundamental role for electrical heating during genesis and maintenance of tornadoes.

Further details of severe storm circulations should become available as more cases are studied (hopefully with close-range aluminum chaff releases to enhance return below cloud base near tornadoes) and as networks of two or three Doppler radars are established. Doppler data in real time is becoming available and should eventually lead to improved tornado warnings.

Doppler Radar

An emerging tool for severe thunderstorm research is the meteorological Doppler radar. This specialized radar measures the Doppler frequency shift caused by the component of radar target movement that is parallel to the radar beam. Two Dopplers spaced on a baseline 20 km or more apart and operated in unison are required to map two-dimensional flow patterns, and three radars at the vertices of a triangle should detail three-dimensional flows.[2, 77, 97]

Recently at the National Severe Storms Laboratory the first Doppler velocity measurements within a radar hook echo were obtained.[11] A high-resolution pulsed Doppler radar with a range of 115 km and a maximum unambiguous velocity spread of $\pm$ 34 m/sec was used. Frequency shift information was recorded in separate sampling volumes or "range gates", each one a cylinder about 300 m in radius and 150 m in length, aligned along the radar beam. The velocity spectrum is obtainable for each volume sample. Figure 16.9 shows Doppler velocities (sample volume means) measured in a squall line with this radar. The indentation in the radar echo locates warm air flowing into an updraft which the Doppler velocities suggest is rotating cyclonically. Some recent cases show more symmetric implied circulations extending up to 8 km.

Direct Probing

Recently, three different methods to directly probe tornadoes with instrument packages have been advanced. A group from Purdue University proposes shooting instrumented rockets through tornadoes.[1] Because of the need for highly sophisticated, rapidly responding instruments and the harsh tornado environment, the group decided to first measure pressure, temperature, humidity, and wind in waterspouts, using a specially designed drag body which a light aircraft tows diametrically through the vortices.[1] Film documentation and flares activated on the sea surface yield supplementary information about the flow. Preliminary results indicate that the temperature peaks in the annulus of rising motions surrounding the funnel (the spray sheath at lower levels) and reaches a minimum near the axis; however, dynamic corrections still must be applied. Other evidence suggests that the annulus of rising motions is warm and very humid and that the flow in the core is downward.[19]

B. J. Morgan[87] proposes ground interceptions of tornadoes using a flatbed trailer and an armored vehicle system. The flatbed trailer would be driven to within visual range of a tornado on the basis of storm observations by the passengers, aided by radio link with a radar site. Thereafter the tornado would be tracked visually from the mobile

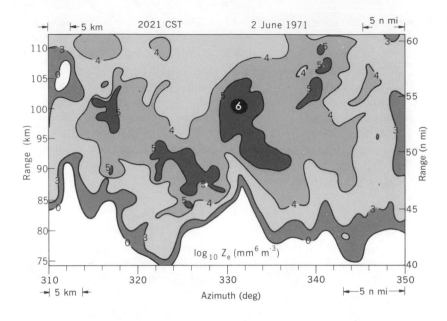

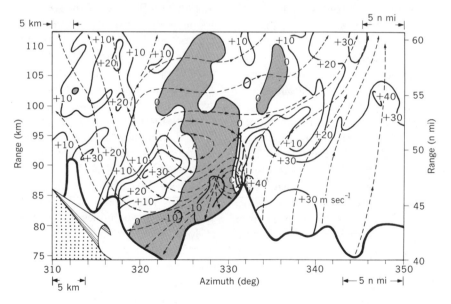

Figure 16.9 Doppler radar measurements made in the squall line between 2020:23 and 2021:56 CST on June 2, 1971. Top, reflectivity pattern in units of the logarithm of the radar reflectivity factor (Z_e). Elevation angle is 1° (height above terrain at the bottom of the figure is 1.6 km and at the top is 2.6 km). Bottom, Doppler velocities (solid lines) and inferred streamlines (dashed lines) relative to the moving storm. Doppler velocities approaching the radar are negative (shaded areas) and receding are positive. The bottom left corner reveals the density of Doppler velocity and reflectivity measurements (every 0.5° in azimuth and 0.6 km in range) (reprinted by permission, from Brown et al.[11]).

station which, aided by navigation techniques, would close in to a strategic position. Then the armored vehicle would be unloaded and driven on an intercept course through the tornado. Instruments on the tank would measure the important meteorological variables. Morgan estimates that one or two tornadoes could be intercepted per year, if there were no obstructions to desired travel directions. However, various constraints including the need to remain on existing roads should greatly reduce the intercept probability. The feasibility of approaching within close range of tornadoes in Oklahoma was demonstrated in 1972 and 1973 with conventional vehicles. Eight rotating cloud systems, three of which produced tornadoes, were tracked and photographically documented from automobiles.

Grant[52] outlines an experimental procedure in which a research airplane drops an instrumented, self-inflating, constant-volume balloon on the ground ahead of a tornado. As the tornado approaches, the balloon inflates and spirals inward toward the core (as well as upwards) because of the radial pressure gradient. If the balloon launches itself close enough to the tornado, then (according to a simple line-vortex tornado model) it arrives at the axis before reaching its float altitude—otherwise, the balloon revolves around the core at constant altitude. The method seems promising and might be used in trial field experiments involving waterspouts.

Waterspouts

Waterspouts appear to embody tornadoes' essential properties, are much more easily approached, and are very frequent near the Florida Keys (volunteer observers reported 300 during the 1969 season). Therefore they are far more amenable to scientific studies than tornadoes.

J. H. Golden[51] comprehensively studied Florida Keys waterspouts. He found that most sightings are multiple and that they tend to occur over very shallow water (less than 12 ft deep) in building cumulus congestus cloud lines with one or two incipient showers. At the time of waterspout formation, the maximum and average cloud top heights may be only 17,000 and 12,000 ft. Strong baroclinicity and vertical wind shear, characteristics of the synoptic conditions under which tornadoes occur, are absent over southern Florida during the waterspout season (June–August). Furthermore, the waterspout environment is also less unstable with a less pronounced dry layer at intermediate levels. Favorable waterspout conditions are high surface temperature ($\sim$87°F) and dew point ($\sim$74°F)[20], conditional instability and the imminent passage of a weak trough line in the lower mean flow (easterlies or westerlies). The shallow coastal water (only 3 or 4 ft deep over wide areas) is propitious as, during the late afternoon, solar heating can raise its temperature to as high as 90°F. In the region of the cloud lines, aircraft measurements indicate mesoscale convergence and cyclonic vorticity, plus a superadiabatic lapse rate in the lower half of the subcloud layer.

In addition, Golden found the following statistical results. The diurnal frequency distribution of waterspouts peaks around 1530–1830 EST with a secondary maximum around 1000 EST. The maximum lifetime observed was 49 min, and the mean was about 10 min. Funnel diameters at the lower tip ranged from 5 to 100 m. There was a strong positive correlation between size (funnel diameter and length) and duration, with the longer-lived ones having condensation funnels reaching all the way to the sea surface. About 10 percent of waterspouts rotate anticyclonically. In some cases anticyclonic and cyclonic waterspouts were present simultaneously in the same cloud line.

Golden observed the following life cycle. The first visible sign of a vortex is a dark spot on the sea surface. A short funnel pendant from the clouds may be present initially or it may develop later in the dark spot stage. Dark spots may occur in groups of two or more with one often dominating the others, which decay. The lifetime of the dark spot stage is from 1 to 22 min and many never evolve further. Tracers dropped on the sea surface indicate that the dark spots are caused by rotation imposed from above. The second stage, not always present, is the formation of spiral patterns on the sea (see Figure 16.7c). Frequently, only one major dark band (scale 500 to 3000 ft) emanates from a nearby shower. The cause of the surface darkening is not known. Flare observations show that the spiral bands depict lines of confluence and difluence on the ocean surface and that regions of flow reversal exist along the surface. The next stage begins as the wind, increasing beyond a critical value ($\sim$22.5 m/sec), throws up a ring of spray from the surface. The funnel increases in size, becomes tilted, and begins moving along the surface as it comes under the influence of the wind shift line associated with the cool air outflow from a neighboring shower. At the same time the spiral tightens. The mature stage lasts 2 to 17 min and is characterized by peak winds, increasing tilt, and maximum forward speed of 5 to 15 knots (30 knots in extreme cases). The waterspout moves generally along a gently curved path. The spray ring evolves into a spray vortex and the funnel sometimes has a double-walled structure (Figure 16.7c). Often, the waterspout is not visible throughout from cloud to sea surface. The decay stage lasts 1 to 3 min as the rain-cooled air

finally overtakes the waterspout. By now the spiral pattern has disappeared and the funnel becomes distorted and finally retracts. During the decay stage a spiral rain curtain sometimes is seen, lasts about 5 min, and occasionally is intense enough to give a hook echo on radar.

Golden has noted several similarities between the life cycles of waterspouts and tornadoes. Perhaps the dust devils observed by Bates[6] and the rotating tail-clouds could be the tornadic counterparts (with different flow visualization) of the dark spots and spiral bands.

The hollow appearance of many waterspouts and the spray's initial annular shape suggest that the axial flow may be downward.[107] The spray sheath also indicates that the highest vertical velocities occur outside the visible funnel[50] (the shape of the sheath is not believed to be accountable by centrifugal forces). From carefully tracking spray particles in a movie taken looking down on top of a spray vortex about 50 ft in depth, Golden derives a tangential wind profile that is a good fit to a combined Rankine profile and obtains a maximum tangential velocity of 144 mph at 12 m for the Lower Matecumbe Key waterspout. These figures correspond to a circulation* of 5 $\times$ 10³ m²/sec. A very large waterspout in 1969 also had a circulation of this magnitude as estimated from time lapse photography of a small funnel revolving around the large one. Comparison with circulations for tornadoes (Table 16.2) shows us that the largest waterspouts have circulations comparable with the weakest tornadoes.

The effect of a change in the lower surface from water to land has been observed quite frequently. E. M. Brooks[9] noted that waterspouts often dissipate on reaching a shoreline. Golden[49] observed a

* Circulation is $2\pi \times$ angular momentum for an axisymmetric vortex and is only weakly dependent on radial distance outside the core.

waterspout which made landfall, and soon afterward weakened and resembled a large dust devil. The visible funnel expanded, became very hollow, and gradually retracted into the parent cloud; after moving off the opposite shore of the island, however, the waterspout, which had retained its vortical identity, reformed.

EXPERIMENTAL MODELS

Concentrated vortices have been generated and studied in the laboratory by many investigators. The rotation introduced into the system by rotating a tank of water[80, 81, 116, 117] or an outer screen, if air is the working fluid,[4, 129, 137] is concentrated into a vortex by establishing meridional circulations in the fluid. The driving mechanism is provided by extracting fluid from the top or bottom of the apparatus,[80, 81, 129, 137], by the drag of rising gas bubbles,[117] or by heating the bottom plate.[4]

Correct modeling of any natural phenomenon requires appropriate selection of nondimensional ratios involving length scales, inertia, rotation, and diffusion. When these ratios are the same for different scale flows, similarity is said to exist. In complex flows with several parameters it is usually impossible to achieve perfect similarity, and the experiment is then designed to satisfy similarity for the more important nondimensional numbers only. For example, the Rossby number, or ratio of inertial to Coriolis forces, is very significant in determining the character of the flow. At the low Rossby numbers (compared to one) which characterize Turner and Lilly's experiment, the Coriolis forces impose such a strong constraint on the lateral flow that the return flow occurs at a radius less than that of the tank. Long's experiment is performed at high Rossby number[75] and hence is relatively unconstrained by the background rotation. Leslie thus concludes that Long's experiment is the more relevant to torna-

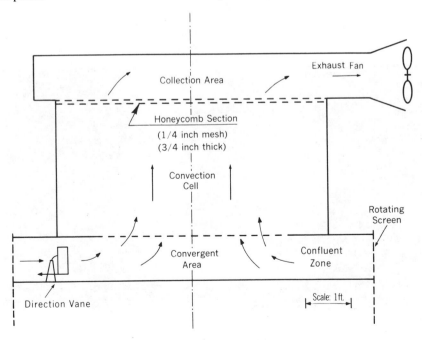

Figure 16.10   Vertical section of Ward's laboratory device.

Figure 16.11 Laboratory vortex enlargement associated with transition from laminar to turbulent flow.

Figure 16.12 Laboratory vortex pair. Compare with figure 16.7b.

does, which are high Rossby number phenomena.[88]

We should consider also how well the various experiments model the physical processes which take place in the atmosphere. Rotating tank experiments suffer from the presence of the lateral wall, which of course is not present in the atmosphere. Driving the meridional flow by forces acting in a narrow neighborhood of the axis[80, 81, 117, 137] compels the vertical velocity on the axis to be of predetermined sign and does not model well the atmospheric case of much wider updraft than vortex. Only Turner and Lilly attempt to model the release of buoyancy by condensation in the atmosphere. They do this by introducing suitable nuclei into carbonated water and causing effervescence to occur. Barcilon's experiment is the most relevant to dust devils as the convection is induced by surface heating at the lower boundary.

Ward[129] probably comes closest to modeling real tornadoes. A diagram of his apparatus is shown in Figure 16.10. An exhaust fan at the top creates convective flow. Air flows into a confluent zone through a rotating screen before entering the convective chamber where it rises and exits the apparatus through a fine mesh honeycomb. The intent is to model a rotating thunderstorm updraft with its low-level inflow concentrating angular momentum near the axis. The honeycomb removes the rotation from the outflow, thereby divorcing the vortex from the exhaust hole by distributing the pressure deficit relatively evenly over the top of the chamber and exposing the top of the vortex to overlying nonrotating flow (as is the case in the atmosphere). Because of this annulment of the central pressure deficit at the top, the axial flow is downward through the honeycomb. This downflow ends at an axial stagnation point, below which the vortex core is laminar

and upflowing. Above the stagnation point, the core is turbulent and bulges out (see Figure 16.11). This phenomenon is known as vortex breakdown.[59]

Another unique feature of this experiment is the use of aspect ratios (updraft diameter to inflow depth) greater than 1. The inflow is confined to the lowest layers in order to model thunderstorm convection. The converging layer of moist air which supplies an updraft is typically 1 mi in depth, and the diameter of the updraft may be 5 mi or more. Only at high aspect ratios can some interesting atmospheric phenomena (discussed below) be simulated.

The angle of inflow, θ, (measured with respect to the radial) at the screen measures the relative strengths of forced convection and circulation. As the inflow angles progressively increase, the stagnation point descends down the axis until it reaches the surface; the radius, r_1, of the turbulent core increases rapidly; and at aspect ratios greater than unity a vortex pair develops (Figure 16.12). The vortices form on opposite sides of the parent vortex near the radius of maximum tangential speed, and the pair rotates around the central axis at about a quarter of that speed since each is in the velocity field of the other. By further increasing the inflow angle, three or four vortices can be obtained.

Ward shows that when the updraft diameter, $2r_0$, is greater than the inflow depth, h, inertial effects associated with large changes in inward radial momentum are important. Deceleration of the inflow explains a high-pressure ring surrounding the core. The associated adverse pressure gradient causes local separation of the inflow from the surface and formation of a very shallow ring of outflow. Barograph recordings of tornado passages often show such pressure peaks (Figure 16.13). With increasing inflow

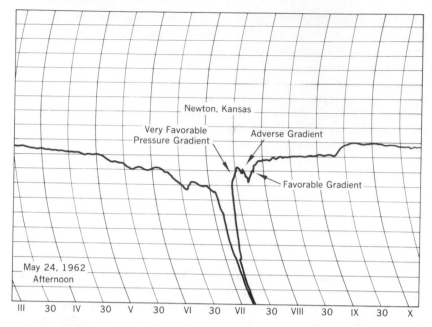

Figure 16.13 Surface pressure record for tornado cyclone of May 24, 1962. Abscissa is time; ordinate pressure from 28 in Hg in intervals of 0.1 in.

angle, the ring disappears as the flow becomes more cyclostrophic. Further analysis of Ward's data shows that the nondimensional core radius, r_1/r_0, is highly dependent on the ratio of tangential velocity at the edge of the updraft to the mean updraft speed above the inflow layer.[22]

S. J. Ying and C. C. Chang[137] have made detailed boundary layer measurements along the lower surface of their vortex chamber. Pressure is almost constant across the boundary layer (as is the case in practically all flows) so that the frictional forces turn the flow inward across the isobars, and produce the

Table 16.4   Resulting Effects of Ground Roughness on Tornadoes

Parameter	No Roughness	Roughness
Core diameter		Larger
Maximal velocity		Smaller
Axial velocity	Slow, possibly downward	Maximal, upward
Mean core updraft		Larger
Axial depression Δp		Smaller
Surface damage		Slighter (decreasing explosive effects and wind velocities)
Appearance	Smooth (waterspouts)	Turbulent (tornadoes)

Source: Dessens.[28]

strongest inflow near the surface. It is the boundary layer air which enters and rises in the core.

J. Dessens[28] has investigated the effects of changing surface roughness in a laboratory model. His results (summarized in Table 16.4) may explain Golden's observation of a waterspout weakening during passage over an island. Thus differing degrees of atmospheric instability, rather than contrasting surface roughness, may be the reason for Midwest tornadoes being more severe than waterspouts. However, Fulks[45] and Leslie[75] have shown that in flows where the tangential and meridional components are strongly coupled (as opposed to more weakly rotating flows which provide forced convection of angular momentum without significant modification of the basic meridional flow), the influence of ground friction may be all-important in establishing enough horizontal convergence for intense vortices to form. Over a rough surface the boundary layer is relatively thick and the tangential component of friction locally reduces centrifugal force and induces the inflow. Over a smooth surface, the layer is much thinner, cyclostrophic balance is disrupted to a lesser extent, and consequently the inflow is weaker. Thus effects of surface roughness are still not clearly resolved.

MATHEMATICAL MODELING

Concept of Vorticity

Fluid motion can be broken down into fields of linear translation, deformation, and rotation which describe effects of the motion on localized fluid blobs (particles). Of most importance to us is the vorticity, the vector curl of the velocity field, or the rate of spin of fluid particles. For a fluid in solid body rotation at rate Ω, the vorticity is constant and equal to 2Ω. Half of this spin is caused by circular

orbits, the other half is caused by radial shear in the velocity field. The two components exactly cancel for a potential vortex ($v \propto \frac{1}{r}$), which is therefore an example of an irrotational (vorticity free) flow. A combined Rankine vortex, which roughly describes the tangential velocity fields of atmospheric vortices, thus consists of a core of concentrated vorticity in an irrotational environment.

Lines in the fluid which are tangential at every point to the vorticity vector are called vortex lines. A vortex tube is the surface containing all the vortex lines that intersect a closed curve, and the enclosed fluid is called a vortex filament. Circulation is the integral of velocity tangent to a closed circuit and is related to vorticity by the result that it is equal to the integral of vorticity across any surface having the circuit as boundary. For a combined Rankine vortex, the circulation is merely 2π times the angular momentum and so is constant outside the core.

It follows from the above properties[105] that in a frictionless fluid the strength of a vortex tube (defined as the circulation around it or vorticity flux along it) is constant along its length, and that vortex tubes cannot intersect or end within the fluid. If in addition the density is uniform, the vortex tubes move with the fluid and always consist of the same particles so that the circulation around a closed fluid curve is constant. This attachment of fluid and vortex lines results in a resistance to motions perpendicular to the vortex lines at high background rotation rates. Viscous and turbulent diffusion is a relatively slow process and normally does not affect the above properties very much although it does provide a limit to the size of vorticity gradients that can be established and maintained.[89]

B. Morton[88] summarizes the ways in which vorticity can be produced or amplified in a fluid, namely, by

1. Stretching of vortex tubes: vorticity is amplified in convergence areas by this means. Since vortex tubes are carried with the flow, their orientation may be changed appreciably.

2. Viscous stresses at solid boundaries: vorticity, generated at the ground, diffuses upward. Vortex layers also form at the boundaries of solid bodies, and may separate and be convected downstream as wakes or roll up to form trailing vortices.

3. Solenoids: vorticity is created by solenoidal circulations which are due to intersecting surfaces of constant density and pressure in the fluid. The lighter (denser) air moves toward low (high) pressure so as to bring the two types of surfaces into coincidence.

Possible Vorticity Sources for Tornado-Genesis

The concentrated vorticity of tornadoes must be caused by either local generation or local amplification from a background of weak vorticity. First consider the latter. The vertical vorticity of the earth itself in midlatitudes and the relative vorticity of cyclones are both around 10^{-4}/sec. Ignoring loss of angular momentum to the ground, air must be drawn into the center of a convergence area from a radius of 10 km to form a tornado with circulation 5×10^4 m²/sec from a large-scale background vorticity of 2×10^{-4}/sec. Assuming constant horizontal convergence of 10^{-3}/sec (a typical value) within the 10-km circle, we conclude that it takes (very) roughly 3 hr to establish a concentrated vortex by this process. Ward[125] showed experimentally that an updraft can be induced to rotate by interaction with a simulated cold air outflow in a cyclonically rotating tank. The presence of the updraft produces a bulge on the outflow boundary, and Coriolis acceleration causes the bulge to pass to the

right of the buoyant column. Thus the left or cyclonic-shear flank is brought under the updraft. A preferred orientation for tornadic activity is for the downdraft to be northwest of the main updraft. This brings the cyclonic-shear flank beneath the southwest side of the parent storm, which is the area where the most rapid convective development is expected to take place. Evidence for this process occurring in the atmosphere was presented in a case study of a tornadic storm by Ward.[128] Note that the Rossby number ($\equiv v_{max}/\Omega r_m$), which may be interpreted as a measure of the strength of the relative local vertical vorticity compared to the background vertical vorticity, 2Ω, of the earth's rotation, is of the order of 10^4 for tornadoes and 10^2 for tornado cyclones so that, although the earth's rotation can serve as a vorticity source, it cannot exercise any lateral constraints on the flow.

Local generation of horizontal vorticity occurs at the earth's surface. Barnes[5] suggested that horizontal convergence might tilt horizontal vortex tubes into the vertical and stretch them. The horizontal vorticity associated with vertical wind shear is roughly 10^{-2}/sec, about two orders of magnitude less than the vertical vorticity of a tornado core. Drawing up loops of horizontal vortex tubes results in an erect vortex pair with opposite rotation senses. Barnes argues that net cyclonic rotation generally results if the low-level inflow is to the right of the storm's path, as is generally the case.

Bates's observation of a line of dust devils preceding a tornado suggests a horizontal shearing instability with the vertical sheet of vorticity at a wind-shift line rolling up into individual vortices, which are then stretched by convection. Cyclonic shear across the gust front is necessary to explain the predominance of cyclonic tornadoes and has been observed in three storms by Ragette.[99] The presence of cyclonic and anticyclonic waterspouts in the same cloudline could be explained by postulating that cyclonic and anticyclonic shear exists across different segments of the wind-shift line.

Some Analytic Solutions

Theoretical treatments of vortices generally have been limited to axisymmetric, laminar flows in basically incompressible fluids. Owing to lack of a better formulation, turbulent flows are treated as if they were laminar except for an upward change of several orders of magnitude in the coefficient of viscosity to take increased diffusion into account. Adiabatic changes of volume in the atmosphere are allowed for by substituting potential temperature and potential density for temperature and density in the equations. To simplify the equations, the Boussinesq approximation (that variations in potential density and potential temperature are relatively small and may be neglected except in the buoyancy term) is made. These assumptions limit us to the lowest 3 km of the atmosphere and to winds less than about 250 mph. Release of latent heat and other effects caused by the presence of water substance are added complications which, although important, are not included here. Only axisymmetric vortex flows about vertical axes are considered.

The equations of motion, continuity, and heat transfer in cylindrical polar coordinates (r, θ, z) with velocity components (u, v, w) are thus

$$\frac{\partial u}{\partial t} + u\frac{\partial u}{\partial r} + w\frac{\partial u}{\partial z} - \frac{v^2}{r} - 2\Omega v = -\frac{1}{\rho}\frac{\partial p}{\partial r}$$

$$+ \nu\left(\frac{\partial^2 u}{\partial r^2} + \frac{1}{r}\frac{\partial u}{\partial r} - \frac{u}{r^2} + \frac{\partial^2 u}{\partial z^2}\right) \quad (16.15)$$

$$\frac{\partial v}{\partial t} + u\frac{\partial v}{\partial r} + w\frac{\partial v}{\partial z} + \frac{uv}{r} + 2\Omega u$$

$$= \nu\left(\frac{\partial^2 v}{\partial r^2} + \frac{1}{r}\frac{\partial v}{\partial r} - \frac{v}{r^2} + \frac{\partial^2 v}{\partial z^2}\right), \quad (16.16)$$

$$\frac{\partial w}{\partial t} + u\frac{\partial w}{\partial r} + w\frac{\partial w}{\partial z} = -\frac{1}{\rho}\frac{\partial p}{\partial z}$$

$$+ \nu\left(\frac{\partial^2 w}{\partial r^2} + \frac{1}{r}\frac{\partial w}{\partial r} + \frac{\partial^2 w}{\partial z^2}\right) + \frac{gT'}{T_a}, \quad (16.17)$$

$$\frac{1}{r}\frac{\partial}{\partial r}(ru) + \frac{\partial w}{\partial z} = 0, \quad (16.18)$$

$$\left(\frac{\partial}{\partial t} + u\frac{\partial}{\partial r} + w\frac{\partial}{\partial z}\right)\frac{T'}{T_a} + \frac{\partial \ln T_a}{\partial z}w$$

$$= \kappa\left(\frac{\partial^2}{\partial r^2} + \frac{1}{r}\frac{\partial}{\partial r} + \frac{\partial^2}{\partial z^2}\right)\frac{T'}{T_a}, \quad (16.19)$$

where Ω is the vertical component of background rotation (the earth's in the case of atmospheric vortices), $p + \rho\Omega^2 r^2/2 - g\rho_a z$ is pressure, $\rho_a(z)$ and $T_a(z)$ are ambient values of density and temperature T' is temperature deviation from T_a, and ν and κ are the coefficients of viscosity and thermal diffusivity. The following exact solutions of these equations do not satisfy realistic boundary conditions but are nonetheless enlightening.

First consider a steady flow $(\partial/\partial t \equiv 0)$ in a homogenous fluid $(T = T_a = $ constant) with no background rotation $(\Omega = 0)$. The simplest vortex type solution to the equations is the Burgers-Rott vortex[14, 104]

$$u = -ar,$$

$$v = \frac{\Gamma}{2\pi r}\left[1 - \exp\left(\frac{-ar^2}{2\nu}\right)\right],$$

$$w = 2az,$$

$$p(r,z) = p(0,0) + \rho\int_0^r \frac{v^2}{r}\,dr$$

$$- \frac{\rho a^2}{2}(r^2 + 4z^2), \quad a > 0$$

$$(16.20)$$

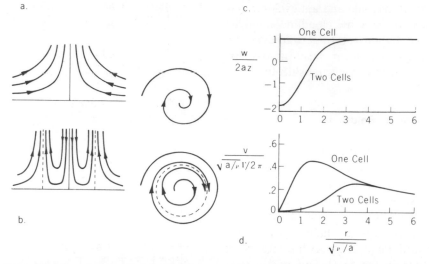

Figure 16.14 Sketches of projections of streamlines onto a vertical plane through the axis and a horizontal plane for (a) the "one-celled" Burgers' vortex; (b) the "two-celled" Sullivan vortex and velocity distributions; (c) axial velocity; (d) tangential velocity as functions of radius (Sullivan[112]).

where Γ is the circulation at infinity, and $2a$ is the (constant) horizontal convergence. The flow is driven by infinite excess pressure at $r = \infty$ or by infinite suction at $z = \infty$. Since u and v are not functions of z, the streamlines in r-z planes are completely independent of circulation. This decoupling of the meridional and swirl flow is seldom achieved in reality, except in the limit of very small circulations. The controlling influence of a ground boundary layer on the flow is not represented because only no-stress boundary conditions ($\partial u/\partial z = \partial v/\partial z = w = 0$) are satisfied at $z = 0$. Eddy viscous forces act only in the tangential direction and establish a core which rotates approximately as a solid body. The maximum tangential velocity, $v_{max} = 0.72/2\pi r_m$, occurs at a radius $r_m = 1.12\sqrt{2\nu/a}$. The pressure deficit at the axis due to circulation is $\Delta p_\Gamma = 1.68\rho v^2_{max}$.

R. D. Sullivan[112] obtained a similar solution but of a "two cell" structure in which inner and outer flows are separated by the surface $r_1 = 2.38\sqrt{2\nu/a}$ and the vertical velocity reverses sign at $r_0 = 1.48\sqrt{2\nu/a}$. The meridional flow is given by

$$u = -ar + \left(\frac{6\nu}{r}\right)\left\{1 - \exp\left(\frac{-ar^2}{2\nu}\right)\right\}$$

$$w = 2az\left\{1 - 3\exp\left(\frac{-ar^2}{2\nu}\right)\right\} \quad (16.21)$$

One- and two-cell solutions are compared in Figure 16.14. In a one-cell vortex the flow spirals in toward the axis and out along it. In the two-cell case, fluid spirals inward and upward in the outer cell, downward near the axis, and out and upward near the outer edge of the inner cell. Interestingly, the two flows are driven by the same axial pressure gradient, $-4\rho a^2 z$. For the same values of ambient convergence and circulation, the one-cell vortex has a smaller core, higher tangential speeds, and lower central pressure. L. N. Gutman[54] and H. L. Kuo[71] found solutions for vortices driven by the release of latent heat in a conditionally unstable saturated atmosphere. Kuo's equations have the same form as Equations 16.15 and 16.19 with a modified equivalent potential temperature substituted for temperature to take the release of latent heat into account. With assumptions of equal constant values of eddy viscosity and thermal diffusivity, exact similarity of the T and rw fields, and neglect of the vertical gradient of perturbation pressure, $\partial p/\partial z = 0$ in Equation 16.17, Kuo obtains specific one- and two-cell solutions. The role of a in the Burgers-Rott and Sullivan vortices is now assumed by $\beta \equiv (-g\partial\,(\ln T_a)/\partial z)^{1/2}$. The two-cell solution is

$$u = \frac{-\beta r}{2} + \left(\frac{4\nu}{r}\right)\left\{1 - \exp\left(\frac{-\beta r^2}{2\nu}\right)\right\},$$

$$w = \beta z\left\{1 - 2\exp\left(\frac{-\beta r^2}{4\nu}\right)\right\},$$

$$\frac{T'}{T_a} = \left(\frac{\beta^2}{g}\right)z\left\{1 - 2\exp\left(\frac{-\beta r^2}{4\nu}\right)\right\}.$$

$$(16.22)$$

Comparison of Equations 16.21 and 16.22 shows that the large-scale convergence is the same for $\beta \equiv 2a$. Note that with this equivalence the two solutions have similar properties, but Kuo's vortex is more concentrated ($r_o = 0.69\,\{4\nu/\beta\}^{1/2}$, $r_1 = 1.59\,\{4\nu/\beta\}^{1/2}$). Kuo's solution yields a maximum tangential velocity $v_{max} = 0.85\,\Gamma/2\pi r_m$ at a radius $r_m = 2.06\,(4\nu/\beta)^{1/2}$, and a circulation-associated central pressure drop $\Delta p_\Gamma = 1.14\rho v^2_{max}$. For Kuo's one-cell solution $v_{max} = 0.68\Gamma/2\pi r_m$, $r_m = 1.22\,(4\nu/\beta^{1/2}$, and $\Delta p_\Gamma = 1.74\rho v^2_{max}$, and comparison with the Burgers-Rott results shows only small changes in the numerical coefficients. Kuo later modified his two-cell solution to satisfy a zero vertical velocity criterion at large radii, and showed that the solution at inner radii was changed very little.[72]

Obtaining numerical values for the above quantities is a dangerous game owing to the nature of turbulence. Al-

though the other parameters can be estimated fairly accurately, the eddy viscosity is known to only an order of magnitude and should be a function of height and radius rather than a constant. However, we will insert $\Gamma = 4.7 \times 10^4$ m²/sec (as observed for the 1953 Cleveland tornado), $\rho = 1.14$ kg/m³, $\beta = 1.5 \times 10^{-2}$/sec corresponding to a strongly unstable 8°C/km lapse rate of equivalent potential temperature in the lowest 2 to 3 km, and $\nu = 10$m²/sec (probably correct to at least an order of magnitude). The results are $v_{max} = 181$ and 134 mph, $r_m = 63$ and 106 m, $\Delta p_\Gamma = 130$ and 47 mb for the one- and two-cell cases, respectively.

In addition to previously made criticisms of cylindrical vortex solutions, Kuo's two-cell vortex has another drawback; the downdraft in the center is cold. As Lilly[79] points out, this is unlikely because there is no way to produce a cold downdraft in a conditionally unstable atmosphere except by evaporation, and a concentration of liquid water sufficient to produce such a downdraft would quickly be centrifuged out of the vortex. A cold core vortex might also be unstable to outward displacements.

Note that all the above solutions predict that the radius of maximum tangential velocity is independent of circulation. This is a direct result of uncoupled meridional and swirl flows and conflicts with Ward's experimental observations.

The vertical vorticity in these solutions is strongly concentrated in the core and the outer flow is almost irrotational. A steady balance is achieved locally between the stretching of vortex tubes due to horizontal convergence, advection of vorticity, and diffusion down the vorticity gradient. There is no tilting of vortex tubes due to the cylindrical similarity. Another basic feature of these solutions is that any uniform rotation in the ambient (as introduced by Kuo) is exactly annulled and the final solution still refers to a vortex core in an irrotational environment.[88]

Structure of Vortices

Theoretical reasoning gives some insight into the structure of vortices. For instance, Morton[88] performed an order of magnitude evaluation of the terms in the governing equations for a long thin vortex which spreads laterally with height. He demonstrated that vertical gradients of the flow variables are important and result in coupling of the swirl and meridional components of motion through the pressure field. In addition, he deduced that the vertical velocity should be of the same order as tangential velocity. Interestingly enough, there are reports of strong vertical velocities in tornadoes. Hoecker[62] determined a maximum vertical velocity of 68 m/sec on the axis at a height of 40 m from movies of the Dallas tornado.

Inflow into a tornado occurs primarily in a thin layer along the lower boundary. Mathematical solutions for the boundary layer flow are extremely hard to obtain even with the assumption of a simplified upper flow. A well-known class of problems in fluid dynamics is the interaction of homogenous flows of the type $u = 0$, $v \propto r^n$ ($-1 \leq n \leq 1$) with a stationary rigid lower surface, which acts as a centrifugal pump by inducing radial convergence. The unspecified vertical velocity in the external flow is deduced from mass continuity. The Bodewadt problem ($n = 1$, i.e., solid body rotation) has a fairly simple solution. The thickness of the boundary layer, δ, is roughly $8 (\nu/\Omega)^{1/2}$ and is independent of radial distance. Applying this result to a tornado core rotating at 1 rad/sec with an assumed eddy viscosity of 10 m²/sec, we obtain $\delta = 25$ m. The other extreme ($n = -1$) is

the matching of a boundary layer to a potential vortex. Some recent work on this problem shows that for a lower surface of radius a (finite for computational reasons), the boundary layer has a double structure with an inner layer next to the surface with thickness of order $(\nu/\Gamma)^{1/2}r$, in which the flow is primarily radial, and an outer layer of order $(\nu/\Gamma)^{1/2} a$ thick, of predominantly inviscid nature in which the flow recovers to the external potential vortex.[15] The boundary layer erupts upward very close to the axis. By using spherical rather than cylindrical polar coordinates, Serrin was able to solve for the axial flow.[107] He showed that only three types of motion are possible, namely, inflow along the ground with upflow along the axis, surface inflow and axial downdraft with compensating outflow along a conical surface whose apex is at the foot of the vortex, and surface outflow with axial downflow. The last motion is clearly contrary to observations, and the first kind is theoretically possible only for $\Gamma/4\pi\nu < 2.86$, a condition not likely to be fulfilled in tornadoes except for eddy viscosities one to two orders of magnitude larger than those currently assumed or for weak circulations. This work thus predicts that central downdrafts commonly occur in tornadoes. Serrin also points out that the second kind of motion provides an explanation other than centrifugal force for the shape of the dust and spray cascades frequently observed at the foot of tornadoes and waterspouts.

Kuo[73] took the turbulent nature of the boundary layer and the vortex core into account. His results show weakly descending motion in the outer region, relatively strong ascending motion in the inner region, with a sharp maximum upward motion occurring inside the radius of maximum tangential wind. On the axis the flow is weakly upward. Kuo's velocity profiles are in qualitative agreement with the model vortex measurements of Ying and Chang.[137]

The direction of the flow along tornado axes remains unresolved. The above theories consider neither thermal effects nor vertical gradients in the free flow, both of which should have a profound influence on the flow in the core. There may even be stagnation points aloft, separating axial upflows and downflows.

Numerical Models

Numerical weather prediction is practiced mostly on the large scale with hemispheric coverage on a grid point spacing of about 400 km.[108] These models provide guidance to the thunderstorm forecaster by predicting the general patterns of horizontal winds, moisture, temperature, vorticity, and vertical currents. Limited area models with a mesh size of 200 km[47] and planetary boundary layer models[55] are now becoming operational, but improvements in accuracy will be impeded by the wide spacing ($\sim$400 km) of upper air data provided by radiosonde stations and by boundary conditions which are taken from a larger model. Statistical models that forecast directly the parameters known to be important to thunderstorm and tornado development are also coming into operational use and provide somewhat more detailed spatial distributions over the United States than have been previously available.[69]

Clearly the above models are unable to resolve a single thunderstorm. They are also incapable of representing the effect on large-scale weather of redistribution of temperature, moisture, and momentum by thunderstorm outbreaks, because this interaction is still not well known. Modeling a single rain shower is in itself a task which is barely feasible with present-day technology because of the

intricacies of deep compressible convection and the complex interactions of dynamics, thermodynamics, and cloud physics. The most comprehensive of today's shower models duplicate several features of natural rain[90, 93, 113] but are two-dimensional with line or axial symmetry; thus interactions with a strongly sheared environment are not accounted for. A model of a three-dimensional thunderstorm is being developed at the National Center for Atmospheric Research. To simulate a 5-hr storm will take 20 hr on one of the world's largest computers, and even then many important features of real storms will not be reproduced.

The core of a tornado is two orders of magnitude smaller in horizontal dimensions than a thunderstorm, and although a thunderstorm model may show areas of enhanced vorticity in which tornadoes are likely to form, it cannot be expected to resolve individual tornadoes. Clearly a model with more limited horizontal area must be used to reveal the structure of tornadoes, but defining suitable initial and boundary conditions is a problem because we have limited physical insight into how the two scales (tornado and thunderstorm) interact.

Numerical models of vortices have been restricted thus far to axisymmetric simulations of laboratory experiments (in cases where the boundary conditions are well defined) or shallow convective processes in a swirling atmosphere. The numerical scheme used to solve the differential equations is very important; it must be conservative in a gross sense because otherwise artificial generation of kinetic energy and vorticity invalidates the solutions.

Experiments with closed circulations are suitable for modeling because of the simple boundary conditions. Leslie[75] has numerically simulated Lilly and Turner's experiment but at a higher Rossby number, so that the return flow is controlled by the tank geometry rather than the lateral constraints imposed by the Coriolis force. To simulate the drag of rising gas bubbles on water, Leslie artificially prescribed a constant buoyancy force on the upper one-third of the axis. The vortex grows downward from the top to the bottom of the tank as cyclostrophic balance becomes established at consecutively lower levels; as the vortex starts interacting with the bottom plate, accelerated amplification occurs because of restrictions on the inflow and the associated surface pressure fall. In the final steady state, radial inflow is confined almost entirely to the lower boundary layer (in contrast to similar flows without rotation). Concentrated vortices form only for a limited range of values of applied force and rotation, and their cores have radii about one-tenth, and angular velocities about thirty times, those of the tank. Substituting a free upper surface for a rigid top results in a slightly weaker vortex, and is attributed to smaller disruption of the interior cyclostrophic balance by a yielding surface.

Vortex Breakdown and Instabilities

Mathematical analyses can show whether basic flows are stable to the infinitesimal perturbations which are bound to be present in any physical system. Because of mathematical complexities, the only swirling flows which have been treated in general have been of the type $u \equiv 0$, $v \equiv v(r)$, $w \equiv w(r)$, and $T' \equiv 0$. For this class of flows, integral techniques have been used to show that certain conditions are sufficient for stability. Proving sufficient conditions for instability is much more difficult, and usually each basic flow has to be considered separately. It is generally assumed that the disturbances are inviscid. Viscosity has a stabilizing influence, except possibly in boundary layers, affects shorter wavelength disturbances the most, and should not invalidate the stability criteria given later. More general vortex flows, such as those discussed earlier, should be subject to the same types of instabilities, as well as possibly others.

There is a strong analogy between the role played by the gravitational force in stratified (i.e., density varying with height) flows and the role of centrifugal force in rotating flows. When the flow speed in a channel is super-critical, that is, exceeds the speed of the fastest (horizontally) moving gravity wave, disturbances are unable to propagate upstream; incompatible upstream and downstream boundary conditions are matched in steady flow by sudden reduction to a conjugate subcritical flow speed and corresponding increase in depth. This phenomenon is known as hydraulic jump and, like a shock wave, is not a classical instability. A similar transition, vortex breakdown, can occur in rotating flows when the updraft speed exceeds the speed of the fastest downward propagating inertia wave (inertia waves are discussed below). Vortex breakdowns are characterized by an abrupt enlargement of the vortex core and, unlike shocks and jumps, have stagnation points and reversed velocities along the axis. They may be viewed alternatively as occurring when the axial pressure gradient becomes large enough to drive secondary backflows. Generally, breakdowns are present whenever the maximum tangential velocity is roughly equal to or greater than the updraft speed, and the core is turbulent above the stagnation point. Sometimes the conjugate flow has a spiral, rather than axisymmetric, character.

Turbulence in a parallel shear flow is suppressed by stable stratification (e.g., the boundary layer near the ground at night). The Richardson number, $g \{\partial(\ln T_a)/\partial z \}/ \{\partial u/\partial z \}^2$, is the crucial parameter; if it is no less than $\frac{1}{4}$, the flow is stable. Howard and Gupta[63] found an analogous criterion for axisymmetric stability in rotational flows, namely, $\{\partial M^2/\partial r \}/r^3 \{\partial w/\partial r \}^2 \geqq \frac{1}{4}$ where $M \equiv vr$ is angular momentum. Thus jets are stabilized by circulation which increases monotonically with radius. In the case where there is no shear in the vertical velocity, the criterion reduces to $\partial M^2/\partial r \geq 0$, a result first obtained by Rayleigh.[100] The latter can be given a simple physical interpretation. Consider a fluid parcel which is displaced radially outward while conserving its angular momentum. If M^2 increases outward in the environment, which is in cyclostrophic balance, the inward pressure gradient force acting on the particle exceeds the centrifugal force and forces the particle back toward its original position. In actuality, the particle oscillates about its original position until viscous forces damp out the motion. These radial oscillations in a vortex are the aforementioned inertia waves. $M^2 =$ constant, which applies in the outer portion of a combined Rankine vortex, represents a state of neutral stability. In a vortex core M^2 increases rapidly with radius. Turbulent vortex cores are thus probably caused by the destabilizing influence of large shears of vertical velocity.

Purely angular disturbances (no axial variance) may also be unstable. A sufficient condition for stability (also due to Rayleigh) is that the vertical component of vorticity be a monotonic function of radius. The Burgers-Rott vortex is definitely stable to this type of disturbance, but a glance at Figure 16.14d indicates that the stability of two-cell vortices is not so definite. The inner cell acquires swirl only through viscous coupling with the outer cell, and the core rotates differentially with a vorticity maximum

away from the axis. An extremely idealized model of this type of flow is a cylindrical sheet of vorticity in an otherwise irrotational flow. Vortex sheets are inherently unstable and wrap up into rows of straight line vortices; therefore, it is reasonable to suppose that unstable angular disturbances evolve into the multiple vortices observed in Ward's experiment.

No general stability criteria exist for three-dimensional (helical) disturbances, and each flow has to be analyzed separately.

Boundary layer instabilities also should be considered. The interaction of a uniform flow U and a parallel flat plate gives rise to a boundary layer which becomes turbulent at a distance x from the leading edge given by $Re = Ux/v \gtrsim 4 \times 10^5$ (Re is the Reynolds number for the flow). This criterion cannot be simply applied to the bottom boundary layers of swirling convergent flows because of the radial pressure gradient force. A favorable pressure gradient (i.e., one which acts in the direction of the flow) reduces turbulence and makes the boundary layer thinner, whereas an unfavorable one acts in the opposite sense (see N. B. Ward[130] for a more detailed discussion). Tornadic pressure traces show mostly favorable gradients. However, a region of adverse gradient is associated with a high-pressure ring, such as the one in Figure 16.13, and should give rise to a local thickening of the boundary layer and increase in turbulence. The boundary layer is turbulent in both Ward's and Chang's models.

Faller[29, 30] in experiments with rotating fluids with radial inflow, found horizontal roll vortices, resembling the spiral bands in hurricanes, in the boundary layer. The bands were oriented about 15° to the left of the horizontal flow above the (Ekman) boundary layer.

OUTLOOK FOR MODIFICATION

Up to this point we have explained what is currently known about tornadoes and what kind of projects are underway to increase our understanding. Vortices, in general, are poorly understood even though they are present in numerous engineering problems. In the atmosphere, concentrated vortices appear to depend on unstable stratification and enhanced vorticity for their creation and maintenance. This view is supported by observations of massive vortices in large fires and near erupting volcanoes.

Too little is known about tornadoes at this stage to attempt to modify them[67]; even the wind speeds are not accurately known. However, we close by enumerating possibilities.

Towering cumulus clouds over forest fires, volcanoes, and atomic bomb blasts suggest that competing meteorological events might be triggered at strategic locations to rob a tornadic storm of its needed inflow. Arrays of huge jet engines or an oil burning device such as the meteotron, developed by H. and J. Dessens in France, might be used.[26, 27] The meteotron had a peak output equivalent to $7 \times 10^5 \, kW$ and was used to stimulate rainfall, but it was successful only under particularly favorable conditions, and the operational costs were excessive. Since strong fire vortices are observed under conditions of strong atmospheric instability, there is a potential for the triggered events themselves becoming severe. Vul'fson and Levin[121] report another approach: the dispersal of convective clouds by the downwash created by flying modified jet aircraft through them. They claim that during all of nine tests, clouds up to 15,000 ft thick dispersed 5 min after such flights. It is hard to imagine a tornadic storm being much affected by this method. Changing the characteristics of the earth's surface

such as the albedo and the availability of water for evaporation is a remote possibility. Variations of several degrees in temperature and several percent in relative humidity could be induced in the surface layer.[74] In fact, Atkinson[3] in England and Chagnon and associates[17] in the United States have shown that man's building and engineering activities inadvertently may be increasing the incidence of summer thunderstorms and showers over cities.

Another proposal is based on the idea that vortices tend to weaken over rougher surfaces because of reduction of net low-level inflow. This theory is supported by the laboratory results of Dessens[28] and observations of waterspouts striking shore. Waterspouts and tornadoes often retract into the clouds on encountering cliffs. The base is arrested while the upper part continues forward; the vortex is stretched longitudinally into a rope shape and soon dissipates.[56] Sometimes the density of buildings appears to reduce structural damage at ground level.[56] Surrounding forests or artificial mounds and ridges might protect cities somewhat, but other observations indicate that intense tornadoes would penetrate such defenses. For example, tornadoes have left behind swaths of uprooted trees in forests[13] and stayed on the ground while passing over mounds[46] or crossing 150-ft valleys.[44] Hardy[56] describes a waterspout that surmounted 100-ft cliffs and became a destructive tornado.

Modification by cloud seeding appears to be a cheaper and easier method if it works in the desired sense. One-dimensional cumulus cloud models are used operationally to predict the results of seeding[109, 132] but do not include rotational effects. Briefly, seeding with silver iodide releases latent heat by inducing freezing of supercooled water. Also it leads to accelerated growth of precipitation particles, but the heavier hydrometeors are centrifuged outward and hence would not exert much of a drag on the rising air close to a tornado. It is reasonable therefore to suppose that initially seeding would increase low-level convergence and hence might intensify a tornado, although Henderson[60] reported seeding weak tornadoes in Texas at cloud base without any noticeable visual effects. On the other hand, seeding could shorten the life of a tornado if the storm's cold air outflow became stronger and overtook the vortex sooner, thereby cutting off the inflow. It might also prove beneficial to seed a neighboring cell that is upstream relative to the low level inflow. Hopefully, the seeded cloud would develop rapidly and by competing for warm, moist air would reduce the inflow into the rotating updraft, thus weakening it.

Any efforts to modify a severe storm with potential or actual tornadoes obviously will have to be carried out with extreme caution, and it would be wise to experiment first with waterspouts, which are less of a threat. Actual modification attempts on menacing tornadoes are probably several years away. In the meantime, we should seek improved building codes and construction practices, and continue research into the actual morphology of convective vortices.

REFERENCES

1. Agee, E. M., Tornado project activities, Purdue University, *Bull. Am. Meteorol. Soc.* **50,** 806, 1969; Part II, **51,** 951, 1970; Part III, **52,** 575, 1971.

2. Armijo, L., A theory for the determination of wind and precipitation velocities with Doppler radars, *J. Atmos. Sci.* **26,** 570–573, 1969.

3. Atkinson, B. W., A preliminary examination of the possible effect of London's urban area on the distribution of thunder rainfall 1951–1960, *Trans. Papers Inst. Brit. Geogr. Publ.* **44,** 97–118, 1968.

4. Barcilon, A. I., A theoretical and experimental model for a dust devil, *J. Atmos. Sci.* **24,** 453–466, 1967.

5. Barnes, S. L., On the source of thunderstorm rotation, *NOAA Tech. Memo* ERLTM-NSSL *38,* National Severe Storms Laboratory, Norman, Okla., 1968, 28 pp.

6. Bates, F. C., A theory and model of the tornado, *Proceedings of the International Conference on Cloud Physics, Montreal, Canada, Aug., 1968.*

7. Bigler, S. G., Tornado damage surveys, *Texas A & M University Department of Oceanography and Meteorology Report,* 1957.

8. Booker, C. A., On transmission towers destroyed by the Worcester, Massachusetts tornado of June 9, 1953, *Bull. Am. Meteorol. Soc.* **35,** 225, 229, 1954.

9. Brooks, E. M., Tornadoes and related phenomena, *Compendium of Meteorology,* American Meteorological Society, Boston, 1951, pp. 673–680.

10. Brook, M., Electric currents accompanying tornadic activity, *Science* **57,** 1434–1436, 1967.

11. Brown, R. A., W. C. Bumgarner, K. C. Crawford, and D. Sirmans, Preliminary Doppler measurements in a developing radar hook echo, *Bull. Am. Meteorol. Soc.* **52,** 1186–1188, 1971.

12. Browning, K. A., Airflow and precipitation trajectories within severe local storms which travel to the right of the winds, *J. Atmos. Sci.* **21,** 634–639, 1964.

13. Budney, L. J., Unique damage patterns caused by a tornado in dense woodlands, *Weatherwise* **86,** 75–77, 1965.

14. Burgers, J. M., A mathematical model illustrating the theory of turbulence, *Advan. Appl. Mechan.* **1,** 197–199, 1948.

15. Burgraff, O. R., K. Stewartson, and R. Belcher, Boundary layer induced by a potential vortex, *Phys. Fluids* **14,** 1821–1833, 1971.

16. Carr, J. A., A preliminary report on the tornadoes of March 21–22, 1952. *Monthly Weather Rev.* **80,** 50–58, 1952.

17. Chagnon, S. A., F. A. Huff, and R. G. Semonin, METROMEX: An investigation of inadvertent weather modification, *Bull. Am. Meteorol. Soc.* **52,** 958–967, 1968.

18. Chollet, R., Waterspouts, *Mariners Weather Log* **2,** 152–156, 1958.

19. Church, C. R. and C. M. Ehresman, A brief

report on the Purdue waterspout research program, *Purdue Tornado Project Report,* 1971.

20. Clemens, G. H., Detailed radar observations and recent climatology of waterspouts in the Key West area, *Proceedings of the Sixth Conference on Severe Local Storms,* American Meteorological Society, 1969, pp. 172–175.

21. Cressman, G. P., Killer storms, *Bull. Am. Meteorol. Soc.* **50,** 850–855, 1969.

22. Davies-Jones, R. P., The dependence of core radius on swirl ratio in a tornado simulator, *J. Atmos. Sci.,* **30,** 1427–1430, 1973.

23. Dergarabedian, P. and F. Fendell, On estimation of maximum wind speeds in tornadoes and hurricanes, *J. Astronaut. Sci.* **17,** 218–236, 1970.

24. Dergarabedian, P. and F. Fendell, Estimation of maximum wind speeds in tornadoes, *Tellus* **22,** 511–515, 1970.

25. Dergarabedian, P. and F. Fendell, A method for rapid estimation of maximum tangential wind speed in tornadoes, *Monthly Weather Rev.* **99,** 143–145, 1971.

26. Dessens, H., A project for a formation of cumulonimbus by artificial convection, *Geophys. Monogr.* **5** (AGU Publ. No. 746), 1960.

27. Dessens, J., Man-made tornadoes, *Nature* **193,** 13, 1962.

28. Dessens, J., Influence of ground roughness on tornadoes: a laboratory simulation, *J. Appl. Meteorol.* **11,** 72–75, 1972.

29. Faller, A. J., An experimental analogy to and proposed explanation of hurricane spiral bands, *Proceedings of Second Technical Conference on Hurricanes, Boston,* American Meteorological Society, 1961, pp. 307–313.

30. Faller, A. J., An experimental study of the instability of the laminar Ekman boundary layer, *J. Fluid Mech.* **15,** 560–576, 1963.

31. Fankhauser, J. C., Thunderstorm-environment interactions determined from aircraft and radar observations, *Monthly Weather Rev.,* **99,** 171–192, 1971.

32. Finley, J. P., Character of six hundred tornadoes, *Professional Papers of the Signal Service,* No. VII, Washington Office of the Chief Signal Officer, 1882.

33. Flora, S. D., *Tornadoes of the United States,* University of Oklahoma Press, 1954, 194 pp.

34. Frankenfield, H. C., The tornado of May 27 at St. Louis, Mo., *Monthly Weather Rev.* **24,** 77–81, 1896.

35. Fritz, C. E., Disaster, *Contemporary Social Problems,* Harcourt, Brace and World, New York, 1961, pp. 651–694.

36. Fujita, T., A detailed analysis of the Fargo tornadoes of June 20, 1957, *Tech. Rep. No. 5,* Severe Local Storms Project, University of Chicago, 1959, pp. 1–29.

37. Fujita, T., Analytical mesometeorology: a review, *Meteorol. Monograph* **5**(27), 77–125, 1963.

38. Fujita, T., Formation and steering mechanisms of tornado cyclones and associated hook echoes, *Monthly Weather Rev.* **93,** 67–78, 1965.

39. Fujita, T., Estimated wind speeds of the Palm Sunday tornadoes, *SMRP Research Paper* **53,** The University of Chicago, 1967, 25 pp.

40. Fujita, T., The Lubbock tornadoes: A study of suction spots, *Weatherwise* **23,** 161–173, 1970.

41. Fujita, T., Proposed mechanism of suction spots accompanied by tornadoes, *Proceedings of the Seventh Conference on Severe Local Storms,* American Meteorological Society, 1971, pp. 208–213.

42. Fujita, T., D. L. Bradbury, and P. G. Black, Estimation of tornado wind speeds from characteristic ground marks, *University of Chicago SMRP Research Paper* **69,** 1967, 19 pp.

43. Fujita, T., D. L. Bradbury, and C. F. Van Thullenar, Palm Sunday tornadoes of April 11, 1965, *Monthly Weather Rev.* **98,** 29–69, 1970.

44. Fujita, T., J. J. Tecson, and L. A. Schaal, Preliminary results of tornado watch experiment 1971, *Proceedings of the Seventh Conference on Severe Local Storms,* American Meteorological Society, 1971, pp. 255–261.

45. Fulks, J. R., On the mechanics of the tornado, *NOAA Tech. Memo. ERLTM-NSSL* **4,** National Severe Storms Laboratory, Norman, Oklahoma, 1962, 33 pp.

46. Galway, J. G., The Topeka tornado of 8 June 1966, *Weatherwise* **19,** 144–149 and 160, 1966.

47. Gerrity, J. P. and R. D. McPherson, Development of a limited-area fine-mesh prediction model, *Monthly Weather Rev.* **97,** 665–669, 1969.

48. Glaser, A. H., An observational deduction of the structure of a tornado vortex, *Proceedings of the First Conference on Cumulus Convection,* Aerophysics Laboratory, AF Cambridge Research Center, Portsmouth,

New Hampshire, May 1959, Pergamon Press, 1960, pp. 157–166.

49. Golden, J. H., Waterspouts at Lower Matecumbe Key, Florida, September 2, 1967, *Weather* **23,** 103–114, 1968.

50. Golden, J. H., Waterspouts and tornadoes over South Florida, *Monthly Weather Rev.* **99,** 146–153, 1971.

51. Golden, J. H., *Life Cycle of the Florida Keys Waterspout as the Result of Five Interacting Scales of Motion,* Ph.D. Thesis, Florida State University, Tallahassee, Fla., 1973, 371 pp.

52. Grant, F. C., Proposed technique for launching instrumented balloons into tornadoes, *NASA Report TN D-6503,* 1971.

53. Gunn, R., Electric field intensity at the ground under active thunderstorms and tornadoes, *J. Meteor.* **13,** 269–273, 1956.

54. Gutman, L. N., Theoretical model of a waterspout, *Bull. Acad. Sci. USSR (Geophys. Ser.),* Pergamon Press translation, New York, Vol. 1, 1957, pp. 87–103.

55. Hadeen, K. D., Air Force global weather central boundary layer model, *AFGWC Tech. Memo. 70-5,* Air Weather Service Offutt Air Force Base, Nebraska, 1970.

56. Hardy, R. N., The Cyprus waterspouts and tornadoes of 22 December 1969, *Meteorol. Mag.* **100,** 74–82, 1971.

57. Harkness, J., The tornado at Little Rock, Arkansas, October 2, 1894, *Monthly Weather Rev.* **22,** 413–414, 1894.

58. Harrison, H. T., Notes on certain tornado and squall line features, *Meteor. Circular No. 36,* United Air Lines, 1952, 13 pp.

59. Harvey, J. K., Some observations of the vortex breakdown phenomenon, *J. Fluid. Mech.* **14,** 585–592, 1962.

60. Henderson, T. J., and W. J. Carley, The airborne seeding of six tornadoes, *Preprints 3rd Conference on Weather Modification, Rapid City, South Dakota,* American Meteorological Society, 1972, pp. 241–244.

61. Hoecker, W. H., History and measurement of the two major Scottsbluff tornadoes of 27 June 1955, *Bull. Am. Meteorol. Soc.* **40,** 117–133, 1959.

62. Hoecker, W. H., Wind speed and air flow patterns in the Dallas tornado of April 2, 1957, *Monthly Weather Rev.* **88,** 167–180, 1960.

63. Howard, L. N., and A. S. Gupta, On the hydrodynamic and hydromagnetic stability of swirling flows, *J. Fluid. Mech.* **14,** 463–476, 1962.

64. Jones, H. L., A sferic method of tornado identification and tracking, *Bull. Am. Meteorol. Soc.* **32,** 380–385, 1951.

65. Kessler, E., Purposes and program of the U.S. Weather Bureau National Severe Storms Laboratory, *Trans. Am. Geophys. Union* **46,** 389–397, 1965.

66. Kessler, E., Tornadoes, *Bull. Am. Meteorol. Soc.,* **51,** 926–936, 1970.

67. Kessler, E., On tornadoes and their modification, *Mass. Inst. Technol. Technol. Rev.* **74,** 6, May 1972, pp. 48–55.

68. Kinzer, G. D., and B. Morgan, Location and movement of lightning centers associated with a tornadic storm, Paper presented at the AGU-AMS Meeting in Washington, D.C., 8–11 April 1968.

69. Klein, W. H., The forecast research program of the Techniques Development Laboratory, *Bull. Am. Meteorol. Soc.* **51,** 133–142, 1970.

70. Koscielski, A., The Black Hills' tornado of 23 July 1967, *Proceedings of the Fifth Conference on Severe Local Storms, St. Louis, Missouri,* American Meteorological Society, Oct. 1967, pp. 226–228.

71. Kuo, H. L., On the dynamics of convective atmospheric vortices, *J. Atmos. Sci.* **23,** 25–42, 1966.

72. Kuo, H. L., Note on the similarity solutions of the vortex equations in an unstably stratified atmosphere, *J. Atmos. Sci.* **24,** 95–97, 1967.

73. Kuo, H. L., Axisymmetric flows in the boundary layer of a maintained vortex, *J. Atmos. Sci.* **28,** 20–41, 1971.

74. Landsberg, H. E. and T. N. Maisel, Micrometeorological observations in an area of urban growth, *Boundary-Layer Meteorol.* **2,** 365–370, 1972.

75. Leslie, L. M., The development of concentrated vortices: A numerical study, *J. Fluid. Mech.* **48,** 1–21, 1971.

76. Lewis, W. and P. J. Perkins, Recorded pressure distribution in the outer portion of a tornado vortex, *Monthly Weather Rev.* **81,** 379–385, 1953.

77. Lhermitte, R. M., Application of pulse Doppler radar technique to meteorology, *Bull. Am. Meteorol. Soc.* **47,** 703–711, 1966.

78. *Lightning,* PI660024, Public Information Office, Environmental Science Services Administration, Supt. of Documents, Washington, D.C., 1969.

79. Lilly, D. K., Tornado dynamics, *NCAR Manuscript 69-117,* National Center for At-

mospheric Research, Boulder, Colorado, 1969.

80. Long, R. R., Sources and sinks at the axis of a rotating liquid, *Quart. J. Mech. Appl. Math.* **9,** 385–393, 1956.

81. Long, R. R., Vortex motion in a viscous fluid, *J. Meteorol.* **15,** 108–112, 1958.

82. McLaughlin, J. M., Design of nuclear power plants for tornadoes, *Proceedings of the Conference on Tornado Phenomenology and Related Protective Design Measures,* University of Wisconsin, April 1970.

83. Mehta, K. C., J. R. McDonald, J. E. Minor, and A. J. Sanger, Response of structural systems to the Lubbock storm, *Texas Tech Univ. Storm Res. Rept. 03,* 1971.

84. Melaragno, M. G., *Tornado Forces and Their Effects on Buildings,* Kansas State University, Manhattan, Kansas, 1968, 51 pp.

85. Miller, R. C., Notes on analysis and severe storm forecasting procedures of the military weather warning center, *Air Weather Service Tech. Rept.* **200,** U.S. Air Force, 1967.

86. Moore, H. E., *Tornadoes over Texas, a Study of Waco and San Angelo in Disaster,* University of Texas Press, Austin, 1958, 334 pp.

87. Morgan, B. J., Tornado detection, tracking and interception, *Univ. Notre Dame, Coll. Eng. Rept.,* 1972.

88. Morton, B., Geophysical vortices, *Progress in Aeronautical Sciences,* Vol. 7, Pergamon Press, New York, 1966, pp. 145–193.

89. Morton, B. R., The physics of fire whirls, *G.F.D.L. Paper* **19,** Monash University, Clayton, Victoria, Australia, 1969.

90. Murray, F. W., Numerical models of a tropical cumulus cloud with bilateral and axial symmetry, *Monthly Weather Rev.* **98,** 14–28, 1970.

91. Newton, C. W., Severe convective storms, *Advan. Geophys.* **12,** 257–308, 1967.

92. Newton, C. W. and J. C. Fankhauser, On the movement of convective storms, with emphasis on size discrimination in relation to water budget requirements, *J. Appl. Meteorol.* **3,** 651–668, 1964.

93. Orville, H. D. and L. J. Sloan, A numerical simulation of the life history of a rainstorm, *J. Atmos. Sci.* **27,** 1148–1159, 1970.

94. Outram, T. S., Storm of August 20, 1904, *Monthly Weather Rev.* **32,** 365–366, 1904.

95. Palmen, E. and C. W. Newton, *Atmospheric Circulation Systems,* Academic Press, New York, 1969, 603 pp.

96. Pautz, M. E., Severe local storm occurrences

1955–1967, *ESSA Tech. Memo. WBTM FCST 12,* Office of Meteorological Operations, Silver Spring, Md., 1969.

97. Peace, R. L. and R. A. Brown, Comparison of single and double Doppler radar velocity measurements in convective storms, *Proceedings of the Thirteenth Radar Meteorological Conference,* American Meteorological Society, Boston, 1968, pp. 464–473.

98. Pearson, A. D. and L. F. Wilson, The tornado season of 1971, *Weatherwise* **25,** 20–25, 1972.

99. Ragette, G., Mesoscale circulations associated with Alberta hailstorms, *Monthly Weather Rev.* **101,** 150–159, 1973.

100. Rayleigh, J. W. S., On the dynamics of revolving fluids. *Proc. Roy. Soc. London Ser. A,* **93,** 148–154, 1916.

101. Reber, C. M., The South Platte Valley tornadoes of June 7, 1953, *Bull. Am. Meteorol. Soc.* **35,** 191–197, 1954.

102. Reynolds, G. W., A practical look at tornado forces, *Proceedings of the Conference on Tornado Phenomenology and Related Protective Design Measures,* University of Wisconsin, April 1970.

103. Rossow, V., Observations of waterspouts and their parent clouds, *NASA Technical Note D-5854,* National Aeronautics and Space Administration, Washington, D.C., 1970, 63 pp.

104. Rott, N., On the viscous core of a line vortex, *Z. Angew. Math. Physik* **96,** 543–553, 1958.

105. Scorer, R. S., *Natural Aerodynamics,* Pergamon Press, New York, 1958, 312 pp.

106. Segner, E. P., Estimates of minimum wind forces causing structural damage, U.S. Weather Bureau *Research Paper No. 41,* The Tornadoes at Dallas, Texas, April 2, 1957, 169–175, 1960.

107. Serrin, J., The swirling vortex, *Phil. Trans. Roy. Soc. London A,* **271,** 325–360, 1972.

108. Shuman, F. and J. B. Hovermale, An operational six-layer primitive equation model, *J. Appl. Meteorol.* **7,** 525–547, 1968.

109. Simpson, J. and V. Wiggert, Models of precipitating cumulus towers. *Monthly Weather Rev.* **97,** 471–489, 1969.

110. Skaggs, R. H., Analysis and regionalization of the diurnal distribution of tornadoes in the United States. *Monthly Weather Rev.* **97,** 103–115, 1969.

111. Somes, N. F., R. D. Dikken, and T. H. Boone, Lubbock tornado—a survey of building damage in an urban area, *NBS*

Report 10254, National Bureau of Standards, U.S. Dept. of Commerce, 1970, 69 pp.

112. Sullivan, R. D., A two-cell vortex solution of the Navier-Stokes equations, *J. Aerospace Sci.* **26**, 767–768, 1959.

113. Takeda, T., Numerical simulation of a precipitating convective cloud: the formation of a long lasting cloud, *J. Atmos. Sci.* **28**, 350–376, 1971.

114. Taylor, W. L., Review of electromagnetic radiation data from severe storms in Oklahoma during April 1970. *NOAA Tech. Memo. ERL WPL-6*, Wave Propagation Laboratory, Boulder, Colo. 1971, 80 pp.

115. Tepper, M. and W. E. Eggert, Tornado proximity traces, *Bull. Am. Meteorol. Soc.* **37**, 152–159, 1956.

116. Turner, J. S., The constraints imposed on tornado-like vortices by the top and bottom boundary conditions, *J. Fluid Mech.* **25**, 377–400, 1966.

117. Turner, J. S. and D. Lilly, The carbonated water tornado vortex, *J. Atmos. Sci.* **20**, 468–471, 1963.

118. U.S. Weather Bureau, Average number of tornadoes by months, *L. S. 6208*, revised 1962, U.S. Weather Bureau, 1962, 1 p.

119. Van Tassel, E. L., The North Platte Valley tornado outbreak of June 27, 1955, *Monthly Weather Rev.* **83**, 255–264, 1955.

120. Vonnegut, B., Electrical theory of tornadoes, *J. Geophys. Res.* **65**, 203–212, 1960.

121. Vul'fson, N. I. and L. M. Levin, Dispersion of developing cumulus clouds by artificially created down currents, *Dokl. Akad. Nauk SSSR* **181**, 855–857, 1968, (translated from Russian).

122. Wagner, N. K., A brief description and analysis of the Austin, Texas, tornado of May 10, 1959, *Bull. Am. Meteorol. Soc.* **41**, 14–17, 1960.

123. Waite, P. J. and C. E. Lamoureux, Corn striations in the Charles City tornado in Iowa, *Weatherwise* **22**, 55–59, 1969.

124. Ward, N. B., Radar and surface observations of tornadoes on May 4, 1961, *Proceedings of the Ninth Weather Radar Conference, Kansas City, Missouri*, American Meteorological Society, Oct. 1961, pp. 175–180.

125. Ward, N. B., The effect of low level wind shear on the formation of atmospheric vortices, *Proceedings of the Second Conference on Severe Storms, Norman, Oklahoma*, American Meteorological Society, Feb. 1962.

126. Ward, N. B., The Newton, Kansas tornado cyclone of May 24, 1962, *Proceedings of the Eleventh Weather Radar Conference, Boulder, Colorado*, American Meteorological Society, Sept. 1964, pp. 410–415.

127. Ward, N. B., An effect of rotation on a buoyant convective column, *Proceedings of the Fifth Conference on Severe Local Storms, St. Louis, Missouri*, American Meteorological Society, Oct. 1967, pp. 368–373.

128. Ward, N. B., Rotational characteristics of a tornado cyclone, *Proceedings of the Thirteenth Radar Meteorology Conference, Montreal, Canada*, American Meteorological Society, Aug. 1968, pp. 94–97.

129. Ward, N. B., The exploration of certain features of tornado dynamics using a laboratory model, *J. Atmos. Sci.*, **29**, 1194–1204, 1972.

130. Ward, N. B., A note on the effects of pressure gradients on fluid flow with atmospheric applications, *J. Atmos. Sci.* **29**, 982–984, 1972.

131. Ward, N. B. and A. B. Arnett, Some relations between surface wind fields and radar echoes, *Proceedings of the Third Conference on Severe Storms, Champaign, Illinois*, American Meteorological Society, Oct. 1963.

132. Weinstein, A. I. and P. B. MacCready, An isolated cumulus cloud modification project, *J. Appl. Meteorol.* **8**, 936–947, 1969.

133. Weller, N. and P. J. Waite, The Weller method: Tornado detection by television, *Preprints of the Sixth Conference on Severe Local Storms*, American Meteorological Society, Boston, 1969, pp. 169–171.

134. Wilkins, E. M., The role of electrical phenomena associated with tornadoes, *J. Geophys. Res.* **69**, 2435–2447, 1964.

135. Winston, J. S., Forecasting tornadoes and severe thunderstorms, *Forecasting Guide No. 1*, U.S. Weather Bureau, Kansas City, Mo., 1956, 34 pp.

136. Wolford, L. V., Tornado occurences in the United States. *Technical Paper No. 20*, Rev. Ed., U.S. Weather Bureau, Washington, D.C., 1960.

137. Ying, S. J. and C. C. Chang, Exploratory model study of tornado-like vortex dynamics, *J. Atmos. Sci.* **27**, 3–14, 1970.

138. Zegel, F. H., Lightning deaths in the United States: a seven year survey from 1959 to 1965, *Weatherwise* **20**, 168–173, 179, 1967.

17 | Lightning Modification

GEORGE DAWSON
DONALD M. FUQUAY
HEINZ W. KASEMIR

An Introduction to Atmospheric Electricity

GEORGE DAWSON

Atmospheric electricity is the study of all those phenomena of an electrical nature that occur in the atmosphere, though the term has usually been restricted to the troposphere. The subject therefore includes such diverse branches as the production of free charges (ions) in the lower atmosphere by the action of cosmic rays and radioactivity, the electrical conductivity of the atmosphere, fair and stormy weather atmospheric potential gradients and the potential of the ionosphere, the rather special processes in electrified clouds such as thunderclouds, and the production and properties of lightning discharges. It is the purpose of this chapter to show that these apparently unrelated areas of study are not grouped

together under the heading of atmospheric electricity for convenience only; they are all intimately connected with the complete atmospheric electrical circuit. We try here to point out the circuit elements of the atmosphere and show their interrelationship.

There will be some differences between the simplified global circuit presented here and the single thunderstorm circuit described by Kasemir.

Current is flowing in the atmospheric circuit and shows no sign of dying out, despite a time constant for decay of only minutes. That at least one of the circuit elements must therefore be an electrical generator was first proposed by Wilson in 1920, and since then it has been almost unanimously accepted that the generators are the thunderstorms taking place somewhere all over the world. The earth-atmosphere system is conceived of as a spherical capacitor whose inner conductor is the earth, whose dielectric is the air, and whose outer conductor is an ill-defined conducting layer called the electrosphere.[31] For the purposes of this chapter, we can equate the electrosphere with the ionosphere, though the difference will become apparent. As a dielectric, the air is unusual in that it is lossy (i.e., partially conducting), but not uniformly so. Its lossiness increases toward the

George Dawson is an Associate Professor in the Institute of Atmospheric Physics, The University of Arizona, Tucson, Arizona. Donald M. Fuquay is Chief Research Meteorologist at the Intermountain Forest and Range Experiment Station, Northern Forest Fire Laboratory, U.S. Department of Agriculture, Missoula, Montana. Heinz W. Kasemir is attached to the Atmospheric Physics and Chemistry Laboratory, National Oceanic and Atmospheric Administration, Boulder, Colorado.

outer layer until, when it is sufficiently conducting, it *becomes* the outer layer at a few tens of kilometers altitude. Embedded in the dielectric are the generators, the thunderstorms (also lossy), which separate charge continuously, keeping the inner electrode negatively charged and the outer electrode at a positive potential. The thunderstorms have the properties of *current* generators; however, at least as important are the potentials developed at the generator ends. It is the electric fields produced by thunderstorms that are responsible for most of their properties.

A thunderstorm does not reach from earth to electrosphere, so there are parts of the circuit joining the generators to the electrodes. The circuit to ground at the base of the thunderstorm is by far the most important passive element, and the most complex. The air at this altitude has low conductivity, that is, can transport very little charge, but under the influence of the thunderstorm electric field, a large point-discharge current is generated from the ground. In this chapter, this is considered to be the principal electric function of a thunderstorm. Some of the ionic current from the ground is lost by being captured by precipitation particles and returned to earth. The remainder is unable to load the thunderstorm generator appreciably, so spark breakdown of the poorly conducting air takes place, to give the spectacular cloud-to-ground lightning discharge. The charge carried by lightning, however, is only a small fraction of that from point discharge.

At the top of the thundercloud, the circuit element is the relatively conducting air.

The atmospheric circuit is completed by a "load" across the electrodes—in this case, the slightly conducting air over all those parts of the world experiencing fair weather. A schematic diagram of the complete circuit is shown in Figure 17.1.

The global electrical circuit seems to have reached an equilibrium state. The number of thunderstorms, the potential of the electrosphere, the current density of the air in fair weather, etc., do not seem to be changing with time. Clearly this equilibrium should not be disturbed; the consequent changes in weather and climate would be difficult to predict. In the following pages, we shall look in more de-

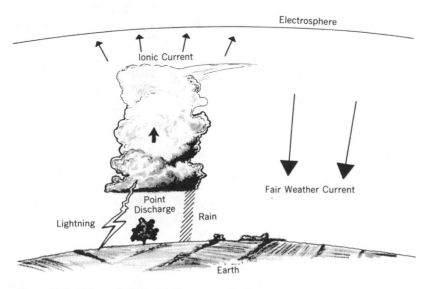

Figure 17.1 Schematic diagram of complete circuit.

tail at the factors which affect the equilibrium, to better understand their relative importance for its stability.

THE GENERATORS

Output

At any instant, there are about 2000 thunderstorms taking place worldwide. The number is not constant but has, in principle, two diurnal maxima corresponding to afternoon over equatorial West Africa and equatorial South America, where thunderstorm activity is particularly strong. In fact, these usually merge into one broad maximum between 1400 and 1900 GMT. An average thunderstorm seems to separate charge at the rate of a few amperes, some of which is lost by internal leakage. The polarity of the current seems to be always the same. The upper part of the cloud becomes positively charged, the lower negatively. The charging current builds up huge potential differences from top to bottom of the thundercloud. The potential of the cloud base has been estimated to be about -3×10^8 V, with a positive potential of similar magnitude being quoted for the cloud top. The net charge contained in the lower part of the cloud is calculated to be typically -30 C, though whether this is in fact a small difference between inner and screening charges an order of magnitude larger and of opposite sign is a subject of active interest.[14, 30,] Slightly smaller positive charges are calculated for the cloud top.

Charging Mechanisms

People have for many years tried to determine "the charging mechanism" of these thunderclouds and have been embarrassed by their own success. Not one, but several mechanisms seem able to account for the magnitude and polarity of the charging. In fact, rather strangely, none has been found that would give appreciable charging of the wrong polarity. Of course, several mechanisms are required. There is no reason to believe that all thunderstorms are charged by the same mechanism, or even that any *one* storm is so charged for the duration of its life. The range of thunderstorm types is surprisingly large, from cold clouds, that is, those having a cloud base higher than the 0°C isotherm, to warm clouds, that is, those lying totally below the 0°C isotherm, from continental to maritime.

Most thunderstorm theories separate the charging process into two parts: first, a microscopic charge separation in some organized fashion on hydrometeors of different sizes, followed by a large-scale separation of these charged particles under gravity. This gravitational separation mechanism has recently been questioned by S. A. Colgate[5] in an interesting and provocative paper; however, it would seem premature to discard gravitational separation on the evidence presented so far. Most theories differ, then, only in their microscopic charge separation mechanism. We will briefly review a few of the most favored processes.

1. Thermoelectric charging: when two ice surfaces of similar impurity content come into contact then separate, as in a bouncing collision between two ice crystals, then charge will move from one surface to the other if there is a temperature difference between the surfaces. The warmer ice surface is left negatively charged; the colder surface has an equal positive charge. For a given temperature difference, the magnitude of the charge is determined by the fundamental efficiency of the process, and the area and time of contact (plus other

secondary factors). Simple theoretical estimates of thermoelectric potential for the ice system give a value of about -2 mV/°C. Time of contact has two effects: if too short, not enough charge can be transferred; if too long, thermal conductivity reduces the temperature difference. The optimum contact time seems to be between 10^{-3} and 10^{-2} sec.

The mechanism of charge transfer is believed to be the migration of the lattice and orientational defects, $(OH_3)^+$, $(OH)^-$, vacant and doubly occupied hydrogen bonds, respectively. The mobility of the positive defects is greater than that of the negative. Since the number of defects rapidly increases with temperature, defects produced in the warmer ice migrate toward the colder ice, becoming annihilated on the way. Because of the difference in mobilities, the warm ice becomes negative, the cold ice positive.

In the thunderstorm, the various thermoelectric mechanisms differ only in the specific way in which the temperature difference is produced. The most important ways are probably frictional heating and latent heat. For example, in asymmetric rubbing collisions, where one surface slides over another, heat will be deposited with different densities, and temperature differences will result. Similarly, if the surfaces are of different roughness, the smoother surface slides on the high spots of the rougher surface, causing greater density of heat and a negative charge on the latter. This process is therefore a likely mechanism for rough hail and graupel particles falling through ice crystals. The effects of roughness were well shown by the experiments of J. Latham and A. H. Miller.[21]

The other major source of temperature difference is the latent heat of freezing of supercooled droplets. A hail or graupel pellet falling through a mixed cloud will accrete supercooled droplets which freeze on impact and keep its temperature above

ambient and above the temperature of other smaller ice crystals with which it also collides. This process was proposed and studied by Reynolds, Brook, and Gourley in 1957.[57] In both these processes, the pellet becomes strongly negatively charged, the crystals positive.

There are also other secondary contributions to temperature differences, e.g., thermal lag, which will not be discussed further.

2. Freezing potentials: rain and cloud drops are dilute solutions of salts, and when dilute solutions freeze, they produce freezing potentials; that is, the ice becomes charged (usually negatively) with respect to the water. The mechanism is the preferential incorporation of some of the anions (negative ions) of the salt into the ice lattice. The only exception is the action of ammonium salts, which makes the ice strongly positive. The potentials developed across the ice/water interface depend on freezing rate and solution concentration but reach maximum values at concentrations of about 10^{-4} M. Typical salt concentrations in continental rainwater are 10^{-6} to 10^{-4} M and at the coast up to 10^{-3} M. This potentially strong mechanism was discussed by Workman and Reynolds in 1950[45] and has since been studied by Prupacher, Steinberger, and Wang.[33] The amount of charge separated by this mechanism can be very great. There has, however, been difficulty in incorporating it into a complete electrification model.

3. Various other ice and water charging mechanisms: there is a considerable amount of evidence that surfaces undergoing riming with supercooled drops become charged by a mechanism that has never been fully explained. All experimenters except one found the riming surface to become negatively charged.

A related mechanism is the one studied

at length by Latham and Mason,[20] whereby supercooled cloud droplets with radii between 15 and 55 μ freeze and eject numerous positively charged splinters upon impact with a hail or graupel pellet. These results have not been confirmed by other workers.

Slight charging of ice has been observed upon evaporation and also upon melting. This latter effect (the Dinger-Gunn effect) has most recently been studied by J. C. Drake.[7]

4. Inductive charging: there is a whole class of inductive charging processes whereby precipitation particles undergo bouncing collisions while polarized by the external electric field of the atmosphere or the cloud. Subsequent separation of the particles enhances the external field. All possible collisions between ice and water particles can be considered, though the efficiency of the process clearly depends on the fraction of bouncing collisions—a number that has not been determined with any accuracy for water drop collisions, and would obviously depend rather critically on the field. An interesting study on water/ice interactions has recently been completed by Aufdermaur and Johnson.[1]

There is a basic objection, on the global scale, to inductive charging as the only major thunderstorm mechanism. Since the charging results from the electrical field of the atmosphere, that is, the potential difference between earth and electrosphere, and this latter is itself a *consequence* of thunderstorm charging, then this is a positive feedback situation, and equilibrium is not easily obtained.

5. Convective charging: Vonnegut[44] has described a charging mechanism in which positive space charge near the ground is drawn into the cloud updraft and transported to the upper part of the cloud. Negative ions above and around the cloud are attracted to, and become at-tached to, the outside of the cloud, and are transported by subsidence to the lower parts of the cloud. The model predicts a self-intensification of the charge until breakdown occurs. Most objections to the theory have been based on the need for organized rapid subsidence at the boundary of the cloud, for which there is little evidence.

6. Evaluation: it is not possible in a limited space to properly evaluate these mechanisms; they can all contribute to thunderstorm electrification. Inductive processes will always be present; with warm clouds, there will also be some convective charging and ion capture. For all other thunderclouds, the various thermoelectric mechanisms seem to be very powerful. Certainly large charging effects have been observed when ice hydrometeors, ice crystals, and supercooled droplets coexist in a cloud.[22] The minimum charging rate for any acceptable mechanism is usually taken as 1 C/km³ min.

Lossiness of the Generator

The major electrical losses in the thundercloud should be caused by ionic conduction, turbulent transport, and intracloud lightning discharges. The ionic conductivity inside a cloud is normally considered to be between 3 and 20 times smaller than in the free air at the same altitude. However, under intense thunderstorm electric fields, corona from hydrometeors should increase the conductivity significantly. The loss due to turbulent transport of charge is much less field dependent, and as pointed out by Imyanitov for many years, should be appreciable. Intracloud discharges are a major source of loss; cloud strokes are between three and five times more numerous than ground strokes, depending on latitude.[32]

THE GENERATOR—GROUND CIRCUIT

Lightning Flashes to Ground

When certain conditions are fulfilled, the lower part of a thundercloud partially discharges to ground via a lightning stroke. These conditions have never been completely defined. The mechanism by which a cloud-to-ground lightning stroke is produced has been the subject of several experimental and theoretical papers,[6, 23] but as of this time, the mechanism is not fully understood. Once started, however, the progress of the visible discharge is well documented. A diffuse, faintly luminous, so-called stepped-leader channel grows down from the cloud toward the ground, branching occasionally, and appearing to propagate in a stepped fashion, that is, a period of rapid advance, typically 50 m in 1 microsecond, followed by a pause of typically 50 μsec. During the pause period, the leader is not noticeably luminous. During each advance, only the new step addition is very luminous, the remainder faint. The average speed to ground of the leader is between 10^5 and 10^6 m/sec. Each leader deposits about 5 C of charge along its path. When the leader reaches 10 or 20 m from the ground, it is met by the upward-going stroke which travels up the leader channel and discharges it to ground. This is the very bright return stroke which is observed in lightning. The average speed of a return-stroke wave front is about 4×10^7 m/sec; peak currents of 1 to 4×10^4 A are reached in about 1 to 3 μsec, giving a rate of rise of current of about 10^{10} A/sec. Peak lightning temperatures are observed to be about 30,000°K. The stroke may be followed by typically 3 or 4 leader/return-stroke sequences along the same path at intervals of, say, 30 msec. The whole episode constitutes the visible lightning flash. The amount of charge brought to ground by a flash is thus of the order of 20 C. In a mature thunderstorm, flashes occur at intervals of between 20 sec and 1 min, indicating a generator charging current that is dissipated by lightning of the order of 1 A.

Point Discharge

Measurements indicate that the primary function of a thunderstorm is not the production of lightning but the generation of positive point discharge from the ground below, as a result of the thunderstorm electric field. The charge given off in this way appears to be at least five times that from lightning. It should also be noted that so far the measurements have been sparse, difficult, somewhat unreliable, and confined to land. The method consists of isolating a "representative" piece of ground, tree, or a metal point, and measuring the current from it. The major question is always how representative the sample was. The subsequent fate of the ions produced from the ground is important for the electrical circuit. We shall see that their lifetime before collection by aerosol particles is only about 30 sec over land, during which time they may have risen only some 30 m. Thereafter, their movement is controlled entirely by wind and negligibly by electrical forces. Even with a vertical wind speed as high as 2 m/sec between cloud and ground, it would take at least 1500 sec, that is, 25 min for them to reach cloud base. Most ions are therefore not going to be taken up by the generator, but instead will be swept away by the wind and eventually deposited as space charge. As a result, they are not a significant load on the generator. It is not necessary for these ions to reach the electrosphere to affect the global circuit, though they would have to change the current density between the resulting space charge and the electrosphere. The residence time of the excess charge could be a week or two.

Charge on Rain

A charge equivalent to between 20 and 30 percent of that from point discharge is returned to the ground on rain; the process may indeed be solely the collection of ions by raindrops polarized in the thunderstorm field, though some discrepancies remain.

THE GENERATOR–ELECTROSPHERE CIRCUIT

Above the thundercloud, the air is relatively conducting, there are no competing processes, and the circuit is very simple. If current conservation holds, the current through the generator can best be measured above the cloud. A measurement by Gish and Wait[11] gave a current of about 0.5 A; Stergis, Rein, and Kangas[38] obtained 1.3 A. Gish and Wait found the conductivity above the cloud to be the same as elsewhere at the same altitude.

These values may be misleading; current continuity need not hold, since the circuit contains capacitance. As a result, charge can be stored and leaked at different rates. Also, a knowledge of the current through the generator may not be enough. It is not clear how much, if any, of the all-important point discharge current goes by this path. The results above the cloud are in agreement with the contribution from cloud-to-ground lightning, though this may be coincidental.

THE LOAD–THE FAIR WEATHER ATMOSPHERE

By making measurements of the atmosphere's electrical conductivity as a function of altitude, it is possible to evaluate approximately a columnar resistance, that is, the resistance of a column of atmosphere of cross-section 1 m² stretching from the ground to the electrosphere. The value is about 10^{17} Ω/m^2. This gives a global resistance of the atmosphere of about 200 Ω. The potential difference between earth and electrosphere across this load is about 3×10^5 V.

The conductivity of the atmosphere results from the presence of free ions, since electrons attach to form negative ions in times of the order of 10^{-9} sec. The initial ions are caused by ionizing agents, cosmic rays, and radioactivity, and rapidly cluster to form complex small ions. These small ions can then be lost by mutual neutralization, by combination with aerosols to form large ions, and by neutralization with large ions of opposite sign. The mobility of small ions, that is, their speed in unit electric field, is typically 1 cm/sec per V/cm; for large ions it is typically 10^{-3} cm/sec per V/cm. As a result, only small ions contribute to the conductivity of the air. Their equilibrium number density is determined by their rates of production and loss. Some typical values near the ground are: generation rate of ion pairs, 10/cm³ sec; equilibrium concentration of small ions, 10^2 to $10^3/cm^3$; concentration of large ions, 10^3 to $10^4/cm^3$; and lifetime of a small ion, 10 to nearly 100 sec. The upper small ion number, the lower large ion concentration, and the longer lifetime values are for cleaner air. There is a roughly inverse relation between conductivity and aerosol content. Above the mixing layer, the conductivity starts to increase rapidly, since the aerosol content of the air is low and, at greater altitudes, the ion production by cosmic rays goes up. Near the cloud base, at 3 km, say, the concentration of small ions should be about $10^3/cm^3$ and their lifetime about 500 sec.

The current density through the fair weather atmosphere over the globe follows the rate of generation of charge by thunderstorms. Since the conductivity increases with altitude, the atmospheric potential gradient must decrease, indicating the presence in the atmosphere of positive space charges, that is, there is an excess of positive ions in the atmosphere. An excess of positive ions is also produced near the ground by other processes that we shall not discuss.

The fair weather field at the ground is associated with a bound charge of about 2×10^{-9} C/m². Since the fair weather current density is about 2×10^{-12} A/m², the charge on the earth would all decay at constant current in about 16 min, showing the need for continuous thunderstorm activity.

SUMMARY OF GLOBAL ELECTRICAL BUDGET

Estimates have been made by several authors of the various contributions to the electrical budget of the earth. One that balances in C/km² year is[12]:

Lightning	−20
Point discharge	−100
Precipitation	+30
Fair weather conductivity	+90
Total	0

A more recent estimate at Kew, England, by J. A. Chalmers[4] gives a total unaccounted for of −249 C/km² year, mostly produced by point discharge. It was hypothesized that a similar excess positive charge was deposited in desert or polar regions, where measurements have not been made.

LIGHTNING MODIFICATION

The major reasons for reducing lightning are to decrease forest fires and to prevent damage and reduce danger to expensive facilities, such as very large passenger aircraft or spacecraft. At first sight, the requirements for these cases appear different; in fact, they have much in common. In principle, forest fires could be stopped by eliminating those cloud-to-ground strokes having continuing currents. Since the cause of these currents is unknown, however, it may in practice be easier to reduce *all* lightning strokes.

There are four possibilities for reducing the lightning hazard:

1. Decrease the basic efficiency of the charging mechanisms.
2. Increase the lossiness of the generators.
3. Block the triggering mechanism that initiates the stroke.
4. Discharge the cloud artificially at a controlled rate without waiting for lightning.

Although lightning modification on a global scale is not contemplated at this time, it is as well to determine at an early stage which modification procedure would disturb the electrical budget of the earth least.

As has been pointed out, the contribution of cloud-to-ground lightning to the overall budget is much smaller than that of point discharge from the ground below thunderclouds. The best modification procedures would therefore maintain thundercloud fields at the ground while substituting for the lightning stroke another current of similar magnitude which passed through the thundercloud generator. Procedure 4 above is therefore superior in principle. Kasemir's chaff-seeding experiment below cloud (though close to cloud base) seems a reasonable approach to this ideal, though there are still some questions regarding ion lifetime and trajectory. His rocket experiments also come in category 4, though they offer less hope for modification over either large or congested airspace areas. Kasemir's proposed chaff seeding project within cloud would come under category 2, and Fuquay's cloud seeding experiment under category 1.

There is always the possibility of unwanted side effects, for example, rainfall reduction, resulting from lightning modification experiments. Again, experiments in categories 3 and 4 would be least likely to produce such side effects, since the cloud is being minimally affected. The only cloud changes would be electrical, and it is presently considered (without much evidence) that electrical forces do not have a significant effect on the growth of precipitation in the critical early stages. With cloud-seeding experiments, one might expect successful lightning suppression to be accompanied by some rainfall changes.

Lightning modification methods that have not been discussed are those in category 3. Most theories of stroke initiation rely on the presence near the cloud base of a small pocket of positive charge, which some consider to be the characteristic feature of a thundercloud. Basic investigations of the initiation process and the origin of this charge under various meteorological conditions seem badly needed.

Lightning Damage and Lightning Modification Caused by Cloud Seeding

DONALD M. FUQUAY

DAMAGE CAUSED BY LIGHTNING

From the dawn of history, lightning has influenced the activities and thinking of man. Schonland,[36] Viemeister,[43] and Uman[42] give many examples from history where lightning has altered events in wars and entered into political decisions of early empires. Today, more than ever, lightning continues as an ever-present threat to the lives and creations of mankind. It is estimated that some 2000 thunderstorms are in progress someplace on the earth's surface at any given time. These storms are producing cloud-to-ground lightning flashes at the rate of over 100/sec. In the United States, in an average year, lightning kills some 600 persons and injures about 1500. For this country, the death toll is greater for lightning than for other severe weather phenomena including tornadoes and hurricanes. Another area of concern is lightning effects on forests where natural resources of ever-increasing value are lost.[40] Lightning-caused forest fires contribute substantially to this loss. The annual cost of controlling lightning fires in the United States approaches $100 million. The total property losses caused by lightning in the United States can only be estimated, but is believed to be on the order of several hundred million dollars per year.

The invention of the lightning rod by Benjamin Franklin provided protection of structures, buildings, barns, and farmhouses. Only in recent years has the thought of introducing protection by preventing lightning at its source or by modifying lightning in a beneficial way advanced from hope or speculation to a distinct possibility.

There are two general situations where lightning protection is needed: one is protecting an object or activity covering only a small area; the second is providing protection over several hundred square miles for an extended period of time. In the first case, a lightning rod system on a building serves quite well to protect the structure. However, there are many other activities of a temporary nature that are vulnerable to lightning discharges but where conventional systems are impractical or inadequate. For example, while routine spacecraft launchings may be delayed to avoid sending them through electrified clouds, a space rescue mission cannot be delayed because the launching must take place at a given time. In this case, it is essential that some method be found to protect this rescue craft from the effects of lightning.

Examples of transient activities requiring lightning protection are refueling or fuel transfer operations, the handling of high explosives, assembling of nuclear devices, and others. All these activities need some type of lightning protection because the operation cannot be secured in the time that a storm can develop and produce lightning. A prime example of the need for large area protection from lightning over extended periods is for the prevention of lightning-caused fires in the forested areas of the world.

In this section we review the principles and status of three approaches to lightning modification: lightning modification using cloud seeding with freezing nuclei to alter the dynamic and microphysical processes affecting the lightning discharge; lightning suppression by chaff seeding to increase the conductivity in the cloud and keep the electric field below the lightning-igniting level; and artificially triggering lightning discharges to discharge a small section of the cloud for a limited period of time. The first two projects are directed to the prevention of lightning-caused forest fires and the third project to provide a fieldfree corridor through electrified clouds for the safe passage of a space vehicle. All three projects are still in a research stage; however, enough progress has been made to provide a sound foundation for future development and an assessment of expected benefits and limitations.

LIGHTNING MODIFICATION BY CLOUD SEEDING

Lightning-Caused Forest Fires

A lightning problem covering a large area is that of the ignition of forest fires. The problem is particularly acute in the western United States, including Alaska, and other forested parts of the world. Lightning ignites, on the average, some 10,000 reported forest fires in the United States per year. In some years the number approaches 15,000. These fires cause damage to commercial timber, watersheds, scenic beauty, and other forest resources. In addition to loss of human lives, lightning fires constitute a growing threat to homes, businesses, and recreational areas. The total number of lightning-caused fires is only one facet of this major problem. Several hundred fires can occur in a period of just a day. In one instance almost 1500 lightning fires were ignited in a two-state area in a period of 10 days. Control of these outbreaks of lightning fires is one of the major problems in carrying out sound forest management practices. At present, our best hope for reducing or solving this critical problem is application of weather modification techniques to reduce the severity of the lightning-fire load.

How could weather modification alleviate the lightning-fire problem? The most obvious approach is to reduce the number of cloud-to-ground discharges, particularly during critical fire periods. This could be accomplished by decreasing the charging rate within an incipient lightning storm or by encouraging the cloud system to dissipate through internal conduction currents and flashes. Another possibility is to alter characteristics of discharges which may favor forest fuel ignition.

Another possible approach has been used by the Bureau of Land Management in Alaska. They have tried for several summers to put out forest fires by using weather modification techniques to produce rains to put out the fires.

The probability of sustained forest ignition from any source is strongly influenced by environmental conditions. The time most favorable for ignition and spread of fires is after a long period of hot, dry weather. Low humidities and drying winds condition the fuel to the point where it can be ignited readily and fires spread rapidly. Any form of weather modification that produces rain, increases ambient relative humidities, and decreases the drying effects of direct solar radiation and wind would be beneficial. Since the moisture content of light, flash fuels are very sensitive to changes in relative humidity, even a small increase in ambient humidity results in a rapid increase in fuel moisture and a reduction of the probability of ignition and initial fire spread. For example, raising ambient relative humidity by only 15 percent can change the fire situation from "dangerous" to "easy." Thus initiating even a very light rain during critical fire conditions would be very beneficial.

Continuing on this same line of reasoning, should we expect all results from lightning modification for fire suppression to be beneficial? Could there be undesirable side effects? We have already seen that dynamic seeding of summer cumulus accelerates vertical growth and precipitation formation. The result can be lightning that would not normally occur. Severe downdrafts often accompany the small thunderstorms of the Rocky Mountain region when intensive evaporation occurs beneath the high cloud bases. Inducing small amounts of precipitation could enhance these downdrafts. The resulting erratic surface winds could seriously endanger firefighters and complicate the fire situation.[2] Also, a reduction in rainfall in some cases may be the

consequence of lightning modification efforts. Seeding should also affect the hail process, as is made evident by the strong association between hail and lightning occurrence in mountain thunderstorms.[3] Until more is known about the effects of seeding on incipient thunderstorms, unexpected and adverse effects must be considered, although improved numerical models that accurately predict cloud development and the effects of seeding should minimize the risk of unexpected events. In the case of experimenting near manned forest fires, a cautious and prudent approach is essential for the safety of firefighters.

It is assumed that, if the total number of cloud-to-ground discharges could be reduced by cloud seeding, there should be a corresponding decrease in the number of forest fire ignitions. Although probably true, this is certainly an oversimplification of the lightning-fire ignition problem. Other factors influence ignition, such as the type of discharge and its current duration, the nature of the storm as it influences discharge type and surface weather conditions, the terrain-fuel complex, and antecedent weather influencing fuel moisture. The nature of discharges causing forest fuel ignition has been observed and measured.[10] In agreement with early laboratory experiments with long sparks, it was found that forest fuel ignition is caused by hybrid cloud-to-ground discharges having long-continuing current phases exceeding 40 msec duration. Further, the probability of ignition is proportional to the duration of the continuing current phase. Thus it appears that lightning modification efforts to prevent forest fires must aim at reducing the total potential for lightning-caused ignition. This includes considering the effects of modification efforts on lightning, surface weather, and the nature and state of the fuel complex.

Forming plausible working hypotheses

for lightning modification experiments has been one of the most difficult phases of lightning modification research. Seeding has been shown to affect the dynamic characteristics of cumulus and the supercooled droplet-ice particle distributions within storms. Most of the available theories on cloud electrification processes suggest that these changes induced by seeding should affect the cloud electrification. However, none of the theories are sufficiently developed to provide valid estimates of the possible changes, [4, 25] On the other hand, additional data on storms are needed to extend our knowledge on the nature of the storms. The ability to induce glaciation in growing cumulus is in itself a valuable tool for studying electrification processes and testing available theories. Although there is much to be done in basic studies of cloud electrification, there are sound physical principles on which to build lightning modification hypotheses for testing. Within-cloud measurements by Latham and Stow[22] indicate a strong positive correlation existing between electrification and the coexistence of rimed aggregates of ice crystals, appreciable quantities of supercooled water droplets, and ice crystals. Their observations within clouds show that the strongest fields and hydrometeor charges are associated with the coexistence of all three types of particles. Lesser electrification occurs when the concentration of either supercooled droplets or ice crystals are diminished. Stow[39] hypothesizes that massive glaciation of growing cumulus by cloud seeding would reduce the supercooled water content, resulting in less electrification of the cloud. Another approach to lightning modification is alteration of the microphysical structure within the cloud. Laboratory studies of ice crystals and water drops in strong electric fields indicate that breakdown potentials are reduced and conduction currents increased

when water drops are converted to ice crystals. Fuquay[8] hypothesizes that glaciation induced by seeding should increase the within-cloud conduction currents and alter the initiation and nature of cloud-to-ground discharges.

Lightning Modification Research

Only a very limited amount of lightning modification work has been done in the United States. In this section, we describe a series of lightning modification experiments performed in the northern Rockies by the Forest Service of the U.S. Department of Agriculture. The first systematic program of lightning modification was conducted in western Montana in the summers of 1960 and 1961.[2] This 2-year pilot experiment was designed to test the effect of seeding on lightning frequency and to evaluate lightning observation methods and cloud seeding techniques in mountainous areas. Thirty-eight percent fewer cloud-to-ground discharges were recorded on seed days than on days when clouds were not seeded. Intracloud and total lightning were less by 8 and 21 percent, respectively, on seed days during the 2-year period. A nonparametric statistical test showed that if seeding had no effect, differences of this magnitude would occur about one time in four.

A second experiment lasting through three summer seasons was begun in 1965. The objectives were to gain additional information on mountain thunderstorms and to build a body of observations of lightning from both seeded and not-seeded storms. These data were then to be used to build working hypotheses and models for testing in future experiments. Appropriate statistical tests were included in the experiment as a basis for evaluating differences attributable to treatment.

Before considering the results of these experiments, let us look at lightning measurements and how they are interpreted. A lightning discharge can be identified and its charge and current change calculated from continuous measurements of the electrostatic field throughout the total discharge. In addition, much can be inferred about the charge transferred from measurements of electric field strength before and after the lightning flash. Needed are records of the electric field in sufficient detail to show the fine structure of the discharge with a time resolution of only a few milliseconds. A complete review of lightning measurement technology and terminology is given by M. A. Uman.[41]

Next we look at some examples of lightning measurements. In this report, any discharge that does not reach the earth is defined as intracloud, and those between cloud and earth, cloud-to-ground discharges. The cloud-to-ground discharges containing continuing-current intervals are referred to as hybrid discharges or flashes and all others are called discrete discharges or flashes. The very luminous component of a discharge is called a return stroke or stroke. Figure 17.2 is a composite of an oscillogram and photograph of a discrete cloud-to-ground discharge. The upper trace shows luminosity and the middle and lower traces indicate slow and fast electric field changes. This lightning discharge had three return strokes. The slow rises at the beginning of the R_1 and R_2 strokes are stepped leaders and their spacing indicates that there were two channels to ground. The two channels are also evident in the lightning-triggered photograph.

An oscillogram of a typical hybrid cloud-to-ground discharge is shown in Figure 17.3. This hybrid discharge is similar to the discrete discharge except that luminosity and field change continue for about 210 msec after the R_2 stroke.

Reference to this type of discharge as causing most lightning fires in the northern Rocky Mountains has already been made. Possible modification of the hybrid discharge by seeding is discussed later in this report. For comparison of characteristics, a typical intracloud discharge is shown in Figure 17.4.

The 1965–1967 cloud seeding experiment was conducted in a mountainous area west of the Continental Divide near Missoula, Montana. The seeding trials took place over the Missoula Valley and the surrounding mountains. Both ground-based and airborne silver iodide (AgI) generator units were used in the cloud seeding trials (Figure 17.5).

All the generators used in this experiment produce silver iodide particles by burning a 2-percent solution of silver iodide by weight and sodium iodide dissolved in acetone. Calibration indicated an output of about 7×10^{14} nuclei/g of AgI (5×10^{13} N/sec) effective at $-20°$C. The consumption rate of AgI during seeding operations averaged about 2.2 kg/hr with periodic maximums of 4.2 kg/hr. On an average we used 13.4 kg of AgI per seed day.

Aircraft penetrations of cumulus clouds provided direct evidence of the degree of glaciation attributable to seeding. The principal instrument used in these measurements was a continuous particle collector which produced replicas of cloud particles on transparent 16-mm film. The difference between water drops and ice particles can be detected easily by this instrument. The most striking evidence of glaciation was obtained by several penetrations of adjacent seeded and not-seeded cumulus clouds. In one case, observations taken a few minutes apart show an untreated cloud consisted primarily of supercooled water droplets with a very low concentration of graupel pellets, while a nearby moderately sized

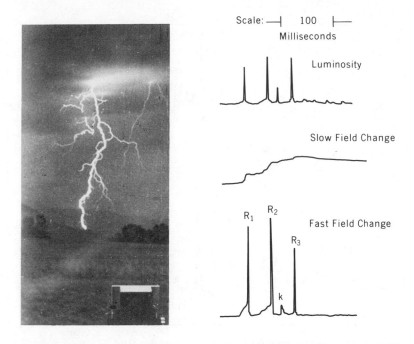

Figure 17.2 Photograph and oscillograph traces for a typical discrete cloud-to-ground discharge.

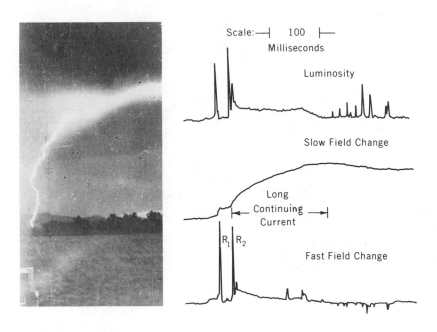

Figure 17.3 Photograph and oscillograph traces for a typical hybrid cloud-to-ground discharge.

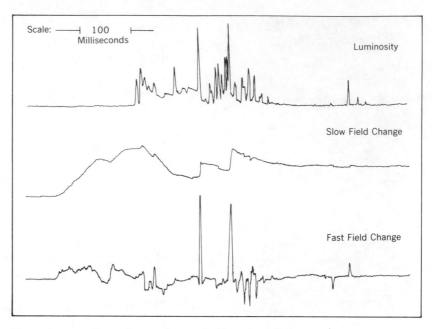

Figure 17.4 Oscillograph traces for a typical intracloud discharge.

Individual Storm of Aug. 14, 1966

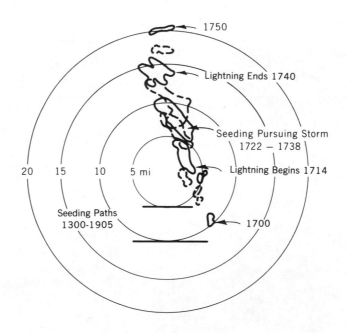

Wind 170-180° 33kts.

Figure 17.5 Illustration of radar echoes and seeding patterns for individual storms.

seeded cloud was essentially completely glaciated.[24] These cloud penetrations, along with many observations of cloud glaciations from the ground, indicate that extensive glaciation by cloud seeding is possible. In some cases, complete glaciation can be induced in growing cumuli at within-cloud temperatures as warm as $-7°C$.[24]

Results

Next we look at some results of the 1965–1967 experiments. The experiment yielded a sample of 12 seeded and 14 unseeded storms whose life history was thoroughly documented by visual, photographic, radar, and lightning-sensing observations. Since these 26 storms existed as distinct entities, they are called *individual storms*.

Analysis of data on the basis of the life cycle of the 26 individual thunderstorms gave the following results at the given level of significance for a two-tailed rank-sum test.[9]

1. Sixty-six percent fewer cloud-to-ground discharges, 50 percent fewer intracloud discharges, and 54 percent less total storm lightning occurred during seeded storms than during the not-seeded storms ($p = 0.17$, 0.08, and 0.10, respectively).

2. The maximum cloud-to-ground flash rate was less for seeded storms: Over a 5-min interval, the maximum rate averaged 8.8 for not-seeded storms and 5.0 for seeded storms ($p = 0.20$); for 15-min intervals, the maximum rate for not-seeded storms averaged 17.7 and 9.1 for seeded storms ($p = 0.16$).

3. The mean duration of lightning activity for the not-seeded and seeded storms was 101 and 64 min, respectively ($p = 0.086$). Lightning duration of the not-seeded storms ranged from 10 to 217 min, while that of seeded storms ranged from 21 to 99 min.

4. There was no difference in the average number of return strokes per discrete discharge (4.1 not-seeded versus 4.0 seeded, $p = 0.928$); however, a significant difference was found for hybrid discharges (5.6 not-seeded versus 3.8 seeded, $p = 0.004$).

5. The average duration of discrete discharges (period between first and last return stroke) decreased from 235 msec for not-seeded storms to 182 msec for seeded storms ($p = 0.004$).

6. The average duration of continuing current in hybrid discharges decreased from 187 msec for not-seeded storms to 115 msec for seeded storms ($p = 0.003$).

Most pertinent to the discussion of lightning-fire prevention is the distribution of hybrid stroke durations from the not-seeded and seeded storms. The distribution of hybrid stroke durations from not-seeded and seeded storms is shown in Figures 17.6a and 17.6b. As the reported statistic implies, the distribution in the seeded storms is shifted toward a larger percentage of short-duration strokes and a substantial reduction in mean stroke duration. The distribution of hybrid stroke duration for fire starters is shown in Figure 17.6c. This sample, although very small, clearly implies that the probability of ignition increases with increasing duration of the hybrid stroke component. These results show that for the sample available, the duration of hybrid discharges was considerably shortened by cloud seeding. If seeding does cause a substantial reduction of the continuing-current portions of the hybrid discharge, this lessens the ability of an individual discharge to ignite fuels. This modification of the nature of the discharge may be much more important than any change in the total amount of lightning produced by a storm.

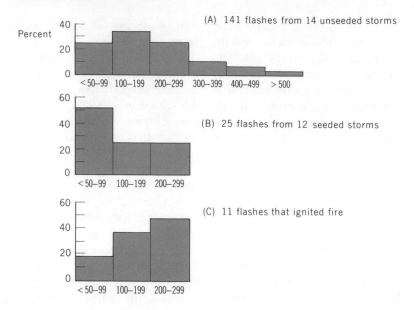

Figure 17.6 Distribution of long-continuing current durations for (*a*) hybrid flashes during 14 individual storms, (*b*) hybrid flashes during 12 seeded individual storms, and (*c*) flashes known to have ignited forest fires.

Progress in Lightning Modification

The experiments in lightning modification by cloud seeding reported here have yielded results strongly suggesting that, in some instances, lightning can be modified in a beneficial manner. Further, these very limited experiments have yielded distributions of lightning events and their characteristics for a test group of storms. This body of data is being used to develop lightning modification hypotheses. Progress has been made in increasing the power of lightning modification experiments by the identification of significant covariates such as the association betwen lightning occurrence and storm size, updraft values, precipitation rates, hail, and others. However, the apparent success of these early experiments in lightning modification for fire prevention, and the great advancements in the entire field of weather modification, should not obscure the magnitude of the research effort that is yet to be done to clearly identify and quantify the degree and applicability of lightning modification to the lightning-fire problem.

Lightning Suppression by Chaff Seeding and Triggered Lightning

HEINZ W. KASEMIR

LIGHTNING SUPPRESSION BY CHAFF SEEDING

The Physical Concept

The purpose of seeding a thunderstorm with chaff is to inhibit lightning discharges. The physical concept of this method is to increase the conductivity of the cloud by ionizing the air by corona discharge on the chaff fibers so that the electric field is kept below the value necessary to ignite lightning. The principal idea of chaff seeding can be easily demonstrated by a laboratory experiment. A metallic sphere of about 0.5-

m diameter is placed over a grounded plate with an air gap of 10 to 20 cm between sphere and ground. If the sphere is charged to several 100 kV—let's say, by a Van de Graff generator—sparks will flash over between the sphere and the grounded plate. If we now bring into the air gap one chaff fiber of about 5 cm length, sparkover will stop immediately. The chaff fiber is attached to the end of a long thin Teflon rod, that is, it is well insulated and doesn't touch either the sphere or the plate. If we remove the chaff fiber from the air gap the sparks will flash over again. In this experiment the charged sphere represents the thunderstorm, the plate is the earth's surface, and the sparks are cloud-to-ground lightning. The chaff fiber shows the effect of chaff seeding. The effect can readily be explained. Corona discharge at the two ends of the fiber produce a large number of ions which flow in a wide stream from the upper end of the fiber to the sphere and from the lower end to the plate. This ion flow increases the conductivity of the air between sphere and plate and the resulting current—if not actually shorting the sphere to ground—is such a load on the Van de Graff generator that

the voltage at the sphere drops below the flashover voltage.

This experiment shows us that there are three important conditions that must be fulfilled if chaff seeding of a thunderstorm is to be effective in suppressing lightning discharges. These conditions are (1) that the volume of air in a storm or between the storm and ground can be made conductive by the corona discharge on the chaff fibers; (2) that the electric field necessary to produce lightning is higher than that to produce corona discharge; and (3) that the current induced by corona discharge will load down the thunderstorm generator so that the electric field remains below the lightning igniting value.

Let us construct an electrical circuit diagram which is equivalent to a thunderstorm (see Figure 17.7.) On the left side of Figure 17.7, the generator symbols G_1 and G_2 represent the positive and negative charge generation in the top and the base of a storm which would result in a charge distribution in the cloud, as shown on the right side of the figure. G_1 and G_2 are assumed to be constant current generators related to the microphysical processes in the cloud

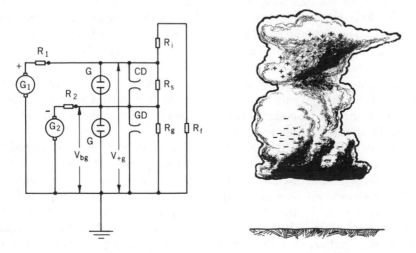

Figure 17.7 Left, circuit diagram of a thunderstorm. Right, charge distribution in a thunderstorm.

which is steadily producing positive and negative ions. The resistors R_1 and R_2 associated with these generators control the supply current, that is, the charge production of the cloud. The potential from cloud bottom to ground V_{bg} due to G_2 and the potential from cloud top to ground V_{tg} due to G_1 both vary with the strength of the current source and also with the resistances between the respective terminal and ground. Three different means are provided in the circuit diagram for the dissipation of charge. These are, from left to right in Figure 17.7, the two glow lamps G, the two spark gaps CD and GD, the ohmic resistors R_s, and the series R_i, R_f, R_g. The current flow through the ohmic resistors represents the conduction current of a thunderstorm. R_s is the shunt resistor between the hot terminals of the two generators; it is determined by the conductivity of the column of air between the upper positive and the lower negative charge inside the storm. R_g is the resistance of the air column between the base of the cloud and the ground, and R_i the resistance of the air column between the top of the storm and the ionosphere. This branch of the circuit is closed by R_f which represents the resistance between the ionosphere and the earth in the fair weather areas. With the exception of R_f, the resistors have comparatively high values (on the order of hundreds of megohms) so that the charge produced by the generators cannot leak away very fast. As a consequence, the voltage at the terminals builds up to high values until a spark is ignited through the spark gaps CD or GD. These spark gaps picture the lightning discharges in a real thunderstorm, CD representing the cloud discharge and GD the ground discharge. The glow lamps of the circuit diagram have no natural equivalent in the thunderstorm. They represent the artificially introduced corona discharge of the chaff fibers. If the voltage across the glow lamps reaches their breakdown value, the lamp will ignite and keep the voltage very effectively at this value. Any further increase in the current output of the generators will be shunted through the glow lamps. If the glow lamps have a lower ignition voltage than the spark gaps, no flashover at the spark gaps will occur.

It is generally assumed that the average thunderstorm produces a current output of about 3 A, which is equally divided between the three load circuits. Between the two main charge centers inside the storm through the shunt resistor R_s, 1 A flows as conduction current. From a technical point of view, this current may be considered a leakage current across faulty insulation between the poles of the generator. It is the purpose of chaff seeding to increase this current by increasing the conductivity between the generator poles so much that the voltage between the poles is reduced below the lightning igniting voltage. It will be the topic of the next section to discuss in detail how the conductivity is increased by the corona discharge on the chaff fibers.

Another part of the current output flows from the positive pole in the top of the storm to the ionosphere through the resistor R_i, spreads out in the ionosphere over the whole globe, flows down to earth through the resistor R_f, and returns to the negative pole of the thunderstorm generator through the resistor R_g. This is the contribution of the storm to the atmospheric electric fair weather field. The current flow in this branch — the global circuit — is assumed to be on the order of 1 A.

The last part of the thunderstorm charge production is dissipated in lightning discharges. If we assume that the average thunderstorm produces one lightning discharge every 30 sec and that 30 C is discharged by each lightning, 1-A continuous-supply current is required to feed the lightning activity. More severe storms may generate lightning discharges every 5 or 10 sec. With the same amount of coulomb discharged, the lightning supply current would be on the order of 3 to 6 A. As has been mentioned, the thunderstorm is in a technical sense, a current generator with a given current output; therefore, if lightning activity is to be suppressed, the lightning supply current of 1 to 6 A has to be channeled through another circuit, which in our case is the two glow lamps in parallel with the resistors R_s and R_g. To absorb this current without increase in the terminal voltage these resistors have to be decreased to $\frac{1}{2}$ to $\frac{1}{5}$ of their natural value by chaff seeding. For proper operation of the circuit the ignition voltage of the glow lamp should be less than the flashover voltage of the spark gap, and the glow lamp should be capable of absorbing the spark gap supply current without a substantial increase in the generator voltage.

Applied to the thunderstorm problem this means that the field to start corona discharge on the chaff fibers should be lower than the lightning-igniting field and that the conductivity of the chaff seeded volume should be increased to such an extent that the lightning supply current can be absorbed without a significant increase of the field. The first condition is met easily because the corona onset field of a chaff fiber 10 to 100 μ thick and 5 to 10 cm long is of the order of 30 kV/m, whereas the lightning-igniting field is about 500 kV/m, that is, more than 10 times as large. The fulfillment of the second condition is not so easily asserted and will be discussed in the next section.

The Effect of Corona Discharge at Chaff Fibers on the Conductivity of the Cloud

The conductivity of the air stems from the ionization of air molecules by cosmic rays and radioactive emanation of the ground. The ion production is given by the ionization constant q, which is of the order of 5 to 20 $\times$ 10^6 ion pairs/sec m^3. One single chaff fiber with a corona current of 1 μA will produce 6 $\times$ 10^{12} ion pairs/sec; that is, it would be equivalent to the natural ion production of a volume of 3 $\times$ 10^5 m^3 or of a cube of 68-m length, width, and height. If we require that the ion production in the cube should be increased three times, then the chaff fiber with 1-μA corona current would provide the required number of ion pairs for a cube with a side length of about 46 m. A cloud volume of 4 $\times$ 4 $\times$ 4 km^3 = 64 $\times$ 10^9 m^3 contains 65 $\times$ 10^4 such cubes. This means that about half a million chaff fibers would be enough to increase the ion production in the cloud by a factor of 3. Implicit in this estimate is the assumption that a field of about 70 kV/m exists to produce the 1-μA corona current and that the chaff fibers are evenly distributed throughout the volume.

In estimating the increase of conductivity by an increase of ion production we have to take into account the loss by recombination and attachment. As the negative ion stream released by the chaff fiber is moving upward and the positive ion stream is moving downward, ions of opposite polarity do not meet; consequently, there is no ion loss by recombination of small ions. Only if the upward moving negative ion stream meets the downward moving positive ion stream produced by another chaff fiber floating at a higher level will there be recombination of small ions. However, this recombination or intermixing will be beneficial for a continuous current flow, as will be discussed later.

A heavy loss of small ions will occur in the first few seconds by attachment to the cloud droplets until the droplets are charged to capacity and reject further attachment of ions of the same sign. An area of negative space charge will form above the chaff fiber and one of positive space charge below it. This will have two

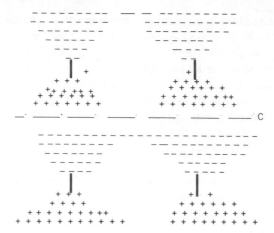

Figure 17.8   Positive and negative space charge distribution of an array of chaff fibers.

effects. First the presence of a negative space charge in the negative ion stream will deflect the ions to both sides until the hitherto uncharged cloud droplets at the sides become also negatively charged. This will result in a kind of trumpet-shaped space charge, as depicted in Figure 17.8. These negative and positive space charges forming above and below the single chaff fiber will generate an electric field, opposed to the existing field, which produces the corona discharge. If the space charge can accumulate unchecked or their opposing field is not cancelled by other means, then the corona discharge will be quenched by its own space-charge generation.

However, this result is applicable only to a single chaff fiber or to the uppermost or lowest layer of the seeded area. If we have two layers of chaff fibers, as in Figure 17.8, the fields of the inner positive and negative space charge layers will approximately cancel at the locus of the chaff fiber. The extrapolation to more than two layers of chaff fibers is evident. Such a volume filled with positive and negative space charge pockets resembles very strongly that of a dielectric material polarized by an external field. The local fields of the internal dipoles cancel each other by the sheer number of dipoles, and

the reduction of the primary field inside the dielectric can be interpreted as the effect of the surface charge at the boundary of the dielectric material.

In the case of chaff seeding for the purpose of lightning suppression, we require in addition to the field reducing effect that an enhanced current flow is carried through the seeded area. This would necessitate that at the contact area—marked by a dashed line C in Figure 17.8—between the positive space charge of the upper chaff fiber and the negative space charge of the lower chaff fiber a strong recombination between the opposite charged cloud droplets takes place, or that this contact area is penetrated by a sufficient number of positive small ions coming down from the upper chaff fiber and of negative small ions coming up from the lower chaff fiber. Even if the small ions recombine rapidly with cloud droplets or small ions of opposite polarity they reduce the space-charge pockets and stimulate a continuous corona discharge. For a continuous current flow through the seeded area it is necessary only that each chaff fiber carry the current through its own region of influence.

A numerical analysis to determine the size of the region of influence is extremely difficult. Probably the best way to solve this problem is to carry out measurements in a large cloud chamber equipped with a plate condenser, which is capable of generating fields in the order of 100 kV/m; however, there are three effects which tend to reduce the little pockets of space charge, thereby helping to prevent the quenching of the corona discharge, and to support the current flow. The effects are turbulent mixing, recombination of positively and negatively charged cloud droplets, and washout by precipitation. The enhanced coagulation of oppositely charged cloud droplets may even lead to precipitation enhancement. This would apply also to raindrops falling through the seeded area. The raindrop picks up a number of highly charged cloud droplets passing through, say, the upper negative space charge pocket. It becomes negatively charged itself. Entering the positive space charge pocket below, the raindrop now coagulates with the positive charged cloud droplets, loses its negative charge, and becomes positively charged. This process continues alternately until the raindrop leaves the seeded area. Besides reducing the space-charge pockets this also increases the growth rate of the raindrop.

We assume that an enhanced current flow can be carried through the seeded area. Given a sufficient overlapping of the region of influence of the chaff fibers, the large number of small ions liberated by corona discharge, the recombination of charged cloud droplets, the effect of turbulent mixing, and the washout of space-charge pockets by precipitation are lumped together in one material parameter, namely, the conductivity inside the seeded area, with the resulting effect that the conductivity inside the seeded area is increased by corona discharge. Accepting the assumption that the conductivity inside the seeded area is increased by chaff seeding, the problem of the field concentration at the surface of a conductive body in an electric field

comes immediately to mind. If we also generate by chaff seeding regions of field concentration, this method of lightning suppression may inadvertently turn out to be a method of lightning triggering. It is therefore quite important to have quantitative answers to the following questions: (1) how does the field concentration at the boundary depend on the conductivity ratio inside to outside of the seeded area; (2) how does it depend on the geometrical shape of the seeded area; and (3) how much is it changed if we are dealing with the inhomogeneous field of a thunderstorm generated by space charges instead of the general assumption of a homogeneous field.

The solution of the third problem is more difficult and has been worked out and discussed.[18] The solution to problems 1 and 2 is given in the next section.

The Field Concentration at the Boundary of the Chaff Seeded Area

A body of a given shape with the constant conductivity λ_2 is embedded into an environment of the constant conductivity λ_1. The whole system is exposed to a homogeneous electric field F_1 in the direction of the z axis of a Cartesian coordinate system $x, y,$ and z. Find the field F_2 inside the body and the maximum field F_m at the boundary of the body. The body shall be represented either by an infinitely long elliptical cylinder or by a spheroid. The solution shall be given for the cylinder for the following cases: (ac) the small axis of the elliptical cross-section is in the z direction, (bc) both axes are equal, in which case the elliptical cylinder becomes a circular cylinder, and (cc) the large axis of the elliptical cross-section is in the z direction; and for the spheroid, for (as) a flat disc, with the short axis in the z direction, (bs) a sphere, and (cs) a prolonged spheroid, with the long axis in the z direction.

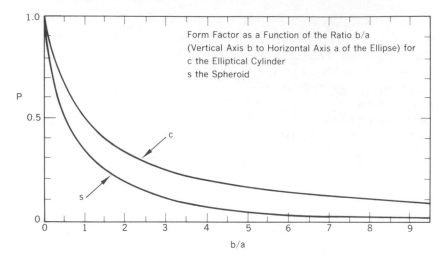

Figure 17.9   Form factor for ρ as a function of the ratio b/a.

The cases ac to bc may be encountered if the chaff is dispersed continuously from an airplane and the cases as to cs if the chaff is dispersed discontinuously in little packages from an airplane, dropsonde, or rocket.

The potential function of cases ac to cc can be obtained from *Static and Dynamic Electricity* by Smyth[37] if the dielectric problem is transformed into a current-flow problem by substituting the conductivities λ_1 and λ_2 for the two dielectric constants ξ_1 and ξ_2. The solution to problems as to cs is given by Kasemir.[13]

From these potential functions a very simple equation has been obtained for the maximum field concentration F_m at the boundary of the seeded area, as well as the attenuated field F_2 inside the seeded area. These equations are valid for all the problems stated earlier and are

$$\frac{F_2}{F_1} = \frac{\lambda_1/\lambda_2}{\lambda_1/\lambda_2 + (1 - \lambda_1/\lambda_2)\rho} \quad (17.1)$$

and

$$\frac{F_m}{F_1} = \frac{1}{\lambda_1/\lambda_2 + (1 - \lambda_1/\lambda_2)\rho}. \quad (17.2)$$

From Equations 17.1 and 17.2 also follows

$$F_2 = \frac{\lambda_1 F_m}{\lambda_2}, \quad (17.3)$$

where F_1 is the original field outside the seeded area, F_2 is the field inside the seeded area, F_m is the maximum field at the boundary, λ_1 is the conductivity outside the seeded area, λ_2 is the conductivity inside the seeded area, and ρ is the form factor, which depends only on the geometrical shape of the seeded area (Figure 17.9). With the horizontal half-axis a and the vertical half-axis b of either the elliptical cylinder or the spheroid, ρ is given for the elliptical cylinder by

$$\rho_c = \frac{a}{a + b} \quad (17.4)$$

and for the spheroid by

$$\rho_s = \frac{a^2}{a^2 - b^2}\left(1 - \frac{b}{\sqrt{a^2 - b^2}}\right.$$
$$\left. \cdot \tan^{-1}\frac{\sqrt{a^2 - b^2}}{b}\right). \quad (17.5)$$

For $a \to \infty$ the elliptical cylinder, as well as the spheroid, degenerate into a horizontal infinite layer. Note that ρ_c and ρ_s approach 1. For $\rho = 1$ it follows from Equation 17.1

$$\frac{F_2}{F_1} = \frac{\lambda_1}{\lambda_2}. \qquad (17.6)$$

As F_1 and λ_1 are given parameters the field F_2—according to Equation 17.6—is inversely proportional to λ_2. This is a well-known result and follows from the assumption of a continuous current flow. If $i_1 = \lambda_1 F_1$ is the current density outside and $i_2 = \lambda_2 F_2$ inside the seeded area we obtain from Equation 17.6

$$i_2 = i_1. \qquad (17.7)$$

If we bring a conductor into a homogeneous field there is usually a field concentration at the boundary. However, in this case inserting $\rho = 1$ into Equation 17.2 we obtain

$$F_m = F_1. \qquad (17.8)$$

The maximum field F_m is equal to the original field F_1 outside the seeded area; that is, there is no field augmentation at the boundary.

If we go to the other extreme and let $b \to \infty$ then the elliptical cylinder deteriorates into a vertical infinite layer and the spheroid into an infinitely long circular cylinder. In this case ρ_s and ρ_c approach zero. From Equations 17.1 and 17.2, we obtain

$$F_2 = F_1 \qquad (17.9)$$

and

$$\frac{F_m}{F_1} = \frac{\lambda_2}{\lambda_1}. \qquad (17.10)$$

These are somewhat unexpected results. Equation 17.9 says that the field inside the seeded area is the same as outside the seeded area no matter how much we increase the conductivity inside the seeded area. This means that continuous seeding from a dropsonde, where we may generate a body of seeded area like a

prolonged spheroid, would have almost no field-reducing effect. Furthermore, we learn from Equation 17.10 that the field concentration on the ends of the prolonged spheroid is proportional to the inside-outside conductivity ratio of the seeded area. Both effects are favorable for generating lightnings and adverse to lightning suppression; therefore, the manner in which a cloud is seeded will have some effect on the outcome, that is, if lightning is suppressed or prematurely triggered.

Equation 17.2 answers completely questions 1 and 2 stated at the beginning of this section. The maximum field concentration F_m/F_1 at the boundary of the seeded area and the field attenuation F_2/F_1 inside the seeded area is shown in Figure 17.10 as a function of the ratio b/a (vertical to horizontal dimension) for the cylindrical as well as the spheroidal case.

To discuss the influence of the conductivity λ_2 of the seeded area on the field concentration we go first to the extreme case that $\lambda_2 = \infty$. This would be equivalent to the case of a metallic conductor in an electrostatic field. According to Equation 17.2 the field concentration factor is $F_m/F_1 = 1/\rho$. This means that the reciprocal of the form factor is identical with the electrostatic field concentration factor. For the sphere, for instance, $\rho = \frac{1}{3}$ and $F_m/F_1 = 3$. If λ_2 is not infinite, but a finite multiple of λ_1, the field concentration factor decreases. In the case of the sphere and for $\lambda_2 = 2\lambda_1$ it drops to 1.5. Finally, for $\lambda_2 = \lambda_1$ the field concentration factor is 1, as it should be. Question 2, about the influence of the geometrical shape on the field concentration factor, is even easier to answer because the maximum "electrostatic" field concentration factor for all shapes can be obtained from $1/\rho$, as shown in Figure 17.9. In general the smaller the ratio b/a (vertical to hori-

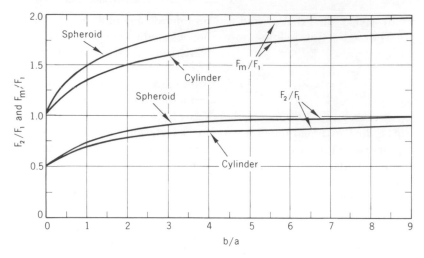

Figure 17.10 Field concentration F_m/F_1 at the boundary of the seeded area. Field attenuation F_2/F_1 inside the seeded area.

zontal dimension of the seeded area), the smaller the field concentration factor, with the cylindrical shape more favorable than the spheroidal shape. The electrostatic field concentration factor of the horizontal circular cylinder is 2 as compared to the sphere, where it is 3. Assuming again an inside conductivity $\lambda_2 = 2\lambda_1$, the value is reduced to 1.33 as compared to the sphere value of 1.5. This means that line seeding from an airplane flying horizontally through or below the cloud will produce the most desirable shape of the seeded area.

Using as a reference a 1973 paper by Kasemir,[19] we can answer question 3— how much is the field concentration factor changed if we are dealing with the inhomogeneous field of a thunderstorm instead of a generally assumed homogeneous field—as follows. The seeded area has an equalizing effect on the field distribution of a thunderstorm. It will decrease the maximum field value between the two main charge centers and will increase the outside field at the upper and lower boundary of the seeded area, analogous to the maximum field

concentration in a homogeneous field. However, this value is lower than in the case of the homogeneous field and if the seeded area is large enough to reach the poles of the thunderstorm generator, it is even lower than the maximum field of the not-seeded storm.[18]

Field Tests with Chaff Seeding

Field tests of chaff seeding underneath thunderstorms have been carried out at Flagstaff, Arizona in 1965 and 1966. For these tests a C-47 airplane was equipped with the following instruments: (a) two field mills measuring the three components of the electric field; (b) two chaff dispensers; (c) one corona discharge indicator; and (d) sensors for different meteorological and airplane parameters, including a strip chart and tape recorder.

Each field mill records two components of the atmospheric electric field and automatically eliminates the influence of the airplane charge on the measurement.[18]

The chaff dispenser contains the chaff not as fibers cut to a certain length and pressed a million apiece into little packages, but as a long strand of conductive fibers wound up on a reel. Ten reels constitute one dispenser unit housed in one wing tank. During operation the 10 strands are forced out through 10 guide holes at great speed, and before leaving the tank completely are chopped by a helical chopper into fibers of a preset length. This design has several features that are crucial for lightning suppression. (1) The chaff is emitted continuously. Bird nesting, that is, bunching together in clumps of several hundred or thousand fibers, is completely eliminated. (2) The chaff is distributed more evenly behind the airplane because a continuous stream of fibers and not individual packages emerges from the airplane. (3) It is possible to experiment with different lengths of fibers, the length depending on the speed of the chopper, which is easily adjustable.

The corona discharge indicator measures the electromagnetic emission of the chaff fibers as soon as they emerge from the chaff dispenser. This range should be limited to 50 to 100 m and the indication selective to corona discharge on the chaff only; that is, corona dis-

charge on the airplane itself should not be indicated. The last point is difficult to establish and needs a more detailed study. Otherwise the instrument seems to be working properly. One essential point has already been confirmed namely, that corona discharge occurs on chaff fibers if the electric field in the atmosphere surpasses a threshold value of about 25kV/m, which is in agreement with the laboratory tests.

The airplane would fly below developing thunderstorms or shower clouds and hunt for areas with electric fields of about 30 kV/m or more. If such an area was found and the field pattern had been established by several passes through this area, chaff would be ejected on two to four test runs. The electromagnetic emission of the corona discharge and the electric field were recorded during consequent passes back and forth through the seeded area until either corona discharge or the strong electric field disappeared. Figures 17.11 and 17.12 are typical examples of such flight records.

Figure 17.11 shows the corona discharge on the upper trace and the vertical field component on the lower trace. Seeding has been marked on the record

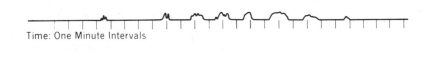

Time: One Minute Intervals

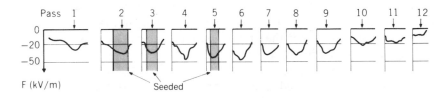

Corona Discharge Generated by Chaff Seeding
2 August 1966

Figure 17.11   Corona discharge generated by chaff seeding.

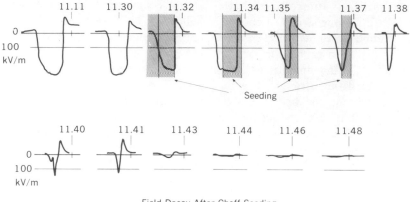

Field Decay After Chaff Seeding
1 August 1966

Figure 17.12 Field decay after seeding.

as shaded areas. Eleven passes have been made below the storm. On the second, third, and fifth passes, chaff was dispensed. On the second and fourth passes, the trace of corona discharge emission was small and irregular. At the third pass the plane missed the previous seeded area completely, and no corona discharge was recorded. But it seems that the chaff fibers spread out very rapidly and after 5 to 10 min the whole area was solidly filled with corona discharge emissions until the fields dropped below 20 kV/m.

Figure 17.12 shows the decay of a strong electric field after chaff seeding. It may be pointed out that between the first and the second pass about 20 min elapsed because the area was lost and could not be relocated any earlier. This proved to be fortunate because it shows that during this time the field remained at its high value of about 300 kV/m. After the area was found again chaff seeding began at the third and following passes. The decay of the field could be recognized 3 min after chaff seeding started. Ten minutes thereafter the field had completely collapsed.

These experiments have established that corona discharge is generated if chaff fibers are dispersed in the electric field of thunderstorms exceeding values of 30 kV/m.

It seems highly probable that the decay of strong electric fields is caused or accelerated by corona current produced by the chaff fibers.

There are a number of open questions which must be answered before chaff seeding becomes an effective tool for lightning suppression. The most important task is to prove experimentally that the effect of chaff seeding on the conductivity inside the cloud is as stated above. This implies that chaff can be dispersed in a reasonable time through a large volume of the cloud. If we visualize the thin line of chaff ejected from the airplane on one seeding run, the fast field decay shown in Figure 17.12 is indeed remarkable.

TRIGGERED LIGHTNING

The two lightning strikes to Apollo 12 shortly after launch in 1968 prompted more systematic research into the problem of rocket triggered lightning discharges. This investigation may even lead to the possibility of temporarily discharging electrified clouds by triggered lightning. The pioneer work of this research was done by M. M. Newman[26-28] with J. R. Stahman and J. D. Robb,[29] who successfully triggered lightning discharges with rockets fired into a thundercloud from a ship. The rockets were connected to the ship by a trailing wire. This research has been continued by this author with L. R. Ruhnke and others.[15-17, 19, 35]

It is not realized in general that the airborne vehicles themselves cause the lightning strikes if they enter regions with high electric fields. It is very seldom that they are hit accidentally by natural lightning produced by the storm. The probability of lightning striking an airplane during the flight through the cloud is much too small to account for the number of hits. Furthermore, Apollo 12 was struck twice in a cloud which did not produce natural lightning. There are several other arguments in favor of the trigger hypothesis; however, the most convincing one is this research itself. If lightning is caused by the airborne vehicle, it should be possible to trigger lightning by test rockets under the right conditions any time. If lightning strikes to airborne vehicles are accidental hits by natural lightning when about 1000 to 10,000 rockets are fired into a storm, only one may be struck by natural lightning. The experimental results, which are discussed later, show quite clearly that the trigger hypothesis is the correct one.

From calculations given in detail[17] it can be deduced (1) that the cloud field, which caused the lightning strikes during the Apollo 12 launch, was only on the order of 10 kV/m, or (2) that the small "folding fin aircraft rocket Mighty Mouse" will trigger a lightning if fired

into a cloud field of about 100 kV/m. Our next step then is to carry out an experimental test to prove or disprove the theory.

Experiments

The theory predicts that lightning can be triggered even with a small inexpensive rocket such as the 2.75 folding fin aircraft rocket Mighty Mouse (Figure 17.13). This is somewhat surprising because it was always thought that large rockets or rockets with long trailing wires are necessary to trigger lightning discharges. The price to pay for the cheap triggering method is the necessity of comparatively high fields on the order of 100 kV/m. Such fields occur only in thunderstorms, and even then are restricted to small areas and to certain time periods in the life cycle of the storm. This imposes two problems on the experimental procedure. First, how to locate the high field areas, and second, how to distinguish a triggered from a natural lightning. The first problem was solved by using an airplane equipped with a field recording system[18] to monitor electric field underneath thunderstorms in the rocket range. If the field reached the theoretically determined threshold value, rockets were fired into this area of the cloud. To distinguish a triggered from a natural lightning several criteria were applied. The distance of the lightning from the rocket launcher as determined by the time interval between lightning field change and thunder arrival should be inside the rocket range. Lightnings triggered by consecutively fired rockets should have the same distance from the launcher and/or occur at approximately the same time interval after launch or both. In addition to these criteria a statistical analysis has been carried out to determine the probability of

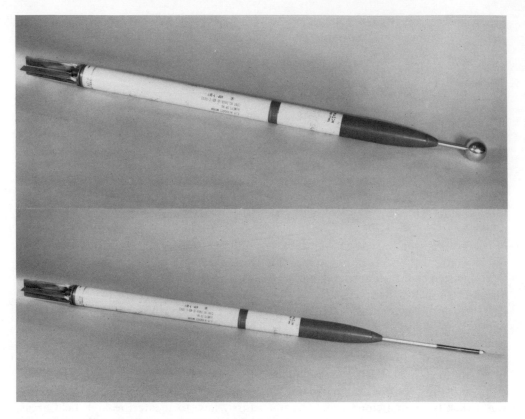

Figure 17.13   Mighty Mouse rocket with trigger payload.

the occurrence of a natural lightning at the time of the triggered lightning.

Field tests to trigger lightning discharges have been carried out at the Langmuir Observatory of the Institute of Mining and Technology at Socorro, New Mexico, with trigger rockets (Mighty Mouse, Figure 17.13) topped with metal spheres of 5, 7, or 10 cm diameter, referred to in Table 17.1 as S (small), M (medium), and L (large); a field mill and a wire antenna at the ground to record the cloud and the lightning field; and an airplane equipped with cylindrical field mills[18] (Figure 17.14) to record the three field components in the air. The airplane checked the field underneath clouds in the target area and transmitted the field

values to the launch site. When the field surpassed 100 kV/m—or at the last firing, 60 kV/m—the airplane left the target area and rockets were fired into the peak field region. After the firing the airplane was recalled to check the discharging effect of the triggered lightning on the cloud field. The distance of the lightning strike from the ground station was determined by the lightning-thunder time interval assuming the speed of sound to be 330 m/sec.

Table 17.1 shows the result of 15 rockets fired into storms at Socorro with the following parameters listed in the columns from left to right: (1) date and time; (2) diameter of the trigger sphere, L = 10 cm, M = 7 cm, S = 5 cm; (3) time interval between launch and lightning (sec); (4) distance between lightning and ground station (km); (5) lightning field (kV/m); (6) cloud field (kV/m); (7) azimuth of launcher (degree); (8) elevation of launcher (degree); (9) airplane field before and after launch (kV/m); and (10) test number. The rockets were fired in five small groups marked I to V, of two, three, or four rockets each aimed to answer specified questions. In the case of groups III and IV, where the field underneath the cloud was measured by the airplane before and after rocket launch, the reduction of the field from 150 to 50 kV/m in group III and from 60 to 40 kV/m in group IV is evident. Group V served as a control test to group IV. The launcher was turned 20° from the location of the high-field area. No lightning was triggered during this misalignment, as could be expected. However, the field at the ground remained the same during both of these tests. This shows that the field at the ground did not indicate localized high-field areas in the cloud and therefore cannot be used to determine either the lightning potential of a cloud or the area in which lightning is most likely to occur.

Another remarkable feature of test group IV is that the three triggered lightning discharges occurred practically the same time after the respective launch. This indicates that the lightnings were caused by the rocket penetrating a high-field area because an occurrence of natural lightning with such an accurate timing after three consecutive launches is improbable. Furthermore, it shows that the cloud area was not completely discharged by the first lightning but that enough charge remained to produce a second and third lightning discharge. However, the third lightning is rather weak, as can be recognized by the small lightning field at the ground (column 5).

We close this section with a brief

Figure 17.14 Airplane with top and nose field mill.

Table 17.1 Triggered Lightning Data from Field Test at Socorro, August 1970

Date, Time	L = 10 cm M = 7 cm S = 5 cm	Time Interval (sec)	Distance Lightn. Station (km)	Lightning Field Ground (kV/m)	Cloud Field Ground (kV/m)	Launch Azimuth (deg)	Launch Evelation (deg)	Cloud Field Airplane (kV/m)	Group
Aug. 8									
17 h 15.38	M	54	5.6	4	−25	—	—	—	
17 h 19.09	S	—	—	—	−25	—	—	—	I
17 h 31.12	M	—	—	—	−25	—	—	—	
Aug. 13									
15 h 42.51	M	53	5.3	12.5	−7.5	196	080	—	
15 h 44.52	L	59	6.9	3.5	−6	196	080	—	II

Date	Time		45	3.6	20	−9	309	079	Before 150	
Aug. 13										
	18 h 55.04	L	45	3.6	20	−9	309	079	Before 150	III
	18 h 56.49	M	27	3.0	10	−11	309	079	—	
	18 h 58.35	L	38	3.6	12	−11	309	079	—	
	19 h 00.42	M	—	—	—	−11	309	079	After 50	
Aug. 14										
	15 h 26.46	M	61.0	—	1	−1	325	080	Before 60	IV
	15 h 27.14	L	59.5	—	5.5	−1.2	325	080	—	
	15 h 27.43	M	60.5	—	0.5	−1	325	080	After 40	
Aug. 14										
	15 h 29.04	M	—	—	—	−1	305	080	—	V
	15 h 29.36	L	—	—	—	−1	305	080	—	
	15 h 30.04	M	—	—	—	−1	305	080	—	

assessment of the research aimed at discharging electrified clouds by triggered lightning discharges.

The theory given in my 1971 paper[17] seems to be broad enough to cover the essential physical facts of the triggered lightning problem, and its predictions agree with the experiments carried out so far. The instruments are adequate to measure the necessary parameters. The field tests must be considered to be in the preliminary stage.[17, 35] There is little doubt that lightning can be triggered even by such a small rocket as the folding fin aircraft rocket Mighty Mouse. Further tests are necessary to determine the electric parameters of the triggered lightning and its effect on the cloud charge. This leads into the problem of discharging the cloud, which is presently of a more speculative nature. Problems to be solved here, for instance, are the charge distribution in the cloud during its life cycle, the amount of charge destroyed by the individual lightning, and the regenerative power of the charge-producing mechanism in the cloud. The preliminary field test at Socorro indicates that the cloud field can be reduced by triggered lightning in a small area of the cloud to such an extent that further lightnings are not ignited by subsequent rockets, but the region discharged and the time it stays discharged are still open questions.

REFERENCES

1. Aufdermaur, A. N., and D. A. Johnson, Charge separation due to riming in an electric field, *Quart. J. Royal Meteorol. Soc.* **98**, 369–382, 1972.

2. Barrows, J. S., Weather modification and the prevention of lightning-caused forest fires, *Human Dimensions of Weather Modification,* W. R. Sewell, Ed., Dept. Geog. Research Paper 105, Univ. Chicago, Ill., 1966, pp. 169–182.

3. Baughman, R. G., and D. M. Fuquay, Hail and lightning occurrence in mountain thunderstorms, *J. Appl. Meteorol.* **9**(4), 657–660, 1970.

4. Chalmers, J. A., *Atmospheric Electricity,* 2nd ed., Pergamon Press, New York, 1967, 515 pp.

5. Colgate, S. A., Differential charge transport in thunderstorm clouds, *J. Geophys. Res.* **77**, 4511–4517, 1972.

6. Dawson, G. A., and D. G. Duff, Initiation of cloud-to-ground lightning strokes, *J. Geophys. Res.* **75**, 5858–5867, 1970.

7. Drake, J. C., Electrification accompanying the melting of ice particles, *Quart. J. Royal Meteorol. Soc.* **94**, 176–191, 1968.

8. Fuquay, D. M., Weather modification and forest fires, *Ground Level Climatology,* American Association for the Advancement of Science, 309–325, 1967.

9. Fuquay, D. M., and R. G. Baughman, Project Skyfire—Lightning Research, Final Report to National Science Foundation, Grant No. GP-2617, 1969, 59 pp.

10. Fuquay, D. M., A. R. Taylor, R. G. Hawe, and C. W. Schmid, Jr., Lightning discharges that caused forest fires, *J. Geophys. Res.* **77**(12) 2156–2158, 1972.

11. Gish, O. H., and G. R. Wait, Thunderstorms and the earth's general electrification, *J. Geophys. Res.* **55**, 473–484, 1950.

12. Israel, H. Bemerkung zum Energieumsatz im Gewitter, *Geofis, Pura Appl.* **24**, 3–11, 1953.

13. Kasemir, H. W., Zur Strömungstheorie des luftelektrischen Feldes II, *Arch. Meteorol. Geophys. Bioklimatol.* **5**(1), 56–70, 1952.

14. Kasemir, H. W., The thundercloud, in *Proceedings of the Third International Conference on Atmospheric Space Electricity,* S. Coroniti, Ed., Elsevier, New York, 1963.

15. Kasemir, H. W., Lightning hazard to rockets during launch I, *ESSA Tech. Rept., ERL 143-APCL 11,* 1969.

16. Kasemir, H. W. Lightning hazard to rockets during launch II, *ESSA Tech. Rept., ERL 144-APCL 12,* 1970.

17. Kasemir, H. W., Basic theory and pilot experiments to the problem of triggering lightning discharges by rockets, *NOAA Tech. Memo. ERL-APCL-12,* 1971.

18. Kasemir, H. W., The cylindrical field mill, *Met. Rundschau* **25**(2); 33–38, 1972.

19. Kasemir, H. W., Lightning suppression by

chaff seeding. NOAA Tech. Rept. *ERL 284-APCL 30,* 1973, 32 pp.

20. Latham, Jr., and B. J. Mason, Generation of electric charge associated with the formation of soft hail in thunderclouds, *Proc. Roy. Soc. A,* **260,** 537–549, 1961.

21. Latham, J., and A. H. Miller, The role of ice specimen geometry and impact velocity in the Reynolds-Brook theory of thunderstorm electrification, *J. Atmos. Sci.* **22,** 505–508, 1965.

22. Latham, J., and C. D. Stow, Airborne studies of the electrical properties of large convective clouds, *Quart. J. Royal Meteorol. Soc.* **95,** 486–500, 1969.

23. Loeb, L. B., The mechanisms of stepped and dart leaders in cloud-to-ground lightning strokes, *J. Geophys. Res.* **71,** 4711–4721, 1966.

24. MacCready, P. B. Jr., and R. G. Baughman, The glaciation of an AgI-seeded cumulus cloud, *J. Appl. Meteorol.* **7**(1), 132–135, 1968.

25. Mason, B. J., *The Physics of Clouds,* 2nd ed., Oxford Univ. Press, London, 1971, 671 pp.

26. Newman, M. M., Use of triggered lightning to study the discharge process in the channel, *Problems of Atmospheric and Space Electricity,* American Elsevier, New York, 1965, pp. 482–490.

27. Newman, M. M., Triggered lightning stroke at very close range, *J. Geophys. Res.* **72**(18), 4761–4764, 1969.

28. Newman, M. M., Lightning discharge simulation and triggered lightning, *Planetary Electrodynamics,* Vol. 2, Gordon and Breach, 1969, pp.213–219.

29. Newman, M. M., J. R. Stahmann, and J. D. Robb, Experimental study of triggered natural lightning discharges, Final Report DS-67-3, Lightning and Transients Research Institute, Minneapolis, Minn., 1967.

30. Phillips, B. B., Charge distribution in a quasi-static thundercloud model, *Monthly Weather Rev.* **95,** 847–853, 1967.

31. Pierce, E. T., Thunder and lightning, *Shell Aviation News,* 9–13, 1958.

32. Pierce, E. T., Latitudinal variation of lightning parameters, *J. Appl. Meteorol.* **9,** 194–195, 1970.

33. Prupacher, H. R., E. H. Steinberger, and T. L. Wang, On the electrical effects which accompany the spontaneous growth of ice in supercooled, aqueous solutions, *Planetary Electrodynamics,* Vol. 1, S. Coroniti and J. Hughes, Eds., Gordon and Breach, 1969, pp. 283–306.

34. Reynolds, S. E., M. Brook, and M. F. Gourley, Thunderstorm charge separation, *J. Meteorol.* **14,** 426–436, 1957.

35. Ruhnke, L. R., A rocket-borne instrument to measure electric fields inside electrified clouds, *NOAA Tech. Rept. ERL 206-APCL 20,* 1971.

36. Schonland, B. F. J., *The Flight of Thunderbolts,* 2nd ed., Clarendon Press, Oxford, 1964, 182 pp.

37. Smyth, W. R., *Static and Dynamic Electricity,* McGraw-Hill, New York, 1950.

38. Stergis, C. G., G. C. Rein, and T. Kangas, Electric field measurements above thunderstorms, *J. Atmos. Terr. Phys.* **11,** 83–90, 1957.

39. Stow, C. D., On the prevention of lightning, *Bull. Am. Meteorol. Soc.* **50**(7), 514–520, 1969.

40. Taylor, A. R., Lightning effects on the forest complex, *Proc. Ann. Tall Timbers Fire Ecology Conf.* **9,** 127–150, 1969.

41. Uman, M. A., *Lightning,* McGraw-Hill, New York, 1969, 264 pp.

42. Uman, M. A., *Understanding Lightning,* Bek Technical Publications, Inc., Carnegie, Pa., 1971, 166 pp.

43. Viemeister, P. E., *The Lightning Book,* Doubleday, Garden City, N. Y., 1961, 316 pp.

44. Vonnegut, B., Some facts and speculations concerning the origin and role of thunderstorm electricity, *Meteorol. Monogr.* **5,** 224–241, 1963.

45. Workman, E. J., and S. E. Reynolds, Electrical phenomena occurring during the freezing of dilute aqueous solutions and their possible relationship to thunderstorm electricity, *Phys. Rev.* **78,** 254–259, 1950.

F

CLIMATIC
CHANGE

Why do glaciers start? Why do they end and where are we now? It has been suggested recently that we are near the end of an interglacial period. Studies of climatic changes are in their infancy. We know that there have been four episodes of glaciation in the recent past covering a period of about 1,000,000 years. A conference at Brown University in January 1972 discussed this problem and the majority of the participants concluded that:

"The global environments of the last several millennia are in sharp contrast with climates that existed during most of the past million years. Warm intervals like the present one have been short-lived and the natural end of our warm epoch is undoubtedly near when considered on a geological time scale. Global cooling and related rapid changes of environment, substantially exceeding the fluctuations experienced by man in historical times, must be expected within the next few millennia or even centuries. In man's quest to utilize global resources, and to produce an adequate supply of food, global climatic change constitutes a first-order environmental hazard which must be thoroughly understood well in advance of the first global indications of deteriorating climate. Interdisciplinary attacks on these problems must be internationally organized and encourage to develop at a rate substantially exceeding the present pace."

We may be facing a climatic catastrophe within a few hundred years, or we may not. Anyway, it is useful to see where we are now in our knowledge of climatic change and man's effect on it, and our ability to model climatic change.

Machta's and Landsberg's chapters show that we can measure some effect on the climate both locally near cities and also globally. The changes observed now are not large, but extrapolations of trends do suggest large problems in the future. Smagorinsky's work on climate modeling is probably the best in the world, but it is still quite primitive in that it really has little predictive ability yet.

18 | Global Atmospheric Modeling and the Numerical Simulation of Climate

JOSEPH SMAGORINSKY

Intriguing fragmentary evidence from geological sources has been accumulating regarding variations in climate of the past. Major glacial variations can be traced back 800 million years, but since the continental distribution was presumably significantly different even as recently as 50 million years before the present (Y.B.P.), continental distribution must be taken as a major factor. It appears that the Antarctic ice sheet formed about 10 million Y.B.P. and that North Atlantic glaciation began about 6 million Y.B.P. Paleoclimatologists tell us that the last major climatic cycle began in the geological quaternary beginning 130,000 Y.B.P. During this interval, a major period of 90,000 years can be detected with minor periods of 11 to 20,000 years in duration. There seems to be little clear evidence that the Arctic sea ice has ever disappeared.

It is tempting to ascribe long-term climatic change to some external cause,

Joseph Smagorinsky is Director of Geophysical Fluid Dynamics Laboratory, NOAA/ERL, Princeton University, Princeton, N. J.

such as variations in the solar radiation. After all, the most dramatic short-term change in climate is that of the seasons. There could also conceivably be terrestrial causes, of course, that are still essentially independent of the atmosphere. Change in atmospheric opacity due to a severe volcanic eruption is a good example. But clearly, how climate will react must depend to some extent on conditions within the atmosphere itself—for example, the atmosphere's buoyant properties and its motions will determine the geographic redistribution of the volcanic material and its residence time in the atmosphere. Of course, there are other more subtle interactions which will govern the effects of suspended volcanic material on climate. Furthermore, many man-made activities may also have climatic consequences: those due to industrialization and transport vehicles yielding unnatural sources of gases and particulates and heat.

One should not forget, however, that because the atmosphere is a very complicated physical system, even without external stimuli it may undergo natural variations of significant magnitude. An important factor is that the temperature regime of our planet, even before life began on it, has been in the range that water substance in its three forms—vapor, liquid, and solid—can coexist in the atmosphere and at its lower boundary. The possibility of changes among these thermodynamic phases adds an element of complexity to the system that profoundly influences the nature of terrestrial climate, its modes of natural variability, and its stability to external stimuli. Hence to understand our climate we must also know how the oceans act as a thermal flywheel, and how snow and ice cover react back upon the atmosphere.

The most powerful tools for understanding the evolution of climate and the causes of its variations were forged within the past 20 years. The advent of high-speed computers has made it possible to determine the evolution of a climate consistent with a given set of postulated governing physical laws, whose mathematical statement is said to constitute a numerical model. It is the extension and development of these techniques that will ultimately permit the simulation of the climatic impact of a variety of hypothetical situations. The purpose of this chapter is to expose the physical basis and nature of these methods, their current level of simulation capability, and to identify problems that must yet be solved before such models can become an effective means for understanding climatic changes of the past and for reliable predictions of the future.

When we speak of climate, we usually think of average characteristics of the atmosphere which we experience where we are, that is, at the earth's surface and at a particular geographic location. We probably would include a measure of the average variability from day to night and from summer to winter. Such characteristics would be the temperature, the precipitation, the snow cover, the cloudiness, the wind speed, and the wind direction. These are the conditions that have always been important to man and have governed his evolution. His ability to survive and thrive within these immediate environmental conditions has been the critical element in his harmonious relationship with the other components of the earth's biosphere. Of course, technological progress has made it possible for man more easily to endure large variants in climate. Structures, temperature control, and water management have thus allowed him to live and conduct his affairs in the deep tropics, in the subtropical deserts, and in the polar regions almost as easily as he can in midlatitudes. But as we are rapidly learning, many of these same benefits of technology may in turn inadvertently make it more difficult for the ecological balance to be maintained within tolerable limits. The ways that this might happen are many—some direct and others insidious. Within recent years these concerns have received the collective attention and scrutiny of scientists throughout the world.[22, 31, 41]

Our first important realization is that the weather and climate that we experience within the 2 m of the earth's surface where we live are the net result of very complex physical, chemical, and dynamical processes taking place in the remaining 99.98 percent of the atmosphere above us. Climate cannot be discussed rationally without a knowledge and understanding of the total workings of the atmosphere and, as will be seen, the hydrosphere and cryosphere as well.

It is perhaps trivial to be reminded that the atmosphere and the oceans constitute a *thin* viscous fluid film adhering to the rotating planet earth (for example, 99 percent of the atmosphere is confined to $1/200$ of an earth radius—relatively comparable to the skin of an apple). As for any planetary atmosphere, the composition of the film is determined by the geological characteristics of the crust and the escape properties of the gaseous components as influenced by the gravitational field of the planet, the solar distance, whether much heat is escaping from the interior of the planet, and its age.

The fluid envelope of the inner planets moves relative to the planet in response to uneven heating of the sun's rays, since their obliqueness varies with latitude and importantly in some cases with longitude (such as for the slowly rotating Venus). On the other hand, some of the outer planetary atmospheres, such as those of Jupiter and Saturn, may derive their primary driving energy from heating from within—that is, they may be "star-like."

For earth we know empirically that its atmosphere and oceans sustain a considerable variety of phenomena, which may be characterized by their time and space scales. But because of the highly interactive character of the excited response, only limited questions regarding it can be answered through analytical mathematical techniques. It was the advent of fast computers and thence of numerical modeling that provided a means for dealing comprehensively with the complex system of interacting processes. An historical account of the development of the use of numerical methods for the simulation of large-scale meteorological processes and their role in the development of a theory for the general circulation and climate can be found elsewhere.[38,39] An excellent book on the nature of the atmospheric general circulation has been written by E. N. Lorenz.[14]

Simply, the object of atmospheric modeling is to approach the problem of simulating the large-scale atmospheric behavior through a consistent application of the governing fundamental physical laws and boundary conditions; it is significant that the macrostructure of these laws has been known for over a century. What had been lacking, and still is the main subject for research, is an understanding of the interactions of the macroscales of primary interest (i.e., greater than several hundred kilometers) with processes of lesser dimensions, such as radiative transfer, turbulent fluxes, and the processes within clouds and by clouds. These relationships must ultimately be expressed as functions of the macroscale variables, such as the average wind, temperature, and humidity in elemental volumes having horizontal dimensions of, say, 100 km and vertical dimensions of tens of meters to several kilometers, depending on the specific process and the altitude at which its interaction is to be determined. The dimensions of the volume are a measure of the *computational resolution*; what happens on scales smaller than that of the volume is known as *subgrid scale processes*; the simplified prescription of how they are related to the macroscale is known as the *parameterizations*; and the empirical constants that enter into the prescription are known as *parameters*, as are the macroscale dependent variables that are involved. The subject of parameterization will come up a number of times in what follows.

Tests of provisional parameterizations can, except for very simple cases, be made only by numerically integrating the mathematically closed (that is, consistently well-posed) system of laws, parameterizations, boundary conditions, and initial conditions. As will be seen, the magnitude of such calculations requires exceedingly fast and large computers, a factor which has inhibited and will continue to inhibit very general testing programs free from serious mathematical degeneracies.

One's strategy is to construct a hierarchy of models in which one systematically relaxes empirical parametric constraints, thereby permitting the model characteristics and evolutions to be more and more self-determining in terms of very fundamental (i.e., more generally valid) statements of the governing physics. The ultimate model of such a hierarchy could be conceived as one appropriate to any of the planetary atmospheres. One does ask that each of the models in such a hierarchy represent a physically reliable system which is similar to, but inevitably simpler than, the real geophysical medium. It is therefore of interest to study the similarities as well as the differences between the model simulations and real atmospheric evolutions. This may not always be possible, especially when one deals with a parameter range sufficiently removed from observed experience. Such a class of problems arises in studying climatic change.

BASIC PARAMETERS OF A MODEL

In any model, depending upon the parameterizations incorporated in the formulation—that is, the modeling approximations—a number of empirically determined numbers must be specified.

Planetary Data

These are clearly external and assume the planet to be nearly spherical.

a The radius.

g The effective acceleration of gravity, which depends on the mass of the planet, and takes account of the centrifugal acceleration and its effect on the slight oblateness of the planet. If the planetary atmosphere is thin compared to a, then slight variations of gravity with altitude and latitude may be ignored.

Ω The angular velocity of the planet will determine the Coriolis forces, that is, the apparent forces acting on atmospheric motions relative to the planet. It will also determine the length of day and night, which is important for the way that solar radiation is absorbed by the atmosphere and the planet's surface. One can conceive of problems where Ω can be altered by what happens in the atmosphere itself. For example, if the polar ice cap were to grow systematically due to predominant snow, the angular momentum of the planet relative to atmosphere could be slightly altered. However, we will not consider such a possibility.

Orbital   The geometry of the planet's annual orbit about the sun and the inclination of the planet's axis of rotation to the plane of the ecliptic.

$S_\infty(\lambda)$ The quantity and quality of the sun's radiation reaching the outer limits of the planetary atmosphere. By quality is meant how much energy is received at different wavelengths, λ, in other words, the shape of the solar spectrum. Various gases react differently to radiation which is multichromatic.

Internal, Bulk Molecular, and Static Properties of the Atmosphere

Here we assume that we already know the constituents of the atmosphere, bypassing, for the most part, the problem of accounting for its evolution. There are some planets whose atmospheric composition is not well-known yet, but recent remote and direct probing experiments are rapidly removing the uncertainty, especially for the near planets.

p_* The total mass of all gaseous constituents expressed as the average force per unit area of the planet's surface—the mean surface pressure.

c_p and c_v The effective specific heats at constant pressure and at constant volume of the mixture of gases.

The most massive constituents of the planetary atmosphere determine p_*, c_p, and c_v. For earth, N_2 and O_2 account for about 98 percent of the total mass (and over 99 percent of the total volume). However, the trace gases may be quite important for their radiative impact. For our planet, the radiatively active gases are CO_2, O_3, and H_2O, as are particulates such as clouds, haze, and dust. The other trace gases that have been found in our atmosphere—argon, neon, helium, and krypton—are relatively inert radiatively, just as are N_2 and O_2. There are transitory catalytic by-products in the stratosphere accompanying photochemical changes in O_3, which is only mentioned here in passing. For Mars and Venus, the radiatively active constituent, that is CO_2 is also the most massive.

When the time scale of variability of the radiatively active trace gases is comparable to the time span of interest and when this variability has sufficiently large amplitude, then their distribution must be internally determined, that is, predicted rather than specified. For example, for time spans beyond 1 or 2 days, variations in water vapor, as it affects the release of latent heat upon condensing, must be predicted. The predicted motions determine the transport of water vapor, the changes in its distribution, and when and where it will cool to the saturation or freezing point. On the other hand, the radiative consequences of water vapor and of liquid water (cloud) are slower acting and become of consequence only over somewhat longer periods—perhaps

one or more weeks. The same is probably true for variations of atmospheric O_3, which result from transport by the air motion and photochemical interactions— an incompletely understood process at present. CO_2 is so well mixed and relatively unchanging by natural processes that its variations over decades may only be of importance. With possibly strong artificial sources due to fossil fuel combustion, a predictive model would require not only an accounting of the transport properties of the medium in dispersing CO_2 from localized sources, but also a biospheric and oceanic buffering mechanism.

Boundary Properties that Depend on Internal Variations

Examples are the surface roughness (z_*), the reflectivity or albedo (A_*), and the thermal conductivity of the earth's surface (K_*). In the first approximation one can, as has been the custom in simple models, specify an average value determined from observed data. In the limit one can, for short-period evolutions, ignore them entirely, but not for problems of climatic scope. For sophisticated models, z_* for a sea surface will depend on the wind in the lower atmosphere and A_* will be materially altered by snow cover. K_* will vary at sea because of wind stirring and even on land will depend on the water content of the soil which is the net result of precipitation, evaporation, and the water storage capacity of the soil.

Scale Assertions in the Choice of Computational Resolution

The computational resolution, that is, the size of the elemental volume (Δx, Δy, Δz) used in approximating the differential equations and in calculating solutions evolving in elemental time increments Δt, should be determined by the possible solutions themselves. One hopes that the scale spectrum of the solutions is not too broad and uniform; otherwise an inordinately large span of scales must be covered or resolved—that is, very small volumes and time increments must be used. The practical difficulties are apparent. On the other hand, if the dominant phenomenological scales are confined to a limited span of the spectrum, the interactions with the smaller scales of lesser amplitude can be approximated by parameterizations rather than treated explicitly. In the case of our own atmosphere, we have observational data to guide us in the choice of Δx, Δy, and Δz. As it turns out, Δt may not be dictated by the phenomenologically dominant modes, but computationally by the fastest propagating mode allowed by the difference analogue of the differential equations of the model, even if that mode does not carry much energy.

In the absence of empirical guidance, a first guess as to the scale of the primary modes can be made from a theoretical scale analysis of the governing differential equations. Otherwise, trial calculations must be made with varying resolutions to determine the maximum volume that can still yield the dominant solutions with tolerable distortion over a substantial span of the spectrum; that is, it would determine where the mathematical zone of convergence lies. For a spectrum as in the earth's atmosphere, which can span 11 orders of magnitude from planetary scale to cloud droplet size, there may be more than one dominant mode; that is, there are a number of scale-distinct phenomena, each of fairly limited size range, such as extratropical cyclones, hurricanes (or tropical cyclones), and cumulus clouds.

The choice of the elemental volume size must therefore be considered as an external parameter of the model since by itself it can influence the outcome of a numerical integration by distorting scale interactions or by excluding some entirely.

THE GOVERNING LAWS

The Physical Equations

We shall qualitatively consider the system of physical laws that together constitute a fully consistent set of mathematical equations to predict the time variations of the primary dependent variables (the macroscale variables) at each point (really a finite volume, as we discussed) in the three-dimensional atmosphere. As we shall see, these variables are the two horizontal wind components (which define a two-dimensional vector $\mathbf{v}$), the temperature (θ), and the humidity (q). Furthermore, at the lower boundary of each finite column, one must also predict variations in the surface pressure (p_*). From a knowledge of these primary dependent variables, at any one time it is then possible to determine such subsidiary variables as the density (ρ) and a measure of the vertical velocity (ω). Actually, ω is the change of pressure in an elemental volume moving with the total three-dimensional wind, and is approximately proportional to the downward component of the air motion.

For a system to be solvable in terms of these primary variables, it is first necessary to assume a priori that, for the macroscales of interest, the vertical particle accelerations are small compared to the acceleration of gravity. This *hydrostatic* approximation is valid for horizontal macroscales larger than a few kilometers, which certainly includes the scales we are concerned with, that is, of hundreds of kilometers and larger. Such a simplification of the differential (and difference) equations eliminates (or "filters") from the final solutions those corresponding to the vertical propagation of sound waves, which, relative to the other macroscale modes, carry relatively little energy and are of high frequency. This permits one to use time increments (Δt) in the numerical integration of the order of minutes rather than seconds.

The conservation equations for the horizontal *momentum* (the equations of motion) express the fact that an elemental volume will accelerate horizontally relative to the planet depending on the imbalance between the Coriolis force, the horizontal pressure gradient, and subgrid scale diffusive processes (which in general are three-dimensional). A volume next to the earth's surface may be decelerated by turbulent stresses at the boundary.

If we assume air (even moist air) to behave as a perfect gas, then Charles' and Boyle's laws together give a simple algebraic relation between the average

pressure, temperature, and density in an elemental volume. This is the equation of *state* and involves the empirical constants c_p and c_v, the specific heats of air at constant pressure and at constant volume, respectively.

The horizontal momentum equations and those of state and hydrostatic balance are schematically combined in Figure 18.1. This set can predict changes in $\mathbf{v}$ provided we can also predict θ and p_*, and determine ω. The law expressing the indestructibility of *mass*, the equation of continuity, provides the means of determining p_* and ω if we know $\mathbf{v}$.

To predict changes in θ we impose the requirement that the energy of an elemental volume can change only if mechanical work is done on it or if heat is transmitted through its boundaries. This is the first law of thermodynamics of the *thermal energy* equation. To calculate

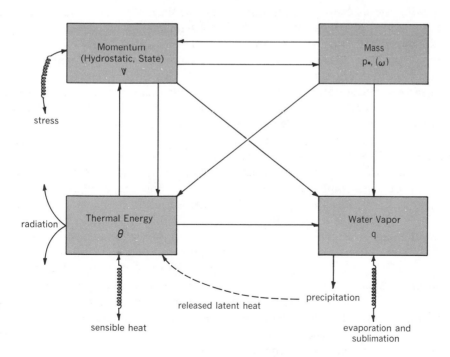

Figure 18.1 A schematic diagram of the relationship of the physical conservation laws which together define the variations of the primary macroscale variables in an elemental volume. Also shown are external sources and sinks of momentum, heat, and water vapor.

such changes, one first of all needs to know v, p_*, and ω. Subgrid scale diffusive processes may also alter the energy in an elemental volume. If water vapor changes phase, for example, condensation and then precipitation occur, then the released latent heat must be accounted for. It may also be possible that energy may be radiated into or out of the volume. The scattering, absorption, or transmission will in general depend on the changing content of water vapor and of liquid water particles in the volume. For our present purposes, as discussed earlier, we assume that O_3 and CO_2 are known and quasiconstant in the volume. Finally, an elemental volume next to the planetary surface could exchange sensible heat through turbulence.

The remaining predictive law needed is that for q, the *water vapor* conservation equation. For this, we say that the water vapor content of an elemental volume moving with the macroscale motion can change (analogously to energy) only by internal subgrid scale diffusion, turbulent transfer from the planetary surface (evaporation or sublimation), or water phase changes (precipitation). To calculate these one must know v, p_*, ω, and θ.

In practice one does not deal explicitly with moving elemental volumes (a "Lagrangian" framework), but one can more conveniently make the equivalent calculation for a geographically and temporally fixed volume (a "Eulerian" framework) through a simple identity from the calculus (the "chain" rule).

Subgrid Scale Diffusion

The major transports of momentum, heat, and water vapor generally are accomplished by the macroscale motions, that is, those that are explicitly resolved. Diffusion by motion of scales smaller than an elemental volume may be im-

portant (as will be seen in Figure 18.8) and must be parameterized. This then requires the specification of some additional empirical parameters.

Of special importance are the small-scale modes which can be excited when the macroscale becomes buoyantly unstable locally. In the atmosphere this is visually observed as cumulus clouds which have dimensions of several kilometers. Families of such clouds can occupy areas of hundreds or thousands of kilometers and, therefore, can have a profound effect back on the macroscale. The excitation of cumulus convection is relatively rapid, of the order of a few hours, and we know that, as a result, large vertical transport of heat, water vapor, and, quite likely, momentum occur with the net result that the macroscale is restored to a buoyantly stable state. This process, as we know from observation, is at least of intermittent importance in extratropical latitudes, but is a primary mechanism in the tropics. The parameterization of cumulus convection is, therefore, a critical element in modeling the earth's atmosphere. Unfortunately, a generally accepted parameterization has yet to be devised; in part, the problem is that there are inadequate observations of the details of convection, so that conclusive verification of candidate parameterization hypotheses is not possible.

Of similar importance is the role of small-scale diffusive processes which may occur within the atmosphere without any accompanying condensation or cloud. There is indirect evidence that such processes may be responsible for as much as half of the total energy dissipated by the macroscale motions. The remainder is lost within the planetary boundary layer, which is about 1 km in depth, being thicker at equatorial latitudes. Provisional parameterizations of internal clear air diffusion and dissipation are used with some success. However, definitive determination in this also awaits the results of detailed observational experiment and theoretical studies.

Surface Balance Conditions

The interaction of the atmosphere with its lower boundary is of critical importance. For example, through surface stresses the atmosphere loses about half its energy. Furthermore, the atmosphere is largely transparent to the sun's radiation. Most of this radiation is absorbed by the planetary surface and then fed back up by infrared radiation and through turbulent transfer of heat in sensible and in latent form. The details of this upward return of heat and how it depends on characteristics of the planetary surface, which may be slowly or rapidly variable depending in part on the atmosphere itself, provide some of the most interesting and challenging modeling problems with particular relevance to questions of the stability of climatic regimes.

In the case of large oceanic expanses, the wind stress together with heat transfer and freshwater exchange (precipitation and evaporation) are ultimately responsible for driving the ocean circulation. The uppermost layer of the ocean responds most rapidly to the interfacial exchanges with a relaxation time of the order of several weeks. On the other hand, the deep ocean circulation is the result of atmospheric coupling over hundreds or even thousands of years.

The oceans themselves are responsible for transporting a great deal of heat poleward. It may be as much as half the magnitude of the atmospheric transport (as will be seen in Figure 18.26). Therefore, the response of the earth's entire fluid envelope to the driving latitudinal heating gradient of the sun's radiation must be considered. A truly climatic model, therefore, cannot be bounded by the earth's surface but must extend into the ocean down to its bottom. Only one such joint model, albeit simple, has been constructed and used for climatic experiments.[16] The results will be discussed at length later.

We now consider those elements that enter into the nature and degree of the interaction between the atmosphere and its lower boundary.

Momentum. The turbulent stress exerted on the lower boundary by the atmosphere would be expected to be of direct importance to a boundary which is mobile, that is, a large body of water. We exclude from consideration the possibility that over long periods of time the stress component in the direction of the earth's rotation (east-west) may exert a net torque which could alter the earth's rotation rate—for the latitudinal distribution of surface easterly and westerly surface winds tend to balance out such that on the average there is no net angular momentum exchange between the atmosphere and the planet.

The stress, however, will exchange momentum with the atmosphere locally, and the surface wind will also be a factor in determining the contribution of turbulent exchange processes in the heat and water vapor balances at the earth-atmosphere interface. We now discuss these balances.

Heat. The limiting cases that one can conceive of are as follows: (*a*) a continental surface with zero heat capacity admitting no conduction of heat into the ground. The surface temperature would, therefore, be determined from a balance of the net radiation at the earth's surface and the turbulent flux of sensible and latent (due to evaporation) heat. Since diurnal variability of surface temperature is in part determined by the daily cycle of solar radiation and by the molecular conduction of heat by the soil, the diurnal cycle could not be included in

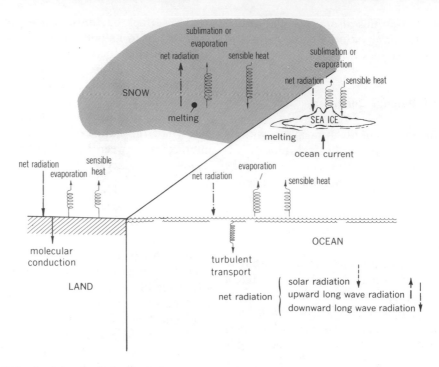

Figure 18.2 A schematic of the physical processes entering into the heat balance at four types of earth's surface: bare land, snow-covered land, open sea, and ice-covered sea.

this limiting case; (*b*) an oceanic surface with infinite heat capacity, the other extreme. This then would be an ocean which transmits heat vertically so rapidly (presumably by turbulent processes) that its surface temperature is unaffected by radiative or turbulent heat fluxes at the interface. This is a reasonable approximation for period of less than 1 or 2 weeks. In this limiting case the sea surface temperature must be specified as an external parameter. Many useful climatic simulation experiments of limited scope have thus been performed by using the observed distribution of sea surface temperature.

We now proceed to the intermediate cases, namely, those of continents with small but finite heat capacity due to molecular conduction, and oceans with large but finite heat capacity because of their vertical and lateral mobility. This

then requires a predictive framework for the surface temperature. Effectively, the boundary of the predictive system is moved to some point below the earth's surface. We further break down the possible classes of lower surface by distinguishing between the presence or absence of snow (or ice) cover; these will also be predicted.

Referring to Figure 18.2, we speak of net radiation at the lower boundary as the sum of downward solar radiation and the upward and downward long-wave (or infrared) radiation.

A snowfree land surface has a temperature determined in the manner discussed earlier. The main difference for snow cover is that molecular conduction becomes negligible, but the heat that goes into the melting of snow becomes a factor in the heat balance.

The ocean surface heat balance in the

presence of sea ice is controlled by the exchange of heat between ocean and atmosphere through molecular conduction in the ice. It is of particular importance that pack ice forms a significant shield which reduces the free interaction of the ocean and atmosphere. Cracks in the sea ice are a complicating factor.

It must already be apparent that to make the above calculations of the heat balance over the variety of boundary surfaces, the water content and snow or ice cover must also be predicted.

Hydrology and Cryology.   As in the case of the heat balance, one can conceive of limiting cases in the hydrologic interaction of the earth's surface with the atmosphere. In particular, one can assume that the lower boundary provides only an evaporative source and a precipitation sink: (a) for a continental boundary one must

specify empirically the soil moisture available for evaporation as a boundary condition; (b) over an ocean boundary, consistent with the corresponding limiting condition for the heat balance, one need only specify the sea surface temperature which determines the amount of water available for evaporation at 100 percent relative humidity.

The real situation lies somewhere between these two limiting situations and for the intermediate cases one must predict surface variations of water substance (see Figure 18.3).

Over a snowfree continental surface, one can prescribe the evaporation rate as a function of soil moisture, which must be predicted. For various soils one can assume a water storage capacity beyond which the excess of precipitation over evaporation will run off in the water shed determined by the topography, eventually

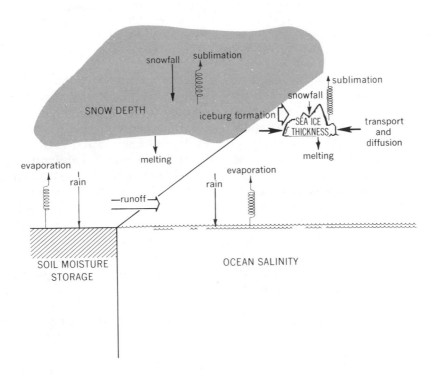

Figure 18.3   A schematic of the physical processes entering into the hydrologic balance at four types of earth's surface: bare land, snow-covered land, open sea, and ice-covered sea.

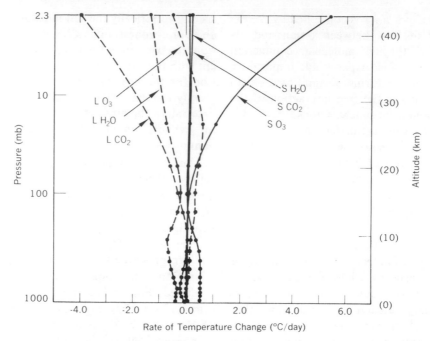

Figure 18.4   Vertical distributions of the rate of temperature change of the ($H_2O + CO_2 + O_3$) atmosphere in pure radiative equilibrium due to various absorbers. LH_2O, LCO_2, and LO_3 show the rate of temperature change due to long-wave radiation of water vapor, CO_2 (the effect of H_2O overlapping was included), and O_3. SH_2O, SCO_2, and SO_3 show the rate of temperature change due to the absorption of solar radiation by water vapor, CO_2, and O_3 (reprinted by permission, from Manabe and Strickler[20]).

reaching the sea. This runoff will contribute fresh coastal surface water to the oceans, which with the local ocean rain-evaporation imbalance will determine the local alteration of the surface salinity. Since ocean salinity and temperature determine the density of seawater, the surface hydrologic imbalance contributes to determining the oceanic motion field.

Returning to continental surfaces, but now snow covered, the difference between snowfall and sublimation, as modified by snow melt, which alters the local soil moisture, determines the snow depth and, where it goes to zero, the snow perimeter. Icebergs form from coastal breakoffs of continental snow (and ice) cover. As we have seen, sea ice may materially alter the thermal characteristics of the ocean, and so its thickness must be predicted in a general model which one would hope

could be applicable to the study of the onset of ice ages, namely, the systematic equatorward growth of the snow perimeter over continents, and of sea ice. The sea ice thickness will change as a net result of snowfall, sublimation, and melting. Where the sea ice is mobile, one must keep track of the ice floe's movements by the upper ocean currents of large scales (transport) and of small scales (diffusion). Melting will also alter the salinity and hence the density stratification of the upper layers of the ocean.

THE NATURE OF THE ENERGY SOURCE

The earth's fluid envelope, and the atmosphere in particular, is driven by the sun's radiation. For our purposes we can ignore the effects of longitudinal and time variations of solar radiation due to the earth's daily rotation (the diurnal variability); the quantities we shall deal with will have been suitably averaged. Of primary importance is the latitudinal variation of the incident solar radiation due to the spherical shape of the Earth. The annual variation, although of obvious importance for understanding the seasonal components of climate, can for some purposes be averaged out for simplicity. Even in discussing seasonal problems, the ellipticity of the Earth's orbit is secondary to the effect of the inclination of the polar axis to the plane of the ecliptic.

What happens to a given amount of solar radiation incident upon the outer reaches of the atmosphere as it passes downward depends on the distribution of the radiatively active trace gases (H_2O, CO_2, and O_3) and their absorptive properties. Furthermore, aerosols and clouds may scatter and reflect some of the radiation.

When the atmosphere or the ground radiate, however, it is at much lower temperatures than that of the sun, of the order of 300°K (Kelvin) rather than 6000°K. This is from another part of the electromagnetic spectrum, the infrared (IR), and the absorptive properties of H_2O, CO_2, and O_3 are quite different.

Consider an atmosphere with the *observed* average vertical distribution of H_2O, CO_2, and O_3. One can then calculate the temperature change at different altitudes due to absorption of solar radiation by each of the constituents, and by long-wave (or infrared) radiation. How-

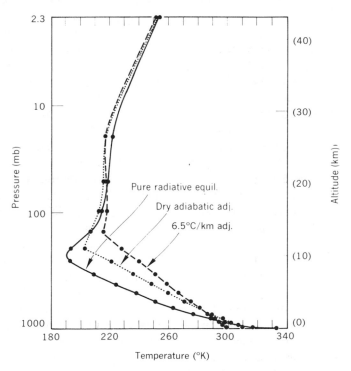

Figure 18.5 The solid curve is the equilibrium vertical temperature distribution resulting from radiative processes only. The dashed and dotted curves result from requiring, in addition, the adjustment of the static stability by convective overturning to a minimum lapse rate of 6.5°C/km (moist) and 10°C/km (dry), respectively.

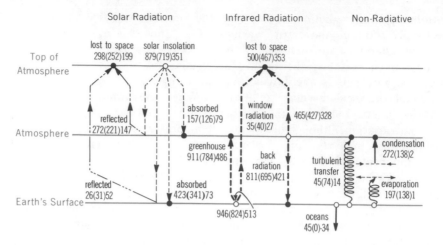

Figure 18.6 Schematic diagram of the processes contributing to the heat balance. The three number sequences give heating rate values in ly/day: at 5° latitude, the global mean (in parentheses), and at 85° latitude (based on Smagorinsky[34]).

ever, the *net* temperature change of all contributions tends to be zero, namely, that there is radiative equilibrium. Such a requirement[20] yields component profiles such as those in Figure 18.4. Note that the absorption by O_3 of the ultraviolet radiation in the 1800–2900 A band contributes to keeping the stratosphere warm, while CO_2 and H_2O long-wave radiations act to cool the stratosphere. In the troposphere, however, H_2O heating by solar radiation and cooling by the long-wave radiation in the 5 to 8 μ band and beyond 19 μ are the dominant contributions to the balance, but long-wave cooling by CO_2 in the 12 to 18 μ band is not negligible.

The resulting vertical temperature profile for pure radiative equilibrium is shown in Figure 18.5. Its slope in the lowest 10 km can be shown to be buoyantly unstable and could not remain so for long. As we mentioned earlier, convective overturning and an upward transfer of heat and water vapor would result. For a typical moist atmosphere, the resulting lapse rate could be about

6.5°C/km, with a net heating of the middle and upper troposphere and cooling at the earth's surface.

Let us consider an atmospheric column as a whole and ask how the annual mean solar radiation is disposed of. Globally, there must be radiative equilibrium or else the mean temperature of the atmosphere would systematically change. On the other hand, we would not necessarily expect local thermal equilibrium, that is, a balance among all processes contributing to the temperature change, by radiation, by condensation, etc. It is the meridional gradient of this imbalance that does in fact provide the driving impetus for the atmosphere.

On the left side of Figure 18.6 we see the disposition of the solar radiation (insolation). Of the 719 ly/day (Langley = cal/cm²) impinging at the top of the atmosphere averaged globally, 221 are reflected back by the atmosphere and clouds and 31 by the earth's surface. This is 1/4 the radiation at the subsolar point (the solar constant). Recent rocket- and balloon-borne observations suggest 698 ly/day, which coincides with a very early estimate. In these determinations the combined atmosphere-earth reflectivity (the planetary albedo) was assumed to be 36 percent. More recently, satellite measurements suggest 30 percent. Only 126 ly/day is absorbed by the atmosphere (mainly by stratospheric O_3) and almost half (341) of the original is absorbed by the ground. The latitudinal gradient of the solar radiation absorbed by the earth is quite large, however, ranging from 423 at 5° latitude to 73 at 85° latitude.

In the center of the figure we see that the ground radiates in the infrared on the average 824 ly/day, which is virtually all absorbed by the atmosphere—the "greenhouse" effect. Only a small amount escapes to space—the radiation

in the "window" between 8 and 12 μ. The atmosphere also radiates in the infrared, upward and downward. A substantial part goes to space, but the majority is returned to the earth's surface, the "back" radiation.

But the atmosphere's heat balance is not exclusively determined by radiative processes. Since there should be no heat accumulation at the earth's surface in the mean, the net radiation must be balanced (see the right side of Figure 18.6) by the sum of the upward turbulent transfer of sensible and latent heat. Although the ocean itself is globally in thermal equilibrium, it absorbs heat at low latitudes and gives up heat to the atmosphere at high latitudes. The ocean, therefore, must respond as a complementary poleward

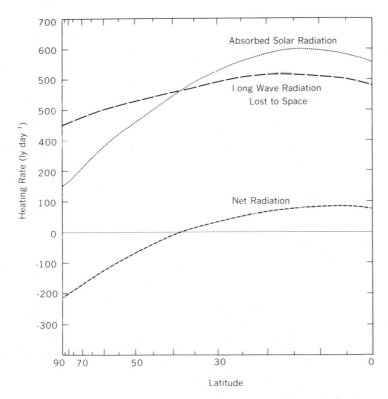

Figure 18.7 Latitudinal profiles of the annual mean components of radiation and heating responsible for driving the atmosphere.

heat transfer agent. The latent heat in the form of water vapor in general will undergo considerable lateral movement before it may condense and be realized sensibly to heat the atmosphere.

In Figure 18.7 are plotted the latitudinal profiles of the absorbed solar radiation (that is, the solar insolation less that lost to space) and the long-wave radiation lost to space, both with the same mean value of 467 ly/day. Hence the net radiation (or radiation excess), which is also shown, represents global radiative equilibrium, but possesses a latitudinal gradient. This is what is responsible for driving the combined atmosphere-ocean system. The atmosphere responds essentially to this gradient, but is somewhat modified by modes excited by seasonal and diurnal variation, which we have ignored.

THE SCOPE OF THE ATMOSPHERIC RESPONSE

We now briefly discuss some aspects of the atmospheric response to the net radiation gradient as it may mold our perspective in modeling the atmosphere and in conducting simulation experiments. Much will be drawn from observation, but, as will be seen, many types of observations do not yet exist and so one has to resort to qualitative and often intuitive estimates; theoretical studies have proved particularly useful. In the final analysis comprehensive observational efforts, such as those planned for the Global Atmospheric Research Program (GARP) and the subsequent routine World Weather Watch (WWW) network, will provide the basic information. A popular account of the broad properties of the observed general circulation can be found in articles by A. H. Oort.[25-27] A definitive determination of the global structure of the general cir-

culation and of climate and their nuances is an essential guide to continued refinements in the building of a hierarchy of atmospheric models. With an ability to establish reliably the atmosphere's natural state and variability and to simulate them, a firm foundation will have been laid upon which to probe the reasons for subtle changes in climatic regime and to predict the consequences of human activities.

The Scale Characteristics of Atmospheric Phenomena

The potential energy continually being created by the solar radiation does not become transformed to kinetic energy of atmospheric motions in a simple way. For example, a weather map does not show disturbances of all sizes. If so, one would have kinetic energy more or less uniformly distributed over many scales—a "white noise" spectrum. Instead, one usually sees disturbances in midlatitudes of the dimensions of about half the United States (extratropical cyclones), in which one can find fronts which can be quite long, but are no more than 50 to 100 km in the transverse dimension. Similarly, we are now learning that in equatorial latitudes only a limited number of distinct "animals" can be found in the tropical menagerie. This tendency toward discreteness in the atmospheric energy spectrum, a propensity for a line structure, reflects the fact that scale-selective instabilities and forced modes are responsible for the orderly transition of energy available in its "potential" state to that of limited variety of phenomena. An understanding of each of these hydrothermodynamic excitation mechanisms provides the constituent theoretical building blocks of dynamic meteorology. A parallel exists in physical oceanography.

The time-space domain for typical atmospheric phenomena is shown in Figure 18.8. This is typical for the free atmosphere (away from the lower boundary). Note that it covers a 10 order of magnitude span of horizontal dimension, but does not include the size of cloud water droplets, about 1 mm. There is a major energy peak at around 3000 to 6000 km corresponding to a zonal wave number of 5 to 8 (the number of waves in the wind or temperature field around a latitude circle). These extratropical cyclones are the result of "baroclinic instability," the major energy converting mechanism in extratropical latitudes. There are secondary peaks corresponding to tropical cyclones (hurricanes), fronts, cumulus convection, tornadoes, clouds, and clear air turbulence. They are generally intermittent and sparse phenomena and thus may not appear in a spectrum taken at any one time.

The fact that the spectrum is not quite linelike is the result of nonlinear spectral transfers of energy which tend to diffuse peaks, a process ultimately responsible for carrying energy down to the molecular scales (an energy "cascade") where it is transformed to heat by viscous forces, typical of most fluid systems. However, peculiar to quasi-two-dimensional fluids, such as the large-scale atmosphere ($> 10^3$ km), energy can "decascade" toward larger modes so as to maintain the jet stream.

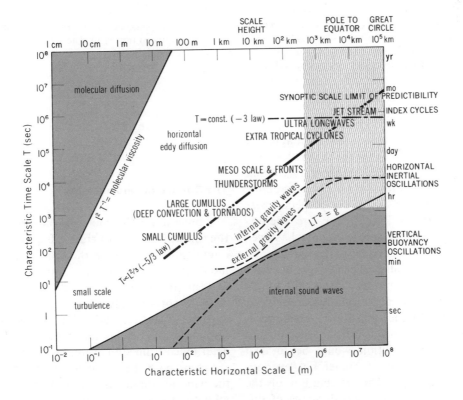

Figure 18.8 A space-time domain for characteristic atmospheric phenomena. The unstippled region encompasses most of the kinetic energy containing phenomena, with the predominance of extratropical cyclones, ultralong waves, and the jet stream. The crosshatched area denotes scales and phenomena typically resolved by general circulation models, that is, the macroscale (based on a figure by H. Fortag as modified by K. Ooyama).

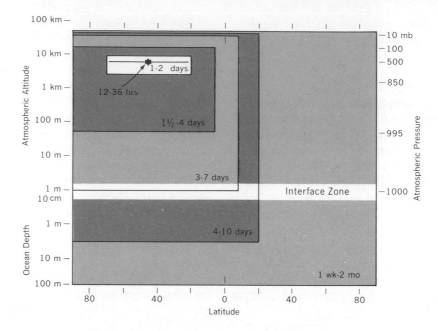

Figure 18.9 A schematic diagram of the domain of initial data dependence for a prediction point in midtroposphere at midlatitudes (denoted by a star) as a function of forecast time span. The atmospheric and ocean elevations are given on a logarithmic scale, increasing upward and downward, respectively. The stippled area is the interface zone. (Smagorinsky[37]).

Usually, a numerical model can at most resolve a two order of magnitude span of scales of this spectrum. Hence a global model with a horizontal grid size of 100 km can resolve wavelengths of 400 km and larger on to the planetary size (about 40,000 km). These are the explicitly treated macroscales. As has been pointed out a number of times, interactions with smaller scales must be dealt with by other means. For some simple problems they can be ignored entirely, or more usually they must be parameterized. Where greater detail is desired locally, an intermediate solution would be to deal with a subregion of higher but overlapping resolution. The calculation of the interaction between two domains of differing resolution, one "nested" in the other, is not yet a solved problem, technically. It has applicability to a number of modeling problems, such as a hurricane embedded in the planetary flow.

The Domain of Dependence

The precision with which one must model the complicated myriad of physical interactions depends on the relative magnitude of the interactions. Furthermore, certain processes and mechanisms become important only over longer periods of time. The "relaxation" times of the responses are one class of properties to be determined from simulation experiments.

Let us pose the problem of predicting the wind at a point in midtroposphere at midlatitudes (Figure 18.9).

For short periods, of the order of 12 to 36 hr, the atmosphere behaves inertially, so that the slow-acting sources and sinks of energy can be ignored. Because of the earth's high rotation rate, it can be shown both theoretically and experimentally that the atmosphere may be considered to be strongly coupled in the vertical, and we need only worry about redistributions of the kinetic energy already inherent in the initial conditions. A model which embodies these physical constraints, that is, a "barotropic" model, has proved very useful for short-range prediction—it needs data at but one level.

Over slightly longer periods, say, for 1 to 2 days, there may be occasional new sources of kinetic energy transformed from the potential energy inherent in the initial conditions. This will be reflected by small departures from a fully vertically coupled barotropic flow and would occur in connection with the intensification of large-scale extratropical cyclones. As a result, there will also be small phase differences between the motion wave disturbances and those of temperature (as will be seen in Figure 18.14). We need information at a minimum of two levels to be able to sense the potential energy available for transformation and we need a simple "baroclinic" model to account for the dynamics of cyclone intensification—that is, the "baroclinic instability" process.

For still longer periods, from 1½ to 4 days, the energy latent in the water vapor already in the atmosphere at the initial instant becomes of importance. As a result of such latent energy transformations in midlatitudes, the rate of development of cyclones is intensified and their scale reduced. Furthermore, dissipation of energy can no longer be ignored. The observational demands for 4-day predictions are thus increased. We now need water vapor data, and we need all the variables: wind, temperature, and pressure extending into the planetary boundary layer and into the deep tropics.

By the time we get to time spans of 3 to 7 days, completely external sources of energy gain significance, such as exchanges of heat and water vapor with the lower boundary. Data is now needed down to the earth's surface, into the lower stratosphere, and probably into the other hemisphere.

Longer time spans are less well understood and thus what we can say is more speculative. For periods of 4 to 10 days duration, the development of extratropical disturbances already evident in the initial conditions will have transported heat from equator to pole so efficiently that second-generation disturbances cannot be adequately accounted for without radiative exchange processes acting to resupply the potential energy. The relative thermal inertia of the oceans may begin to wane as the surface layers of the ocean react to a systematic heat exchange with the atmosphere. The domain of influence thus extends from the middle stratosphere, where ultraviolet absorption by ozone takes place, into the surface layers of the ocean, and probably well into the other hemisphere.

For a week and longer, data from the entire global atmosphere and probably down to the seasonal thermocline of the oceans are necessary. Indeed, for such time spans and longer into seasons, it will be unavoidable that dynamical models for the entire fluid envelope of the earth be employed. One need only recall that, although there is virtually no radiative gradient between pole and equator in summer, the atmosphere nevertheless is driven by the temperature gradient in the oceans that had been established by a succession of many winters.

Hence to account for the variability of the wind at midtroposphere at midlatitudes requires an enlarging domain of dependence with time which influences both the number of degrees of freedom in the total model, and the amount and kind of data one needs to specify initial conditions for a prediction.

CONDUCTING NUMERICAL EXPERIMENTS

The Design of Numerical Experiments

There have customarily been two types of experiments. Formally, they appear to be alike; that is, one starts with the properly posed set of initial conditions to which is applied the discretized or "differenced" form of the set of governing partial differential equations (as described earlier, we actually deal with a finite number of elemental macroscale volumes, not an infinite number of points) and the boundary conditions.

Transient Evolutions. These experiments are designed to determine the nature and limits of deterministic predictability of the atmosphere. To study transient details of the evolutions, one makes calculations from real initial conditions simulating changes over several days or weeks with slightly different theoretical models and then compares them with each other and with reality. In this way one can learn which processes are of importance, mechanistically why, and the rate at which they influence the large-scale atmosphere.

We refer to these as *prediction experiments.*

Asymptotic Calculation to a Statistical Equilibrium. One starts with trivial initial conditions which possess as few of the observed properties of the atmosphere as possible. Then one makes an infinite prediction (that is, long compared to the dissipation time of the total system) so that the initial conditions are forgotten. One seeks a statistical equilibrium and studies its phenomenological and mechanistic properties, for example, the characteristic time and space scales, the energy cycle, the mechanistic partitioning which leads to a heat, momentum, water balance, and the sensitivity of these

characteristics to the external parameters: the rotation rate, the solar radiation, etc. One can then compare the similarities and differences between the theoretical model simulation and the real atmosphere.

We refer to these as *general circulation experiments.*

It should be noted that we speak of a "statistical equilibrium." The reason is that fluctuations with characteristic times of several days (corresponding, as we will see, to the lifetime of baroclinic disturbances) must occur to maintain the system energetically. A system in true equilibrium could not be maintained.

It is the general circulation type of experiment that we are concerned with in the study of climate.

Computational Aspects

Resolution and Mapping. We have seen already that there is a need to span about two orders of magnitude in the horizontal scale to deal with the large-scale atmospheric phenomena containing the major portion of energy. Since at least four mesh widths are required per wavelength, we thus need a horizontal computational grid with a mesh width of about 100 km.

The number of vertical subdivisions in the computational grid is much less than the horizontal because of the small ratio of the vertical atmospheric scales to the horizontal. The distance between the gridpoints in the vertical may vary from between about 50 m in the boundary layer to several kilometers in the free atmosphere.

As we have seen, the fact that the scales of primary interest lie within the available resolution does not obviate the possibility that significant interactions must be accounted for with scales that have been truncated. These scales may be ignored for short periods of the order of a day, but for longer periods they become increasingly important both as a source and as a sink of energy for the macroscale (that is, the scales explicitly resolvable). The subgrid scale processes, or more precisely subresolution processes, cannot in practice be dealt with explicitly, but one must seek to express their statistical effect on the macroscale, while in turn they are determined in terms of the macroscale parameters.

The general concept of parameterization may be demonstrated schematically. In Figure 18.10, the atmospheric energy density distribution (or spectrum) is represented as a function of the horizontal wave number $k \propto l^{-1}$, where l is a horizontal distance. The computational domain has dimensions l_0, while l_1 denotes the dimension of the smallest scale which can barely be resolved if the mesh width is l_2 ($l_1 = 4l_2$), which corresponds to Δx or Δy, or generally, Δ. Physical processes which have a dimension less than l_1 thus need to be parameterized. The stippled area, representing the portion of the energy which must be parameterized, extends from l_1 to the smallest values of l for which energy interaction with the larger scales is significant. As indicated earlier, (see Figure 18.8), for a global domain $l_0 \sim 4 \times 10^4$ km, $l_1 \sim 400$ km, and $l_2 \sim 100$ km.

The early practice of numerical simulation in meteorology was to use conformal mapping methods in subglobal domains. Hence a stereographic projection has been employed for hemispheric domains (see Figure 18.11) and a Mercator was used for zonal (latitudinal) channel domains. Nonconformal methods are necessary to cover the globe and a variety are being used.

A considerable effort is now going into adopting truncated spherical spectral representation as an alternative discretizing method to grid mapping.

Computational Load. The number of points in a horizontal coordinate surface

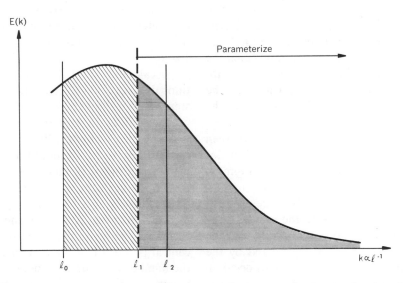

Figure 18.10 The curve represents the variation of atmospheric energy density as a function of the wave number $k \propto l^{-1}$. The hatched area is the explicitly resolved part of the spectrum, the macroscale. The stippled area indicates the portion of the energy spectrum within which the physical processes cannot be resolved by a domain with the dimension l_0 and the mesh width l_2.

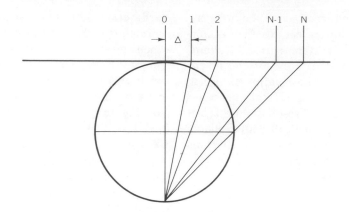

N	Distance On Earth Corresponding To Δ			Approximate Number of Points in Hemisphere
	Pole	45°	Equator	
5	2560	2160	1280 km	80
10	1280	1080	640	320
20	640	540	320	1250
40	320	270	160	5000

Figure 18.11 Polar stereographic projection. N is the number of horizontal grid intervals between the equator and pole (Smagorinsky et al.[40]).

is A/Δ^2, where A is the area of the domain and Δ is the average horizontal grid size. For K vertical coordinate levels, the total number of points is then KA/Δ^2. If there are n variables per point, then the total variables at any one time are nKA/Δ^2, which is a measure of the number of degrees of freedom in the model.

The maximum time increment between successive stages in the numerical integration Δt for the integration to remain stable in a computational sense for so-called "explicit" methods is determined by the speed (relative to the earth) of the fastest wave mode admissible by the system of equations $C + U$, and by the minimum grid length which can occur in the mapping chosen, $\Delta_{\min}$. In most modern large-scale models, C is the phase speed of external gravity waves, about 300 m/sec, and U is the maximum wind speed which can occur, say, 200 m/sec.

"Implicit" methods may provide a means for much longer time steps, and although the calculation is more complex there may be a net speedup by a factor of 5 or 10.

To calculate from the governing equations the incremental changes over a time step Δt in each of the n variables at each of the points in the three-dimensional mesh requires about 200 to 1000 computer operations (denoted by O), depending on the physical complexity of the model. Typical operations are a multiply or a subtract, or a fetch of a number (or "word") from a particular memory location to an arithmetic register in the central processor of the computer. Hence the number of computer operations to advance all the variables in the entire domain one time step is $n\,KAO/\Delta^2$. If the simulated elapsed time of an experiment is τ, then the number of time steps per experiment is $\tau/\Delta t = \tau(C + U)/\Delta_{\min}$.

We then have that the number of computer operations to perform an experiment is $n\,KAO\,\tau\,(C + U)/\Delta^2\Delta_{min}$.

We are now in a position to estimate how long it will take a given computer to do a given experiment with a particular model. We only require a measure of the average rate at which a computer can process a typical mix of instructions including the use of peripheral storage devices, such as magnetic drums or disks—the "throughput" rate. The units customarily used are millions of instructions per second: MIPS. In Figure 18.12 we see an historical summary of computers dating back to 1953, the beginning of the modern electronic computer era. Shown are top-of-the-line computers at the time of their commercial debut. The MIPS rating is not very precise and depending on the criteria of determination may easily vary by a factor of two in either direction. Of particular importance is that there has been a relatively uniform ex-

ponential growth of power with a doubling about every 1½ years. Present-day computers are about 10 MIPS, almost 1000 times faster than in 1953. Within the next few years we can expect to see "fifth generation" computers capable of about 100 MIPS. Many computer designers feel that an ultimate limit of 10^4 MIPS will be determined by the speed of light and the maximum degree of miniaturization of electronic components and their interconnections. Parallel (rather than sequential) modes of calculation should provide further extensibility as is already becoming evident.

If S is the throughput speed of the computer being used, then the computer time to do a single experiment is

$$\frac{nKA(C + U)}{\Delta^2\Delta_{min}}\frac{\tau O}{S}.$$

Of special significance is that the grid size occurs to the third power in the denominator. Generally, the vertical reso-

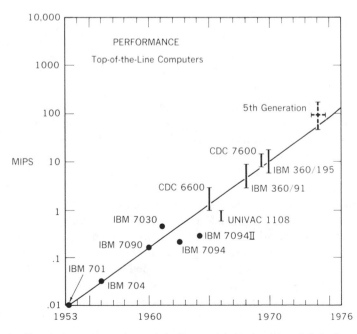

Figure 18.12 An historical summary of top-of-the-line computers in terms of their throughput capability (in millions of instructions per second, MIPS) and the year of their commercial debut. Depending upon the problem and method of measurement, the speed can easily differ by a factor of two in either direction.

lution increases as we increase the horizontal resolution, so that resolution increase becomes by far the dominant factor in determining the computational load. Hence with no other changes in models, the increase of computer power by 10^4 in little more than 20 years could only accommodate about a factor of 10 increase in resolution in each of the three space dimensions!

To estimate the computer time needed to do a currently typical experiment we shall choose a global ($A = 4\pi a^2$, in which the earth's radius is $a = 6400$ km), 11(= K)-level model with 4 (= n) variables per point (two horizontal wind components, temperature, and humidity). The average horizontal grid size is 200 km, which we will take for Δ and $\Delta_{\min}$. We shall consider a 500-day (= τ) simulation and assume that O is 500. For an IBM 360/91 type of computer S is 5 MIPS; therefore,

$$\frac{4 \times 11 \times 4\pi \times (6400)^2 \text{ km}^2 \times 500 \text{ m/sec}}{(200)^3 \text{ km}^3}$$

$$\times \frac{500 \text{ days} \times 500}{5 \times 10^6/\text{sec}} = 56 \text{ days},$$

which is about 10 times faster than real time, a typical ratio for experiments of this type. Prediction experiments, simulating several weeks, can use higher resolution for models of similar physical complexity. It is not uncommon to compute only twice as fast as real time in experimental predictions, although for operational predictions the models and their resolution must be adjusted to available computer power such that it does not take much more than an hour to produce a 24-hr prediction.

It is obvious that long climatic simulations for 10's or even 10,000's of years cannot be done in this usual way, even with the very fastest computers imaginable.

EXPERIMENTS IN THE SIMULATION OF CLIMATE

We concern ourselves exclusively with the first class of experiment, which John von Neumann in 1960 termed the "infinite prediction."[44] Presumably, the final state is independent of how one starts out and is wholly controlled by the boundary conditions and the governing physical laws. That such a unique "statistical equilibrium" exists has been seriously questioned by Lorenz.[14] This is supported by experimental evidence with the rotating laboratory annulus. In highly interactive systems in which many processes interplay, one cannot rule out the possibility that there may be more than one discrete climatic state which will apparently satisfy all the external conditions. With this reservation in mind, we now proceed to ignore this possibility in what follows.

The Breakdown of the Zonally Symmetric Circulation

If the earth's surface were homogeneous and the planet were not rotating, the imposition of a latitudinal heating gradient would result in a single circulation cell with an upward limb at the equator and a downward limb at the pole, much the same as in a tea kettle heated at the rim. The upper air flow poleward and the lower flow equatorward both contribute toward relieving the radiative heating gradient. It is called an energetically direct cell because it is the result of a transformation of potential to kinetic energy. This is also referred to as a Hadley cell in tribute to the eighteenth-century scientist who first deduced its existence.

When the earth rotates, the situation is altered in that the Coriolis force turns the wind to the right relative to the earth in the Northwestern Hemisphere and to the left in the Southern Hemisphere. Therefore, in the Northern Hemisphere the upper poleward current assumes a strong eastward component (westerly wind) and the lower equatorward current assumes a westward component (easterly wind). For a given heating gradient, the turning of the wind tends to reduce the efficiency with which the single cell can transport heat poleward. For a sufficiently large rotation rate, a balance cannot be effected and the meridional

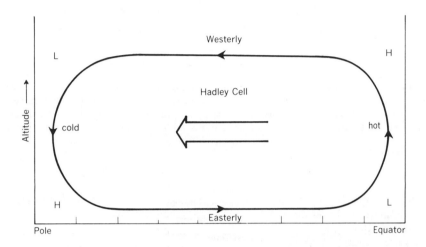

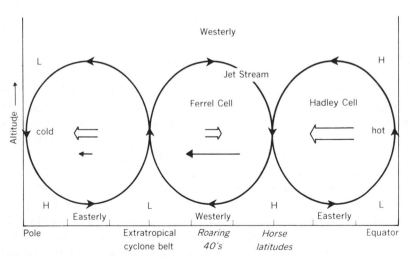

Figure 18.13   The mean meridional circulation for a horizontally zonally symmetric circulation (top) and for a zonal circulation which has become unstable and developed wave modes along latitude circles (bottom). *H* and *L* denote high- and low-pressure extrema relative to a given altitude. The horizontal arrows denote the direction and relative magnitude of heat transport from equator to pole, the open arrows are due to the mean meridional circulation, and the solid arrows are due to the wave motion.

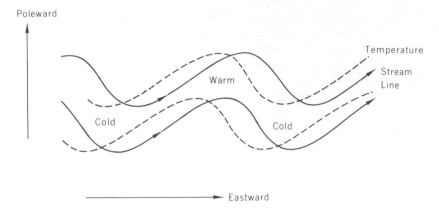

Figure 18.14 A typical baroclinic wave in mid-atmosphere traveling eastward along a midlatitude circle—it is intimately connected with extratropical cyclones at the earth's surface. The fact that the temperature wave lags westward behind the streamline wave means that cold air is being transported equatorward and warm air poleward. Hence the zonally asymmetric eddies are responsible for a systematic poleward heat transfer.

temperature gradient continues to increase, while remaining circularly symmetric about the pole (top of Figure 18.13). There comes a point, however, when small meridional displacements of a particular east-west size can become dynamically unstable and grow. The development of such zonal asymmetries is known as *baroclinic instability* and was discovered theoretically in 1947 by Charney[4] and independently in 1949 by Eady,[7] although it was known observationally well before. Even the earlier numerical models of the general circulation with only two levels of vertical resolution were capable of simulating the breakdown of the symmetric polar vortex.[23, 29, 34] The processes involved are essential to the energy balance of the earth's atmospheric circulation which, in the final analysis, is responsible for climate.

It is worthwhile here to describe in further detail how the evolution takes place. One can show theoretically that disturbances with the dimensions of five to eight waves about the pole will grow fastest. At the surface, the polar anticyclone (high pressure) develops nodules or

waves which grow and ultimately become detached finding a final residence in the subtropics—the subtropical high-pressure belt. Between the two, at 50 or 60° latitude, is a belt of low pressure where the extratropical cyclones dwell. As this occurs, the air motion seeks to balance the new pressure distribution because of the Coriolis force—the geostrophic wind. As a result (bottom of Figure 18.13), the westerlies aloft intrude into the lower atmosphere, lying at the earth's surface between the subtropical high pressure (at the "horse latitudes") and the extratropical cyclone belt. The belt of "roaring 40s" of the Southern Hemisphere is particularly pronounced because it is less interrupted by continents than in the Northern Hemisphere and because of the snow covered Antarctic Continent. Weak surface easterlies still exist between the extratropical belt and the polar high and also equatorward of the subtropical anticyclone. Aloft, the wave disturbances force a convergence of angular momentum, tending to sharpen and intensify the jet stream in the westerlies.

The mean meridional circulation is no

longer simple. The single equator-to-pole direct cell, is contracted equatorward, so that the poleward descending limb coincides with the axis of the surface subtropical anticyclone. In these latitudes baroclinic instability is virtually nonexistent because of the weak Coriolis force, so that the direct mean meridional circulation (the Hadley cell) is still the most efficient means for effecting a poleward heat transfer. In midlatitudes, the unstable wave motion is the most efficient means for the poleward heat transfer (Figure 18.14). In fact one finds observationally, theoretically, and numerically that a weak energetically indirect mean meridional circulation is excited (the Ferrel cell), transfering a small amount of heat equatorward with a corresponding transformation of kinetic to potential energy, partially compensating the potential to kinetic energy transformation by the baroclinic wave instability.

The numerical models of the latter 1950s had only two levels, providing just sufficient degrees of freedom to permit baroclinic instability to operate and to account for the dominant energetic features of the atmospheric response to the radiative heating gradient energy source and the frictional energy sink. The net effect of the complex radiative components was represented by a simple heating function. To explain further the observed details required still more complex models.

The Vertical Structure

The next major step was to simulate the evolution of the vertical structure of the atmosphere.[9, 13, 36] As an example, we consider a situation which represented the vertical structure by nine discrete levels and calculated the radiative exchanges as a function of an arbitrary distribution of

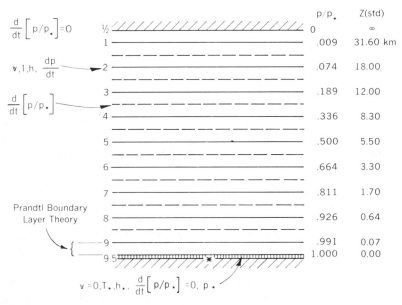

Figure 18.15 Vertical distribution of horizontal coordinate surfaces and their approximate heights. Notations are made as to where the model variables are predicted. h is the relative humidity, an alternative measure of water vapor content. $d[]/dt$ is the rate of change of a property in a moving atmospheric elemental volume (Smagorinsky et al.[36]).

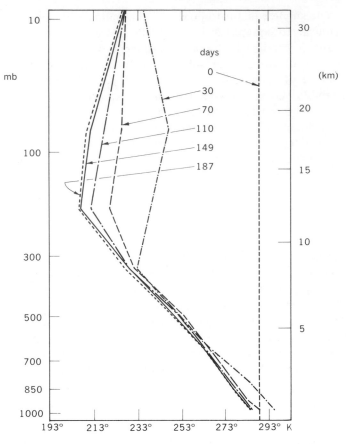

Figure 18.16   Variations with time of the vertical distribution of temperature of the model atmosphere at 45° latitude (after Manabe et al.[19]).

radiatively active gases and cloud, which would have to be either predicted or prespecified.[36] Since one was still limiting attention to the annual mean radiation and excluding seasonal forcing, it was sufficient to deal with a hemispheric domain. The nine levels (Figure 18.15) were unevenly distributed with elevation to optimize the resolution needed to account properly for the processes thought to be of importance. The uppermost level was placed at about 32 km, the lowest elevation at which one could still take account of the major absorption by ozone of the ultraviolet portion of the solar radiation. At least three levels were kept

at or above 12 km so that the temperature lapse rate of the stratosphere, if it were correctly simulated, could be described. The lowest level was set at 70 m, the upper limit of the surface boundary (or Prandtl) layer, in which turbulent exchange processes could be assumed to be independent of the earth's rotation and to follow rather simple parameterization laws. There were two levels of resolution below 1 km, the minimum required to take account, even crudely, of the interaction between Coriolis forces and turbulent viscous forces which are of comparable magnitude in determining the structure of the planetary (or Ekman) boundary layer, which has a key role in linking the dynamics of the free atmosphere to turbulent exchanges of heat, momentum, and water vapor with the lower boundary. The remaining intermediate levels were the minimum thought to be necessary to resolve the dynamical coupling in the free troposphere and to define the tropopause.

It is only because of the strong influence of the earth's rotation that there is so much coherence in the vertical, requiring so little resolution. In near equatorial latitudes, where the Coriolis force is weak, more vertical structure can result with a corresponding need for increased computational resolution.

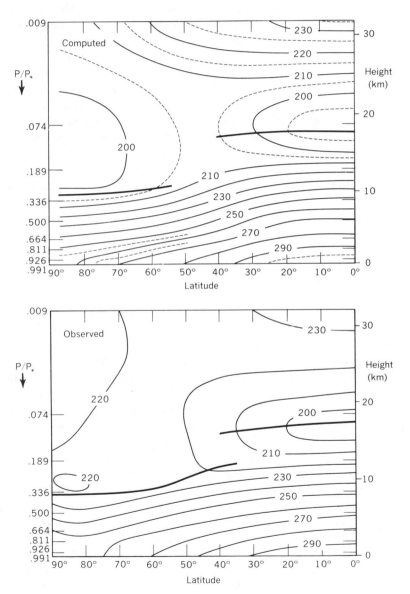

Figure 18.17 The latitude-height distribution of the zonal mean temperature, computed and observed. The tropopause is indicated by a heavy line broken in middle latitudes (Smagorinsky et al.[36]).

With a prescribed latitudinal and height distribution of CO_2, O_3, H_2O, and clouds, a calculation was performed from rest and isothermal conditions (the "trivial" initial conditions). Figure 18.16 shows that the evolution of the thermal structure of the troposphere is rapid, attaining its essential characteristics within a month or two. On the other hand the stratosphere takes longer. The equilibrium state compared with observation from equator to pole is shown in Figure 18.17. In this calculation it was assumed that the lower boundary was a uniform land surface with no heat capacity at all. If it were taken to be an ocean with very large heat capacity instead, then the total system would take much longer to come into statistical equilibrium, hundreds or thousands of years, depending on what we assumed the initial temperature to be. This is discussed below.

This experiment established that one can simulate the thermal and dynamic necessity for a stratosphere provided the mean distribution of water vapor, ozone, and cloud is known. CO_2 is assumed to be spatially uniform and well mixed; only the mean concentration need be specified. A more fundamental experiment would be for the model itself to *predict* their evolution together with that of the thermal and wind structure. It is only recently that this has been done for water vapor.

Another very important result came from this simulation, namely, an elucidation of the means by which the stratospheric kinetic energy is maintained. It has been known observationally that the lower stratosphere transports heat poleward toward higher temperature, a process which transforms kinetic to potential energy. Further losses of kinetic energy result from frictional dissipation. The simulation indeed reproduced the countergradient heat transport and showed, moreover, that the lower

stratospheric kinetic energy is maintained through pressure interactions from the troposphere.

The Dynamical Influence of Water Vapor

We have already discussed the important radiative role of water vapor even in the stratosphere, where the humidity is very low. But because of the copious sources of water vapor the atmospheric relative humidity is quite high in the troposphere. Even the rather small vertical velocities associated with large-scale weather systems can be as large as 10 cm/sec. If sustained for a day at that rate, they can result in a vertical displacement of about 9 km, virtually the full depth of the troposphere. If the rising parcel were dry, then it would cool by adiabatic expansion at a rate of almost 10°C/km. Since it would be increasingly cooler than its environment, which has a typical temperature stratification (or lapse rate) of 6.5°C/km, the parcel's upward course is subjected to increased resistance by the buoyant stability of its environment.

But if it started out near the earth's surface at, say, 50 percent relative humidity (Figure 18.18), the dry adiabatic cooling would bring the parcel to saturation after about 1 km rise within several hours. Thereafter, the condensing water vapor would release latent heat as precipitation occurs from the parcel, thus slowing down the rate of temperature decrease to about 5 or 6°C/km, almost half its previous rate. But also the parcel's upward speed may increase by a factor of 2 to 10 depending on a variety of circumstances including the reduced resistance of the environment's stability. Hence the availability of water vapor for condensation makes the parcel less stable to upward displacements with respect to its environment than if the parcel were completely dry.

The thermodynamic consequences of released latent heat are, therefore, much more rapid than the direct radiative consequences. For example, the reduced effective static stability of a moist atmosphere (that is, in those places where significant amounts of condensation are occurring) has a profound dynamical influence, because baroclinic instability critically depends on the magnitude of the effective static stability (Figure 18.19). The result is that the most baroclinically unstable disturbances (given by the minimum of each curve) are shorter and the resulting meridional temperature gradient (which is proportional to the vertical wind shear du/dz) is weaker. Numerical experiments with dry and moist general circulation models verify this.[19] Figure 18.19 also shows that an $N = 20$

($\Delta = 540$ km) stereographic grid just barely resolves moist baroclinic instability, while $N = 10$ does not.

Furthermore, since heat is being transferred in both sensible and latent form, the total heat transfer requirements can be satisfied by smaller amplitude baroclinic waves than for a dry atmosphere.

The moist simulation started from dry initial conditions, but the lower boundary was assumed to have a constant source of water vapor available for evaporation while having the thermal properties of a land surface. Effectively, the hemispheric domain of the earth was assumed to be a swamp. It was, therefore, of interest that the resulting water vapor distribution in the statistical equilibrium in this simulation was remarkably similar to that which has been observed for the annual

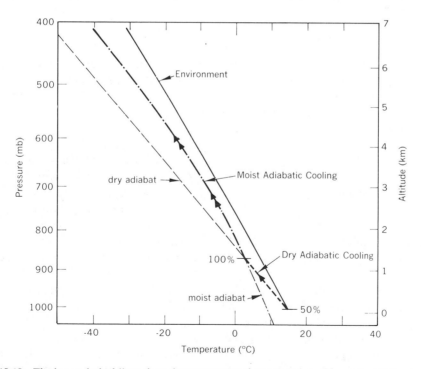

Figure 18.18 The heavy dashed lines show the temperature that a parcel would assume relative to its environment (solid line) as it is forced upward from its initial position at the earth's surface. If it initially possessed 50 percent relative humidity, it would first rise dry adiabatically until it cooled to saturation, and thereafter would cool at a slower rate, because of heat released during condensation, that is, moist adiabatically.

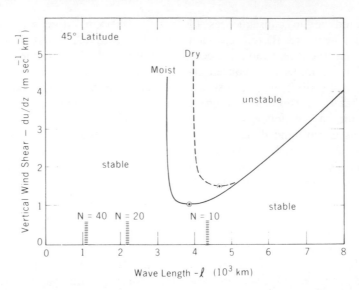

Figure 18.19 The dependence of baroclinic instability on the wavelength of the disturbance ($l = 2\pi a/k$, where k is the wave number) and the vertical wind shear (du/dz) which is proportional to the meridional temperature gradient. The "dry" curve is the neutral criterion for a dry atmosphere, while the "moist" curve reflects effect on the baroclinic stability criterion of the condensation of moisture. The most unstable waves are denoted by stars on minima of each curve. The points $N = 10, 20,$ and 40 show the scales resolvable by stereographic grids of different resolution.

mean. Furthermore, the simulated meridional distributions of precipitation and evaporation predicted the tendency for deserts to form in the subtropical belt because evaporation exceeded the available precipitation.

Effects of Surface Asymmetries

We have seen that a symmetric flow is dynamically unstable, breaking down into traveling wave disturbances which transfer heat poleward and are responsible for transforming potential energy to kinetic energy, thus maintaining the general circulation of the atmosphere. These are connected with extratropical storms and, therefore, are responsible for typical weather variations in middle latitudes.

If these transient waves had random geographic position (phase), we might expect that if we were to look at an average over a long enough period, the waves would nullify each other and the averaged flow would be zonally symmetric. That this is not actually so can be seen from Figure 18.20, which is an average over many Januarys and many Julys of the surface pressure and the surface temperature in the Northern Hemisphere. In fact, the amplitude of the normal zonal asymmetries in the pressure and temperature is quite large, especially in the wintertime. Most noteworthy in January are the two large surface pressure minima in the North Pacific and North Atlantic, that is, the Aleutian and Icelandic lows, and the massive high-pressure area (anticyclone) over Asia. The latter reverses to a low-pressure area in July, a manifestation of the monsoonal transition with season. The subtropical surface anticyclones, more pronounced in summer than in winter, would exist as a continuous belt in the mean even if the traveling waves nullified each other. What is significant is that in actuality, the belt is interrupted at certain longitudes, particularly at continents, resulting in cells of high pressure centered in the western Pacific and Atlantic Oceans.

Even on a year-to-year basis, the monthly mean surface pressure distribution looks quite similar to the normals, so that the quasistationary disturbances are quite definitely geographically fixed. One would suspect asymmetries of the lower boundary to be responsible, and therefore, the quasistationary patterns to be different in each hemisphere. Indeed, early theoretical studies suggested that the large mountain masses can kinematically excite quasistationary disturbances that are deep and have almost the same phase at all levels (that is, they are quasibarotropic), whereas the difference in thermal properties between oceans and continents can create quasistationary disturbances with maximum amplitude in the lower atmosphere and which can tilt westward or eastward with elevation (baroclinic).

These theoretical results were first verified by the numerical model simulations of Mintz in 1965 (see Figure 18.21). Recent calculations with more sophisticated models admitting an interactive hydrologic cycle clearly delineate the

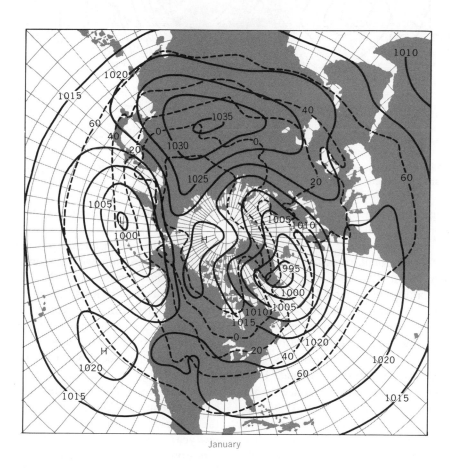

January

Figure 18.20a Normal January sea level pressure (solid lines) in millibars and temperature in degrees Fahrenheit (dashed lines) (after Smagorinsky[33]).

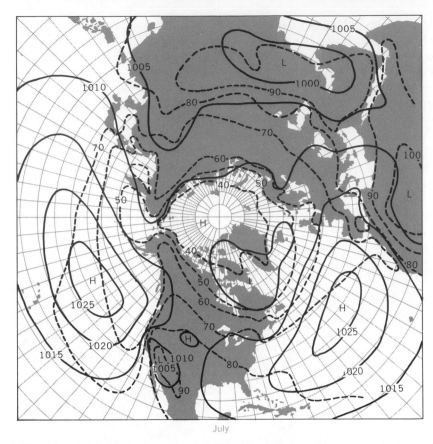

July

Figure 18.20b Normal July sea level pressure (solid lines) in millibars and temperature in degrees Fahrenheit (dashed lines) (after Smagorinsky[33]).

profound influence of the world's mountains on the global climatic precipitation regime.[17] Figures 18.22 and 18.23 show the results of calculations for a perpetual January, both with and without mountains, against the observed January precipitation. Figure 18.24 shows the computed soil moisture, another element in the hydrologic cycle. The prediction of the world's arid regions is reasonably clear: the Sahara, Southwest Africa, India, Western Australia, Mexico, and Southern South America. It is of particular interest that the predicted snow line in the Northern Hemisphere is too far south, a consequence of imposing January solar radiation conditions perpetually in this simulation experiment.

The Influence of an Interactive Ocean

We have already discussed the fact that beyond a week or two, the ocean can react to the atmosphere. The depth of reaction increases with time interval, so that by definition, the seasonal thermocline, at about 100 m, is the reacting depth to seasonal variations of the atmosphere. The deep oceans owe their characteristics to the action of the atmosphere over 100 or even 1000 years.

An attempt to build a joint model of the atmosphere and ocean was made a few years ago using a physically complex atmospheric model with nine coordinate levels and the most advanced six-level ocean model available at that time.[16]

There was a particular problem to avoid doing a 100-year simulation of the joint evolution of the circulations of the ocean and atmosphere in the short time steps demanded for computational stability of the atmospheric model. This was dealt with rather ingeniously by first calcu-

lating with the atmospheric model for a year and then applying a smoothing filter which removes the fluctuations of less than a week; that is, those with periods shorter than the characteristic energy cycle of individual atmospheric disturbances. The resulting statistical prop-

Figure 18.21 Surface pressure reduced to sea level, for Northern Hemisphere winter and Southern Hemisphere summer, in millibars (the broken lines are intermediate 2½-mb isobars). The upper figure is the 30-day mean (from Day 256 to 285) computed in the numerical experiment with mountains. The lower figure shows the observed normal January sea level pressure (reprinted by permission, from Mintz[23]).

Mean Precipitation Rate

With Mountains

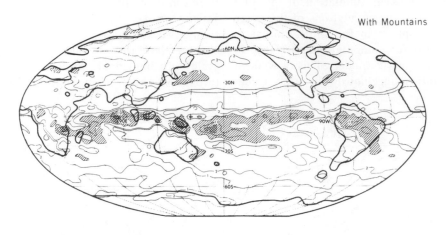

☐ < .1 cm/day
☐ .1 to .5 cm/day (Contours .2 cm/day)
▨ .5 to 1.5 cm/day (Contours at 1.0 cm/day)
▦ 1.5 cm/day

Without Mountains

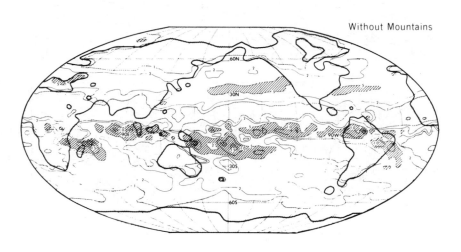

Figure 18.22 Distribution of simulated January mean rate of precipitation. Upper half, the mountain model. Lower half, the mountainless model (reprinted by permission, from Manabe and Holloway[17]).

Mean Observed Precipitation Rate

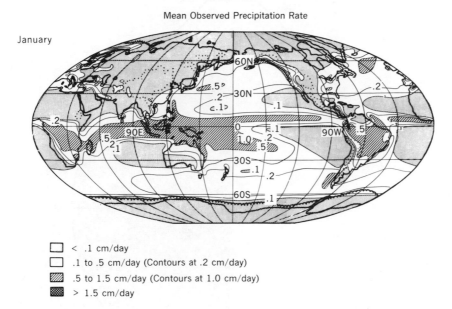

□ < .1 cm/day
□ .1 to .5 cm/day (Contours at .2 cm/day)
▨ .5 to 1.5 cm/day (Contours at 1.0 cm/day)
▩ > 1.5 cm/day

Figure 18.23 Distribution of the observed January mean rate of precipitation (reprinted by permission, from Möller[24]).

Mean Computed Soil Moisture

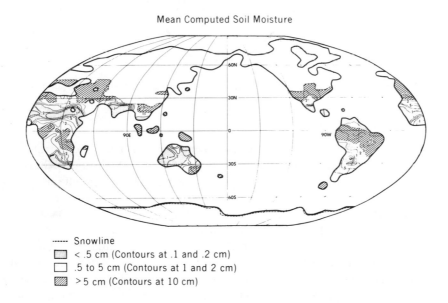

----- Snowline
□ < .5 cm (Contours at .1 and .2 cm)
□ .5 to 5 cm (Contours at 1 and 2 cm)
▨ > 5 cm (Contours at 10 cm)

Figure 18.24 Computed continental soil moisture with the mountain model. Dotted shading delineates areas having less than 0.5 cm of soil moisture; slashed shading, more than 5 cm. Thick dashed lines indicate the boundary of snow cover. Soil moisture under snow cover is not shown (after Manabe and Holloway[17]).

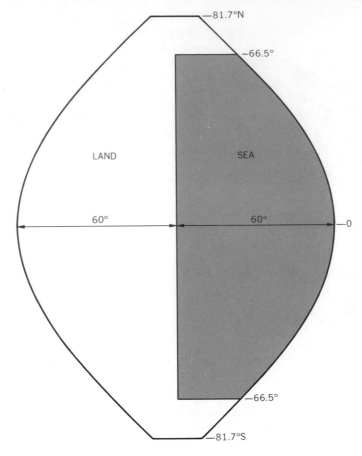

Figure 18.25 Ocean-continent configuration of the ocean-atmosphere coupled model (after Manabe and Bryan[16]).

erties of the atmospheric surface stress, heat flux (radiative and turbulent), and water substance (rainfall, runoff, evaporation, sublimation, snowfall, and iceberg formation) were then applied as boundary conditions to drive the oceans for 100 years in 166.7-min time steps. The resulting sea surface temperature and ice cover (its thickness is predicted) were thereupon used to influence the transient variability of the atmosphere in 10-min time steps, and so on.

To further ease the computational demands, an idealized geography was used that was sixfold symmetric (Figure 18.25), with three equally spaced con-

tinents and a polar continent in each hemisphere. Although there were assumed to be no mountains, the snow cover and surface hydrology were self-determined. For the latter, a north-south continental divide was invoked midway. In these calculations the annual mean solar radiation was imposed, and for the purposes of radiative calculations H_2O, O_3, and cloud distribution were prescribed functions of latitude and elevation while the CO_2 was assumed to be uniform. For the release of latent heat and all other elements of the hydrologic cycle, water vapor was completely self-determined.

In Figure 18.26 we can see how well

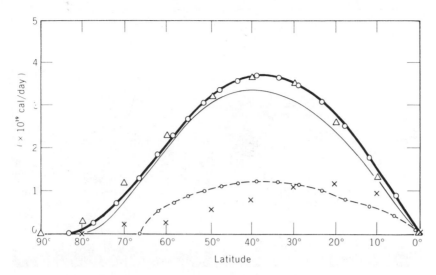

Figure 18.26 Latitudinal distribution of the rate of the poleward transport of energy by the joint atmosphere-ocean model is shown by a heavy solid line; the corresponding quantity for the actual ocean-atmosphere system is given by solid triangles; the rate of the poleward transport of heat contributed by the model ocean is shown by a dashed line; the corresponding quantity for the actual ocean is shown by ×'s (average of the two hemispheres); for comparison, the poleward transport of energy by the atmospheric-model atmosphere, without a participating ocean, is given by a light line (after Manabe[15]).

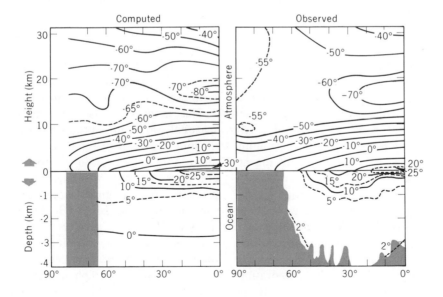

Figure 18.27 Zonal mean temperature of the joint ocean-atmosphere system—left-hand side. The distributions of the two hemispheres are averaged. The right-hand side shows the observed distribution in the Northern Hemisphere. The atmospheric part represents the zonally averaged annual mean temperature. The oceanic part is based on a cross-section for the western North Atlantic (after Manabe and Bryan[16]).

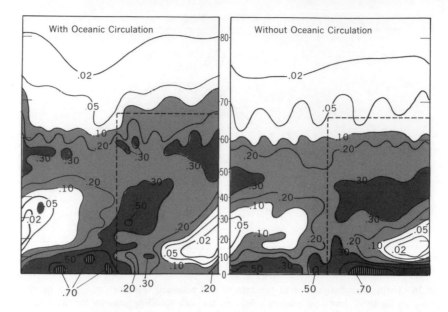

Figure 18.28 Area-distribution of the time mean rate of precipitation (rainfall and snowfall) of the joint model (cm/day) on the Mercator map projection. The area of the ocean is indicated by the box in the lower right side. The distributions of the two hemispheres are averaged (after Manabe and Bryan[16]).

the partitioning of the poleward heat transfer by the atmosphere and oceans is predicted, particularly that the oceans are responsible for one-third the total transport. It is significant that when no ocean participation is permitted, the atmosphere alone becomes more active and accomplishes the necessary poleward heat transfer by itself, which is also characteristic of what happens when water vapor transfer is suppressed. The resulting zonally averaged, quasiequilibrium temperature distribution in the atmosphere and ocean is shown in Figure 18.27. The distribution of precipitation is shown in the right side of Figure 18.28. For comparison, the left side shows the result of a model without a reactive mobile ocean—it acts as a swamp, only providing a source of water for evaporation. The most striking difference is to be found near the equator. In the model without ocean circulation the precipitation maximum occurs over the equatorial "ocean" where the greatest

evaporative supply of water is to be found. However, when the ocean can react, a cold equatorial surface current (corresponding to the observed Cromwell current) is formed which inhibits atmospheric convection relative to the continent. The maximum precipitation thus occurs over the equatorial continent, reminiscent of the Amazon basin.

This model has also been used for a seasonal simulation in the statistical equilibrium. However, it is still far too inadequate for studies of long-term climatic changes. Severe weaknesses in our present understanding of ocean dynamics present a limiting factor in building ocean models. In part the problem is, in contrast to the atmosphere, that the ocean transports are significantly accomplished by mesoscale phenomena having characteristic horizontal scales of a few hundred kilometers. The dynamical nature of these motions and their systematic transport properties are not well-known. Furthermore, the small-scale vertical turbulent transfer mechanisms in the ocean are but imperfectly understood. These modeling deficiencies are now under intensive study by oceanographers.

MODELS FOR LONG-TERM PREDICTION

The atmosphere, just as the oceans and the rivers, is a finite reservoir for man-made waste products. Because the media are fluid, they can also transport foreign matter, as well as natural constituents, from their sources to far-off places. Also, since our ability to artificially produce chemicals and energy is beginning to rival nature quantitatively, the composition of the atmosphere and the forces acting upon it *may* significantly be altered.

An ability to predict reliably the nature and magnitude of such atmospheric behavior is, therefore, of paramount importance in calculating risk. The precision required is great since the biosphere's tolerance for environmental aberrations is evidently not very large. Preventative measures to protect the biosphere and the environment from man-made pollutants will for practical reasons generally not be a question of complete cessation of certain activities. Instead, it will require

decisions of prudent moderation. For example, where is the least harmful place for a particular industry to be concentrated, or how many SST's can be safely tolerated and along which routes? It is for this reason that rather precise and reliable predictions will be needed of the environmental impact of a given human activity. Crude or premature estimates can be very misleading in providing guidance for such far-reaching decisions and may be far more damaging than no estimate at all.

In principle, numerical models can provide a predictive capability in two classes of problems:

- A calculation of the large-scale, long-term dispersive and storage characteristics of the combined atmosphere-hydrosphere system.
- A determination of secondary interactions which could significantly influence the structure and variability of the combined system, that is, which could result in climatic change.

Dispersion

The present level of modeling capability of the atmosphere and oceans probably

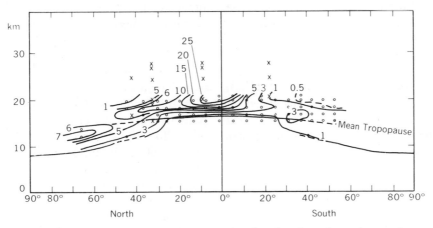

Figure 18.29 The measured W[185] distribution in the atmosphere based on observations made over North and South America (by R. J. List and K. Telegadas; see Hunt and Manabe[8]).

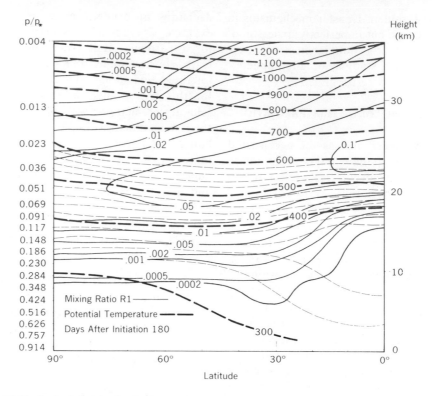

Figure 18.30 Latitude-height distributions of the calculated zonal mean tracer concentration 180 days after initiation (units: g/g) (after Hunt and Manabe[8]).

will allow meaningful estimates to be made of the residence times and the dispersion rates and patterns of inert material. Recent experience has shown that verbal arguments may be wholly misleading,[8] for example, in the case of the dispersion of radioactive tungsten released in the lower equatorial stratosphere, the observed slow rate, which was well simulated by model calculation (see Figures 18.29 and 18.30), was the result of an almost balanced opposition of the transports by the mean meridional circulation and the large-scale quasihorizontal eddies.

Critical Factors in Climate Variation

The second question, will there be an appreciable modification of climate, is a delicate one. It requires models that are capable of simulating the natural climate and its variability in sufficient detail, for example, as it was 100 years ago before the industrial revolution. A decade ago modeling finesse was clearly inadequate to answer most questions. However, as we have seen in the preceding section, a great deal of progress has been made, although an unqualified potential is far from having been achieved.

In this section we try to judge the compatibility of some of the questions we would like to ask with the limitations of our modeling ability. To do so, we draw upon and elaborate on some of the material developed in earlier sections.

It is important to be aware of the fact that many, if not most, of the inadvertent climate change questions we pose represent probable variations comparable to the noise level of the natural variability of the atmosphere and the uncertainty of current modeling techniques. For example, in a 5-year period there was at least a 20 percent variability in the poleward heat flux by the large-scale eddying motion (Figure 18.31). Figure 18.32 shows the great sensitivity of the calculated poleward heat flux in models in which only computational resolution or the role of water vapor has been altered.

We have seen that over the past 15 years global circulation models have become exceedingly sophisticated. The dynamical interactions have attained a high level of similitude to reality, although some severe problems still remain. However, the climatic equilibrium is strongly influenced, if not dominated, by many subsidiary processes whose interaction with the atmospheric dynamics must be modeled with sufficient fidelity to account for the evolution of climate and its possible variants.

Oceanic response. Looking a bit more closely upon the oceans, one can detect a mechanism for an asymmetry in the thermal response of the upper layers of the ocean. We first note that because the ocean has a much larger heat capacity (that is, thermal inertia) than the at-

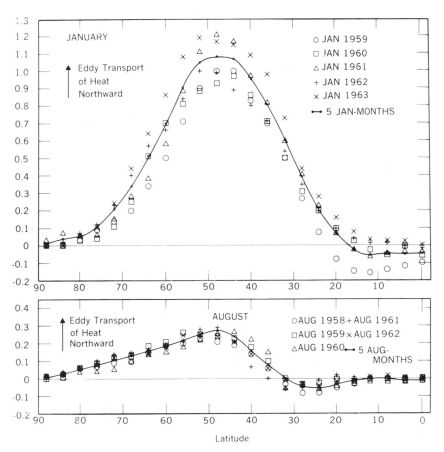

Figure 18.31 Observed vertically integrated poleward eddy transport of sensible heat for five different years for the months of January and August. The solid curve in each is the 5-year mean. Units are 10^{20}cal/day (based on Oort and Rasmusson[28]).

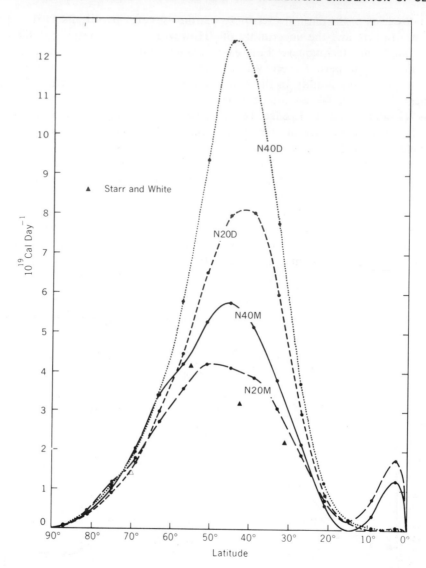

Figure 18.32 Vertically integrated poleward eddy transport of sensible heat. Observed annual mean values by Starr and White are given by solid triangles. The curves give the results of four different model simulations in which the horizontal resolution ("N20" and "N40" are low and high resolution, respectively) and the role of water vapor ("M" is for a model with moisture thermodynamically active; "D," moisture is absent) were varied (after Manabe et al.[18]).

mosphere, it resists temperature changes. The existing main thermocline is buoyantly stable down to about 800 m. Heat removed at the ocean surface produces buoyant instability which can be relieved only through mixing deep enough to form an isothermal or neutrally stratified layer. On the other hand, sur- face heating is stabilizing and remains confined to the uppermost layer. Thus a surface temperature decrease requires a greater heat exchange across the air-sea interface and is more difficult to accom- plish than a surface temperature increase (Figure 18.33).

In the limiting case, when we start with

a well-mixed isothermal layer in the upper ocean (Figure 18.33), for example, as is the case at high latitudes, the skewness of response becomes extreme. The removal of heat at the ocean surface yields buoyant instability which must mix heat through a very deep layer.

In either case, wind stirring of the upper layers of the sea will mechanically promote the vertical transport of heat, especially when surface heating tries to create large vertical temperature gradients.

The effects of salinity of the ocean on its buoyant stability is considerably less, especially at lower latitudes. In midlatitudes and to a lesser extent in the tropics, where precipitation clearly exceeds evaporation, the freshening of the surface waters contributes to stabilizing the upper layers; on the other hand, in the subtropics, where there is a minimum of precipitation, we would expect the opposite.

Moreover, the ocean's poleward heat transfer will tend to reduce an imposed pole to equator heating gradient in the atmosphere. However, the partitioning of the heat transport between the oceans and the atmosphere would obviously be altered if there were substantially more or less sea ice, attenuating or enhancing the thermal interaction between the atmosphere and ocean. The presence of sea

ice will, of course, prevent wind stirring of the uppermost layers of the ocean.

Hence the existence of oceans over three-fourths of the earth's surface can have a curiously stabilizing influence on climatic change, one that is, as yet, incompletely understood and may vastly alter estimates of the climatic impact of a given human impulse.

Snow Cover. The earth is only partially glaciated. But even the variation from an ice age to an interglacial period represents a relatively small latitudinal shift of the snow perimeter. It is very easy (in fact, too easy, as was evidenced in connection with Figure 18.24), in simulation experiments, to produce an ice age. A prime mechanism derives from the annual variability of solar radiation, that is, the balance between snowfall in winter and thaw in summer. And it is only recently that the simulation of seasonal variation has been attempted with models. Also, one should keep in mind that the Antarctic polar cap underlain by a continental mass will respond differently than will the Arctic cap which overrides relatively warm waters. The intermediate case is to be found as the Antarctic snowpack extends northward out over the adjacent oceans as winter advances, almost doubling the effective size of the continent.

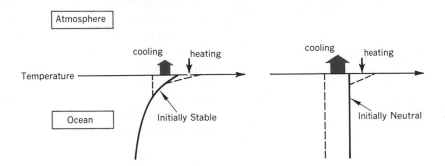

Figure 18.33 The depth of response of the ocean and the amount of heat exchange with the atmosphere necessary for a *given* sea surface temperature increase or decrease. The left figure is for a stably stratified upper ocean, and the right figure is for neutral stratification. The role of salinity has been ignored.

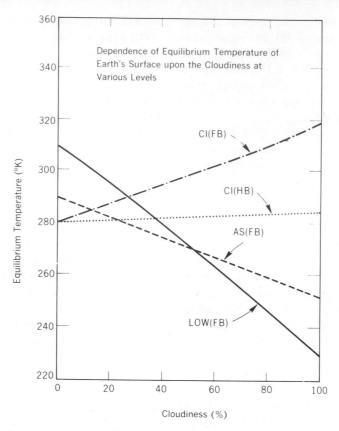

Figure 18.34 Radiative convective equilibrium temperature at the earth's surface as a function of cloudiness (cirrus, altostratus, low cloud). FB and HB refer to full black and half black, respectively (after Manabe and Wetherald[21]).

Aerosols. Except for cloud studies, aerosols have been ignored in modeling—not only the effects of their variability but even their systematic influence on the radiation balance. Conclusions concerning climatic effects due to changes of the aerosol distribution, quality, and amount due to human activity are, therefore, not an immediate nor an obvious expectation. Atmospheric sampling and laboratory optical studies will be needed.

Qualitatively, one would expect that aerosols would have two different effects:

- An increase in albedo and, therefore, a decrease of the net downward solar radiation at the top of the atmosphere.

- An increase in absorption of solar radiation and, therefore, a decrease in the net outgoing radiation.

The first would result in a cooling of the atmosphere-earth-ocean system and the second would give a heating. The sense of the balance among these contributions would depend on the specific optical properties of the aerosol and probably the altitude of its occurrence, as will be evident when we discuss the effects of clouds.

Clouds. Even that preponderant atmosphere aerosol-like constituent, cloud, has been dealt with cavalierly. Cloud amount in most models is not predicted explicitly (for an exception, see Kasahara and Washington[10]), although water vapor is predicted in some models. The reason is that cloud storage of water is a negligible element in the water balance and is difficult to predict with the necessary accuracy for radiative purposes.

Attempts have been made empirically to determine stratiform cloud amount from the water vapor distribution—but it has not proved acceptably satisfactory as yet—especially since corresponding statistics for cumulus clouds have not been determined, which may be particularly critical in the tropics. However, the average effects of cloud based on the observed vertical and latitudinal variation on the radiative transfer is now taken into account.

Some simple studies to assess cloud-radiative coupling do suggest a sensitivity, but only in the absence of the myriad of other stabilizing interactions.[21] Basically, clouds should behave as aerosols, as we just discussed. But in the case of clouds, it has been possible to do some simple calculations for a radiative-convective equilibrium. These show (Figure 18.34) that for low clouds the high albedo (about 0.7) and the high emission temperature yield a sharp decrease of equilibrium surface temperature with increasing cloudiness. On the other hand, for fully black high cirrus clouds (that is, acting as "black" radiators, emitters, or absorbers), the low albedo (0.2) and low emission temperature give an increase of surface temperature with increasing cloud amount. However, observations indicate that cirrus clouds behave more as if they are "half-black" radiators, in which case, they transmit more infrared radiation upward for the same albedo, thereby reducing the sensitivity of surface temperature to cloud amount. Direct radiometer measurements indicate that under cirrus in the tropics there is warming and in midlatitudes cooling.[5].

It is clear that the net effects of clouds are yet to be determined in different climatic regimes.

Typical Instabilities

Of all the processes in the atmosphere-hydrosphere-cryosphere system which interact to determine climate and its variations, it is always possible to select a few which in combination, but *to the exclusion of all others*, can conspire to yield a climatic instability through a cyclic chain of positive feedbacks. Two such feedback chains that are often cited are the greenhouse instability and the snow cover instability.

Greenhouse Instability This feedback chain can be reasoned purely in terms of the interaction between the infrared radiation field, the water vapor content, and the convection in the presence of buoyant instability. It must be assumed that the solar radiation is constant, that cloud amount and CO_2 are not affected, and that there are no atmospheric dynamics or ocean interactions.

One can reason, referring to the left side of Figure 18.35, that

- an increase in temperature throughout the troposphere ($1 \rightarrow 2$)
- results in an increase in water vapor content, q (3),
- which increases in the absorption of infrared (IR) radiation from below (the "greenhouse") (4),
- which increases the in situ temperature and, therefore, the back-IR radiation,
- which yields an increase of temperature in the lower troposphere and earth-ocean boundary below (5),
- which gives rise to convective instability (6),
- which distributes the heat throughout the troposphere (7),

but which *reduces* the magnitude of the

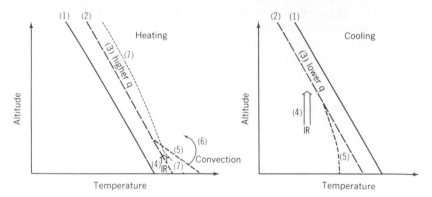

Figure 18.35 Stages in the greenhouse instability due to initial heating (left) and initial cooling (right). Refer to the text.

surface temperature increase (a negative feedback).

The reverse situation, that of cooling (the right side of Figure 18.35), would give an even larger surface temperature change, but now a decrease, since the negative feedback of convective instability plays a lesser role. This is quite analogous to the situation in the surface layers of the ocean discussed earlier. We see then that the atmosphere resists an increase of temperature at the earth's surface, but that oceans resist a decrease there even more so.

Snow Cover Instability This positive feedback involves only the interaction between snow cover, its effect on reflecting solar radiation, and the net effect on the resulting temperature regime. The solar radiation is constant, while trace gases and cloud amount are assumed to remain unaltered. Again, atmospheric dynamics and ocean interaction are ignored.

Reasoning, as before, we have that

- an increase in atmospheric temperature
- decreases the area of snow cover,
- which decreases the albedo of the earth's surface and, therefore, increases the absorbed solar radiation,

- which further increases the atmosphere-earth-ocean temperature at the latitude of reduced snow perimeter,
- which results in a poleward movement of the zone of maximum baroclinic activity, that is, of storminess.

This argument is generally reversible for a temperature decrease.

Keeping the qualitative sense of these instabilities in mind, we shall now go on to consider the results of some actual calculations and numerical experiments.

Limited Experiments

A Solar Constant Increase of 2 Percent. We first of all note that a 6 percent/year global increase of energy production would in the span of a century, yield the equivalent of a 2 percent increase in the absorbed solar energy.[41] We shall consider the results of three different models.

First, consider a one-dimensional model in which we fix the absolute humidity. In this calculation the radiative-convective interaction is the only operative mechanism. One finds a 1.2°C surface temperature increase, which is essentially the Boltzmann infrared temperature change.

Now let us consider a model in which we fix the relative humidity. This allows a simple water vapor interaction together with radiative-convective interaction and is also one-dimensional. Now one gets a 2.4°C surface temperature increase, which is larger than before because of the greenhouse instability.

Finally, if we use a three-dimensional dynamical model with idealized geography (as yet an unpublished work of Syukuro Manabe), we can allow complex water vapor interactions. There is no heat storage by continents or ocean, but "swamp water" in the oceanic locations is available for evaporation. This model predicts snow cover and soil-moisture changes over continents, and sea-ice changes over the ocean. The albedo criteria for bare soil, snow, sea, and sea ice

are empirically specified. Figure 18.36 shows the incremental change of the statistical equilibria between a simulation with a solar constant 2 percent larger than the normal and that with the normal value. We find that this integration resulted in a new stable equilibrium state with a 2 to 4°C surface temperature increase equatorward of 60° latitude and a 4 to 10°C surface temperature increase poleward of 60° owing to the positive feedback of the snow cover and the lack of a convective redistribution upward near the pole where the static stability is high. There is 3.1°C global average surface temperature increase. We see that the snow-albedo feedback results in a further destabilization. The 4°C increase in the extrapolar upper troposphere is due to heat released during condensation. It is

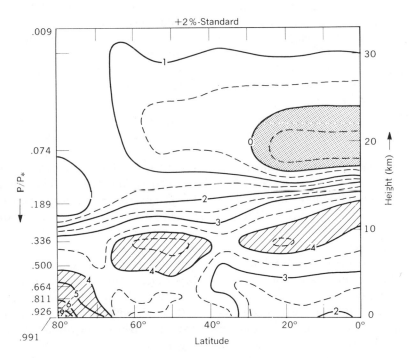

Figure 18.36 The latitude-height distribution of the zonally averaged temperature difference (°C) between the statistical equilibrium of a general circulation simulation with a solar constant 2 percent larger than standard and with the standard solar constant. This is a fully three-dimensional model with idealized geography (after Manabe—to be published).

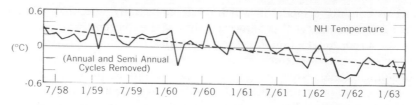

Figure 18.37 The Northern Hemisphere temperature during the period May, 1958 through April, 1963. The values are the mass weighted averages between the surface and about 18 km height with the mean and the annual and semi-annual cycles removed (after Starr and Oort[42]).

also found that snow cover is reduced as predicted by the instability argument.[1, 32]

Another calculation by Manabe showed that a 2 percent decrease in the solar constant also results in a new stable state with a larger ice cap. However, it appears that a large decrease in the solar constant, that is, greater than 4 percent, may result in unstable growth of the ice perimeter because convective instability is not operating to distribute the cooling upward. M. I. Budyko predicted the stability of an initially "white" earth.[2]

We must keep in mind, however, that these calculations ignore the annual variability of the solar radiation and interactive changes of cloud cover. Furthermore, by neglecting ocean interaction (heat storage and lateral mobility), we have eliminated a strong stabilizing influence to large temperature changes.

It is of interest to note that the best estimates from observations indicate an actual decrease of atmospheric temperature in recent years (Figure 18.37). This means that the sequence we just described could not be the dominating one in contemporary reality.

Actually, the effect of heating sources may be much greater, but for another reason. Horizontal gradients associated with nonuniform sources of heat will drive subsidiary circulations. For example, a megalopolis heating strip at the east coast of the United States could result in a shift of the effect of Gulf Stream slightly westward, thus broadening it. An expected consequence would be a westward shift of wintertime storm tracks along the East Coast with more precipitation on the coast.

A CO_2 Increase by a Factor of 2. The observed increase of CO_2 this century has been 20 to 25 percent. It is estimated that approximately this much has also been buffered by the oceans.[41] In the following calculations it has been assumed that the solar constant and the distributions and amounts of cloud and ozone do not change. It is assumed that the CO_2 concentration is doubled from 300 to 600 ppm.

Again, let us consider a one-dimensional model unit with fixed relative humidity. As before, this permits simple water vapor interaction with the radiative-convective interaction. The result is a 1.9°C surface temperature increase. The increased CO_2 increases the upward and downward infrared radiation, the latter decreasing the loss at the earth's surface—a greenhouselike effect. The water vapor interaction accentuates the heating.

A three-dimensional model, as before, with idealized geography (also, as yet, an unpublished work of Manabe) yields (see Figure 18.38) a 2.9°C global average surface temperature rise with the main increase poleward of 60° latitude, as before. On the other hand, the stratospheric temperature decreases. Anticipating a surface temperature rise,

Budyko,[3] using a model with very crude dynamics, predicted the clearing of the Arctic ice through CO_2 pollution. But again, we must remember that annual variability, ocean, and cloud reaction have not been accounted for.

For example, the presence of a horizontally stagnant ocean can be used to reason a further destabilizing complication. If an increase of atmospheric CO_2 increases the surface temperature of the sea, it will also reduce the ocean's capacity to buffer CO_2. Hence a given CO_2 source rate of increase would yield a greater rate increase of atmospheric CO_2 concentration and thence surface temperature.

Parameterization of Synoptic Scale Dynamics

We have seen that the most important phenomenological reaction of the earth's atmosphere in midlatitudes to solar radiation is connected with baroclinic instability. Extratropical climate, therefore, is critically dependent on the sum total of all episodes of cyclogenesis in their poleward transport of heat, water vapor, and angular momentum. On the face of it, because of the short characteristic time scale of the energy cycle of extratropical cyclone families, that is, 1 to 3 weeks, combined with the requirement for mathematical accuracy, it is necessary to do calculations in very short time increments even if one wishes to determine only the average consequences. It is already reasonably well established that for a turbulent fluid, such as our atmosphere, the details of these calculations must be deterministically limited, so that an individual extratropical storm is predictable in position and intensity for only a few weeks. Yet many of the systematic or statistical properties of such storms

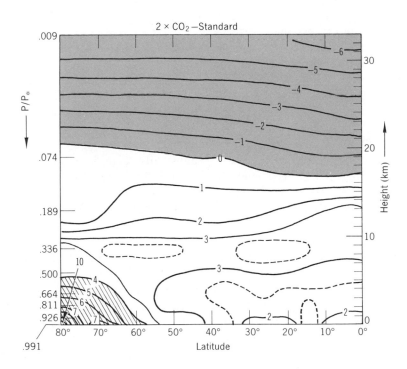

Figure 18.38 Same as Figure 18.36 except that the CO_2 concentration is increased from 300 to 600 ppm, but with standard solar constant (after Manabe—to be published).

should be predictable for much longer, and indeed constitute deterministic climatic variation.

From the computing load discussion it is clear that even computers of 10^4 MIPS capacity will not permit very long experiments appropriate to climatic time scales. If we modify the example given in the section, "Conducting Numerical Experiments," to a model with twice the horizontal and vertical resolution to be applied to a 10,000-year simulation, then it would take 10 full years to do the calculation with a computer 2000 times faster than the IBM 360/91. There is, therefore, a very compelling reason to ask whether some alternative can be found to calculating a many-thousand-year simulation in time steps of a few minutes. What this amounts to is finding a means for parameterizing the transport and energetic properties of synoptic scale disturbances so as to avoid a detailed calculation. This quest is similar to the one in the field of turbulence theory, in which one seeks to express the statistical properties of the ensemble of turbulent elements in terms of properties of the mean flow.

Although the problem in this form is new, attempts to construct "turbulence" theories of the atmospheric general circulation, or *gross-austausch* formulations, date back to A. Defant's work in 1921.[6] Considerable progress has been made in recent years,[11, 35, 43, 46] but there appear to be some inherent difficulties in application to the study of climatic change. Nevertheless, even the seasonal problem has been given attention[12, 45] while efforts to include effects of continentality variations with longitude are rudimentary.[30] So far, attempts have succeeded only with simple models, with effectively only two levels in the vertical and in which the operative physical mechanisms excluded most of the processes and interactions apparently important for the evolution and variability of climate, for example, water

vapor phase changes and oceanic coupling. Hence, as yet, we seem to be at a plateau of progress in the development of means to parameterize the synoptic scale dynamics and thus compress the number of calculations necessary to simulate long-term climatic excursions. Nevertheless, a great deal of attention is being given to the search for closure conditions, with insight being drawn from observational, analytic, and numerical studies of the atmospheric general circulation. It must be said at this point, however, that we have no hard proof that in principle such a closure condition must exist. It may be that the general strategy which has been taken by most investigators is not sufficiently flexible to include all those physical interactions that are relevant to climatic change. In which case an alternative method, similar to that used by Manabe and Bryan to couple an atmospheric and oceanic calculation, may prove fruitful.

Concluding Remarks

It is clear that we do not fully know the real potentialities and limitations of our present modeling capability, and it will not be known until we do a sufficient number of experiments with the most physically complete models and with high-enough resolution to be in the mathematically convergent range.

However, even the most pessimistic prophesies of doom allow a century of grace.

The acceleration of a deliberate and scientifically sound simulation capability is the prime objective of the Global Atmospheric Research Program (GARP) which is now underway. Under this program, the necessary technical tools will be developed, the critical basic atmospheric and oceanic data will be acquired, and the existing modeling limitations will be diminished. There is no question that modeling simulation techniques will provide the most reliable key to predicting natural and unnatural climatic change.

It does warn us, however, that if current physically comprehensive models are inadequate to answer some of our questions, then certainly we should be wary of basing broad national or international decisions on hand-waving arguments or back-of-the envelope calculations.

ACKNOWLEDGMENTS

The author gratefully acknowledges the critical comments on a draft manuscript by G. S. Benton, C. E. Leith, A. H. Oort, J. D. Mahlman, and especially, S. Manabe.

REFERENCES

1. Budyko, M. I., The effect of solar radiation variations on the climate of the earth, *Tellus* **21,** 611–619, 1969.

2. Budyko, M. I., *Climate and Life,* Hydrometeorological Publishing House of the USSR, Leningrad, 1971.

3. Budyko, M. I., *Man's Impact on Climate,* Hydrometeorological Publishing House of the USSR, Leningrad, 1972, (in Russian).

4. Charney, J. G., The dynamics of long waves in a baroclinic westerly current, *J. Meteorol.* **4**(5), 135–162, 1947.

5. Cox, S. K., Cirrus clouds and the climate, *J. Atmos. Sci.* **28**(8), 1513–1515, 1971.

6. Defant, A., Die zirkulation der atmosphare in der gemassigten breiten der erde, *Geograf. Ann.* **3**(3), 209–266, 1921.

7. Eady, E. T., Long waves and cyclone waves, *Tellus* **1**(3), 35–52, 1949.

8. Hunt, B. G. and S. Manabe, Experiments with a stratospheric general circulation model: 2. large-scale diffusion of tracers in the stratosphere, *Monthly Weather Rev.* **96**(8), 503–539, 1968.

9. Kasahara, A. and W. M. Washington, NCAR global general circulation model of the atmosphere, *Monthly Weather Rev.* **95**(7), 389–402, 1967.

10. Kasahara, A. and W. M. Washington, General circulation experiments with a six-layer NCAR model, including orography, cloudiness and surface temperature calculations, *J. Atmos. Sci.* **28**(5), 657–701, 1971.

11. Kurihara, Y., A statistical-dynamical model of the general circulation of the atmosphere, *J. Atmos. Sci.* **27,** 847–870, 1970.

12. Kurihara, Y., Experiments on the seasonal variation of the general circulation in a statistical-dynamical model, to be published in *J. Atmos. Sci.,* 1973.

13. Leith, C. E., Numerical simulation of the earth's atmosphere, *Methods Comput. Phys.* **4,** 1–28, 1965.

14. Lorenz, E. N., *The General Circulation of the Atmosphere,* World Meteorological Organization No. 218.TP.115, 1967, 161 pp.

15. Manabe, S., Climate and the ocean circulation: 2. The atmospheric circulation and the effect of heat transfer by ocean currents, *Monthly Weather Rev.* **97**(11), 775–805, 1969.

16. Manabe, S. and K. Bryan, Climate calculations with a combined ocean-atmosphere model, *J. Atmos. Sci.* **26**(4), 786–789, 1969.

17. Manabe, S. and J. L. Holloway, Jr., Simulation of the hydrologic cycle of the global atmospheric circulation by a mathematical model, *Proceedings of the Reading Symposium, International Association of Scientific Hydrology, World Water Balance,* 1970, pp. 387–400.

18. Manabe, S., J. Smagorinsky, J. L. Holloway, Jr., and H. M. Stone, Simulated climatology of a general circulation model with a hydrologic cycle: 3. Effects of increased horizontal computational resolution, *Monthly Weather Rev.* **98**(3), 175–212, 1970.

19. Manabe, S., J. Smagorinsky, and R. F. Strickler, Simulated climatology of a general circulation model with a hydrologic cycle, *Monthly Weather Rev.* **93**(12), 769–798, 1965.

20. Manabe, S. and R. F. Strickler, Thermal equilibrium of the atmosphere with a convective adjustment, *J. Atmos. Sci.* **21**(4), 361–385, 1964.

21. Manabe, S. and R. T. Wetherald, Thermal equilibrium of the atmosphere with a given distribution of relative humidity, *J. Atmos. Sci.* **24**(3), 241–259, 1967.

22. Matthews, W. H., W. W. Kellogg, and G. D. Robinson, Eds., *Man's Impact on the Climate,* The MIT Press, Cambridge, Mass., 594 pp.

23. Mintz, Y., Very long-term global integration of the primitive equations of atmospheric motion, *WMO-IUGG Symposium on Research and Development Aspects of Long-Range Forecasting,* World Metoerological Organization Tech. Note No. 66, 1965, pp. 141–155.

24. Möller, F., Vierteljahrskarten des niederschlags fur die ganze erde, (Quarterly charts of rainfall for the whole earth), *Petermanns Geographische Mitteilungen* (Justus Perthes, Gotha, East Germany) **95**(1), 1–7, 1951.

25. Oort, A. H., The energy cycle of the earth, *Sci. Am.* **223**(3), 54–63, 1970.

26. Oort, A. H., The atmospheric circulation: the weather machine of the earth, *Sci. Teacher* **38**(9), 12–16, 1971.

27. Oort, A. H., The present climate and its anomalies, *Yale Sci.* **46**(5), 19–23, 1972.

28. Oort, A. H. and E. M. Rasmusson, Atmospheric circulation statistics, *NOAA Professional Paper 5,* U. S. Government Printing Office, Washington, D. C., 1971, 323 pp.

29. Phillips, N. A., The general circulation of the atmosphere: a numerical experiment, *Quart. J. Royal Meteorol. Soc.* **82,** 123–164, 1956.

30. Saltzman, B. and A. D. Vernekar, An equilibrium solution for the axially-symmetric component of the earth's macro-climate, *J. Geophys. Res.* **76,** 1498–1524, 1971.

31. SCEP, Report of the study of critical environmental problems, *Man's Impact on the Global Environment,* The MIT Press, Cambridge, Mass., 1970, 319 pp.

32. Sellers, W. D., A global climatic model based on the energy balance of the earth-atmosphere system, *J. Appl. Meteorol.* **8,** 392 pp., 1969.

33. Smagorinsky, J., The dynamical influence of large-scale heat sources and sinks on the quasi-stationary mean motions of the atmosphere, *Quart. J. Royal Meteorol. Soc.* **79**(341), 342–366, 1953.

34. Smagorinsky, J., General circulation experiments with the primitive equations. 1. The basic experiment, *Monthly Weather Rev.* **91**(3), 99–164, 1963.

35. Smagorinsky, J., Some aspects of the general circulation, *Quart. J. Royal Meteorol. Soc.* **90**(383), 1–14, 1964.

36. Smagorinsky, J., S. Manabe, and J. L. Holloway, Jr., Numerical results from a nine-level general circulation model of the atmosphere, *Monthly Weather Rev.* **93**(12), 727–768, 1965.

37. Smagorinsky, J., The role of numerical modeling, *Bull. Meteorol. Soc.* **48**(2), 89–93, 1967.

38. Smagorinsky, J., Numerical simulation of the global atmosphere, *The Global Circulation of the Atmosphere,* Proceedings of conference on the global circulation of the atmosphere, London, England, August 1969, G. A. Corby, Ed., 1970, pp. 24–41.

39. Smagorinsky, J., The general circulation of the Atmosphere, *Meteorological Challenges: A History,* D. P. McIntyre, Ed., Information Canada: Ottawa, 1972, pp. 3–41.

40. Smagorinsky, J., R. F. Strickler, W. E. Sangster, S. Manabe, J. L. Holloway, and G. D. Hembree, Prediction Experiments with a general circulation model, *Proceedings of the IAMAP/WMO International Symposium on Dynamics of Large Scale Processes in the Atmosphere—June 23-30, 1965* (Moscow, U.S.S.R.), 1967, pp. 70–134.

41. SMIC, Report of the study of man's impact on climate, *Inadvertent Climate Modification,* MIT Press, Cambridge, Mass., 1971, 308 pp.

42. Starr, V. P. and A. H. Oort, *Nature* **242,** 310–313, 1973.

43. Thompson, P. D., Prognostic equations for the mean motions of simple fluid systems and their relation to the theory of large-scale atmospheric turbulence, *Tellus* **6**(2), 150–164, 1954.

44. vonNeumann, J., Some remarks on the problem of forecasting climatic fluctuations, *Dynamics of Climate,* R. L. Pfeffer, Ed., Pergamon Press, New York, 1960, pp. 9 11.

45. Wiin-Nielsen, A., A theoretical study of the annual variation of atmospheric energy, *Tellus* **22,** 1–16, 1970.

46. Williams, G. P. and D. R. Davies, A mean motion model of the general circulation, *Quart. J. Royal Meteorol. Soc.* **91,** 471–489, 1965.

19 | Inadvertent Large-Scale Weather Modification

LESTER MACHTA

KOSTA TELEGADAS

The composition of the earth's atmosphere began concerning mankind as soon as its chemical constituents were even vaguely known. For example, a need for oxygen to support life has led to many doomsday prophecies predicting the depletion of this vital element by natural geological processes or through inadvertent human intervention. But the *depletion* of atmospheric oxygen is not the only potential hazard to life[41]; every 1 percent increase in atmospheric oxygen concentration doubles the probability of forest fires, and if oxygen concentration reaches 25 percent rather than the present 21 percent, *all of the forests will be burned,* while there are valid reasons for expecting a constant oxygen concentration, still mankind must watch its values.

Equally frightening are some of the forecasts, rarely documented, that the increase in atmospheric carbon dioxide through the "greenhouse" warming will melt the glaciers and ice caps and inundate the vital coastal plains of the world. Indeed maps have been prepared

Kosta Telegadas is a Research Meteorologist in the Air Resources Laboratory, National Oceanic and Atmospheric Administration, in Silver Springs, Maryland, where Lester Machta is the Director.

of the world's land distribution if and when the ocean surface rises about 250 m.[36] The picture is frightening, but there appears at this time no evidence that such atmospheric warming and sea level increases will occur in the foreseeable future.

The weather and climate of the earth changes on virtually all time scales. Fluctuations from minutes to weeks are categorized as weather, those greater than a few years as climate. Recorded climate observed by instruments began in the latter part of the 17th century, but climatic variations dating back thousands to millions of years have been deduced from indirect evidence. The network of weather observing stations has grown with time; hemispheric mean surface temperatures reflecting possible large-scale changes date back to the latter part of the 19th century.

Prior to the past 50 to 100 years, changing climate was almost certainly unrelated to human activity, and contrary to some laymen's opinions, climate changes do not necessarily depend on man's disturbance of the earth or its atmosphere; however, in recent and probably in future years, man's impact becomes progressively more evident in pollution and land alteration. A major scientific challenge to the atmospheric scientist is the separation of natural from man-made climate modifications. This chapter treats those aspects of changing air composition caused by pollution from human activity which may influence the climate, but aside from local phenomena such as the urban heat island, one cannot unequivocably attribute any climatic change to man.

Climate includes a large variety of parameters: temperature, rainfall and snowfall, wind, sunshine, humidity, and many others. Changes in any of these can prove to be important to society; in some cases even small changes can be significant. For example, the annual temperature at Washington, D.C., during 1958–1967 was only 0.17°C cooler than in the 1940–1949 period, but since winter temperatures are near the freezing point, the snowfall almost doubled; the winter temperature cooled by 1°C. Similar or larger increases in the snowfall along the United States' middle Atlantic states occurred during the same periods. Since this chapter deals with large-scale changes, it is necessary to confine those changes to weather elements commonly observed in a routine, systematic fashion and that are not subject to purely local anomalies. In practice, this effectively limits the available weather parameters to temperature, at least until future research reveals others.

This chapter treats primarily the evidence for changes in air composition with only minimal remarks on their climatic consequences; the reader is referred to Chapter 18 for a further elaboration on modeling climate changes. Unfortunately, this latter step of converting an air quality change to a climate change can now be treated in only a qualitative or at best semiquantitative fashion, contrasting with the evidence for changing air composition which, although imperfect, rests on a sounder quantitative basis.

Each of the parameters discussed, carbon dioxide, particles, clouds, land use, and heat, may contribute to climatic change. If the climate changes, it is because of the net effect of all these and other contributions, despite the artificial treatment of each in isolation. The last section of this chapter reviews briefly evidence of past large-scale temperature trends and a review of efforts to predict the warming and cooling variations.

CARBON DIOXIDE

The atmospheric constituent most often mentioned in expressing concern that man may be altering the climate is carbon dioxide, CO_2. Its concentration is increasing with time, by perhaps 10 percent since the start of the industrialization in the latter half of the last century; qualitatively, the many consequences of increased CO_2 are well established, and projections of fossil fuel consumption may be estimated with some confidence. It is likely that atmospheric CO_2 will increase in the foreseeable future. If the lower atmosphere does indeed warm because of increased CO_2, a number of potential effects have been suggested: melting of ice caps, rise in sea level and its consequences, and warming of the seawater.

Other nonclimatic effects may be attributed to changes in atmospheric concentrations of CO_2 such as increased acidity of fresh waters, which is likely to be unfavorable although the effects are likely to be very small, and an increase in the rate of photosynthesis.[59] This latter aspect will be incorporated in a predictive model of future CO_2 since, in principle, more carbon will be tied up in the biosphere. The increased photosynthesis may be beneficial to mankind. Implicit in the growth of atmospheric CO_2 is the consumption of fossil fuels; the depletion of fossil fuel reserves may prove more detrimental to mankind than the production of CO_2.

Source

There is a significant natural seasonal source of atmospheric CO_2 when organic matter decays and living organisms respire, which is balanced by a photosynthetic sink in other seasons resulting in no or little net annual change. Although the validity of the assumed exact balance is open to question, no firm information is at hand to disprove it. Man-made production of CO_2 or certain changing natural processes are the most likely source of the long-term increase in atmospheric CO_2. Numerical values for the fossil fuel consumption are given later. The burning of limestone for cement production, fluxing stone, or other artificial CO_2 sources is only about 2 percent of the total fossil fuel consumption.

Oceanic warming releases CO_2 to the air. It is estimated that the average ocean temperature rose by not more than 0.05°C over the past century based on a 10-cm rise in sea level.[59] This implies about a 0.5°C temperature rise in the upper 400 m of the ocean, resulting in a nearly 3 percent increase in the surface ocean partial pressure of CO_2. After equilibration between the surface oceans and the air, the atmospheric CO_2 concentration would be about 2.5 percent higher.

Changes in land use alter the amount of humus in the soil. Most of the humus resides on the floor of forests and grassland. When these lands are cultivated, the humus may be more quickly oxidized to carbon dioxide. Less than 5 percent of the atmospheric CO_2 might originate from this source, based on a 50 percent increase in farmlands since the middle of the 19th century.

Less than 4 percent of the marine carbon reservoir consists of organic matter. A decrease by 1 percent in the marine organic carbon raises the partial pressure of the oceans and the concentration in the air by less than 0.5 percent. Uncertainties in the marine biosphere and oceanic temperature variations (which affect the relative oxidation versus photosynthesis rate) could permit a several percent change in atmospheric CO_2 in the last century without detection in the oceans.

The deep waters of the oceans hold about 15 to 20 percent more CO_2 than they would if they were in equilibrium with the current atmospheric content. This excess CO_2 results from the sinking of dead organic matter from the surface waters to greater depths and its subsequent oxidation to produce CO_2. This biological-gravitational pump thus maintains the relatively low CO_2 concentrations in the surface waters of the oceans. If, for some reason, the pump ceased to function, atmospheric CO_2 would increase fivefold. Variations in the effectiveness of the pump could have occurred without detection during the past century, causing notable changes in atmospheric CO_2.

The volume of the oceans partially controls the atmospheric CO_2 content. While variations in the ocean volume during the Ice Age varied by about 5 percent, in recent times it has been much smaller and the consequent effect on the atmospheric CO_2 is negligible.

Volcanic gases have been the principal sources of atmospheric CO_2 but estimates based on current volcanic activity place the present volcanic contributions a hundredfold smaller than the consumption of fossil fuels.

It is estimated that the precipitation of calcium and magnesium carbonate to the ocean floor and the chemical weathering of limestone and dolomite on land might alter the atmospheric CO_2, but very much slower than changes caused by the combustion of fossil fuels.

The only source of atmospheric CO_2 comparable in magnitude to the burning of fossil fuels known to have been in progress in the past decades (or centuries) is the decrease in soil humus caused by encroachment of cultivation on forest or grasslands. A decreased content of dissolved organic matter in the surface waters or a reduction of the CO_2 content of the deep oceans can have equal or larger effects on the air than fossil fuel CO_2, but it is unlikely that either has been significant in the last century.

Instrumentation

Certain stringent criteria must be imposed on a program designed to detect long-term trends in background, global atmospheric CO_2 concentrations. One must select a site remote from local sources and sinks. This isolation has been achieved in two ways; first by finding or establishing an observatory as far from human activity and vegetation as possible. In practice, no perfect location can be found, so that the data almost always contain undesirable noise. A selection of acceptable periods of record by wind direction or other factor is often required to deal with truly clean air. Second, observations may be obtained in the free air aboard aircraft or balloons. For example, Bischof and Bolin have successfully sampled air from the ventilation system of commercial aircraft making sure there has been no cabin recirculation.[7] Keeling, Harris, and Wilkins have collected whole air samples aboard U.S. Air Weather Service reconnaissance aircraft.[33] Aircraft samples are normally returned to a laboratory for analysis, although Bischof[5] and Georgii and Jost[20] also carried a portable analyzer aboard

an aircraft for in situ analyses. Collections must be performed with care; for example, samples containing traces of human breath will yield unwanted high concentrations.

The measurement of background levels of atmospheric CO_2 to detect long-term trends has been performed with nondispersive infrared analyzers. This instrument, described by Smith, measures the energy loss from a beam of infrared radiation passing through the air sample.[67] To avoid interference by water vapor absorption, the gas sample is thoroughly dried before analysis. The instrument is not absolute, requiring the intercalibration with gas from a "working reference" standard. The absolute standard has been prepared manometrically by Dr. C. David Keeling of Scripps Institution of Oceanography in California. The precision of an individual flask sample, one standard deviation, is approximately ± 0.3 ppm (parts of CO_2 per million parts of dry air by volume), while that of samples drawn directly through analyzers is slightly less than 0.3 ppm. The absolute accuracy of the CO_2 standard is known to about 3 ppm but all data in this chapter reported after 1957 have been referred to Keeling's standard.

Historical Data

G. S. Callendar examined the measurements of CO_2 from 1866 to 1956 and argued that the secular increase of CO_2 beginning in about 1900 was due to the burning of fossil fuel by man.[13, 14] He further concluded that most of the fossil fuel CO_2 remained in the atmosphere and would raise the temperature of the earth since it would increase the absorption of outgoing radiation. Figure 19.1 shows the data selected by Callendar between 1866 and 1956. The curve provides an estimate by Revelle and Suess[60] of the fossil fuel production, assuming all CO_2 remains

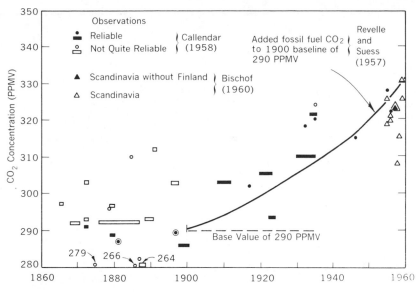

Figure 19.1   The secular increase of CO_2 according to Callendar.[14] The solid line represents the expected increase by fossil fuel consumption since 1900 if all CO_2 remained airborne (adapted from Junge[29]).

airborne, which Callendar used as evidence for the observed increase in the total atmospheric CO_2. Callendar used a base of 290 ppm and his points, it was claimed, are in agreement with the estimated fossil fuel production of CO_2. Also shown in Figure 19.1 are a series of Scandinavian measurements taken in the late 1950s.[4]

In a detailed statistical analysis of all the data Bray concluded that the increase in measured CO_2 was most likely from industrial activities and emphasized the need for further sampling at selected locations.[10]

R. Revelle and H. Suess and many others deduced the growth of fossil fuel CO_2 in the air from the decrease of the carbon-14 to carbon-12 ratio, the Suess effect.[60] The CO_2 created by the burning of fossil fuels contains no carbon-14. The decrease in the ratio of carbon-14 to carbon-12 therefore measures the CO_2 increase if the number of carbon-14 atoms in the atmosphere is unchanged. By 1954 the decrease in the ratio of carbon-14 to carbon-12 was only about 2

to 3 percent, rather than approximately 10 percent as expected from the fossil fuel curve in Figure 19.1; thus there appeared to be an inconsistency between the observed carbon dioxide increase as found by Callendar and the dilution of carbon-14 in the atmosphere. In our model, the Suess effect is predicted to be about 3½ percent.

The growth of carbon dioxide in the atmosphere since the early 1960s is sufficiently large that annual increments can be detected with good instrumentation at observatories removed from pollution sources. The dilution of carbon-14 by fossil fuel carbon-12 is no longer useful as a means of estimating the growth of atmospheric CO_2 because of the contribution of weapon test carbon-14.

Modern Data

The evidence for a trend in atmospheric CO_2 has been well established for the past decade or more.[30] Figure 19.2 presents the available history of atmospheric

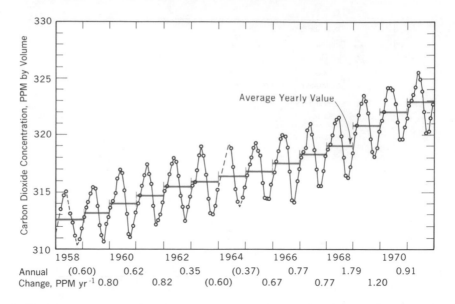

Figure 19.2   Mean monthly carbon dioxide concentrations at Mauna Loa. Annual changes in parentheses are based on incomplete record (after Keeting et al.[31]).

carbon dioxide at the U.S. Geophysical Monitoring Observatory, Mauna Loa, near the summit of a volcanic mountain (more than 3000 m) at 19° N on the island of Hawaii.[31] The continuous infrared analyzer record is carefully analyzed by Keeling and, insofar as possible, local influences are removed. The mean monthly values appear as open circles. These are joined by straight lines for continuous data points and by a dashed line where data are missing. The average annual carbon dioxide concentration is denoted by the horizontal dashed line.

Three types of fluctuations are apparent in the record. The first is a seasonal variation with an amplitude of about 6 ppm, having a maximum in April and a minimum in October. The decrease represents the excess photosynthetic uptake of carbon dioxide over decay and respiration during the Northern Hemisphere summer; the increase is mainly due to the opposite.

The second variation is a general up-

ward trend with time. Between 1958 and 1970, this averaged about 0.75 ppm/year. If all the fossil fuel CO_2 remained within the atmosphere, then the growth would be about 1.44 ppm/year; thus almost 50 percent of the man-made CO_2 has remained airborne during this period, and the other 50 percent has gone into either the sea or the biosphere.

The third type of variation is a change in the annual growth rate. This may be noted from the numbers along the bottom of Figure 19.2; parentheses indicate that the growth rate was estimated since some of the average monthly values were missing. The growth rate between 1962 and 1965 was less than average while the growth rate since 1968 exceeds the average. No reason for the change in the annual growth rate can be given. Insofar as is known, it is not due to any abrupt change in the consumption of fossil fuels; however, relatively small changes in either the sea surface temperatures, which alter the oceanic uptake and release of

CO_2, or the biospheric uptake or decay, if they occurred, can account for the change in the atmospheric growth rate. The trends at some other "modern" stations appear in Figures 19.3 and 19.4 for comparison with the Mauna Loa record.

In figure 19.3, the Scandinavian aircraft data[8] and the South Pole data[32] are derived primarily from occasional flask samples rather than continuous analyzers as at Mauna Loa or Point Barrow.[34] The Mauna Loa 12-month running mean derived from Figure 19.2 is also shown in Figure 19.3. It is not surprising that the South Pole points (triangles) generally, but not always, fall below the Mauna Loa curve, while the Scandinavian (aboard aircraft) and northern Alaskan (Point Barrow) points lie above the Mauna Loa curve, since the main source of fossil fuel CO_2 originates in temperate and high northern latitudes. The long-term trend found at Mauna Loa is confirmed by data at other stations, as is the more rapid annual increase after about 1968.

Figure 19.4, on the other hand, presents station results in the north temperate and polar latitudes only in recent years. The Point Barrow change from 1965 to 1966 was small (actually negative as noted by the number −0.2 ppm/year). Mauna Loa between 1965 and 1966 had returned to an annual growth rate of 0.67 ppm. The flasks collected in 1971

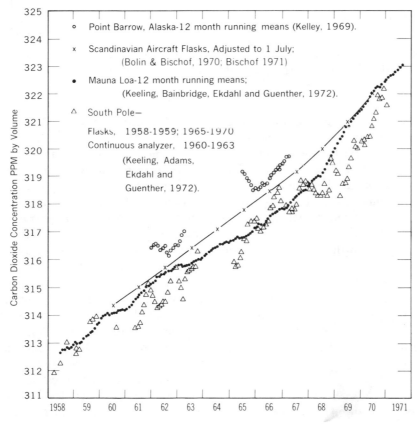

Figure 19.3 Comparison of the atmospheric carbon dioxide concentration at Mauna Loa with those observed at other locations.

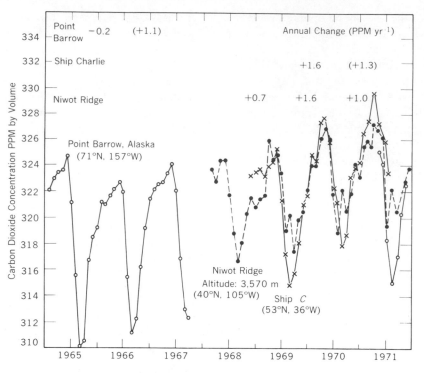

Figure 19.4 Some recent measurements of atmospheric CO_2 mean monthly values (source of data, Kelley,[34] T. E. Harris[24]). Annual changes in parentheses are based on incomplete record.

at Point Barrow indicate that the average growth rate must have been about 0.5 ppm/year since 1967. Niwot Ridge, on the Continental Divide in Colorado, U.S.A., and Ocean Weather Ship "C" were both sampled with flasks and from 1969 to 1971 reported growth rates very similar to Mauna Loa. Parentheses indicate that the growth rate was estimated from incomplete records.

It is suggested that all the CO_2 data at clean air locations confirm the general long-term growth found at Mauna Loa and, further, tend to support a substantial increase in the annual growth rate since about 1968.

Predictions of Future CO_2 Concentration

The transfer of fossil fuel CO_2 from the atmosphere to and from other reser-

voirs has been the subject of many papers.[9, 60, 69] Most of these employ the concept of homogeneous reservoirs requiring the transfer rate between those which exchange with one another. The main uncertainty in this approach stems from uncertainties in the exchange rates rather than the reservoir size. Predictions of future atmospheric CO_2 concentrations demand knowledge of fossil fuel inputs as well as its partitioning among the several reservoirs. Economists have projected a continued 4 percent annual growth in consumption to the year 1980, followed by a decrease to $3\frac{1}{2}$ percent until the year 2000.[73] The amounts of CO_2 expressed in grams of carbon in CO_2 since 1860 based on United Nations' estimates of fossil fuel production appear in Table 19.1. These numbers are based on assumed conversions of the fuels through combustion to CO_2; the amount of con-

Table 19.1 Annual Release of Fossil Fuel Carbon Dioxide to the Atmosphere

Year	10^{16} g C	Year	10^{16} g C	Year	10^{16} g C	Year	10^{16} g C
1860	0.009	1900	0.052	1940	0.130	1980	0.573
61	0.010	01	0.054	41	0.134	81	0.593
62	0.010	02	0.055	42	0.133	82	0.613
63	0.011	03	0.061	43	0.136	83	0.635
64	0.012	04	0.061	44	0.135	84	0.657
65	0.012	05	0.065	45	0.120	85	0.680
66	0.013	06	0.070	46	0.127	86	0.704
67	0.014	07	0.077	47	0.142	87	0.728
68	0.014	08	0.074	48	0.152	88	0.754
69	0.014	09	0.077	49	0.147	89	0.780
70	0.015	10	0.080	50	0.161	90	0.808
71	0.016	11	0.082	51	0.174	91	0.856
72	0.018	12	0.087	52	0.177	92	0.865
73	0.019	13	0.093	53	0.181	93	0.896
74	0.018	14	0.084	54	0.186	94	0.927
75	0.019	15	0.083	55	0.203	95	0.959
76	0.019	16	0.089	56	0.215	96	0.993
77	0.020	17	0.095	57	0.225	97	1.028
78	0.020	18	0.093	58	0.233	98	1.064
79	0.021	19	0.083	59	0.247	99	1.101
80	0.023	20	0.096	60	0.261		
81	0.024	21	0.083	61	0.257		
82	0.026	22	0.089	62	0.271		
83	0.028	23	0.101	63	0.286		
84	0.028	24	0.100	64	0.303		
85	0.028	25	0.101	65	0.316		
86	0.028	26	0.101	66	0.333		
87	0.030	27	0.110	67	0.338		
88	0.032	28	0.109	68	0.359		
89	0.033	29	0.117	69	0.374		
90	0.035	30	0.108	70	0.389		
91	0.037	31	0.097	71	0.404		
92	0.037	32	0.087	72	0.420		
93	0.036	33	0.092	73	0.437		
94	0.038	34	0.100	74	0.455		
95	0.040	35	0.103	75	0.473		
96	0.041	36	0.115	76	0.492		
97	0.043	37	0.123	77	0.511		
98	0.045	38	0.117	78	0.532		
99	0.050	39	0.123	79	0.553		

Source: 1860–1968, Keeling.[30] 1969–1979, estimated annual growth rate of 4%. 1980–1999, estimated annual growth rate of $3\frac{1}{2}$%.

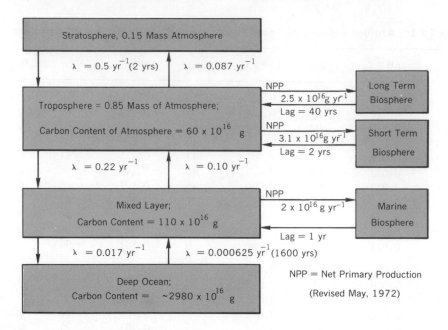

Figure 19.5 The model of the three reservoirs of CO_2. The quantity λ denotes the fraction of the CO_2 in one reservoir being transferred to an adjacent reservoir according to the sense of the arrow.

sumed fossil fuels as well as in the conversion to CO_2 is subject to considerable uncertainty so that errors may range up to several tens of percent.

About 50 percent of the fossil fuel CO_2 has remained airborne from 1860 to 1970. A predictive model must verify this airborne fraction. Alternately, this 50 percent airborne fraction might also be applied to future fossil fuel CO_2 outputs in the future. If the CO_2 from 1970 to 2000 is mixed with the entire mass of the atmosphere (5.14×10^{21} g), the additional fossil fuel CO_2 would increase the atmospheric CO_2 by 100 ppm. But since only 50 percent remains airborne, about 50 ppm would add to the 1970 readings of about 320 ppm to suggest a prediction of 370 ppm in the year 2000.

An example of a box model of CO_2 reservoirs and exchanges is given in Figure 19.5 with both the mass of carbon and the exchange rates.[43, 45] For the atmosphere and oceans, first-order kinetics

has been assumed as is usual in box models. A fraction of the CO_2 in a box, given by λ, transfers in the direction of the arrow each year. The transfer rate from the stratosphere to troposphere is based on an assumed mean residence time ($= 1/\lambda$) of 2 years estimated from nuclear bomb debris. The reverse transfer rate from troposphere to stratosphere is based on the assumption that in an equilibrium state an equal number of CO_2 molecules move across the tropopause in each direction. The transfer between the mixed and deep layers of the oceans assumes a mean residence time of the deep ocean of 1600 years from carbon-14 dating. For predictions tens of years in the future the air-sea exchange is the most crucial transfer, and this has been estimated in Figure 19.5 from the atmospheric behavior of bomb carbon-14. Transfer to and from the biosphere assumes that a given amount of CO_2 is taken up for photosynthesis each year, the net primary production.[73] The biosphere is broken down into two parts, the first with a short (2-year) delay time and the second with a 40-year delay before decay of the organic matter converts the carbon to atmospheric CO_2.

Modeling of the carbon cycle should cope with two special problems. First, for that part of the biosphere which is not water or nutrient limited (assumed to be half of the world's forests), larger concentrations of atmospheric CO_2 will enhance photosynthesis; G. Woodwell* in 1972 suggested a 5 percent photosynthesis increase, on the average, for each 10 percent increase in CO_2. The second problem is the buffering action of the oceans; Revelle and Suess suggest a buffering factor of about 10 due to the acidifying action of CO_2 added to the ocean.[60] If, for illustration, the oceans and atmosphere have equal amounts of carbon, then 11 units of fossil fuel CO_2 at equilibrium will partition 10 in the air and one in the water rather than the expected equal partition. Figure 19.6 shows the result of the introduction of the values of Table 19.1 in the model of Figure 19.5.

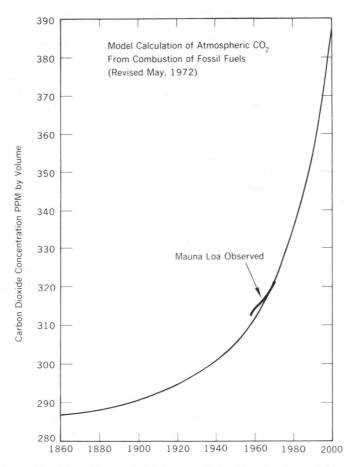

Figure 19.6   The predicted time history of global atmospheric CO_2. The observed CO_2 concentrations at Mauna Loa are due to Keeling et al.[31]).

* Private communication, 1972.

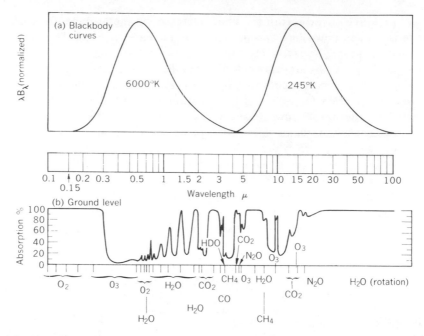

Figure 19.7 Absorption of various gaseous constituents of the atmosphere. (*a*) Blackbody emission for 6000°K (solar) and 245°K (the atmosphere). (*b*) Atmospheric absorption spectra for a solar beam reaching ground level (reprinted by permission, from Goody and Robinson,[21] Robinson[61]).

The forecast for the year 2000 is about 385 ppm, slightly higher than the forecast assuming a 50 percent airborne fraction. The model is anchored to 313 ppm in 1958. The concentration in 1860 predicted by the model is about 290 ppm, in the range of numbers near that date in Figure 19.1.

The combustion of carbon to form CO_2 withdraws oxygen (O_2) from the air. Since O_2 is so vital to life, there have been frequent concerns and doomsday predictions based on the loss of atmospheric O_2. Man cannot quickly deplete O_2 from the air on a global scale by any activity. Even if O_2 were not renewed, it would take thousands to tens of thousands of years for the concentration to halve because of natural processes. The removal of O_2 during combustion of fossil fuels has and will be very small. For example, between 1910 and 1967 the calculated decrease in O_2 due to the formation of CO_2 amounts

to 0.005 percent by volume.[46] Comparison of O_2 measurements made in 1911 by Benedict[3] with those of Machta and Hughes[46] in 1969 shows no detectable change to the limit of accuracy of the two sets of data. Further, when all recoverable fossil fuels are consumed, it can be shown that the O_2 concentration could at most drop from 20.945 percent to 20.8 percent by volume. This decrease, while undesirable, would not be catastrophic.

It is more difficult to assess the climatic impact of an increase in atmospheric CO_2 than to predict the amount of the increase. The radiative properties of CO_2 are well established; Figure 19.7 shows the bands at which the CO_2 and other atmospheric gases absorb and reemit radiant energy. Most of the absorption takes place at long wavelengths rather than in the shorter-wave solar spectrum. This accounts for the greenhouse effect of CO_2, since solar radiation passes through the atmosphere largely unattenuated, but the atmosphere absorbs and reemits more long-wave terrestrial radiation when more CO_2 is present. The result is a net warming of the lower atmosphere, which receives both the solar and returned long-wave radiation, but a cooling of the upper atmosphere, which suffers greater atmospheric energy loss by radiation.

The difficulty in predicting the climatic change due to CO_2, as well as other atmospheric properties, results from the uncertain response of the atmosphere to thermal changes induced by the radiative effects. It is possible that the circulation, temperature, and moisture patterns might adjust in a way to drastically modify the simple lower-level warming.

A simple approach to the prediction of temperature changes resulting from increased CO_2 was undertaken by Manabe and Wetherald using a one-dimensional (vertical) model.[48] The new CO_2 warms the lower atmosphere which can then hold more moisture. A fixed relative humidity attempts to cope with the increasing moisture. The vertical distribution of relative humidity fits climatology. The dynamics of the atmosphere are bypassed by assuming convective adjustment; the vertical temperature structure is returned to a reasonable lapse rate (6.5°C/km) when radiative temperature changes produce superadiabatic profiles. At high altitudes, above about 20 km, the temperature decrease exceeds the ground level increase. Figure 19.8 shows the change of ground

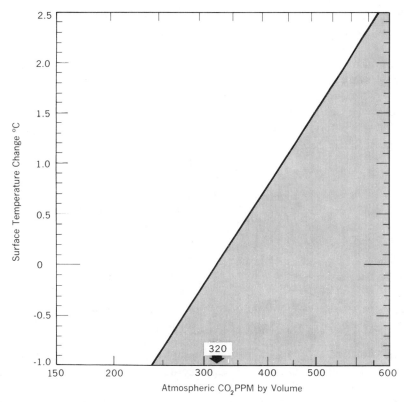

Figure 19.8 Surface air temperature change due to increased atmospheric carbon dioxide with average cloudiness assuming a convective-radiation equilibrium and a fixed relative humidity (reprinted by permission, from Manabe and Wetherald[48]).

level temperature with changing CO_2 content of the atmosphere. The lower atmosphere is predicted to warm by about 0.5°C when the CO_2 concentration reaches 385 ppm in the year 2000.

S. Manabe has modeled the climatic effects of increasing CO_2 concentration in a simplified three-dimensional model.[47] The results show a somewhat larger warming than from the global averaged one-dimensional model; at high altitudes the warming is about twice as great as the hemisphere average because the higher temperatures force a retreat of the arctic ice and snow.

Climatic predictions based on incomplete models are fraught with uncertainties and possible feedback mechanisms may be treated inadequately. For example, all models predicting CO_2 changes employ a climatological cloud cover that cannot adjust to the increased CO_2. If the warmer lower atmosphere increases lower cloudliness through increased natural evaporation, this might negate the greenhouse warming; an increase of low cloudiness of only 0.6 percent could also decrease the low-level temperature by 0.5°C.[74] Alternately, the warming of the lower atmosphere may also warm the ocean surface, releasing more CO_2 to the air. This might have a positive feedback in that the enhanced CO_2 could speed up the warming and so forth.

PARTICLES

Atmospheric particles, in contrast to CO_2, make themselves evident to an observer in a variety of phenomena; dust layers, restrictions to visibility, dirty windshields, and so forth. Dust in the air may be of either natural or man-made origin. On a global basis, by far the greater fraction is of natural origin. Particle character and concentrations depend on proximity to source and sink regions. Particles influence weather and climate in two ways: first, by absorbing, radiating, and scattering radiant energy; and second, by acting as nuclei for cloud elements.

In this discussion, the terms "particles" and "dust" will be used interchangeably. The term "aerosol" is also often used. It is defined as a colloidal system in which the dispersed phase is composed of either solid or liquid particles and in which the dispersion medium is some gas, usually air. Since the present discussion deals with the solid or liquid particles only, the term aerosol will be avoided.

The nomenclature and roles played by natural particles as a function of size appear in Figure 19.9,[29] and particle numbers in three environments may be found in Figure 19.10.[74] These figures show that particles with radii of about 0.1 to 1 μ are both very abundant and important in atmospheric optics since their dimensions are of the same order as the wavelength of visible sunlight (0.4 to 0.7 μ). Particles in the same size range also act as condensation and freezing nuclei if they possess correct hygroscopic properties. Particles tend to grow with time by coagulation and attachment so that aged particles tend to have a larger size than newly created ones. Figure 19.10 indicates that the continents are the main source for particles principally because continental air masses have more smaller particles than do maritime air masses. Sea spray creates relatively few particles (about 10/cm³) in the size range between about 0.5 and 20 μ. The background total particle concentrations at very clean air locations amount to about 700/cm³ in the Northern Hemisphere. Pollution remote from the sources themselves is evident in particle size distributions mainly below radii of 0.1 μ. There is still doubt whether the particle number increases or decreases with decreasing radii below 0.01 μ; that is, whether curves

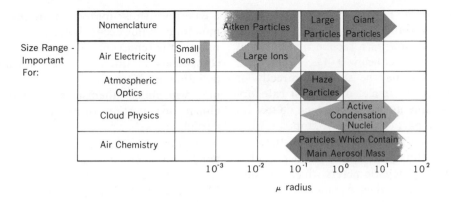

Size Range - Important For:								
Nomenclature			Aitken Particles			Large Particles	Giant Particles	
Air Electricity	Small Ions		Large Ions					
Atmospheric Optics					Haze Particles			
Cloud Physics						Active Condensation Nuclei		
Air Chemistry					Particles Which Contain Main Aerosol Mass			

$$10^{-3} \quad 10^{-2} \quad 10^{-1} \quad 10^{0} \quad 10^{1} \quad 10^{2}$$

μ radius

Figure 19.9 Nomenclature of natural aerosols and the importance of particle sizes for various fields of meteorology (reprinted by permission, from Junge[29]).

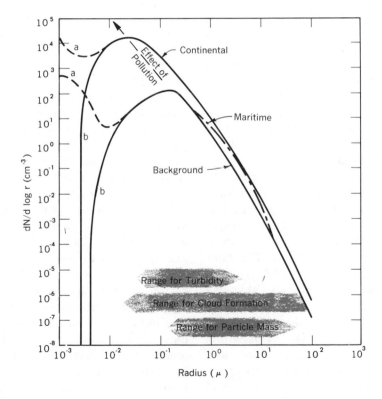

Figure 19.10 Typical comprehensive size distributions for the principal tropospheric regimes and the size ranges important for turbidity, cloud formation, and mass concentration of particles. Curves *a* and *b* refer to possible variation of the size distribution with and without continuous production of very small particles. The arrow indicates the effect of pollution on the location of the maximum of the size distribution (reprinted by permission, from Wilson et al.[74]).

a or *b* in Figure 19.10 are correct. Another source of very small particles would account for a second maximum of particle number below 0.01 μ.

Particles may be sampled directly or indirectly. Examples of the first category of instruments are those which collect individual particles on impactors or filter papers or those which detect particles by light scattering from individual particles. Integrating nephelometers which measure the total scattering, pyranometers which measure normal incidence or direct solar radiation, electrical conductivity instruments, and Aitken nuclei counters exemplify indirect instrumentation.

Sources and Sinks of Particles

Particles originate from both natural and man-made sources. Table 19.2 lists the present estimates in these two categories, as well as a future prediction for the year 2000 A.D. of the man-made contribution. A large fraction of the smaller natural and most of the man-made particles result from indirect production, that is, from gases emitted into the air which are converted, mainly in the presence of sunlight, to sulfates and other particles. The subdivision of particles between sizes smaller and larger than 5 μ separates those more likely to play a role in weather and climate ($< 5\ \mu$) from those less likely to do so ($>5\ \mu$). The table shows the larger mass of particles to be of natural origin. The figures in Table 19.2 should be accepted with reservations. In part, the uncertainty in the values is due to large year-to-year variability of natural phenomena like forest fires and volcanoes. But more important there is a very large uncertainty in the estimates of both the current as well as the projected particulate emissions.

Future estimates of man-made particles depend sensitively on new control measures; thus many countries are mounting major technological efforts to reduce the emission of gaseous sulfur compounds in order to abate local air pollution problems. Such control measures will reflect themselves in reduced global concentrations of indirectly formed particles as well.

The meteorological role of particles depends on their optical and chemical characteristics as well as their sizes. Many commonly found natural and man-made substances have been tested for their nucleating properties. One of the more common indirectly formed particles is ammonium sulfate, which is a good nucleating agent.[53] Radioactive particles will influence electrical properties, but in amounts released by the nuclear industry this effect should be negligible. Some changes in air conductivity have, however, been attributed to radioactive fallout from nuclear tests.[26] Most important are the optical properties of particles, their absorption and scattering; unfortunately, very little is known about these particle characteristics. This will prove to be a vital missing link in the argument concerning climatic change, since the ratio of absorption to scattering determines whether a dust layer in the lower troposphere warms or cools the air.

Fine particles move with the air for all intents; a spherical particle with a radius of 1 μ and density 1.0 g/cm³, for example, settles through the lower tropospheric air at a rate of less than 0.02 cm/sec, appreciably smaller than most air motions. Particles larger than about 10 μ in radius will sediment out of the air fairly rapidly; however, a particle 10 μ in radius will sink at a speed of 1.3 cm/sec. In a matter of days, a point source of material will be dispersed over a horizontal area of 10⁴ km² or more, and as illustrated by the dispersal of radioactive debris from nuclear tests, over a hemisphere in a matter of about a month. In the

Table 19.2 Estimated Global Direct and Converted Particle Production (in 10^6 Metric Tons Per Year)

Due to Natural Phenoma	Particle Diameter	
	$>5\mu$	$<5\mu$
Direct particle production[b]		
Sea salt	500	500
Windblown dust	250	250
Forest fires	30	5
Meteoric debris	10	0
Volcanoes	?	25[a]
Total	790+	780
Particles formed from gases		
Sulfates	85	335
Hydrocarbons	0	75
Nitrates	15	60
Total	100	470

Due to Human Activities	1968		2000
	$>5\mu$	$<5\mu$	$<5\mu$
Direct particle production[b]			
Transportation	0.4	1.8	
Stationary sources (fuel combustion)	33.8	9.6	
Industrial processes	44.0	12.4	
Solid waste disposal	2.0	0.4	
Miscellaneous	23.4	5.4	
Total	103.6	29.6	100
Particles formed from gases[b]			
Converted sulfates	20	200	450
Converted nitrates	5	35	80
Converted hydrocarbons	0	15	50
Total	25	250	580

Source: Peterson and Junge.[56]

[a] Volcanic emissions are highly variable from year to year.

[b] These emissions (for both 1968 and 2000) use the amount of emission control available in 1968.

first few days, the vertical extent may be confined to the lower several kilometers of the troposphere from a ground level source, but a month later almost the entire troposphere becomes filled. In general, because of the zonal nature of the winds, transport will take place faster around circles of latitude than in a north-south direction.

Figure 19.10 shows that most particles are smaller than 1 μ and consequently will not be greatly influenced by gravitational settling; it is fortunate, therefore, that other removal processes are highly effective for fine particles. The most important mechanism is precipitation scavenging. The efficiency of this removal process depends on several particle and precipitation parameters. Among the more important is the particle size. Very small particles ($< 0.1 \mu$) are effectively transferred to cloud or precipitation elements by Brownian motions, while larger particles are impacted on the falling precipitation elements; however, theory suggests that particles in the size range of 0.1 to 1 μ have relatively low scavenging efficiency by precipitation and this may partly account for the maximum in the background particle size concentration in this size range (Figure 19.10).[22] Ultimately, particles that do not wash out quickly can coagulate or become attached to larger particles with removal efficiency increased over those in the submicron range. Certain particles can act as the nuclei of precipitation elements. They are removed from the atmosphere when precipitation deposits on the ground.

Impaction on obstructing surfaces such as buildings and vegetation also removes particles from the air. As with falling precipitation elements, this impaction process, which depends on particle inertia, is also more effective for particles larger than 1 μ. Brownian motion may also remove particles smaller than about 0.1 μ onto the ground surfaces.

The complexities of removal processes have been bypassed by employing the concept of a mean residence time of particles in the air; that is, the time for the concentration to reduce to $1/e$ (about 37 percent) of its original content. The values for such times quoted below have been derived from analogies to marked particles such as radioactive particles. The generalization of the results to other particles, although an assumption, is probably correct. Using particles from the ground marked with natural radioactivity (radon daughters), the residence time is about 3 to 8 days in temperate latitudes. Other particles introduced directly into the upper troposphere by nuclear tests suggest a residence time of 20 to 30 days. Precipitating clouds and precipitation occur more often in the lower several kilometers than aloft. Particles injected into the upper troposphere require time to mix down to the rainbearing layers, while ground level emissions already lie in the precipitation scavenging region. The proximity to the altitude of scavenging accounts for most of the differences in residence times between ground and upper tropospheric injections. Precipitation, the main mode of particle removal, is variable in time and space, and this probably accounts for most of the residence time variability for particles injected into the same altitude.

Trends in Particle Concentrations

One possible means of identifying man-made contributions to atmospheric particles lies in the detection of trends. With the assumption that naturally formed particles are in equilibrium, emissions and removal balance one another. For sources such as volcanoes this assumption is clearly invalid. Nevertheless, major volcanic eruptions, being sporadic in nature, are unlikely to produce a long-term increase in atmospheric particles that would confuse increasing man-made outputs.

There exist very few data to shed light on global trends in atmospheric dust loadings. Most dust measurements are taken at ground level within built-up areas and reflect purely local conditions. Within the United States, for example, many cities exhibit a downward trend reflecting the local imposition of control measures.[42]

The industrial temperate latitudes contain frequent and widespread precipitation so that one expects an irregular upward trend to parallel a growth

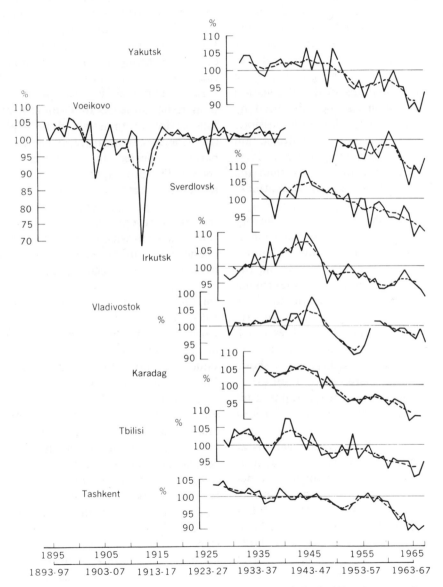

Figure 19.11 Normal incident solar radiation as percent of normal. Dashed line: 5-year running mean (reprinted by permission, from Pivovarova[57]).

in the sources of particles. But the precipitation scavenging, combined with predominantly zonal rather than meridional air movements, tends to confine particles near their latitudes of origin.

Airborne dust concentrations can be inferred from normal incident solar radiation measurements. By integrating the net effect of particle attenuation of solar radiation throughout the vertical extent of the atmosphere, these observations minimize local influences. Fortunately, this measure of atmospheric dust has been observed for a long time at numerous stations over the globe. On the other hand, variations in the attenuation of solar radiation can be influenced by gases as well as particles. It is possible, but not likely, that increasing attenuation of solar radiation with time reflects changes in the gaseous rather than the particulate composition of the atmosphere.

Figure 19.11 shows a long-term history of the normal incident solar radiation at eight observatories in the Soviet Union.[57] All stations show a decrease in the amount of solar radiation at the ground or an increase in atmospheric turbidity after the 1940s. All stations but one are located within or near large cities which have grown in the past century; Karadag is isolated from a large city, and it shows the same large increase in turbidity as found at the other seven stations after 1940. All the Soviet stations lie north of 40°N.

Table 19.3 lists four other locations at which long-term records of normal incident solar radiation also display a decrease in solar radiation in time reflecting an increase in atmospheric turbidity.[28] Except for Davos, in eastern Switzerland, the locations are all very close to large industrial areas and even Davos, in central Europe, is not very remote from heavy industry and population centers.

Many other temperate latitude observatories in the U.S.A., Japan and elsewhere display the same general decrease in normal incident solar radiation as in Figure 19.11 and Table 19.3.

Figure 19.12 gives a similar, but shorter, history of normal incident solar radiation at Mauna Loa Observatory on the Island of Hawaii (19°N).[18] The station initiated its solar radiation program in 1957, but since then, there has been no net decrease in the normal incident radiation. The sharp drop seen in 1963 followed the stratospheric injection of dust by a major volcanic eruption on the Indonesian Island of Bali, Mt. Agung. The recovery to the pre-1963 solar radiation levels took about 6 years. This is longer than would be expected from the 2-year mean residence time for particles injected in the lower stratosphere. The prolonged reduction in solar radiation, longer than expected from the Mt. Agung injection, was probably caused by several subsequent large volcanic eruptions which added to the stratospheric dust burden. The names of these possible volcanic injections are also shown in Figure 19.12. Since about 1970 the solar attenuation is no greater than before 1963. Data in 1971, not shown in Figure 19.12, confirm this statement. Stations in the Southern Hemisphere show the same recovery after 1963 as seen at Mauna Loa Observatory.

Another indirect measure of fine atmospheric particles is given by atmospheric electrical conductivity in 1967 with similar measurements in the same general area as far back as 1907.[15] The comparison in the North Atlantic Ocean reveals a marked increase in conductivity while the comparison in the Pacific Ocean of the Southern Hemisphere showed no change. The conductivity measurements reflect only changes in particle concentration near the sea surface.

The above data suggest that most, if not all, of the temperate latitude locations

Table 19.3   Long-Term Atmospheric Turbidity Trends at Various Stations

Station	First Period	$\bar{\beta}_1{}^a$	Second Period	$\bar{\beta}_2{}^a$	$\Delta\beta$	Secular Increase Change (%)	Increase Change per Decade (%)
Davos, Switzerland	1914–1926	0.024	1957–1959	0.043	0.019	78	~20
Washington, D.C., USA	1903–1907	0.098	1962–1966	0.154	0.056	57	~10
Mexico City, Mexico	1911–1928	0.081	1957–1962	0.169	0.088	111	~25
Jerusalem, Israel	1930–1934	0.037	1961–1968	0.048	0.011	34	~10

Source: Joseph and Manes.[28]
^{a} $\bar{\beta}$ = average Ångström turbidity coefficient.

have suffered increasing particle concentrations since about the 1940s and in some cases earlier. On the other hand, south of about 30°N, only stations near major cities or industrial areas (Mexico City, for example) have increased their dust concentrations in recent decades; others show no change. On a global scale, then, most of the northern quarter and small parts of the rest of the globe have upward trends, but elsewhere there has been no detectable change in dust con-

centrations. This description fits the expected distribution from a source in the north temperature latitudes and a relatively short mean residence time for the dust in the air.

Particles and Climate—Radiation

Particles both absorb and scatter incident radiation. Too little information about particles exists with which to make

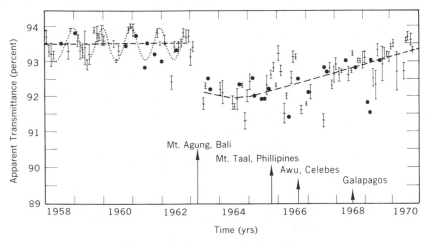

Figure 19.12 Temporal variations of the monthly means of transmission factors of solar radiation. Arrows and names identify large volcanic eruptions (reprinted by permission, copyright 1971 by the American Association for the Advancement of Science[18]).

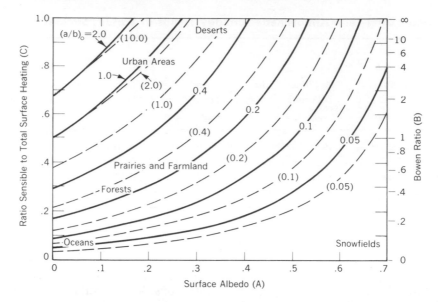

Figure 19.13 Solution of (a/b) as a function of the relevant properties of the earth's surface for two values of tropospheric particle distribution (solid curves, all dust in lower troposphere; dashed curves, three-fourths of dust in the lower troposphere). Different surface types belong to different regions of the figures as indicated (reprinted by permission from Mitchell[50]).

definitive quantitative statements on the relative roles of these two processes. Thus it will not be possible to predict with certainity even the sign of the ground level temperature change that will accompany an increase in atmospheric particle concentration. However, in some special cases, the consequences of enhanced particle concentrations are relatively well-known.

Many investigators have calculated the influence of particles on the atmospheric radiation balance making certain assumptions.[58] The particle size distribution for background conditions, as shown in Figure 19.10, the vertical distribution of particle concentration deduced by Elterman,[19] and certain assumed values for the absorption and scattering coefficients and surface albedos have been incorporated into a model of short-wave solar radiative transfer. The result shows that the atmosphere as a whole will cool; more solar energy will be reflected back

to space because of the dust. This situation most often comes to mind in a qualitative assessment of enhanced particle concentration, namely, atmospheric cooling.

J. M. Mitchell in 1971 analyzed the particle and other requirements for the lower atmosphere to warm or cool.[51] Part of Mitchell's analysis is in Figure 19.13, which provides a series of values for critical particle absorption to backscattering ratios. When particles have an absorption (a) to backscatter (b) ratio greater than the critical numbers on the sloping lines, the lower atmosphere will warm; when less than the critical value the lower atmosphere will cool. The other parameters in this graph are the surface albedo (a clear sky is assumed) as the abscissa, the ratio of heat warming the ground and lower atmosphere to the total incident heat, part of which evaporates moisture, and two crude two-layer particle concentration distributions with height (the solid and dashed lines). The solid lines apply to a case of all particles in the near ground layer of the troposphere while the dashed lines are applicable to a case with only three-fourths of the particles in the near ground layer. Certain characteristic surfaces such as oceans, snowfields, and deserts appear in the figure. For oceans, for example, most of the incident solar radiation is absorbed; hence the surface albedo is very low, but a large fraction of the absorbed energy is used for evaporation rather than heating the water. Mitchell considers that some particles are believed to have ratios of absorption to backscattering greater than 2, and these will exceed the critical ratios for all types of surfaces. The lower atmosphere would mostly warm rather than cool in this circumstance.

Mitchell has also tried to assess the influence of particles in a partly cloudy sky, making several assumptions about the height of the particle layer above or below the clouds and their effect in altering the albedo of the clouds. The results show that the critical absorption to backscattering ratio becomes smaller when cloud albedo interactions are included, implying a greater likelihood that particles will warm rather than cool the lower atmosphere.

In the infrared region of the spectrum, it is possible that particles can provide a "greenhouse" effect similar to that noted for CO_2 in which the lower atmosphere warms but the upper atmosphere cools; however, except for very heavy industrial pollution or at high latitudes, the attenuation of solar radiation by particles almost always exceeds the radiative effects in the infrared region of the spectrum.[74]

Particles and Climate—Clouds and Precipitation

A second role for particles in possible weather or climate modification is their interaction with natural clouds and

precipitation. Particles in clouds can alter the optical properties of water droplets so that the absorption in the visible part of the solar spectrum may be more significant than for pure water. More important, however, are the nucleating effects of particles and consequent changes in precipitation and cloud character.

There are three types of nucleating particles: cloud condensation nuclei around which cloud droplets form at relative humidities just over 100 percent, sublimation nuclei which directly initiate ice crystals from the vapor state, and ice or freezing nuclei which convert water droplets to ice at temperatures below $0°C$, depending on the efficiency of the nuclei. It is believed that most ice crystal clouds have, if only momentarily, started in the liquid water stage; direct sublimation is rare or absent. The number of cloud droplets has been shown to be directly related to the number of cloud condensation nuclei, but the number of ice nuclei is usually less than the number of ice crystals contained in clouds.[53]

If man-made particles can act as nuclei, they may have several possible influences on clouds and subsequent precipitation. More condensation nuclei create more cloud droplets and reduce their average size by using up the available water vapor. The addition of more ice nuclei can accelerate the glaciation (conversion of water to ice cloud). Both these possibilities may alter the character of the cloud system and subsequent precipitation. For example, the glaciation process, in temperate latitudes, precedes precipitation.

There is little doubt that man's activities can greatly increase the numbers of cloud condensation nuclei, for example, by the burning of sugar cane or from steel manufacturing plants.[25, 70] But once removed by many hundreds of miles from the source of the nuclei, excess above a variable background is more difficult to detect. On a global scale the direct addition of nuclei by human activities amounts to only a few percent or less of the total condensation nuclei. The contribution of particles from man-made gaseous emissions is more difficult to assess. But, since 1968 when systematic measurements were begun in Australia, no trend has been observed (to detect a trend would require a 10 percent/year change). Although most nuclei appear to originate over continents, there is no correlation with human population so that natural sources predominate. The mass of condensation nuclei produced each year, about 4×10^5 tons, is a very small part of the present direct or indirect annual particle formation rate; thus there are already more airborne particles on a global scale than are needed as cloud condensation nuclei.

In terms of precipitation inhibition or enhancement from man-made particles, the situation is not clear since plausible arguments may be offered on both sides. Added cloud condensation nuclei produce more, smaller, and perhaps more uniform cloud droplets than would otherwise occur and will, in theory at least, reduce precipitation. On the other hand, ice nuclei from pollution sources can alter precipitation in either way, depending on the existence of natural ice nuclei.

As already noted, the cloud systems of the globe play a vital role on the radiation budget of the atmosphere mainly in reducing the solar radiation available to drive the atmospheric circulation systems. More condensation nuclei, as noted above, may increase the number of cloud droplets. Other factors being equal, a cloud with more numerous droplets both absorbs and reflects more solar energy, resulting in less transmission to lower levels. The radiative consequences of changing the character of clouds can be significant on a global scale should man-made cloud nuclei increase significantly.

Long-wave radiation is also influenced by the number of cloud droplets. In this case, however, the infrared radiation from below will be intercepted and reradiated back to the lower atmosphere more effectively when there are more numerous cloud droplets.

CLOUDINESS

The global cloud cover plays a vital role in the earth's radiative budget in reflecting solar energy back to space, in adsorbing both solar and long-wave radiation, and in emitting its energy downward and outward into space. Changes in cloud cover will alter the heat balance of the atmosphere. In the preceding section the possible effects of particles on cloud droplets were discussed; in this section, the effect on clouds of the direct input of moisture by human activities will be evaluated. On a global or even a regional scale, the very small amounts of moisture which man adds by his land practices, combustion of fossil fuels, mining of subterranean water, and so forth, are negligible in comparison with nature; they are, at most, minute fractions of the natural global evaporation. Locally, such as downwind of a man-made reservoir or deforested area, changes in evaporation may possibly alter the cloud cover. But on a regional scale, only one form of injection for increasing cloudiness suggests itself; aircraft formation of contrails.

Contrail Formation

H. Appleman has specified the environmental conditions under which contrails are likely to form.[1] Both heat and moisture come out of the aircraft exhaust; the former tends to reduce the relative humidity in air while the added moisture brings it closer to saturation, if unsaturated. Figure 19.14 shows that a layer of air in the upper troposphere where present-day subsonic jet aircraft fly is most prone to contrail formation. Further, although statistics are unavailable, this layer probably has a higher average relative humidity than the middle troposphere below or stratosphere above; however, for contrails to be effective in climate modification, they must persist at least many tens of minutes. It can be shown that the amount of moisture exhausted behind a large four-engine jet aircraft when diluted into an equivalent cross-sectional area of a circle of say, 1-km radius, produces a concentration of water vapor of no more, and usually much less, than 1 percent of that required to saturate the air.[44] This means that the air must be at least 99 percent saturated for the added moisture from the aircraft to produce persistent saturation after it is diluted by the aircraft wake and normal atmospheric turbulence. Considering the relatively high frequency of contrail observations near air lanes, explanations for their persistence other than near-saturated air are necessary because large volumes of

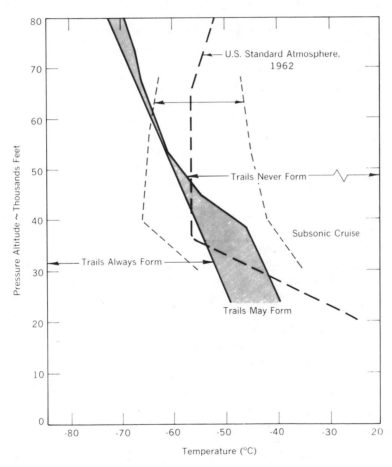

Figure 19.14 Temperature and pressure altitude criteria for contrail formation. The U.S. standard atmosphere and its range are shown by the dashed lines (adapted from Appleman[1]).

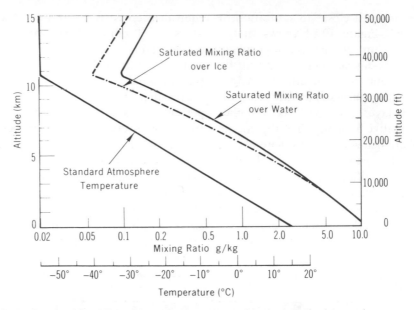

Figure 19.15 Saturated humidity mixing ratio over water and ice in a standard atmosphere.

the upper atmosphere without clouds are not this close to saturation. Beckwith reports that contrails formed and lasted less than 5 min during 38 percent of his observations from a commercial jet liner, while 25 percent were persistent (sighted for more than 5 min).[2]

Two other explanations for the persistence of contrails suggest themselves. First, it is common knowledge among meteorologists that high thin clouds may not be detected by ground observers. The formation of contrails in an apparently clear sky may really be a thickening in an existing thin, and to the ground, invisible cloud deck. Many casual observations when natural cirrus are present confirm that some contrails represent a thickening of existing high clouds. Beckwith reports that only 7 of 25 percent persistent contrail sightings occurred in clear or scattered cloud skies at or near aircraft altitudes. But contrails also form in clear skies and a second explanation is frequently advanced for this case.[35] Most,

if not all, natural ice crystal clouds (and clouds in the 10 to 13 km layer at which jet aircraft cruise contain ice crystals) form through the freezing of water droplets. Thus the air must be saturated with respect to liquid water, if only momentarily. On the other hand, after the water droplet forms and transforms to an ice crystal it will survive and even grow when the relative humidity, with respect to water, is less than 100 percent, so long as the air is saturated with respect to ice. Figure 19.15 shows the moisture required for saturation with respect to water and ice for the United States standard atmosphere. The region between the two curves constitutes an area in which contrails can persist once formed.

The final aspect of contrail formation, its growth into large sheets, is not explained readily. The larger ice crystal particles settle to lower altitudes and can be spread by shearing winds in addition to turbulent diffusion. Knollenberg has shown that the contrail ice crystals extract much more moisture from the air than can be accounted for from the engine exhaust.[35] Moreover, some multiplicative mechanism such as splintering of the large crystals is required to create the very many crystals that may occur when the contrails seemingly cover a much larger area than that of a single trail. This unknown mechanism apparently, at times, produces an overcast. Montefinale discusses the growth of ice crystal clouds.[53]

Trends in High Cloudiness

Although there is little doubt concerning the reality of artificial contrails formed behind aircraft, one can properly question whether the increase in cloudiness due to contrails is significant in amount. In part these doubts arise because persistent contrails may be rare compared to natural high clouds or because atmospheric conditions conducive to the formation of persistent contrails might subsequently produce natural clouds anyhow. In an attempt to assess the contribution of contrails to high cloudiness, the trends in cirrus clouds have been examined from 1949 to 1970. During this period air traffic grew rapidly and increasing jet aircraft flying in the 10 to 13 km altitude range began to cover much of the United States skies.[44]

The trends in high cloud amounts from routine National Weather Service records of visual cloud observations suffer from many serious shortcomings. First, only those observations with no or very few low or middle clouds permit sighting of high clouds from the ground. The data below, therefore, limit themselves to these conditions of zero or a few tenths of low and middle clouds. The bias introduced by these restrictions is unknown but it is not likely to account for a trend with time in high clouds. The second and more serious aspect of using the routine

observations of cloudiness is their poor quality. There is minimal interest in accurate, high cloud observations for either precipitation forecasting or for aviation. Even with effort, it is often difficult to distinguish between middle and high clouds, to correctly estimate the number of tenths of high clouds even in the absence of low or middle clouds, and in some cases to even detect the presence of high clouds except at sunrise and sunset or from aircraft. It is assumed that the errors introduced by inaccurate observations will exhibit no trend but will more likely superimpose random variability on a time history of high cloudiness. There is also an uncertainty in the interpretation of trends of high cloudiness amounts because of possible natural fluctuations.

Figure 19.16 shows the time history of high cloud amounts of seven U.S. stations using the three hourly observations at National Weather Service stations during which the low and/or middle cloud amounts were less than or equal to 0.3. The figure shows that after 1965 when jet flights became more numerous the average annual high cloudiness was also greater. The seven stations individually reflect the same upward trend; statistically, the post-1965 cloudiness amounts are significantly greater than those of the pre-1958 period. But the year-to-year association between jet fuel consumption and cloudiness changes is not good. This poor relationship may be due either to the fact that total jet fuel consumption is not a good indicator of the number of flights in the troposphere where contrails form, or to the variability introduced by poor quality observations, if the increase in cloudiness is indeed caused by contrails. The mere existence of greater cloudiness during the period of active jet aircraft operations does not suffice to implicate the jet aircraft as the cause of the increase. It is possible, as noted above, that changing atmospheric

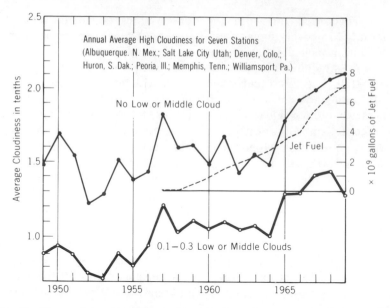

Figure 19.16 The history of the annual average high cloudiness for seven stations with zero low or middle clouds (upper curve) and 0.1 to 0.3 low or middle clouds (lower curve). The dashed line shows the growth of jet fuel consumption by U.S. domestic commercial jet aircraft.

circulation, moisture, and thermal patterns can also induce variations in the amount of high clouds. Thus more definitive studies to prove a connection, if any, between the significant cloud increases and jet aircraft remain to be undertaken.

Climatic Effects of Increased High Cloudiness

The most likely effect of increasing high cloudiness will be a disruption of the natural radiative balance of the atmosphere. The earth's albedo will probably increase, and the effective radiating temperature of the earth's atmosphere will drop. Since clouds also absorb and reemit long-wave radiation, this reduced transmitted solar radiation does not necessarily imply a decrease in temperature near the ground. Rather the ground level temperature effect depends on the amount of solar energy transmit-

ted by the cloud cover, the blackbody character of the clouds to infrared radiation, and the temperature of the cloud, which controls how much long-wave radiation is radiated downward. Figure 19.17 illustrates, for two cases of blackbody radiation, an example of the calculated response of the ground level temperature to changing cloud cover.[48] It may be seen that for high clouds, this combination of assumptions acts more like a "greenhouse" since ground level temperatures warm with increasing cloudiness. On the other hand, Kuhn has measured the long-wave radiative balance below a contrail sheet at 13°N over water and calculated the probable solar radiation balance below the cloud and estimates that this cloud sheet produced a 7 percent decrease in total radiant energy[37]; the solar decrease in this case exceeded the infrared gain. Cox has analyzed actual values of infrared emissivities of cirrus clouds in tropical and temperate latitudes and concludes that low-altitude

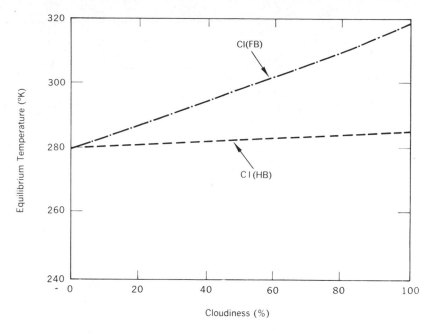

Figure 19.17 Radiative-convective equilibrium temperature at the earth's surface as a function of cirrus cloudiness. FB and HB refer to full black and half black, respectively (reprinted by permission, from Manabe and Wetherald[48]).

warming would occur in the tropics and cooling in the temperate latitudes.[16] If correct, added artificial cirrus clouds would tend to increase the existing, natural north-south temperature gradient at ground level. Nicodemus and McQuigg have shown with the help of a simulation model that the daily maximum temperature in summertime over a continental area can be reduced by as much as 2°C if enough contrail cloudiness is created to decrease the percent sunshine by 15 to 35 percent.[55]

The evidence for falling ice crystals from contrails supports the speculation that they may alter the properties of certain clouds into which they settle.[35] Specifically, ice crystals from cirrus clouds falling into subcooled water clouds below have been observed to initiate the precipitation mechanism. But the significance of the ice crystal seeding mechanism is unevaluated; for example, it may turn out that the upper parts of any natural cloud lying below contrails will almost always be cold enough to have their own abundant ice crystals.

LAND USAGE AND HEAT

Two other aspects of human activities are often cited as possible bases for inadvertent climate modification: land usage and direct heat insertion into the atmosphere. The changing deployment of land from its natural cover and state to urbanization and a cultivated agriculture alters the local albedo, heat capacity, fraction of solar radiation used for evaporation (the Bowen ratio), and surface roughness. The combustion of fossil fuels adds heat directly to the lower atmosphere; even natural heat sources may have their timing and geographical distribution altered by human intervention.

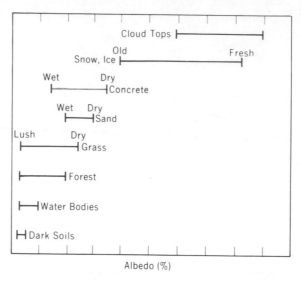

Figure 19.18 The albedo (percent reflectivity) of various surfaces (after Slade[66]).

Figure 19.18 shows the estimates of the albedo of many surfaces. The albedo for clouds and snow is higher than for ground or ocean surfaces. The albedo over ocean surfaces is very low (less than 0.05) when the sun is within about 45° of the vertical even when sun plus sky radiation is included. Gravel and concrete (light-colored) roads generally have albedos greater than natural, wooded land settings, while asphalt (dark) roads may have the same or slightly lower albedos. Urbanized areas also have higher albedo values than their natural surroundings. But it is likely that man can produce more significant changes of the global albedo by directly or indirectly modifying the cloud or ice and snow cover than by localized changes in albedo from land usage; in many cases, the altered surface may have about the same albedo as was formerly present.

The ratio of incident radiation used to heat the surface and lower atmosphere to that used for evaporation can be dramatically changed by substitution of man-made for natural surfaces, replacement of a forest by a city, for example. The largest areas changed by man involve the conversion of forests into agriculture. When either coniferous or deciduous forests are converted to wet arable land in temperate and high latitudes, the amount of evaporation will increase sizably. When converted to relatively dry types of arable land, the change will be small; however, in the temperate latitudes most of the replacement of forests to agriculture has already taken place. In some areas marginal farmland is being returned to forest. The most likely deforestation in the future will take place in the tropics where, in many regions, there can be a large decrease in the rate of evapotranspiration. In such areas, large additions of latent heat will be replaced by sensible heat.

Evaporation rates may also be altered by irrigation and artifitial lakes and reservoirs. It is estimated that about 2 million km² or about 1 ½ percent of the area of the continents is under irrigation,[11] while the estimated 300 artificial lakes and reservoirs cover about 300,000 km² of the earth's surface.[12]

Human activities other than land

Table 19.4 Sources and Magnitude of Certain Energy Releases

Description of Energy Release	Magnitude (W/m²)
Net solar radiation at earth's surface (dependent on latitude)[a]	~100
Urban industrial area estimate[b]	12
1968 energy production distributed evenly over all continents[c]	0.040
1968 energy production distributed evenly over whole globe[c]	0.012
Annual net photosynthetic production of the continental vegetation cover[d]	0.13
Global average heat flux from the earth's interior[e]	0.062

Source: Wilson et al.[74]
[a] Budyko.[11]
[b] Assuming 75 percent of the energy consumption to be concentrated in a total industrial area of the globe of 0.5×10^6 km².
[c] This figure was obtained using the 1968 energy production value of 5.9×10^6 MW from Table 19.5.
[d] Lieth.[40]
[e] Higher locally by a factor of 10 to 50 in geothermal areas.

19.4 compares this estimate with the solar energy. The numbers are derived from a world energy production of 5.9×10^6 MW for 1968. The global average man-made sources of heat are thus almost five orders of magnitude smaller than that received from the sun. In local urbanized and industrial areas, especially in winter when solar radiation is minimal, there may be more man-made heat than that from the sun. Part of the well-known urban heat island (in which built-up areas are warmer than the countryside) results from the artificial heat sources; another part arises from the greater heat capacity of the concrete, steel, and other structures and surfaces. Table 19.5 presents estimates of future heat injections into the air. By the year 2000 it is expected that the rate of heat addition will be approximately four times that in 1968. Predictions of even larger energy needs after 2000 can be extrapolated using the 4.5 percent growth rate employed in Table 19.5 between 1980 and 2000, but many generations are still required to approach an equality with solar energy on a global scale.

W. M. Washington has run a three-dimensional general circulation model with

management will also affect the relative amount of solar energy used for evaporation compared to that used to heat the ground and nearby air; thus swamp drainage can reduce evaporation, but river engineering and flood control may either increase or decrease evaporation.

It is most likely that only the deforestation of the tropics may affect global climate.[54] Other changes in land usage do not appear to be important on a global or large regional scale.

The direct insertion of heat by fossil fuel combustion can be estimated from the consumption of fossil fuels. Table

Table 19.5 World Energy Consumption (10^{12} W)

	1968	Estimated 1980	Estimated 2000
Coal	2.14	2.30	3.33
Oil	2.72	5.60	12.23
Natural gas	1.06	2.24	5.00
Nuclear	0.02	0.71	5.74
Total	5.94	10.85	26.30

Source: Shurr.[63]

an amount of heat one order of magnitude beyond that expected in 2000; 3×10^{14} W were geographically distributed in the same fashion as the current population.[71] The departures from a control run without the additional man-made heating were found to be no different from a randomly distributed case or for that matter a case with a negative heat addition. One may conclude that the present state of modeling is inadequate to detect with confidence the global climatic consequences of added heat or that very much larger amounts of heat than that used are needed to produce detectable climatic anomalies.

CLIMATIC PREDICTION

Climatic changes may be due either to natural or inadvertent man-made causes. It is not possible to see a trend in climate as proof of man's influences. This section describes the observed trends in global climate since about 1850 and some efforts to predict these trends in more modern times.

Observed Trends in Surface Temperature

There are many stations reporting surface temperatures in the networks of national weather services in contrast to the very limited number of locations at which the history of air quality measurements is available. Despite this greater number of temperature reports, global trends must still be accepted with reservation for the reasons discussed below.

Zonal mean temperatures are derived from an interpolation between stations; the zonal averages are weighted by area for global (or near global) mean values.[49] The first and most crucial problem involves the amount of station data; thus only after 1880 is there at least one station in every 10° latitude band between 60°S and 80°N. By 1900, only 26 stations contributed to the Southern Hemisphere mean temperatures. The second problem is that virtually no stations exist over much of the world oceans so that interpolation is necessary to estimate the temperature in these large regions. Third, many stations are located near large urban areas which have local heat island effects that introduce trends at the stations of local rather than general significance. Despite these very formidable difficulties, the major trends in global temperatures derived from this procedure and data are believed to be statistically real.

Figure 19.19 shows the global trends in surface temperature between 1870 and 1968. Several features are evident. The temperature has generally risen between 1880 and about 1940, after which it has fallen. The amplitude of the changes is larger in the Northern than Southern Hemisphere, and in the Northern Hemisphere, larger in winter than summer. The warming from 1880 to 1940 was almost twice as large in temperate and high latitudes (25° to 60°N) as in the tropics (25°N to 25°S).[49] Figure 19.20 shows the temperature changes in the decade of the 1960s from those in the period 1931–1960. The entire cooling in the Northern Hemisphere occurred north of about 50°N. The cooling in high latitudes in the 1960s, again, took place mostly in the winter half of the year.

No satisfactory explanation can be offered for the dominance of the Northern Hemisphere high-latitude winter in global temperature changes. It has been suggested that the presence of strong, persistent temperature inversions will amplify any temperature effect, but an analysis of radiosonde data at Point Barrow, Alaska in winter suggests similar mean monthly changes throughout the troposphere. Starr and Oort have shown an even more marked cooling for the Northern Hemisphere from 1012.5 to 73 mb from 1958 to 1963 than is found at ground level.[68]

H. H. Lamb's Review of Prediction of Natural Climatic Changes[39]

There are a number of natural phenomena which have formed the basis for climatic predictions during the decades since about 1950.

1. Astronomical variations in the ellipticity of the earth's orbit, distance from the sun, and tilt of the polar axis occur on much too slow a time scale to contribute to climatic changes over a period of tens of years.

2. Variations in the energy output of the sun possibly caused by tidal effect of neighboring planets and solar disturbances including sunspots. Quantitative estimates of climatic changes from increased or reduced solar output have some basis, but precision of measurements or solar output is inadequate to show changes smaller than 1 percent.

3. Variations in the tidal forces of the sun-moon combination either directly or

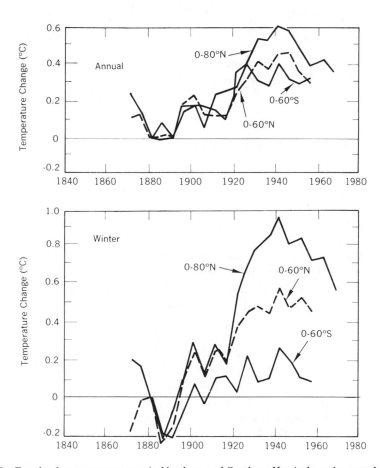

Figure 19.19 Trends of mean temperature in Northern and Southern Hemisphere, by pentads, 1870–1968, shown as successive 5-year means relative to the 1880–1884 mean (reprinted by permission, from Mitchell[52]).

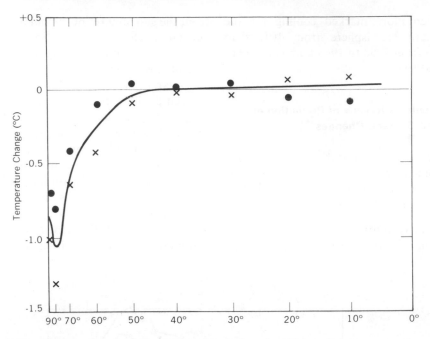

Figure 19.20   Mean temperature in the Northern Hemisphere for 10° latitude belts, expressed as a deviation from the average of 1931–1960. The solid line gives the change in the decade 1961–1970, the circle the changes in 1961–1965, and the crosses in 1966–1970 (reprinted by permission, from Wilson et al.[74]).

indirectly through their influence on the seas and oceans.

4. Variations in the ocean circulation and sea surface temperatures.

5. Changes in atmospheric transparency caused by volcanic dust and natural variations in water vapor and cloudiness. There are some supporting evidence and a large number of studies relating volcanic dust to subsequent temperature cooling, especially after the Krakatoa eruption in 1883.[38, 50]

6. Suggestions that the atmosphere-ocean system contains natural long-period oscillations. These might reflect themselves in a large variety of meteorological parameters such as circulation, rainfall, temperature, and cloudiness patterns. Many scientists have sought and always found periodicities of climate or weather from a few to thousands of years without necessarily explaining their origin.

Actual Forecasts

1. R. Scherhag deduced and extrapolated periodicities of circulation in the atmosphere and in 1939 predicted that the European winters would become progressively more severe than they were in the 1930s.[62] This forecast became especially well-known because the next three European winters during World War II were the coldest of the past 50 years or more. The trend for severe winters has continued up to 1970 so that the forecast may be considered successful.

2. H. C. Willett in 1951 based his climatic prediction on an extrapolation of cyclic trends tentatively associated with sunspot activity.[72] The forecast called for a decrease in temperatures over much of the world, especially over the Spitzbergen-Greenland-Iceland area and, to a lesser extent, over northern Europe and the eastern United States. Exceptions to

the downward temperature trend were predicted for the northwestern United States and especially the eastern Mediterranean. The recession of the glaciers should have been reversed in the next 20 years. The minimum temperatures should occur during the period 1960–1965. The forecast of the sunspot cycle proved to be almost entirely in error. Most of the temperature forecasts were correct, on the other hand, although the minimum temperatures did not occur in the first half of the 1960s as expected.

3. A. A. Girs in 1966 based his forecast prepared in the late 1950s and early 1960s on an extrapolation of circulation patterns having periodicities of 9 to 29 years as derived from past history.[23] More westerly wind patterns were expected to dominate until the mid-1980s or 1990; however, Lamb has found that the westerly wind patterns have in fact diminished over the British Isles and by 1968–1971 reached their smallest frequency in the past 110 years.

4. G. Siren in 1961 performed a periodogram analysis of the growth of pine trees in the northern forest limit of Finland between 1181 and 1960.[65] The forecast extrapolates the six, and especially the three, largest amplitude periodicities (73, 93, and 204 years) which have been shown to correlate with average summer temperatures. The summer temperatures at high latitudes were warmer in 1930–1940 than in 1900–1910; in northern Finland the warming was 2.7°C. The extrapolation called for a marked deterioration of climate between 1960 and 1975, followed by a small variation around the new climate up to the year 2000. Up to 1970, the forecast has been verified. This prediction included neither meteorological nor sunspot data; the extrapolation derives solely from the response of trees to the climate permitting periodicity analysis for long periods in the past.

5. B. L. Dzerdzeevskij in 1962 predicted the character of the meridional circulation.[17] Meridionality or the intensity of the north-south wind components was expected to continue and become a maximum in the mid-1960s, followed by a trend toward more zonal flow. Later he postponed the time of the reversal to zonal flow until possibly 1975, since meridional conditions persisted. The zonal character of the flow was predicted to continue for about 20 years. The forecasts of continued meridional flow have been verified up to 1971.

6. C. J. E. Schuurmans in 1969 based his forecast on particle radiation from solar flares which make the circulation of the middle latitudes more meridional.[64] His reasoning predicts lower mean temperatures in western Europe north of 50°N in the quarter century following 1962 than in the quarter century preceding that date. For the first 9 years, the forecast appears to be correct.

7. S. J. Johnsen, W. Dansgaard, and H. B. Clausen in 1970 have based their extrapolation of the climate after 1970 on the oxygen and hydrogen isotope ratio in ice and snow of a core bored in northwestern Greenland.[27] The extrapolation of a Fourier analysis of the ratios and an assumed relationship between the isotope ratio and temperature leads to a forecast of continued cooling for the next 10 to 20 years followed by a warming trend reaching a new climatic optimum about 2015.

The above forecasts almost universally lack a real justification based on meteorological reasoning. A surprisingly large number correctly foreshadowed the cooling that occurred after about 1950, and as in the case of Willett's prediction, even suggested the correct areas of cooling. It is possible that many of the forecasts were directly or indirectly influenced by the cooling trend in progress at the time of the forecast.

Despite the considerable measure of success in many of the above and other forecasts, the future climate based presumably on natural events cannot be predicted reliably at this time. As a corollary, it follows that man-made changes in climate, if any, probably cannot be distinguished from those due to nature. Making this distinction represents a major challenge to the scientist.

SUMMARY AND CONCLUSIONS

Meteorologists, it is believed, understand the likely means by which man's activities may inadvertently modify the weather or climate: CO_2, particles, cloudiness, ozone, albedo, and direct heat input. Predictions of future inputs of pollutants, although vital, fail to supply all the information needed to predict the air concentrations; CO_2 added to the atmosphere by the combustion of fossil fuels, for example, enters the oceans and biosphere as well as remaining airborne. A full appreciation of changes of air composition requires knowledge of the life cycle of each constituent. But even more critical is the present imperfect capability to convert a change in the atmospheric character into a climatic modification.

Nature competes or in some cases overwhelms man-made inputs; it is believed that most of the atmospheric dust, for example, originates independently of man's presence on the earth. But even more significant is the natural fluctuation of climate that has been in progress far longer than man's capability to alter the climate. Efforts to predict these changes, although superficially successful, display a striking ignorance of the true cause of climatic variability; most predictions are based on extrapolation of some presumed periodicity. Since society is prone to blame human intervention as

the cause of adverse climatic changes, it is doubly critical that scientists be capable of estimating natural fluctuations.

The following specific points raised in this chapter deserve emphasis:

1. The growth of CO_2 in the atmosphere at a rate of about 0.7 ppm/year has been established during the period 1958-1970. The increase is attributed to the combustion of fossil fuel, about 50 percent of which has remained airborne with the remainder presumably entering the biosphere and oceans. Projections of future CO_2 consumption inserted into a model of CO_2 exchanges between air, water, and biosphere lead to a prediction of a CO_2 concentrations in the year 2000 of about 385 ppm. Very crude calculations of the effect of this increase suggest a global ground level temperature rise of about 0.5°C.

2. Using the attenuation of solar radiation as a measure of atmospheric turbidity, almost all observatories north of about 30°N reflect an increase in dust loadings since about 1945. Only stations near major cities south of 30°N show a similar trend; others such as Mauna Loa indicate no increase in atmospheric turbidity attributable to man-made particles. Although increased particle loadings may result in several possible climatic effects, the most frequently associated consequence is a warming or cooling of the lower atmosphere, which depends on the uncertain character of the particles.

3. Cloudiness plays a crucial role in controlling climate. Aircraft contrails increase high clouds, at least superficially. Examination of the conventional cloud observations at weather stations display an increase in high cloudiness after the appearance of jet aircraft. But this association does not prove or disprove that the increase in high clouds is due to aircraft operations.

4. Changing the land usage may also

contribute to climatic change but the net effect is uncertain. Current models of climatic change are unable to detect global alterations from even the amount of heat an order of magnitude greater than that expected in the year 2000 from the natural, current climatic variability.

5. Natural climatic variability cannot now be predicted with confidence even though some past statistical or even quasitheoretical attempts to perform these have proved to be moderately successful.

ACKNOWLEDGMENT

Partial support of this work by the Division of Biomedical and Environmental Research, U.S. Atomic Energy Commission, is gratefully acknowledged.

REFERENCES

1. Appleman, H., The formation of exhaust condensation trails by jet aircraft, *Bull. Am. Meteor. Soc.* **34,** 14–20, 1953.

2. Beckwith, W. B., Future patterns of aircraft operations and fuel burnouts with remarks on contrail formation over the United States, *International Conference on Aerospace and Aeronautical Meteorology of the American Meteorological Society*, Washington, D.C., 1972, pp. 422–426.

3. Benedict, F. G., Composition of the atmosphere with special reference to its oxygen content, *Carnegie Inst. Wash. Publ.* **166,** 1912.

4. Bischof, W., Periodical variations of the atmospheric CO_2-content in Scandinavia, *Tellus* **12,**216–226, 1960.

5. Bischof, W., Carbon dioxide measurements from aircraft, *Tellus* **22,**545–549, 1970.

6. Bischof, W., Summary of recent carbon-dioxide measurements in the atmosphere, Summary letter 110/1813, University of Stockholm, Inst. of Meteorology, 1971.

7. Bischof, W., and B. Bolin, Space and time variations of the CO_2 content of the troposphere and lower stratosphere, *Tellus* **18,**155–159, 1966.

8. Bolin, B., and W. Bischof, Variations in the carbon dioxide content of the atmosphere of the Northern Hemisphere, *Tellus* **22,** 431–442, 1970.

9. Bolin, B., and E. Ericksson, Changes in the carbon dioxide content of the atmosphere and sea due to fossil fuel combustion, *Rossby Memorial Volume,* Rockefeller Institute Press, New York, 1959, pp. 130–142.

10. Bray, J. B., An analysis of the possible recent change in atmospheric carbon dioxide concentration, *Tellus* **11,**220:230, 1959.

11. Budyko, M. I., *The Heat Budget of the Earth,* Hydrological Publishing House, Leningrad; 1963.

12. Budyko, M. I., *Climate and Life*, Hydrological Publishing House, Leningrad, 1971.

13. Callendar, G. S., The artificial production of carbon dioxide and its influence on temperature, *Quart. J. Royal Meteorol. Soc.* **64,** 223–240, 1938.

14. Callendar, G. S., On the amount of carbon dioxide in the atmosphere, *Tellus* **10,**243–248, 1958.

15. Cobb, W. E., and W. E. Wells, The electrical conductivity of oceanic air and its correlation to global atmospheric pollution, *J. Atmos. Sci.* **27,** 814–819, 1970.

16. Cox, S. K., Cirrus clouds and the climate, *J. Atmos. Sci.* **28,** 1513–1514, 1971.

17. Dzerdzeevskij, B. L., Long-term variability of the general circulation of the atmosphere as a basis for forecasting climate, *Proceedings of a Conference on the General Circulation of the Atmosphere, Leningrad, Gidromet. Izdat,* 1962 (in Russian).

18. Ellis, H. T., and R. F. Pueschel, Solar radiation, absence of air pollution trends at Mauna Loa, *Science* **172,**845–846, 1971.

19. Elterman, L., Rayleigh and extinction coefficients to 50 km for the region .27 μ to .55 μ, *Appl. Optics* **3,**1139–1147, 1964.

20. Georgii, H. W., and D. Jost, Concentration of CO_2 in the upper troposphere and lower stratosphere, *Nature* **221,**1040, 1969.

21. Goody, R. M., and G. D. Robinson, Radiation in the troposphere and lower stratosphere, *Quart. J. Royal Meteorol. Soc.* **77,**151–185, 1951.

22. Greenfield, S. M., Rain scavenging of radioactive particulate matter from the atmosphere, *J. Meteorol.* **14,** 115–125, 1957.

23. Girs, A. A., On peculiarities of the arctic

meteorological regime in different stages of the circulation epoch of 1949–1964, *Polar Meteorology, WMO Tech. Note* **87,**454–477, 1966.

24. Harris, T. E., personal communication, 1972.

25. Hobbs, W. E., L. F. Radke, and S. E. Schumway, Cloud condensation nuclei from industrial sources and their apparent influence on precipitation in Washington State, *J. Atmos. Sci.* **27,**81–89, 1970.

26. Huzita, A., Effect of radioactive fallout upon the electrical conductivity of the lower atmosphere, *International Conference on the Universal Aspects of Atmospheric Electricity, 4th Tokyo, 1968,* S. C. Coroniti and James Hughes, Eds., Gordon and Breach, New York, Vol. 1, 1969, pp. 49–57.

27. Johnsen, S. J., W. Dansgaard, and H. B. Clausen, Climatic oscillations 1200–2000 AD, *Nature* **227,**482–483, 1970.

28. Joseph, J. H., and W. Manes, Secular and seasonal variations of atmospheric turbidity at Jerusalem, *J. Appl. Meteorol.* **10,** 453–462, 1971.

29. Junge, C. E., *Air Chemistry and Radioactivity*, Academic Press, New York, 1963.

30. Keeling, C. D., Industrial production of carbon dioxide from fossil fuels and limestone, *Tellus* **25,** 174–198, 1973.

31. Keeling, C. D., A. E. Bainbridge, C. A. Ekdahl, P. Guenther, and J. F. S. Chin, Atmospheric carbon dioxide variation at Mauna Loa Observatory, Hawaii, 1958–1970, *Tellus* **25** (5), 1973, in press.

32. Keeling, C. D., J. A. Adams, C. A. Ekdahl, and P. R. Guenther, Atmospheric carbon dioxide variations at the South Pole: 1957–1970, *Tellus* **25** (5), 1973, in press.

33. Keeling, C. D., T. B. Harris, and E. M. Wilkins, Concentration of atmospheric carbon dioxide at 500 and 700 millibars, *J. Geophys. Res.* **73,** 4511–4528, 1968.

34. Kelley, J. J., Jr., An analysis of carbon dioxide in the arctic atmosphere near Barrow, Alaska, 1961 to 1967, *Sci. Rept. Dept. of Atmospheric Sci.,* University of Washington, 1969.

35. Knollenberg, R. G., Measurement of the growth of the ice budget in a persisting contrail, *J. Atmos. Sci.* **29,** 1367–1374, 1972.

36. Kopec, R. J., Global climate change and the impact of a maximum sea level on coastal settlement, *J. Geography,* 541–550, 1971.

37. Kuhn, P. M., Airborne observations of contrail effects on the thermal radiation budget, *J. Atmos. Sci.* **27,** 937–942, 1970.

38. Lamb, H. H., Volcanic dust in the atmosphere; with a chronology and assessment of its meteorological significance, *Phil. Trans. Roy. Soc. Ser. A,* **266,** 425–534, 1970.

39. Lamb, H. H., Climate forecasting: a brief survey of the position and forecasts issued up to 1971. Preliminary report of WMO CoSAMC Working Group on Climatic Fluctuations, 1971.

40. Lieth, H., Versuch einer kartographischen Darstellung der Pflanzendecke auf der Erde, *Geographisches Taschenbuch 1964/65, Steiner, Wiesbaden, 1964, pp. 72–80.*

41. *Lovelock, J. E., and J. P. Lodge, Jr., Oxygen in the contemporary atmosphere, Atmospheric Environment,* Vol. 6, Pergamon Press, Great Britain, 1972, pp. 575–578.

42. Ludwig, J. H., G. B. Morgan, and T. B. Mc-Mullen, Trends in urban air quality, *Trans. Am. Geophy. Union* **51,** 468–475, 1970.

43. Machta, L., The role of the oceans and biosphere in the carbon dioxide cycle, *Nobel 20 Symposium,* The Changing Chemistry of the Oceans, Gothenburg, Sweden, Aug. 1971, D. Dryssen and D. Jagner, Eds., Almquist and Wiksell, Stockholm, 1972, pp. 121–145.

44. Machta, L., Global effects of contaminations in the upper atmosphere, presented at the American Inst. of Chemical Engineers 64th Annual Meeting, Nov. 28–Dec. 2, 1971, San Francisco, Calif.

45. Machta, L., Prediction of CO_2 in the atmosphere, *Brookhaven Symposium on Biology, No. 24,* Carbon and the Biosphere, Upton, N.Y., May 1972, G. M. Woodwell and E. V. Pecan, Eds., U.S.A.E.C. Conference 720510, Oak Ridge, Tennessee 1973, pp. 21–31.

46. Machta, L. and E. Hughes, Atmospheric oxygen in 1967 to 1970, *Science* **168,** 1582–1584, 1970.

47. Manabe, S., Private communication, 1972.

48. Manabe, S. and R. T. Wetherald, Thermal equilibrium of the atmosphere with a given distribution of relative humidity, *J. Atmos. Sci.* **24,** 241–259, 1967.

49. Mitchell, J. M., Jr., Changes of mean temperature since 1870, *Ann. N.Y. Acad. Sci.* **95,** 235–250, 1961.

50. Mitchell, J. M., Jr., A preliminary evaluation of atmospheric pollution as a cause of the

global temperature fluctuation of the past century. *Global Effects of Environmental Pollution,* S. F. Singer, Ed., (Springer-Verlag, New York, 1970, pp. 139–155.

51. Mitchell, J. M., Jr., The effect of atmospheric aerosols on climate with special reference to temperature near the earth's surface, *J. Appl. Meteorol.* **10,** 703–714, 1971.

52. Mitchell, J. M., Jr., Air pollution and climatic change, presented at the American Inst. of Chemical Engineers 64th Annual Meeting, Nov. 28–Dec. 2, 1971, San Francisco, Calif.

53. Montefinale, A. C., T. Montefinale, and H. M. Papee, Recent advances in the chemistry and properties of atmospheric nucleants: a review, *Pure Appl. Geophys. (PAGEOPH)* **91,** 1–234, 1971.

54. Newell, R. E., The Amazon forest and atmospheric general circulation. *Man's Impact on the Climate,* W. H. Matthews, W. W. Kellogg, and G. D. Robinson, Eds., M.I.T. Press, Cambridge, Mass., 1971, pp. 457–460.

55. Nicodemus, M. L. and J. D. McQuigg, A simulation model for studying possible modification of surface temperature, *J. Appl. Meteorol.* **8,** 199–204, 1969.

56. Peterson, J. T. and C. E. Junge, Sources of particulate matter in the atmosphere, *Man's Impact on the Climate,* W. H. Matthews, W. W. Kellogg, and G. D. Robinson, Eds. M.I.T. Press, Cambridge, Mass., 1971, pp. 310–320.

57. Pivovarova, Z. I., Study of the regime of atmosphere transparency, Proceedings of the WMO/IUGE Symposium in Bergen, August 1968, *WMO Tech. Note 104,* Geneva, Switzerland, 1970, pp. 181–185.

58. Rasool, S. I. and S. H. Schneider, Atmospheric carbon dioxide and aerosols: effects of large increases on global climate, *Science* **173,** 138–141, 1971.

59. *Restoring the Quality of Our Environment,* U.S. Government, Appendix Y4, Report of the Environmental Pollution Panel, President's Science Advisory Committee, Washington, D.C., 1965.

60. Revelle, R. and H. Suess, Carbon dioxide exchange between atmosphere and ocean and the question of an increase of atmospheric CO_2 during the past decades, *Tellus* **9,** 18–27, 1957.

61. Robinson, G. D., Meteorological aspects of radiation, *Advances in Geophysics,* Vol. 14, Academic Press, New York, 1970, pp. 285–306.

62. Scherhag, R., Warming of the Arctic, *Ann. Hydrograhi Maritimen Meteorol.* **67,** 57–67, 1939.

63. Schurr, S. H., Ed., *Energy, Economic Growth and the Environment*, published for Resources for the Future, Inc., by the John Hopkins University Press, 1972.

64. Schuurmans, C. J. E., The influence of solar flares on the tropospheric circulation, *Mededelingen en Verhandelingen,* Koninklijk Nederlands Meteorologisch Instituut, Vol. 92, 1969, 122 pp.

65. Siren, G., Skogsgranstallen som indikator for klimatfluctuation-erna i norra Fennoskandien under historisk tid, *Communications Institute Forestalis Fenniae,* 54, Helsingfors, 1961.

66. Slade, D. H., *Meteorology and Atomic Energy 1968,* U.S. Atomic Energy Commission/Division of Tech. Info., Oak Ridge, Tenn., 1968, p. 15.

67. Smith, V. N., A recording infrared analyzer, *Instruments* **26,** 421–427, 1953.

68. Starr, V. P. and A. H. Oort, Five-year climatic trend for the northern hemisphere, *Nature* **242,** 310–313, 1973.

69. Walton, A., M. Ergin, and D. C. Harkness, Carbon 14 concentrations in the atmosphere and carbon dioxide exchange rates, *J. Geophys. Res.* **75,** 3089–3098, 1970.

70. Warner, J. and S. Twomey, The production of cloud nuclei by cane fires and the effects on cloud droplet concentration, *J. Atmos. Sci.* **24,** 704–706, 1967.

71. Washington, W. M., Numerical climate-change experiments: the effect of man's production of thermal energy, submitted to *J. Appl. Meteor.,* 1972.

72. Willett, H. C., Extrapolation of sunspot-climate relationships, *J. Meteorol.* **8,** 1–6, 1951.

73. Wilson, C. L. and W. H. Mattews, Eds., *Report of the Study of Man's Impact on the Global Environment,* M.I.T. Press, Cambridge, Mass., 1970, 319 pp.

74. Wilson, C. L., W. H. Mattews, W. W. Kellogg, and G. D. Robinson, Eds., *Report of the Study of Man's Impact on Climate,* M.I.T. Press, Cambridge, Mass., 1971, 308 pp.

20 | Inadvertent Atmospheric Modification Through Urbanization

HELMUT LANDSBERG

The fact that there are changes in the atmospheric environment of towns was recognized early. It was the sense of smell rather than scientific analysis that signaled the change. The air of towns was often foul and stale. Poor sanitation and combustion processes made air pollution known to the ancients as well as to the modern world.

It was not long after the start of systematic atmospheric observations with instruments that other alterations became obvious. As early as 1818 Luke Howard clearly established that the city of London had higher temperatures than the nearby countryside, which then was not as far removed from the town center as now. He also offered an explanation by citing the large amount of heat given off by coal-fired domestic heating and manufacturing processes.

Toward the beginning of this century changes of other elements in urban areas, such as wind, rainfall, and particulate

level found some documentation. Monographs on weather and climate of individual cities that appeared in increasing number usually included some reference to the city-country contrasts, and the 1930 decade until the outbreak of World War II saw a remarkable effort in the study of urban mesometeorology. It started with the introduction of motor vehicles as instrument carriers for traverses across cities to ascertain variations of meteorological elements in metropolitan areas. A general review of the available findings and a comprehensive literature survey were given by Kratzer.[61] In 1970 Chandler compiled another very useful classified bibliography.[16]

After World War II the pace of research accelerated when a series of catastrophic episodes showed that weather and pollution interacted to cause deaths and illness. The most dramatic event of this type was the *killer smog* of December, 1952 in London, England. In the course of 4 days 4000 excess deaths occurred, and the problem was lifted from the obscurity of scientific research to a public issue.

This concern about pollution and also

Helmut Landsberg is a professor at the Institute for Fluid Dynamics and Applied Mathematics at the University of Maryland, College Park, Maryland.

the fear that man may have started irreversible trends in the evolution of atmospheric parameters has stimulated an intense research effort in metropolitan meteorology. The most conspicuous development in recent years has been an attempt to formulate physical and numerical models of the observed phenomena. Although this is only in the incipient stages, it promises to lift the questions out of an entirely empirical framework into the realm of predictable cause-and-effect relationships. An attempt will be made here to relate some of the relevant facts and conclusions.

URBAN EFFECTS ON RADIATION AND ENERGY BALANCE

One may appropriately divide the effects acting on the radiative balance in urban areas into those caused by man-made alterations of the surface and those caused by interceptors brought into the atmosphere by elements from human activities. These can be assessed by measuring the changes brought about in the various heat fluxes. The net heat gain (+) or heat loss (−) of the surface can be expressed as follows:

$$Q_N = Q_I(1 - a) + Q_{L\downarrow} - Q_{L\uparrow}$$
$$= \pm Q_S \pm Q_H \pm Q_E + Q_P, \quad (20.1)$$

where

Q_N net energy balance
Q_I incoming short-wave radiation
a albedo of surface (reflectivity)
$Q_{L\downarrow}$ long-wave heat flux downward (atmospheric infrared radiation)
$Q_{L\uparrow}$ long-wave heat flux upward ($Q_{L\uparrow} = \epsilon\sigma T^4$, where ϵ emissivity of surface, σ Stefan-Boltzmann constant, T absolute temperature of surface)
Q_S heat flux in or out of surface of the ground

Q_H sensible heat transfer between atmosphere and ground
Q_E heat loss by evaporation from surface or gain by condensation (latent heat)
Q_P heat production or heat rejection (from combustion processes, air conditioning, metabolism)

Even a cursory glance at the elements of the energy balance shows immediately that many of the factors are measurably changed in the urban environment compared with rural regions. The incoming short-wave radiation can be affected by the pollution pall over cities; that will be discussed in a subsequent section. The albedo is usually smaller in urban areas, with dark roofs and asphalted surfaces, than that of actively growing and even dried vegetation. A notable difference develops particularly in winter when a highly reflective snow cover exists in the open farm spaces but little or no snow remains in the city, because of either removal or melting. Kung[62] has measured some of these differences from the air, and 10 to 30 percent lower albedos in urban areas give a feel for the magnitude of the difference.

Long-wave radiation of the atmosphere downward is not appreciably different under most circumstances although occasionally a high atmospheric dust content over the city might add to this component. However, the long-wave flux upward is profoundly affected by urbanization even though there is little difference in the emissivity of the various ground surfaces. The principal change stems from the higher absolute temperatures these surfaces have in the city. The immediate reason for that is the greater absorption of incoming shortwave radiation because of lower albedo by the city surfaces and the considerably greater heat conductivity and heat capacity of the surfaces common in cities (concrete, asphalt, stone, etc.) compared with plant-covered or natural soils, which are poor heat conductors and have relatively low heat capacity. The ensuing surface temperature differences are measured readily by infrared measuring devices. A number of such observations from low-flying aircraft have been published by Lorenz. An example from one of his papers illustrates the case.[74] On a nearly cloudless day helicopter flights were made at 94 m, from 1000 to 1600 hours, across terrain with mixed surface. Air temperatures ranged from 19 to 25°C, temperatures of a forested area from 22 to 26°C, and of highway surfaces from 29 to 37°C. Although flight restrictions prevented actual measurements over towns, other traverses[73] took place over man-made surfaces at an airport that should approximate what can be expected in urban environments. In that case the forests were generally 1°C cooler than air temperature, and a concrete apron was 8°C, an asphalt surface was 20°C, and hangar roofs were 17°C warmer than air temperature. Even under overcast conditions the buildings of a small village showed an infrared temperature 3°C higher than the surroundings. At the surface many infrared measurements have been made. They suffer a bit from the fact that the field of view subtended by the instrument is very small and hence individual values can have substantial sampling errors. Here only a few observations of Kessler[58] at Bonn, Germany are quoted. They are typical for what may be expected on clear days. The daily maximum of an asphalt street surface was 46.6°C, the minimum 14.5°C. On the same day the maximum of a lawn surface was 32.5°C, the minimum 9°C; the shelter temperatures were maximum 23.1°C, the minimum 11.9°C.

The other elements in the energy balance also change. The flux through the surface, Q_S, is essentially a function of vertical temperature distribution in the soil, building material, or plant cover, and the heat conductivity. The latter is notoriously low in loose soil and organic litter which constitute the surface in natural or cultivated vegetation. This contrasts with relatively good heat conductivity of many surfaces in cities, such as roadbeds, parking lots, bricks, and stone surfaces. Similarly, the factor Q_E is profoundly affected by the change from rural to urban environment. The principal reason again is the change in the surface, which radically alters the availability of water. Therefore, a much higher rate of evaporation can be maintained from soil and plants in the rural area which also store substantial amounts of precipitated water. In contrast, urban areas usually have at least 50 percent impermeable surfaces which do not retain water. Any precipitation runs off and is drained away as rapidly as possible, a process that not only affects the evaporative heat flux but also leads to flash floods in urbanized areas. The latter effect can only be mentioned in passing as urban hydrological changes are not a main theme for this discussion. In addition to minimal direct evaporation from the urban sector, there is a notable reduction in evapotranspiration because of the drastic reduction in plant cover that would at least seasonally contribute to this flux. Although not too much quantitative information on Q_H is available, this factor is usually considerably smaller than the other fluxes and the differences between rural and urban areas are minor.

A great deal of speculation has been devoted to the factor Q_P, the heat production by man and his activity. The metabolic contribution from human beings and animals is negligible but heat rejection from industry, domestic heating, air conditioning, and internal combustion

Table 20.1   Relative Values of Urban Heat Production[a]

Heat Source	Heat Share (%)	Weighted Heat Share Through Day (%)			
		0800	1300	2000	Night
Stationary	66.6	71	64	71	41
Mobile	33.3	69	45	25	12
Human and animal metabolism	0.1	0.05	0.2	0.1	0.02
Total	100	140	110	97	53
Cal/cm² min	0.037	0.052	0.041	0.037	0.02

Source: Bach.[10]

[a] Based on Cincinnati conditions in summer.

engines of cars must be considered. Table 20.1 gives an appreciation of the relative magnitude of various contributions, based on summer conditions in Cincinnati, Ohio.

This table shows that in summer the direct heat production Q_P is not a major contribution to the energy balance. Bulk studies in other cities have also shown that this factor is not the overwhelming element, even in metropolitan heat balances, that has been projected by some. Q_P is usually given as percentage of the incoming solar radiation Q_I. Because some higher-latitude localities receive little radiation from sun and sky in winter because of low solar elevation and much cloud cover, Q_I can have a rather small value and thus Q_P represents a larger fraction. But even for as large a city as Montreal it does not exceed 33 percent and averages only about 20 percent.[35]

In New York City yearly heat production is estimated at 2.8 × 10¹⁷ cal.[12] Manhattan winter heat release is 0.28 cal/cm² min, in summer 0.06 cal/cm² min. Seasonally and in the restricted area this can be an appreciable factor in the heat balance.

Actually the changes in heat balance caused by the alterations of the surface, especially in the fluxes, Q_S and Q_E, are those mainly giving the urban areas its thermal characteristics. This is readily demonstrated by comparative measurements over a typical paved surface of the environment in cities and a large vegetated surface. Such a case is illustrated by the values of Table 20.2, obtained in the newly urbanized area of Columbia, Maryland, during a clear day in summer. Q_P was not a factor in the case.

It is well to remember here that the heat balance considerations are considerably more complex than can be inferred from observations over horizontal surfaces. The urban surface is highly serrated by streets with buildings, with vertical walls and sloping roofs variously exposed to solar radiation. These surfaces often have temperatures far in excess of the air temperature in daytime.[64] They cool off only slowly after sunset and, instead of radiating fully toward the sky, part of the absorbed energy is radiated back and forth between walls on opposite sides of the street and the roadbed.

The time development of the radiation

Table 20.2 Diurnal Variation of Elements of Heat Balance at Columbia, Maryland over Vegetated and Paved Surfaces under Calm Clear Conditions

Surface type	Time of day	T_A (°C)	T_S (°C)	Q_I	Q_N	$Q_{L\downarrow}$	$Q_{L\uparrow}$	Q_E	Q_H	Q_S
				Heat Factors (cal/cm² min)						
Vegetation	00	12.7	15.5	—	−0.10	0.41	0.54	—	—	−0.13
Parking lot		15.0	21.5	—	−0.12	0.41	0.61	—	—	−0.20
Vegetation	05	12.2	11.0	—	−0.09	0.40	0.50	—	—	−0.10
Parking lot		12.7	18.0	—	−0.13	0.41	0.57	—	—	−0.16
Vegetation	12	24.7	32.0	1.20	—	0.43	0.67	0.30	0.12	0.24
Parking lot		24.7	47.5	1.23	—	0.43	0.85	0.00	0.10	0.64

Source: Landsberg and Maisel.[66]
Note: T_A air temperature at 2 m; T_S surface temperature.

flux in the city pall has been considered by Atwater,[9] who describes it by the differential equation

$$\frac{\partial T}{\partial t} = -\frac{1}{\rho c_p} \left| \frac{\partial Q_I (1-a)}{\partial z} \right.$$

$$\left. + \frac{\partial (Q_{L\uparrow} - Q_{L\downarrow})}{\partial z} \right|, \quad (20.2)$$

where T is the radiative temperature, ρ the air density, c_p the specific heat of air at constant pressure, z the height above ground, and t time.

Numerical solution of this equation shows that a low-level aerosol cloud can intensify inversion formation after sunset when only the long-wave fluxes operate. In daytime it can lead to elevated temperatures in the dusty air layers, a fact that had already been established by Roach[92] from aircraft observations in the London area.

This dust layer over cities in turn affects the Q_I at the surface materially. This is highly dependent on wavelength, in that shorter wavelengths are more affected than the longer wavelengths; thus the ultraviolet ($\lambda < 400$ nm) is often completely filtered out. But the effect is also notable in comparisons of all-wave global radiation measurements between neighboring urban and rural stations. East[34] reports from Montreal that the attenuation of total radiation runs parallel to particle contamination. On an average, Q_I is 9 percent less in the city than in the rural environs, varying from 4 percent on clear days to 18 percent on overcast days, and from 3 to 4 percent in summer to 12 to 15 percent in winter.

In recent years, because of its simplicity, many measurements have been made with the Volz sunphotometer, which permits an estimate of the depletion of the solar beam (normal incidence) at the single wavelength of 500 nm. This permits calculation of a turbidity factor B, defined by the following relation:

$$J_{(500)} = \frac{J_{0(500)}}{R_0{}^2} \exp - a\left[\left(\frac{p}{p_0}\right) + B\right] m,$$

$$(20.3)$$

where

$J_{(500)}$ measured solar radiation intensity normal to beam at 500 nm

$J_{0(500)}$ extraterrestrial solar radiation intensity normal to beam at 500 nm

R_0 distance to sun in astronomical units

a a constant (~ 0.0674)

m optical air mass (secant of angular zenith distance of sun)

Measurements in relatively uncontaminated air in the United States have shown low values of $B \approx 0.02$. These can be taken as background values.[39] Surveys at 26 localities in the United States show for the "cleanest" towns values of ~ 0.07, with a winter minimum of 0.04 and a summer maximum of 0.09. Industrial centers have much larger values, as exemplified by Baltimore, Maryland, which has an average around 0.2, a winter minimum of 0.13, and a summer maximum around 0.3. Extreme values on individual days can, of course, run still higher.

In this context one cannot overlook the effects on the more commonly measured element of sunshine duration, which also is reduced in urban areas; however, because other factors also affect the duration of sunshine, it will be discussed in a later section.

THE URBAN TEMPERATURE FIELD

Not only in terms of its early discovery but also as a target for investigation, the urban temperature changes have stood in the center of attention of urban mesometeorology. There are probably as many papers on this subject in the literature dealing with the urban atmosphere as on all other themes combined. For this reason only a selection are cited here to illustrate the principal facts.

Various techniques are in use to establish these facts. Most commonly used in the past were comparisons of measurements at fixed stations in the city and at nearby rural stations. This procedure always left the doubt that even without the city, micrometeorological settings caused by topography might have caused all or part of the observed differences. Mobile surveys helped in assessing the topographic aspects better, and reduction of observed temperatures by assumed lapse rates to a common level has helped in isolating the urban influence. These surveys also established that city temperature patterns often stand out boldly from a fairly uniform rural temperature field. The city nearly always is warmer than its surroundings, a phenomenon that has been given the graphically descriptive label "urban heat island." The drawback of mobile surveys is their occasional, and hence selective, character. In a few cities substantial urban observing networks have recently been organized which overcome this shortcoming. The extension of the mobile surveys to the third dimension by use of helicopters has been most welcome and, in spite of their sporadic nature, has added materially to our knowledge.

The last technique has been the statistical analysis of long series of records for differences in trends inside and outside the city, to establish temperature increases with time as the cities grow. One point has to be made here, namely, that the heat island is an intermittent condition, largely similar to the local temperature patterns that give rise to land and lake breezes or mountain and valley breezes. All these local happenings in the planetary boundary layer can take place only when the general synoptic weather conditions are favorable. They are essentially "good weather" phenomena, occurring notably when pressure gradients are weak and skies are clear. Actually, under these conditions one can show that a single block of buildings, sur-

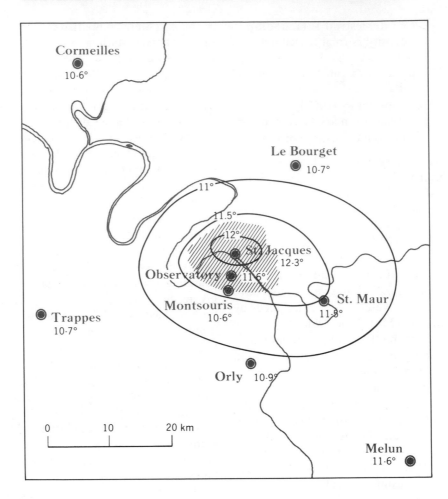

Figure 20.1 Typical metropolitan heat island, shown by the mean annual isotherms of the Paris region (after Dettwiller[29]). Temperatures in °C.

rounded by grass and vegetation, will become a miniature heat island, showing its maximum development on calm, clear nights 2 to 3 hr after sunset. At that time the infrared temperature differences between paved and grassy surfaces is about 7.5°C with an air temperature difference at 2 m of about 1°C.[65] The areal extent and intensity of the heat island indeed exhibits a trend, as observations in the growing town of Columbia, Maryland have shown. There on clear evenings, when the town had 2000 inhabitants, the largest temperature differences to the environs were +1 to +3°C, values that climbed to +2 to 4.5°C 2 years later when the town had a population of 16, 000.[64]

Others have looked at mean monthly and annual temperature trends. Central London had in winter, compared with other stations in southeastern England, a rise of 1°C in the mean daily minimum temperature and 0.5°C in the mean maximum between the 1920s and 1950s. By the 1960s the annual average had risen by 1°C, larger in winter by about 0.4°C and less by the same amount in spring and fall.[82] In a worldwide critical survey of city temperatures Dronia estimated for the 90-year interval 1871–1960 the average city rise at 0.7°C.[32]

Perhaps the most illustrative case history involves Paris.[29] There the time-dependent changes could be followed not only in the air but also in a 28-m deep cellar, where records have been kept since the 18th century. The mean annual air temperature in central Paris lies 1.7°C above the environs. This can be seen from Figure 20.1, which shows the Paris heat island. The region is so devoid of topography and open to the westerly weather regime from the Atlantic that this notable inhomogeneity cannot be attributed to other than man-made causes. In the cellar there was for more than 100 years no measurable temperature change. But in the last 70 to 80 years the subterranean temperature has risen about 1.5°C, a remarkable heating effect if one considers that the normal annual temperature wave becomes unmeasurable at about 10 m depth. Dettwiller's analysis goes much further. For the air temperature he showed an increase of 1.1°C/century and a strong increase for the mean daily minima of 1.9°C/century with only a slight rise of 0.6°C for the maxima.

Even in smaller towns trends, often only seasonal, sometimes become apparent. A comparison of Sacramento,

Figure 20.2 Nocturnal heat island in a small town (Corvallis, Oregon). Isotherms in °C (reprinted by permission, from Hutcheon et al.[54]; photo by Western Ways, courtesy of Dr. W. P. Lowry).

Figure 20.3 Mosaic of infrared aerial photographs taken over the developing town of Columbia, Maryland. White areas have high, gray areas moderate, and black areas low surface temperatures. Note solid heat island over business districts, notable warmth of roads, and isolated warm spots identifying houses in residential section.

California with nearby rural Davis, showed a rise of 1°C in the June to September period for the city. This might be in part attributable to large values of heat rejection in that city with 2×10^8 kg cal/hr from air conditioning on hot days. But this is notably outdistanced by the 10^{12} kg cal/hr released from the 360,000 automobiles used in the city.

Lest there be any misconception that the urban heat island is a typically middle- and high-latitude phenomenon, there is accumulating ample evidence of similar conditions in the tropics.

Similarly, there is now ample evidence that under suitable synoptic conditions even relatively small towns develop notable heat islands. Almost every case has its own peculiarities though. These often furnish insight into small-scale meteorological patterns. Not all of them are always immediately explainable.

Of particular interest has always been the maximum value of the heat island which generally requires very calm clear nights following a sunny day. A typical example of such a case is Corvallis, Oregon, a town of 21,000 population, covering 1350 km² with 130 m elevation differences (see Figure 20.2). The isotherms show a maximum heat island difference of $+5°C$.[54] Ordinary micrometeorological doctrine predicts that under such synoptic conditions the lowest temperatures are found in the lowest areas of the terrain and concave land forms, but human occupation of such areas can counteract the gravity flow of cold air and create selective warm spots.

Usually there is a microclimatic patchwork of cooler and warmer areas in the urban mesoclimatic heat island. Parklands, especially with trees, stand out as cool spots but even smaller-scale differentiations, presumably influenced by different radiative qualities, abound. This is particularly evident from infrared aerial photographs which depict surface temperatures (Figure 20.3). In these, paved and built-up areas and even roofs of individual houses stand out as warm spots, whereas lawns, water surfaces, and trees are cool. This may not necessarily find expression in air temperatures because air motions readily equalize these smallest-scale differences.

Some other general observations apply to large cities. One of them is the crowding of nocturnal isotherms at the edge of the built-up areas in large cities, such as London.[15] The pattern has all the earmarks of a miniature cold front. The extent to which the nocturnal rural-urban temperature difference can reach in anticyclonic weather, with a stagnant polar air mass, is illustrated by statistics of Heuseler[44] for Berlin, Germany. In 14 percent of the cases the difference was 6.1 to 8.0°C, in 57 percent 8.1 to 10.1°C, in 18 percent 10.1 to 12.1°C, and in 11 percent 12.1 to 13.3°C. The last value was the highest rural-urban difference observed, well of the order of magnitude of a major cold front.

The magnitude of the difference has been related to the intensity of the ground inversion that often builds up in the rural area at night while temperature lapse continues over the city. Ludwig and Kealoha[75] give the following empirically derived relations:

For urban areas with
$$\begin{cases} <500{,}000 \text{ population:} \\ 500{,}000\text{--}2 \text{ million:} \\ >2 \text{ million:} \end{cases}$$

$$\Delta T = 1.3 - 6.78\gamma$$
$$\Delta T = 1.7 - 7.24\gamma$$
$$\Delta T = 2.3 - -14.8\gamma \quad (20.4)$$

where γ is the lapse rate in lowest 100 m, in °C; in inversions γ is negative.

Although population as a nonphysical parameter may well be a suitable correlate for the magnitude of the heat is-

land, the main governing meteorological element is the wind speed near the surface. Should the general synoptic situation create a brisk or strong wind, the heat island effect is weakened or eliminated. Oke and Hannel[87] tried to develop a relation for the interplay of city size (represented by population number P) and the critical wind speed (U_{crit} in m/sec) for development of a heat island: $U_{crit} = 3.4 \log P - 11.6$. In a later study, Oke[86] empirically relates the magnitude of the temperature difference (urban-rural) to these same variables and states that it approximates

$$\Delta T \simeq \frac{P^{1/4}}{4\bar{u}^{1/2}}. \quad (20.5)$$

This statistical relation for a large number of cities explains about 70 percent of the variance. It should be noted here that there is a limit to the temperature difference that can be maintained because buoyancy factors will cause the warm air to rise and to be replaced by cooler country air.

SOME SECONDARY EFFECTS OF THE URBAN HEAT ISLAND

The heat island has a number of consequences. One of them is the lowering of relative humidity values at the 2-m level compared to the rural area. This lowering, for equal vapor pressures, is simply a consequence of the definition of this parameter. Actually, the higher temperatures are not the only influence, because the vapor pressures are also a fraction lower in many instances because of the reduction in Q_E following rapid runoff and lack of vegetation. On the whole, humidity decreases of 2 to 6 percent in cities are common. This may not be at all the case at somewhat higher levels, where urban industrial combustion

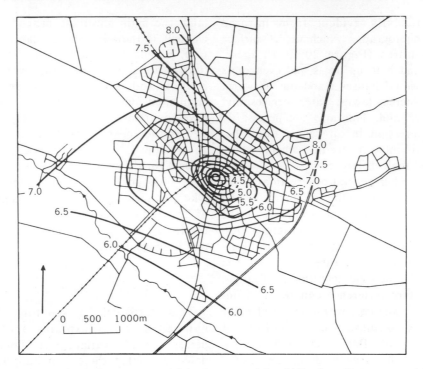

Figure 20.4 Snow depth (in cm) measured in the town of Lund, Sweden, after a recent fall. Note systematic increase from urban center to rural environs (reprinted by permission, from Lindqist[71]).

and cooling towers using water may discharge substantial amounts of water vapor into the atmosphere.

A very consistent urban effect is the reduction in frequency and duration of snow covers. Much snow falls under synoptic conditions when surface temperatures are close to the freezing point. Under those circumstances heat flux from pavements and man-made sources can melt snow on impact or gradually, if the total amount is moderate. A very notable example is shown in Figures 20.4 and 20.5 taken from a study by Lindquist in Lund, Sweden.[71] One shows the snow depth on a winter day in 1965, with an almost concentric decrease from 8 to 3 cm into the center of town. The isolines look exactly as the expected isotherms of a heat island, except for the reversed gradient. Figure 20.5 shows an aerial photograph of the

town and its surroundings. A monotonous white snow blanket surrounds the dark core of the city. Even if a few patches of snow remain here and there in the urban area, the contrast is striking. On such occasions snow removal may be partly responsible, but the fact that the roof areas are also free of snow shows that heat from the fluxes Q_S and Q_P are the energies at work under such circumstances.

A synoptic climatology of the snow conditions in the New York City area has been presented by Grillo and Spar.[43] They used observations from 33 stations in the metropolitan area for two winters. In the central city, snow probability under synoptic conditions favoring precipitation is only around 27 to 35 percent compared with 40 to 45 percent in outlying and rural areas, giving adequate consideration to the climatic gradients existing in the region. In lower Manhattan less than 30 percent snow probability exists.

Another typical heat island effect is the lower value of degree-days below a given threshold of air temperature. In the United States this is generally set at 18°C (65°F) for the mean daily temperature. Below this value, space heating starts and the fuel consumption is approximately proportional to the degree-day value. This applies particularly to cumulative values over a month or a heating season. Turner[108] showed for St. Louis that on weekdays the fuel need for space heating showed a correlation coefficient of .96 with the mean daily temperature. A comparison in the United States, using a 30-year period, shows that there is a degree-day difference between airports and cities of 5 to 15 percent, indicating a beneficial effect of the heat island on fuel economy.

Similar degree-day concepts can be applied to values above a certain threshold, usually geared to plant growth. These are generally counted cumulatively above 5°C. The higher values in the urban areas lead to earlier leafing and flowering of plants. The difference between urban rural areas, for the same species, may be 1 to 2 weeks. Observations of Eriksen[38]

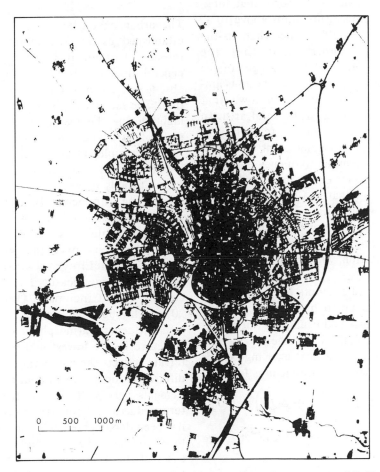

Figure 20.5 Aerial photograph of Lund, Sweden, and surroundings after recent snowfall. Note dark town area (snow melted) and highly reflective snow cover outside (reprinted by permission, from Lindqist[71]).

for Kiel, Germany, may serve as an example. Hourly heat sums above 5°C, starting March 1, yield the following values for various dates:

	March 15	April 1	April 15	May 1
City center	1500	2250	3550	6050
Edge of city	1300	1900	3150	5400
Percent of city surplus	15	18	13	12

This gives the city a phenological difference of about 10 days in spring. In autumn the first freezes are similarly delayed and the growing season is prolonged. Davitaya[27] reports that long-term daily temperature sums above 10°C in Moscow lie 250° higher than in the surrounding country and that the frostfree period is 30 days longer in the city than outside. This often permits imported species to survive in town that would not get along in the rural area. In areas where mean winter temperatures are close to the freezing point there is a notable change in the frequency of freeze-thaw cycles between urban and rural areas, a factor that has an important influence on weathering.

Other effects of the urban heat island concern changes in the local wind field and precipitation, which will be discussed in later sections.

A number of attempts have been made to model the urban heat island. This is not an easy task and usually requires simplifying assumptions. In the absence of advection and latent heat exchange Oke[85] depicted the time development as follows

$$\frac{\Delta \bar{T}}{\Delta t} = \frac{1}{\rho c_p}\left(\Delta\left(\bar{Q}_H \frac{\Delta T}{\Delta z}\right) + \Delta Q_N\right), \quad (20.6)$$

where the quantities with an overbar are means and

T temperature
t time
ρ density of air
c_p specific heat of air at constant pressure
Q_H convective heat transfer
z height
Q_N net radiative heat fluxes

It is difficult to establish the height for which the vertical temperature gradient enters and, obviously, the premises cannot be met because, even in the rare case of absence of advection, the urban-rural temperature difference would induce a wind field.

THE URBAN WIND FIELD

The urban heat island is, of course, not established as a static temperature difference. Instead it interacts with the general flow of air and if such flow is absent sets its own circulation into motion. The latter case is most common at night during the maximum development of the heat island. The nocturnal pattern is quite simple. Over the city there is a slight vertical motion because the air stays unstably or neutrally stratified with the heating from below. In the rural areas a ground inversion forms with generally another stable layer aloft at a few hundred meters over the whole area. Clark and McElroy[22] showed a cross-section of this, measured from a helicopter, over Cincinnati, Ohio (Figure 20.6). Although there is some rolling terrain, the general atmospheric temperature structure in the lowest 400 m above the terrain reflects the mesoscale atmospheric differentiation brought about by large settlements very well. This isothermal pattern induces a "country breeze" which has been documented as a converging surface flow in several localities. Notable cases of this type have been summarized for clear nights over Frankfurt, Germany by Stummer,[104] as shown in Figure 20.7.

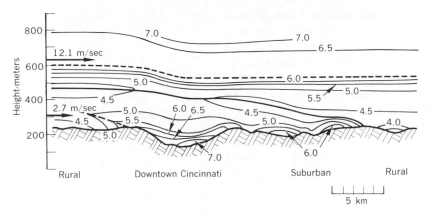

Figure 20.6 Cross-section of vertical temperature distribution (°C) over Cincinnati, Ohio, and environs during night with well-developed urban heat island (reprinted by permission, from Clarke and McElroy[22]).

Even the heat island developed over a runway complex will show a measureable convergence from which Maisel[76] determined the vertical speed of the rising air current to have an approximate magnitude of 10 cm/sec. Angell, Pack, and Hoecker[4] obtained from tetroon flights (radar-tracked constant-pressure balloons) over New York City somewhat larger values, consistent with the much larger heat island there. Their median values for nocturnal releases (0300 to 0900 hours) were about 20 cm/sec and for daytime releases (0900 to 1500 hours) about 40 cm/sec. The lift imparted during the day in mid-Manhattan was from 300 to 800 m, within the obviously unstable layer created by the heat island.

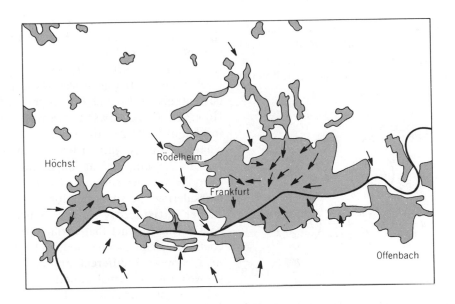

Figure 20.7 Surface wind flow, indicated by arrows, on calm clear nights in Frankfurt am Main (after Stummer[104]).

Of course, intense individual heat sources will also cause rising currents, such as a plume of an incinerator, which made one tetroon rise about 100 m. Incidentally, this convective flow exercises a mitigating effect upon urban effluents by lifting them aloft.

Even an isolated industrial complex can induce its own wind field as Schmidt and Boer showed for a refinery district in the Netherlands.[95] They noted wind convergence toward the area and measured upward speeds of 15 cm/sec with downdrafts in the surroundings up to 25 to 30 cm/sec. All this fits vertical soundings of temperature made by Duckworth and Sandberg over San Francisco, Palo Alto, and San Jose, California and surroundings.[33] There they found nocturnally nearly isothermal conditions over the city and a very steep surface inversion over the rural sector. The soundings intersected at about 25 m (crossover effect). Below that level the urban atmosphere was warmer and above considerably colder than the environment, obviously a structure conducive to a circulation equalizing the differences.

Several attempts have been made to produce a numerical time-dependent model of the heat island circulation. Delage and Taylor[28] used the following prognostic equations:

$$\frac{\partial u}{\partial t} = -u\frac{\partial u}{\partial x} - w\frac{\partial u}{\partial z} - \frac{1}{\rho}\frac{\partial p}{\partial z}$$

$$+ \frac{\partial}{\partial z}\left(K_z\frac{\partial u}{\partial z}\right) + Kx_M\frac{\partial^2 u}{\partial x^2} \quad (20.7)$$

$$\frac{\partial \theta}{\partial t} = -u\frac{\partial \theta}{\partial z} - w\frac{\partial \theta}{\partial z} + \frac{\partial}{\partial z}\left(Kz\frac{\partial \theta}{\partial z}\right)$$

$$+ Kx_H\frac{\partial^2 \theta}{\partial z^2} \quad (20.8)$$

with vertical wind speeds calculated from the horizontal wind field using the continuity condition

$$\rho\frac{\partial u}{\partial x} + \frac{\partial}{\partial z}(\rho w) = 0 \quad (20.9)$$

with the following meaning for the symbols

x,y,z Cartesian coordinates
t time
u,w horizontal and vertical wind speed
ρ density of air
p pressure
θ potential temperature
Kx_H, Kx_M horizontal eddy transfer coefficients for heat and momentum
K_z vertical eddy transfer coefficient

Using appropriate boundary conditions these differential equations can be solved and the time development of wind and temperature field be calculated. Figure 20.8 shows such a solution after $t = 90$ min for an urban rural temperature difference of 2.5°C, a low-level vertical gradient

$\partial\theta/\partial z = 2^0C/km^{-1}$, setting $Kz = Kx_H$

$$= 50 \text{ m}^2/\text{sec and } Kx_M = 0.$$

The solution, which depicts one wing of the circulation with the heat island at zero of the x-axis, gives a rather realistic, if simplified, structure. It shows the low-level inflow to the heat island, the rising current above it, and the upper return flow. It also shows a rural ground inversion from 2.5 to 5 km from the center, a slightly warmer layer at the surface to 2.5 km, and an upper inversion over the warmer surface zone at 600 m. Considering the simplifications, this is a rather satisfactory result, especially considering the fact that variable transfer coefficients can also be used. Vukovich,[112] in a somewhat different approach, also showed for nocturnal conditions a two-cell system with low-level convergence and high-level divergence when the external wind is slow. That model also indicates that with increasing general winds

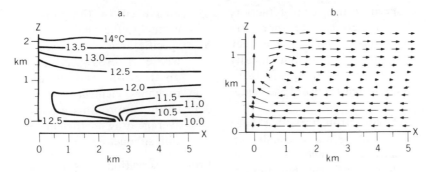

Figure 20.8 Calculated temperature (*a*) and wind field (*b*) induced by heat island at $x = 0$, after 90 min under limitations indicated in the text (reprinted by permission, from Delage and Taylor[28]).

the pattern shifts and the position of maximum air rise is shifted downwind, and some subsidence may take place over the heat island center. In essence this model suggests a city plume, a factor which may be of some importance in formation of precipitation, yet to be discussed.

Obviously, however, the city's wind flow is not only determined by the thermal contrasts, inside and outside. In fact, these circumstances are primarily operative when general winds are weak. There is also the difference of flow in the streets and above the roof as well as the changes in the vertical wind profile caused by the increased aerodynamic roughness of the built-up area. Relatively little work has been done on the wind in the streets because the measurements there are so far removed from the usual meteorogical standards requiring obstacle-free exposure of anemometers. Only data collected for other special purposes, such as air pollution studies, are available. The wind speed variations according to the angle that a street forms with the wind direction are plain. Motion is speeded in the street channels parallel to the wind direction and inhibited in those at an angle to that direction. Also, the middle of the street usually shows higher speeds than the sections close to buildings. Dirmhirn and Sauberer[31] indi-

cate that if flow in the middle of the street is set as 100 percent, the windward sidewalk may experience 90 percent of that speed and the lee sidewalk as little as 45 percent. They also noted the effect of trees which, if in leaf, reduce the speed by 20 to 30 percent, and in parks even up to 50 percent.

Most studies report mean wind speed reductions of 20 to 30 percent in cities, compared with nearby airports for standard exposures of anemometers. A more elaborate study by Bornstein for New York City compared winds measured upwind and downwind at 32 and 16 km distance from the reasonably well exposed station at Central Park.[13] In daytime his results suggested a relatively steady decrease across this profile, with a 10 percent decrease in midtown and 20 percent at the most leeward station. The area 10 km upwind and downwind is rather well urbanized, and the reduction can be explained well by the increased friction. However, the fact that the wind speed at a further distance of 16 km downwind has not recovered remains unexplained. At night, wind speeds 32 km upwind and downwind are essentially equal but less than those at the other stations, which are 10 to 20 percent higher. This is explained by the low inversion in the countryside, with weak winds and the convergence flow to the city, with a maximum speed at the edge of the city where the horizontal temperature gradient is steepest. This has not been verified elsewhere but is dynamically plausible also because of the gradual surface roughness increase toward the center of the city.

The aerodynamic roughness has been determined in various ways, usually indirectly. Ideally one would like to determine a drag coefficient but present methods are not suitable for measuring this in a heterogeneous area. The usual procedure is to obtain wind speeds at various heights above the surface and determine the roughness parameter z_0 from these measurements, assuming generally neutral stability in the surface-near air layer, to about 200 to 300 m. This is not too bad an assumption over the heat island. Under those conditions the vertical wind profile is given by

$$\bar{u} = \frac{u^*}{k} \ln\left(\frac{z}{z_0}\right), \qquad (20.10)$$

where

$\bar{u}$ mean wind speed at height z
u^* friction velocity, $= \sqrt{\tau/\rho}$, τ shear stress, ρ air density
k von Kármán's constant
z_0 roughness length

Measurements over urban areas have variously yielded values of 1 to 5 m for z_0.

Lettau[69] proposed a method by which one can approximate the value of the roughness parameter as a function of the geometric properties of the obstacles to the wind:

$$z_0 = 0.5h^* \frac{s}{S}, \qquad (20.11)$$

where

h^* is the average vertical height or "effective obstacle height"
s silhouette area of the average obstacle "seen" by the wind,

measured in the vertical plane normal to the wind
S specific area or let area $S = A/n$, where A cross-sectional area of obstruction and n number of roughness elements
0.5   is the approximate value of the average drag coefficient for obstacle with silhouette s.

From observations this relation can be simplified (perhaps oversimplified) to

$$z_0 = 0.058h^{*1.19},$$

The values here change notably if the cross-sections of the obscuring objects are considerably higher than the open spaces between them.

In a slightly different approach, Davenport prefers to relate the wind speed near the surface to the gradient wind, loosely defined as the level where local frictional influences disappear and the synoptic pressure field determines the wind speed. This relation can be expressed as a power law:

$$\bar{u} = v_G \left(\frac{z}{z_G}\right)^\alpha, \qquad (20.12)$$

where

$\bar{u}$ mean wind speed near the surface
v_G gradient wind speed
z, z_G heights of surface and gradient wind measurement

Davenport's values for α and z_G in various environments are given in Table 20.3.

These average wind profiles are shown in Figure 20.9 which impressively depicts not only the slowing down of surface winds over the city but also the height to which it affects the wind field. Some earlier work showed that in crossing large urban areas the motion of fronts at the surface can be deformed and slowed down by an average of 25 percent of their speed.

Table 20.3 Parameters of Vertical Wind Profile in Various Environments

Environment	Exponent α of Power Relation	Gradient Wind Level, z_G (m)
Flat open country	0.16	270
Suburban area	0.28	390
Urban centers	0.40	420

Source: Davenport.[25]

Such static average patterns, however, fail to portray one of the important facets of urban air motion, namely, its turbulent structure. Heurtier tried to describe this condition by presenting a profile across the Paris metropolitan area of the gustiness factor. This is defined as the difference of the highest and lowest wind gust speed in an hour divided by the hourly mean speed. His line was 73 km long, reaching well upwind and downwind from Paris. For cases of flow along this NW–SE line he could show that the gustiness factor is 3.6 times larger at a station (26 m height) over the center of Paris than at the upwind rural station (10 m). This drops to 2.2 times the upwind value at 56 m and has smoothed out at the height of the Eiffel Tower (317 m) to only 0.8 times the upwind value. This procedure does not permit any judgment on how much of the added turbulence can be attributed to thermal and how much to mechanical causes. Davenport approached the problem in a more sophisticated way for London, Ontario by determining the powerspectrum from records taken on a microwave tower in the center of town (49 m). He found a smooth horizontal spectrum with a single peak of 1.6 $\times$ 10^{-3} waves/m ($\sim$a 650-m wavelength), for a roughness parameter z_0 of $\sim$2 m, obviously a shorter value than over a smooth terrain. But such numbers are only an approximate indication of conditions which are strongly dependent on wind speed and stability.

Some of the peculiarities of urban flow can be conveniently studied in the wind tunnel by use of scale models. Cermak using dynamic similarity tested a model of the city of Denver (scale 1:400) and obtained moving pictures of smoke flow in-

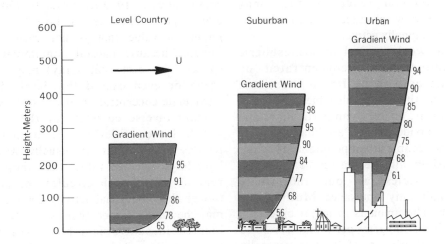

Figure 20.9 Average vertical wind speed profiles over surfaces of varying roughness, using power relation (reprinted by permission, from Davenport[25]).

dicating very complex patterns of turbulence induced by the city. In extension of this work Chaudhry and Cermak[20] were able to demonstrate experimentally the deformation of the vertical wind profile, corroborating the studies of Davenport.

URBAN POLLUTION

The urban atmosphere is continually being loaded with effluents resulting from human activity. A wide variety of substances are added. Some of them are just end products of combustion processes such as water vapor and carbon dioxide. These are also present in nature and may have only long-range implications for atmospheric changes. But a large number of pollutants which are either rare or absent in the natural atmosphere enter the urban air space and interact with meteorological processes. In this connection those compounds that may have no or only a remote relation to these processes, such as carbon monoxide, shall not be dealt with here. One should, however, be aware of the possibilities that even these substances may play a not unimportant role in urban chemical and photochemical reactions that may produce new products which can be meteorologically active.

Past work on urban atmospheric contaminants has concentrated on substances with health implications. Many of these have also meteorological side effects, among them particulates and sulfur dioxide (SO_2). Particulates are the most noticeable urban pollutant. They are injected in all size classes. The largest dust particles fall out and create essentially only a nuisance. Much of it is fly ash. The amounts in major cities can reach 30 to 100 g/m² per month (or 30 to 100 tons/km²). This fallout is generally rapid for sizes above 10 to 20 μm diameter and occurs very close to the sources.

The reduction from city to suburban or nearby rural areas is rapid. Steinhauser[101] reports for Vienna 40 to 50 percent in the windward suburbs of the amounts noted in the center of the city.

Meteorologically more important than gross dust is that fraction of the aerosol that stays in suspension. Most of these particles are less than 5 μm in diameter, the vast majority of them, by number, in the size range of the Aitken condensation nuclei (diameter ≤ 0.1 μm). These particles, which in clean uncontaminated air range from 10^2 to 10^3/cm³, generally number 10^5 to 10^6/cm³ in urban atmospheres. Although usually only a small fraction of them play an active role in condensation processes, they have a large influence on meteorological factors. They can be conveniently determined and are a handy measure of relative levels of contamination in time and space. Another commonly used measure of particulate pollution level is the dust loading expressed in μg/m³. This is, of course, heavily influenced by the larger size classes present in the aerosol. In industrially heavily polluted atmospheres this can reach values of 1000 μg/m³, although the average values in United States cities in 1970 lie closer to 150 to 200 μg/m³. The desired target is 60 μg/m³, a value that is not easily attainable because natural contamination (pollen, soil dust, salt spray) may also reach or even exceed this level. The particulate concentration at the surface is a close inverse correlate of the wind speed. Hence cities in climatological regions where winds are usually strong have lower average values than those in regions of sluggish circulation, even though their industrial activity and traffic may be much larger than some towns located in subtropical high pressure regions or protected mountain valleys.[64]

Considerable information is also available on SO_2 concentration because of the adverse health effects of this gas. Unfortunately, its chemical life history is not entirely understood as yet, although a portion of it generally ends up as sulfuric acid before being eliminated from the atmosphere.[3] Most of the SO_2 comes from sulfur-containing fossil fuels. In 1970 about 30 million tons/year were emitted in the United States, mostly in or near urban areas. Concentrations in urban air average about 0.05 to 0.1 ppm in United States urban centers but in high pollution episodes values can reach or exceed 1 ppm. In the context of this discussion interest concentrates on the fact that sulfuric acid is highly hygroscopic and forms droplets with atmospheric water vapor long before saturation is reached. The small droplets certainly begin to grow at 70 percent relative humidity. SO_2 concentrations show much the same behavior as particulates. They are very dependent on wind speeds.

Many contaminants have a notable annual variation, generally with higher values in winter than in summer. This is a function of greater use of low-grade fuels for space heating and, in some areas, high frequency of low-level inversions. Parry[89] gives for Reading, England a ratio of 5:1 between summer and winter in suburban and residential sectors but 35:1 in industrial sectors, for particulate loading. The weekly and diurnal variation is also pronounced. Weekends generally show less contamination and the morning hours of weekdays have a pronounced maximum at 7 to 9 a.m. This is the time of traffic rush hour and intensification of domestic and industrial activity, with vestiges of nocturnal inversion conditions left. The minimum is in the early afternoon (1300 to 1500 hours) when convection is apt to be strongest and a secondary maximum (1700 to 2000 hours) coincides with evening rush hour and reestablishment of the surface-near inversion.[105]

One additional anthropogenic contaminant deserves singling out. This is the lead particulate produced presently in profusion by motor vehicles using leaded fuels. For example, in San Diego at a sampling site passed by about 175,000 cars daily, lead in particulate form constitutes about 2 to 5 percent of the suspended material. Weekly averages were as high as 8 $\mu g/m^3$.[21] The public health aspect alone would make this a pollutant deserving close attention. But Schaefer[93] showed that in the presence of iodine vapor these lead particles will form freezing nuclei active on supercooled water of -3 to $-20°C$. It had, of course, been known long before that lead iodide could act as an excellent seeding agent for water clouds at subfreezing temperatures. Schaefer[94] showed that smoke from bituminous coal or burning wood could be a source of the reactant for lead from car exhausts. Iodine is also always present in maritime air, especially near the coastlines. He also pointed out that a high correlation existed between Aitken nuclei and lead iodide freezing nuclei. Schaefer's work was done at eastern locations; other data from Denver and Palo Alto show that no geographic dependence is present. Moyers, Zoller, and Duce[84] found for the Boston urban atmosphere an average atmospheric lead content of 3 $\mu g/m^3$ (1 percent of the suspended material) and an iodine concentration 14 ng/m^3. The measurements at various lead concentrations suggest that the lead scavenges the gaseous iodine, although this is a relatively slow chemical reaction. From a meteorological point of view only a few highly active freezing nuclei per liter may be needed to have an effect in supercooled clouds.

In this and other contexts it is important to know how the aerosol behaves in and around urban areas. Several observational techniques can give such information. In early experiments, near

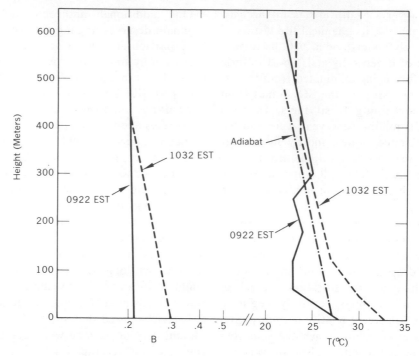

Figure 20.10 Turbidity and temperature sounding near and above Cincinnati in anticyclonic weather during a summer day (left section shows turbidity factor B, with logarithmic abscissa; right section shows temperature. 0932-hr values were obtained upwind in rural area; 1032-hr values were measured over the city) (reprinted by permission, from McCormick and Kurfis[78]).

Leipzig, Germany, using a dust collector in low-flying aircraft, Löbner[72] could show that for large particulates (2 to 30 μm diameter) a very rapid decrease takes place from areas of dense settlement to the suburbs. He showed that, aside from a maximum concentration at the surface, there is a distinct secondary maximum at the average height of stacks (40 to 50 m), with a rapid decrease above, which he attributed to the higher wind speeds, promoting dispersion. In recent years more sophisticated equipment has permitted better insight. Using a Volz sunphotometer, McCormick and Kurfis[78] obtained some soundings over Cincinnati. One of their results is reproduced here in Figure 20.10. It represents both the turbidity factor B and the temperature stratification over the city and the nearby countryside upwind. Over the city an ex-

ponential decrease of particulates is shown through the mixing layer, reaching the rural values at about 450 m height. McCormick and Baulch[77] had already shown that up to 200 m one could express the vertical distribution of the turbidity factor B by a simple exponential function of the height z: $B(z) = B_0 \exp - 0.00346z$, where B_0 is the surface turbidity value. The particulate loading (in μm^{-3}) is obtained by multiplying the turbidity value by a factor of about 10^3.

Ahlquist and Charlson[2] used an integrating nephelometer over Seattle to investigate the dust dome. They measured a particulate-depending light scattering coefficient of $\sim 4 \times 10^{-4}$/m near the surface and showed that this decreased to one-tenth of this value above 500 m. Lidar also offers great promise to get at the vertical distribution of pollutants.

This equipment has the potential to map aerosol clouds three-dimensionally. Uthe[109] followed the diurnal development of a pollutant cloud through a summer day over St. Louis with a ruby laser (0.69 μm), using the backscatter from particulates for analysis. At 0730 hours a dense aerosol layer existed in a layer from the surface to 250 m, with an inversion at the level and stable stratification to 1.5 km. Convection beginning at 1215 hours raised the aerosol layer at a rate of 125 m/10 min interval to about 2 km, a level to which an adiabatic lapse rate had been established. Cloud formation took place and precipitation occurred, creating a layer of 750 m without an aerosol return of the laser beam. Such observations do not permit conclusions if the aerosol was involved in the cloud and precipitation formation but, if multiplied, might offer hints in that direction.

A large number of attempts has been made to model the interaction of urban pollutants with the atmospheric parameters. Usually the purpose has been to get an approximation of the concentration at the surface; however, many useful things can be learned about the behavior of city plumes; hence a brief summary of the more common approaches to this problem seems to be worthwhile. Extensive discussion of the problem can be found in the proceedings of a symposium devoted to this theme.[102]

In its simplest form one can restrict the analysis to an atmosphere with a shallow adiabatically stratified layer beneath an upper inversion. This yields a simple box model[26] showing the time-dependent buildup of a contaminant:

$$dV = Pdt - v\frac{V}{V_a}dt, \quad (20.13)$$

where

V is volume of contaminant
V_a total volume of air through mixing layer over city area

P production rate
v ventilation rate through mixing layer
t time

A critical element in such a model is the depth of the mixing layer. This is to a considerable extent a function of the urban heat island. This quantity can also be well approximated by a model developed by Leahy and Friend[68]

$$h^2(x) = \frac{2}{c_p\rho u(\Gamma - \gamma)}\int_{x_1}^{x} Q_s dx, \quad (20.14)$$

where

$h(x)$ mixing depth in direction x (direction of flow)
Q_s sensible heat flux through bottom of air column
u wind speed in lower layer
ρ air density
c_p specific heat at constant pressure
$\Gamma - \gamma$ difference between adiabatic lapse rate and lapse rate in overlying stable layer

This can be further developed by substituting long-wave and other heat fluxes in the following manner:

$$h^2(x) = \frac{2}{c_p\rho u(\Gamma - \gamma)}\int_{x_1}^{x} [Q_{s_c} + Q_P + Q_{s_r} - \sigma(T_{0_c}^4 - T_{0_r}^4)]dx, \quad (20.15)$$

where

Q_{s_r}, Q_{s_r} the surface heat flux in the city and rural area
Q_P anthropogenically produced heat flux
T_{0_c}, T_{0_r} absolute temperatures in the city and rural area at the surface

The vertical distribution of large urban particulates in surface-near layer has also been estimated based on a simple exponential law of decrease from the source which is quite realistic.[96]

$$\chi(z) = \chi_1 \exp - vK_z(z_1 - z), \quad (20.16)$$

where

$\chi(z)$ concentration at height z
χ_1 concentration near surface at height z_1
v Stokes' terminal velocity of particles
K_z vertical diffusivity

But the majority of attempts have been made to model the urban plume analogous to point sources, using a Gaussian dispersion.[107] This started with very simplified assumptions on uniform distribution of emission sources and uniform rate of emission, single emission heights, and uniform pollutant distribution through the mixing layer, and hence permitted no conclusions on the plume development downwind.

In a rather formalistic way Egan and Mahoney[36] developed the numerical solution of a differential equation for a diffusion model for an urban emission puff in a time-dependent two-dimensional section:

$$\frac{\partial \chi}{\partial t} = -u \frac{\partial \chi}{\partial x} \frac{\partial}{\partial z} K_z \frac{\partial \chi}{\partial z} + Q, \quad (20.17)$$

where

χ pollutant concentration
t time
u wind speed
K_z vertical diffusivity
Q added material
x, z horizontal and vertical coordinate

Calculations show the vertical exponential dropoff, but no tests against observations were made. None of these models take the all-important question of residence time of pollutants into account. This concept includes time-dependent processes of elimination, whether mechanical or chemical. Should meteorological processes, such as condensation, also interfere, the modeling task becomes indeed very formidable. In his Gaussian diffusion model of area pollutants,

Slade[98] at least considers a "half-life" factor for pollutants, but in the absence of observed values for this parameter for most pollutants one has to make assumptions that again lead only to order-of-magnitude concentration values.

Semiempirical methods seem to be quite helpful in approximating a solution. These involve a release of tracer material upwind of an urban area and have been done for the relatively simple topographic setting of Fort Wayne, Indiana by Hilst and Bowne.[46] These experiments showed that the increased roughness of the city and thermal instability created by the heat island caused 30 to 50 percent more mixing of the aerosol over the city than in nearby rural areas. More importantly, it showed the development of an urban plume with a notable enhancement of concentration aloft to the lee of the city.

SOME METEOROLOGICAL EFFECTS OF URBAN POLLUTION

Most of the obvious atmospheric effects of urban pollutants are caused by the particulates. They deplete the solar and sky radiation in all wave lengths, as has already been pointed out. This depletion of the solar beam takes place in a very shallow layer. The average reduction in urban areas can be set at 8 to 30 percent depending on size of the city, topographic setting, and ventilation rate. On clear days when the Volz sunphotometer can be used to observe the rate of change in stagnation situations, an increase in optical thickness of the atmosphere of 10 percent per day was noted in Central Europe.[111] On cloudy winter days things can become so bad that daylight illumination in London goes as low as 2 klx.[42]*

The relation between solar radiation and particulate loading is very im-

* 1 lux = 1 lm/m² = 0.0929 ft-candles.

pressively shown in a table of Monteith[83] for the center of London and suburban Kew. It presents a time series which straddled the period when smoke controls were put into effect (Table 20.4). This shows not only the notable difference between city and suburb but also depicts the gradual improvement particularly noticeable in the ratio of solar intensity to sunshine hours in the city. At dust loads of 200 to 300 $\mu g/m^3$, the radiation loss in the city is 20 to 30 percent compared with surroundings. In the seven years the normal incidence radiation in the city increased from 21 to 26 mw/cm² hr and is now nearly the same as at Kew. The attenuation at both places amounts to about 1 percent per 10 $\mu g/m^3$ particulate. Using sunshine data for London from 1958 to 1967 and comparing them with the 30-year interval 1931 to 1960 Jenkins[56] also checks on the effectiveness of smoke abatement. He finds little change during the warm season (April to September) but notable improvements during the heating season, as follows:

	Oct.	Nov.	Dec.
Sunshine hour percent increase	15	30	60

	Jan.	Feb.	Mar.
Sunshine hour percent increase	45	25	15

This indicates the rapid reversibility of meteorological effects of particulate pollution.

The light-attenuating properties of the aerosol have also a notable impact on the horizontal visibility. It has long been known that meteorological visual range v, especially in daytime is inversely proportional to an attenuation or extinction coefficient,[79] assuming the normal visual threshhold contrast $\epsilon = 0.02$:

$$V = \frac{3.912}{\sigma} \text{ (km)},$$

where σ ranges from 0.03/km in very clean air to about 0.5/km in dirty air. The attenuation is directly related to the cross-section of the particles (or πr^2 in

Table 20-4   Relation between Solar Radiation, Sunshine, and Particulate Loading in London and Kew

	Kew				London			
Year	Solar Radiation (mW-hr/ cm²)	Sun- shine (hr)	Ratio	Particu- lates ($\mu g/m^3$)	Solar Radiation (mW-hr/ cm²)	Sun- shine (hr)	Ratio	Particu- lates ($\mu g/m^3$)
1957	109	4.29	25.5	90	90	3.32	20.9	343
1958	86	3.80	22.7	94	69	3.34	20.6	234
1959	130	5.07	25.6	93	120	4.76	25.2	222
1960	102	4.13	24.7	70	84	3.93	21.4	160
1961	105	4.40	23.9	72	91	4.25	21.4	123
1962	107	4.08	26.2	73	100	3.97	25.2	103
1963	99	3.94	25.1	74	99	3.72	26.6	105

Source: Monteith.[83]

[a] Table footnote.

Table 20.5   Comparison of Frequencies of Low Visibilities (< 1000 m) at 0900 Hrs at Manchester Center and Airport (Average Number of Days per Month, 1967–1968)

Location	I	II	III	IV	V	VI	VII	VIII	IX	X	XI	XII
						Month						
Airport	2.9	1.3	0.6	0.8	0	0	0.1	0.1	0.9	1.6	2.8	2.4
Center	6.1	4.1	4.1	2.4	0.3	0.6	0.1	0.6	1.8	3.0	5.6	6.0

Source: Collier.[23]

case of small spherical particles) and the number per unit volume.

Many investigations, especially for earlier years, have shown a deterioration of visual ranges in the city compared with rural areas. For a while this led to an increase of cases with very low visibilities and a large reduction in the cases of long-distance visibilities. It also led to the introduction of a new term "smog" into the vocabulary and other descriptive phrases, such as London fog. The latter was already graphically described by Howard.[51] But in recent years the same reversal of the trend already noted for sunshine has also been observed for very low visibilities. This does not mean that such phenomena have been eliminated. Mörick[81] reports a case for Budapest where during a late morning hour in March, 1959 visibility in the center of town fell to 80 m, with particulate concentration 5000 $\mu g/m^3$, while at the airport visibility was 2000 m. Usually the low visibilities in cities are noted in the early morning hours at the time of maximum output of particulates with a confining inversion over the city still in place. Collier[23] has published statistics for Manchester, England that are typical for this situation and could be duplicated for many other localities. The figures are shown in Table 20.5.

This shows again that these low visibilities in that climatic region are a distinct cold-season phenomenon when maximum particulate production and meteorological confinement coincide.

In London before the great cleanup campaign prompted by the 1952 catastrophe, fog was nearly twice as frequent as in the rural areas of southeastern England. Some values taken from Shellard[97] show this clearly. They represent the average number of cases per year of visibilities less than 1 km at the observational hours 0300, 0900, 1500, and 2100 for the decade 1947–1956.

Kingsway, London center	Kew, suburban	London airport	South-eastern England, 7 rural stations
940	633	562	494

Comparable data for the ensuing period have not been published but some other available information shows more recent trends. The number of cases of visibility less than 500 m has indeed decreased,[56] but it is not clearly established that this is a result of smoke abatement. In Los Angeles, midday visibilities show in recent years fewer days with visibilities < 2 km, but a deterioration in the class of good visibilities > 50 km, so that a class of mediocre visibilities in the 5 to 10 km range has become considerably inflated.[57] One might well raise the question here if the shift in major pollutant sources from domestic and industrial use of coal with lots of fly ash to oil and to the output of motor vehicles has brought about this shift. There is evidence that the contribution from car exhausts with large amounts of particulates and aerosols produced by secondary reactions in the atmosphere of oxides of nitrogen, which also come from cars, may now be the major cause of reduced visibilities.[14]

It also seems that not only the great metropolitan centers have suffered from deteriorating visibilities. Cities of intermediate size are also being afflicted, especially in the daylight hours.[80] Akron, Ohio, Memphis, Tennessee, and Lexington, Kentucky showed this trend for visibilities less than 10 km, which increased over 10 percent in frequency. This is shown in Figure 20.11 for 1600 hours, indicating that in the interval 1966–1969 the frequency of poor visibilities below 10 km was at least twice as high as those places than in the preceding interval 1962–1965. Man-made contaminants are the most likely source of this change.

The role of suspended particulates in impairing visibility has never been in doubt, and the physical theory governing the relations is well understood. The corresponding effect of hygroscopic aerosols has been much less explored and remains essentially qualitative. Human activity brings many hygroscopic substances into

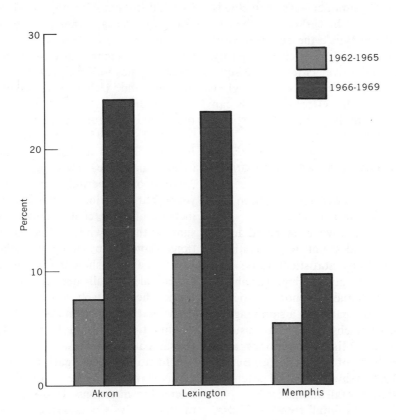

Figure 20.11   Frequency changes in visibilities < 10 km between 1962–1965 and 1966–1969 at three stations (after Miller et al.[80]).

the atmosphere but it is generally agreed upon that sulfuric acid resulting from oxidation and subsequent hydrolysis of SO_2 is the most effective. There have been some estimates of visibility reduction in relation to fuel consumption, but these are very crude. Thomas reports that visibility at the Los Angeles airport shows a linear relation to fuel oil use on days with low wind speeds.[106] He estimates that for each 2.5×10^3 liters of fuel oil used per day, visibility goes down by 1 km. For a better assessment it would be better to use measured values of SO_2 and H_2SO_4 as well as ambient relative humidities.

The urban heat island and production of hygroscopic nuclei, and possibly the release of water vapor from stacks and cooling towers, combine to create more convective clouds over towns than over rural areas. On summer days such clouds may appear 1 to 2 hr earlier over the city than elsewhere. Hence the climatological records show a shift from clear to partly cloudy conditions over cities during summer. In winter, at least in higher latitudes, no notable effect on clouds can be extracted from the available statistics.

URBAN EFFECTS ON PRECIPITATION

A great deal of attention has been devoted to the urban effects of precipitation. These have been noted for a number of decades but were relatively hard to verify by statistical tests. The reason for this is the very high variability of rain amounts and the poor qualities of the ordinary rain gauge as a sampling device. Decades of observations are usually needed to establish differences at a reasonable level of significance. But as early as 1929 Ashworth called attention to some urban rainfall anomalies.

An analysis of rainfall trends in Tulsa, Oklahoma, a city which grew from an Indian village of a few hundred inhabitants in 1900 to a metropolis of several hundred thousand in five decades, also showed a gradual increase of precipitation totals compared with its surroundings.[63] This increase was about 10 percent in summer and 5 percent in winter. Different mechanisms for the increases in the warm season and the cold season were suggested, with the heat island playing a major role in the former, and nucleation in the latter.

A great deal of work since has, by and large, confirmed at least the notion of increased precipitation in urban areas. We cite here only some of the studies. In a review Changnon[18] arrives at values of 5 percent urban increases in annual precipitation for both Urbana and Chicago. In Urbana the colder half of the year showed a 12 percent increase in snowfall compared with the airport. He also reported an annual increase of 7 percent for St. Louis. At all these cities he noted also increases in thunderstorm frequency (Chicago, 6 percent; St. Louis, 11 percent; and Urbana, 7 percent). For another climatic region of the world, Barrett[11] analyzed rainfall in a 70-year interval (1890–1959) near Manchester, England. This is a highly urbanized and industrialized area. He compares the first 35 years of the period with the second 35 years. Three stations in the center had an increase of 15 percent, somewhat more in summer than in winter. Nine outlying stations showed no change. Ericksen[37] in Kiel was able to show that in an area with a general rainfall decrease toward the Baltic, the city itself was an isohyetal island with an annual increase in precipitation totals of 10 percent, nearly the same in all seasons.

Additional investigations in the St. Louis area[52] have raised further interesting questions. If one extends the rainfall analysis considerably beyond the urban confines, a notable anomaly in summer rainfall appears 15 km southeast of St. Louis, essentially to the lee, considering the prevailing winds. The conditions are reflected in the anomaly pattern shown in Figure 20.12. The time trend of the rainfall ratio of the urban area to Centerville, the lee side station most affected, shows: 1941, 1.0; 1953, 1.05; 1960, 1.1; and 1965, 1.2. Heavy summer rains > 50 mm in 24 hr show similar patterns with 43 cases at Centerville but only 25 in the city during a 20-year period. The airport, to the west of the city, during a similar time span had 344 thunderstorms that also affected the city, but single thunderstorms affected the airport only 67 times against 105 in the city. The question whether or not the leeward rainfall increases have man-made causes is now the target of a large re-

search project "Metromex" that may furnish the answers in a few years.

The occurrence of instability showers and their possible growth into thunderstorms over an urban area is, of course, not really surprising. The heat island and its accompanying unstable temperature stratification in the lower layers and ascending air currents can impart enough impetus to a slow-moving moist air mass to create a storm cell. The tetroon experiments with several hundred meters of added lift over New York City indicate values that could readily lead to thunderstorms in potentially unstable air. Individual case studies lend support to this hypothesis.

Parry[88] cites an occurrence for Reading, England. It is a prototype for

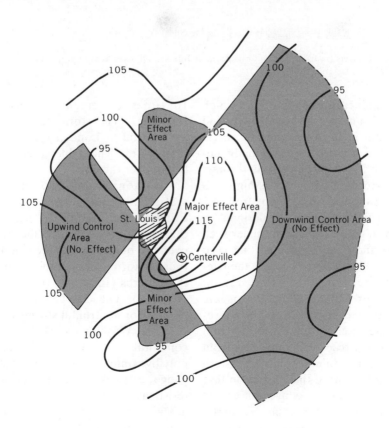

Figure 20.12 Summer rainfall departures (%) in the St. Louis area from that in a nearby undisturbed control area (reprinted by permission, from Huff et al.[52]).

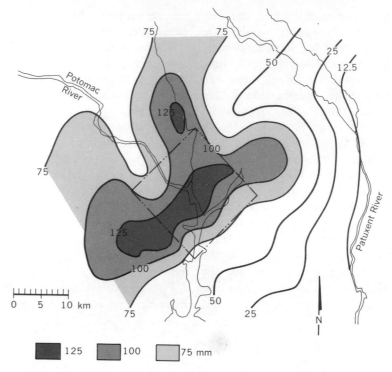

Figure 20.13   Isohyets (mm) of "urban" rainstorm of July 9, 1970 over Washington, D.C.

such an induction of rainfall, which yielded 34.5 mm in 2 hr over the city and only 2.5 mm in the rural area. It was a nonfrontal storm with conditionally unstable air to 2 km moving slowly through the area. Moisture content was high to 3 km, and instability was present from 4.5 to 8 km height. Only a small input is needed to initiate a rainstorm under such conditions and Parry suggested that even a 1°C heat island might become the triggering force. In the same region Atkinson[8] presented some mesoscale analyses for the greater London metropolis. In one case, not too dissimilar from Parry's the city received 68 mm rain from a thunderstorm while all of southeastern England had no more than 3 mm on that particular day. Surface convergence over the city was definitely present. Similarly, in another case, convergence rose from 0 to 20 × 10⁻⁵/sec and London had about

10 times as much rain as the surroundings. The storm formed about 30 km upwind of London with radar echoes to 6.5 km. These increased to 9.5 km height over the city.

Atkinson[7] could also show through statistical analysis that in a decade London had about 20 percent more thunderstorms than the suburbs and that over the built-up area precipitation from thunderstorms was notably higher than in the suburbs (1000 mm versus 600 mm in 10 years). The effect seems to be most notable for nonfrontal storms in summer.

In the Washington, D.C., area similar conditions have been spotted but analyzed only by comparisons within the dense rain gauge network.[53] These storms characteristically dump around 75 to 100 mm in a few hours over the metropolitan area, in many instances three to four times as much as in the rural area. A typical case is shown in Figure 20.13. The topographic setting alone would lead one to expect increased values to the west where elevations are 160 m higher than in the downtown area; however, we are confronted here with one of those unanswerable scientific questions: what would have happened if the city weren't there? Only a great many detailed case studies can answer this, and then only as a physicostatistical inference.

At present this very problem has to be faced also in the assessment of possible effects of anthropogenic pollutants on urban precipitation. In case of precipitation induced by the heat island, there is at least a very plausible physical model. For the pollutants our ideas are much less specific. The evidence is entirely circumstantial. The suspicion was voiced early by Ashworth[6] when he discovered an apparent influence of the day of the week on precipitation in Rochdale, England. In that industrial town weekdays had, on the average, 20 percent more precipitation than Sundays, compared to a nearby station where the difference was small. Ashworth attributed this to smoke and heat produced by factories. Since then others have found similar things. Recently Dettwiller[29] showed a weekly variation of precipitation for Paris, reproduced here in Figure 20.14. In the 8-year interval 1960–1967 there appears to be a gradual increase from Monday to Friday and a sharp drop in precipitation totals on weekends. Average for weekdays was 1.93 mm but only 1.47 mm on weekends, a decrease of 24 percent. The difference is

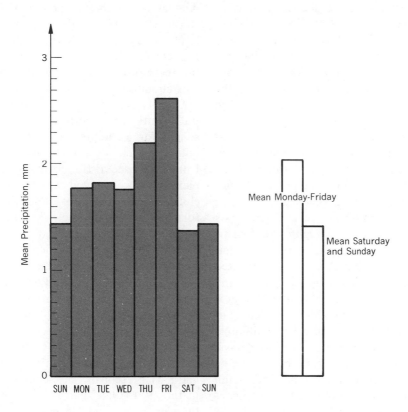

Figure 20.14   Average precipitation by day-of-week in Paris, France, 1960–1967 (reprinted by permission, from Dettwiller[30]).

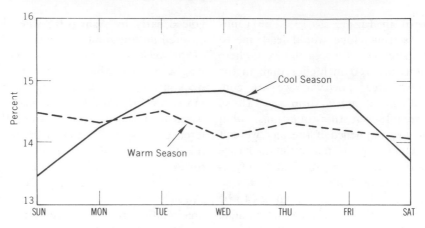

Figure 20.15 Average percent of total precipitation for each day of the week at 22 stations in the eastern 1912–1961 (reprinted by permission, from Frederick[40]).

statistically significant at the 95 percent confidence level. Four other towns in northern France showed weekday-weekend precipitation decreases from 14 to 32 percent. A study in the United States by Frederick[40] also found a weekly cycle. He used the much longer interval of 50 years (1912–1961) with observations from 22 stations in the eastern part of the country; however, the cycle shows only in winter data. The summer values essentially show only random fluctuations. The winter data, on the other hand, both in daily amounts and ranks of the days of the week show the preponderance of weekdays over weekends at better than 95 percent significance level. Frederick's findings are shown in Figure 20.15. This work seems to fit the hypothesis that urban precipitation in summer is affected in a random fashion by the heat island but that the winter excess on weekdays is stimulated, in some fashion, by seeding of supercooled clouds with urban pollutants that are more prevalent on workdays. How tenuous such hypotheses are is shown by a study of Lawrence of London conditions.[67] He finds a fairly significant weekly cycle for that city over a 20-year span with a correlation of the difference of Thursday (highest value) to Sunday

(lowest value) with the urban daily sunshine duration divided by wind speed during the warm months May to July. And although the maximum of Sunday sunshine and large value for the heat island development might well be dependent on low pollutant concentrations, the reasoning to rainfall effects becomes very circuitous.

Direct evidence for influences of pollutants on precipitation has been sought for an appreciable time. Stout[103] noted low clouds above and downwind from the Tuscola, Illinois industrial complex. Once he noted a cumulonimbus cloud reaching to the 7-km level, with formation of a funnel. Several times radar echoes were noted from these clouds. More than coincidence to the effluents was suggested. He also mentioned a rainfall anomaly at LaPorte, Indiana. That island of excessive rainfall has since become a classic case of scientific controversy. It was placed into this category by a paper by Changnon[17] who reported that the little town of LaPorte, Indiana had a 31 percent increase of precipitation in the 1951–1965 span, 38 percent more thunderstorms, and 246 percent more hail than surrounding stations. The town is 48 km east of the large industrial complex around South Chicago, Illinois-Whiting, Indiana, where steel production had increased manifold.

Sceptics disagreed with Changnon's acceptance of the basic observations at LaPorte,[48] which they contended contained observer errors and held that his interpretations were synoptically not defensible. A recording rain gauge installed in 1963 failed to show the increases allegedly observed in the previous decades. In a reply Changnon[19] defended his analysis and tried to bolster his argument with noninstrumental observations at LaPorte. Hidore[45] entered the fray with an analysis of runoff data, 1926–1960, of the Kankakee River supporting the precipitation anomaly at LaPorte, located in the upper watershed of the river. This was followed by an exchange between him and Holzman,[49] who argued that observations at the river-gauging stations fail to verify the LaPorte anomaly. Finally, Ashby and Fritts[5] made an attempt to check the anomaly by tree-ring analysis with other trees in the Chicago and Indiana area as controls. They excluded the "suspect" LaPorte precipitation record from the analysis. A large part of the variance, 59 percent, in white oak growth near LaPorte could be related to general climatic fluctuations. But the procedure neither proved nor disproved the LaPorte anomaly. It suggests that some factor in the LaPorte area became increasingly more limiting to tree growth than climate since 1940, and that factor might well have been air pollution directly affecting the tree growth. The case must be put to rest here. One can only remark that it is generally a difficult task to prove anything pro and con about rainfall trends with a single rain gauge.

On a much larger scale Hobbs et al.[47] attempted to show changes in rainfall regime in the whole state of Washington between 1929–1946 and 1947–1966. Particularly notable were new islands of higher precipitation totals which these authors attribute to increased cloud condensation nuclei from industrial plants. They claim that the rapid rise of the pulp and paper industry since World War II in the state is a cause, and estimate the cloud condensation nuclei production to be as high as 10^{19}/sec. The cloud condensation nuclei are not identical with the Aitken nuclei but include only those which act as nuclei at saturation or very low supersaturation. Yet it is not at all clear that large numbers of these nuclei, while easily leading to cloud formation, necessarily help in the formation of rain. They might do this if they are highly hygroscopic and contribute to the growth of the small droplets of > 20 μm diameter. On the other hand, an oversupply of these nuclei might well lead to a large number of very small droplets per unit volume, a circumstance that might stabilize the cloud and inhibit rainfall.

There is no doubt that large cities contribute greatly to nuclei concentrations not only at the surface but also aloft. From airborne measurements, upwind and downwind of Denver, Colorado, Squires[100] obtained a production of 10^5 Aitken nuclei/cm² sec in the city area. About 5×10^3 of these act as cloud condensation nuclei at 0.5 percent supersaturation. This, for any meteorological event, is an oversupply. The critical question of what substances are present and how they might affect condensation, coalescence, and freezing processes has not yet been answered.

Although most of the previously reported urban precipitation studies have stressed increases—for one reason or another—in or near the urban area, one should not overlook at least one case of contrary behavior (again with the reservations necessary for single-rain gauge evidence). That was reported by Spar and Ronberg,[99] who noted a decrease in precipitation at the Central Park Station in the middle of Manhattan, New York. In contrast to regional trends at 10 surrounding stations, a 35-year record there (1926–1961) showed a decrease of 5 mm/year. If real, is this a case of overnucleation or overseeding? The answer is open. We have only the work of Schaefer,[93, 94] which cogently points out that city air contains highly effective freezing nuclei, but a superabundance of these and other nuclei that may travel dozens or even hundreds of kilometers and could well create cloud effects far downwind. Some have occasionally been observed downwind from power plants with a combination of stacks emitting, among other pollutants, SO_2 and cooling towers supplying large amounts of water vapor. These have not only caused occasional local fogs but cloud plumes tens of kilometers long.

Presently, until many mysteries of atmospheric chemistry and cloud physics are clarified, the most convincing approach to inadvertent man-made cloud modification and rainfall is still the case study where events happen in urban areas that are not to be expected by synoptic or climatological circumstances. A few such cases have been documented but there are undoubtedly many that have escaped attention.

There are now several convincing cases in the literature, however. The first was reported by J. V. Kienle[59] for the heavily industrialized area of Mannheim-Ludwigshafen, Germany. There snowfall was observed on January 28 and 30, 1948, in a synoptic situation when chances for such an occurrence were very low. Winds were calm and with surface air temperatures of $-4°C$ and an intense inversion at 500 m with upper temperatures above freezing, a dense ground fog and stratus formed over the cities with skies elsewhere in the area completely clear. Light snow fell from the fog for about 4 hr each on the 2 days. Accumulation was only about 6 mm. The fog and cloud formation was evidently promoted by urban effluents and the snow resulted very probably from seeding of the supercooled fog with suitable, anthropogenic nuclei.

Culkowski[24] observed on December 22, 1960 an anomalous snowfall downwind from the cooling towers of the gaseous diffusion plant 15 km southwest of Oak Ridge, Tennessee. This snow fell for 4 hr (0800 to 1200 hours) 5 to 8 km from the towers which at the time used 7.6×10^7 liter/day of water, most of which was evaporated into the air. This water condensed and formed low stratus at a 100-m height and a few scattered cumulus clouds. No other clouds were present outside the affected area. These clouds must have been definitely supercooled with temperatures at 0500 of $-13.9°C$, dew point $-16.1°C$; and at 1200 of $-8.3°C$, dew point $-11.7°C$. The wind was southwest at 3 m/sec. Culkowski suspected that a ferromanganese plant at Rockwood, Tennessee, 29 km upwind from the gaseous diffusion plant, may have furnished the freezing nuclei. A snow depth of 2.5 mm was noted on an area of 6.5 km² with an estimated water equivalent 1.6×10^6 liters or about 12 percent of the water vapor output of the cooling towers in the 4-hr interval.

Another light snowfall, apparently induced by effluents from a coal-fired power plant, was observed by Agee[1] on January 11, 1971 at Lafayette, Indiana. It occurred in a supercooled ground fog ($-12.8°C$). A total of 6 mm accumulated on the ground downwind from the stacks. Agee suggests that aluminum oxide (Al_2O_3) found in the power plant effluent, forming freezing nuclei active at $-6.5°C$, could have been the inducing agent.

Occasionally a whole city may act as the source of freezing nuclei, or so it would seem. Potter[90] noted that the 1959–1960 cold season snowfall in the

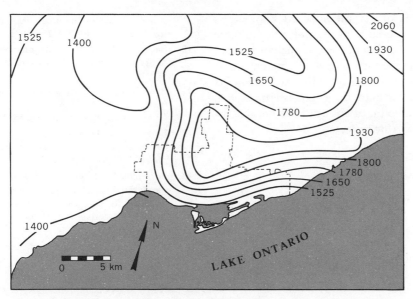

Figure 20.16   Seasonal depth of snowfall (cm) in the Toronto, Canada area in the 1959–60 season, showing a "suspicious" city plume (reprinted by permission, from Potter[90]).

Toronto, Canada region was markedly different from the normal climatic pattern. That pattern shows the usual increase of total snow northward, inland from the lake with a southwest to northeast gradient at the west end of the city, with increases to the north and northeast from 100 to 140 cm. In the anomalous year heaviest snowfall (211 cm) was in the center of the city and a well-developed "plume" extended from west (152 cm) to east (203 cm), as shown in Figure 20.16. Here again, it is difficult to "prove" with scientific specificity that the urban heat island and effluents were the causes for this snowfall distribution, but one can at least suspect such an influence.

CONCLUSIONS

It has been established beyond reasonable doubt that urban agglomerations cause measurable changes in the atmosphere immediately adjacent to them. Temperatures are increased, low-level lapse rates steepened, horizontal winds slowed, and updrafts induced. Turbulence and cloud formation are increased, summer rainfall is enhanced, and possibly, some winter snowfall is stimulated. Snow on the ground is diminished and so are surface-near humidities. Most apparent is the increase in pollutants from one to several orders of magnitude. They reduce solar radiation intensity, eliminate all the short-wave and a substantial portion of the long-wave ultraviolet, and shorten sunshine duration. Their effect on cloud formation and rainfall over and in the vicinity of the cities is still somewhat uncertain but evidence points to occasional cases of stimulation of precipitation and perhaps some rare cases of inhibition.

Use of urban meteorological models is making some progress and permits assessment of time-dependent developments of low-level atmospheric parameters.

Although city pollutant plumes have occasionally been followed for long distances of several hundred kilometers, there is presently no sign of other notable effects of city influence on meteorological variables beyond a few kilometers or at most several tens of kilometers. But as cities grow into large conurbations one can foresee that they will have notable regional weather effects.

ACKNOWLEDGMENT

Work on which this review is based has been sponsored, in part, by the National Science Foundation under Grant No. GA-29304x.

REFERENCES

1. Agee, E. M., An artificially induced local snowstorm. *Bull. Am. Meteorol. Soc.,* **52,** 557–560, 1971.

2. Ahlquist, N. C., and R. J. Charlson, Measurement of the vertical and horizontal profile of aerosol concentration in urban air with the integrating nephelometer, *Environ. Sci. Technol.,* **2,** 363–366, 1968.

3. Altshuller, A. P., Composition and reactions of pollutants in community atmospheres, Urban Climatology, *World Meteorol. Org. Tech. Note 108,* 179–193, 1970.

4. Angell, J. K., D. H., Pack, and W. H. Hoecker, Urban influence on nighttime air flow estimated from tetroon flights, *J. Appl. Meteorol.* **10,** 194–204, 1969.

5. Ashby, W. C. and H. C. Fritts, Tree growth, air pollution, and climate near LaPorte, Ind., *Bull. Am. Meteorol. Soc.,* **53,** 246–251, 1972.

6. Ashworth, J. R., The influence of smoke and hot gases from factory chimneys on rainfall, *Quart J. Royal Meteorol. Soc.* **55,** 341–350, 1929.

7. Atkinson, B. W., A further examination of the urban maximum of thunder rainfall in London, 1951–1960, *Inst. Brit. Geogr. Trans. Papers Publ.* **48,** 97–117, 1969.

8. Atkinson, B. W., The effect of an urban area on the precipitation from a moving thunderstorm, *J. Appl. Meteorol.* **10,** 47–55, 1971.

9. Atwater, M. A., The radiation budget for polluted layers of the urban environment, *J. Appl. Meteorol.* **10,** 205–214, 1971.

10. Bach, W., An urban circulation model. *Arch. Meteorol. Geophys. Bioklimatol. Ser. B* **18,** 155–168, 1970.

11. Barrett, E. C., Local variations in rainfall trends in the Manchester region, *Inst. Brit. Geogr. Trans. Papers Publ.* **35,** 55–71, 1964.

12. Bornstein, R. D., Observations of the urban heat island effect in New York City, *J. Appl. Meteorol.* **7,** 575–585, 1968.

13. Bornstein, R. D., Observed urban-rural wind speed differences in New York City, *EOS (TAGU)* **50,** 626, 1969. Paper presented at AGU Fall Meeting, Dec. 15–18, 1969.

14. Buchan, W. E. and R. J. Carlson, Urban haze: the extent of automotive contribution, *Science* **159,** 192–194, 1968.

15. Chandler, T. J., The climate of London, Hutchison, London, 1965, 292 pp.

16. Chandler, T. J., Selected bibliography on urban climate, *World Meteorol. Org. No. 276, TP 155,* Geneva, 1970, 383 pp.

17. Changnon, S. A., Jr., The LaPorte weather anomaly—fact or fiction? *Bull. Am. Meteorol. Soc.* **49,** 4–11, 1968.

18. Changnon, S. A., Jr., Recent studies of urban effects on precipitation in the United States, Urban Climatology, *World Meteorol. Org. Tech. Note 108,* 325–341, 1970.

19. Changnon, S. A., Jr., Reply, *Bull. Am. Meteorol. Soc.* **51,** 337–342, 1970.

20. Chaudhry, F. H. and J. E. Cermak, Simulation of flow and diffusion over an urban complex, *Conference on Air Pollution Meteorology, Raleigh, N.C., April 5–9, 1971,* American Meteorological Society, 1971, pp. 126–131.

21. Chow, T. J. and J. L. Earl, Lead aerosols in the atmosphere: increasing concentrations, *Science* **169,** 577–580, 1970.

22. Clarke, J. F. and J. L. McElroy, Experimental studies of the nocturnal urban boundary layer, in "Urban Climatology", *World Meteorol. Org. Tech. Note* **108,** 108–112, 1970.

23. Collier, C. G., Fog at Manchester, *Weather* **25,** 25–29, 1970.

24. Culkowski, W. M., An anomalous snow at Oak Ridge, Tennessee, *Monthly Weather Rev.* **90,** 194–196, 1962.

25. Davenport, A. B., The relationship of wind structure to wind loading, *Proceedings of Conference on Wind Effects on Structures,*

Ntl. Phys. Lab. Her M. Stat. Off. London Sympos., 16, Vol. I, 1965, pp. 53–102.

26. Davis, F. K., The air over Philadelphia, in "Air over Cities", *U.S. Public Health Serv. SEC Tech. Rep't. 62–5,* 115–126, 1961.

27. Davitaya, F. F., Principles and methods of agricultural evaluation climates, in "Agrometeorological Problems", abbrev. reports to 2nd session CAg Met., World Meteorol. Org., Moscow, 1958.

28. Delage, Y. and P. A. Taylor, Numerical studies of heat island circulations, *Boundary Layer Meteorol.* **1**, 201–226, 1970.

29. Dettwiller, J., Deep soil temperature trends and urban effects at Paris, *J. Appl. Meteorol.* **9**, 178–180, 1970.

30. Dettwiller, J., Incidence possible de l'activité industrielle sur les précipitations à Paris, in Urban Meteorology, *World Meteorol. Org. Tech. Note 108,* 361–362, 1970.

31. Dirmhirn, I. and F. Sauberer, Das Strassenklima von Wien, Chap. IV in Pt. III of Steinhauser et al., *Klima and Bioklima von Wein,* 1959, pp. 122–135.

32. Dronia, H., Der Städteinfluss auf den weltweiten Temperaturtrend, *Meteorol. Abh.* **74**, (4), 1–68, 1967.

33. Duckworth, F. S. and J. S. Sandberg, The effect of cities upon horizontal and vertical temperature gradients, *Bull. Am. Meteorol. Soc.* **35**, 198–207, 1954.

34. East, C., Comparison du rayonement solaire en ville et la campagne. *Cahiers Geogr. Quebec* **12**(25), 81–89, 1968.

35. East, C., Chaleur urbaine a Montréal, *Atmosphere* **9**, 112–122, 1971.

36. Egan, B. A. and J. R. Mahoney, Numerical modeling of advection and diffusion of urban area source pollutants, *J. Appl. Meteorol.* **11**, 312–322, 1972.

37. Eriksen, W., Beiträge zum Stadtklima von Kiel, *Schrift. J. Geogr. Inst. Univ. Kiel* **22**(1), 1–218, 1964.

38. Eriksen, W., Das Stadtklima, seine Stellung in der Klimatologie und Beiträge zu einer witterungsklimatologisckhen Betrachtungsweise *Erdkunde* **18**, 257–266, 1964.

39. Fischer, William H. and G. E. Sturdy, The contribution of a city to atmospheric turbidity and the turbidity background, *Atmos. Environ.* **5**(7), 561–563, 7 refs., 1971.

40. Frederick, R. H., Preliminary results of a study of precipitation by day-of-the-week over Eastern United States, *Second National Conference on Weather Modification,* American Meteorological Society, 1970, pp. 209–214.

41. Garnett, A., A survey of air pollution in Sheffield under characteristic winter anticyclonic conditions, *Int. J. Air Water Pollut.* **7**, 963–968, 1963.

42. Gildersleeves, P. B., A contribution to the problem of day-darkness over London, *Meteorol. Mag.* **91**, 356–369, 1962.

43. Grillo, J. N. and J. Spar, Rain-snow mesoclimatology of the New York metropolitan area, *J. Appl. Meteorol.* **10**, 56–61, 1971.

44. Heuseler, H., Extreme Temperatur Differenzen Stadt-Land in Strahlungsnächten, *Umschau* **65**(2), 60–61, 1965.

45. Hidore, J. F., The effects of accidental weather modification on the flow of the Kankakee River, *Bull. Am. Meteorol. Soc.* **52**, 99–103, 1971.

46. Hilst, G. R. and N. E. Bowne, Diffusion of aerosols released upwind of an urban complex, *Environ. Sci. Technol.* **5**, 327–333, 1971.

47. Hobbs, P. V. and L. F. Radke, and S. W. Shumway, Cloud condensation nuclei from industrial sources and their apparent influence on precipitation in Washington State, *J. Atmos. Sci.* **27**, 81–89, 1970.

48. Holzman, B. G., La Porte precipitation fallacy, *Science* **171**, 847, 1971.

49. Holzman, B. G., More on the La Porte fallacy (with reply by J. F. Hidore), *Bull. Am. Meteorol. Soc.* **52**, 572–574, 1971.

50. Horvath, H., A comparison of natural and urban aerosol distribution measured with the aerosol spectrometer, *Environ. Sci. Technol.* **1**, 651–655, 1967.

51. Howard, L., *The Climate of London,* Vol. I, London, 1818.

52. Huff, F. A., S. A., Changnon, Jr., and T. A. Lewis, Climatological assessment of urban effects on precipitation, *Conference on Air Pollution Meteorology, Raleigh, N.C., April 5–9, 1971,* American Meteorological Society, 98–103, 1971.

53. Hull, B. B., Once-in-hundred-year rainstorm, Washington, D.C., 4 September 1939, *Weatherwise* **10**, 128–131, 139, 1957.

54. Hutcheon, R. I., R. H. Johnson, W. P. Lowry, C. M. Black, and D. Hadley, Observations of the urban heat island in a small city, *Bull. Am. Meteorol. Soc.* **48**, 7–9, 1967.

55. Jenkins, I., Increase in averages of sunshine in central London, in Urban Climates, *World Meteorol. Org. Tech. Note 108*, 292–294, 1970.

56. Jenkins, I., Decrease in the frequency of fog in central London. *Meteorol. Mag* **100**(1192), 317–322, 1971.

57. Keith, Ralph W., Downtown Los Angeles noon visibility trends, in *Conference on Air Pollution Meteorology, April 5–9, 1971, Raleigh, N.C.*, Preprints, American Meteorological Society, Boston, 1971, pp. 85–91.

58. Kessler, A., Über den Tagesgang von Oberflächentemperaturen in der Bonner Innenstadt an einem sommerlichen Strahlungstag, *Erdkunde* **25**, 13–20, 1971.

59. Kienle, J. v., Ein stadtgebundener Schneefall in Mannheim, *Meteorol. Rundschau* **5**, 132–133, 1952.

60. Kimosita, T. and T. Sonda, Change of runoff due to urbanization, *UNESCO, International Symposium on Floods and their Computation*, preprint, 1967, 8 pp.

61. Kratzer, A., *Das Stadtklima, Die Wissenschaft*, Vol. 90, 2nd ed., Vieweg und Sohn, Braunschweig, 1956, 184 pp.

62. Kung, E. C., R. A. Bryson, and D. H. Lenchow, Study of continental surface albedo on the basis of flight measurements, *Monthly Weather Rev.* **92**, 543–564, 1964.

63. Landsberg, H. E., The climate of towns, in W. L. Thomas, Ed., *Man's Role in Changing the Face of the Earth*, University of Chicago Press, 1956, pp. 584–603.

64. Landsberg, H. E., Air pollution and urban climate, in *Biometeorology, Proceedings of the 3rd International Biometereology Congress*, **2**, Pt. 2, 1963, pp. 648–656.

65. Landsberg, H. E., Micrometeorological temperature differentiation through urbanization, in Urban Climates, *World Meteorol. Org. Tech. Note 108*, 129–136, 1970.

66. Landsberg, H. E. and T. N. Maisel, Micrometeorological observations in an area of urban growth, *Boundary Layer Meteorol.* **2**, 365–370, 1972.

67. Lawrence, E. N., Day-of-the-week variation in weather, *Weather* **26**, 386–391, 1971.

68. Leahy, D. M. and J. P. Friend, A model for predicting the depth of the mixing layer over an urban heat island with applications to New York City, *J. Appl. Meteorol.* **10**, 1162–1173, 1971.

69. Lettau, H., Note on aerodynamic roughness-element description, *J. Appl. Meteorol.* **8**, 828–832, 1969.

70. Lewis, J. E., Jr., F. W. Nicholas, S. M. Scales, and C. A. Woollum, Some effects of urban morphology on street level temperatures at Washington, D.C., *Wash. Acad. Sci.* **81**, 258–265, 1971.

71. Lindquist, S., Studies on the local climate in Lund and its environs, *Lund Studies Geogr. Ser. A*, **42**, 79–93, 1968.

72. Löbner, A., Horizontale und vertikale Staubverteilung einer Grosstadt, *Veröffentl Geophys. Inst. Univ. Leipzig Ser. 2*, **7**(2), 1–99, 1935.

73. Lorenz, D., Messungen der Bodenoberflächentemperatur vom Hubschrauber aus, *Ber. Deut. Wetterd* **11** (82), 1–29, 1962.

74. Lorenz, D., Temperature measurements of natural surfaces using infrared radiometers, *Appl. Optics* **7**, 1705–1710, 1968.

75. Ludwig, F. L. and J. H. S. Kealoha, Urban climatological studies, *Stanford Res. Inst. Final Rep., SRI Proj. MU 6300-140*, 1–195, 1968.

76. Maisel, T. N., Measurements of the horizontal convergence caused by the heat island of a paved airfield, *Inst. Fluid Dyn. Appl. Math. Tech. Note BN 735*, 1–9, 1972.

77. McCormick, R. A. and D. M. Baulch, The variation with height of the dust loading over a city as determined from the atmospheric turbidity, *J. Air Pollut. Control Assoc.* **12**, 492–496, 1962.

78. McCormick, R. A. and K. R. Kurfis, Vertical diffusion of aerosols over a city, *Quart J. Royal Meteorol. Soc.* **92**, 392–396, 1966.

79. Middleton, J. F. K., *Visibility in Meteorology*, 2nd ed., University of Toronto Press, 1941, 165 pp.

80. Miller, M. E., N. L. Canfield, T. A. Ritter, and C. R. Weaver, Visibility changes in Ohio, Kentucky, and Tennessee from 1962 to 1969, *Monthly Weather Rev.* **100**, 67–71, 1972.

81. Mörick, J., Probleme der Verhütung von Luftverunreinigungen in Ungarn, *Angew. Meteorol.* **4**, 274–279, 1963.

82. Moffitt, B. J., The effects of urbanization on mean temperatures at Kew Observatory, *Weather* **27**, 121–129, 1972.

83. Monteith, J. L., Local differences in the attenuation of solar radiation over Britain, *Quart. J. Royal Meteorol. Ser.* **92**, 254–262, 1966.

84. Moyers, J. L., W. H. Zoller, and R. A. Duce,

Gaseous iodine measurements and their relationship to particulate lead in a polluted atmosphere, *J. Atmos. Sci.* **28**, 95–98, 1971.

85. Oke, T. R., Towards a more rational understanding of the urban heat island, *McGill Univ. Climatol. Bull.* **5**, 1–20, 1929.

86. Oke, T. R., City size and the urban heat island, *Atmos. Environ.*, **7**(8), Aug. 1973, pp. 769–779.

87. Oke, T. R. and F. G. Hannell, The form of the urban heat island in Hamilton, Canada, in Urban Climatology, *World Meteorol. Org. Tech. Note 108*, 113–126, 1970.

88. Parry, M., An "urban rainstorm" in the Reading area, *Weather* **11**, 41–48, 1956.

89. Parry, M., Sources of Reading's air pollution, in "Urban Climates," *World Meteorol. Org. Tech. Note* **108**, 295–305, 1970.

90. Potter, J. G., Areal snowfall in Metropolitan Toronto, *Meteorol. Branch, Dept. Transport, CiR-3431 TEC-342*, 1961, 9 pp.

91. Richter, B. L., Verlauf von Temperatur und der Feuchte an einer Reihe won heiteren Tagen des Sommers 1968 in der Innenstadt und in den Aussenbezirken West-Berlin, *Meteorol. Abh.* **110**,(3), 1–52, 1969.

92. Roach, W. T., Some aircraft observations of fluxes of solar radiation in the atmosphere, *Quart. J. Royal Meteorol. Soc.* **87**, 346–363, 1961.

93. Schaefer, V. J., Ice nuclei form auto exhaust and iodide vapor, *Science* **154**, 1555–1557, 1966.

94. Schaefer, V. J., Ice nuclei from auto exhaust and organic vapors, *J. Appl. Meteorol.* **7**, 148–149, 1968.

95. Schmidt, F. H. and J. H. Boer, Local circulation around an industrial area, *Ber. Deut. Wetterd.* **12**(91), 28–31, 1963.

96. Sheleikhovskii, G. V., Smoke pollution of towns, English translation (1961)—Israel Program F. Scient. Transl. f. NSF-USDoc., 1949, 203 pp.

97. Shellard, H. C., The frequency of fog in the London area compared with that in rural areas of East Anglia and southeast England, *Meteorol. Mag.* **88**, 321–323, 1959.

98. Slade, D. H., Modeling air pollution in the Washington, D.C. to Boston megalopolis, *Science* **157**, 1304–1307, 1967.

99. Spar, J. and P. Ronberg, Note on an apparent trend in annual precipitation at New York City, *Monthly Weather Rev.* **96**, 169–171, 1968.

100. Squires, P., An estimate of the anthropogenic production of cloud nuclei, J. Rech. Atmos. 297–308, 1966.

101. Steinhauser, F., Ergebnisse mehrjähiger Beobachtungen der Staubablagerung in Österreich, *Wetter und Leben* **23**, 89–102, 1971.

102. Stern, Arthur, C., Ed., *Proceedings of Symposium on multiple-source urban diffusion, Air Pollution Control Office, Publ. No. AP-86*, Washington, D.C., 1970.

103. Stout, E., Some observations of cloud initiation in industrial areas, in "Air over Cities" *U.S. Public Health Serv. SEC Tech. Rept. 62-5*, 147–152, 1961.

104. Stummer, G., Klimatische Untersuchungen in Frankfurt am Main und seinen Vororten, *Ber. Meteorol.-Geophys. Inst., Frankfurt*, **5**, 1939.

105. Summers, P. W., Smoke concentrations in Montreal related to local meteorological factors, in Air over Cities, *U.S. Public Health Serv. SEC Tech. Rept. 62-5*, 1961, pp. 89–112.

106. Thomas, M. D., Sulfur dioxide, sulfuric acid aerosol and visibility in Los Angeles, *Int. J. Air Water Pollut.* **6**, 443–454, 1962.

107. Turner, D. B., A diffusion model for an urban area, *J. Appl. Meteorol.* **3**, 83–91, 1964.

108. Turner, D. B., The diurnal and day-to-day variations of fuel usage for space heating in St. Louis, Missouri, *Atmos. Environ.* **2**, 339–351, 1967.

109. Uthe, E. E., Wider observations of the urban aerosol structure, *Bull. Am. Meteorol. Soc.* **53**, 358–360, 1972.

110. Vilkner, H., Die Nachttemperatur am Erdboden einer Stadt, *Z. Meteorol.* **15**, 141–147, 1961.

111. Volz, F. E., Some results of turbidity networks, *Tellus* **21**, 626–630, 1969.

112. Vukovich, F. M., Theoretical analysis of the effect of mean wind and stability on a heat island circulation characteristic of an urban complex, *Monthly Weather Rev.* **99**, 919–926, 1971.

G
OTHER
PROBLEMS

If cloud seeding produces rain to the benefit of many farmers, should they pay for the added water, and, if so, how? Who should decide if a weather modification project should be allowed to go ahead? Who should decide whether a hurricane approaching Florida should be seeded or not? If such a hurricane is seeded, should it become labeled "Property of U.S. Government"? Even if the damage from the storm is reduced, some property will probably be lost. How should the government respond to the very probable

lawsuits resulting from this damage? It may turn out that it is easier to develop ways to modify weather beneficially than to develop ways to apply this knowledge. If seeding a cloud produces severe weather, then people who suffer damage should probably be able to collect damages from the seeder. The problems of proving causality here are not trivial.

Chapters 21 and 22 cover these social and legal problems and are written by specialists actively working on these problems.

21 | Weather Modification Litigation and Statutes

RAY J. DAVIS

THE LAWSUITS

Dan Ming, Henry Hooker, and the Judicial Process

During Arizona pioneer days Henry Hooker controlled a range of nearly 30 mi². One fellow cattleman who was no admirer of Hooker was "Big Dan" Ming. During a severe drought in 1885 it was suggested at a cattlemen's meeting that they should follow the example of their Indian neighbors and pray for rain. Dan was called upon. After having the men remove their hats, Big Dan proceeded:

Oh Lord, I'm about to round you up for a good plain talking. Now, Lord, I ain't like these fellows who come bothering you every day. This is the first time I ever tackled you for anything, and if you will only grant this, I promise never to bother you again. We want rain, Good Lord, and we want it bad; we ask you to send us some. But if you can't or don't want to send us some, then for Christ's sake don't make it rain up around Hooker's or Leitch's ranges, but treat us all alike. Amen.

The judiciary's goal in dealing with litigation over artificial rainmaking is to treat all like cases alike.

Ray J. Davis is a Professor in the College of Law, University of Arizona, Tucson, Arizona.

When faced with the responsibility of deciding lawsuits, judges look initially to constitutions and statutes for authoritative rules. Statutes dealing directly with cloud seeding concern themselves primarily with the relationship of weather modifiers to the government, not with their interaction with other individuals; the latter is the setting for most weather modification litigation. Judges must therefore look to the common law, their own past rulings, and prior decisions of other judges. Their first step is to examine the precedents and compare them with the case at hand. The doctrine of *stare decisis*, following precedents from similar cases, is the foundation of the common law system.

When lawsuits involving use of weather technology first reached the courts in the 1950s there were of course no precedents directly on point. Lacking authoritative weather modification statutes and cases, the judges looked for analogous situations and then adapted to weather modification the rules which had earlier been applied to supposedly similar cases. The judges, however, did not agree as to which analogies were most apt. They arrived at differing results.

By molding existing law to new situations the judiciary in effect makes new law. Judges must and do legislate, "but they can do so only interstitially; they are confined from molar to molecular motions." They deal with precedents and reach decisions only upon these facts established in the case before them. Consequently there are gaps in the spectrum of judge-made law areas where cases have not yet arisen in which legal norms can be created. Such is the case with weather technology litigation. Less than a dozen lawsuits have been litigated and in only half of them have there been written opinions filed by the judges that other courts can examine to find precedents.

So we have little authority in the present common law bearing directly upon cloud seeding. It is not only scattered but also contradictory. In future cases, though, judges will examine the precedents in an effort to satisfy the desires of litigants who, like Dan Ming, seek uniformity of treatment. Accordingly the few cases we have may well form the basis of future weather modification law.

Whose Cloud Was That You Rustled?

Over 300 years ago an English judge wrote that a man's land extends from Hell to Heaven. So long as claims concerning interference with this pie-shaped domain involved only overhanging eaves, tree limbs, and shooting across the land, the courts had no real trouble applying the rule. The advent of travel by aircraft through space unusable by the landowner caused judges to reevaluate their pronouncements about ownership of the atmosphere. Now the aircraft operator is liable to the landowner only if his flight interferes substantially with the use and enjoyment of the land beneath.

Cloud seeding presented the courts with another new fact pattern, a so-called "unprovided case." The air flight cases

were not analogous because passage extracts nothing from the skies; precipitation enhancement is designed to accomplish that. The old Heaven to Hell ownership theory had been developed centuries before anyone regarded clouds as a resource; it could be further modified in view of the new technological capabilities. Judges have indicated their attitudes about ownership of atmospheric waters in three of the lawsuits that have been tried. One case affirms individual ownership in the landowner, another refers to cloud moisture as common property, and a third dismisses the whole notion of property rights in clouds.

The defendants in the *Southwest Weather Research* litigation were employed by a group of west Texas farmers to suppress hail by cloud seeding. Ranchers, asserting that the seeding destroyed rain clouds over their property, obtained an injunction in a local court commanding the defendants to refrain from seeding "over plaintiffs' lands and in the area of plaintiffs' lands" pending a final hearing. The defendants appealed to the Texas Court of Civil Appeals, which handed down its judgment in 1958.

The appellants protested the issuance of the injunction on several grounds. They asserted that they had every right to seed to protect their crops from hail, that the ranchers had no right to prevent them from flying over the ranchers' lands, that no one owns the clouds unless it be the state, and that the injunction was too broad in its terms. The key issue was whether the landowners were entitled to the precipitation that would naturally fall from the clouds. Or, to put it another way, were the seeders rustling the ranchers' clouds? The court ruled that they were, and stated:

We believe that under our system of government the landowner is entitled to such precipitation as Nature deigns to bestow. We believe that the landowner is entitled, therefore and thereby, to such rainfall as may come from clouds over his property that Nature, in her caprice, may provide. It follows, therefore, that this enjoyment of or entitlement to the benefits of Nature should be protected by the courts if interfered with improperly and unlawfully.

The farmers and their cloud seeder did prevail on one issue: the injunction was too sweeping in its terms. The ban on seeding in the area of the plaintiffs' lands was lifted. The Texas Supreme Court upheld this action by the intermediate appellate court and construed the injunction as limited to only those cloud seeding activities conducted over lands that the plaintiffs owned and that directly affected such lands. In Texas the landowner's atmospheric rights do not extend beyond the vertical boundaries of his land.

In 1968 a Pennsylvania judge considered the Texas decision and looked at an earlier New York case, the one denying individual rights in the moisture of the skies, and proceeded to adopt a position part way between them. It was his opinion that "clouds and the moisture in the clouds, like air and sunshine, are part of space and are common property belonging to everyone who will benefit from what occurs naturally in those clouds." He then, however, went on to lean in the direction of the *Southwest Weather Research* cases by affirming that "every landowner has a property right in the clouds and in the water in them."

Ownership claims to the waters of the rivers of the skies can be analogized to assertions of rights in surface and ground waters. A landowner would argue that he is a riparian, someone whose property abuts the stream. In most American states, water rights have depended upon ownership of such riparian lands. Surface water law, though, is applied to "watercourses." Atmospheric waters do not flow in channels and are not contained by banks. The analogy does not fit.

The "prior appropriation" doctrine of water rights which applies to surface and ground waters in most western states would, if applied to atmospheric waters, tend to support the claims of weather modifiers to ownership of water harvested from clouds treated by them. According to that doctrine, priority of rights is based upon priority of time of diversion. The weather modifier who acted first would acquire a vested water right. The doctrine of prior appropriation, though, was invented during the nineteenth century when courts wished to encourage exploitation of natural resources. Today's temper is quite different. What was good law for Gold Rush times will not necessarily be extended into the latter half of the twentieth century.

Ten Million New Yorkers Need Pure Water

In 1950 a trial court in New York City handed down the first written opinion on the subject of scientific weather modification. In that case, *Slutsky v. City of New York*, the owners of a resort in the Catskills sought an injunction to restrain the city from engaging in efforts to induce rain artificially. They asserted that actual or threatened rainfall would harm the resort business and that runoff from augmented rains would damage lands along streams in the area. The city presented witnesses who testified that drought conditions coupled with the delay of its water development projects during World War II had created a water emergency. The judge denied the motion for a temporary injunction.

In the opinion, the court first dismissed any notion of vested property rights in the clouds or the moisture in them, then it stated that there had been no proof by the plaintiffs that they would in fact be harmed, and finally it expressed its view on the right of the city to embark upon precipitation enhancement.

This court must balance the conflicting interests between a remote possibility of inconvenience to plaintiffs' resort and its guests with the problem of maintaining and supplying the inhabitants of the City of New York and surrounding areas, with a population of about 10 million inhabitants, with an adequate supply of pure and wholesome water. The relief which plaintiffs ask is opposed to the general welfare and public good; and the dangers which plaintiffs apprehend are purely speculative. This court will not protect a possible private injury at the expense of a positive public advantage.

The city had the right to seed clouds.

It is important to the outcome of *Slutsky* that the judge determined that there were no individual cloud property rights. If landowners hold title to clouds above their properties, the right to seed such clouds would be theirs. It is possible that courts adopting this position might permit someone else to engage in cloud seeding to protect their property from storms or droughts even though in doing so they will alter the clouds belonging to the holders of property under them. Such, though, was not the fate of the Texas farmers who wanted to protect their croplands from hail damage. They were barred from using weather management to protect their crops because the ranchers asserted that the seeding was depriving them of moisture that otherwise would have fallen naturally from their clouds.

Judicial adoption of a cloud ownership concept which limited a proprietor's rights would permit some artificial nucleation efforts. Thus the judge in the 1968 Pennsylvania case reached the conclusion that whatever property right the landowner has in the clouds and the water in them, it cannot be an unqualified right. His opinion stated that "weather modification activities undertaken in the public interest, as opposed to private interests, and under the direction and control of governmental authority should and must be permitted." This view has the merit of permitting proper weather modification activities to be undertaken. It does not, however, prevent persons from recovering damages for their losses when their interests might in fact have been harmed. This would include landowners who have been deprived of precipitation which would have fallen but for the cloud seeding.

Honest, Judge, I Didn't Do Anything

An essential element in a lawsuit seeking judicial relief from unwanted consequences of weather management is that there be some reasonable connection between the activities of the weather modifier and the damage which the plaintiff assertedly has suffered or will suffer. The plaintiff must convince the jury or the judge, if the court is deciding the facts, that the weather modification was a necessary antecedent to the injury. Under the rule now generally being applied by the courts, "the defendant's conduct is a cause of the event if it was a material element and a substantial factor in bringing it about." Defendants can say, "Honest, Judge, I didn't do anything." Plaintiffs must persuade the trier of facts that the conduct of the defendant was a substantial factor in causing them harm. They have been almost uniformly unsuccessful in bearing this burden. This, more than anything else, accounts for the dearth of lawsuits involving weather modification. People cannot establish that they have been or will be injured by cloud seeding.

The Pennsylvania case previously referred to is an illustration of the difficulty in proving causation. It is titled *Pennsyl-*

vania Natural Weather Association v. Blue Ridge Weather Modification Association. The defendant weather modifier contracted with the defendant association to operate on its behalf a hail modification project in the area where West Virginia, Maryland, and Pennsylvania come together. The plaintiff, an organization of individuals from the vicinity who were opposed to cloud seeding, sought injunctive relief to prohibit further seeding. The trial judge was basically sympathetic to the plaintiff's case. He found a property interest of landowners in clouds and he noted that possible harm can result from weather modification activities. He nevertheless denied the relief requested. The Natural Weather Association did not prove causation.

The plaintiff association, according to the judge, did no more than show the possibility of future harm. He noted that:

No particular witness for plaintiff was able to relate the activities of defendants to damage suffered by him. That there was a drought in the entire area is not disputed, but the drought began before, and continued after, the weather modification activities complained of here. Nowhere in the record is there any evidence of "over seeding" clouds by defendants to reduce precipitation. In fact, defendant's evidence is to the effect that the hail suppression increased precipitation.

Similarly in the *Slutsky* case the New York court relied upon the proofs of New York City that the production of artificial rain would not interfere with the resort owners' business, and it branded the assertions by the plaintiffs of possible damage as "exaggerated." According to the opinion, "the factual situation fails to demonstrate any possible irreparable injury to plaintiffs."

Proof of causal relationship in a weather modification lawsuit will inevitably require testimony by expert witnesses based upon adequate data relating to what would have happened under existing meteorological conditions had the defendant not acted, what the defendant weather modifier did, and what thereupon took place. This sort of information may not be available to the plaintiffs and their attorneys in sufficient quantity to give the experts adequate foundation for arriving at an opinion on causation. Moreover, the experts themselves might not agree upon conclusions reached by one another based upon the same data.

This somber picture, looking at it from the point of view of plantiffs, might not always be so bleak. An increase in instrumentation will yield more data. Rules of court procedure permit plaintiffs to get access to data in the files of defendants. And experts seem not to be so badly split on fundamental issues relating to weather technology as has been the case heretofore.

The *Southwest Weather Research* litigation provides another ray of hope for claimants. In that case there was a sharp difference of opinion between the experts about the probable effect of a hail suppression program like that of the defendants. Three of the ranchers and other lay witnesses testified that they had visually observed the destruction of potential rain clouds by the defendants' use of their equipment. The appellate court stated that:

The trial court apparently, as reflected by his findings included in the judgment, believed the testimony of the lay witnesses and that part of the expert testimony in harmony with his judgement. This he had a right to do as the trier of facts.

It should be noted, however, that when an appellate court determines that a finding of fact made in the lower court was unreasonable, the decision below will be reversed. The testimony of the ranchers, crucial in the outcome of the litigation, might well have been regarded as so unreliable as to render findings of causation based in part upon it unreasonable.

Even If I Did Something I Was Careful

Adams v. California was a lawsuit filed on behalf of victims of the December, 1955 flood of Yuba City, California to obtain compensation for the losses they suffered when levees intended to protect the city gave way. They had employed legal counsel who inquired into the manner in which the levee system had been built and operated. In the course of their investigations, counsel discovered that the Pacific Gas and Electric Company had employed a cloud seeder, North American Weather Consultants, who had conducted seeding in the drainage area upstream from Yuba City during part of the storm responsible for the flooding. The attorneys filed suit against the state for its alleged failures in design, construction, and maintenance of the levees. They also sued Pacific Gas and Electric and North American Weather Consultants. They claimed that there was "negligent maintenance and operation of the rainmaking equipment" and that artificial rainmaking was an "ultrahazardous activity." These allegations were essential to their complaint. Complainants in court have the burden not only of alleging and proving causation, but also of asserting and establishing a liability theory.

Lawyers who have tried weather modification cases and legal scholars who have analyzed the legal ramifications of artificial nucleation have looked to four different theories of liability: trespass, negligence, strict liability, and nuisance.

Trespass to property is committed when someone intentionally enters land in the possession of another person or causes a thing to do so. Consent of the possessor of the land is a defense. In *Adams v. California* the flood victims had no basis for claiming that the weather modifiers had entered their lands without obtaining consent to do so. It is customary for weather modifiers to obtain consent of landowners before setting up generators or instruments on their property. The Yuba City landholders were outside the immediate area of the defendants' activities, so there had been no entry by the defendants or their employees. But water allegedly produced by them had flooded the lands. Courts have held that someone whose property has been harmed by water intentionally released may maintain a legal action for recovery of damages. Here, however, there would have been difficulty in showing that the defendants intentionally released floodwaters; in any event, plaintiffs' counsel chose not to rely on a trespass theory.

Adams and the other claimants did, though, assert that the weather modifiers had been negligent. Negligence is carelessness. The standard of conduct to which a person must conform to avoid being negligent is that of a reasonable man under like circumstances. The plaintiffs, to win their case on the negligence theory, had to prove that the weather modifiers fell below the standard of the profession, that is, that they had failed to act as reasonable modifiers would have acted under the circumstances. Experts were called to testify on this issue. The defense used Vincent Schaefer. He was a most impressive witness. Proof of failure to adhere to the reasonable man standard was hard to come by in the Yuba City flood lawsuit.

The other theory upon which the plaintiffs tried their suit was strict liability. They asserted that the cloud seeding by the defendants constituted an ultrahazardous activity. If they could have established that claim, they would not have needed to prove negligence or any other sort of fault by the defendants. Courts have ruled than an activity which is ultrahazardous or abnormally dangerous subjects its perpetrator to liability for harm caused by it, even though he has exercised the utmost care to prevent such harm.

Is cloud seeding an abnormally dangerous activity? Courts consider the following factors in responding to such a question:

(a) Whether the activity involves a high degree of risk of some harm to the person, land or chattels of others;

(b) Whether the gravity of the harm which may result from it is likely to be great;

(c) Whether the risk cannot be eliminated by the exercise of reasonable care;

(d) Whether the activity is not a matter of common usage;

(e) Whether the activity is inappropriate to the place where it is carried on; and

(f) The value of the activity to the community.

Neither the judge in *Adams v. California* nor any other judge has thus far characterized a weather modification effort as abnormally dangerous. There is, however, no reason to believe that all types of weather modification projects would be regarded in the same manner by courts. Hurricane modification may well someday be regarded as so dangerous that its practitioners will be answerable for damages caused by their activities even though they have done everything in their power to prevent harm. Fog dispersal may be regarded as not constituting such an abnormally dangerous activity.

The final liability theory, nuisance, is one which has been regarded by many writers as potentially the most useful in weather modification cases. Conduct constitutes a private nuisance when it invades an owner's interest in the use and enjoyment of his land, and such invasion is intentional and unreasonable, negligent or reckless or regarded as an abnormally dangerous activity. Weather modification

intentionally carried on would be regarded as unreasonable enough to constitute a nuisance, unless the utility of the modifer's conduct outweighs the gravity of the harm it engenders. A nuisance case then resolves itself upon a consideration of the balance of interests. Most weather modification is intended to result in a favorable cost benefit ratio. Heretofore economic considerations have weighed heavily in most nuisance cases. That is one reason why the law of private nuisance has not been successful in coping with pollution problems. There now is a shift toward giving more consideration to aesthetic, ecological, and other nonpecuniary factors. The law of nuisance may yet become an important factor in weather modification litigation.

THE STATE STATUTES

Commissioner Knight and the Tasmanian Parliament

Sir Alan Knight, the Hydroelectric Commissioner of Tasmania, is a very influential and highly respected public official of that Australian state. Tasmania has no weather modification control act, and the Hydroelectric Commission has not sought to have one enacted. The commission, which has been a partner of an agency of the Australian federal government in a cloud seeding program over the mountains of western Tasmania, could benefit from a law regulating weather modification, granting it a legally protected interest in water generated by its cloud seeding and protecting it and its employees from lawsuits related to their weather engineering activities. The commission has, however, earlier obtained legislation exempting it from being enjoined from carrying out its programs. Commissioner Knight felt that proposals to the state parliament for specific legislation on weather modification might result in a law that would obstruct their cloud seeding project. This "let's let well enough alone" attitude probably accounts for the lack of legislation elsewhere too.

Two facts of life about the legal order are applicable here. First, law is enacted only at the instigation of those interested; and secondly, under the common law, what the law does not forbid it allows. Statutes that in one way or another mention weather modification have been enacted in 60 percent of the American states. It is noteworthy that this list includes those states in which most weather modification has been practiced. There also is a correlation between the extent of legislative activity and the extent of weather modification activity. Recent legislative activity in Colorado was generated by persons who believe that cloud seeding would have an impact upon them. In Illinois draft legislation has been prepared under the sponsorship of the organization most likely to conduct weather modification in the state. Persons with something to gain are those who obtain enactment of weather modification laws.

There Ought To Be A Law Against That

Weather modifiers have with some people the image of an impotent rapist. On the one hand they are regarded as impotent to bring about any real change in the weather. "It would have happened anyway." That is the motto of such unbelievers. On the other hand weather modifiers have been regarded in some quarters as potent enough to rape the atmosphere and bring about droughts and floods. Legislators in the tri-state Maryland, Pennsylvania, and West Virginia area have been shown the impotent rapist image by opponents of weather modification; they have reacted by enacting laws designed to impede weather modification.

In the absence of such statutes, cloud seeding was allowed. The effect of the legislation has been to shut down weather modification in the area.

The Maryland legislature has responded most vigorously to the "there ought to be a law against that" mood. It has enacted several short-term prohibitions against "any form of cloud seeding or any other artificial form of weather modification." Prior to expiration of the term of each statutory proscription, the law has been reenacted to extend the ban. The last one expired in 1971. No one has yet been prosecuted for kindling fires in Maryland and that is at least arguably an "artificial form of weather modification." Neither has anyone been tried for cloud seeding in violation of the law. It is most unlikely that anyone has done any seeding in the state since the law was passed. It has achieved its purpose.

It is interesting to note that both Pennsylvania and West Virginia have cooperated with Maryland's suppression of weather management by providing in their laws that nothing in them shall be interpreted to authorize any person to carry out a cloud seeding operation in those states in order to seed in "another state where such cloud seeding is prohibited."

Short of an absolute prohibition of weather modification it is possible to discourage modifiers to such an extent by restrictive legislative provisions that they will not operate. The statutes of Pennsylvania and West Virginia have the effect of imposing strict liability on modifiers who have caused droughts, "heavy downpours or storms which cause damage to lands. . . ." These laws nowhere require claimants to prove fault, but permit recovery if causation can be established. Setting high financial responsibility bonding or insurance requirements is another legislative technique to hinder weather modification. A Colorado proposal, engineered by persons who are suspicious of weather modification, would have set licensing fees at $500/year. Jacking up the costs of operators is hardly the sort of thing that encourages their activities. The legislature established a $100 fee.

No Person Shall be Liable for Any Loss Caused

Legislative assessment of weather modification technology in most states has been favorable. For example, statutes in some jurisdictions have created a permit system which has the effect of providing cloud seeders with franchises to operate without interference within given geographical areas; appropriations have been made for support of weather alterations; and some state laws set up specialized governmental units and empower them to raise funds to pay for cloud seeding. In a few jurisdictions lawmakers have altered the law of liability to protect weather modifiers from all or some potential court-ordered injunctions or judgments for damages.

The enactment offering the most protection to cloud seeders is the *Victoria Rainmaking Control Act of 1967*. That law had its genesis in drought conditions which hit Australia during the mid-1960s. The states, including Victoria, called upon the federal government of the commonwealth to assist them by conducting drought relief rainmaking. Federal scientists were for a while diverted from their role as experimenters in order to seed clouds for the states. The commonwealth government, though, decided that it would be more appropriate for its personnel to give scientific advice and for the states to do their own weather modification. A committee of officials from the civil service was appointed to recommend what the government of Victoria should do. The committee suggested, among

other things, that the state conduct cloud seeding, but that legislation provide for immunity of the crown and of state employees from liability.

The parliament of Victoria responded favorably to the committee suggestions. The statute that was enacted included the following section:

No person carrying out rain-making operations authorized by the Minister under this Act shall in any way be liable in respect of any loss or damage caused by or arising out of precipitation . . . in consequence of the rain-making operation so carried out.

The intent of the law and the manner in which it has been administered restricts grants of authorization to seed to state employees. This section gives them immunity from liability. Prior interpretation of a general statute on liability of the crown had ruled that the state would not be liable for damages unless its employee was liable. So in effect the law on rainmaking also immunized Victoria from liability.

In the 1972 revision of its weather control law, Colorado stipulated that officers and employees of the state and its subdivisions and their agencies would be "immune from liability resulting from any weather modification operations approved or conducted by them" under the regulatory law. The immunity does not by its terms apply to the state or its entities; neither does it protect private weather modifiers.

It is noteworthy that while Victoria and Colorado have by legislation provided for immunity, in most American states the defense of sovereign immunity of the states has been rapidly eroding. At one time the states could not be held responsible in their own courts for the harm caused by their employees. That has changed. The final case in the series of lawsuits which led to the abolition of sovereign immunity in California was decided by that state's supreme court while the trial of *Adams v. California* was taking place. When the supreme court abolished the immunity, the trial court in *Adams* was bound by its action. Even though Pacific Gas and Electric Company and North American Weather Consultants were not held liable to the Yuba City flood victims, the state was. California settled its obligation for $6.3 million.

A less drastic type of legislative tinkering with the law of liability is a part of the Texas weather control law. Upon the recommendation of counsel for the weather modifiers in the *Adams* case, Texas officials proposed to the legislature a bill intended to bar the courts from considering weather modification as an ultrahazardous activity. The law reads:

[A]n operation conducted under the license and permit requirements of this chapter is not an ultrahazardous activity which makes the participants subject to liability without fault.

In Texas, although modifers can still be held for harm caused by their carelessness or other fault and for committing a nuisance, they are not liable without such proof of fault.

We, The Undersigned Electors Request

A recurring problem with weather modification is raising adequate funds to pay for operations. Individual electrical power generating companies usually have the kind of resources necessary to support cloud seeding to augment stream flow. Individual farmers do not have enough spare money to finance precipitation enhancement or hail suppression. It is not surprising, then, that several utilities have a history of many seasons of sponsorship of seeding, and that individual farmers or loosely knit farmers' groups have tended to get involved in cloud treatment on a very hit-and-miss basis, one season on,

and then several off. Also it is not surprising that there have been pressures in the legislatures of the states to devise laws that will make financing easier.

The approach taken in the Dakotas, Nebraska, and Texas has been to pass enabling legislation that will provide for the creation of special districts which will have power to raise money to pay for cloud seeding. These laws have not been uniformly effective. The Nebraska legislation was struck down by the state's courts because it did not give a voice to all the property owners in the proposed weather control districts on whether a district should be created or not. The Texas law has not been used often. In North Dakota under the original law an election was held in the proposed district to ascertain whether the voters would approve setting it up and obligating themselves to provide its financial support. Urban interests were able to outvote the rural voters. There has been success in South Dakota where multi-county districts operate with coordination by a state director and the state regulatory agency.

North Dakota has now amended its weather modification authority law. Such authorities are now created by the county commissioners upon petition of the electorate. The petition format contains the following paragraph:

We, the undersigned qualified electors of (name of county) . . . by this initiated petition request that the . . . board . . . create by resolution a . . . weather modification authority and to appoint . . . the commissioners for the (name of county) weather modification authority . . . .

This approach should be more successful than an election, except for the requirement than 51 percent of the qualified electors of the county sign the petition.

Rather than setting up special districts, legislative authorization could be extended to existing units of government giving them power to engage in cloud seeding. California authorizes any agency with power over water supply to engage in precipitation enhancement. New York grants municipalities such power. The county commissioners in Nevada can enter into contracts with governmental and private organizations engaged in weather modification. Minnesota law names nine counties that can each spend up to $5000/year on payments to weather control companies.

Direct appropriations from the state legislature to units of government for weather modification purposes are meaningful manifestations of legislative encouragement of cloud seeding. Drought conditions in the arid southwest have caused some legislatures to loosen their purse strings and provide funds for precipitation enhancement efforts.

An Open Public Hearing In The Area To Be Affected

Lawmakers have decided upon a system of licensing professions. Before anyone can be a doctor, dentist, accountant, beautician, or lawyer he must obtain a license. Such a licensing system is intended to protect the public against persons who are dishonest or incompetent. Weather modification statutes in many states provide for professional weather modifier licenses.

Laws have delegated power to agencies to issue permits authorizing activities in certain areas. Telephone companies, bus lines, natural gas suppliers, owners of cattle grazing on public land, and power plants operate under the terms of permits. They are subject to regulation, but in return get a franchise excluding all or some competition within the area covered by their permit. A number of states have laws that require operational permits for all or some kinds of weather management activities.

The more recently enacted state laws, such as those of Montana and Texas, require both professional licenses and operational permits. According to the Montana law permits can be issued only:

(1) if the applicant is licensed pursuant to this act;

(2) if sufficient notice of intention is published and proof of publication is filed as required in . . . this act;

(3) if an applicant furnishes proof of financial responsibility in an amount to be determined by the board as required in . . . this act;

(4) if the fee for the permit is paid as required in . . . this act;

(5) if the weather modification and control activities to be conducted are determined by the board to be for the general welfare and the public good;

(6) A public hearing may be held in the area to be affected by the issuance of the permit. . . .

In other words, to get a permit in Montana a modifier must have a license, publish notice of intent to conduct an operation, prove he is financially responsible, pay his fee, persuade the board that his project is in the public interest, and may need to be at a public hearing.

A number of states insist upon notice to the public by publication about an impending project. The South African law requires publication in the government gazette. This is motivated by the philosophy that the people have a right to know what is going on. Legal notices, though, are not the sort of light reading that the normal newspaper subscriber indulges in. That kind of notice may be more "legal" than real.

The Montana public hearing requirement, a provision also found in the Massachusetts and Colorado laws, goes beyond mere notice. It affords the public a chance to participate in the decision-making process. The law does not bind the board to use, in arriving at its decision on a permit application, any of the information brought to it at the hearing. But Montana practice has already shown that adverse comments about a project at a hearing can be very persuasive with the administrators.

Public hearings are normal statutory prerequisites to issuance and renewal of franchises. They do, however, provide a forum which may be misused by persons who are not well informed about cloud seeding or who may for personal reasons, rather than in the public interest, seek to block projects. That may account for the fact that hearing requirements have not been more widely imposed.

Most laws that have licensing and permit features do not spell out the procedures to be followed in issuance of such documents. They sometimes, like the Quebec act, delegate power to administrators to promulgate regulations and develop forms which will set forth the procedure. Some laws, particularly a few of the earlier statutes, have very little procedure spelled out in them because they ask very little of applicants. The Idaho law is such an enactment. It merely requires persons who conduct operations in Idaho to produce rainfall artificially to register with the state agriculture department, and to file logs with the department.

But Only Registered Professional Engineers

Assuming that a state does license weather modifiers, what sort of qualifications should be possessed by licensees? Professional licensing provisions in other fields have been ofttimes cluttered by demands made of applicants that are irrele-vant to the issues of competency and integrity. Weather modification has its example also of the questionable prerequisite for licensing. Under the Wyoming law only a "registered professional engineer" could be licensed. Many highly competent weather modifiers are not registered as professional engineers either in Wyoming or elsewhere; indeed most cloud seeders are not so registered; nor is there any good reason why they should be. Wyoming saw the light and in 1971 changed its statute to eliminate the professional engineer requirement.

Competency in meteorology is the most common requirement for obtaining a professional license, with some laws clearly listing it as a licensing condition, and others merely asking applicants to indicate the extent to which they possess such qualifications. Some laws give licensing agencies express power to determine whether applicants have the skills necessary for licenses.

Not all laws are as explicit as is that of Montana in setting forth the preconditions of obtaining a permit. Where licenses as well as permits are required, the permit seeker would have to measure up to whatever is needed to get a license. And in some states the application forms for permits request the kind of information that would give the agency issuing permits a reading on whether grant of the application would be in the public interest or not. When one project would contaminate another one because of their proximity, the agency may choose not to issue the permit. A form which required the applicant to indicate where he would operate and the manner in which he would carry on the project would give the government the kind of information it would need to make an informed decision about whether to grant the permit.

Praise For Our Sister The Water

"Water is our humble servant and precious treasure. Held in trust high on the mountain it runs down hill to give us light and heat and power. It fills our lakes and reservoirs and brings life to the thirsty land. Continually a mist rises to fall again as rain and never a drop is lost. Sustaining and cleansing all things water returns to its virgin purity." Praise for our sister the water.

Water is also the stuff of which legal battles are made. Many lawyers have made good livings litigating water rights. Lack of legislation clearly delineating such rights has been a contributing factor to the confusion that often ends in court fights. Weather modification statutes do not address themselves to the question of ownership of runoff made possible through cloud seeding. Either the laws must be amended or the costly process of litigation will attempt to provide answers.

The recently revised Colorado law declares that *any* moisture which falls naturally or is artificially induced to fall is the property of the people of the state and will be administered under existing Colorado water law. Colorado follows the prior appropriation doctrine. A modifier's claim would depend upon the priority of appropriation.

More and more states are adopting a permit system for distribution of ground and surface water resources. The legislative charters of the agencies administering these laws may be read broadly enough to authorize them to allocate runoff from weather modification undertakings. In some states the weather modification control agency is the same governmental body as the water resources department, but where that is not the case there exists a potential conflict of jurisdiction between officials controlling natural waters and those giving weather modification permits. Who will allocate runoff from cloud seeding? Perhaps without legislation spelling out the answer the job will have to be done in court.

THE FEDERAL BILLS

Men Must Turn Square Corners When They Deal With the Government

Federal money has paid for a very significant portion of the scientific inquiry into cloud physics, laboratory experiments involving weather modification, and field studies to test engineering applications of hypotheses about cloud seeding. Although in-house work accounts for part of these expenditures, much of the money has been spent to hire contractors to do the experiments, studies, and tests.

"Men must turn square corners when they deal with the government." Clauses in federal procurement contracts that weather modifiers who want government business must sign, deal not only with performance of the contract, but also with matters not related to nongovernmental contracts. By federal law, clauses are inserted that relate to governmental social and economic policy, such as wages and hours of employees, the use of small business concerns, and equal employment opportunity. Congress has also required inclusion in procurement contracts of clauses designed to protect the federal treasury. These are some of the "square corners" that must be turned.

Federal contractors, which include many of the most experienced weather management companies, are to some extent indirectly regulated as to their nongovernmental business. A pertinent illustration is reporting to the National Science Foundation on cloud seeding activities. When the Foundation lost the authority to require weather modifiers to report, it tried to maintain continuity in the records by asking for voluntary reports. Organizations holding government contracts and some other operators saw the wisdom of reporting; other modifiers did not. The power of the federal purse was a factor in this result.

Who Will Bell the Cat?

Apart from appropriations statutes and laws on reporting, the federal government has talked a lot about weather modification law but has done nothing. Students of the legislative process have attributed this to several factors: legislative inertia, doubt as to the efficacy of cloud seeding, lack of any really pressing need for federal legislation, and interdepartmental rivalries. Desire for leadership of the federal weather modification effort made more than one department seek to have power to bell the weather modification cat. Proposals from the agencies and their congressional allies all spoke "the language of pluralism, in which one agency is more equal than the other."

Federal legal activity can be sketched quickly. In 1953 Congress enacted a bill setting up the Advisory Committee on Weather Control with a charter calling for it to evaluate cloud seeding. In 1957 the National Science Foundation was granted authority to support weather modification research, to obtain information, and to report annually to the government. The Foundation at first gathered data from cloud seeders on a voluntary basis, then in 1966 it promulgated regulations requiring reporting. In 1968 its power to demand reporting was repealed by Congress. Just before Christmas, 1971, President Nixon signed into law an act which gives the Department of Commerce the power to require reporting. It does not cover the federal government, its employees, and its

contractors. Other persons may not engage in weather modification activity in the United States unless they submit "such reports with respect thereto, in such form and containing such information, as the Secretary may by rule prescribe."

Significantly Affecting the Quality of the Human Environment

Although only appropriation laws and the reporting statute deal directly with weather modification, several federal enactments have an indirect impact upon it. Probably the most important of these statutes is the National Environmental Policy Act of 1969. It imposes two principal requirements upon federal agencies and projects funded with federal money: utilization of interdisciplinary planning and filing environmental impact statements.

Section 102 (2) A of the Environmental Policy act demands that all agencies of the federal government:

[U]tilize a systematic, interdisciplinary approach which will insure the integrated use of the natural and social sciences and the environmental design arts in planning and in decisionmaking which may have an impact on man's environment.

Planning for weather modification projects, according to this law, must not be the exclusive province of physical scientists, engineers, and administrators. Now ecologists, biologists, sociologists, and other types of professionals must be included. This provision has not been strictly enforced yet, but foresighted administrators are starting to comply with its mandate.

The federal courts have been enforcing subsection C of section 102 (2) of the National Environmental Policy Act. It requires all agencies of the federal government to include an environmental impact statement "in every recommendation or report on proposals for legislation and other major Federal actions significantly affecting the quality of the human environment." Two statutory interpretation questions are presented in ascertaining whether impact statements must be prepared in connection with weather modification projects: what is a "major" federal action and does it "significantly" affect environmental quality. Laboratory studies and possibly field experiments would appear not to be covered. On the other hand pilot projects and clearly operational weather modification would be included.

Environmental impact statements, according to the law, must be "detailed." Judges have taken this part of the statute to heart. A number of important federal projects have been stalled because courts determined that the impact statements filed were inadequate. The leading illustration is the pipeline across Alaska. Statements must include:

(i) the environmental impact of the proposed action,
(ii) any adverse environmental effects which cannot be avoided should the proposal be implemented,
(iii) alternatives to the proposed action,
(iv) the relationship between local short-term uses of man's environment and the maintenance and enhancement of long-term productivity; and
(v) any irreversible and irretrievable commitments of resources which would be involved in the proposed action should it be implemented.

In preparing impact statements agencies are required to consult other federal organizations that have jurisdiction over or expertise with regard to any environmental impact involved. Their comments and the statements must be filed with the Council on Environmental Quality and circulated among appropriate agencies and to certain officials, including the president.

The National Environmental Policy Act has given a new legal weapon to environmentalists. Until fairly recently these groups were unable to use the judicial system effectively to achieve their aims. They did not have that sort of economic interest that courts recognized as a basis for "standing to sue," the right to enter courts as plaintiffs. The law in this area has been liberalized so that persons with nonpecuniary interests can sue. A second difficulty formerly blocking environmentalists was that federal agencies were given wide discretion by Congress to carry out their programs and were, at least in most instances, not forced by law to consider environmental factors. The statute now requires them to do so. If agencies do not comply, environmentalists have something to point to in court in order to get an injunction blocking the project until there is compliance.

There is nothing in the Environmental Policy Act that would prevent a weather modification project from being operated even if its sponsors admit in their impact statement that there are or might be adverse ecological consequences caused by the cloud seeding. Subsection A merely requires inclusion of those types of planners who are knowledgeable in environmental matters; subsection C merely demands that a complete impact statement be prepared. Groups opposed to a particular federal project might be able to delay it if those persons in charge had not complied with the statute to the satisfaction of the judiciary. But, unless delay itself prevents a season of seeding, the law does not block implementation of federal projects.

There are statutes dealing with each of the federal departments that engage in weather modification. They of course must comply with them. Some of these laws applying to the functions of an agency also have an impact on persons doing business with such agency. An example of such a law which would bear upon weather modification by the Bureau of Reclamation is the Excess-Land Law. It reads:

No right to the use of water for land in private ownership shall be sold for a tract exceeding 160 acres to any one landowner.

No landowner with a tract over the 160-acre limit can buy water generated by the Bureau's cloud seeding projects.

THE UNANSWERED QUESTIONS

How Can a State Control Activities Affecting Another?

The township board of supervisors of Ayr Township in Fulton County, Pennsylvania enacted an ordinance in 1964 making it a summary offense for any person to seed clouds. A weather modifier operated a ground-based generator in the township to suppress hail on orchards nearby in West Virginia and Maryland. He was arrested, charged with violation of the ordinance, tried before a justice of the peace, found guilty, and fined $100.

The modifier appealed his conviction to the Fulton County Court of Common Pleas. The case was designated as *Pennsylvania ex rel. Township of Ayr v. Fulk*. It was heard along with the *Pennsylvania Natural Weather Association case*. Opinions in both cases were handed down on the same day early in 1968.

A major contention of the defendant Fulk was that, as applied to him, the township ordinance was unconstitutional because it imposed an undue burden on interstate commerce. The so-called "Commerce Clause" of the federal constitution gives an express grant of power of Congress "to regulate commerce . . . among the several states." It has been interpreted by the courts not

only as an affirmative grant of power to the federal government, but also as a restriction upon the power of the states to regulate interstate commerce. States can regulate commerce among the states only insofar as their actions interfere with the national economy no more than may be justified by the importance of the resulting protection of local interests. This limitation is expressed in terms of a constitutional ban upon imposing "an undue burden on interstate commerce."

The judge in the *Fulk* case affirmed the conviction of the defendant. He sought to overcome the constitutional arguments by asserting:

[I]t is only when state legislation invades the domain belonging exclusively to Congress and imposes real obstacles or burdens upon such commerce that it cannot be allowed to operate . . . . It is obvious that the ordinance now before us was never intended to regulate interstate commerce, rather it was to protect the health and welfare of the people of the township. Further, we are not at all convinced that weather modification as such is "commerce" . . . .

The judge's position is subject to several criticisms. The intent of the township supervisors is not determinative of the case. Even though they may not have meant to regulate interstate commerce, they did regulate an activity which was designed to have an interstate impact. The issues are whether that activity and its impact constituted "commerce" and whether the regulation interfered with national interests more than the intended local protection justified. From the earliest days of the nation federal courts have read the word "commerce" very broadly, so broadly that most judges would not have questioned whether the activity involved in *Fulk* was "commerce." Where the operation was not being conducted to affect Pennsylvania interests, but rather to suppress hail in Maryland and West Virginia, it seems that no local interest was really being protected and that a national interest was being adversely affected. The ordinance as applied to the project in question should have been declared unconstitutional.

The Ayr Township ordinance now has been superseded by the present Pennsylvania law. It, like the West Virginia law, bans weather modification in the state which is designed to affect the weather in a state which does not permit cloud seeding. These two enactments are intended to help Maryland free itself from weather modification. Constitutional validity of such provisions might also be challenged. They are not as constitutionally dubious as the Ayr Township ordinance. There one state was barring an activity which at that time was lawful in the neighboring states. Here two states were prohibiting an activity which was unlawful in Maryland. That kind of cooperative legislation is clearly distinguishable from what the township undertook to do.

Colorado, New Mexico, and Utah also have interstate provisions in their statutes. New Mexico provides:

Weather control or cloud modification operations may not be carried on in New Mexico for the purpose of affecting weather in any other state which prohibits such operations, or which probibits operations in that state for the benefit of New Mexico or its inhabitants.

This statute applies to instances other than cases like the Maryland prohibition of weather modification; it also bars someone from trying to alter the weather in a state which will not let people in its jurisdiction attempt to change weather to benefit New Mexico. That would include a state which would not allow cloud seeding to cause added snowfall within that state which upon melting would augment downstream water flow in New Mexico, as well as a state which would not permit cloud seeding designed to enhance precipitation in New Mexico.

This is a sort of "least favored nation" law—if your law will not let people help us in your state, then our law will not let our territory be used to help you.

For policy reasons, as well as constitutional questions, there is ample justification to regulate weather modification on the federal level. Even if state efforts to deal with extraterritorial weather modification impact are constitutional, that approach is not necessarily sound. Weather and state boundaries have very little in common. It makes good sense to have a National Weather Service, rather than a collection of state weather bureaus. It also is sensible to deal with weather management as an interstate problem which can best be regulated on a national basis.

As has been indicated previously there is very little federal law of weather modification. There are some devices within our federal system which have been used to resolve interstate problems and which might be employed to cope with the interstate ramifications of weather modification. These are resort to the Supreme Court in a lawsuit between states, concluding interstate compacts, and enactment of federal legislation.

The federal constitution and the judicial code of the United States permit states to file lawsuits against each other in the Supreme Court of the United States and regulate the procedure to be followed in such litigation. It is usual for the Court to refer such a suit to a master, often a sitting or retired federal district court judge, to take evidence, make findings of fact and reach conclusions of law. The master's report is then reviewed by the Court which renders its decision. This process has been used by the states in obtaining apportionment of water rights in interstate streams. The Supreme Court has developed the doctrine of equitable apportionment. In *Nebraska v. Wyoming*, a 1945 decision allocating the waters of an interstate stream, the Court gave an illustrative, but not exhaustive, list of the factors they had looked to in adjusting the interests of the states:

Priority of appropriation is the guiding principle. But physical and climatic conditions, the consumptive use of water in the several sections of the river, the character and rate of return flows, the extent of established uses, the availability of storage water, the practical effect of wasteful uses on downstream areas, the damage to upstream areas if a limitation is imposed on the former—these are all relevant factors.

No decree in an equitable apportionment case has as yet purported to allocate supplemental water in an interstate stream which has been generated by weather modification. So by their express terms the decrees are inapplicable. But language in some decrees might be interpreted to include water harvested by cloud seeding. For example, the term "imported water" used in some decrees might serve as the basis for a claim to water flows augumented by precipitation enhancement.

The more important aspect of use of the original jurisdiction of the Supreme Court is that states feeling aggrieved by weather modification conducted by another state might invoke such jurisdiction and ask to have the Court adjudicate their claims. It is likely, if that ever happens, that the Court will look to the equitable apportionment cases for guidance in deciding the matter.

There is a prohibition in the constitution against states entering into treaties with one another without the consent of Congress. States have entered into numerous "interstate compacts" with each other and have been given congressional consent to do so. Interstate authorities have been created to build and operate transportation facilities. For example, the Port of New York Authority was set up by compact between New York and New Jersey, with congressional approval.

The interstate compact device might prove to be a useful structure for regulation of weather modification on a regional basis, operation of projects, and allocation of interstate weather resources. No real effort has thus far been made to set up an interstate regulatory agency by use of the compact device. There are, however, compacts creating authorities and granting them regulatory power in more than one state. For example, zoning in the Lake Tahoe region is now to some extent controlled by a California-Nevada interstate authority. During 1971 there were suggestions that an interstate agency be created to carry out weather modification operations in the lower Great Plains states. Some already existing multistate organizations may have power to perform cloud seeding. An illustration would be fog suppression performed by an agency which operates airports. No compact by its express terms purports to allocate weather resources among the signatory states; but arguments have been advanced that compacts allocating waters of interstate streams, like equitable apportionment decrees, might be interpreted to encompass division of water harvested from precipitation modification; and interstate compacts which allocate atmospheric water resources are feasible.

Federal legislation is the most likely ultimate solution to interstate weather modification problems. Congress has constitutional power to regulate weather modification, to make appropriations to finance cloud seeding operations, and to divide the benefits of weather technology among the various states. It has appropriated money and it has enacted a reporting statute which is a sort of regulatory measure. Congress, though, has not expressly allocated atmospheric waters.

The Colorado River Basin Project Act has the indirect effect of dividing the benefits of Colorado River water augmentation among basin states. By treaty the United States must annually deliver to Mexico a million and a half acrefeet of water. Congress has stipulated that satisfaction of Mexican rights

constitutes a national obligation which shall be the first obligation of any water augmentation project . . . Accordingly, the States . . . shall be relieved from all obligations which may have been imposed upon them by . . . the Colorado River Compact so long as . . . means are available and in operation which augment the water supply of the Colorado River system in such quantity as to satisfy the requirements of the Mexican Water Treaty together with any losses of water associated with the performance of that treaty.

Thus augmentation by weather engineering will make water available to the states that otherwise they could not have used. To that extent the national legislature has allocated the benefits of stream flow augmentation in the basin.

Will This Merge Each Nation's Affairs With Those of Every Other?

Weather does not need a passport to cross an international boundary, nor do the effects of weather modification. It has been asserted that use of technological capabilities here will tend to merge the affairs of each nation with those of every other country. As yet this has not happened. That is just as well. There is very little international law applicable to weather control.

Cooperation among scientists interested in weather and weather modification has been good. There is rather free exchange of meteorological data, and nations possessing weather technological capabilities have shared their understanding of cloud seeding techniques with other countries. Australian scientists have trained cloud seeders from many countries, and some American commercial firms have contracts on an international basis.

There is, however, no international agency that possesses regulatory power over transnational weather modification impacts. A draft protocol on weather modification offered by the World Peace Through Law Center, an international nongovernmental organization of lawyers and jurists, would lodge such authority in the World Meteorological Organization. Other scholars have offered other scientific international agencies as the recipients of regulatory power. In a world fraught with much more serious international complications than those that might be attributable to weather modification activities, creation of an international administrative authority is not a matter of the highest priority.

What then can be done under existing international law about the case in which activities in one nation bring about adverse effects in another country? There are two possibly analogous cases: the *Trail Smelter* arbitration between the United States and Canada which was finally decided in 1941, and the *Lake Lannoux* arbitration which was concluded in 1957 between France and Spain. The first was an air pollution case, the second a water rights decision.

In the *Trail Smelter* case fumes from a smelter in British Columbia were causing damage in the state of Washington. Canada and the United States submitted the case to an arbitration tribunal which concluded that:

Under the principles of international law . . . no State has the right to use or permit the use of its territory in such a manner as to cause injury by fumes in or to the territory of another or the properties or persons therein, when the case is of serious consequences and the injury is established by clear and convincing evidence.

The general principle here enunciated would apply to injuries across an international boundary caused by weather modification activities.

The *Lake Lannoux* case dealt with a French proposal to divert waters of the Carol River which flows from the lake and drop them to the Ariege River in order to use the fall for power generation. The French further proposed to replenish the Carol River downstream before it entered Spain. The Spanish insisted that the project not be started without their consent. The arbitration tribunal to which the two nations submitted the dispute found that neither under an existing treaty between the two countries nor as a matter of general law did France require Spanish consent. There would be no change in the flow of the river. The upstream riparian "has the right to prefer the solution offered by its own project, provided it takes the interests of the lower state into consideration in a reasonable manner." Likewise an upwind country could carry out weather modification projects without permission from a downwind nation, but it would have to consider the impact of its operations on the other country.

REFERENCES

Cases

Adams v. California, No. 10112 (Superior Court, Sutter County, California, April 6, 1964).

Atmospherics, Inc. v. Ten Eyck, Civil Action (D. Ct. Alamosa County, Colo., April 4, 1973).

Auvil Orchard Co. v. Weather Modification, Inc., No. 19268 (Superior Court, Chelan County, Washington, 1956).

Pennsylvania ex rel. Township of Ayr v. Fulk, No. 53 (Court of Common Pleas, Fulton County, Pennsylvania, Feb. 28, 1968).

Pennsylvania Natural Weather Association v. Blue Ridge Weather Modification Association, 44 Pa. D. & C. 2d 749 (1968).

Slutsky vs. City of New York, 197 Misc. 730, 97 N.Y.S. 2d 238 (Supreme Court, Trial Division, 1950).

Southwest Weather Research, Inc. v. Rounsaville, 320 S.W.2d 211, and Southwest Weather Research, Inc. v. Duncan, 319 S.W.2d 940 (Texas Court of Civil Appeals, 1958), both affirmed under the name Southwest Weather Research, Inc. v. Jones, 160 Tex. 104, 327 S.W.2d 417 (1959).

Samples v. Irving P. Krick, Inc., Civil Nos. 6212, 6223 and 6224 (U.S. District Court, Western District, Oklahoma, 1954).

Shawcroft v. Department of Natural Resources (Alamosa County, Colorado, Sept. 20, 1972).

Summerville v. North Platte Valley Weather Control Dist., 170 Nebr. 46, 101 N.W. 2d 748 (1960).

Weather Engineering Corporation of America v. United States, No. 343–72 (Court of Claims, filed Sept. 8, 1972).

Books, Reports, and Articles

Ray Jay Davis, *The Legal Implications of Atmospheric Water Resources Development and Management*, Report to Bureau of Reclamation, Oct. 1968.

Ray Jay Davis, The law of precipitation enhancement in Victoria, *Land Water Law Rev.* **7,**1, 1972.

Dean Mann, The Yuba City flood: a case study of weather modification litigation, *Bull. Am. Meteorol. Soc.* **49,** 690, 1968.

Howard Taubenfeld, *Weather Modification Law, Controls, Operations,* Report to the National Science Foundation Special Committee on Weather Modification, 1966.

Howard Taubenfeld, Ed., *Weather Modification and the Law*, Oceana Publications, 1968.

Howard Taubenfeld, Ed. *Controlling the Weather: A Study of Law and Regulatory Processes*, Dunellen, 1970.

22 | Sociological Aspects of Weather Modification

J. EUGENE HAAS

THE SOCIOLOGICAL SIGNIFICANCE OF WEATHER MODIFICATION

Man is a social being. What he experiences, how he perceives his external environment, and how he tries to cope with it are all influenced by the culture and social structure in which he operates. From the human perspective the atmosphere is both benign and harmful; it is also capricious.

Since the beginning of human history it is clear that man has tried a variety of methods for coping with perceived atmospheric hazards and inconveniences by devising protective technologies: buildings, heating and cooling devices, dams and levees, drought-resistant crops, drainage and irrigation systems, and forecasting techniques, to name a few. These technologies are all *adjustment* mechanisms designed to allow man to adapt more profitably and comfortably to atmospheric conditions as they occur naturally. In recent decades man has added a new tool to his kit—a *manipulative* mechanism designed to alter directly unwanted conditions of the atmosphere. To passive adjustment technologies he has added a frontal attack technology, scientifically based, planned weather modification concepts and techniques. This frontal attack became possible after science and engineering had developed as institutional forms and thought systems in modern societies.

As with many, but certainly not all, new technologies, scientifically based, planned weather modification started in scientific and engineering institutions and gradually moved into other areas of the social system.[11] It has been adopted in limited measure by some electric power company managers, municipal and state water resource managers, forest resource and range land management officials, farm and ranch owners and operators, and air transportation managers. The term adoption here refers to investing money in operational weather modification efforts. Indications of its acceptance in U.S. Governmental institutions may be seen in the work of the Interdepartmental Committee for Atmospheric Sciences (ICAS) in the Executive Office of the President, the Office of Atmospheric Water Resources in the Bu-

J. Eugene Haas is Professor of Sociology and Head of the Research Program on Technology, Environment, and Man in the Institute of Behavioral Science, University of Colorado.

reau of Reclamation, the Office of Weather Modification in the National Oceanic and Atmospheric Administration (NOAA), and at the state level, the South Dakota Weather Control Commission and the Texas Water Development Board. The U.S. Congress provided, perhaps inadvertently, for the use of weather modification across a broad range of atmospheric conditions when it passed the Disaster Relief Act of 1970 (PL 91-606).

The principal justification for both experimental and operational use of this technology is for precipitation augmentation (see Table 22.1). Compared to the demand, water has been in short supply in some areas of the United States for many decades. In most of those areas, however, the population has been sparse. Now the burgeoning population growth in the Southwest, along the Pacific Coast, and in some other areas is bringing new, insistent demands that weather modification be used to produce additional water resources. Earthbound water technologies (dams, reservoirs, transmountain tunnels) no longer seem adequate to meet the projected demands for water.[18] The per capita consumption of water is increasing at the same time that semi-arid and arid regions of the country are experiencing rapid population growth. No serious attempts are being made to deter population increases in those areas already having a water deficit. It is generally assumed that the only viable solution is to search for more water. Thus the proponents of weather modification have a strong ally in the demographic forces already underway.[4]

In order to provide background for the subsequent sections of this chapter, let's look briefly at several different types of weather modification projects with emphasis on the social setting in which they are or have taken place.

Florida Rain Augmentation Program, April–July, 1971

For several years before 1971, scientists of the Experimental Meteorological Lab (EML), NOAA, had been conducting cloud seeding experiments from aircraft over the Gulf of Mexico and over southern Florida outside of the Miami metropolitan area. On one occasion the intended target area was modified after strong disapproval was voiced by agriculturalists in the area. The meteorological and related aspects of this project are discussed in Chapter 6.

During the last half of 1970 and early 1971, rainfall in central and southern Florida was much below normal to the point that there was apparently consensus that drought conditions existed or were very near by March, 1971. Upon the recommendation of the Central and Southern Florida Flood Control District and after extended informal discussions with NOAA scientists and administrators, both Governor Kirk in late 1970 and Governor Askew in early 1971 officially asked the EML, NOAA, to conduct a cloud seeding program over a 4800 mi² target area in Central Florida during April and May. A special network of rain gauges was installed in this target area. The criteria to be used in conducting the cloud seeding program were arrived at by mutual agreeement between EML and the Central and Southern Florida Flood Control District. While the cloud seeding program was originally experimental in format, a major objective was to produce additional rainfall for drought relief purposes. The costs of the effort were shared, with NOAA paying the largest share.

The program began in early April but by May 1, 1971, very few cloud seeding opportunities appeared over the central Florida target area. Therefore, early in May a section of southern Florida was

Table 22.1 Areas in Which Some Type of Weather Modification Field Effort Was Conducted During Some Period of 1971 or 1972, U.S.A.[a]

	Precipitation		Fog		Hail		Lightning		Other	
	G	NG	G	NG	G	NG	G	NG	G	NG
Alabama			1							
Alaska	1		2	1						
Arizona	1									
Arkansas		1								
California	5	1	2							
Colorado	2	2[b]			1	1[b]				
Florida	3						1		1	
Hawaii	1									
Idaho				1						
Iowa		1								
Kansas	1									
Louisiana		1								
Michigan		1								
Montana	2									
Nebraska				1						
Nevada	1									
New Mexico	1	1								
New York	1									
North Dakota	1	1				1				
Oklahoma	4	2								
Oregon				1						
South Dakota	5[b]	1			2[b]					
Texas	3	1[b]				1[b]				1
Utah	1			1						
Washington		3	1							
West Virginia			2							
Wyoming	2									

[a] The following definitions were used in assembling these data. G = funding by some governmental agency or unit such as a county or municipality—funding is directly from tax revenues. NG = all other sources of funding. Area = a geographical entity being treated as a single unit for purposes of conducting the weather modification effort. For example, there may be nine counties being treated as a single district for purposes of operating in the South Dakota State Weather Modification Program. This is counted as one area rather than nine. Where a project involves more than one state it is listed under the state in which the largest part of the seeding operation takes place. These figures were gathered from a variety of unofficial sources. They probably represent an underestimation of weather modification activity.

[b] Where a program has both rain augmentation and hail suppression as objectives it is listed under both headings.

designated as a target area also. It was called the "secondary" target, with the original target then called the "primary" target area. Most of the cloud seeding was conducted over the secondary area, sans rain gauges, the planned randomization of seeding days was dropped, and the program was extended 6 weeks beyond the June 1 planned termination date.

Initial research by the author and his associates indicated that not all agriculturalists in the target area were being negatively affected by the drought. Indeed, the melon and tomato growers prefer to have no rainfall at all just prior to and during the harvest season, which in this case happened to coincide with the April–May rain augmentation program period. Success in growing tomatoes in Florida depends on the use of wells for irrigation so that moisture levels can be carefully controlled. Both the tomato plant and the fruit are easily and quickly damaged by rain. Melons are somewhat less vulnerable to small amounts of rain at harvest but can be severely damaged by heavy rainfall. There were approximately 3500 acres of tomatoes and 300 to 500 acres of melons being grown in the target area.

Tomato farmers contract with the few processing corporations in the area. Since both the farmer and the corporation invest in the production of the crop, both stand to lose if the crop is damaged or lost completely. Dairy farmers gain very little from spring rains since the cattle are fed carefully controlled supplements. Heavy rains create some disease and general sanitation problems for the dairymen. Beef cattle and citrus growers, the two largest agricultural enterprises in the area, were being hurt by the drought and welcomed anything close to normal spring rainfall. There is very little tourism in the target area during April and May. Most of the towns in the target area appeared to have either stable or declining populations, and thus building construction was not a significant economic activity.

The setting, then, was one in which the vast majority of persons and investment groups would gain from normal or slightly above normal spring rainfall. The few who could be hurt could lose heavily from as little as 2 or 3 in. of rain if it fell during certain critical time periods. There was a clear, latent conflict of interest among the diverse economic activities in the target area. The most significant considerations became whether or not there would be appreciable rainfall during April and May, whether or not those standing to lose would conclude that the rain was, in part, produced by cloud seeding, and the timing and location of the rainfall.

Those individuals residing or operating agricultural enterprises in the target area were reinterviewed in late June, 1971. Tomato growers expressed more negative views whereas persons in most other occupational categories expressed more favorable views than had been the case before the start of the cloud seeding program. Tomato growers were significantly more apt to attribute losses they knew of to the cloud seeding than were other groups. Also, tomato growers were less apt than other groups to attribute any benefit whatsoever to the program, less likely to believe that anyone benefited. One large corporation with investments in tomato production threatened to sue to recover alleged losses produced by the cloud seeding program but did not follow through with the threat. Sixty-one percent of the tomato growers believed that they had suffered direct personal losses attributable to the cloud seeding effort, a figure more than double that of any other group interviewed. None of the growers took any legal action. One small claims settlement was made to a person who claimed that hail from cloud seeding had damaged his car windshield. Relations between the Central and Southern Florida Flood Control District and EML, NOAA became strained over public relations issues.

This was the first drought relief effort with large-scale federal support in the United States. Since that time there have been similar efforts in Texas, Arizona, and Oklahoma.

San Luis Valley, Colorado, Augmentation and Suppression Program

During the late 1960s, many growers of Moravian barley in the San Luis Valley organized to hire a commercial weather modification firm to assist in the production of premium-quality barley grown under contract for a brewery. The intent was to produce additional rainfall during the early growing season, suppress hail during the midsummer season, and to suppress rainfall during the final weeks in which the barley crop was maturing. After the initial year of operation, a special-purpose insurance corporation was formed with the insurance premiums being used to hire the weather modification firm and to guarantee policyholders a minimum per acre income in the event that their Moravian barley crop was damaged by hail. All policyholders were barley growers under contract to the brewery.

A vocal opposition group formed claiming that the cloud seeding had reduced the amount of rainfall in the area. They were principally cattle growers and lettuce and potato farmers. They contacted the Governor and other state officials. In 1971 their anger was partially cooled for a time by two events. First was the action of a state legislative committee

in holding public hearings preparatory to writing a new weather modification law for consideration by the 1972 Colorado Legislature. Perhaps the new law would correct what they saw as significant grievances. Second, hail damage was so extensive that the insurance company was unable to pay the claims and declared bankruptcy!

But that did not end the cloud seeding effort. In 1972 the same barley growers reorganized and hired a different commercial weather modification firm. The head of this firm was more experienced than the previous operator and also appeared to be much better in dealing with public relations problems.

A new state weather modification law went into effect in midsummer. The leader of the opposition in prior years was appointed a member of the Advisory Committee to the state official who made the decisions on requests for weather modification permits. New leaders emerged in the opposition, and they started using a new name for their group. The opposition, feeling that 1972 was the worst drought in many years and that cloud seeding had made it even worse, appeared in large numbers at a public hearing on the request for a permit under the new law. The hearing examiner following the meeting recommended that the request be denied. The Advisory Committee quickly recommended that the permit be granted and it was. An attorney for the opposition filed an appeal in court challenging the decision of the state official granting the permit. Several weeks later the weather modifier's trailer was bombed. As of mid-November, 1972 there had been no arrests for the bombing, but a straw vote on the November 7 ballot showed a 4 to 1 vote against the use of weather modification in the valley.

Were the sponsoring growers of

Moravian barley getting any real change in precipitation or hail damage for their investment? Were other farmers and ranchers suffering losses from the cloud seeding activity?

In the absence of any hard data the answers were simply matters of opinion.

The year 1973 shaped up as another one of confrontation.

South Dakota State Weather Modification Program

Weather modification has been tried periodically in South Dakota in at least a half dozen areas almost every year since the mid-1950s. Both hail suppression and rain augmentation have been tried. Most of the efforts were locally sponsored on a year-to-year basis either by county tax funds or by donations from interested farmers and ranchers. Almost all these programs were intended to be operational rather than experimental. Since 1965 the Institute of Atmospheric Sciences, South Dakota School of Mines and Technology, has conducted one or more experimental projects every season.

In 1971 the South Dakota Legislature passed legislation putting the state squarely in the weather modification business. This included authorizing the state Weather Control Commission to develop a statewide plan for operational hail suppression and rain augmentation programs. The plans that were developed and put into operation starting May, 1972 contained the following features.

- Elected commissioners of each county could choose whether the county would participate or not.
- When enough contiguous countries requested participation they would be considered a possible district or area for inclusion in a program.
- The state and the counties shared the cost of the operation on a 75 percent-25 percent basis.
- The number of areas to be included in any given year depended on both the number of contiguous counties requesting participation and the funding level provided by the state legislature.
- Each operational district had a local advisory committee which could decide when soil conditions were such that cloud seeding for rain augmentation should *not* take place.

In 1972 more counties officially requested participation than could be served given the limited level of funding by the 1972 legislature. Operations for both hail suppression and rain augmentation were conducted during the period of May 1 to September 1. Full operations were conducted in two districts composed of 21 counties in the southeastern part of the state. Supplemental assistance was provided to several counties conducting their own programs in the western part of the state.

Evaluation of the effectiveness of the program was conducted with funding from the State of South Dakota, NOAA, and the Bureau of Reclamation.

By late 1972 two-thirds of the counties in the state had already requested participation in the 1973 program.

The National Hail Research Experiment, 1971 and 1972

Northeastern Colorado is often referred to as "hail alley." During the 1950s and 1960s there were several short-lived operational hail suppression programs in the area. Meteorologists from nearby Colorado State University at Fort Collins, Colorado, conducted research on hail suppression periodically during the

1960s. They also mailed questionnaires to gather data on hailfall throughout the area.

In 1971 the National Hail Research Experiment (NHRE) began installing equipment and running some initial tests of the various parts of *the system*. A few tests of the cloud seeding technique for hail suppression were carried out. The National Center for Atmospheric Research (NCAR) in Boulder, Colorado, has the management responsibility for the 5-year research project which is conducted under National Science Foundation (NSF) support and supervision. Other participating organizations included several universities, a commercial weather modification firm, and NOAA.

Social research was conducted through interviews with leaders of potentially interested groups and organizations in the area and with a random sample of adults in the experimental or target area, a surrounding or marginal area, and a selected control area. Interviews were conducted both before and after the hail season in 1971 and 1972. The findings for 1971 showed that approximately two-thirds of those residing in the experimental area favored NHRE. They believed that NHRE offers potential benefit through reduced hail damage and that scientific research in general is desirable. A consistent strong belief was expressed that local residents should have a major role in decisions regarding weather modification programs in their area, but they also anticipated that opportunity for such participation would not take place. By the end of the summer none of the opponents or proponents of NHRE had taken any action. This behavior was consistent with what they had said they would or wouldn't do. At the end of the season almost no one in the experimental area attributed either losses or benefits to the NHRE activities. Potential interest groups and organizations had shown little interest in the NHRE program.

Prior to the 1972 season NCAR established a Citizen's Council on Hail Research. It was composed of individuals representing various occupations who lived in or near the target area. This group had a monthly meeting with the NHRE scientists and administrators to discuss some phase of the NHRE effort. Members of the Council understood that they were not participating in a decision-making body. Despite considerable mention of the Council in the news media almost none of the respondents interviewed at the end of the 1972 season knew of the Council's existence.

The NHRE was fully operational during the 1972 season, but it had little immediate social impact.

VARIOUS PERSPECTIVES ON WEATHER MODIFICATION

Now let us try to briefly summarize the orientation toward weather modification of various classes of participants or onlookers.

It is a social fact that most human beings individually and collectively in groups and organizations seem to operate primarily on the basis of perceived self-interest. This is not to imply condemnation or praise, but simply to point to what *is* in the social world. Both the proponents and opponents of weather modification reflect this self-interest orientation.

Federal Elected Officials

A central aim of elected officials is to get reelected to their current or "better" positions; thus any President, U.S. Senator, or Representative tends to look on weather modification (and any other issue) as a potential means to the end of reelection. This can be seen in the fact that Federal Legislators whose constituents in general or home-based powerful interest groups are perceived to favor weather modification have been supporters of federally financed weather modification efforts.[14] Such projects have the potential for a two-pronged payoff for the elected official—the federal expenditures bring additional income into the state or district, and the project itself offers evidence that the politician is supporting a scientific effort to solve a significant problem for his constituents. Most such projects are described for public consumption as scientific "experiments." The federally funded programs of emergency drought relief in Texas, Arizona, and Oklahoma were sometimes also described as operational programs.

This point should not be overstressed, however. Apparently elected officials see weather modification as a relatively minor vehicle to carry them to reelection. To our knowledge few Senators or Congressmen have ever conducted a public opinion survey in the home area to ascertain what proportion of the voters favor the application of weather modification technology either experimentally or operationally. They do see to it where possible, however, that the first announcement of a federally funded weather modification program comes from their offices, but that is standard political procedure.

Within limits the President also has something to gain and little to lose from backing weather modification. When the Governor of a state, pointing to the provisions of the 1970 Disaster Relief Act, which authorizes the expenditure of federal funds to lessen the impact of an impending disaster such as drought, requests federal assistance through the Federal Disaster Assistance Adminis-

tration, it is to the President's advantage to approve the request. Never mind if the majority of atmospheric scientists in the country believe that such emergency drought relief cloud seeding efforts are premature. Most voters and powerful politicians are not scientists. If it is believed that an emergency rainmaking effort will win more votes than it will lose, the effort will be authorized even when there are very severe budget restraints. Reelection is the name of the game. But, again, it should be remembered that the use of such decision criteria is not unique to weather modification nor are the criteria used by members of only one political party. It is part and parcel of the political process.

Federal Agencies

The primary driving force in any organization is for increased autonomy, security, and prestige for the organization. Despite an almost unlimited range of motivations behind the actions of individual members within an organization, when viewed as a totality the organization seems to be almost single mindedly pursuing policies which, if unconstrained by external forces, will lead to improvements in the degree of autonomy, security, and prestige granted the organization by outside power centers.[10] This generalization applies to organizations of all types. Federal agencies are no exception.

It is true that the Congressionally authorized statement of a federal agency's "mission" does serve as the general definition of the limits of its domain. However, such mission statements typically include enough unspecific language so that agency officials can find adequate justification for a variety of interpretations of the mission. This makes it possible for the agency to find a mission justification for almost any program which its top officials seriously wish to conduct. The constraints on the range of programs an agency can pursue come far more from the political clout of its competitors than from its legally defined mission.

The perspective of any federal agency toward any weather modification program may be understood by examining how the program or a proposed program appears to contribute to the autonomy, security, and prestige of the agency. A weather modification program which is *only* scientifically important will play second fiddle to another program which offers *both* a potential scientific payoff and the promise of increased security through added appropriations. The decision to beef up support for one weather modification program rather than another may be justified in official explanations as due to its higher potential for successful accomplishment of its objectives or its inherent scientific merit or whatever. A significant but often unstated reason for the decision rests in the belief of the top agency officials that the favored program offers a greater potential for ensuring the health of the organization. In short, the most important (though not the only) question is, what will the program do for the agency as a whole? How much "sex appeal" does it have?

In any federal agency weather modification programs must compete with other programs for support. The expansion of weather modification efforts, therefore, usually must come at the expense of other programs in the agency and vice versa. The growth or decline of weather modification programs in any agency then depends not alone on their scientific merit or their anticipated contribution to agency income but also on the characteristics of competing programs within the agency.

State and Local Governmental Units

Relatively few state and local governmental units have significant involvement with weather modification efforts. A few state agencies such as the Texas Water Development Board have gone beyond the routine licensing and permit-granting functions,[1] but only the South Dakota Weather Control Commission is in a position where the success or failure of weather modification programs is directly related to the survival and prestige of the organization itself. The Illinois State Water Survey may move to a point in the mid-1970s where the success of its summer rain augmentation is important to its prestige.

Some municipalities and water conservancy districts in California have had weather modification as an integral part of their operations for up to 20 years. In most cases, however, the *results* of these weather modification efforts are unknown to the public and indeed very nearly unknown altogether because no routine evaluations of seeding effects are conducted. These programs seem to run on continued faith and inertia, along with the modest input of local tax revenues.

Research Oriented Versus Operations Oriented Groups

One of the most interesting and initially puzzling aspects of weather modification for the nonatmospheric scientist is the very high level of dissent among weather modification scientists and practitioners regarding the state of the art. In private conversations and in professional meetings one can hear expressions of sharp disagreement regarding a whole host of issues ranging from seeding agents and techniques to evaluation of measurement data.

Generally speaking, scientists employed in federal laboratories and academic scientists see weather modification as offering enormous promise for the future but claim that the only reasonable approach now is to conduct more and better experiments or "demonstration" projects. It is in their interest to have such a view accepted by the public and governmental officials and legislators. For the most part commercial seeders hold that they have developed and refined techniques for the various types of weather modification activities in which they engage. It is in their interest to have such a view accepted by the public and potential consumer groups, both governmental and private.

A few of the commercial firms concentrate on research projects, but a majority practice the state-of-the-art as they understand it; that is, they seek contracts to run operational efforts. They believe that they can do what they are paid to do—increase rain and snow and decrease hail. But even those that contend that the technology has already developed to the operational phase are willing to sign contracts for experimental work. Within the same state in 1972 one commercial operator conducted an operational hail suppression program financed by farmers in one location and simultaneously in another part of the state participated in an experimental hail suppression program financed by federal funds.

Organized Consumers

Organized consumers are sets of individuals, corporations, and governmental units who specifically hire weather modifiers to produce a change in the weather. They intend to profit in some manner directly from the hoped-for change in the weather (e.g., more precipitation, less hail). They have little or no interest in supporting financially attempts to increase knowledge about weather modification processes. Although it is perhaps inaccurate to say that they are all true believers in weather modification, they do seem to have a perspective that stops just short of blind faith. The typical expression is that the costs of an operational program are relatively low and the potential returns so high that it is silly not to take the risk. But these consumers seldom know with even minimal certainty what it is they are getting for their money because evaluation techniques are not usually incorporated into the program. (Some electric power companies have run evaluations for brief periods.) Thus in the absence of benefit data it seems that the consumer must have a high degree of faith unless he is satisfied with the social and psychological rewards he sees coming to him as the result of his financial participation. Oral tradition has it that during the 1950s and 1960s farmers and ranchers in the Dakotas would hire weather modifiers after a "dry" year or two and would not renew the contract after a year or two of normal rainfall. If true, this action suggests a belief in persistence in seasonal weather patterns combined with a belief in the efficacy of weather modification techniques.

It should be noted that the typical citizen whose taxes are used without his knowledge for funding either experimental or operational programs is not a consumer in the sense being discussed here. He does not knowingly make a decision to help finance the weather modification effort. He is essentially a nonparticipant in the process and he is likely to be unaware that a weather modification effort is being conducted in his area even though his taxes paid to the Internal Revenue Service are helping to pay for it.* This fact of noninvolvement is a major condition which has made government-funded field research programs possible.

Other Interest Groups

In addition to the organized consumers, weather modifiers, and their governmental sponsors, who is interested in weather modification? Our research results suggest that other interested groups are relatively few and far between. Those that do exist are primarily composed of individuals and corporations who believe that their economic interests are or may be threatened by a weather modification project. Only rarely have "environmental" groups shown any active interest. Those who perceive the possibility of economic benefit to themselves are much less likely to articulate that view and to take any action to support a weather modification program. A December, 1971 survey of all organizations and groups in South Dakota thought to be potentially interested in weather modification brought 60 replies. Although 25 percent of those responding had the topic of weather modification on a meeting agenda sometime during 1971, only 13 percent passed a resolution of some type or took an official position regarding weather modification. None contributed money either in support of or in opposition to weather modification, and none anticipated doing so in the following 6 months. This relative inaction should be interpreted in light of the findings presented in Table 22.2. If the officers of the responding interest groups

* In one area in Utah, for example, more than three-fourths of the residents interviewed in the target area had not even heard of the weather modification effort at the end of the first year of the program.

Table 22.2 Official's Estimated Value of Weather Modification to Members of South Dakota Interest Groups, 1971 ($N = 60$)

"What is your estimate of the probable value of weather modification to *most* of your members as regards the following items?"

Response (%)	Hail Reduction	Increase in Moisture
Very valuable	21	70
Moderately valuable	50	20
Slightly valuable	21	7
Of no particular value	8	3

are correct in their estimates of the probable value of weather modification to most members, we have further confirmation for the tendency of interest groups to sit idly by even when their officers recognize the potential benefits which organization members could reap from weather modification. Lack of information about *how* to take effective action might be a partial explanation for this organizational inactivity, a sense of powerlessness may be involved, or in the case of South Dakota, it may well be that the interest group leaders may believe that "someone else" is going to look after their interests and therefore no action is necessary.

Where most potential interest groups do not express their views or take any action toward a weather modification project, the few groups which do act generally have a very significant impact. Where there is mostly silence, even a little noise is very noticeable! Where previously established corporations or farmer or rancher associations decide to oppose a weather modification project their views receive more than courteous attention, especially where a project is publicly fi-

Table 22.3 Attitudes Toward Experimental Cloud Seeding Programs in General (Prior to Start of Local Project)

"It is a good idea to experiment with cloud seeding so that we can find out if it really does work."

Response (%)	Colorado (N = 168)	South Dakota (N = 182)	N.Y.-Mont.-Utah (N = 240)
Agree	75	87	89
Undecided	8	5	4
Disagree	17	7	6

nanced. Politicians may not "jump" when an individual complaint comes in but they very nearly do so when an established group claiming a large membership expresses concern over a weather modification project.

The Unorganized: Approvers, Disapprovers, and the Disinterested

One of the most consistent findings from survey research in six states covering five different types of weather modification projects is that most persons know very little about weather modification in general or about the local project. Despite the lack of information the majority in most areas have a generally favorable orientation toward weather modification. Illustrative findings may be seen in Tables 22.3 through 22.5.

When the "undecideds" are deleted from considerations, the findings show an overwhelming majority approving of weather modification in general and a very significant majority approving the proposed local weather modification project. In general, a higher proportion of

persons anticipate personal economic benefit than losses from the local weather modification project. But despite that favorable orientation, a significant proportion of respondents express misgivings about the impact of cloud seeding on the balance of nature and the congruence of cloud seeding programs with "God's plan for man and the weather" (see Table 22.6 and 22.7). Clearly some people give assent to weather modification while maintaining a negative "religio-natural" orientation toward it. They appear to be logically inconsistent in their views toward weather modification.

The best explanation for this apparent inconsistency is that these are matters of low salience for such persons. They know relatively little and care little about weather modification. Repeatedly we have found that respondents have very little factual knowledge about weather modification in general or about the local project. The majority state explicitly that they are not well informed in these matters (see Table 22.8). For most respondents prior to the start of their local proj-

Table 22.4 Attitudes Toward Local Weather Modification Project (Prior to Start of Local Project)

"Suppose that a cloud seeding program was proposed for your community or nearby, a program designed to (intent of local project). How would you feel about this idea?"

Response (%)	Colorado (N = 168)	South Dakota (N = 182)	N.Y.-Mont.-Utah (N = 240)
Favor	52	68	45
Undecided	33	18	32
Oppose	15	14	21

Table 22.5 Views[a] of Possible Personal Benefits or Losses Resulting from Local Weather Modification Project (Prior to Start of Local Project)

Response (%)	Florida (N = 101) (Rain Augmentation)	Colorado (N = 168) (Hail Suppression)	South Dakota (N = 182)	
			(Hail Suppression)	(Rain Augmentation)
Will benefit me	58	82	80	74
Neither loss nor benefit	8	12	17	20
Will cause me loss	25	3	1	2
Undecided, other or qualified response	9	3	2	3

[a] Florida sample included only persons engaged in agricultural activity whereas the Colorado and South Dakota samples included both rural and small town residents. Questions were phrased as follows:
Florida (Regarding cloud seeding during April and May): "Do you think that you will benefit from the cloud seeding or do you think that you will have losses?"
Colorado: "If scientists were able to suppress hail—reduce damage from hail—would it most likely be of economic benefit or harm to you, or would it probably make no difference?"
South Dakota: "If a cloud seeding program were able to (suppress hail, reduce hail damage, increase moisture), would you say it would probably be of economic benefit to you, harmful to you, or make no difference to you?"

ect these are hypothetical questions. They are not issues to which much thought has been given. Recognition of this cognitive dissonance* may come later, but initially the incongruence in their attitudes is not troublesome.

When those who have expressed opposition to the proposed local weather modification effort are asked why they take that view it becomes clear that they believe in the effectiveness of weather modification technology. They believe, as in the central Florida case and apparently in the San Luis Valley in Colorado, that cloud seeding will actually produce some change in precipitation and that changed

* For a discussion of resolution of cognitive dissonance regarding weather modification, see Haas, Boggs, and Bonner.[7]

Table 22.6 Views[a] of Effect of Cloud Seeding on the Balance of Nature (Prior to Start of Local Project)

"Even when carefully controlled, cloud seeding programs are very likely to upset the balance of nature."

Response (%)	Colorado (N = 168)	Florida (N = 101)	South Dakota (N = 181)
Agree	54	30	37
Undecided	19	38	22
Disagree	27	32	41

[a] Florida sample included only persons engaged in agricultural activity, whereas the Colorado and South Dakota samples included both rural and small-town residents.

Table 22.7   Views[a] of Cloud Seeding As It Relates to God's Plans for Man and the Weather (Prior to Start of Local Project)

"Cloud seeding probably violates God's plans for man and the weather."

Response (%)	Colorado ($N = 168$)	Florida ($N = 101$)	South Dakota ($N = 182$)
Agree	46	30	38
Undecided	9	12	12
Disagree	45	58	50

[a] Florida sample included only persons engaged in agricultural activity whereas the Colorado and South Dakota samples included both rural and small-town residents.

set of weather events will affect them negatively. Their verbalized response is a clear statement of concern over the impact of the cloud seeding on their personal economic self-interest.

But here again expressed attitude is one thing and action based on that attitude is quite another matter. In northeast Colorado (1971) less than one-third of those expressing a negative attitude toward the project said they would take any action at all to support that view. By the close of the first season no one had taken any action. The initial disapprovers did not feel that they had been harmed in any way by the cloud seeding.

In central Florida (1971) the disapprovers were almost exclusively tomato growers who felt that additional rainfall at harvest time, which coincided with the planned cloud seeding period, would cause significant losses. They felt that they should have been consulted but they weren't. Almost to a man they said they would back their views with action but they were very pessimistic about their chances of affecting any change. They felt frustrated and helpless. As in northeast

Colorado, almost no one mentioned the possibility of legal action. As it turned out their anticipations were correct. Most of them did take action, principally talking to other tomato growers, but it was ineffective just as they thought it would be. Furthermore, 61 percent of the tomato growers were convinced that they had suffered losses from the cloud seeding and 89 percent believed that other persons had suffered losses. They were bitter but no one initiated court action.

As with other public issues the unorganized, so long as they stay unorganized, have little impact on public policy matters.

WEATHER MODIFICATION AND CULTURAL LAG

There is a view, now widely mentioned among thoughtful observers of modern

Table 22.8   Perceived Level of Knowledge About the Local Weather Modification Project (Prior to Start of Project)

"How well informed do you feel you are about the (proposed local weather modification project)?"

	Response (%)		
	Not at All	Slightly	Fairly/ Very Well
Colorado ($N = 168$)			
Target area	49	46	5
Adjacent area	76	22	2
Control area	76	24	—
South Dakota ($N = 182$)			
Southeast area	66	34	0
South-central area	45	45	10
Southwest area	35	60	5

society, that there is a tendency for the material or technical components of culture to change more rapidly than the related nonmaterial aspects of the same culture. In 1922 William Ogburn noted that the technological aspects of culture tend to move ahead rapidly in their rate of change while norms, beliefs, values, and patterns of social organization change much more slowly. The consequence of such differing rates of change is a phenomenon he termed *cultural lag*.[13] For example, the factory system was introduced with an ensuing sharp increase in worker injuries and deaths, but it was not until some years later that safety devices and accident compensation were required by law. Automobiles came into general use long before any emission control devices were required.

Is there cultural lag with regard to the application of weather modification technology? On the face of it one might be inclined to answer in the affirmative. But let us examine the concept more closely. A lag is said to exist when a technology has come into use without "appropriate" adjustments in the normative views, including the law, and practices of the society. But what is "appropriate" is a value decision. And who shall decide what the appropriate adjustments should be? Should those engaged in weather modification be the spokesmen for society? Shall the consumers who knowingly hire the weather modifiers serve as societal arbiters? Or are those appropriate adjustment decisions to be left to the legislative process with the courts as the interpreter of legislative intent?

However one may answer these questions it is a fact that cloud seeding has repeatedly been practiced in the absence of any significant societal adjustments whatsoever. Although some state legislatures have slowly taken note of weather modification, the adjustments stipulated by their legislative acts have for the most part been minimal. And the U.S. Congress, while appropriating funds for weather modification field experiments and demonstration projects for more than a decade, failed to enact a comprehensive and readily enforceable weather modification report law. In short, if one believes that social and legal adjustments relevant to cloud seeding should take place at all, then he is forced to the conclusion that such adjustments have indeed lagged behind the application of cloud seeding technology.[7]

POLITICAL AND ECONOMIC IMPLICATIONS

Political and economic considerations usually blend together in the public decision-making process. Members of Congress and federal agency officials advocating the expenditures of tax revenues for weather modification frequently use economic arguments to support their favorite program proposals. The benefit cost numbers game has become a standard part of the political process. Every agency has its favorite set of numbers purporting to show that the *potential* benefits from a proposed weather modification program greatly exceed the costs thereof.[2] The fact that these numbers represent estimates usually put together by noneconomists without benefit of adequate data has not been seriously challenged in the literature. Indeed, in one case where a group of economists working under government contract reportedly concluded that the benefit cost ratio of a proposed weather modification program was approximately one to one, their report was sent back to them for further consideration![17] This is not to say that benefit cost ratios may not turn out to be very favorable, but rather that frequently political considerations appear

Table 22.9 Attitudes Toward the Involvement of Federal Scientists in Local Cloud Seeding Programs (Prior to Start of the Local Program)

"Federal government scientists should not be involved with cloud seeding programs in (name of state)."

Response (%)	Florida (N = 101)	Colorado (N = 168)	South Dakota (N = 182)
Agree	30	35	26
Undecided	14	20	20
Disagree	56	45	54

to have taken precedence over scholarship in the computing of the ratios. Furthermore, it may be noted that the benefit cost numbers game was being played in the political arena long before there was federal financing of weather modification programs.

A second area of interest in the political realm deals with federal-state relations. In general, it may be said that federal agencies have sought and gained the approval of top state officials before embarking on a weather modification effort *in* the state. But given the uncertainty regarding extended area effects of cloud seeding the same cannot be said for clearance in advance with officials of adjacent states which could conceivably be affected by the cloud seeding effort. This potential problem may be seen in its most complex form in considering attempts at hurricane modification. Officials of how many states, if any, should be consulted when a hurricane has developed in the Caribbean? A pattern has yet to be established.

On the hunch that citizen views of a weather modification project might be conditioned by antipathy toward perceived federal "intervention" in state affairs the issue was treated explicitly in interviews in three states. The findings are summarized in Table 22.9.

Approximately one-third of the respondents feel that federal scientists should not be involved in cloud seeding programs in their state; however, in the Florida case at least, the view is markedly different when severe drought conditions are considered. When presented with the statement "when there are severe drought problems in Florida, it is all right for scientists of the federal government to try to help." 81 percent expressed agreement.[5]

But the acceptance of the involvement of federal scientists in a local program is not equated with federal *control*, as may be seen in Table 22.10. State control is viewed as preferable.

It appears quite likely that the passage of the 1970 Disaster Relief Act marked a turning point in the practice of weather modification. That Act includes a provision authorizing the expenditure of federal funds to lessen the impact of an impending disaster. In 1971 a new precedent was set. The Governors of Texas, Arizona, and Oklahoma believed

Table 22.10 Attitudes Toward State vs. Federal Control of Weather Modification Experiments (Prior to Start of the Local Program)

"If there are going to be weather modification experiments such as cloud seeding, individual states rather than the Federal government should control and conduct them."

Response (%)	Colorado (N = 168)	South Dakota (N = 182)
Agree	57	59
Undecided	11	17
Disagree	32	24

that there was an impending disaster from drought. They approached the President's Office of Emergency Preparedness requesting that the relevant provisions of the Disaster Relief Act be activated. Federal funds were expended for cloud seeding for rain enhancement through the Bureau of Reclamation with an assist from the U.S. Air Force in the Texas case. These efforts, plus the semioperational drought relief effort of the National Oceanic and Atmospheric Administration in the spring and early summer of 1971 in Florida, put the federal government squarely in the middle of operational programs for the first time.

For public information purposes these drought relief efforts were occasionally referred to as experimental projects, but by the usual scientific standards they were a long way from being experiments.

One does not always have to be a prophet to recognize the start of a trend. Any Governor who becomes convinced that a drought has started in a part of his state is very likely to call on the Federal Disaster Assistance Administration for federal assistance. As in the 1971 Oklahoma case, state officials may not prefer the particular contractors hired by the responsible federal agency, but when federal funds become available the agency with the funds in hand has considerable clout in such negotiations.

In late 1971 Congress passed the Weather Modification Reporting Law (PL 92–205) which authorized the Secretary of Commerce to establish reporting procedures for all nonfederally sponsored weather modification projects. The Weather Modification Association (WMA), many of whose members represent commercial weather modification firms, did not urge its members to express their views when the legislation was being considered in the House and Senate. After the measure because law, however, and drafts of proposed reporting requirements became known to WMA members, there was great consternation. The anticipated reporting requirements were the focus of much heated discussion among WMA members in early 1972. Two principal concerns were expressed. First was the additional cost involved in keeping the required records. To the degree that the requirements established by the Secretary of Commerce entailed extensive record keeping and perhaps additional instrumentation intended to secure data on physical effects, the requirements, it was argued, could threaten the financial solvency of some firms.

A second concern centered on the amount of time permitted between notifying the Secretary of Commerce of the intention to begin a cloud seeding project and the actual starting date of the project. Commercial operators report that it is not unusual for the period from the signing of a contract to the beginning of seeding to be less than 1 week. An early draft of proposed regulations called for a 3-day advance notice. Such a requirement could be the death blow to any number of potential projects, the commercial weather modifiers contended.

The WMA at its February 25, 1972, meeting passed a resolution that, in effect, called on the Secretary of Commerce to alter the proposed reporting regulations to make them less stringent. Specifically the resolution called for, among other things, deletion of any requirement for reporting in advance on anticipated weather modification efforts, simplified record keeping requirements and formal written notice, and an opportunity to comply before any punitive action against the weather modifiers is taken.[12]

The rules and regulations on "Maintaining Records and Submitting

Reports on Weather Modification Activity" to be administered by the Administrator of NOAA became effective on November 1, 1972. The regulations require that the Administrator of NOAA receive 10 days in advance a written report of intent to start a weather modification project. Exceptions to the 10-day requirement may be made provided that the "tardy" report include an explanation deemed adequate by the Administrator. A chronological record of activities must be kept, preferably in the form of a daily log including starting and ending time for activity each day, position of each aircraft or location of each item of weather modification apparatus during each mission, and rates of dispersal and total amounts of each seeding agent used. For each airborne weather modification run records must be maintained on altitude, air speed, release points of modification agents, temperature at release altitude, etc. All measurements made of precipitation in target and control areas and any "unusual results" must also be recorded. These and other records must be maintained for 5 years after completion of the activity to which they relate.[3]

The 1971 reporting legislation passed by Congress does not require federal agencies to report weather modification activity; however, President Nixon, by a 1972 Executive Order, acted to make the reporting requirements uniform. Presumably information on "secret" federally funded weather modification projects, should they exist, will still be secure from public disclosure.[15] Certain information on nonfederal projects may also be withheld from public disclosure depending on the judgment of the Administrator of NOAA:[3]

(c) Persons reporting weather modification projects or related activities shall specifically identify all information that they consider not to be subject to public disclosure under terms of Public Law 92–205 and provide reasons in support thereof. A determination as to whether or not reported information is subject to public dissemination shall be made by the Administrator.

HOW SHALL THE INTERESTS OF THE UNORGANIZED BE PROTECTED?

As indicated in the discussion of cultural lag, there has been little institutional adjustment accompanying the introduction and practice of the new technology of weather modification. On the assumption that the unorganized citizens on whom the technology is being imposed might have views on potential new decision-making procedures, this issue was treated in interviews in several states. Earlier research in New York, Montana, and Utah showed that the preponderant preference was for decision-making regarding a proposed weather modification project to be at the local level. But the respondents also thought that in fact such decisions would probably be made at "higher" levels.[7] In Colorado and South Dakota we asked respondents. "Who do you think *should* make the decision whether or not a cloud seeding program should be allowed in your area?" and "Who do you think *will* make the decision whether or not a cloud seeding program should be allowed in your area?". The findings in Table 22.11 indicate that more than half of those expressing an opinion believe that local residents or local government officials should be involved in that type of decision. The largest plurality hold that it should be the *sole* decision of local residents.

But as in the earlier research findings these citizens don't think that it will happen that way! The proportion answering "don't know" increases and the responses tend to be more evenly distributed across

Table 22.11   Citizen Views of Who _Should_ and Who _Will_ Make the Decision Regarding a Local Cloud Seeding Project (Prior to Start of Local Program)

Response (%)	Colorado (N = 168)		South Dakota (N = 182)	
	Should	Will	Should	Will
Local residents	58	16	36	7
Local government	4	2	7	13
County and state government	a	a	9	15
State government	8	14	7	21
State and federal government	7	15	6	8
Federal government	7	18	1	8
Scientists	7	13	7	1
Other, including combinations[b]	5	8	24	7
Don't know	4	14	3	20

[a] Not included in Colorado survey.

[b] Includes 6% who said, "farmers and ranchers" without specifying area of residence.

the answer categories; however, South Dakotans see the state government as playing the dominant role, whereas Coloradoans are more likely to anticipate federal involvement in the basic decision. This is in keeping with what had already been decided by the state and federal governments before the interviews.

Since the dominant view is that there should be local involvement in decision making, we attempted to present to each respondent several decision-making procedures or plans which could be used in the future.* Each respondent was asked to consider carefully all the alternatives and then to select the one viewed as "most satisfactory" and the one viewed as "least satisfactory." The rank order of expressed preferences may be seen in Table 22.12.

In general, across the three states the most satisfactory mechanism is seen as "a referendum submitted to the vote of all owners and operators of agricultural land in the affected area" (agricultural referendum). The most frequent expla-

nation given for selecting this plan was that these are the people who may be affected so they should decide. It is equally

* In the Florida study, the alternative procedures were described in some detail. Example: "A referendum would be submitted to the vote of _all_ citizens living in the affected area. It would be a 'yes' or 'no' vote on the right of the flood control district or other state water management agency to operate a cloud seeding program. The proposal to be voted on would state the _criteria_ to be followed in the operation of the cloud seeding program; e.g., 'cloud seeding shall not begin unless the water level has become critical.'" After using such detailed descriptions in the interviews, we observed that most respondents seemed to be responding to only one part of the procedure being described—_who_ was involved in the decision making. Subsequently, in the Colorado research, we simplified the description of the alternative procedures. The lead-in statement asked the respondent to note that each alternative would involve both a decision on whether or not a cloud seeding program should be permitted and, if so, a decision regarding the _guidelines_ to be followed in conducting the program. Following that each alternative plan was stated very concisely; for example, "The decision should be made by a referendum submitted to a vote of _all citizens_ living in the affected area."

Table 22.12   Rank Order of Citizen Preferences Regarding Decision-Making Procedures to be Used in Deciding About A Weather Modification Effort (Prior to Start of Local Program)

	Florida[a] (N = 101)		Colorado (N = 168)		South Dakota (N = 179)	
	Most Satisfactory	Least Satisfactory	Most Satisfactory	Least Satisfactory	Most Satisfactory	Least Satisfactory
All-citizen referendum	4	1	2	3	1	2
Agriculture referendum	1	4	1	4	2.5	3
Elected control board	2	5	3	5	2.5	4
Appointed advisory board	3	2	4	2	4	5
Scientists	5	3	5	1	5	1

[a] All respondents in sample were engaged in agricultural activity of some type.

clear that these respondents would prefer any of the named political procedures to decision making by scientists *alone*.* The view seems to be that scientists may not understand the needs and problems of the agriculturalists in the affected area. An appointed advisory board with the responsibility to advise scientists in the conduct of a program was seen as somewhat more acceptable.

The response to several items in our interview schedule show that the majority have a positive orientation toward science and scientific experimentation in general, but these citizens also hold to a generalized belief that local residents or their local officials, or both, should play an important part in making decisions about local weather modification programs. When it comes to this type of decision, they would prefer to trust their own judgment over that of scientists.

* In Colorado and South Dakota the description was, "The decision should be left entirely up to the scientists conducting the program." In Florida the choice, "the way it was done this time," was described as one in which the Governor asked the federal scientists to conduct a cloud seeding program, with the location, time, and so forth being decided by mutual agreement.

Whether these citizens are concerned about the power of the organized and special interest groups or not, they express preferences for institutional mechanisms that allow the unorganized some protection from such groups through the use of the vote. Although they have some indirect voice in such matters now in that they elect state officials who establish general policies for weather modification projects, special referenda would assure the opportunity for more direct influence on weather modification efforts which could have a direct effect on them.

HOW SHALL THE INTERESTS OF THE MINORITY IN A WEATHER MODIFICATION AREA BE PROTECTED?

It is rare when scientists, public officials, and weather modifiers planning a weather modification project *know* in any meaningful sense how the citizens in the target area view a proposed effort. It is not unusual, however, for such planners to *claim* that they know "how people around here feel." When asked how they know that their claims are correct they will refer to a few casual, nonrandom conversations that they have had with local residents. From these they will contend that they can make empirically based generalizations about what the people are thinking. Repeatedly, we have found that even those atmospheric scientists who are very rigorous in the design and conduct of weather modification experiments find it convenient to ignore many of the most basic of scientific procedures in reaching conclusions about social facts.

Let us assume for a moment that those responsible for planning a weather modification effort do in fact know with an acceptable degree of certainty that a large majority of the target area residents favor a proposed project. What responsibility, if any, do the planners have for the minority who are opposed, because they are worried about potential losses resulting from the cloud seeding? Does the minority have any "rights" in such a setting?

There is no established mechanism or institutional arrangement in our society which protects the rights of such a minority, although there are a few standard ways for such individuals to express their grievances. A person can write a protest letter to the editor of a local newspaper, speak up on a radio "talk show," contact an elected public official, contact someone associated with the proposed weather modification project or, if sufficiently wealthy, appeal for relief in a court of law. In some states the individual has a right to appear and speak at a required public hearing when a weather modification permit is being requested. *Evidence suggests that all these approaches are essentially ineffective unless a large number of persons combine and organize their efforts and take political action prior to the start of the cloud seeding effort.* Realistically, then, a person opposed to a proposed weather modification project must operate in an institutional "no man's land." Small wonder that the opposers seldom try to influence the planning and conduct of a cloud seeding project and even less frequently have any success when they do try.

As Ray Davis (Chapter 21) and Howard Taubenfield[16] have pointed out, getting compensation for damage caused by cloud seeding is also an uncharted sea. Any individual or business firm believing that damage due to cloud seeding has occurred faces an almost insurmountable set of requirements in the search for compensation.

If there are uninsured losses, should compensation be paid? This question assumes that it is possible to establish a cause-effect link between the cloud seeding action and the loss. Apparently it is rare that this can be done to the satisfaction of an atmospheric scientist or in a court of law. But the question is important nevertheless, because if the claims for the effectiveness of cloud seeding techniques are to be taken seriously then the issue is likely to be raised repeatedly in the future.

If compensation is to be made, who should pay? The respondents in Florida and Colorado were asked these two open-ended questions about compensation. Surprisingly, only 46 percent in Florida and 40 percent in Colorado thought that there should be compensation for uninsured losses due to weather modification. And many of these persons qualified their affirmative response. The proportion of respondents who were "unsure" (Florida, 13 percent; Colorado, 30 percent) also suggests that these citizens are at least vaguely aware of some of the many complex problems involved in finding a workable solution to this issue.

The data summarized in Table 22.13 refer to a drought relief effort in Florida and a hail suppression experiment in Colorado. Although both projects were

Table 22.13 Citizen Views of Who Should Pay for Uninsured Losses Resulting From A Weather Modification Project (Prior to Start of Local Program)

Response (%)	Florida (N = 101)	Colorado (N = 168)
Federal government	29	19
State government	20	0
Federal and state governments	31	8
Those conducting program	20	56
Other	0	17

primarily or entirely funded by the federal government, this fact was not mentioned in the interview. An agency of the state government was actively involved in the Florida program while the Colorado project was being conducted by several federal agencies, individual university research groups, and a nonprofit consortium of universities under NSF sponsorship, the National Center for Atmospheric Research. Since this question was asked near the close of the interview the respondent could have known, at least approximately, which agencies or groups were involved in the local program. Basically, the respondents seem to be saying that "whoever is running the show should pay the bills." There was a notable absence of the view that "those who directly benefit from the improved weather should pay for any losses."

It is, perhaps, too much to hope for in the near future, but serious consideration of some administrative mechanism for dealing with damage claims would seem to be in order. Some set of procedures analogous to no-fault insurance might at least deal with the run-of-the-mill, small-claim cases. Or if such an approach is not generally feasible then at least federally funded programs could have procedures patterned after those of the Atomic Energy Commission. If federal tax funds are to be used to conduct cloud seeding efforts designed to benefit some of the citizen taxpayers, why not use some of those funds also to help the minority who have reaped disbenefits? As one man said, "Why should my taxes be used to hurt me?"

WHAT FACTORS ARE ASSOCIATED WITH CONFLICT OVER PLANNED WEATHER MODIFICATION EFFORTS?

There are three basic findings which should be noted here. First, for the vast majority of weather modification projects across the country *no* organized or collective resistance has developed. The few cases where conflict has developed are discussed repeatedly among weather modifiers. Second, weather modifiers express a fear that there is a large measure of underlying hostility or potential hostility toward weather modification among members of the general public. This fear seems to rest in part on the weather modifiers' memory of those few cases where organized protest developed and on the weather modifiers' experience where one or a few worried or disgruntled residents expressed some concern about a cloud seeding program. These negative responses seem to us to be exaggerated in the thinking of the weather modifier. Third, there is a tendency by weather modifiers to be somewhat secretive about field programs. It is generally assumed that under most conditions it is better not to inform the public about what is being planned or is being done, better to let sleeping dogs lie. This view is congruent with the notion that the conduct of social research in a weather modification area is likely to "stir up trouble" for the weather modifier. An experimental test of potential "interview effect" was carried out in northeast Colorado in 1971 and 1972. No measurable effect on the respondents was found.[8, 9]

What is known about the factors associated with conflict over planned weather modification? Not very much. In our effort to examine this issue we have made a distinction between strictly individual action and organized resistance or protest. In almost any area there are likely to be a few disgruntled individuals who protest against something or other with considerable frequency. They in no sense, represent the general public or even specialized interest groups. Such sporadic behavior appears to be almost a constant across communities. Organized resistance or protest is a relatively rare and distinctly different phenomenon. Thus we discuss those conditions associated with the actions of two or more persons working together in the planning or conducting of resistance to a weather modification effort. The focus is on organized response.

Over the past several years we have gathered data on this aspect of weather modification in 25 locations around the country. The generalizations listed below should be viewed as hypotheses which have emerged from the data gathering process and from general sociological theory. They have not been tested as yet in any rigorous sense. They are presented only to give the reader a perspective on the kinds of conditions which *may* be associated systematically with organized resistance to precipitation and hail weather modification.

Factors Hypothesized to be Associated With Conflict

1. (Assuming there has been a past weather modification effort in or near the area under consideration) in the past there has been economic loss (within or near the area) *attributed* by local residents to past weather modification efforts.

2. Objective analysis indicates there is a high probability of some economic loss during the current weather modification effort. The economic loss must be demonstrably related to the weather modification activity.

3. Negative "unusual" weather events or conditions occur during the general time period in which the weather modification activity is going on. These "unusual" events or conditions may be episodic or persistent in character, (e.g., flash flooding or drought).

4. There is evidence of significant antipathy by local power groups and/or community influentials toward the weather modifiers, their sponsors or affiliated groups, and organizations.

5. Objective analysis indicates that within and near the area there are or will be some losers but few, if any, gainers as a consequence of the weather modification effort (weather modifiers and their sponsors are excluded from this consideration).

6. Both the weather modifiers and their principal sponsors are external to the local area.

7. The losers have readily available economic resources or readily utilizable authority or interpersonal influence which could be used in support of losers' complaints.

8. Objective analysis indicates that losers have high potential for mobilization.

9. Prior to the onset of protest or conflict, the behavior of the weather modifiers and/or their associates toward losers or potential losers is characterized as secretive, deceptive, abrasive, threatening, or condenscending.

10. Prior to the onset of protest or

conflict, weather modifiers and/or their associates make no effort to negotiate "protection," compensation mechanisms, and/or trade-offs with the actual or potential losers.

11. Prior to the onset of protest or conflict, weather modifiers and/or their associates make no effort to consult with or make advisory and/or decision-making arrangements that include actual or potential losers.

12. The losers have significant bases for concluding that their "rights" have been or will be violated. ("Significant bases" include such conceptions as long-term tenure on the land, significant investment which may be harmed, large number of persons and/or units involved in that type of investment, etc.)

13. There is evidence of significant antipathy of mass media officials toward the weather modifiers and their associates.

Although it is too early to specify now, it seems likely that some of these hypothesized conditions are necessary but not sufficient to produce conflict. Some conditions probably will need to be weighted more heavily than others when a model is designed to forecast the onset of conflict.

As the number of weather modification efforts around the country increases it should be possible to get approximate tests of these hypotheses. On the other hand, if new approaches are developed for local participation in decision-making and if procedures are perfected so that any bona fide loser can be assured of prompt, fair adjudication of claims for compensation, then organized resistance and ensuing conflict may largely disappear. The most severe test of the adequacy of new institutional adjustments will come when there are a large number of operational programs which encounter a spate of "unusual" negative weather

events such as flash flooding, "freak" snow storms, and protracted dry spells.

ACKNOWLEDGMENTS

The research on which this article is based was conducted under the following grants or contracts: NSF GA-28364 and GA-18724, NOAA Contract No. E-22-89-71 (N), and National Center for Atmospheric Research Contract No. NCAR 173-71. The following made significant contributions to the various research efforts: Suzanne Ageton, Keith S. Boggs, E. J. Bonner, Barbara Farhar, Donald Pfost, Sigmund Krane, and Patricia Trainer.

REFERENCES

1. Carr, John T., Jr., Weather modification in Texas, *J. Weather Modif.* **4**(1), 6–16, 1972.

2. Federal Council for Science and Technology, National Atmospheric Sciences Program, Fiscal Year 1972, *Interdepartmental Committee for Atmospheric Sciences, Report 15*, March 1971.

3. *Federal Register,* **37**(208), 22974–77, Oct. 27, 1972.

4. Haas, J. Eugene, Response to planned weather modification: implications for urban resource management, in *Proceedings of 1970 Western Resources Conference*, 1970, pp. 251–257.

5. Haas, J. Eugene, *Socio-economic Implications of the April-May 1971 Florida Rain Augmentation Program*, Final report submitted to the Environmental Research Laboratories, NOAA, Sept. 1971, 74 pp.

6. Haas, J. Eugene, Keith S. Boggs, and E. J. Bonner, Weather modification and the decision process, *Environment and Behavior,* **3**, 179–189, 1971.

7. Haas, J. Eugene, Keith S. Boggs, and E. J. Bonner, Science, technology and the public: the case of planned weather modification, in William Burch, et al., Eds., *Social Behavior, Natural Resources and the Environment,* Harper and Row, New York, 1972, pp. 151–173.

8. Haas, J. Eugene and Donald Pfost, Social implications of the National Hail Research

Experiment, *1971 Final report to the National Center for Atmospheric Research*, Feb. 1972 (mimeo).

9. Haas, J. Eugene and Sigmund Krane, Social implications of the National Hail Research Experiment, *1972 Final Report to the National Center for Atmospheric Research*, Feb. 1973 (mimeo).

10. Haas, J. Eugene and Thomas E. Drabek, *Complex Organizations: A Sociological Perspective*, Macmillan, New York, 1973.

11. Lambright, W. Henry, Government and technological innovation: weather modification as a case in point, *Public Administration Rev., 72,* 1–10, Jan.-Feb. 1972.

12. Minutes of the February 24–25, 1972 meeting of the *Weather Modification Association*, Appendix III.

13. Ogburn, William F., *Social Change*, Viking, New York, 1922.

14. Oppenheimer, Jack V. and W. Henry Lambright, Technology assessment and weather modification, *Southern California Law Rev., 45,* (2), 570–595, 1972.

15. Shapley, Deborah, Rainmaking: Rumored use over Laos alarms experts, scientists, *Science* **1216,** 175–4040, 1972.

16. Taubenfeld, Howard Ed., *Controlling the Weather: A Study of Law and Regulatory Procedures*, Dunellen, New York, 1970.

17. *Technology Assessment of Winter Orographic Snowpack Augmentation in the Upper Colorado River Basin*, Summary Report, Stanford Research Institute, Menlo Park, Calif., 1972.

18. White, Gilbert F., *Strategies of American Water Management*, University of Michigan Press, Ann Arbor, 1969.

Index

Aburakawa, H., 116
Academy of Sciences, 15, 71
Accelerometers, 154
Accretional growth of ice crystals, 290
Acetone AgI generator, 337, 338
Acetone Solution, 344
Acoustic, Doppler, 190
  shock waves, 32, 525
  sounding system, 190
Adams v. California, 772-73, 776
Adderly, E. E., 24
Additive hypothesis, 212
Adiabatic process, 98, 99, 100, 103, 527, 662
Adjustment decisions, 801
Adjustment Mechanisms, 787
Advection Fog, See Fog
Advisory Committee on Weather Control, 72, 294,
    780
Aeolian (wind) effects, 182
Aerometric Research, Inc., 74
Aerosols, 181, 645, 678, 700
  background, 182
  chemical properties of, 187
  formation of, 182
  natural, 182
Africa, 133, 261, 271, 359
  Congo, 6
  North African Desert, 475
Aganin, M. A., 390
Agee, E. M., 758
AgI seeding, 261. See also Silver iodide
Ageton, S., 810
Aggregation, effect of turbulence on the rate of,
    393
Agricultural referendum, 805
Ahlquist, N. C., 746
Air, as a dielectric, 596
  columns, 102
  density, 201
Airborne measurements, 757
Aircraft, airflow around fuselage, 164
  angle of attack, 153, 164
  attitude of, 153
  contrails, 722
  penetrations of cumulus clouds, 608
  research, 549
  systems for delivering seeding agents, 193
  trailing wire, 143
Air Force, 25, 143, 155

Arnold Test Center, 198
  Cambridge Research Laboratory (AFCRL), 173
Airplane, 420
Airquakes, 5
Air-sea exchange, 696
Air Transport Association, 367
Air Weather Service, 367, 369, 379
Air speed, indicated, 145, 155
Air speed, true, 145, 155
Aitken, J., 6, 7
Aitken nuclei, 7, 97, 745, 757
Aitken nuclei counter, 6, 185, 186, 702
Alaka, M. A., 528
Alaska, 48, 359, 361, 367, 381, 781
Alazan Valley, 419
Alazani rockets, 420, 423
Albedo, 400, 590, 637, 727
  of clouds, 183
  of many surfaces, 716
Albrecht, F., 10
Aleutian lows, 664
Alexander, V., 20
Alleghenie Mountains, 499
Alternative hypothesis, 218
Aluminum, 194
  foil impactor, 165, 168, 179, 473, 483
  oxide, 758
Amazon River Basin, 672
American Association for the Advancement of
    Science, 516
American commercial firms, 785
American Meteorological Society, 13, 25, 33, 34,
    70, 72
  Bulletin, 19, 34, 84, 85
  Monographs, 28
American Society of Civil Engineers, 86
Ammonium iodide, 193
Ammonium nitrate-area-water, 375
Aneroid barometer, 566
Angell, J. K., 739
Angular momentum, 510, 530, 534
Animas River, Colorado, 312
Antarctic, 342, 658, 677
Anthes, R. A., 539
Anti-Hail Service, USSR, 421
Anticyclone, 93, 103
Anthropogenic sources, 182
Apollo 10 astronauts, 239
Appalachian Valleys, 360

Appleman, H., 710
Arctic cap, 677
Arctic Ocean, 399
Areal effect, caused by seeding, 349
Arenberg, D. L., 10
Argentina experiment, 31, 32
Arizona, aerial seeding, 69
  drought relief effort, 791, 794, 802
  experiment, 29, 268
  precipitation amounts, 285
  Program II, 489, 490
  Republic, 16
  Santa Catalina Mountains, 490
  seeding experiments on single cumuli, 261
Army Signal Corps Engineering Laboratories, 24,
    25, 27
Arnett, A. B., 560
Artificial cloud nucleation (ACN), 13, 73
Artificial crystallization, 420
Artificial freezing nuclei, 246
Ashby, W. C., 757
Asia, Southeast, 371
Ashworth, J. R., 755
Asphalt coatins, 273
Astronomical variations, 719
Asymmetric features, 545
Aitkinson, B. W., 590, 754
Atlantic Ocean, 665
Atlantic Research Company, 194
Atmosphere, general circulation of the, 518
  large scale circulation, 400
Atmosphere-ocean system, 720
Atmospheric, carbon dioxide, 691, 692
  cooling, 708
  dust loadings, global trends, 705
  electrical circuit, 596
  energy sources, 573
  opacity, 633
  oxygen, depletion of, 687
  particles, 700
  particles, indirect measure, 606
  processes, 93
  radiation balance, 708
  rights, Texas Landowners, 769
  transparency, caused by volcanic dust, 720
  turbidity, 706, 722
  water management, 326-27
  water, potential of, 351
  waters, ownership of, 768
Atmospheric Physics and Chemistry Laboratory
    (NOAA), 75, 318
Atmospheric Science, 37
Atmospheric Water Resources Management Pro-
    gram, *see* Projects
Atmospheric Water Resources Research Group, 75
Atmospherics, Inc., 47, 77
Atomic Energy Committee (AEC), 201, 808
Atwater, M. A., 730

Aufdermauer, A. N., 600
Australia, 52, 210, 247, 261, 274
  cloud seeding, 23-24, 432-453
  currents, 359
  law, 775
  Western, 666
Australian Academy of Science, 85
Australian experiment, CSIRO, 31, 304
Australian scientists, 275, 785
Australian Weather Bureau, 23
Automatic counting systems, 178
Autonomy, 794

Background effects, 218
Bali, Indonesian Island, 706
Ball Brothers Instruments, 154
Ball variometer, 154
Balloon-borne sondes, 144
Baltic Sea, 752
Baltimore, Maryland, 731
Barbados, BWI, 239, 330
Barbados Oceanographic and Meteorological Ex-
    pedition (BOMEX), 152
Barcilon, A. I., 579
Barger, G. L., 212
Barnes, C. L., 47, 67
Barnes, S. L., 582
Baroclinic instability, 649, 658, 660
Bartlett, J. T., 480
Bates, F. C., 561, 576
Bathymetry, 499
Bathythermographs, 144
Battan, L. J., 26, 31, 74, 76, 490
Battle, J., 47, 67, 85
Baulch, D. M., 746
Bayesian statistics, 220, 221, 222, 266
Beckwith, W. B., 49
Beer Sheva, Israel, 458
Benard cell, 339, 345
Benedict, F. G., 698
Bergeron, T., 8, 45, 52, 118, 282, 288, 352, 391
Bering Sea, 359
Bering Strait, 399
Berlin Airlift Operations, 379
Bottom topography, 499
Bergeron-Findeisen, process, 52, 104, 111, 348
  theory, 8
Bergeron-Wegener, 38
Berkeley Statistical Symposium, Fifth, 309
Berlin, Germany, 735
Berrey, F., 73
Berson, A., 7
Bet D gan, Israel, 463
Bethwaite, F. D., 24
Bias of transformations, 214-215
  methods of correcting, 219
Bibilashvili, N. Sh., 489
Big Bend Farm Project, 48

Bigg-Warner chamber, 186
Biosphere, 673
Bischof, W., 690
Black, J. F., 273
Black, P. G., 571
Blanket clouds, 284
Blanket orographic clouds, 284
Blodgett, K. B., 10, 17, 21, 163
Blue Ridge Weather Modification Association, 82
Blyth, C. R., 208
Bob Marshall Wilderness, 84
Boggs, Keith S., 810
Bollay, E., 47, 67, 79, 87
Bollay, E., Associates, 49, 79
Boltzmann infrared temperature change, 680
BOMEX, *see* Barbados Oceanographic and
        Meteorological Expedition
Bona fide loser in litigation, 810
Bonn, Germany, 728
Bonner, E. J., 810
Bonneville Power Authority, 84, 86
Booker, R., 49, 79, 178
Bornstein, R. D., 741
Boundary layer, 660, 661
   friction of, 531
   instabilities, 589
Boundary Layer Profiler (BLP), 190
Boundary conditions, 538
Boundary layer, 531
Bowen, E. G., 24, 75, 210
Bowen's hypothesis, 24
Bowne, N. E., 748
Boyd, D. W., 221
Bradbury, D. L., 571
Braham, R. R., 26, 29, 30, 38, 63, 74, 76
Bray, J. B., 691
Brazil, 359
Breezes, land and sea, 93
Brewer, W., 79
Bridger Range, Montana, 305, 308
Brier, G. W., 20, 210, 219, 220
Briggs, G. A., 125
British Columbia, 80
Brook, M., 599
Brooks, E. M., 576
Brown, E. N., 159, 160, 177
Brownian motion, 704
Brownlee, K. A., 35, 37, 210
Bruce, M., 79
Bryan, K., 684
Bubble, convective, 328-9
Bubbles, 391
Budyko, M. I., 682, 683
Buffalo, New York, 120-121, 329, 331
Buffalo WSR-57 radar, 329, 339
"Buffer zone," Israel, 463
Buffering factor, buffering action of oceans, 697
Bulgakov, N. A., 390

Bulgaria, 31
Buoyancy, 102, 249
   alteration of, 59
   forces, 527
Buoyant, stability, 662
Buoys, telemetering, 144
Bureau of Land Management, 605
Bureau of Reclamation, 49, 54, 75, 76, 84, 304,
        782, 788, 792, 803
   Pilot project, 308
   reservoirs, 312
Burgers-Rott vortex, 583, 585, 589
Burner seedings, 194
Butcher, S. S., 182

C alpha tests, 212, 213
CCN, *see* Cloud condensation nuclei
CSIRO, *see* Commonwealth Scientific and Indus-
        trial Research Organization
California, 376, 378, 795
   China Lake, 68, 512
   legislation, 775
   ocean current, 775
   randomized seeding program, 68, 306
   seeding convective storms, 247
   urban wind convergence, 740
   watershed seeding, 81
   weather modification in the private sector, 70
   winter orographic weather modification experi-
        ment, 308
California Electric Power Company, 47, 66, 81, 85
California Institute of Technology, 47
California-Nevada Interstate Authority, 784
California State Water Resources Board, 67
California, University of, 106
Callendar, G. S., 690-691
Calusen, H. B., 721
Calvin, L. D., 210, 220
Cambridge Systems Division of E. G. and G., 149
Camera, drop, 199
   dual-image, 169
Canada, 31, 285, 322
   plains, 318
   shore, 327
Canaries currents, 359
Cannon, T. W., 169, 170
Capillary collector, 161
Carbon black, 272
Carbon cycle, 697
Carbon-12, 691
Carbon-14, 691
Carbon-14 to carbon-12, 691
Carbon dioxide (CO_2), atmospheric, 637, 687-690
   box model of reservoirs and exchanges, 696
   climatic effects of increasing concentration, 700
   concentrations, 690
   fossil fuel source, 696
   growth of, 722

long-term increase in atmosphere, 689
man-made production, 689
natural seasonal source, 689
net primary production of, 697
pollution, 683
radiative properties of, 698
secular increase, 690
solid, as a seeding material, 5, 245
sources of atmospheric, 689
transfer between mixed and deep layers, 696
transfer to and from biosphere, 696
trend in atmosphere, 691
Carboniferous fuel, 77
Caribbean Sea, 219, 802
Carpenter, T. H., 220
Carr, J. T., 810
Cascade Range, Washington, 286, 308
Castro, Fidel, 81
Cathode ray tube (CRT), 155
Caucasus Hail Suppression Project (KKPE), USSR,
    31-32, 419, 421
Central America, 75
Central Arizona Project, 309
Central Asian Regional Hydrometeorological Insti-
    tute (SARNIGMI), USSR, 410
Centrifugal force, 501
Chaff seeding, 604, 617
    experiment, 603
    field tests, 620-621
Chafee, D. L., 5
Chain-reaction theory, 17
Chalmers, J. A., 603
Chamberlain, A. R., 78
Chandler, T. J., 726
Chang, C. C., 580, 587, 589
Changnon, S. A., Jr., 556, 590, 752, 757
Chappell, C. F., 291
Charged particles, 5
"Charging mechanism", 598
Charney, J. G., 528, 539, 658
Charles-Boyle equation of state, 524, 639
Charlson, R. J., 182, 184, 746
Chemicals, condensation-retardant, 364
    evaporation inhibiting, 363
Chicago, Illinois, 352
Chicago Midway Laboratory, 26, 27
Chihuahua, Mexico, 69
China Lake, California, 68, 512
Chodes, N., 480
Chotorlishvili, L. S., 416
Cincinnati, Ohio, 729, 738
Circulation, of the atmosphere, 518
    large-scale, 93, 94
    patterns, 721
    transverse, 530, 534, 539
Cirrus clouds, 99
    infrared emissivities, 714
    shields, 523

trends, 713
Citizen's Council of Hail Research, 793
"City plume," 741, 747
Clarke, J. F., 738
Clermont-Ferrand University group, 33
Climate, change, 397, 673, 688
    defined, 398
    large-scale modification, 398
    regional modification, 400
    USSR research in weather, 389
    variation, 674
Climatic, equilibrium, 675
    gradients, 468
    predictions, 719
Climatological changes, 398
Climax, Colorado experiment, 63, 291-314, 463
    Climax I sample, 208, 292, 295, 297-298, 301
    Climax II sample, 208, 295-298
    Climax IIb sample, 298
    data, 297
    experimental area, 294
    model, 313
Cloud (clouds), Atlantic cumuli, 239
    blanket, 284
    bow, 169
    buoyancy, 201
    cap, 61
    cirrus, 99, 713-714
    cold stratus, 25
    condensation of, 93
    continental, 176
    continental cumulus, 490
    convective, 25, 93, 284, 292, 326, 504, 505
    convective stratocumulus layer, 327
    cumulonimbus, 328
    cumulonimbus anvil, 269
    cumulus, 61, 102, 391, 395
    cumulus congestus, 328, 490, 575
    cumulus mediocris, 229, 328
    cumulus, tropical, 522
    diffusion chamber, 184
    dynamics, 31, 233, 527
    feeder, 111
    formation of, 93, 94
    giant cumuli, 229
    glaciation of, 341
    ice, 162
    interactions, 235
    maritime, 176, 183
    microphysical processes, 474
    model, 469, 544
    orographic, 282, 287
    optical effects in, 169
    ownership, 770
    particles, 175
    particle growth, 244
    physics, 9, 233
    pileus, 327

releaser, 118, 121, 122
  sampling in, 162
  science, 233
  seeder, 111, 118, 119
  self-cleaning mechanism of growing, 332
  spender, 119, 121
  simulator, 198-199
  single, 270
  stratus, 14, 21, 25, 327, 395, 399
  summer cumulus, 49, 76
  supersaturation of, 136, 140
  tropical cumuli, 233
  wall, 507, 508
Cloud chamber, experiments, 7
  isothermal, 302
  large, 393
  thermal gradient diffusion, 184
Cloud condensation nuclei (CCN), 7, 97, 99, 110,
    140, 183, 234, 242, 286, 361, 392, 709, 757
  apparatus for measuring, 391
  diffusion chambers, continuous-flow types, 185
  mass produced each year, 710
  spectra, 471, 475, 476
Cloud droplets, 97, 155, 160, 170
  condensation growth of, 392
  crystallization of supercooled, 394
  growth by condensation, 109, 110
  region, 116
  size distribution, 108, 140, 141, 155, 177, 471,
    477, 480
  spectrum of, 141, 162, 177, 199, 474
Cloud modification, extended space and time
    effects of, 274
  history of, 4, 390
  warm, 383
  see also Cloud seeding
Cloud particles, crystallization of, 392, 411
Cloud-physics, 31
  airplane flights, 30
  development of theory and applications, 4
  history of, 4
  laboratory, 26
  meetings on, 85
Cloud Physics Project of the U. S. Weather Bureau,
    18
Cloud seeding, 94, 102, 604, 608
  annual expenditure, 49
  contract extent western, U. S., 50
  federal expenditure, 49
  New York City, 22
  ordinances against, see Congressional Weather
    Modification Bills; Laws
  private sector, 46, 86, 310
  randomized, 458
  South America, 67
  statutes, 767
  to augment water supply, 22
  to clear airport fog, 48-49

  see also Cloud modification
Cloud strokes (lightning), 600
Cloud systems, 326, 327, 331
Cloud top temperatures, 255, 292, 297, 306
Cloud water instrument for measuring, 158, 176
Cloud-to-ground lightning, 602
Coalescence, 286, 362, 488
  early models, 243
  efficiency, 104, 106
  process, 64, 104, 242, 490
  seeding, 242
Coasting phase, 328-29
Cognitive dissonance, 799
Cold-core, 522
Colgate, S. A., 598
Collection coefficiencies, 108
Collection efficiency, 104, 105, 106, 158, 162-167,
    258, 392, 480, 484
Collier, C. G., 750
Collision-coalescence, 9, 17, 94, 104
  growth, 474
  mechanism, 491
  processes, 480-4
Collision and collection, 393, 394
Collision process, 104
Colloidally stable, 365, 474, 484, 490
Columbia, 59
Colorado, freezing nuclei measurement over, 330
  High Altitude Observatory, 291
  law, 309, 310, 775, 779, 783
  legislature, 791
  precipitation profile, 289
  National Hail Research Project, 48
Colorado River basin, 301, 308, 310, 312
  Pilot Project, 308, 309, 784
  Project Act, 309, 310
Colorado Rocky Mountains, 285, 289, 295, 297
Colorado State University, 30, 48, 63, 69, 76, 78,
    87, 173, 288, 292, 295, 296, 302, 309, 310,
    313, 315
Colspan Environmental Systems Company, 194
Columbia, Maryland, 729, 732
"Commerce Clause," 782
Commercial seeders, 796
Committee on Atmospheric Sciences (NAS), 76
Committee on Cloud Physics and Weather Modifi-
    cation, 33, 34
Common law, 767
Commonwealth Scientific and Industrial Research
    Organization (CSIRO) Australia, 23, 26, 64,
    75, 304
Compensation mechanisms, 807, 810
Compensation, uninsured losses, 807
Computational domain, 538
Computational resolution, 635
Computer, experiments, 461
  high-speed digital, 103, 524
  model simulation, 369

simulation studies, 373
studies of cloud particle growth, 244
Concentration factors for particles, 164
Concentration maximum, 476
Concept of vorticity, 581
Condensate, 288, 292
Condensation, 94, 96, 97, 373, 392, 570
    and drop formation, 94
    by evaporation, two situations, 98
    level, 98, 102
    lifting level, 527
    pressure, 98
    process, 233
    with subsequent crystallization, 99
    temperature, 98
Condensation nucleus (CN) counter, 185
Conditional instability of the second kind (CISK),
        528, 530-532
Confidence limits, 207
Conflict over planned weather modification, 808,
        809
Congress, 780
Congressional Weather Control Bills, bills into laws,
        80
    Bill, HR8708, 82
        HR9055, 80
        S5, 72
        S23, 78
        S86, 74
        S152, 75
        S222, 72
        S798, 72
        S943, 75
        S1020, 78
        S1182, 80
        S2826, 80
        S2875, 78
        S2916, 78
    see also Laws
Conrad, G., 201
Contamination, 36, 209
    dynamic, 209
    inadvertent, 465
Continental Divide, 331, 608, 694
Continuous particle collector, 608
Contrails, 710
    cloudiness, 715
    formation, 711, 712
    persistence of, 712
Control area, 209
Convection, 102, 503
    free, 100
    stimulated, 544
Convective, charging, 600
    cloud, mountain induced, 284
    cloud motions, 74
    cloud systems over the Great Lakes, 326
Convergence, 333

horizontal, 102, 103
Converters, 229
Coriolis, acceleration, 582
    force, 501, 659, 636, 661
Cornell Aeronautical Laboratories, 198, 344
Corona, discharge, 613
    indicator, 621
Corpus Christi, Texas, 500
Corvallis, Oregon, 128, 734
Cotton, W. R., 115, 117, 139
Coulier, P. J., 6
Council on Environmental Quality, 781
Covariate predictor, 209
Covariates, 215, 216
Cox, S. K., 714
Cromwell current, 672
Crossover design, 210, 458, 459, 461
Crow, L., 59
Cryosphere, 634
Crystal, 166
    growth of, 167
    habits, 111, 115, 116, 117, 118, 123, 352
Crystals, concentration of single, 330
    forms, 165
    size of, 125
Crystallization, 8, 394
    of the liquid phase, 99
    reaction, 394
Crystallizing reagent. See Silver iodide, 417, 419
Cuba, 81
Culkowski, W. M., 758
Cumuli, Atlantic, 239
Cumulonimbus cloud, 414, 415, 523, 552
    towers, 229, 233
Cumulus clouds, 9, 133, 392, 468, 470, 482, 486,
        489, 510
    aircraft penetrations, 608
    bases, 240
    convection, 640
    continental type, 462, 490
    dendritic growth within, 489
    dynamic multiple seeding, 262
    experiments, 242
    giant, 229
    interactions of, 240
    man-made, 345
    maritime-type, 482
    model of, 59, 237
    modification, of, 232, 240, 241, 242, 279
    numerical models of, 37
    parameterization of, 531, 532, 550
    rain augmentation in, 242
    rain, 234, 414
    seeding of, 14, 29, 48
    summer, 49, 76, 490
    winter, 490
    treating, 18
    towering, 590

tropical, 233
Cundiff, S., 85
Cunningham, R., 173, 178
Current, thunderstorm produced, 614
Current density, 597
Cycloidal suction marks, 571
Cyclones, 93, 102, 103
  extratropical, 497, 648
  tropical, 522
Cyclonic flow, 470

Daily log, 804
"Daily suitability criterion," 262
Dalinsky, P., 462
Dallas project, 73
Damage, lightning, 604
  related to variation of wind speed, 499
  wind, 498
  see also Hurricanes; Tornadoes
Dansgaard, W., 721
Dark-field scattering, 176
Data, analysis, 216, 217, 221
  Centers, 219
  Processing, 218
Dauzère, C., 7
Davenport, A. B., 742
Davis, L. G., 59
Davis, M. H., 104, 105, 480
Davis, R., 807
Davis-Weinstein cumulus model, 340
Davitaya, F. F., 738
Davos, Switzerland, 706
Dawson, G. A., 596
Decelerator, particle, 167
Decision analysis (Bayesian approach), 222
Decision-making procedures, 804, 806
Deep waters, 689
Deep moist layer, 539
Defant, A., 684
Deforestation of the tropics, 717
Delage, Y., 740
Demographic forces, 788
Demonstration projects, 801
Dendrites, 115, 118, 343, 484
Dennis, A. S., 87, 269
Density, at flight level, 145
  true air speed, 155
Denver Research Institute, 194
Department of Agriculture, 5, 34
Department of Commerce, 34, 78, 780. See also
  National Oceanic and Atmospheric Admini-
  stration (NOAA)
Department of Defense, 33, 34, 35, 507
Department of Interior, 34, 35, 72, 78
Department of Transportation, 355
Deposition of ice, 104
Depression, tropical, 540
Dergarabedian, P., 566

Derivatives, local time, 524
  space, 524
Desert Research Institute, 79
Design, context, 208
  crossover, 210, 458, 459, 461
  for experiments in weather modification, 210
  target-control, 208, 209
Dessens, H., 6, 32, 590
Dessens, J., 6, 581
Dettwiller, J., 733
Dewpoint, 165, 196
  hygrometers, 149, 150, 160
  temperature, 98
Diamond dust, 111
Die Atmosphere als Kolloid (Schmauss and Wigand
  1912), 7
Dielectric, 597
Differential equations, ordinary, 524
  partial, 524
Difference techniques, finite, 524
Diffusional growth, 111, 341, 342, 487
Dinger-Gunn effect, 600
Direct heat insertion, 715
Direction finders, radiowave, 155
Dirmhirn, I., 741
Disaster Relief Act, 788, 794, 802, 803
Disdrometer, optical, 176, 179
Dispersion, 673
Dissipation, frictional, 523
Divergence, 329
Division of Atmospheric Resources Management,
  75
Downwash, mining, 369
Dobrowolski, A., 342
Doppler radar, 155, 189, 200, 332, 244, 245, 573
  acoustic, 190
  dual, 189
  meteorological, 573
  velocity measurements, 573
Dossett, C. K., 179
Douglas, L. W., 34
Douglas, W. J., 58
Downdraft, 276
Downwind effect, 63, 274, 462
Drabek, T. E., 811
Drag coefficient, 540
Drizzle drops, 103
Droessler, E. G., 36
Drones, radio controlled, 143
Dronia, H., 733
Droplet, see Cloud droplets
Dropsonde, 144, 151, 152
Drought, 275
Drought relief effort, 791
Droxtals, definition, 361
Dry ice, 21, 58, 194, 245, 335, 344, 366, 391, 393,
  394
  pellets, 356

seeding, 366, 367, 381
 see also Carbon Dioxide
Duce, R. A., 745
Duckworth, F. S., 740
Dunkirk, New York, 344
Duran, B. S., 213
Dussault, J., 125, 126
"Dust," 700
Dust devils, 582
"Dynamic approach," seeding experiments, 246
Dynamic, contamination, 209
 cumulus seeding, multiple, 262
 effects of seeding, 58, 246
 heating, 151, 165
 pressure, 145, 155
 range, 136
 seeding approach, 246
 seeding programs, 38, 241, 261, 271
 processes, 94
 temperature rise, 145
Dynamics of clouds, 31, 233, 527
Dyrenforth, R., 5
Dzerdzeevskij, B. L., 721

Eady, E. T., 658
Earthbound water technologies, 788
East Coast Experiment (Project SCUD), 25
Eastern Georgia, USSR, 416
Eberle, A. M., 34
Echo areas, 339
Economic benefit, 798
Economic interests, 797
Eddy correlation terms, 531
Eddy fluxes, 196
Egan, B. A., 748
Eglin Air Force Base, Florida, 198, 199
Eilat, Israel, 458
Eielson Air Force Base, Alaska, 381
Ekman (or planetary) boundary layer, 589, 660
El'brus-2 rockets, 420, 421, 423
El'brus-3 rockets, 420, 423
Electric field, measuring of, 192
Electrical, conductivity, 706
 effects associated with ice, 172
 noise, 196
 parameters, measurement of, 192
 quantities, atmospheric, 192
Electron microprobe, 188
Electronic light detectors, 176
Electronic noise, 195
Electrosphere, 596, 597, 601
Electrostatic, disdrometer, 173
 noise, 195
 particle counter, 186
 precipitation of fog particles, 362
 probe, 173
Eliassen, A., 528, 539
Elk Mountain, Wyoming, 308

Elliott, R. D., 22, 38, 67, 306
Elliott, S. D., Jr., 506
Elmendorf Air Force Base, Alaska, 367
El Niño, 399
Emergency rainmaking, 794
Emmons, G., 34
Energy, 130
 budgets, 537
 internal, 132
 kinetic, 132
 potential, 132
 requirements, 377
 thermal, 132
 turbulent, 132
Enger, I., 210
England, 355, 356, 378
Entity model, 239, 248
Entrainment, 102, 235, 249, 531
Environment One, Inc., 186
Environmental Protection Agency, 58
Environmental Research Laboratories (NOAA), 75
Environmentalists, 781
Epitazy, 99
Equations, of continuity, 639
 gradient-wind, 530
 hydrodynamical, 524
 linear partial differential, 526
 of motion, 639
 Navier-Stokes, 524
 of state, 524, 639
 primitive, 525
Equitable apportionment, 783
Equatorial South America, 598
Equatorial trough zone, 229, 277
Equilibrium, hydrostatic, 525
 vapor pressure, 96, 97
Eriksen, W., 737, 752
Errors, round-off, 546
 truncation, 525, 546
Espy, J. P., 4, 6
Europe, 359, 367, 381
Evaporation, 504, 529
 inhibiting, 505
 methods of fog dispersal, 362
 two situations of condensation by, 98
 of water, 275
 variations in rapidity of, 637
Everglades, Florida, 270-271
Excess-Land Law, 782
Executive Order, 804
Exhaust heat from jet engines, 379
Experimental design, 36
Experimental Meteorological Laboratories
  (NOAA), 75, 201, 249, 260, 788
Experimental units, selection of, 219
Experiments, computer, 461, 525
 control, 533
 crossover, 458

EG&G, 49, 305
nondeveloping, 543
randomized, 514
sensitivity, 538
Exploding rockets, sound effects of, 32
Explosive growth of seeded clouds, 254
Extended area effects, 63, 274
Extinction measurements, 176
Extratropical cyclones, 497, 648, 649
Extratropical cyclone seeding, 25. *See also* Project
    SCUD
Eyeball evaluation, questionable value of, 219
Eyewall, *see* Hurricane eyewall

Fabre, R., 380
Fairchild Air Force Base, Washington, 369
Fair weather cumulus, 328
Fall velocities of rimed and nonrimed flakes, 343
Faller, A. J., 589
Fankhauser, J. C., 558
Fans, use in modifying patchy ground fog, 362
Faraday cage, 193
Farhar, B., 810
Farm project – Big Bend, Washington, 48
Feder, P., 461
Federal Aviation Administration, 381
Federal, contractors, 779
control of cloud seeding, 802
courts, 81
elected officials, 793
expenditures, 77
government, 283
government and private sector, 70
legislation, 784
Federal Council for Science and Technology, 810
Federal Disaster Assistance Administration, 794,
    803
Fedorov, Ye. K., 423
Feeder cloud, 118, 120
Fendell, F., 566
Ferrel cell, 659
Field, experimentation, 369
meter, radioactive probe electric, 192
mill, 192
-of-motion models, 239
Filtration, 362
Findeisen, W., 8, 9, 45, 391, 482
Finley, J. P., 572
Finnegan, W., 201
Fire method (for stimulating rain), 6
First National Conference on Weather Modifica-
    tion, 85
First Polar Year, 342
Fisher, B., 79
Fisher, R. A., 207, 223
Fitzgerald, D. R., 192
Flagstaff, Arizona, 620
Flake density, 343

Flame emission spectrometer, 188
Flares, dropping, 194
end-burning, 194
pyrotechnic silver iodide, 194
Flatlands projects, 77
Fletcher, N. H., 186, 342
Floating targets, 265, 270
Florida, 128, 518, 788
Central, 788
cloud seeding experiments, 219, 258, 260
cloud simulator facility, 199
cumulus experimentation, 251
drought relief effort, 803
rain augmentation program, 788
state government, 808
thunderstorms, 240
Flueck, J. A., 217, 218
Fog, 94, 184, 355
airport clear of, 365
advection, 357, 361-362
artificial dissipation process, 365
clearing with helicopters, warm, 369
climatology, United States, 360
cost figures, 377
definition, 360
in the United States, 360
ice, 358, 360
London, England, 750
morphology of, 357
physical properties of, 360
prevention, 363, 364
radiation, 357, 359, 361-362
revenue loss caused by, 355
steam, 98, 357
studies, 6
structure, 360
summer sea, 359
supercooled, 358, 361, 366, 367, 381
tropical air, 359
types, 357
upslope, 357, 358
warm, 281, 358, 361, 376
warm-front type, 98
water content, 373, 376
world-wide frequency of days of, 359
Fog dispersal, cost requirements, 377
Fog Intensive Dispersal of (FIDO), 356, 378, 379
general requirements, 364
history of, 356
supercooled, 365, 369
thermal, 378
U. S. Air Force installations, 33
warm, 362, 369, 377, 378
Fog dissipation. *See* Fog dispersal
Fog Facility at State University of New York, 198
Fog modification, 357. *See also* Fog dissipation;
    Fog dispersal
Fog particles, physical removal of, 362

scavenging of, 362
    foil sampler, 178
    folklore, 4
    force, centrifugal, 529
    force, pressure-gradient, 523, 530
Forecasts of severe storms and tornadoes, 557
Forecasting, progress in, 79
Forest fires, 76, 603
    eliminating, 603
    areas and seeding, 48
Forest Service, U. S. Department of Agriculture,
        607
Formvar replicator, 163, 165, 169, 472
Fort Lewis College, Colorado, 315
Fort Wayne, Indiana, 748
Fortag, H., 649
Foss, K., 562
France, 133, 366, 369, 379, 380
    arbitration tribunal, 786
    commercial cloud seeding, 73
    fog dispersion, 194
    fog experiments, 379
    hail suppression, 31
    Ministry of Transportation, 380
    northern precipitation statistics, 756
    Observatory of Puy du Dome, University of
        Clermont-Ferrand, 6, 32
    Orly Airport, 194, 380
    Paris Airport Authority, 369
    Paris, survey of city temperature, 733, 755
    salt-seeding experiment, 244
    Tignes, France, Project, 73
Frankfurt, Germany, 738
Franklin, B., 604
Frazer, A., 172
Frederick, R. H., 756
Fredonia, New York Lakeside Laboratory, 344
Freezing nuclei, 116, 329, 366
Freezing nuclei counter, 330
Freezing potentials, 599
Fremont Pass, Colorado, 295
Friction, 501
Friedman, H., 201
Fritts, H. C., 757
Fronts, 102
Frost, point sonde, 144
    prevention of, 131
    protection, 355
Fuel combustion, 378
Fujita, T. T., 142, 285, 562, 563, 571, 572
Fujiwara, M., 144, 333
Fulks, J. R., 581
Fuquay, D. M., 604, 607
Furman, R. W., 292

Gabriel, K. R., 458-459, 461, 463
Gagin, A., 461, 483, 486-487, 490
Gamma distribution, 211-212

Gartska, W., 86
Gaussian dispersion, 748
Genard cells, 122
General circulation, 635
    experiments, 652
    theory, 635
General Electric Company, 14, 186, 356, 505
General Electric Research Laboratories, 12, 14, 15,
        21, 45, 58, 72, 282
General Electric Research Laboratory Report, 20
General Electric Review 11/52, 12
Generators, 54, 302, 313
    ground, 77
    ground-based silver iodide, 19, 31
    nucleating, 57, 81
    pyrotechnic, 250
    remote radio-controlled, 55
    silver iodide smoke, 22
    silver iodide-sodium iodide-acetone type, 76
    sky-fire needle type, 296
    typical, 53
Gentry, R. C., 31, 511
George, J. J., 34
Georgia, 428
Georgii, H. W., 690
Geostrophic wind, 658
Gerdien ion meter, 193
Germany, 144, 758
Giant nuclei, 110
Giant wet snowflakes, 111
Giant salt particles (NaCl dosage), 418
Gillespie, C. A., 173
Girs, A. A., 721
Gish, O. H., 602
Gissar Valley, Tadjik Soviet Socialist Republic,
        USSR, 426
Glaciation, artificial, 245
    complete, 341, 350
    ice formation, 15
    processes, 483
Glaize, F., 83
Glaser, A. H., 566, 569
Global Atmospheric Research Program (GARP),
        648, 684
Global, climate, 717
    climatic precipitation regime, 666
    electrical circuit, 597
    radiative equilibrium, 648
Goldbeater's skin, 151. See also Hygrometer
Golden, J. H., 553, 575, 576, 581
Gori, E. G., 5
Görög and A. Rovó, 6
Gourley, M. F., 599
Goyer, G. G., 26, 231
Granby, Colorado, 311
Grand Banks area, Newfoundland,
        359
Grant, F. C., 575

Grant, L. O., 30, 52, 54, 63, 69, 76, 87, 173, 186, 208, 210, 287
Graupel (soft hail), 167, 326, 340, 484, 599
  conical, 117, 126
  in convective clouds, 123
  in orographic clouds, 111
  process, 117
Gravitational separation mechanism, 598
Gravity waves, 525
Gray, W. M., 278
Great Lakes Basin, 120, 318, 330, 331, 351
Great Lakes cloud systems, 327
Great Lakes storm, 322
Great Plains, 556
  hail suppression, 48
  Northern, 260
"Greenhouse" effect, 647
Greenhouse instability, 679
Greenland, 359, 721
Grids, nested, 550
  spacing, 525
Gross-austausch formulations, 684
Grossversuch III, Switzerland, 31
Ground-based, heating, 377
  propane systems, 366
  seeders, 394
  systems, 381
Growth, by accretion of supercooled clouds, 394
  diffusional, 111
  regimes, 254
Gucker, F. T., 176
Guilbert, M., 7
Gulf Coast, 554, 555
Gulf of Alaska, 277
Gulf of Mexico, 788
Gulf states, 70
Gulf Stream, 682
Gulf Stream waters, 359
Gunn, R., 64, 572
Gupta, A. S., 589
Gustiness factor, 743
Gutman, L. N., 584
Gyroscopes, 154

Haas, J. E., 810
Hadley cell, 229, 278, 656
Hail, 65, 94, 111, 123, 425, 782
  damage, 397, 411, 424
  clouds, 126
  control, 396
  embryos, 126, 258
  estimated risk of a cloud, 418
  formation, necessary conditions for, 414
  "hail alley," 792
  List theory of growth, 125
  modification method, ZakNIGMI, USSR, 417
  rough, 599
  soft (or graupel), 111, 340

  swathes, 277
Hailstones, 117, 123
  artificial multiplication, 415
  formation, 412
  growth of, by riming, 32
  habits, 126
  embryos, artificial, 415
Hail processes, characteristics of, 411
Hail suppression, 31, 48, 391, 792
  experiments, 32
  experiments in USSR, 394
  operations in USSR, 420-22
Hall, F., 19, 26, 66
Hallett, J., 115, 486, 488
Hallgren, R. E., 342, 343
Halo phenomena, 169, 171
Halstead, M. H., 17
Hannell, F. G., 735
Hanscomb-Field, Massachusetts, 178
Harris, T. E., 690
Hartman, L. M., 38
Haurwitz, B., 34
Haurwitz Committee, 71
Havens, B. S., 19
Hawkins, H. F., 20, 502, 537
Haze layers, 183
Heat, 688
  balance, 642, 730
  global average man-made source, 717
  injections, 717
  of condensation, 93
  plume systems, 378
  required amount of, 377
  spatial distribution of, 378
Heating, of the air, 369
  dynamic, 151
  ground-based, 377
  systems, 379
Heat island, 273, 735
  circulation model, 740
  effect, 737
  maximum difference, 734
  mid-Manhattan, 739
Heat transfer, sensible, 541
Hebrew University, 187, 480
Helicopter, downwash mixing, 33, 369, 382
  sonde, 143
Henderson, T. J., 47, 54, 66
Heuseler, H., 735
Hicks, J. R., 371
Hidore, J. F., 757
High Altitude Observatory, Climax, Colorado, 291
High Mountain Geophysical Institute of the Hydrometeorological Service, USSR, 31
High Mountain Geophysical Institute (VGI), Central Aerological Observatory, USSR, 410
High Plains region, 352
High-pressure areas, 103

Hilst, G. R., 748
Hindman, E., 287
Hirsch, J., 260
Hirsch model, 261
Historical, data, 210
  period, 209
  period of no seeding, 208
  regression method, 220
Hobbs, P. V., 172, 757
Hocking, L. M., 104, 480
Hoecker, W. H., 569, 586, 739
Hoehne, W., 152
Hoke, P., 83
Hokkaido, Island of, 333
Holography, 200
Holroyd, E. W., III, 322, 341, 342
Holzman, B. G., 757
Homogeneous nucleation, 194
Honduras experiments, Langmuir, 17, 38
Hooker, Henry, litigation, 767
Hoopers, materials dispensing, 194
Horizontal resolution, 538
Hosler, C. L., 342, 343
Houghton, H. G., 6, 9, 17, 33, 356
Hot towers, 236
Hot wire instrument, Johnson-Williams (J-W), 158
Howard, L. N., 589, 726, 750
Howard, R. A., 222, 517, 518
Howell, W. E., 22, 59, 65, 67, 71, 72, 76, 77, 78,
  83
Howell, W. E., Associates, 73, 81
Huff, F. A., 38, 39
Huffman, P. J., 171
Hughes, E., 698
Hull, D., 8
Humboldt current, 359
Humidity, 150
  free-air temperature, 145
  measurements, 149, 165
  reduction, 363
  relative, 95, 151
  specific, 290
Humidity-sensitive sensors, carbon-coated, 150
Humphreys, W. J., 4
Hungry Horse Project, 58
Hurricanes, 93, 94, 103, 132, 497-551
  belt of maximum winds, 515
  circularly symmetric models, 532
  clouds, 510, 514
  damages, 497
  development, 527
  early models, 528
  efficiency of mature storm, 538
  energetics, 500, 508
  eye wall, 505, 508, 510, 511, 515, 522-23, 534,
    544, 547
  high wind area, 505
  in Kentucky, 499

  life history, 508
  mature, 498, 534
  modification experiments, 14, 497-551
  natural variability of, 514
  in New England, 499
  in New York, 499
  in Pennsylvania, 499
  physical processes of, 500
  rain, 499
  rain bands, 505
  seeding, 504
  structure, 508
  synoptic scales, 504
  theoretical models, 505, 514
  thermal structure, 534
  United States, 497
  in Virginia, 499
  in West Virginia, 499
Hurricane Agnes, 497, 499
Hurricane Betsy, 497
Hurricane Beulah, 506, 507, 511, 512, 553
Hurricane Camille, 497, 499
Hurricane Celia, 497, 500
Hurricane Daisy, 509, 511, 537
Hurricane Debbie, 506, 507, 511, 512, 516, 517
Hurricane Diane, 499
Hurricane Ella, 507
Hurricane Esther, 506, 507, 509, 511
Hurricane Ginger, 506, 507, 509
Hurricane Hilda, 501
Hydroelectric Commission of Tasmania, 773
Hydrologic balance, 643
Hydrology, 318, 352
Hydrometeor, fall velocities, 189
  foil sampler, 177
Hydrometeorological Service of the Armenian
    Soviet Socialist Republic, USSR, 421
Hydrometeorological Service of the Azerbaijan
    Soviet Socialist Republic, USSR, 421
Hydrosphere, 634
Hydrostatic, approximation, 638
  stability, 100
Hygrometer, dew point, 149, 150
  Lyman-alpha, 151
Hygroscopic, experiments, 243
  nuclei, 7, 9, 97
  particle seeding, 376
  seeding experiments, 244
Hypothesis, additive, 212
  alternative, 218
  null, 211
  testing, 208, 214, 216, 221
Hypsometer, 151

Ice, 366, 709
  clear and opaque, 126
  electrical effects associated with, 172
  fog, 359, 361, 382

fog elimination, 381
  measurement of, 169
  multiplication, 286, 308
  pseudoadiabat, 543
  spectra, 258
Ice crystals, dendritic habit, 165
  growth of, 489
  growth mechanism, (Bergeron), 234
  measurement of, 170-171, 172, 186
  in strong electric fields, 607
  multiplying mechanism, 286, 491
  process, 468, 481
  produced by silver iodide, 275
  replication, 169, 483
  sampling, 162, 167
Ice nuclei, counting, 287, 472
  effective number of, 486
  forming, 393
  freezing, 99, 102, 186, 193, 302
  measurements, 186
  from pollution sources, 710
  primary, 286
  spectra, 99
Ice-particle counter, Mee, 172
Ice particles, hail embryos, 258
  junk ice fragments, 258
  snowflakes, 258
Ice pellets, 416
Icc-phasc sccding, 194, 242
Icicles, 125
Icing measurements, 161
Idaho, 84
Idaho Falls Reactor Test Station, 199
Illinois State Water Survey, 38, 129, 795
Impact statements, 781
Immunity from liability, 775
Impaction, 704
Impactor, aluminum foil, 165, 168
Impactor slide, 173
Imperial College of Science and Technology,
    London, England, 26
Inadvertent climate modification, 715
Increasing high cloudiness, effect of, 714
Independent replication, 295
India, 666
Inductive charging processes, 600
Inertial navigation, 154
Inference, scientific, 223
Inferences, significance tests, 210-211
Inflow, 529, 530
Infrared, analyzers, dispersive, 690
  emissivities, cirrus clouds, 714
  radiation, 642
  radiometry, 200
  remote sensing, 191, 192, 327
  temperature, 732
  thermometers, 200
Initial asymmetry, 548

Initial conditions, 525, 533
Injunction, 81
Instability, static, 527
Institute of Artificial Rain, 390
Institute of Atmospheric Physics, University of
    Arizona, 28
Institute of Atmospheric Sciences, South Dakota
    School of Mines and Technology, 48, 792
Institute of Experimental Meteorology, 390
Institute of Geophysics of the Academy of Sciences
    of the Georgian Soviet Socialist Republic,
    (IGAN), 410, 411, 412, 415, 416, 418, 419,
    420, 428
Institutional aspects of weather modification, 804,
    806, 807
Instruments, 221
  calibration, 197
  classes of, 199
  errors, 218
Instrumentation, priorities in, 138
Integrating nephelometer, 184
Integration, numerical, 524, 528
Intensity of turbulence, 365
Interactive ocean, influence, 666
Interdepartmental Committee for Atmospheric
    Sciences (ICAS), 787
Internal Revenue Service, 796
Internal (or thermal) energy, 132
International Hail Conference, Verona, Italy, 48
International law, 785
Interstate commerce, 782
Interstate compact, 784
"Interview effect," 809
Interviews, hail season, 793
Inverse-radius entrainment law, 249
Inverse transform, 215
Inversion, local, 54
Iodine, 12
Iodoform, 12
Ionosphere, 596
Ion production, 615
Ions, 193
Iran, 81
Irrigation water supplies, 282
Isoecho contours, 197
Isokinetic flow, 160
Isothermal cloud chamber, 302
Isothermal layer, upper ocean, 677
Israel, 31, 454-494
  cloud physics research, 454-494
  experiment, 210, 463, 490
  ice crystal process, 468
  Ice Nucleus Counter, 187
  Jerusalem, 471
  measurement of electrical parameters, 192
  private sector seeding, 75
  rain stimulation, 454-494
  rainy days, 463, 475

seeding, 247, 458, 462
winter cumulus clouds, 490
Meteorological Service, 463, 466, 474
Italy, 31

Jamestown, New York, 344
Japan, 359, 706
Japan, Sea of, 333
Jayaweera, K. O. L. P., 486, 488
Jemez Mountains, New Mexico, 305, 306, 308
Jerusalem, Israel, 471
Jet engine technique for warm fog modification, 379, 380, 391
Jiusto, J. E., 117, 319, 310, 322, 341-343
Johnsen, S. J., 721
Johnson, D. A., 600
Johnson-Williams, Inc., 158
Johnson-Williams (J-W) hot wire instrument, 156, 157, 158, 177
Joint model of atmosphere and ocean, 666
Jonas, P. R., 104, 480
Jones, H. L., 572
Jordan, Kingdom of, 463
Jost, D., 690
Journal of Meteorology, 13, 15, 17
Judiciary's goal in litigation over artificial rainmaking, 767
Junk ice fragments, 258
Junge, C. E., 7
Jupiter, 635

Kahan, A. M., 75
Kangas, T., 602
Kartsivadze, A. I., 416
Kasahara, A., 678
Kasemir, H. W., 596, 603, 604, 618, 620
Kazakh, Northern, 395
Kealoha, J. H. S., 735
Keeling, C. D., 690
Kentucky, seeding project, 73
Kenya, explosives experiment, 32
hail suppression, 31
Kepler, J., 118
Kessler, A., 728
Kessler, E., 558, 569
Kew Observatory, London, England, 7, 603, 649
Khrigian, A. Kh., 108
Khrigian-Mazin formula, 412
Kiel, Germany, 738, 752
Kienle, J. v., 758
Killer smog, 726
Kinetic energy, 132, 510, 538, 549, 648, 659
Kinetics, hailstone ensemble, 416
Kinzer, G. D., 572
Kite-balloons (kytoons), 144
Klett, J. D., 105
Kline, D. B., 220
Klopsteg, P. E., 36

Knight, C. A., 366
Knight, Sir Alan, 773
Knollenberg, R. G., 157, 176
Knollenberg optical array, 157, 169
Kockmond, W. C., 373
Koenig, L. R., 30, 482-84
Köhler, H., 7
Korea, 359
Koscielski, A., 269, 562
Krane, S., 711
Kratzer, A., 726
Kraus, E. B., 23
Krick, Irving P., 46-47, 58, 67, 71, 72, 76
Krick, Irving P., Associates, 73
Kuhn, P. M., 714
Kumai, M., 13
Kung, E. C., 727
Kunkel, B. A., 373
Kuo, H. L., 528, 532-584, 586
Kyle, T., 159, 172
Kytoons (kite-balloons), 144

Lafayette, Indiana, light snowfall, 758
"Langrangian" framework, 640
Lake Erie, 120, 318, 321, 324, 327, 330, 344
Lake Huron, 322
Lake Lannoux arbitration, 786
Lake Ontario, 120-1, 321
Lake Michigan, 352
Lake storms, 327
Lake Tahoe, 308, 784
Lake Tiberias, Israel, 458
Lamb, H. H., 719
Lambrights, W. Henry, 811
Lamoureux, C. E., 572
Landing systems, Category I, 364
Category II, 364
Category IIIa, b, c, 364
Landsberg, H. E., 7, 33
Land usage, 688, 715, 722
Langleben, M. P., 343
Langmuir, I., 9-34, 51, 60, 70-71, 85, 123, 163, 242, 248, 278, 282, 391, 506
Langmuir's, chain reaction, 38, 243
Honduras experiments, 38
hypothesis, 9
new probability theory, 19
Nobel laureate, 45
periodic seeding experiment, 19
results on periodic seeding and weather, 19
Langmuir Observatory, Institute of Mining and Technology, Socorro, New Mexico, 624
Langsdorf, A., 184
Language in some decrees, 784
Lapcheva, V. F., 489
LaPorte, Illinois anomaly, 756, 757
Lasers, 176, 201
Latent heat, 100, 102, 132, 501, 637

of condensation, 501, 509
of fusion, 508, 509
release of, 94
Latham, J., 599, 600, 607
Laverty, Finley, 86
Lavoie, R. L., 322, 325
Law, Adams v. California, 772-773, 776
  Australia, 775
  California legislation, 775
  Causation, 770-771
  Colorado law, 775, 779, 783, 791
  Common law, 767
  county tax funds, 792
  Disaster Relief Act, 788, 794, 802, 803
  Domain, 794
  Excess land law, 782
  Executive Order, 804
  First written opinion on weather modification,
    Slutsky v. City of New York, 769
  Gainers, 809
  Hearing examiner, 791
  Judiciary's goal in litigation, 767
  Lake Lannoux arbitration, 786
  Language, 784
  Liability, 518, 772
  Immunity from liability, 775
  Injunction, 81
  Internation law, 785
  Moratorium Bill, 83
  National Environmental Policty Act, 780, 781
  Nebraska, 776
  Nebraska v. Wyoming, 783
  Negligence, 772
  New Mexico, 783
  New York, 81, 769
  North Dakota legislation, 776
  of liability, 775
  of riparian rights, 60, 769
  Oklahoma law suit, 81
  outlaw cloud seeding by air, 82
  Pennsylvania, 83, 769, 771, 782, 783
  Port of New York Authority Compact, 784
  Procurement contracts, 779
  Professional licenses, 777
  Public law 256, 72
    90-407, 35
    85-510, 36
    85-50, 78
    85-510, 35
  Quebec act, 778
  Simpon bill, 71
  South African, 777
  South Dakota, 776, 792
  Stare Decisis doctrine, 767
  Slutsky v. City of New York, 769
  Supreme Court, 783
  surface water, 769
  Texas, 81, 768, 769, 777, 788, 795

unprovided case, 768
Victoria Rainmaking Control Act, 775
weather modification report law, 801
West Virginia, 783
Wilderness Act, 84
See also Congressional Weather Control Bills
Lawrence, E. N., 756
Lay witnesses, 771
Lead iodide, 12, 246, 393, 394, 418
Leadville, Colorado, 294
Least-squares, method of, 216
Lee, C. W., 116, 118
Lee Wilson Engineering, 194
  and socioeconomic problems, 33
  problems, 78
  ramifications of artificial nucleation, 772
  scholars, 772
  Legal implications, 78. See also Law
Legislation, see Congressional Weather Control
    Bills; Law
Legislation assessment of weather modification,
    775
Leipzig, Germany, 746
Lemon Reservoir, 312
Leopold, B., 17
Leslie, L. M., 577, 581, 587, 588
Lettau, H., 742
Levine, J., 158, 248
Levine, L. M., 590
Lewisburg, West Virginia, 371
Lhermitte, R., 189
Liability, government, 518
Licensing professions, 777
Lidar, 746
  probing techniques, 191
  spectroscopy, 200
Lightning, 94, 192, 596-629
  artificially triggered discharges, 605
  -caused forest fires, 604, 605
  duration, 611
  flashes to ground, 601
  modification, 603
  modification using cloud seeding, 605
  modification work in United States, 607
  protection, 604
  regenerative power of charges producing
    mechanism in cloud, 628
  suppression by chaff seeding, 605, 612
  triggered by fired rockets, 623
Lightning rod, invention of, 604
Lilly, D. K., 569, 570, 579, 585, 587
Limestone, burning of, 689
Lindquist, S., 736
Linear analysis, 526
Linear computational instability, 524
Liquid propane, 194, 366, 393
Liquid water, 159, 161, 176, 191, 196
  accumulation level of, 126

content (LWC), 155, 157, 177
  meter, Johnson-Williams, 156-7
List, R., 117, 126
List theory of hail growth, 125
Little, A. D., Inc., 25, 174
Little, G., 190, 201
Löbner, A., 746
Local participation, 810
Local power groups, 809
Lodge, J. P., 26
Lohse burner, 193
London, England, 726, 749, 754
Long, R. R., 577
Long-wave (infrared) radiation, 642
Lopez, M. E., 67, 81
Loran system, 155
Lorenz, D., 728
Lorenz, E. N., 635, 656
Los Angeles International Airport, 379, 382
Losers, 809
Lossiness generator, 600
Lovasich, J. L., 218
Ludlam, F. H., 288, 489
Lund, Sweden, 736
Lyman-alpha, line, 151, 160
  hygrometer, 151
  total-water instrument, 172

McClellan Air Force Base, California, 376
McCormick, R. A., 746
McDonald, J. E., 26, 76, 112, 248, 563
McElroy, J. L., 738
McQuigg, J. D., 715
McTaggert-Cowan, J. D., 172
McVehil, G. E., 329
MacCready, P. B., Jr., 49, 59
MacDonald, J. F., 36
Mach, W. H., 172
Machta, L., 698
Macroscale, 391, 635
Magnesium, 194
Magnetic noise, 195
Magnus effect, 562
Magono, C., 116, 118, 144, 342
Mahoney, J. R., 748
Maimahd Scientific Instrumentation, 187
Main Water Carrier, Israel, 458
Malkus, J. S., 232, 537, 543
Malone, T., 79
Manabe, S., 644, 660, 681, 682, 684, 699, 700
Man-made particles, 709
Man-made pollutants, 673
Mann-Whitney-Wilcoxon test, 266
Mapping, 652
Marine carbon reservoir, 689
Maritime clouds, 176, 183
Mars, 637
Marshall, J. S., 129

Marshall-Palmer precipitation spectrum, 129, 257
Martin, J. R., 83
Maryland, hail modification projection, 771
  hail suppression, 782
Mason, B. J., 8, 105, 114, 115, 117, 161, 186, 600
Mason, H., 201
Massachusetts Institute of Technology, fog studies,
  6
Massachusetts law, 777
Mass media officials, 810
Mass, virtual, 249
Mastenbrook, H. J., 144
Mathematical modeling, 553, 581
  of cumulus cloud, 413
  See also Numerical modeling
Matheson, J. E., 222, 517, 518
Mauna Loa, Hawaii, 693, 706
Mauna Loa Observatory, atmospheric turbidity,
  722
  normal incident solar radiation, 706
Maynard, R. H., 15
Mean, meridonal circulation, 658, 674
  quasihorizontal eddies, 674
  residence time of particles in air, 704
Mean-square-error (MSE), 218
Measurement, errors, 218
  techniques, 221
Measurement technology and terminology, 608
Mechanical mixing, 369
Mechanism of charge transfer, 599
Medaliyev, Kh. Kh., 417
Mediterranean Sea, 464, 470, 475
Mee ice-particle counter, 172
Mee Industries, 172
Mehta, K. C., 563
Merged complex, 240
Merger, 240
Merrill, R., 48
Mesh, grid, 549
Mesoscale, 672
  circulation, 562
  modeling, 277
  structure of Lake Storms, 327
Metaldehyde, 194
Meteoritic dust, 24
Meteorological stratification, 304
Meteorological visibility, 362
Meteorology Research Inc., 29, 49, 68, 74, 75, 177
Meteotron, 391
"Metromex," 753
Mexican Power and Light Company, 67
Mexico (Chihuahua), aerial seeding, 69
Mexico City, dust concentration, 707
Mexico, world's arid region, 666
Miami, Florida, 128
Microencapsulation, 375
Microphysical, characteristics of cold orographic
  clouds, 285

cloud particles, 283
  processes, 94, 474
  seeding, 38
Microphysics of clouds, 233
Microscopic charge separation, 598
Microstructure, 391
Microwave, emissions, 176
  radar, 190
  radiometer, 176
  refractive index, 152
  refractometer, 152
  region, 191
  sensing, 190
Middle East, 274
Mie, extinction efficiency factor, 177
  scattering laws, 191, 418
Mielke, P. W., Jr., 52, 208, 210, 213
Mighty Mouse rocket, 628
Miller, A. H., 599
Miller, A., 201
Ming, "Big Dan," 767, 768
Minicomputers, 197
Minimum per acre income, 791
Minnesota, 186
Ministry of Agriculture of the Georgian Soviet
    Socialist Republic, USSR, 421
Minor, J. E., 563
Minority, 807
Mintz, Y., 665, 667
Mission statements, 794
Missouri, 76
Mitchell, J. M., 708
Mitchell, V. L., 305
Mixed cloud, 8, 599
Mixed phase, 160
Mobile surveys, 731
Mobile vans, 142
Models, barotropic, 651
  cumulus, 237
  microphysical precipitation, 339
  numerical, 238, 239, 525, 526, 587, 634
  one-dimensional entity, 248
  physical, 216, 220
  real-data, 550
  shower, 587
  statistical, 587
  three-dimensional, 550
  three-dimensional thunderstorm, 587
  treatment effect, 218
Modeling, 138, 270
  a hail cloud, 415
  mesoscale, 277
  physical, 220
  problems, 549, 650
Modification, beneficial, 545
  of climate, 398, 400
  experiment, evaluation of cumulus, 241
  extended space and time effects of

  cloud, 274, 279
  of large-scale meteorological phenomena, 398
  of weather, social and political implications, 94
  of weather systems, 94
  potential, 240, 331
Modified volume, boundaries of the, 337
Moist layers, occurrences of, 331
Molecular sieves, 150
Monarch Pass, Colorado, 297
Monsanto Chemical Company, 165
Montana, 798, 804
  professional licenses, 777
  rainfall analyses, 269
  State University experiments, 305
  Water Board, 84
Montana State University, 305
  experiments, 305
Monte Carlo computer experiments, 464-5
Monte Carlo methods, simulations, 213
Monteith, J. L., 749
Montreal, Canada, 730
Mood, A. M., 213
Moran, P. A. P., 208, 210, 212, 213
Moratorium, bill, 83
  on weather modification, 82
Moravian barley crop, 791
Morgan, B. J., 572, 573, 575
Mörick, J., 750
Morton, B., 569, 581, 586
Morton, K. W., 524
Moscow airport, 369
Moss, M. S., 544
Mosteller, F., 207, 217, 222
Motion, equations of, 323
  radial, 534
  subgrid, 525, 531
"Mountain Skywater," technical movie, 315
Moyers, J. L., 745
MRI Skyfire burner, 193
Mueller, E. A., 129
Multicylinders, 161
Multiplicative effect, 212

National Center for Atmospheric Research (NCAR),
    159, 169, 231, 329, 587, 793, 808
Nakaya, U., 117
Natick Laboratories, Massachusetts, 198
National Academy of Sciences, 10, 14, 36, 63
National Academy of Sciences Committee on At-
    mospheric Science, 37, 76
National Conference on Weather Modification, Cal-
    ifornia, 85
National Environmental Policy Act, 780, 781
National Gazette and Literary Register, 4
National Hail Research Project (Colorado), 48,
    792-793
National Hurricane Research Laboratory (NOAA),
    201, 532-533, 545

National Oceanic and Atmospheric Administration (NOAA), 31, 59, 71, 75, 249, 338, 507, 788, 792, 803, 804

National Science Foundation, 28, 34-36, 49, 53, 68, 295, 779, 780, 793, 808

National Severe Storms Forecast Center (NOAA), 552, 558

National Severe Storms Laboratory (NOAA), 566, 573

National Weather Service, see U. S. Weather Bureau

Natural climatic variability, 723

Natural hailstone embryos, 415

Natural particles, 700

Natural phenomena, number of, 719

Natural Weather Association, 771

Naval Ammunition Depot Indiana, 194

Naval Research Laboratory (NRL), 143, 272

Naval Weapons Center, 57, 68, 194, 250, 506, 512

Navy, 24

Nebraska, 274

Nebraska v. Wyoming, 783

Neel, C. B., 157

Negative ions, 600

Negative "unusual" weather events, 809

Negative views on weather modification, 790

Negev Desert, 463

Neiburger, M., 105

Nelson, R. T., 144

Neumann, J., 459, 461, 487

Neutron activation, 58

Nevada, 308

New England experiments, 304

New England Coast, 360

Newman, M. M., 623

New Mexico, ground seeding test, 69, 70
  periodic seeding test, 69, 70
  State University experiment, 305, 308

New Mexico Institute of Mining and Technology, 18

Newnham, T. D., 156

Newark Sunday Star-Ledger, 76

Newton, C. W., 558

New York, 120-121, 798, 804
  injunction against, 81
  State University of, 34, 172, 344

Neyman, J., 22, 210, 213, 218, 463

Nicodemus, M. L., 714

Noise-collecting surfaces, 195

Nonhygroscopic nuclei, 7

Nonorographic, 73

Nonparametric tests, 213, 461

Nonscientific nature of many commercial weather modification operations, 36

Normal distribution, 211

Normality, 211

North Africa, 75

North American Weather Consultants, 47, 67, 73, 77, 81, 82, 85, 306, 772, 776

North Atlantic, 664

North Carolina, 499

North, D. W., 222, 517, 518

North Dakota, 79
  experimental seeding, 271
  rainfall analysis, 269

North Pacific, 664

Northern Great Plains seeding experiments, 260, 271

Northern Hemisphere, 182, 277, 656, 658, 666, 667, 700, 718

Nucleating, generator, 81
  particles, 709
  properties, 702

Nucleation, 99
  artificial, 73
  efficiency, 187
  homogeneous, 194
  heterogeneous, 366

Nucleant, types of, 57
  smoke, 54

Nuclei, anthropogenic, 329
  artificial freezing, 246
  concentration, 329
  freezing, See also Ice Nuclei, 99, 366, 506, 508, 513
  giant, 184
  measurement studies, 74
  production efficiency, 313
  sea salt, 242
  sublimation, 709

Nuisance, 772, 773

Null hypothesis, 211

Numbers, random, 549

Numerical analysis, 616

Numerical model, 38, 54, 59, 77, 238, 634, 650
  of cumulus dynamics, 37
  of winter lake storm, 322
  See also Mathematical marketing

Numerical weather prediction, 587

Oak Ridge, Tennessee, 758

Objective indicator of natural conditions, 460

Oblako rocket, 421, 423

Obninsk region, USSR, 395

Obolenskiy, V. N., 390

Observations, 549, 690

Observatory of Puy du Dome University of Clermont-Ferrand, France, 6, 32

Ocean, 634
  dynamics, 672
  evaporation, 540
  poleward heat transfer, 677
  variations in circulation, 720

Ocean Weather Ship "C", 694

Office of Atmospheric Water Resources, 787

Office of Emergency Preparedness, 803

Office of Naval Research, 25

Office of Weather Modification, 788
Ogburn, W. F., 801, 811
O'Hanlon, B., 47, 67
Ohio, 70
Ohtake, T., 187
Oke, T. R., 735
Okinawa, 271
Oklahoma, drought relief effort, 791, 794, 802
    frequency of hurricane occurrence, 554
    law suit, 81
Olin Weather Systems IL, 194
Oort, A. H., 648, 718
Ooyama, K., 528, 529, 532, 541, 649
Operational, design, 313
    programs, 366, 796
Optical, array instrument, Knollenberg, 157, 169
    disdrometers, 173, 179
    effects in the cloud, 169
    (lidar) probing techniques, 191
    properties, 702
    scattering instruments, 176, 178
    transmission, 177
Omega system, 155
Ontario, Canada, 318
Ordinances against cloud seeding, *See also* Laws;
        Congressional Weather Modification Bills
Oregon, 84, 128
Organizational period, 533, 540
Organized and special interest groups, 806
Organized consumers, 796
Organized resistance, 809
Orly (Paris airport). *See* France
Orographic cloud, 73, 282, 283, 287, 292
    blanket, 284
    microphysical characteristics of wintertime, 285
    modification, 395
    seeding, 61
    systems, 282
    critical factors, 284
Orographic storms, winter, 76, 208
Orographic upslope flow, 57
Orographic weather modification, winter experi-
        ments, 308
"Orphan anvils," 275
Orville, H. T., 34, 72
Orville Committee, 34, 35
Overseeding, 38, 342, 462, 484
Oxygen, 687
Oxygen and hydrogen isotope ratio in ice and
        snow, 721
Ozone (O_3), 637

Pacific Coast, 360, 788
Pacific Ocean, 665, 706
Pacific Gas and Electric Company, 58, 68, 82, 772,
        776
Pack, D. H., 739
Paleoclimatologists, 633

Palmen, E. H., 523, 531, 538
Palmer, W. M., 129
Panama, 59
Parallx, 169
Parameterization, 525, 635
Paris Airport Authority, Orly Airport, Paris, France,
        369. *See also* France.
Park Range, Colorado, 305, 308
Parry, M., 753, 754
Particles, atmospheric, 700-709
    concentration, 708
    contributing to climatic change, 688
    direct or indirect annual formation rate, 710
    optimum size, 373
    radiation from solar flares, 721
    sampled, 702
    seeding, 382
    size distribution, 708
    wettable, 97
Particle Measuring Systems, 157
Particulates, 637
Partitioning, 220
Parungo-Rhea method, 350
Patten burner, 193
Patrich, J., 473
Patrick Air Force Base, Florida, 143
Pearson, A., 552
Pennsylvania, law, 771, 783
    ordinance cloud seeding, 83
    ex rel. Township of Ayr Fulk, 782
    Natural Weather Association v. Blue Ridge Wea-
        ther Modification Association, 771
    seeding experiment, 261
Pennsylvania Natural Weather Association, 782
Pennsylvania State University, 172, 244, 260, 344
Peppler, W., 9
Period prior to 1946, 4
Periodicities, 20, 720
Pcriodicitics of circulation, 720
Perez-Siliceo, E., 67
Persistence effect, 210, 274
Personal losses, 790
Perspectives on weather modification, 793
Peterssen, S., 25
Pezier, J., 221
Pfost, D., 810
PGI rockets, 423
Phillippine Islands, 271
Philosophy of Storms, 4
Phosphorus pentoxide P_2O_5, 150
Photosynthesis, 688
Physical model. *See* models.
The Physics of Clouds, 8
Picknett, R. G., 105
Pileus clouds, 327
Pilot project concept, 308
Pine trees, growth, 721
Pineapple Research Institute, 17

Pitch angle, 153
Pitot, pressure, 145
  tube, 155
Plane waves, 195
Planetary (or Ekman) boundary layer, 660
Plume, city, 741, 747
  nucleant smoke, 54
Plank, V. G., 370, 371
Plutarch, 4
Point Barrow, 693, 694
Point discharge, 601
Point discharge current, 193
Poisson's formula, 193
Polar air masses, 554
Polar vortex, 658
Poleward heat transfer by atmosphere and oceans,
    672
Political and economic considerations, 801
Pollutants, 183, 673
Pollution of the natural environment, 388
Polycrystal, 116
Polyelectrolytes, 362
Post hoc stratification, 220
Port of New York Authority compact, 784
Portable analyzer, 690
Potassium nitrate, 194
Potter, J. G., 758
Power, calculations, 221
  definition of, 214
  supplies, regulated, 196
Powers, E., 5
Prandtl layer, 660
Precipitation, augmentation, 257, 288, 294, 351-
    352, 425
  charges on, 172
  cloud seeding for, 94
  development, 103, 140, 329
  drops, 97, 149
  duration, 302
  effects, positive, 63
  efficiency (PE), 53, 287-288, 297, 330
  efficiency (PE) of small cumulus clouds, 234
  elements of, 339
  embryos, artificial, 243
  Formval and, 160, 165
  intensity, 302
  in water vapor budget equation, 529
  mechanism, efficient, 330
  particles, 392
  potential, 351
  redistribution of, 340
  remote sensing of, 189
  runoff relationships, 311
  scheme, 257
Precipitation Control Company, 67, 85
"Prediction," 206, 209, 555, 650
  error, 216
  experiments, 652

  from physical models, 221
  meteorological, 215
  of future atmospheric CO_2 concentrations, 694
  point, 650
  skillful, 549
Predictor, covariate, 209
  variable, 209, 216
President's Advisory Committee, 294
Pressure, 151
  dynamic, 155
  gradient, 502
  process, 543
  static, 155
  surface, 501
Price, G., 49
Prism, 118, 120, 361
Prism bundle, 116, 120, 343
Private sector, 70, 80
Probability, 211
  problems of, 206
  statement, 207
  theory, 211
Probing techniques, optical (lidar), 191
Procurement contracts, 779
Problem, initial-value, 524
Professional licenses, 777
Profile, temperature, 190
  wind, 190
Program II, Arizona, 28
Prohibition of weather modification, 774
Projects (Programs), Modification of weather, see
    Weather Modification Projects
Propane, liquid, 366
  protection, 810
Protection technologies, 787
Pseudoadiabatic process, 527
Public, decision-making process, 801
  disclosure, 804
  health service, 58
  hearings, 778, 791
  see also Weather modification, sociological
    aspects
Puerto Rico, 251, 517
Purdue University, 573
"Pseudo-cold front," 558, 561
Psychrometers, sling, 150
Pyrotechnics, 251, 418
  developed, 512
  devices, 57
  flare, 57, 306, 333, 344
  generators, 250, 506
  silver iodide flares, 194
  material, 420

Quate, B., 48, 59
Quebec act, 778

Racetrack patterns, 337

Radar, acoustic, 190
  data, 544
  determination of rainfall rate, 199
  Doppler, 155, 189, 190, 200, 332, 344, 345, 573
  echoes, 329
  hooks, 562
  identification of hail, 418
  long range, 142
  M-33, 344
  parameters of clouds, 411
  sensing, 189, 242, 391
  two-beam system, 391
Radford, W. H., 6, 356
Radiation, 357, 642
  equilibrium, 646
  fogs, *See* Fog
  solar, 102, 275
  to induce sufficient surface level warming, 382
Radiative balance, 714, 727
Radiative equilibrium, 646
Radioactive probe electric field meter, 192
Radioacoustic sounding system, 190
Radio controlled drones, 143
Radio frequency interference (RFI), 195
Radiometer, 327
Radiometer, microwave, 176
Radiosonde, 151, 152
Radio theodolite, 151
Radio wave direction finders, 155
Radio waves, 195
Radke, L., 172, 184
Ragette, G., 582
Rain, areas, 522
  augmentation in cumuli, 242
  augmentation problems, 262
  bands, 508
  clouds, 470
  caused flooding, 498
  data, historical, 462
  Hawaiian, 128
  warm, 104
  see also Precipitation
Rainbow, 169
Raindrops, 97, 103, 155, 161, 162, 178, 180
  concentration of, 140
  impressions, 165
  measurements of, 177
  radar echoes from, 127
  size distribution of, 127, 129, 178, 180, 199
  samplers, 177
  scavenging by, 182
Rainfall, 184
  anomaly, 756
  excessive, 84
  natural fluctuations in, 459
  rates, 523, 537, 547
Rain gauge, as observational instrumentation, 51
Rain-inducement proposal, first, 4

Rainmaking, experiments, methodology of, 208
  emergency, 794
Rainmakers, 22
Rainmakers, commercial, 17
Rainwater instrument, 158
Raman scattering, 191
Randomization, 26, 28, 30, 207, 241, 295, 351
Randomized, area experiment in South Florida, 262
  crossover, 31
  designs, 460
  experiment, 293
  experiment in hail suppression, 31
  seeding, 38, 68, 295
  seeding project, 68
  seeding test, 66
Rankine profile, 576
Rankine vortex, 562, 569, 581, 589
Rapid Project, 267, 268
Rate-of-climb meter, 153-154
Rawinsonde, 151, 344
Rayleigh scattering laws, 418
Reading, England, 745
Reagents, 394
Record keeping, 803
Reducing lightning hazard, 603
Refractive index, microwave, 152
Refractometer, microwave, 152
Registered professional engineer, 778
Regression, coefficients, 216
  equation, 209
  method, 66, 77, 220
Regulation of weather modification, 784
Reichelderfer, F. W., 15, 34, 64, 70, 71, 84
Rein, G. C., 602
Releaser cloud, 118, 121, 122, 352
Remote measurements, 189, 201
Remote radio-controlled generators, 55
Remote sensing, 189
  hydrometer fall velocities, 189
  techniques, 192
  techniques, infrared, 191
Remote sensor, 142, 192
Removal processes, 704
Replicas, 166
Replication, 165, 169, 207
Replicator, Formvar, 163, 169
Representative sample, 163
Rescue operations, 382
Research Flight Facility (NOAA), 38, 150, 201, 344
Research programs, government-funded, field, 797.
  See also Weather Modification Projects
Resin binder, 194
Revelle, R., 670, 690-691
Reverse flow housing, 146
Reynolds number, 589
Reynolds, S. E., 599

Richardson, L. F., 524
Richardson number, 589
Richtmyer, R. D., 524
Riehl, H., 511, 523, 531, 537, 538
Rime, 10
Rimed particles, 290
Riming, 104, 111, 342, 474, 484, 487
  on crystals, 340
  growth, 341
  growth of hailstones, 32
  on snowflakes, 326, 343
Riparian, defined, 769
Riparian rights, laws of, 60
Roach, W. T., 730
Road contractors, 83
Roadways, 382
Robb, J. D., 623
Robustness, 217, 221
Rockets, 144, 194, 242, 603
  Mighty Mouse, 623
Rockwood, Tennessee, 758
Rocky Mountains, 284-286, 302
Roll angle, 153
Ronberg, P., 757
Rose, D. A., 176
Rosemount thermometer, 146, 148
Rosenthal, S. L., 538, 544
Rossby, Carl-Gustaf, 16
Rossby number, 577, 579, 582, 587, 588
Rossow, V., 553, 572
Rotating cone, 160
Rotating cumulonimbi. *See* tornado cyclones, 561
Rotating tank experiments, 579
Rubsam, D. T., 502, 537
Ruby, F. L., 32
Ruhnke, L. R., 623
Ruskin, R. E., 151, 160, 177
Russell, B., 218
Ryan, R., 174, 175
Rye, J., 174

Saginaw Bay, 321
Sahara, 666
Sailplanes, 143
Salience, low, 798
Salinity of the ocean, 677
Salt seeding, 65, 244
Sampling, 162, 165
  in clouds, 162
  error, 217
  problems, 166
Sandberg, J. S., 740
Sanger, A. J., 563
San Juan Mountains, Colorado, 311, 312, 314
San Luis Valley, Colorado, 791, 799
Santa Barbara I, 306
Santa Barbara II, 306
Santa Barbara County field tests, 52

Santa Barbara project, 22, 38, 63, 68, 74, 81
Santa Catalina Mountains, California, 28, 490
Santa Marta, Columbia, 59
Sartor, J. D., 104, 480
Satellite, 103, 144, 324
Saturation, 94
  adiabatic rate, 100
  vapor pressure, 95
Saturn, 635
Sauberer, F., 741
Savannah, Georgia, 506
Scale of weather phenomena, 93
Scale of seeding effects, 60
Scale resolution, 637
Scandinavian aircraft data, 693
Scanning electron microscopy, 200
Scattered light, 171, 172
Scattering, Mie, 191
Scattering, Raman, 191
Schaefer, V. J., 9-22, 45-46, 52, 69, 79, 85, 165,
    206, 245, 278, 282, 288, 356, 391, 758, 772
Scherhag, R., 720
Schleusener, R. A., 33, 48, 79
Schmauss and Wigand (1929), 7
Schonland, B. F. J., 604
Schuurmans, C. J. E., 721
Science (11/15/46), 12, 17, 19
Science Newsletter, 64
"The Scientific Basis of Weather Modification
    Studies" University of Arizona, 35
"Scientific Problems of Weather Modification", 36,
    76
Scott, E. L., 210, 213, 218
Scripps Institute of Oceanography in California,
    690
Sealed instructions, 253
Seasonal variation, 692
Seasonal weather patterns, persistence of, 796
Sea salt nuclei, 242
Sea surface temperature, 670
Secretary of Commerce, 803
Secretary of Interior, 78
Security, 794
Seedability, 36, 216, 254, 307
  of clouds, 508
  criteria, 295, 301
Seeder cloud, 118
Seeding, agent, 344, 366
  aircraft, 63, 76, 242, 265
  burner, 194
  by pilot balloons, 419
  carryover effects, 313
  concentrations, 377
  of cumulus clouds over forest fire areas, 48
  development of strong echoes after, 334
  for hail suppression of, 418
  curtain, 339
  effects, 38, 58, 60-61, 254, 335, 338

environmental considerations, 314
equipment, 302
fog, 48, 49, 394
ground-based, 60, 394
hypothesis, 325, 326, 344, 351
ice phase, 194, 242
larger-scale effects of, 59
of summer convective clouds, 29
of summer cumuli, 65
of extratropical cyclones (New York University), 25
pattern, 349
randomized, 68
reagents, 420
salt, 65, 244
silver iodide, 67
single line hygroscopic particle, 376
snowpack operations, 77
South Dakota, 80
subroutine, 252
tests, 28
USSR experiments in, 394
watershed, 81
wide-area, 377
with a ground generator, 60
Seeding material, dispersal and transport, 308
effectiveness of, 142
model computations, 377
solid CO_2, 5
Seed/no-seed analysis, 310
Seeding particles, electrically charged or neutral, 362
size of, 373
Seeding procedures, economic value, 214
Seeding programs, EG&G, 306
information oriented, 51
randomized, 38
short-term, 77
water oriented, 50-51
Selecting effective reagents, 418
Selection bias, 219
Self-interest orientation, 793
Semi-empirical, laws, 239
methods, 748
Sensitivity, power of, 215
Sensors, carbon-coated humidity-sensitive, 150
Serpolay, R., 33
Settling chamber, 187
Sevan Lake, 396
Severe storm modification, 591
Sewell, W. R., 38
Shafrir, U., 105
Shapley, D., 811
Shawnigan Resins Corporation, 165
Shear, anticyclonic, 583
cyclonic, 583
Sheets, R. C., 509
Shellard, H. C., 750

Shimbursky, E., 459, 463
Shiskin, N. S., 114
Shooting anti-aircraft shells and rockets, 32
Siberia, 395
Siegel, S., 213
Sierra Mountains, California, 286, 308
Sierra Nevada project, 67
Sierra Research Incorporated, 197
Sign test, 213
Signal-to-noise ratio, 197
Significance test, Bayesian, 210, 221
Silver in precipitation, 58, 305
Silver iodide (AgI), as artificial ice nucleus, 366, 393
compound, 344
crystals, 344, 506, 513, 543
delivery systems, 17, 31, 193
inducing glaciation, 246
nuclei, 30, 486
pyrotechnic flares, 194
seeding, 23, 28, 31, 38, 80, 107, 260-261, 275, 335, 339, 394, 417, 418, 512, 608
smokes, 13, 19, 74, 490
sodium iodide-acetone type generators, 76
Vonnegut experiments, 12, 21
Silverman, B. A., 373, 374
Silverthorne, J., 17
Simpson bill, 71
Simpson, G. C., 9, 17
Simpson, J., 31, 59, 75, 201, 221, 507
Simpson, R. H., 506, 510, 543
Simulations, 221
Simulating, a modification experiment, 514
precipitation, 394
See also Numerical modeling
Siren, G., 721
Situ analyses, 690
Size, distribution of cloud and rain droplets, 411
spectra, of what, 177
Skew T-log p diagram, 101
Skyfire burner, MRI, 193
Sky-fire needle-type ground generator, 296
Skyline Conference on the Design and Conduct of Experiments in Weather Modification, 36, 207
Sky, man-made, clear, 345
Slade, D. H., 748
Slide gun, 173
Sling psychrometers, 150
Slutsky v. New York City, 769-771
Smagorinsky, J., 650, 661
Smith, E. J., 23, 24
Smith, E. M., 210
Smith, T., 49
Smith, V. N., 690
Smog layers, 183
Smyth, W. R., 618
Snow cover, 399, 677, 679, 680

Snow crystals, 111
  aggregation of, 343
  efficiency of the production of natural, 339
  habits, 327
  interlocking of, 342
  replicas, 11
Snowfall, redistribution of, 343, 351
Snowflake, 167, 258, 342
  formation of, 342
  giant wet, 111
  properties of, 343
Snow gauges, 344
Snowpack seeding operations, 77
Snow pellets, 117
Snow pillows, 199
Snow shower, man-made, 345, 348
Snowstorms, intense, 318
Snowy Mountain experiments, 23, 304
Snowy Mountains Hydro-Electric Authority
    (SMHEA), 23
Snowy Mountain silver iodide seeding, 23
Social, considerations, 314
  facts, 807
  implications of weather modification, 78, 94
  see also Weather modification, Sociological
    Aspects
Societal adjustments, 801
Society interacts with its natural environment, 387
Socorro, New Mexico, 60, 71
Sodium chloride, 363, 374
Soil moisture, 637
Soil, humus, 690
Solar, constant, 646, 680
  energy, 94
  heating, 285
Solar radiation, 132, 275, 636, 677, 706, 722, 749
Sondes, 151
  balloon-borne, 144
  frost point, 144
  helicopter, 143
Sonic, velocities, 198
Soulage, G., 6
Sounding curve, 101, 102
South African law, 777
South America, 75, 359, 399, 666
Southern California Edison Company, 48, 55
South Carolina, 73
South Dakota, cumulus seeding experiments, 244,
    276
  hailstorm, 234
  legislation, 776, 792
  rainfall observations, 269, 271
  survey of group interested in weather modifica-
    tion, 797
  survey of views, personal benefit or losses, 799,
    804
  thunderstorms, 261, 262
South Dakota School of Mines and Technology,

64, 79, 260, 267, 271
South Dakota Weather Control Commission, 79,
    788, 792, 795
South Dakota Weather Modification Program, 792
Southern Florida Flood Control District, 788
Southern Hemisphere, 657, 658, 667, 706
South Pole data, 693
Southwest Africa, 666
Southwest Weather Research, 771
Southwest Weather Research litigation, 768
Soviet Union, see U.S.S.R.
Space charge density, 193
Spain, 786
Spar, J., 25, 528, 757
Spatial dendrites, 116, 117, 120, 340, 343
Spatial distribution of heat, 378
Spatola, A. A., 371
Specific humidity, 290
Spectral, absorption, 150
  line, 151
Spectrometer, flame, 188
Spectrum of cloud droplets, 162, 480, 481
Spender cloud, 119, 121, 352
Spengler, K. C., 34
Spherical vortex, 249
Spickles, D., 82
Spiral bands, 523, 547
Spiral crystal growth, 116
Spongy stone, 126
Squires, P., 23, 173, 481
Squared rank tests, 213
Square-root transformation, 214
St. Amand, P., 506, 512
St. Croix, Virgin Islands, 244
St. Louis, Missouri, 752
Stability, perturbation, 526
  rotational, 534
  of saturated air, 101
  of unsaturated air, 100
Stahman, J. R., 623
Standard deviation, 211
Standard Weather Bureau stations, 295
Stanford Research Institute, 811
Stare decisis, doctrine of, 767
Starr, V. P., 718
State control, 802
State law. See Laws; Congressional Weather mod-
    ification bills
State University of New York (SUNY), 85, 172,
    198
Static, approach, 246
  pressure, 145, 155
  stability, 663
Statistical, decision theory, 222
  design of weather modification experiments, 36
  methods, 207
  validity, 162
Statistics, 211

Statistics of Time Series, 219
Steam fog, 98
Steel manufacturing plants, 709
Steinhauser, F., 744
Stellar ice crystals, 179
Steris, C. G., 602
Stigler, S. M., 219
Stimulating updrafts, 391
Stochastic, collection, 243
    condensation, theory of, 392
Stockholm, Sweden, 355
Stommel, H., 235
Storm, age, 545
    incipient, 532
    right moving, 562
    surge, 498, 545
    track, 549
Stormfury, Project, *See* Weather Modification
        Projects
Stout, G. E., 129, 756
Stow, C. D., 607
Stratification, post hoc, 220
Stratosphere, 660
Stratosphere to troposphere, 696
Stratus clouds, 14, 395
Stratocumulus cumulogenitus, 122, 327
Stratocumulus layer, 327, 328
Stratus decks, clearing, 21
Stratus and fog, dissipation of cold, 25
Stream flow, 77, 309, 320-321
Strickler, R. F., 67, 644
Strict liability, 772
Stummer, G., 738
Stuve, G., 8
Subcloud layer, *see* Cumulus bases
Subcooled stratus clouds, experiments, 27
Subgrid scale processes, 635
Sublimation, 8
Subsun phenomena, 171
Suess, H., 690, 691, 697
Sugar, burning of, 709
Sulakvelidze, B. K., 31-33, 477, 489
Sullivan, R. D., 584
Sullivan vortices, 585
Sun radiation, 102, 719
Sunshine, 715, 749
Sunspot activity, 721
Super, A. B., 305
Supercooled, clouds, 6
    cloud water droplets, 45
    particles, 14
    water, 506
Supersaturations, 7, 97, 183
Supersaturation in a cloud, 136, 140
Suppressing lightning, 21
Suppressing rainfall, 791
Supreme Court of the United States, 783
Surface, boundary layer, 660

drag, 534
    temperature, global trends, 718
    water law, 769
Sutherland, J. L., 178
Swiss Hail Prevention Experiment, 463
Switzerland, 31, 247
Systems design, 195

Taha, M., 213
Talbot, C., 12
Tangential wind, 534
Target, moving or floating, 210
Target-control, comparison, 310
    design, 209
Targeting, 376
Target-only design, 208
Tasmania, 304, 773
Taubenfeld, H. J., 38, 711, 807
Taunus Observatory, Germany, 7
Taylor, P. A., 740
Taylor, W. L., 572
Technology assessment, 307
Telecommunication links, 199
Telemetering buoys, 144
Temperature, dynamic rise in, 145
    free-air, and humidity, 145
    measurements in flight, 148
    profile, 190
    Rosemount, 148
    virtual, 201
    wet-bulb, 149
Terada, T., 117
Terminal velocities in tropical clouds, 258
Test, C (alpha), 212, 213
    nonparametric, 213
    significance, 210
    squared rank, 213
    t-, 211, 212
    Wilcoxon-Mann-Whitney, 213
    Wilcoxon rank test, 213
Texas, drought, 271, 791, 794, 802
    injunction against hail suppression, 81
    law, 776, 777
    tornadoes in, 553
    weather control law, 776
Texas Court of Civil Appeals, 768
Texas Supreme Court, 769
Texas Water Development Board, 788, 795
Thermal, energy, 377
    equation, 639
    fog, dissipation system, 379, 381, 382
    mountains, 273
    gradient diffusion cloud chamber, 184
Thermistor, 151
Thermocouples, 196
Thermodynamic, diagram, 101
    first law of, 524
Thermodynamically metastable state, 365

Thermodynamik der Atmosphare, 7
Thermoelectric charging, 598
Thermoelectric potential, 599
Thermometer, infrared (IR), 200
  platinum-resistance vortex-type, 473
  Rosemount, 146
  vortex, 146
Thermo-Systems, Inc., 186
Thom, H. C. S., 35, 212
Three-phase process of precipitation formation,
  102, 104, 111
Thunder Bay, Michigan, 345
Thundercloud, electrical losses, 600
Thunderstorms, 93, 132, 553, 597, 598
  circulation, 558
  South Dakota, 261
Thunderstorm Project, 15
Tiberias, Lake, 468
Tidal forces, variations in the, 719
Time, derivatives, local, 524
  steps, 525
Tornadoes, 93, 497
  characteristics, 552, 562
  Charles City, Iowa, 572
  climatology of, 554, 555
  cyclones, 561; *see also* Rotating cumulonibi
  Dallas, Texas, 569
  defined, 558
  described, 562
  diurnal distribution of, 555
  echo, 558
  Fargo, 572
  forecasts, 558
  forms of, 552
  frequency in West Texas, Oklahoma, Kansas, 556
  funnel, 552
  Illinois, 572
  Lubbock, Texas, 563
  modeling, 577
  movement, 553
  Newton, Kansas, tornado cyclone, 566
  Oklahoma tornadic storm, 572
  Palm Sunday, 553, 566, 572
  photographing, 566
  prediction of, 555, 557
  probing with instruments, 573
  reported per year, 555
  season distribution of, 554
  sferics, 572
  storm that devastated Udall, Kansas, 572
  "suction spots," 563
  vortex core, 586
  warning, 558
Toronto, Canada, 329, 759
Toronto, University of, 106
Torque, frictional, 530
Total water, 160
  content, 177
  measurement of, 161, 172
Toxicity, 58
Trace gases, 637
Trade-offs, 810
Trail Smelter arbitration, 785
Trainer, P., 810
Transcaucasia, 417
Transcaucasian Hydrometeorological Institute
  (TsAO). *See* U.S.S.R.
Transform, inverse, 215
Transformation, 214, 215
Transforming observed variable, 211
Travelers Insurance Company, 79
Treatment effect model, 218
Trespass, 772
Triboelectric effects, 196
Tri-counties project, 73
Triggered lightning, 622, 628
Tri-State Natural Weather Association, 83
Tropical, air fogs, 359
  air masses, 554
  clouds: ice spectra, collection efficiencies, termi-
    nal velocities, 258
  cumuli, 233
  cyclone, 500, 522
  disturbances, 503
  drought, 270
  oceans, 275
  storms, 93, 277, 500
Tropical Storm Frieda, 509, 511
Tropical Storm Helene, 509
Tropopause, 235, 661
Troposphere, 661, 679, 696
Tropospheric air chemistry, 182
Tukey, J. W., 207, 217, 222
Turboclair, Orly Airport, Paris, 380
Turbulence, 110, 164, 365, 376, 380
  effect on rate of aggregation, 393
  intensity of, 365
  spectra, 196
Turbulent, energy, 132
  stresses, 639
  vortex cores, 589
Turner, J. S., 579, 587
Tuscola, Illinois, 756
Twomey, S., 188, 476

Ultrasonic sound waves, 362
Uman, M. A., 604
Undersun, 330
Uninsured losses, 807
Union of Soviet Socialist Republics (U.S.S.R.),
  387-431
  Caucasus hail suppression project, 31-32, 419,
    421
  dissipation of fog, 31, 33, 366, 367
  expenditures, 77
  hail suppression and control, 246, 394, 396

instrumentation used, 159, 392
northern Caucasus, 428
observatories, 706
origin and structure of convective currents, 411
research in weather and climate modification,
     389, 396, 398
rockets used in seeding, 144, 194, 242
seeding clouds, 394
three methods of modification, 428
three purposes of use of crystallization of super-
     cooled droplets, 394
Ukrainian Soviet Socialist Republic (Ukraine),
     395, 425
Uzbek, 395
Central Asian Regional Hydrometeorological In-
     stitute, 410
High Mountain Geophysical Institute (VGI), 412,
     415, 416, 418, 420, 421
VGI method of hail prediction, 413
VGI model, 413
Hydrometeorological Service, 390, 410, 413, 421
Institute of Artificial Rain, 390
Transcaucasian Hydrometeorological Institute
     (TsAO), 410, 411, 412, 418, 420, 425, 428
Ukrainian Scientific-Research Hydrometeoro-
     logical Institute, 395
ZakNIGMI Zakavkazskiy gidremeteorological-
     eskiye Institut, 412, 420, 428
United Airlines meteorologist, 49
United Fruit Company, 17
United Nations, 694
United States, Army, 14, 83, 371
     Bureau of Reclamation, 29, 71, 260
     Category II system, 364
     Congress, 788
     Corps of Engineers, 82
     cyclones, 648
     FIDO system of fog dissipation, 356
     fog, 359, 360, 366, 367, 379, 380, 381
     Forest Service, 15, 68, 71, 75
     Meteorological Service, 390
     military organizations, 512
     observatories, 706
     rockets, 144, 194
     salt seeding experiments, 244
     Weather Bureau, 13-26, 34, 35, 64-73, 75, 79,
          295, 325, 506-507, 558, 566, 678, 713, 783
United States Advisory Committee on Weather
     Control. See Orville Committee
United States Air Force, 49, 371, 379, 380, 803
United States Air Weather Service, 690
University of Arizona – seeding tests, 28, 35, 76
University of California, 67, 68
University of Chicago – cloud physics laboratory,
     16, 25, 26, 37, 63, 64, 76
University of Chicago, Chicago Midway Labora-
     tory, 28
University of Clermont-Ferrand, 32, 33

University of Colorado, 315
University of Miami, 189
University of Minnesota, 186
University of Missouri, 186
University of Nevada, 79
University of Sapporo, 144
University of Toulouse, 6
University of Washington, 172, 304
University of Wyoming, 304
Unorganized citizens, 804
"Unprovided case," 768
Updraft, 62
     active, 528
     radius, 153
     suppressing, 392
     velocities, 152, 153, 329
Urban, air motion, 742, 743
     area thermal characteristics, 717, 729, 731, 734,
          738
     chemical reactions, 744
     cloud plumes, 758
     convective clouds, 752
     daylight illumination, 748
     degree-days, 737
     mesometeorology, 726
     mixing layer, 748
     particles, 744, 745, 758
     phenological difference, 738
     plume, 748
     pollution, 744, 747, 748
     precipitation, 752, 754, 755, 757
     radiation loss, 749
     relative humidity, 735
     snowfall, 736, 758, 759
     temperature field, 728, 731
     thunderstorms, 753, 754
     turbidity factor, 730
     vapor pressures, 735
     visibility, 749, 750, 752
     wind field, 738, 739, 740, 741
Urea, 363, 376, 377
Utah, 86, 308, 797, 798, 804
Utah State Experiment, 306
Utah Water Research Laboratory, 306
Uthe, E. E., 747

Vailevskis, M., 210
Vallecito Reservoir, 312
Value decision, 801
Van de Graff generator, 613
Van Tassel, E. L., 572
Vapor, deposition of, 99, 290
     diffusion, growth by, 474
     pressure, 96, 111
     tension, 6, 8
Vardiman, L., 287
Variability, natural, 207, 208, 217, 241, 517
Variable predictor, 209

Variance, 217, 218, 548
Variometer, 154
Venus, 635, 637
Veraart, A. W., 5
Verona, Italy, 48
Vertical, wind profile, 741
  wind shear, 376
Veynberg, B. P., 390
Victoria, B. C., 775
Victoria Rainmaking Control Act, 775
Viemeister, P. E., 604
Vienna, Austria, 744
Virginia, 194
Virtual mass, 249
Viscosity, 588
Viscous stresses, 582
Visibility, 357, 751
  Budapest, Hungary, 750
  deteriorating, 751
  improving, 364
  V-meteorological, 362
Visual range, Budapest Hungary, 750
  Manchester, England, 750
  slant, 191
Visual techniques, 168
Volcanic, Dust, changing atmospheric transparency, 720
  gases, 689
Vonnegut, K., 391
Volz sunphotometer, 730, 746, 748
von Neumann, J., 656
Vonnegut, B., 12, 19, 21, 46, 79, 572, 600
Vortex, breakdown, 588
  circumpolar, 103
  spherical, 249
  thermometer, 146, 473
  numerical model of, 587
Vorticity, 553, 581
  source for tornado-genesis, 582
Vukovich, F. M., 740
Vul'fson, N. I., 590

Wadsworth, G. P., 34
Wagner, A., 7
Wait, G. R., 602
Waite, P. J., 572
Wall clouds, 507, 508
Walser, J. T., 67
Wash, D. T., 24
War and Weather, 5
Ward, N. B., 560, 562, 563, 579, 580, 582, 589
Warm core, 522
Warner, J., 156, 480, 481
Warragamba experiments, 304
Warren, L. F., 5
Wasatch Mountains, Utah, 308
Washington, 69, 84, 308
Washington, D. C., 80, 754

Washington, W. M., 679, 716
Water, Cloud Modification Res. Act, *see* Congressional Weather Modification Bill S86
Water, drops, 167
  evaporation of droplets, 369
  rights, 779
  spray seeding, 243
  storage, 637
  supercooled, 543
  supply deficiencies, 309
  vapor, 637
Water Resources Development Corporation, 47, 67, 72
Water vapor, budget, 529
  convergence, 528
  density, 196
  diffusion, 484
  frictional inflow of, 540
Watershed seeding, 81
Waterspouts, 575-583
Wave clouds, 284
Wave Propagation Laboratory (NOAA), 190, 201
Waves, gravity, 547
Weapon test carbon-14, 691
Weather, control districts, 776
  "Controlling the Weather—A Study of Law and Regulatory Procedures," 38
  man-made, 348
Weather modification, AMS statement on, 84
  draft protocol by the World Peace Through Law Center, 785
  ecological implications of, 78
  growth in private sector, 80
  guideline statistical evaluation, 36
  historical survey, 3, 37
  human aspects of, 38
  inadvertent, 37
  legal aspects, 38, 771
  moratorium on, 82
  permits, 791
  political implications, 94
  potential, 295, 307
  programs perspective, federal, 795
  regulation of, 79
  reporting engagement, 780
  social implications of, 38, 78
  statutes and cases, 767
  techniques, 796
  USSR, *see* USSR
  see also Law; Weather modification Projects; Sociological Aspects.
Weather Modification, Projects (Programs), Arizona, 28
  Australian, 432-450
  Caucasus AntiHail Expedition, USSR, 31-32, 419, 421
  Central Arizona Project, 309
  Cirrus, 14-36, 52, 505

Cloud Physics, U. S. Weather Bureau, 18
Cloud Catcher, 260-262
Dallas, 73
Flatlands, 77
Global Atmospheric Research Program (GARP), 648, 684
Hungry Horse, 58
Israeli, 210, 454-494
Kentucky seeding, 73
Kenya Hail Suppression, 31
National Hail Research, 48, 792-793
Park Range, 54
Rapid, 267, 268
SCUD, 25
short-term, 77
Skyfire, 21, 75
randomized seeding, 68, 306
Skywater, 271, 308
Snowy Mountain, 23, 304
Stormfury, 142, 194, 251, 507, 512, 515, 519
Swiss Hail Prevention, 463
Tasmania, 304
Tri-counties, 73
Thunderstorm, 15
Tignes, France, 73
TsAO, USSR, 410-428
Whitetop, 29, 38, 63, 274, 459, 463, 489, 490
Winter Orographic, 305-306, 308
Weather modification, sociological aspects, benefit-cost ratio, 801
benefits, potential, 801
Citizens Council on Hail Research, 793
citizens preferences, 806
community influentials, 809
compensation, uninsured losses, 807
disapproves, 798
disinterested, the, 798
emperically based generalizations, 807
expert witness, 771
expressed attitude, cloud seeding, 800
factual knowledge, 798
human dimensions, 38, 308
"God's plan for man and the weather," 798
grievances, 807
interest groups and organizations, 793
interviews, hail season, 793
nature, balance of, 798
noninvolvement, 796
nonrandom conversations, 807
opposition, 791
personal losses, 790
political and economic considerations, 801
population growth, 788
population mean, 220
prestige, 794
prior appropriation doctrine of water rights, 769, 782
private sector, 70, 80

procurement contract, 779
professional license, 777, 778
public decision making process, 801
public disclosure, 804
public health service, 58
public hearings, 278, 791
record keeping, 803
regulation, 784
religio-material orientation, 798
social adjustments, 801
social considerations, 314
social facts, 807
statewide plan, 792
Straw vote, 791
underlying hostility, 808
unorganized citizens, 804
Weather modifiers, 85
Weather modifier licenses, 777
Weather or climate modification, 709
Weather radar, remote sensing, 189
three cm, 466, 474
"Weather and Climate Modification Problems and Prospects" 1966, 37
Weather Bureau, see U.S. Weather Bureau
Weather Control, Final Report of the Advisory Committee on, 294
Weather Control Research Association, 79, 85
Weather Engineering, 48, 59
Weather Modification Association (WMA), 80, 85, 803
Weather Modification Company, 47, 48, 67, 73, 82
Weather Modification, President's Advisory Panel on, 283
Weather Science, Inc, 49, 178
Weather Services, Inc., 79, 270
Wegener, A., 7, 8
Weickmann, H. K., 75, 344
Weinstein, A. J., 59
Weller, N., 572
Wells, M. A., 210, 218
West Africa, 598
West Indies, 330
West Virginia, 82, 371, 771, 782
Western Scientific Services, Inc., 49
Wet-bulb temperature, 149
Wetherrald, R. T., 699
Wexler, Harry, 15
Whirling arm facility, 198
Whitby, R. T., 186
White, F. G., 811
Whitetop cloud seeding experiment, 74, 210, 218, 268, 490
Whole air samples, 690
"Widespread Control of Weather by Silver Iodide Seeding," 20
Wieland, W., 184
Wigand, A., 7
Wilcoxon signed rank test, 213

Wilcoxon Statistic One-Sided p-values, 299, 300, 301
Wilcoxon-Mann-Whitney (WMW) nonparmetric test, 213, 461
Wilkins, E. M., 572, 690
Willett, H. C., 34, 720
Williams, J. S., 210
Wilson, C. T. R., 7
Wind, 152, 361, 498
    field, 189, 373
    gradient, 530
    profiles, 190
    shear, 335, 376
    speed, 365, 378
    tangential component of, 530, 534
    tunnel, 198, 302
    waves, 499
Winstein, A. I., 374
Winter Orographic Cloud research experiment, 305-306, 308
Wiresondes, 144
Wolfe Creek Pass, Colorado experiment, 197, 303, 309, 312, 314
Woodwell, G., 697
Working model of a cloud, 412

Workman, E. J., 18
World War II, 235, 248, 356, 726
World Weather Watch (WWW), 648
Wright, H. L., 7
Wu, S., 210
Wurtele, Z. S., 462, 463
Wyoming, 308

X-ray crystallography, 12

Yanagisawa, Z., 333
Yanai, M., 528
Yaw angle, 153
Yaw, R. H., 305
Yenukashvili, I. M., 416
Ying, S. J., 580, 587
Yuba City, California, Adams V. California, 772
    city flood, 81
Yugoslavia, 31

Zakavkazskiy Gidremeteorologicheskiye Institut (ZakNIGMI), 412, 420, 428
Zoller, W. H., 745
Zonal mean temperature, 671
Z-R relationship, 199